International Conference on Digital Solutions for Automotive Industry, Roadway Maintenance and Traffic Control (DS ART 2019)

IOP Conference Series: Materials Science and Engineering
Volume 832

Cholpon-Ata, Kyrgyzstan
1 November 2019

ISBN: 978-1-7138-4725-0
ISSN: 1757-8981

Printed from e-media with permission by:

Curran Associates, Inc.
57 Morehouse Lane
Red Hook, NY 12571

Some format issues inherent in the e-media version may also appear in this print version.

Institute of Physics
Dirac House, Temple Back
Bristol BS1 6BE UK

Phone: 44 1 17 929 7481
Fax: 44 1 17 920 0979

techtracking@iop.org

Additional copies of this publication are available from:

Curran Associates, Inc.
57 Morehouse Lane
Red Hook, NY 12571 USA
Phone: 845-758-0400
Fax: 845-758-2633
Email: curran@proceedings.com
Web: www.proceedings.com

TABLE OF CONTENTS

IOP Conf. Series: Materials Science and Engineering **832** (2020) 011001 doi:10.1088/1757-899X/832/1/011001

PREFACE

International Conference on Digital Solutions for Automotive Industry, Roadway Maintenance and Traffic Control (DS ART 2019)

International Conference on Digital Solutions for Automotive Industry, Roadway Maintenance and Traffic Control was held in Cholpon-Ata, the Kyrgyz Republic, on 1–3 November 2019. DS ART 2019 was organized by Moscow Automobile and Road Construction State Technical University (MADI), Moscow, Russia, with the promotional sponsorship of International Association for Automotive and Road Engineering Education (IAAREE).

The participants came together to discuss the topical issues of automobile and roadway complex, automotive production, maintenance and repair as well as the acute problem of traffic control. The questions of personnel training in the automotive industry were also raised. The need to move towards creating digital solutions in the industry was emphasized.

All the papers submitted to DS ART 2019 were carefully reviewed and proofread by the editors to select the most valuable and significant ones. They cover the range of relevant research on the current state of the automotive industry and innovative ways to solve its problems.

We express our gratitude to all the participants and the organizing committee for their contribution to preparing the DS ART 2019 proceedings.

With the best regards,
Prof. Aleksei A. Solntsev, Chairman of the Organizing Commette

IOP Conf. Series: Materials Science and Engineering **832** (2020) 011002 doi:10.1088/1757-899X/832/1/011002

Peer review statement

All papers published in this volume of *IOP Conference Series: Materials Science and Engineering* have been peer reviewed through processes administered by the proceedings Editors. Reviews were conducted by expert referees to the professional and scientific standards expected of a proceedings journal published by IOP Publishing.

Analysis of some indicators of environmental safety of buses with diesel engines (on the example of Moscow)

A P Boldin

Moscow State Automobile and Road Technical University (MSARTU), 64, Leningradsky Ave. Moscow, 125319, Russia

E-mail: boldin1940@yandex.ru

Abstract. The results of studies performed personally by the author and under his leadership on the current level of environmental safety of city buses of Russian-made ecological class Euro-4 (LiAZ-5292-21) and Euro-5 (LiAZ-5256-13 and LiAZ-529-22) according to the smoke level of the exhaust gases provided by GOST R 52160-2003, and not required in actual operation in terms of the content of nitrogen oxides **NOx** in the exhaust gases of a diesel engine when operating at idle modes. Previous studies on passenger cars and light commercial vehicles with diesel engines (using the operating method developed by us) in accordance with UNECE rules 83.03 (based on urban driving cycles with the results of environmental pollution in g/km) showed an excess of **NOx** emissions from 7 to 15 times, on the basis of which a simplified technology was developed to control the environmental safety of these vehicles by indicators of the content of $\mathbf{CO_2}$ and **NOx** at idle for a conventional engine with a working m volume of 1 liter .. This technique is appropriate to apply for a large vehicle mass, for which **NOx** emissions at idle is much higher. This will provide operational control of environmental safety by **NOx** and increased requirements for manufacturers. In a first approximation, the maximum permissible value of the **NOx** content is justified – no more than **530 ppm**. (For comparison, this indicator for cars is significantly lower – no more than **200 ppm**). It was additionally established that for estimating $\mathbf{CO_2}$ greenhouse gas emissions it is enough (with an error of no more than 10 % to decrease) to use the general indicator of fuel economy – actual fuel consumption: at the same time per **1 kg** of diesel fuel, gasoline, liquefied and compressed natural gas, used in operation, **1,500 g** of $\mathbf{CO_2}$ will be accounted for (conversion to liters is carried out according to the energy density indicators). The studies were carried out as part of comprehensive work to improve the quality of technical servicing of cars, trucks and buses, conducted at the department "Operation of automobile transport and car service" MSARTU (Head of the Department of Doctor of Technical Sciences, Professor A.A. Solntsev).

Keywords: buses, Euro environmental classes, bench tests for compliance with the EB-RO class, emissions of **CO,** $\mathbf{CO_2}$**, CH and NOx**, diagnostics of the ecological safety of diesel engines.

1. Introduction

City buses carry the vast majority of passenger traffic in medium and large cities, including in megacities, such as Moscow, which are characterized by the most stringent requirements for air pollution from car exhaust due to their specific (almost circular diameter up to 40 km according to MRHW) planning and high density construction of multi-storey buildings. The latter leads to the need to limit the use of a classic single or articulated diesel bus for the central districts of the city. Currently, single Russian-made electric buses are already used for individual routes, however, they cannot be widely distributed in the near future, and for non-central regions, as well as for communication with cities and settlements of the Moscow Region, full-size and route buses will be increasingly used buses with diesel engines. It is generally accepted that the exhaust gases of these cars, despite the modern achievements of scientific and technological progress, are much more aggressive compared to gasoline or using cheaper gas fuel. However, to obtain significant power (during bus transportation), the diesel engine remains a priority, which has always been given great attention to the environmental safety of the whole world. At the same time, the current level of solving this problem is still far from complete despite the rather stringent state restrictions [3–5], which is confirmed by numerous foreign studies on passenger cars with diesel engines [6–9], and made by us [10]. As for city buses, with obviously worse environmental safety indicators, here manufacturers use an information sign on the back of a vehicle of an environmental class value (for Moscow it is usually 4th or 5th), which according to Russian laws should be confirmed during routine maintenance by checking exhaust gases for smokiness in the free acceleration mode of the crankshaft without external load (according to GOST R 52160-2003: FUMES OF EXHAUST GASES. Standards and methods for control when assessing the technical condition). At the same time, such an important indicator of the level of exhaust gas pollution as the content of NO_x nitrogen oxides in them, due to technical regulations adopted in the last century, which have long been necessary to revise to improve the environmental safety of diesel engines of heavy trucks and buses, is not used in operation. Therefore, the development of possible options for solving this problem is relevant.

2. Statement of the problem and analysis of possible methods for solving it

Here it is necessary to take into account that traditionally in our country the environmental characteristics of the diesel engine in operation were checked by the parameters of exhaust smoke according to GOST 17.2.2.03-87, which was adjusted in the direction of tightening, and since 2005 has been replaced by GOST R 52160-2003, harmonized with European standards – UNECE Regulations 49 and 83, which contain environmental requirements for vehicles with compression-ignition engines and sparks. According to these rules, in Russia from 1992 to 2016, manufacturers must ensure the release of new cars in accordance with EURO 1–6 standards, and thus ensure the gradual replacement of rolling stock with environmentally friendly cars.

In this regard, UNECE Regulation 83 with its current supplements regarding the certification of cars and light commercial vehicles (weighing up to 3,500 kg) with gasoline and diesel engines, according to the standards of the ecological classes EURO, is of the greatest interest. A general idea of the essence of the EURO standards for cars can be obtained from the Internet materials [4], and in general and a more detailed presentation for cars and heavy vehicles and buses can be found in the Internet resource [5], although there are certain inaccuracies in the list indicators. A theoretical presentation of the features of testing cars and engines for environmental certification is given in the textbook [4]. In ordinary operation, as a rule, there is no clear idea of these standards and methods for their assessment, more so as in Russia the environmental class indicator was reflected in the data sheet only for cars (Russian and foreign) manufactured in 2006 and later. Therefore, for the bulk of passenger cars in operation, it is indicated that this indicator is "not defined", although for foreign models it could be quite high – EURO-3 or more. For this reason, it turned out to be practically impossible to regulate air pollution in the central zones of large cities by allowing cars with a high environmental class to enter there, and the application of this rule for municipal (state) rolling stock was limited only to bureaucratic procedures (for a fee for foreign cars it was confirmed by the basis of the year of release, its regulated environmental class, although in fact as a result of using the resource it could s is much lower). Therefore, there was a need to develop a method for the rapid assessment of the actual level of the ecological class, primarily for gasoline cars, since the modernized GOST R 52033-2003 (Cars with gasoline engines. EMISSIONS OF POLLUTING EMISSIONS WITH EXHAUST GASES. Norms and control methods for assessment technical condition), intended mainly for carburetor feed systems, did not foresee any measurements of NO_X emissions underlying EURO. Similarly, a similar situation was observed in the operational practice of the countries of Western Europe. In addition, the automatic engine control systems for some foreign models were not adapted for emission tests for emissions from CO and CH at elevated frequencies of rotation of the crankshaft (**0.6 nmax**) and gave anomalous values from CO (for example 3 %) with a high environmental class. Therefore, for cars of ecological classes Euro-3 and higher, such control no longer provided high-quality maintenance of exhaust gas aftertreatment systems based on platinum catalysts, primarily regarding NO_X, and required the development of more complex technologies applicable in operation.

We managed to develop such a technology on the basis of tests of a heated car on unloaded running drums in direct transmission at speeds of 40, 60 and 80 km/h, at which the content of carbon monoxide CO, carbon dioxide CO_2, hydrocarbons C_mH_n (abbreviated CH) was measured and nitrogen oxides NO_X using a five-channel non-dispersed gas analyzer infrared type [1]. The results obtained were then adjusted for conditions according to EURO standards (a detailed description of the test technology is given in the textbook [3]). They are based on a four-fold repetition of the so-called European city driving cycle (ECE test), which was used even before using the EURO classification, according to which, at a specialized dynamometer stand with automatic control, the vehicle idling and subsequent accelerations were simulated using I -th gear, I-II-th, and I-II-III-th gears to speeds of 15, 30 and 50 km/h, respectively. Over a period of four test cycles, the car theoretically "traveled" 4,052 km of track, while all exhaust gases were collected in a common tank and diluted with clean air in a ratio of 1: 3 (to exclude vapor condensation). Then, the resulting mixture was analyzed for the content of CO and CO_2 by a non-dispersed infrared gas analyzer, and C_mH_n and NO_X were respectively evaluated by ionization-flame and chemiluminescent methods (not used in ordinary operation). Moreover, according to the research of Doctor of Technical Sciences, Professor V.A. Zvonov, the figures of the C_mH_n infrared gas analyzer used in carburetor engines operations turned out to be half lower than the ionization-flame method (due to the underestimation of non-vaporized fuel droplets when working on enriched mixtures). When testing on EURO since 1993 in accordance with the rules of the EEC No. 83.03, a speed cycle of up to 120 km / h (EUDC test) with a duration of 400 s and a track length of 6.955 km was added to the 4th city cycles. At the same time, for EURO-2, the engine was preheated from the initial temperature of 200 °C for 40 s, for EURO-3, such heating was already excluded, and for EURO-4, tests should have started when the car and engine were cooled to negative temperatures of –70 °C. Tests for cars with gasoline and diesel engines were completely identical, although the standards for NOX diesel emissions were set much less stringent (three times higher), and for CO – more stringent (about four times lower). In Russia, facilities for testing cars can be counted practically on the fingers, and the complexity of the process did not allow the possibility of its use in operation. In Russia, facilities for testing passenger cars can be counted practically on the fingers, and the complexity of the process did not allow the possibility of its use in operation.

At the same time, the operational methodology developed by us for testing cars on unloaded running drums and the recalculation model of the obtained indicators provided the maximum approximation to the certification test 4 × ECE + EUDC "1", and the values of mass emissions of explosives (CO, CH and NOX in g/km) in a first approximation could be compared with the environmental standards of Euro [3–5]. (Here it is necessary to take into account the inevitable error in underestimating the NOX values due to the exclusion of the speed cycle, which could be compensated by its initial low values for an unheated or cold engine, and, accordingly, the underestimation of the indicators for CO and CH for the same reason). However, the final analysis of the diagnostic results obtained in this way for passenger cars with gasoline engines showed a good agreement between the components of CO and CH with the standards of the declared environmental class, although deviations beyond the standards could be observed by NOX indicators for cars of Russian models (2–3 times), due to the insufficient efficiency of the exhaust gas neutralization system, which could be eliminated by certain technical influences [1]. At the same time, for cars of foreign manufacture of ecological classes Euro-2 and higher, there was a "zeroing" of NOX values during tests at idle mode and at 3-speed platforms on unloaded running drums (a similar "zeroing" was recorded , as a rule, also in terms of CO and CH for Euro-4-class cars). At the same time, it was revealed that the increase in the mass of emissions of all three harmful substances (even outside Euro-1) was associated with an unacceptable deterioration in the condition of the cylinder-piston group, and this defect was not detected by the diagnostic parameters customary for operation - the deterioration of vehicle dynamics and increased fuel consumption. After replacing the pistons and rings, the environmental performance of the engine in test modes for testing the car was fully restored, although there were no technological impacts on the neutralizers. Therefore, an excess of the calculated emission values compared to Euro standards is certainly evidence of a deterioration in the state of operation of the engine and the neutralization system as a whole or for individual components.

During logical processing, it is necessary to consider such specifics of technical diagnostics as the need to use the inevitable statistical practice of estimating the maximum permissible value of a diagnostic parameter, the rationale of which is given in the textbook [2]. This is due to the fact that in operation it is impossible to maintain the "ideal" technical condition of the controlled object for a long time, and its deterioration to the level at which its further use by social or technical and economic criteria becomes undesirable or ineffective is allowed. In other words, if the manufacturer manufactures a car of the required environmental class, then in actual use their condition will inevitably be one or two classes worse (controlled emission figures will approximately double [3–5]. The range of permissible changes is determined by the statistical processing the values of the diagnostic parameters for the same type of aggregates, units and mechanisms of a car of various states, usually during the next service, while for informative parameters of its the scattering should be described by a unimodal theoretical distribution law (usually normal or gamma distribution with an offset parameter). The permissible value is then determined by unilaterally or bilaterally limiting the range of variation of the theoretical law with a probability of 0.9 for non-relevant elements (while minimizing the so-called error of the 1st kind – a false malfunction), and with a probability of 0.8 for socially significant elements (while minimizing the so-called error of the second kind – malfunction skipping) [2]. From these positions, the "inaccuracies" of the theoretical or instrumental nature of the technological process of determining the diagnostic parameter are compensated by final statistics – the diagnosis is logically made by comparing the parameter with its maximum permissible value. At the same time, the significance of the test test mode (in this case, on unloaded running drums) is confirmed. In this case, the parameter should be considered informative, in which the a priori (practical) probability of going beyond the range of the permissible value will be at least 25–30 %. Thus, in a real diagnosis of the compliance of a car and an engine with the ecological class of Euro, a situation can be observed when the test (tied to equipment and technology of its application) parameter values can exceed the ecological standards of Euro, but it is enough to objectively identify the need for technical influences to improve the condition within the design limits features of this system and technological capabilities of maintenance and repair. It is necessary to add to this that even with worse parameters compared with the declared values, they generally turn out to be better in comparison with the still used cars of older models (lower environmental class), the compulsory prohibition or restriction on the use of which will undoubtedly lead to increased social tension.

A certain misunderstanding for the operators may be caused by the environmental standards given on the Internet for cars and light commercial vehicles with both gasoline and diesel engines [4, 5]. First, unlike the educational literature [3], there is a substitution of the concept of the content of carbon-hydrogens $\boldsymbol{C_mH_n}$ in the exhaust gases, which is traditional for car ecology, with the content of **HC** hydrocarbons (acid salts of carbonic acid, formed during prolonged transmission of CO_2 through an aqueous solution containing carbonates). The latter has nothing to do with the operation of automobile engines, but for several practitioners it leads to the conclusion that it is impossible to directly control environmental indicators, since the measurement of LS cannot be performed by an infrared gas analyzer. In addition, there are certain discrepancies in the control figures, even considering the comments made, which leads to additional confusion and does not contribute to more complex types of direct control. Based on the comments made, it is advisable to have one agreed table of Euro environmental standards on the Internet, reflecting the specifics of road transport (similar to that presented in the textbook [3] on p. 110), with a more "soft" adjustment depending on the category of cars. In this case, the permissible NOX emission standards for cars (category M) with Euro-4 diesel engines are 0.25 g/km, and for light commercial vehicles (vans) weighing up to 3500 kg (category N1-III & N2) increase to **0.39 g/km**.

Thus, the recalculation principles outlined above were also aimed at obtaining lower calculated values of harmful emissions in relation to the specifics of the tests according to the ECE test, which is convenient for operation due to the reduced likelihood of receiving a negative result during diagnosis. Based on this, and also due to the complexity of the recalculation procedure for real practice, it was concluded that it is possible to universalize the proposed diagnostic technology for all passenger cars with gasoline engines, which may have larger or lower values of their own masses and some differences in the design of power transmission, in particular when using automatic transmissions, which are mainly installed on foreign models. As noted above, the application of the statistical method of substantiating the maximum permissible values allows us to compensate for technical diagnostic errors, which may be small due to the stability of the test test mode on unloaded running drums (slight deviations of the simulated speed practically do not affect the readings recorded by the infrared gas analyzer) . On the other hand, since the Euro environmental classes for cars and small commercial vehicles with diesel engines are determined according to standards similar to gasoline cars (4 × ECE + EUDC combined test), it is appropriate to use both the technology and the allocation model for the diagnosis of diesel engines. It is important to know that the pattern of changing the content of the base harmful substance from the load for the diesel engine as a whole is similar to gasoline in nature (a significant increase in NOX with increasing load), and the engine's reaction to the load change, which was recorded for a gasoline car through the vacuum values in the intake manifold can also be considered similar. In addition, the Euro standards for diesel engines of Euro-3 and higher class [3–5] are no longer identified as an independent indicator of the emission of hydrocarbons CH, unlike gasoline, and the main attention is paid to NOX (CO and CH indicators when checking a conventional infrared gas analyzer, as shown by our measurements, is practically zeroed). At the same time, the technology is based on the correct estimation of mass emissions of harmful substances by a car while simulating its operation at stationary speed modes on unloaded running drums, which is unchanged for passenger cars with gasoline and diesel engines [1]. Therefore, from the standpoint of the systematic approach outlined in the textbook [2], the universalization of the technology described above for operational tests is advisable.

The results of our respective tests for compliance with the declared environmental classes EURO-4–5 of foreign models of cars and minibuses with diesel engines are reflected in the work [10]. First of all, attention was paid to the fact that the fact of a substantial, about two-fold, decrease in the mass of emissions of the non-toxic component of CO_2, which has recently been given increased attention due to the increase in the greenhouse effect of the atmosphere, is practically not covered by the operation, especially for major cities. Data processing according to the above algorithm showed practical compliance with the declared Euro standards for mass emissions of CO (and CH) [3–5], which confirmed the universality of the test and recalculation method. However, according to the calculated mass emission indices of nitrogen oxides, NO_X was found to be significant, from 6 (1.5 g/km against the norm of 0.25) to 35 times (6.74 g/km against the norm of 0.18 for a Volkswagen car-Caddy), exceeding Euro standards. This differed from the data on the four-fold increase in emissions of the Volkswagen-Caddy car given in Article [7] of the permissible values of the Euro standards (0.85 g/km), and indicates a certain lack of efficiency in the operation of the NOX component neutralization system for passenger car diesel engines in general , which is already evident when checking in idle mode (in our measurements, these values were in the range of 120-160 ppm for EURO-5 models, 180–280 ppm for EURO-4 and 630 ppm for Volkswagen-Caddy). Further experiments revealed a certain difficulty in conducting operational tests on the unloaded running drums of modern passenger cars with diesel engines due to the predominant presence of automatic transmissions and all-wheel drive for all 4 wheels, which required complicating the design of the mechanical stand, allowing at least "to hang out" "On the lift one of the axles of the car (preferably the front). However, the corresponding statistical processing of the measurement results for already examined cars with mechanical gearboxes [1], taking into account the practical "zeroing" of the readings of the infrared gas analyzer in CO and CH in all modes, allowed us to obtain an extremely simplified method for diagnosing a car in accordance with the environmental class in the mass of NO_X emissions during measurements only for idle modes, followed by an assessment of the concentrations of NO_X and CO_2 for sites 40, 60 and 80 km/h, according to toryh final evaluation conducted mass emission in km.

Despite the negative level of these indicators, foreign publications, in particular those reflected in the Internet resource [5], confirm the practical unsolved problem of the operational ecology of diesel cars in relation to NO_X emissions (some of them reflect competition in the sales markets [6] and may lead to erroneous conclusions). It is noted that studies conducted by the International Clean Transport Council (ICCT) in October 2014 showed an average seven-fold excess NO_X pollution for EURO-6 vehicles (0.56 g/km instead of 0.08 g/km). At the same time, it is noted that diesel cars (LDDV) in the European Union make up more than 50 % of all new brands and are one of the main sources of NO_X air pollution, and in order to reduce emissions, it has been recommended since 2017 to automakers to carry out certification except for laboratory on dynamometer stands, real road tests using portable emission measurement systems (PEMS). Thus, the EEC technical regulation excluded the possibility of any direct NO_X measurements for passenger diesel cars in operation (using five-channel infrared gas analyzers), which, in our opinion, was and will be the main reason for the multiple exceeding of the content of this most aggressive component compared with EURO standards, and ongoing fraud regarding the actual level of environmental safety for Russian users (in Europe and the USA, a similar process with the name "Dieselgate").

At the same time, foreign publications using certain methods make attempts to assess the consequences of exceeding NO_X emissions of diesel cars (in Europe, their emissions are considered the main reasons for exceeding NO_X limit values in the atmosphere) to public health and nature [7–9]. Most impressive in this regard are the collective data cited in publication [9], although the authors note the conditionality of these estimates regarding the relationship between NO_X and health. They noted that the sharp increase in the share of LDDV sales in the UES since the late 1990s was associated with an increase in the cost of gasoline and a potential reduction in CO_2 greenhouse gas emissions compared to similar gasoline engines. Although the main danger from diesel cars is particulate emissions (more than 400,000 deaths and 15,000 premature deaths were associated with excess concentrations of particulate matter and ozone), it is estimated that about 3.5 % of premature deaths from particulate matter (PM, the amount of which increases with increasing smoke), and 2.3 % of the effects of ozone can be attributed to NO_X emissions from LDDV, and 80 % of deaths could be avoided if diesel emissions corresponded to the level of gasoline cars. For Germany, for 2015, 6,900 premature deaths from excess LDDV emissions were calculated, and another 4,650 premature deaths from the effects of NO_X emissions from heavy trucks and buses. According to [9] estimates, Italy, Germany and France accounted for two-thirds of premature deaths in the EU-28 + from excess NO_X LDDV emissions, with a specific rate (from 2.85 to 4.4 per 100,000 inhabitants) per 40–140 % higher than the EU-28 + average (1.8 cases per 100,000 inhabitants). However, it is noted that for some countries (Belgium, the Netherlands, the United Kingdom) a decrease in NOX emissions led to a decrease in ozone production, and an increase of 70–90 cases of deaths associated with the influence of ozone. Also, a generally insignificant effect of NO_X emissions on the deterioration of freshwater quality (eutrophication) in the Po Valley and in the western parts of France was noted.

As for buses (and heavy trucks), the current technical regulations (in Europe and Russia) practically exclude any kind of control of NO_X emissions in operation, since the environmental certification of these cars, unlike cars, is not carried out using specialized dynamometer stands, and is carried out by testing a new engine mounted on a motor stand in the laboratory (UNECE Regulation R49) [3]. At the same time, 13 modes of the so-called 13-step European cycle specialized in speed, load and time are simulated, and the results for CO, CH, NO_X and PM are given (and normalized) in terms of g/(kW·h), which are difficult to compare with operational tests by the car itself (on running drums or in road conditions). Obviously, for the conditions of large megalopolises in Russia (primarily Moscow), the negative consequences of NO_X emissions can be several times greater than the above average values for the EU-28 + from light diesel engines, and the level of emissions of this component for buses (and especially the heavier trucks, access to which on the MRHW is possible only at night) also have a significantly higher value. This makes it expedient to develop additional methods for estimating NO_X emissions for buses (and heavy trucks) for operation and standardizing them to reduce negative consequences, since there are no other options at present.

3. The results of experimental studies

At the initial stage, we carried out the simplest (for a preliminary assessment of the level of environmental safety) measurements of NO_X (and CO_2) emissions at the minimum engine crankshaft rotational frequencies and smoke free acceleration indicators for 14 LBP (Likinsky Bus Plant) city buses. Such measurements are currently not difficult to carry out even in the field (outside the production and technical bases of motor transport organizations), however, this requires the use of a five-channel infrared gas analyzer for CO, CO_2, CH, NO_X and O_2 (instead of the four-channel one provided by GOST R 52033-2003 for gasoline-new cars). Such gas analyzers are produced (on a four-channel platform) upon the submission of a special application by several Russian manufacturers (the necessary information is available on the Internet).

The measurement results are given in the electronic table of the STATISTICA-12 program (Fig. 1), using the modules of which the necessary processing [2] was carried out. First of all, the presence of a statistical relationship was determined, which reflects the degree of influence of the engine's technical condition in terms of smoke on NO_X emissions, for which the graphic module of the program selected a polynomial model of one-factor regression, shown in Fig. 1 in a separate window with the obtained scatter of experimental data (the model is shown in the upper part of the graph window). Already from the graphical representation, one can see the lack of correlation between these main indicators, and additional mathematical processing confirmed their independence (the Fisher test, reflecting the informativeness of the regression dependence, had an estimate of F=0.6 at a significance level of $p<0.56$, and the determination coefficient, reflecting the degree of variation of the experimental data from the polynomial model was an insignificant value of $R_2=0.1$). Thus, the NOX emission indicator during engine operation at minimum engine speeds is an independent diagnostic parameter, and it must be used to identify the need for maintenance of city buses, in particular, to improve the functioning of the exhaust gas recirculation system, on which NOX emissions, and to test the effectiveness of some operational impacts, such as urea additives in diesel fuel, etc.

A more aggressive level of NOX emissions by buses compared to cars (more than double) can be seen from the histograms of their scatter in Fig. 2 (indicators for cars are taken from work [10]). The scattering of the value of this indicator (when processing the results of the STATISTICA [2] program) can be written by the normal distribution law with an average value of 635 ppm (ppm) and a standard (mean square) deviation of 131 ppm (ppm), with its high the probability of consistency with experimental data (according to the Kolmogorov-Smirnov criterion, the value of D, showing the maximum value of the modulus of the difference between the statistical function obtained from the experiment, the distribution function and the corresponding theoretical function, was 0.112, which sponds to the probability of consistency $R_{sogl}> 0.964$). From fig. 2 this scattering range only covers the emission value for the Volkswagen passenger car, which in the USA fell under the Dieselgate process [6, 8]. However, from the data in Fig. 2, it can be seen that, in general, cars and even heavier trucks and buses are still far from the actual implementation of the declared EURO standards, and for this, manufacturers may require new, more radical and easily controlled in operation, approaches in comparison with applicable now. Therefore, for practical purposes, it is currently reasonable to use the intermediate value of the diagnostic parameter as the lower limit of the 90 % limit of the theoretical distribution of the NOX emission indicator at idle (in the STATISTICA program a 10 % percentile), which amounted to 530–540 ppm (ppm); it is possible to obtain a reduction in environmental pollution of the order of 20 %.

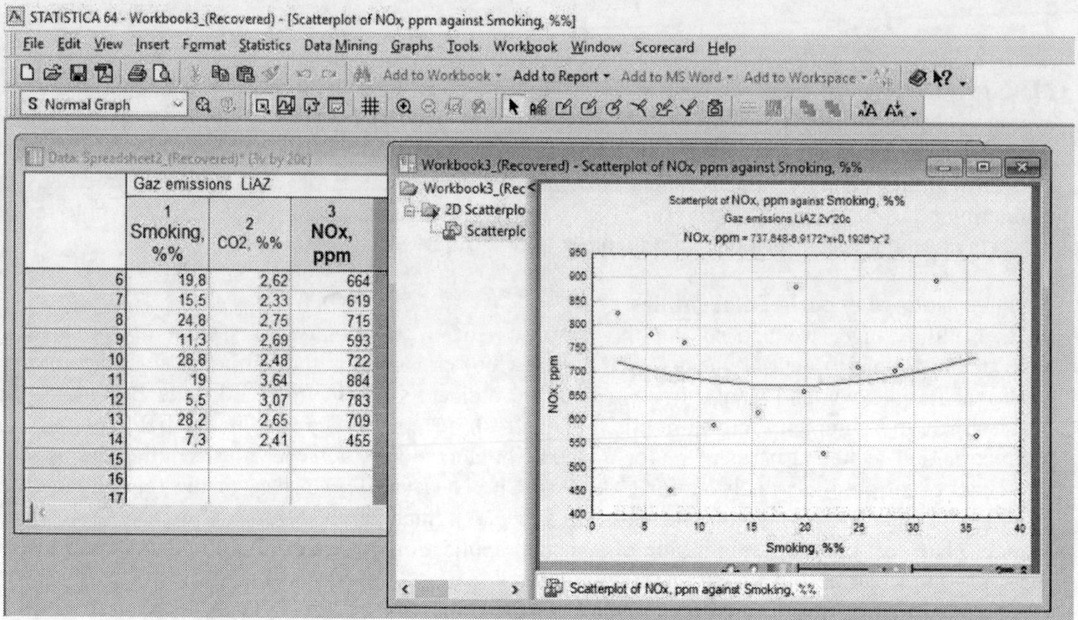

Figure 1. STATISTICA-12 program window with the results of parallel measurement of the content of nitrogen oxides NO_X (and CO_2) in the exhaust gases of 14 city buses LBP (Likinsky Bus Plant) of ecological classes Euro-4-5 when the engines are idling and smoke levels on the acceleration mode of free acceleration of the engine crankshaft without external load, and the obtained schedule of the proposed regression, reflecting the tightness of the relationship between these indicators

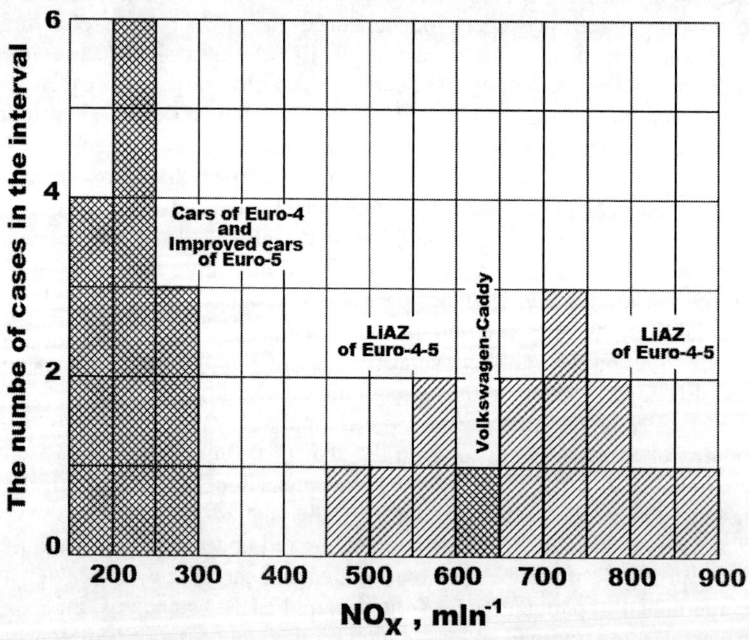

Figure 2. Histograms of the distribution of NOX nitrogen oxide in the exhaust gases of 14 cars and vans with diesel engines of the declared environmental classes Euro-4-5 at the nominal engine speed at idle speed, and for 14 city buses of LBP (Likinsky Bus Plant) declared environmental classes Euro-4-5

IOP Conf. Series: Materials Science and Engineering **832** (2020) 012001 doi:10.1088/1757-899X/832/1/012001

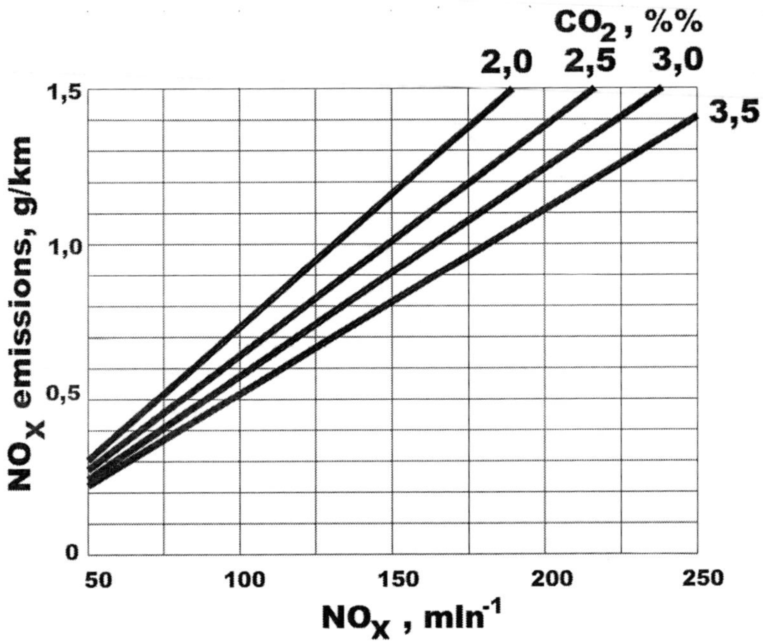

Figure 3. Nomograms for the rapid estimation of the mass of NOX emissions (in g / 1 km for a car with a 1-liter diesel engine, subject to its European urban driving cycle test) depending on the content of NO_X and CO_2 in the exhaust gases when idling [10]

For a physical estimate of the amount of emissions, we can use the nomograms calculated by us to convert the NO_X values at idle to the mass emission indicators when tested according to EURO standards (in g/km) for a conventional passenger car with a 1 liter diesel engine, according to which it is not difficult to carry out practical assessments (Fig. 3 [10]). Since the actual values of NOX emissions for buses are more than three times the range of Fig. 3, when calculating, simultaneously taking into account the working volume of the bus engines (about 6 liters versus 1.3–2.2 liters for cars and vans) it is possible to proportionally increase the results obtained from the nomograms of Fig. 3, the more so since the range of CO_2 indicators for buses (Fig. 1) coincides with the nomograms. (As noted above, the latter were obtained without taking into account the high-speed test section and therefore practically correspond to the specifics of using city buses, for which in the future it will be possible to develop more sophisticated operational monitoring methods on unloaded running drums of an extremely simple design compared to cars) Although as a result of such calculations extremely negative indicators (about 20 g/km) are obtained, it can be argued that with real road measurements, as provided for in the certification of diesel cars in the EU-28 +, they will turn out to be even worse, in view of significantly more tense parameters of traffic flows at the entrance of cars to megacities, especially to Moscow [11].

Another important environmental indicator, which has recently been given increased attention due to climate change, is the problem of reducing greenhouse gas emissions, primarily CO_2, from automobile transport (affected in work [8]), while In our opinion, targets were given unreasonable preference due to the reduction of these emissions. In particular, in recent foreign developments of portable means for continuous monitoring that are built into a car (the ELM327 adapter, connected to the OBDII universal diagnostic connector and transmitting data to mobile phones), CO_2 emissions in g/km are provided (Torque application is installed via the Internet for a small fee), although for a more important NOX indicator, such control is absent due to its complexity. When reducing the CO, CH, and NOX emission indicators when testing on unloaded running drums [1] to the EURO standards, similar calculations were also carried out for CO_2 emissions measured by a five-channel infrared type gas analyzer to estimate the current values of excess air coefficients α for different test steps. At the same time, at 3 sites of constant load mode (at speeds of 40, 60 and 80 km/h in direct transmission), the total CO_2 emissions in g / h and fuel consumption in kg/h were easily determined. The ratio of these two indicators both for gasoline and especially for diesel cars always led to the same value – 1,500 g of CO_2 per 1 kg of spent fuel (with a maximum error of 10 % in the direction of decrease with the maximum enrichment of the mixture for carburetor engines). This was due to an almost insignificant difference in the ratios of carbon content (approximately 85–86 %) and hydrogen (12.5–13 %) for all types of hydrocarbon fuels, including kerosene, as well as compressed (methane) and liquefied (butane -propane) of natural gases, for which this ratio is somewhat different – approximately 82.2 % and 17.85, respectively, but falls within the above 10 % range. Accordingly, in operation it is extremely simple (and clear) to estimate CO_2 emissions by indicators of car efficiency – fuel consumption in liters per 100 kilometers, recalculating taking into account the density (800–845 kg/m^3 for diesel fuel, 770–780 kg/m^3 for gasoline and 550 kg/m^3 for liquefied gas. Thus, the above technology (in conjunction with smoke testing) will provide in operation a fairly complete control of the environmental safety of city buses and stimulate new technical and technological directions for its increase.

4. Conclusion

The carried-out research phase revealed the need to use an additional diagnostic parameter – NO_X emissions – to strengthen control over the level of environmental safety of city buses in operation. A lower level of environmental safety (by 2–3 times) was confirmed for this indicator compared to diesel cars that still do not fit significantly into the EURO standards (practical excess of 7–10 or more times), and the expediency of operational monitoring the technical condition of buses at present by the NOX content in the engine exhaust gas at idle. Based on the statistical approach, the maximum permissible value of this indicator is substantiated – no more than 530 ppm (ppm), the maintenance of which during maintenance will reduce environmental pollution by about 20 %. Based on previously developed calculation models for recalculating the toxicity indicators of gasoline and diesel light-vehicle exhaust gases during their operational tests on unloaded running drums to test modes in accordance with EURO standards, an estimate of the mass of CO_2 greenhouse gas emissions based on car economy indicators is justified – the total fuel consumption in liters when carrying out transportation, or specific consumption per 100 km, taking as a basis the theoretically established value: 1500 g of CO_2 per 1 kg of consumed about fuel. Final calculations must be carried out considering the density of the fuel used: 800–845 kg/m^3 for diesel engines, 770–780 kg/m^3 for gasoline and 550 kg/m^3 for liquefied gas.

References

[1] Boldin A P and Akstnov P V 2017 Vehicle diagnostics for compliance with EURO emissions standards *Vestnic MADI* **1(48)** 27–34
[2] Boldin A P and Maksimov V A 2014 *Foundations of research studies* 2-nd ed, revis and suppl (Moscow: Akademia) 336
[3] Erokhov V I 2013 *Toxicity of modern vehicle: metods and tools of reduction of harmful atmospheric emissions* (Moscow: Forum) pp 447
[4] *Euro emissions standards* Retrieved from: http://www.stroyten.ru.
[5] *Euro emissions standards* Retrieved from: http://www.avtonov.info > normy-evro
[6] Clemente M and Gabbionets C 2017 How Does the Media Frame Corporqte Scandals? The Case of German Newspapers and the Volkswagen Diesel Scandal *J. of Management Inquiry* **26(3)**, 287–302
[7] Chossiere G P, Malina R, Ashok A, Dedoussi I C, Eastham S D, Speth R L and Barrett S R H 2017 Public Health impacts of excess NO_X emissions from Volkswagen diesel passenger vehicles in Germany *Environmental Res. Letters* **12(3)** Retrieved from: http://hdl.handle.net/1721.1/107400.

IOP Conf. Series: Materials Science and Engineering **832** (2020) 012001 doi:10.1088/1757-899X/832/1/012001

[8] Hoekman S K 2016 Comment on "Damagers and Expected Deaths Due to Excess NO_X Emissions from 2009 to 2015 Volkswagen Diesel Vehicles" *Environmental Sci. & Technol.* **50(7)** 4135–6

[9] Jonson J E, Borken-Kleefeld J, Simpson D, Anyiri A, Posch M and Heyes C 2017 Impact of excess NO_X emissions from diesel cars on air quality, public health and eutrophication in Europe *The Lancet Planetary Heart* pp 3–11

[10] Boldin A P, Sarbaev V I and Aksenov P V 2017 Diagnostics of passenger cars and minibuses with diesel engines for compliance with Euro emissions standards *Int. J. of Mechanical Engineer. and Technol.* **8(12)** 933–43

[11] Malinovsky M, Solntsev A, Juravleva A and Macovsky S 2017 Estimation method of monresident vehicle fleet iflow ifluencing road traffic safety in megalopolis *Transportation Res. Proc.* pp 751–5

On the question of the rational behavior of firms in the market of technical services in the field of the autogas system

V A Egorov, E N Bespalchaya, Yu V Panov and T N Egorova

Moscow Automobile and Road State Technical University (MADI), 64, Leningradsky ave., Moscow, 125319, Russia

E-mail: onti@nich.madi.ru

Abstract. The article presents new approaches to formalizing and evaluating the actions of market participants in technical services in the field of installation and maintenance of the autogas system in order to determine the rational boundaries of a number of characteristics of the behavior of these firms-players in the market. Currently, a wide range of gas equipment is produced for installation on cars, which has a wide range of consumer properties. This diversity makes it difficult for consumers to choose these products, being faced with the task of determining the nomenclature of models of autogas kits, their required number depending on various input parameters – the market environment in which the considered service station operates, the power of the service station focused on the gas market segment, etc. To solve the above problems, it is proposed to evaluate the market potential of a particular kit of AGS, goods offered to the consumer-owner of the car, which is an essential characteristic that determines the priority of inclusion (or exclusion) of a set in the gas equipment nomenclature at a particular service station. To determine the predicted effectiveness of certain decisions, the article proposes the formalization of the described market processes and the construction of a mathematical model of the market participants' behavior depending on different influence factors, the behavior of the components included in the complex objective function and the creation of an algorithm and software for finding rational solutions.

Keywords: car, autogas system, market potential, objective function, market behavior, vehicles.

1. Introduction

In recent years, in many countries of the world gas motor fuel (GMF) is gaining more and more popularity. The fleet of gas tank vehicles (GTV) is replenished with vehicles in which the autogas system (AGS) is installed directly at the vehicle manufacturing plant. But in most cases, AGS is installed after the vehicle is put into operation. This type of replenishment of the GTV fleet will be the main one for a long time [2].

A wide range of AGS is produced, which has a big set of consumer properties [1, 4].

Historically, two production strategies have developed on the global AGS market.

1) Plants produce the entire set of equipment.

The following manufacturers can be mentioned as an example:

- BRC, Lovato (Italy), A C (Poland)

In the case of this strategy, the manufacturer makes a profit from each element of the AGS system, however, the plant needs to invest in the development and modernization of equipment, as well as to store and deliver products.

2) Suppliers of sets of AGS place the production of most components at third-party manufacturing plants, and from the resulting components assemble a certified set [3]. The following companies can be mentioned as an example:

- Digitronic, Poletron, LPGtech, Elpigaz.

This strategy allows companies to diversify their risks, which depend both on the company's dependence on one supplier and on the loss of "dead" capital invested in the development and adjustment of the technology for the production of many components, if changes are necessary in the product line.

Often during the life of a product on the market there are periods when ups and downs of quality occur. Plants of the first type, working on a full production cycle, have to pass all periods of release of goods. But suppliers of the second type, such as Digitronic, can create a kit by choosing components from a wide range of elements, selecting them at the peak (on the rise) of quality. Thus elements on decline of quality are excluded from a kit.

Also, there is a tendency in the modern AGS market in which companies and plants merge under the leadership of one large concern. At the same time, companies retain a certain functional independence and do not lose their technical uniqueness (Fig. 1).

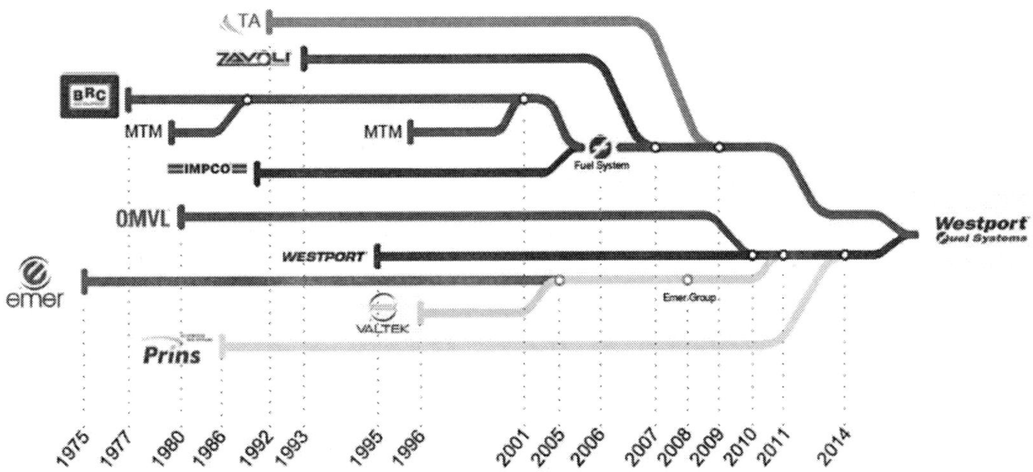

Figure 1. The scheme of merging companies in the Westport concern

As a result, several dozen manufacturers of AGS and its components operate on the European market. This diversity makes it difficult for consumers to choose these products. Under the consumer we can understand a large range of both legal entities and individuals. In general, they can be divided into two groups. The first includes installers (as part of service stations) who buy kits or components from a manufacturer or distributor, perform both installation of gas equipment, and maintenance and repair. The second group includes directly the owners of vehicles converted to work on the GMF. Both of these groups of consumers are customers, although at different levels.

All of the above shows the relevance of the question of determining the nomenclature of models of AGS kits, the required amount depending on various input parameters – the market environment in which the considered automobile service station(ASS) operates(or will operate), on the power of the service station (or service station units focused on the gas market segment), etc. [5].

2. Approach to construction a mathematical model of effective resource management

The purchase or market potential of a particular set of gas equipment, the goods offered to the consumer-owner of the car, is a significant characteristic that determines the priority of including the kit in a set operated by a particular service station, or, conversely, excluded from the supported nomenclature [3]. The analysis of opinions, behavior of market participants gives the chance to predict not only whether or not there will be sales in this direction in the surveyed region, but the economic efficiency of forming one or another package of offers from the service station to customers – car owners [4]. To determine the predicted effectiveness of certain solutions to the above problem, it is necessary to formalize the described market processes and build a mathematical model of market participants' behavior depending on the identified influence factors, the behavior of the components included in the complex objective function and the creation of an algorithm and software for finding rational solutions.

As a result of creating a mathematical model for making a decision on optimizing the AGS nomenclature, which one or another service station supports, a so-called objective function is prepared that takes into account various cost components, both plus and minus.

The objective function determines some financial indicators. Trying to determine the rational set of sets of AGS (products) that will be stored, installed, repaired at the automobile service station (and the ability to work with these sets will be supported), apparently, it is necessary to consider the funds remaining as a result at the service station, i.e. profit. This may be, for example, specific profit per unit of time (day, year ...), and, possibly, per one post. At the same time, it should be borne in mind that the developed rational set of kits will depend not only on the distribution of interests of the customer's market base for those factors that are selected as basic (in our case: reliability, functionality, engine power serviced by AGS, price – are considered as basic), but also on the power of the ASS (or the element of the ASS that deals with AGS). And therefore, obviously, different recommendations can be developed for different capacities of ASS. / customer's gain in costs per 1 km. in this approach can be manifested and taken into account as the degree of the client's desire to switch to gas (and, accordingly, in the probability that the client will do this)/.

What components should be taken into account in the full objective function of the model?

First, this is the income from the installation of AGS sets on automobiles, then – the income from the repair of AGS, from the maintenance of AGS, as well as the expenses spent on the storage of all sets of kits (and their components) at service stations, installation, and periodic training of personnel, associated with a change in the produced nomenclature of AGS, with a change in software, advertising costs, etc.

However, in this article we consider the formation of only the first component-revenues from the installation of AGS. /This is the basic component, since both the current fleet of gas equipment already in operation in the surrounding region will need to be maintained and repaired, and increasingly those cars that we (the service station under consideration, together with competitors, of course) will saturate with gas equipment in the course of our work/.

When determining income, it is necessary to understand from whom we receive it, i.e. to determine the market base of consumers of the considered type of work, which can become our customers.

We accept the following conditions.

1) Our clients are mainly those who live, work, i.e. are mainly located in some adjacent region. Transients, of course, can call in, especially if repairs are necessary, and this is taken into account in the full model.

2) Our (considered) service station is not unique, so there are other stations in the region that operate to intercept our client in the event of our mistakes – insufficient production capacity, reduced quality, etc. Work in a competitive environment, in market conditions, is taken into account by probabilities of various kinds.

3) We accept that all the functional dependencies mentioned below are known to us.

So, we will analyze the "behavior" of the client base in order to determine the "market potential", i.e. the number of customers who can contact us.

Let's assume that there is a probability G0 that the client generally wants to install AGS on his car, and a probability G1 that he wants to install HBO or repair it at our station. These two probabilities depend on the region, on (i.e. the development of the situation in time), on advertising, on the breadth of options presented for the client (the more options we have, the sets of AGS, the more likely the client will reach us, – all other things being equal or almost equal conditions), etc. These probabilities are included in the expressions of the objective function after formalizing each of the components separately.

In the meantime, we consider how the probabilities of a "customer reaction" are formed depending on the parameters characterizing a particular set of AGS.

We assume that the parameter of the supported engine power is not further considered in conjunction with the others, because according to this parameter, customers were initially separated, having cars with different engines and require different (on this parameter) gas equipment.

Let us know the following probability distribution (or, more simply, the distribution of the number of customers) according to their requirements by two parameters, the meaning of which – for any combination (x,y) probability ρ(x,y)*dxdy is the proportion of customers who, starting with this parameter value and higher/better (on each axis), allow themselves to purchase (and install) a kit of AGS (hereinafter – the product, goods). At the same time, we do not consider the cost at this stage, putting in this function the sense that with the worst quality in terms of the parameters the client does not want to purchase the goods (although, maybe, he will take it at a bargain price, see below).

Let this distribution look like this:

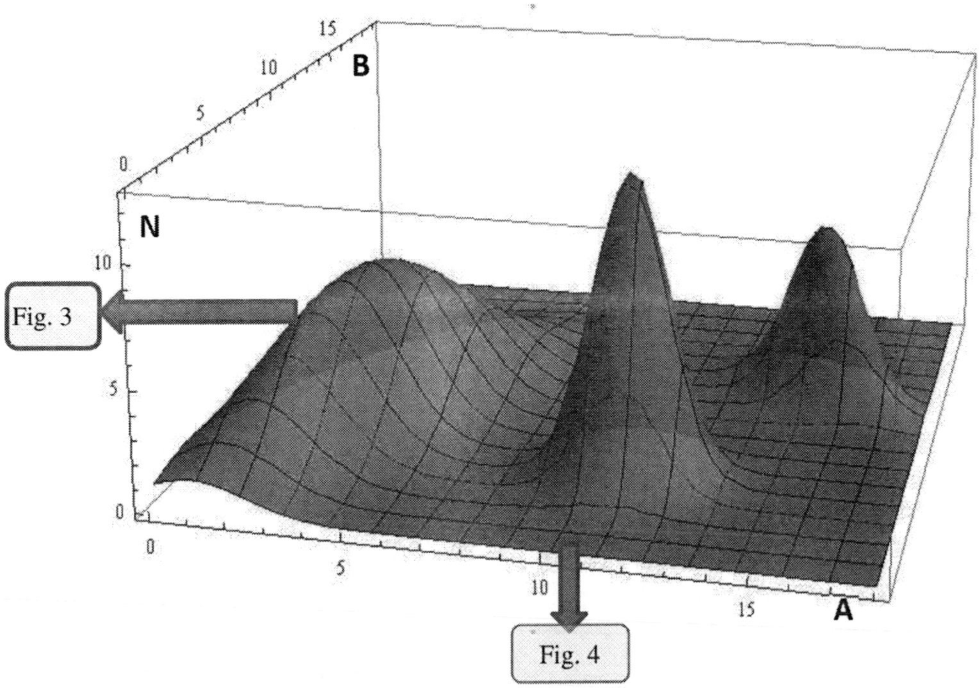

Figure 2. Example of the distribution of customers according to the level of requirements for the quality of goods by 2 parameters

/ Hereinafter the figures – with conditional parameters and conditional values /

Here, along the X axis (in front of us) is parameter A, along the Y axis (away from us) is parameter B, and along the Z axis is probability (number of customers).

Further, for the convenience of visual presentation and explanation, we will consider the dependence on only one factor.

To get distributions separately for each factor, we simply take the integral separately along each axis.

As a result of integration along the X axis, we obtain the distribution of the parameter B (from Y):

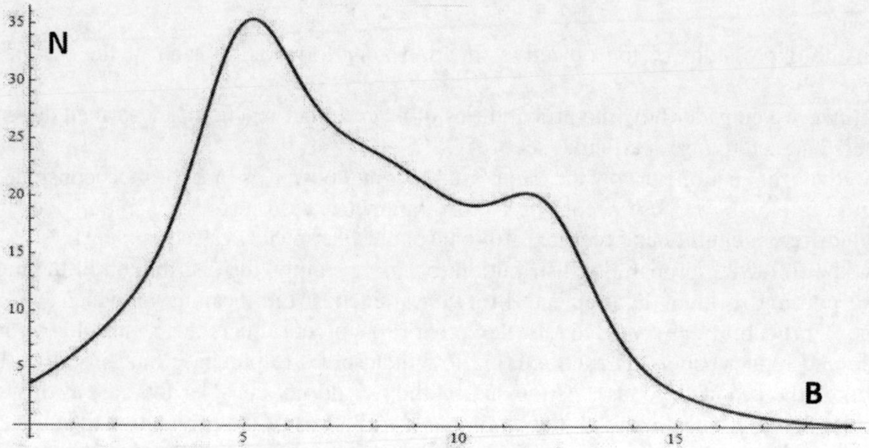

Figure 3. Distribution of customers on the parameter B

As a result of integration along the Y axis, we obtain the distribution of the parameter A (from X):

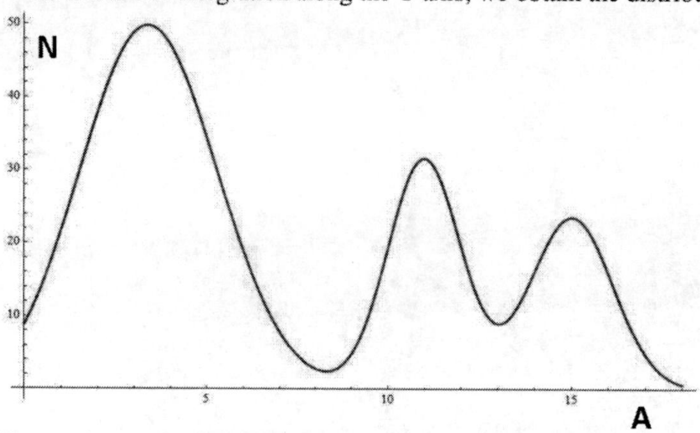

Figure 4. Distribution of customers on the parameter A

With this distribution of clients N(A) we will continue to work – again, for the convenience of visualization (let's assume, for example, that these are the client's requirements for one of the main components of quality – reliability, mean time between failures on a certain scale).

Let's ask a question how to describe client behavior depending on values of the considered parameters?

First, there are claims of the client, or his requirements to goods (further on all figures this axis will be denoted by the letter "W"). The approximate meaning of this – the quality below the W_X I (the client) do not consider. (The variant of distribution together by two parameters of this value is shown in Fig. 2, separately – in Fig. 3, 4). However, this is quite conditional, the customer can say it to himself, in fact, the customer can purchase goods below The W_y border, if the price is low, or in the mood – with a certain probability. So, the probability of making a decision is also affected by the price. In a survey of customers, they can specify as boundary two prices – one at which I will definitely buy goods of such and such quality, and the other, above which I definitely will not buy goods of such and such quality. Such deterministic customer behavior is shown in Fig. 5a. Here on axes: W-claims of the client, K – quality of goods, C – the price of goods. The green surface is the upper allowable price for a customer with W_y claims for a product with K_x quality. The lower violet surface is the boundary of the 100 % acquisition decision. All this is subject to the availability of only one product offered by our (or some other) service station.

IOP Conf. Series: Materials Science and Engineering **832** (2020) 012002 doi:10.1088/1757-899X/832/1/012002

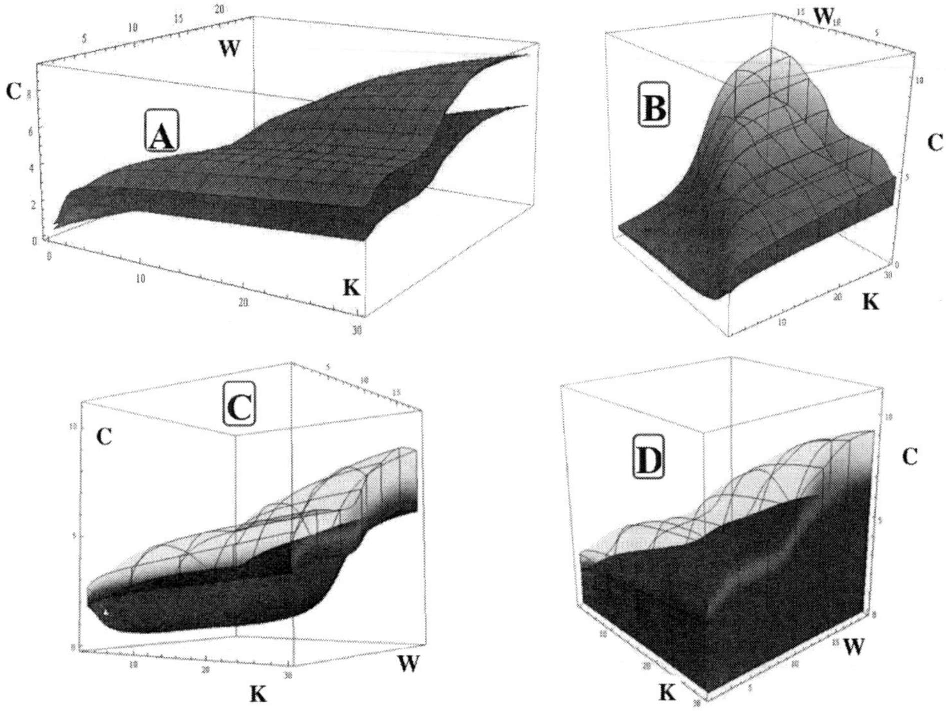

Figure 5. A discrete and probabilistic graphical description of the client's behavior when deciding on the purchase of goods (installation of a kit of gas equipment)

However, as already mentioned, the behavior of the client mass is not strictly determined, and is more correctly described by probabilistic functions. The representation of such a function is given in Fig. 5b, 5c and 5d. These functionals are based on the same functions that are shown in Fig. 5a, but with the distribution of the probability of making a decision between the green and purple surfaces (also above and below them). In Fig. 5b, the probability value is determined by the transparency of the corresponding point (it looks like a more or less faded blue color, the lighter the lower the probability), in Fig. 5c – from another point of view, color across the palette of the visible rainbow – from violet (highest value) to red (lowest value) and with graduation transparency. Figure 5d shows a complete representation of the behavior of the probability of making a decision on the level of claims, on the quality of the product and its price – with the addition of a zone of almost 100 % probability (bottom blue part).

Fig. 6a – 6c show sections of the obtained object P (K,W, C) by different planes. The intersection of planes with a 3D object-graph represents the behavior of the probability of a decision:

a) for a given quality of goods, the dependence of the price boundaries on the level of customer complaints, and the behavior of the probability of making a decision on the combination of the level of claims and the value of the price (cross section by the green plane, Fig. 6a);

b) for a given level of the client's claims, the dependence of price boundaries on the quality of the goods, and the behavior of the probability of decision-making on the combination of the quality of the goods and the value of the price (cross section with a raspberry plane, Fig. 6b);

c) at a given price level, the dependence of the boundaries of acceptable product quality level on customer's claims and behavior of the probability of the decision from a combination of quality of product and level of customer's claims (the yellow section of the plane, Fig. 6c).

/The named dependencies are visible by the behavior of the grid lines and by the slight darkening of the section planes./

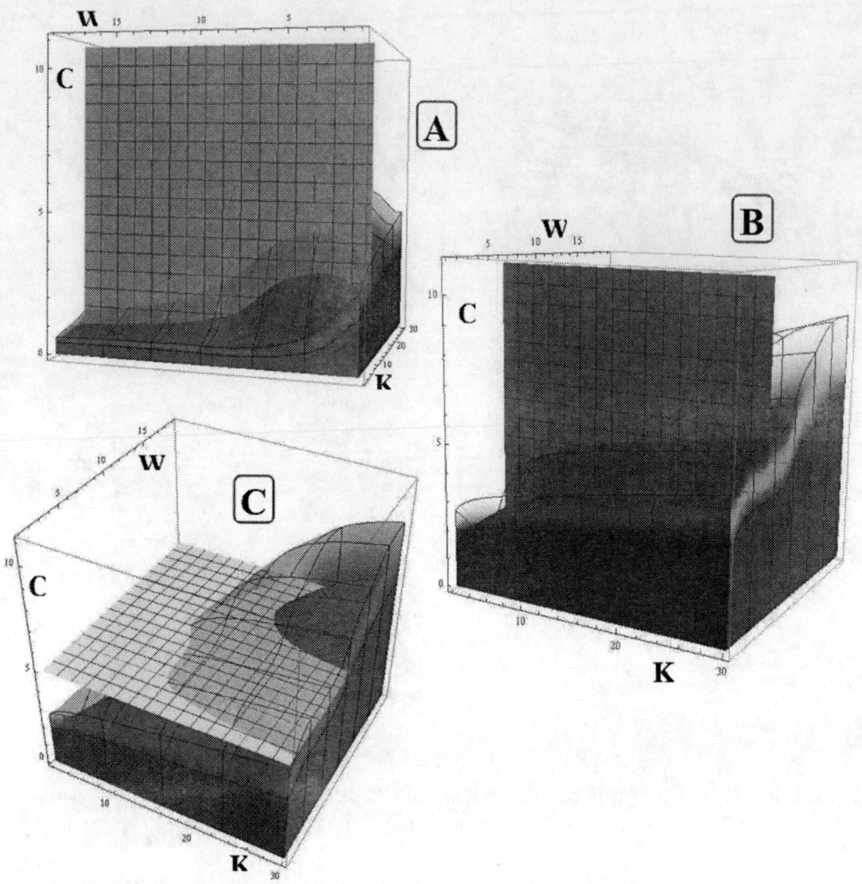

Figure 6. Cross-sections of the function P (K, W, C) by planes of constant quality (A), constant level of quality requirements (B) and constant price (C)

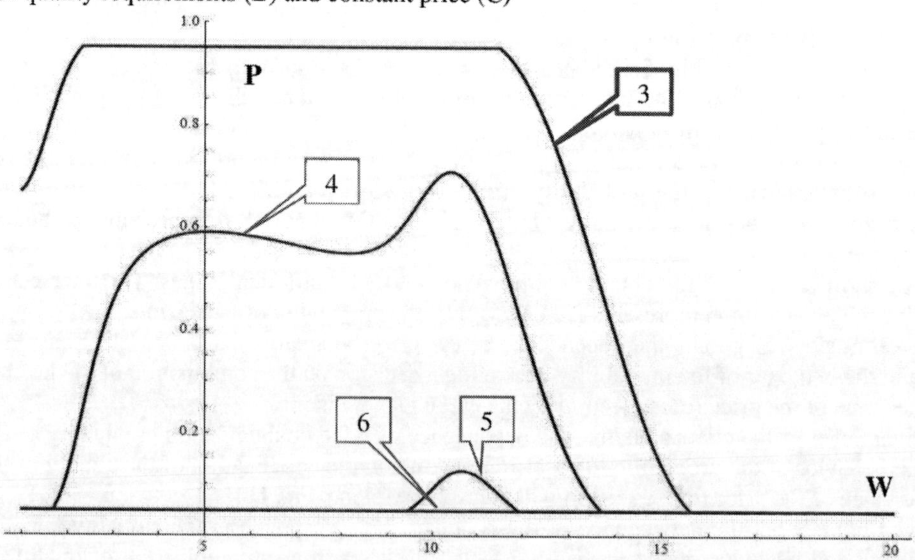

Figure 7. The possible behavior of the probability of making a decision to purchase a product at a fixed level of quality, depending on the level of customer requirements at different fixed prices

IOP Conf. Series: Materials Science and Engineering **832** (2020) 012002 doi:10.1088/1757-899X/832/1/012002

Separately (Fig. 7) determine the behavior of the probability of making a decision to buy a product of const. quality (here K=12 in the selected conditional scale), and with different fixed price of the goods (conditionally from 3 to 6, lower price – the graph above), when the abscissa axis is set to increase the customer's quality requirements, i.e. the minimum quality (or so), which client admits for himself by the parameter under consideration.

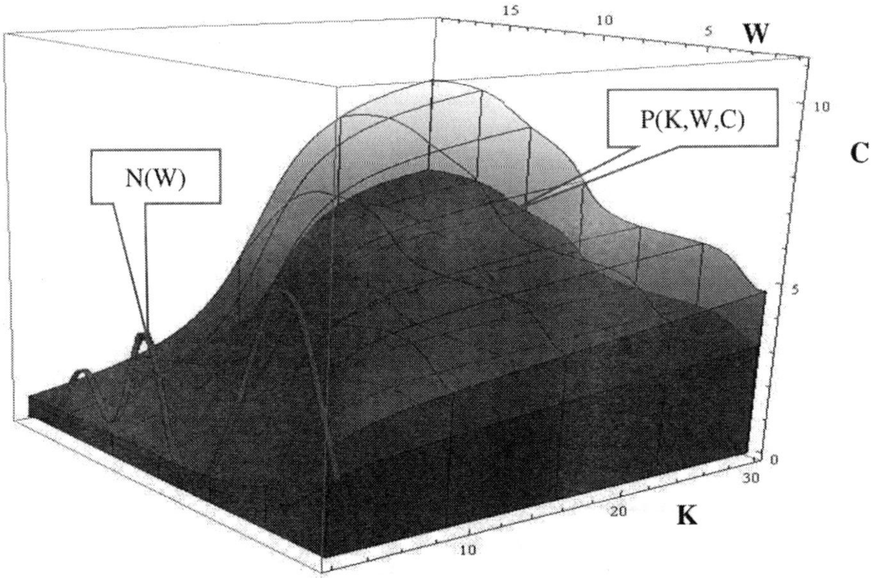

Figure 8. Composition of functionals P(K,W,C) and N (W)

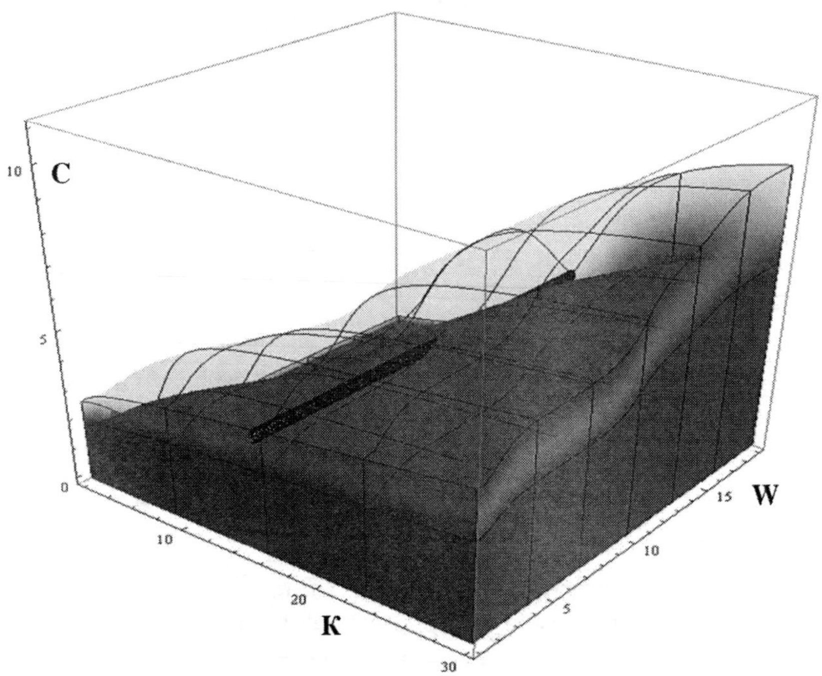

Figure 9. The position of the integration path (determined by the properties of the product) of the function P (C, W, P) with the weight N (W) along the claim axis W

It is evident that for a very low price for a product of this quality (not very bad), they will take it until the customer's claims for quality increase significantly, and even then there is probability that the customer will take the product for a low price. At a higher price, a client with low quality requirements (and, accordingly, with a lower upper and lower price acceptable for himself) will not take such a product at all, but with an increase in quality requirements (and the corresponding acceptable price limits for the client), there is a probability of taking the goods. It rises until the quality requirements from the client exceed the quality of this product (at a specific price), and then falls down with increasing customer requirements. For some price (here – 6) for such price of such quality nobody will buy goods – the probability goes to 0 regardless of the level of customer requirements.

Next, we proceed to determine the market potential of the product, taking into account the distribution of opinions, the behavior of the customer mass, expressed in terms of the probabilities that have just been considered (Figs. 5–7), and the distribution of the mass of customers by the quantity-how many of these or those we have customers by the level of claims (shown in Fig. 4). Together, the functionals P (K, W, C) and N (W) are shown in Fig. 8 in a different color representation. At the same time, the distribution function of clients by the parameter "P" is given by a red ribbon.

Figure 9 gives an idea of a representation of the probability of making a purchase decision, indicating here the axis (rod) that defines a particular product. This is the axis, the fixed coordinates of which are the cost and quality of the goods, along which this probability function is integrated along the "W" axis with a weight of N (W).

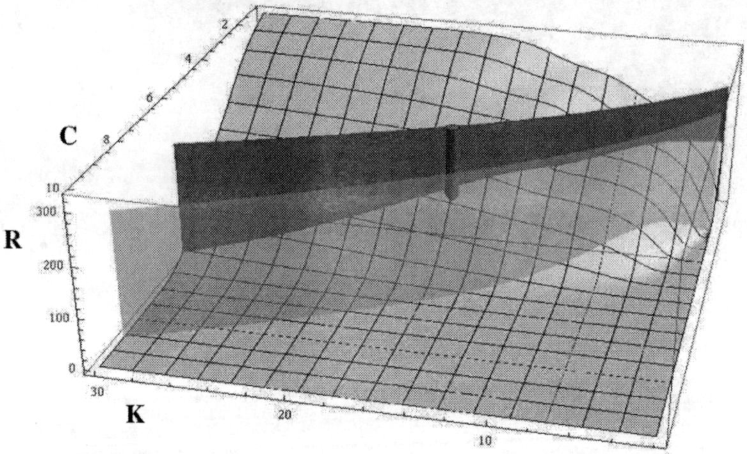

Figure 10. Illustration of the position of the image of a particular product (set of AGS) on the function of the market potential of the product in the region under consideration for the considered station

Carrying out such integration along the axis of the customer's claims level "W" for different combinations of the cost of C and the quality of the goods, we obtain the dependence function shown in Fig. 10. It represents the market potential R (on a conditional scale, as mentioned above) for the goods of value C and quality K, i.e. how many customers from the considered area, the region may well purchase such goods. / It should be said that the potential is an opportunity stretched over time, but using the average time that client makes a decision on the purchase of goods, we can reach the flow of applications for work per unit time, which will allow us to further represent the generated objective function completely in the form of time-specific revenues and costs and optimize the desired solution./

Also in Fig. 10 added (for illustration) approximate boundaries within which a combination of product characteristics {K, C} can be real. Borders are crimson (lower price) and yellow translucent (upper price) color. To all this is added a green label-cylinder, marking the position of a particular product (here the product with the same coordinates as in Fig. 9).

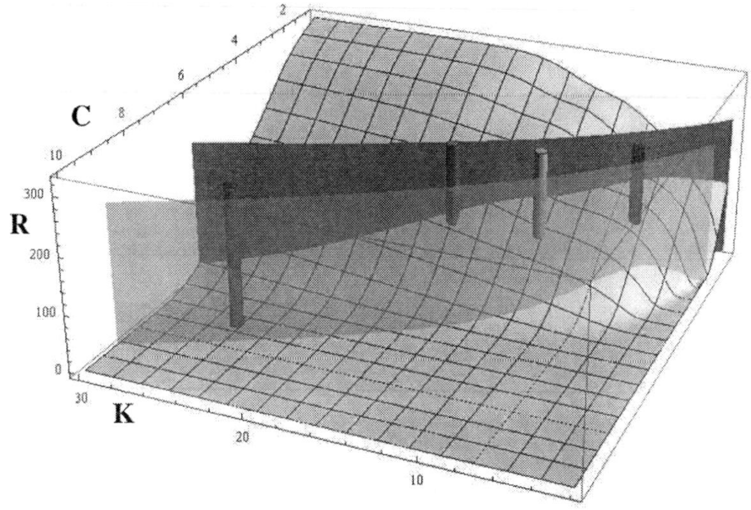

Figure 11. The combination of images of several products on the function of market potential

So, let the service station have several types of goods (kits of AGS) offered to customers. In Fig. 11 they are shown in the same environment as in Fig. 10, several labels-cylinders (given in different colors). However, the total market potential in this case will not be equal to just the sum of the potentials of these four, taken separately.

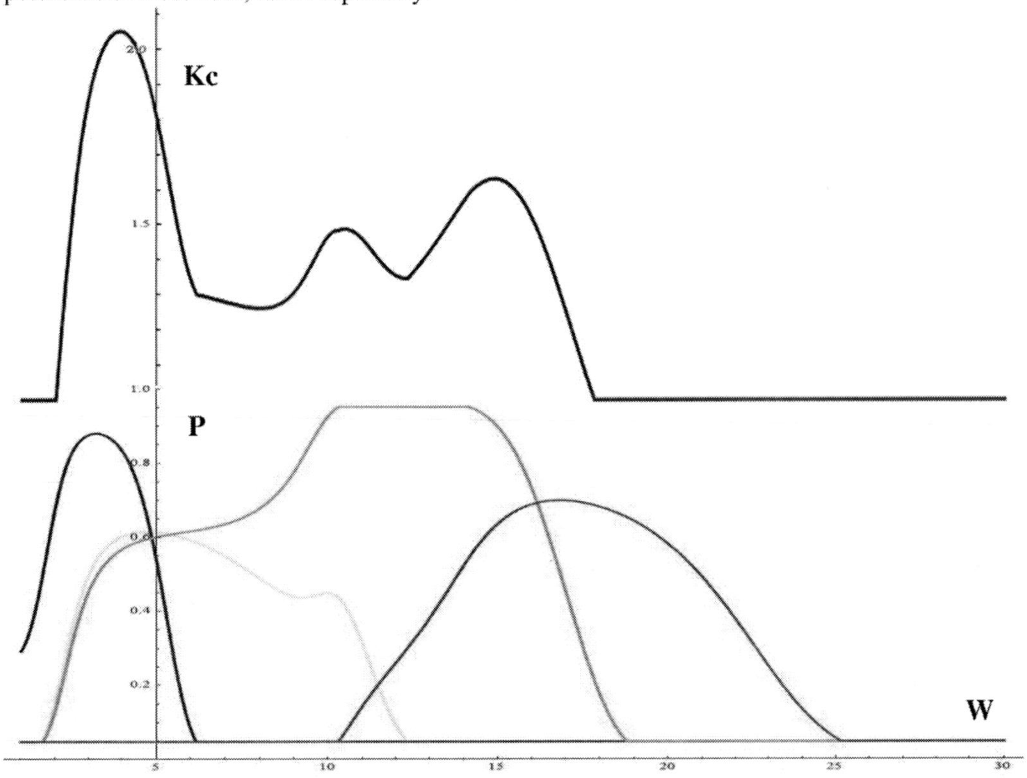

Figure 12. Illustration of a combination of probability functions for several products depending on the level of customer claims and the formed correction factor

We consider the behavior of the probability of a positive decision on the purchase of goods by the customer of the product "No. i" depending on the level of customer claims. For the example of the four products in Fig. 12 graphs of these probabilities are given, shown in the corresponding colors. It is seen that for some values of "W" the sum of these graphs exceeds 1. However, this amount is the probability that the customer will buy something from us. It cannot be higher than 1, because according to the terms of the task, the client will install no more than one set on one of his vehicles.

Where the sum of the probabilities is less than 1, we accept that the interest of customers is distributed between different goods, and thus the probability that some kind of purchase will happen is increased and taken equal to the sum of the individual probabilities

And where the sum of probabilities is greater than 1, we correct all incoming functions in such a way as to reduce the value of the sum of probabilities to 1, i.e. the correction coefficient, by which we will divide the functions included in the consideration, is equal to the sum of these functions if this sum is greater than one, or is equal to 1 if the sum is less than 1. In Fig. 12, the color of the correction factor Kc is shown in black.

The correction performed affects the probability values only for specific goods and for each combination of goods (kits of AGS) it is considered separately, and after correction, the obtained market potentials can be correctly added. In the table 1 the values of the market potential for the considered 4 products before and after the correction are given.

Table 1. Calculation of individual and total market potential for the selected example

data before correction			
quality	cost	market potential expressed in objects	market potential expressed in money
6	3	122.3	366.8
11	3.9	104.3	406.7
16	4.2	116.0	487.3
25	6.9	1.5	10.3
Total		344.1	1271.1
data with correction			
quality	cost	market potential expressed in objects	market potential expressed in money
6	3.0	69,3	207,7
11	3.9	56.0	218.3
16	4.2	64.3	270.0
25	6.9	0.8	5.2
Total		190,4	701,2

Thus, the market potential of 4 goods represented at one service station is higher than that of any one of the components of this group, but lower than the sum of the potentials of any two separate goods (except for the most expensive and therefore with very low potential – within the framework of the above functions).

Going through the various options for sets of goods (kits), we calculate the resulting total market potential, and then the income received from this. When constructing the full model, it is also taken into account that the probability that the client will choose a specific reporting station increases with the number of types of products offered, at the same time, this increases the cost of storage, personnel training, etc. In addition, in the course of a full analysis, it is expected that recommendations will be formed on a rational set of AGS kits depending on the power (size) of the service station. For example, if the station is of low power, then it is not necessary to cover all groups / types of customers – the market potential of a small range of installed AGS, working for any one group, may be quite enough. Thus, depending on the size of the station, you can give different sets of recommendations depended of number and type of sets of goods/kits.

IOP Conf. Series: Materials Science and Engineering **832** (2020) 012002 doi:10.1088/1757-899X/832/1/012002

References

[1] Panov Yu V 2013 *Automotive gas injection systems* Textbook for Universities (Moscow: Third Rome)
[2] *Gassystemeinbauprüfung (GSP)* 2013 Handbuch für GSP – die Schulung der verantwortlichen Personen und Farkräfte 5 vollig uberarbiatete Aũuflage 5. (Bonn, Auflage Akademie des Deutschen Kraftfahrzeuggewerbes GmbH (TAK)) 256 s 1
[3] Panov Yu V, Bespalaya E N, Pronikov A N and Pavelko M A 2017 Analysis of application of technical regulations 018/2011 to make changes in the design with the installation of gas equipment *Coll. of sci. papers on the mater. of the 75th scie.-method. and res. Conf. "Problems of technical operation and auto-service of rolling stock of road transport"* (Moscow: MADI) pp 146–51
[4] Panov Yu V and Bespalaya E N 2018 Research on the significance of the properties of autogas systems for consumers *Coll. of sci. papers on the mater. of the 75th sci.-method. and res. Conf. "Problems of technical operation and auto-service of a rolling stock of road transport"* (Moscow: MADI) pp 162–5
[5] Bespalchaya E N, Zenchenko V A, Panov Yu V and Pavelko M A 2019 Improving the efficiency of the gas service on example of the Gas Point service station *Coll. of sci. papers on the mater. of the 77th sci.-method.l and res. conf. "Problems of technical operation and car service of rolling stock of motor vehicles"* (Moscow: MADI) pp 86–95

IOP Conf. Series: Materials Science and Engineering **832** (2020) 012003 doi:10.1088/1757-899X/832/1/012003

The introduction of unmanned automotive vehicles in Russia

N A Filippova, T E Melnikova and R V Litvinenko

Moscow automobile and road state technical University (MADI), 64 Leningradsky Ave., Moscow, 125319, Russia

E-mail: umen@bk.ru

Abstract. The article presents the modern concept of an unmanned vehicle according to the SAE classification. The features of use of unmanned transport, including its influence on traffic safety, efficiency of the organization of transportations and decrease in fuel consumption are considered. A brief overview of existing Russian developments in the field of unmanned road transport is given. The scheme of movement of an unmanned vehicle through the ice crossing using V2X technology is presented. The authors list the reasons of using unmanned vehicles on public roads.

1. Introduction
In today's world, unmanned technologies in transport are becoming increasingly common. Especially clearly their development is manifested in the automotive sector, where in recent years the spread of drones has increased significantly. Many large companies working in the field of road transport, such as BMW, Volkswagen and Tesla, are actively implementing automated control systems in their products. Unmanned technologies are one of the most markedly progressive areas of scientific and technological progress in road transport, and it is likely that over the next decade they will organically join in everyday life, so their consideration in this article is relevant.

2. The modern concept of an unmanned vehicle
An unmanned vehicle, in its simple definition, is a vehicle equipped with an automatic control system that can move without human intervention. However, at this stage of development of such cars there is the peak of popularity, and each company offers its own set of technical solutions, which leads to a difference in the degree of automation in cars of different manufacturers.

For the convenience of classification of the Community of automotive engineers (Society of Automotive Engineers, SAE) developed the standard for automated traffic control systems [1], which is widespread. The classification according to this standard is presented in table 1.

3. Features of the use of unmanned vehicles
The human factor in driving still plays a key role in creating emergencies. According to the annual statistics of the traffic police for 2018, a significant part of the accident was caused by the driver.[2] On the contrary, an accident with

participation of unmanned vehicles, which occurred due to the fault of the automated system, has not been recorded at this time. In Russia, there is no database on accidents with unmanned vehicles due to the lack of such vehicles on public roads. However, the experience of foreign companies shows that their unmanned vehicles get into an accident due to the fault of other road users.

Table 1. Classification of unmanned vehicles according to the SAE standard

No.	Name	Definition	Degree of automation
		driver controls	
0	manual vehicles	The driver continuously drives the vehicle (vehicle), even if it is improved by warning and intervention systems	no
1	Driver assistance system	The automatic assistant in a particular driving mode performs either steering or acceleration / deceleration using information about the driving environment, assuming that the driver will take over all other aspects of vehicle control	Partial automation
2	Partial automation	In a particular driving mode, one or more driver assistance systems perform both steering and acceleration / deceleration using information about the driving environment, assuming that the driver will take over all other aspects of vehicle control	Partial automation
		Automatic driving system controls the movement	
3	Conditional automation	The system controls all aspects of driving provided that the driver confirms the system's request for intervention	Partial automation
4	High automation	The system controls all aspects of driving, even if the driver does not respond to the system's request for intervention	Partial automation
5	Full automation	The system controls all aspects of driving that the driver could take on, under any road conditions and environmental conditions	Full automation

When transporting goods in columns (platoons) there is a decrease in fuel consumption (figure 1). Scania tests have shown that driving trucks using "platoons" can reduce fuel consumption by up to 12 percent [3].

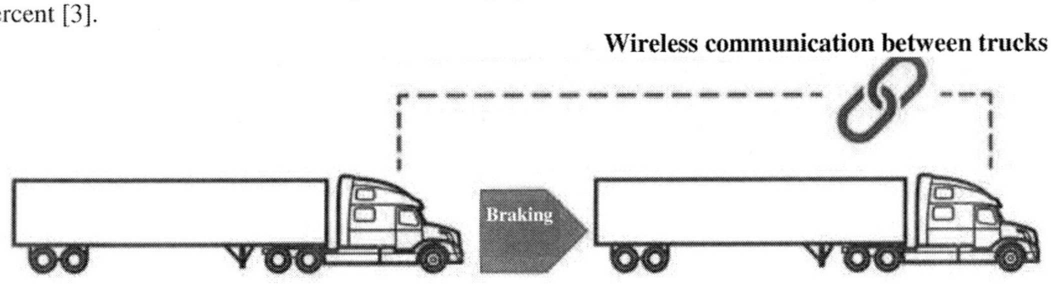

Figure 1. Movement of trucks "platoons»

It is also important to take into account the time that the driver spends in the outfit when transporting goods. According to the mode of work and rest of drivers, the travel time should not exceed more than 8 hours, while the period of continuous driving is 4 hours. The use of unmanned trucks will increase the speed of delivery due to the continuous driving of the vehicle.

It should also be borne in mind that the automated system would be able to identify the optimal route of transportation faster than a person, as it is able to continuously process information from a large number of sources and take it into account in the coordination and management of traffic flows. Evaluation and application of optimal routes will improve traffic and reduce the load on public roads.

4. Development of unmanned vehicles in Russia

Russia is actively developing unmanned road transport, both in the passenger and cargo sectors, which are not only the choice of manufacturing companies, but also supported by the state. Thus, in 2016, the national research initiative "Avtonet" was established and the state plan for the development of the market of electric vehicles and unmanned vehicles until 2035 was approved.

IOP Conf. Series: Materials Science and Engineering **832** (2020) 012003 doi:10.1088/1757-899X/832/1/012003

The most notable successes of Russian drones in the passenger sector in passenger transport. Thus, the company "Yandex" presented its unmanned taxi back in 2018, and at the moment actively continues its test in various conditions. In the test area SKOLKOVO, an unmanned vehicle can be called through the application Yandex.Taxi [4].

In addition to Moscow, Yandex drones also travel to Tatarstan. Innopolis became the first city in Europe where you can use the drone for everyday trips. Despite the automated control of the vehicle, there is a test engineer in the cabin, ready to take control of the vehicle if necessary.

In addition to a taxi in Russia has experience of transporting passengers in driverless buses. In 2018, the manufacturer of trucks "KAMAZ" demonstrated in Kazan, the prototype of an unmanned bus a "SHUTTLE". For unmanned bus has built a separate closed track, with a length of 650 meters. Unfortunately, at the moment these vehicles are not widely used, but the development in this direction continues [5].

Not without attention and freight transport. An unmanned truck was presented at the St. Petersburg international economic forum (SPIEF). The KAMAZ company plans to start using the first unmanned truck for intra-plant transportation in 2019. The Ministry of transport also noted that the use of unmanned trucks, for example, on the route from St. Petersburg to Moscow, will reduce the cost of transportation by 60%. It is assumed that initially unmanned vehicles will be launched on dedicated lanes, and then gradually transferred to such a mode all the way [6].

5. Prospects for the use of unmanned vehicles in the Far North

In addition to the common types of vehicles and developers are engaged in the creation of atypical – for example, all-terrain vehicles. This direction is important for Russia, which has a large area with low traffic. The far North of the Russian Federation and equivalent areas occupy about 70 % of territory of Russia the Importance of the North for economy is primarily determined by the raw potential [7].

Unmanned Rover Snowbus introduced the company Volgabus in June 2019. The all-terrain vehicle is designed for use in all conditions thanks to the large wheels and the 8x8 formula (figure 2).

Figure 2. Snowbus unmanned all-terrain vehicle

It is reported that the "polar bus" can go on snow, ice, ice, creating low pressure on the surface. Also, the all-terrain vehicle can swim on the water, overcome vertical obstacles up to 1.6 meters high and crevices up to 2 meters wide. The all-terrain vehicle is designed to carry passengers, but the presence of this development suggests that companies are interested in creating such vehicles, which in the future will be able to provide year-round transport links in the Northern regions [8].

The Northern regions of the Russian Federation are characterized by off-road, a large number of obstacles in the summer are very frequent cases of heaving roads, the formation of cracks, potholes, bumps and as a result – increased wear of the roadway.

Where the density of the road network is less than 0.3 km per 1 km^2 of the territory, the main accident rate increases sharply – the number of people killed per 10 thousand vehicles [9].

Ice crossings are one of the most dangerous types of seasonal roads, despite the serious measures to ensure the safety of people and goods during their operation, many cars fall under the ice every year. Proof of the seriousness of the security situation on the ice crossings can also be considered presented by the Road research Institute GOST project for ice crossings and winter roads [10].

Human security is one of the priorities in the development of unmanned vehicles. The use of this technology would reduce the level of danger on the roads of the Far North, for example, with the introduction of V2X systems.

The "car-anything" (V2X) concept uses the latest generation of information and communication technologies to implement the omnidirectional vehicle connection in the following configurations: "car-machine" (V2V), "car-infrastructure" (V2I), "car-pedestrian" (V2P) and "car-network" (V2N). This concept is schematically presented in figure 3.

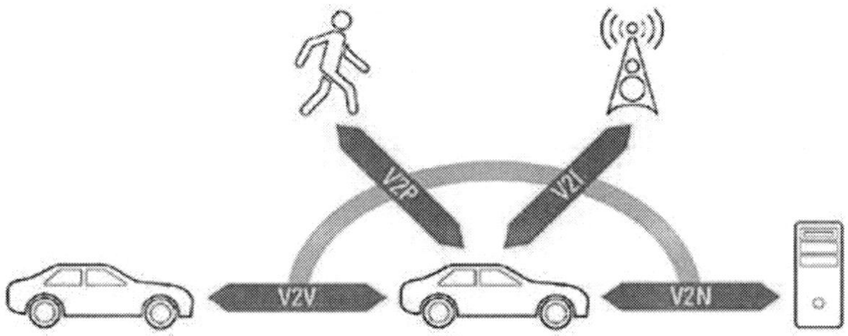

Figure 3. V2x technology

This technology links various elements of transport, such as pedestrians, vehicles, roads and cloud environments, which can contribute to the creation of an intelligent transport system and contribute to the development of new modes of transport and new types of vehicles and transport services in the future. Improving road traffic efficiency following the introduction of such technology can not only save resources and reduce fuel consumption, which is extremely important in the Northern regions with their increased fuel consumption, but also to reduce the frequency of accidents and improve traffic management.

The principle of application of this technology is shown in the figure. A truck equipped with an autopilot is driven on safe sections of the road by the driver, but when driving through the ice crossing, the person leaves the cabin and then the car follows the dangerous part of the path on autopilot, focusing on the signal received from the network, another car or infrastructure element on the other side. After crossing the ice crossing, the driver takes over driving again (figure 4). The strokes indicate the section of the road on which the vehicle is controlled by the autopilot. In this case, the car as the data exchange with the infrastructure for the route, and betrays and receives information from other cars to calculate the safe distance between them when driving.

Despite the actively ongoing development of unmanned road transport and its advantages over equipment controlled by a human driver, in Russia there are a number of factors that prevent the rapid emergence of this technology on public roads. Among them there are significant initial investments caused by the high cost of development and components and imperfection of infrastructure, but the most important of them is the lack of documents regulating this type of transportation.

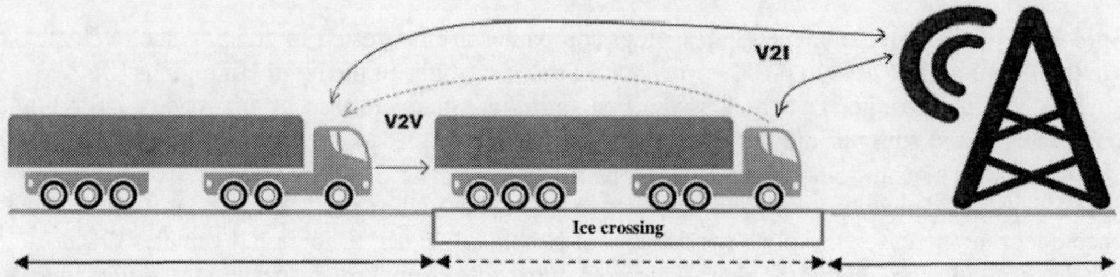

Figure 4. Movement on the ice crossing with the use of V2X

The main question facing legislators is who should be responsible in case of an accident. At the current stage of the introduction of drones in the car is a driver who is able to take control if necessary, which allows you to register his responsibility in the legislation.

However, such changes will only delay the solution of this problem until the time when the machines reach the 5th level of SAE automation, in which the control of driving is completely carried out by the control system without human intervention. If we take into account the growth rate of development of unmanned technologies, we can confidently conclude that will happen in the next decade.

6. Conclusion

Unmanned technology is steadily gaining its niche in the field of road transport. Many companies offer their technical solutions for automation of car control, including in Russia. At the moment, the Russian Federation is developing both passenger and freight transport, but before the appearance of these cars on public roads, it is necessary to solve a number of problems, first of all, to create a legal framework for this type of transportation.

References

[1] Sae J 2016 *Taxonomy and Definitions for Terms Related to On-Road Motor Vehicle Automated Driving Systems*

[2] *Road Safety Indicators.* Retrieved from: http://stat.gibdd.ru/ (date of access 04.08.2019)

[3] *Platooning saves up to 12 percent fuel.* Retrieved from: https://www.scania.com/group/en/platooning-saves-up-to-12-percent-fuel/ (accessed 06.08.2009)

[4] *Drone on the roads Skolkovo.* Retrieved from: https://taxi.yandex.ru/blog/bespilotnik-v-skolkovo (date of access 04.08.2019)

[5] *In Kazan in 2018 launched the first unmanned bus.* Retrieved from: https://www.rbc.ru/tatarstan/freenews/5b1f94569a79472c515b9e1f (date of access 04.08.2019)

[6] *Company "KAMAZ" plans to begin using the first unmanned cargo vehicle for in-plant transportation in 2019.* Retrieved from: https://tass.ru/ekonomika/6434449 (date of access 04.08.2019)

[7] Filippova N A, Efimenko D B and Ledovsky A A 2018 Ensuring the efficiency of transport processes in the Far North *World of transport* **16(4)** 150–9

[8] *Russia has developed an unmanned Rover.* Retrieved from: https://www.ntv.ru/novosti/2198743/ (date of access 07.08.2019)

[9] Reshetnikov A P, M. Ishkov A and Boyarshinov A L 2017 Operational reliability of transport, its impact on the accident in the North *Bulletin of the Irkutsk state technical University* **21(7)** 164–70

[10] *Russia will introduce the GOST for roads, which can be used only in cold time of the year.* Retrieved from: https://www.kommersant.ru/doc/4046400 (date of access 07.08.2019)

IOP Conf. Series: Materials Science and Engineering **832** (2020) 012004 doi:10.1088/1757-899X/832/1/012004

Research of the influence of alternative fuels on diesel engine noise level

A L Iakovenko[1,2], A Y Dunin[1], P V Dushkin[1], E A Savastenko[1] and M G Shatrov[1]

Moscow Automobile and Road Construction State Technical University (MADI), 64, Leningradski Ave., Moscow 124319, Russia

Email: iakovenko_home@mail.ru

Abstract. One of the advantages of using alternative fuels in internal combustion engines is the possibility of reducing noise. The article presents the methodology and some results of the research of structure-borne noise of internal combustion engines related to the working process. The main characteristics of alternative fuels allowing reduction of the noise level by modifying parameters of the heat release law and reducing the pressure growth rate are analyzed. The results of calculated and experimental research of the diesel engine structure-borne noise during its operation on gas and mixed fuels are presented. The research showed that the greatest effect is achieved when converting diesel engine to spark ignition gas engine. For gas-diesel engines, noise reduction of the working process is much lower because a high compression ratio is preserved. The use of mixed fuel revealed a need to correct the injection advance angle to reduce the noise level.

1. Introduction

The problem of vehicle noise is very important for big cities because the number of vehicles on the roads increases from year to year. Noise has harmful effects on the environment and human health. Therefore, it is necessary to improve vehicles not only in terms of reducing fuel consumption and emissions of harmful substances, but also to reduce its noise level.

2. Internal combustion engine as a source of noise

One of the main sources of vehicle noise is its engine. The internal combustion engine is a complex source of noise, as its individual sources have different nature.

In the overall acoustic balance of the vehicle, aerodynamic noise exceeds the structure-borne noise value, but its level can be reduced by installing silencers in the intake and exhaust systems [1].

It is very difficult to reduce structure-borne noise because its sources have different origins. Working process and piston shift are the most important sources of the structure-borne noise.

There are various ways to decrease the engine structure-borne noise, which include improvement of working process and engine design or implementation of special measures for additional noise insulation of the engine. One of the methods of the working process modification is the use of alternative fuels [1]. Noise reduction occurs as a result of decreasing the combustion process toughness, that is, the pressure growth rate in the cylinder.

To assess the impact of alternative fuels on the engine noise level, computational and experimental studies were performed at the Department of Thermal Science and Automotive Engines of MADI. Some results of these studies are presented in this article.

3. Methods of modeling and experimental research of engine noise

Currently, taking into account the development of information technology, all engine research is carried out within a single information space. This approach allows to reduce research time and to improve the quality of the results. At the same time, different methods and software for noise calculation should be used at different stages of the design. At the initial stage, quite simplified models and empirical formulas are used, and at the stage of detailed design development, three-dimensional models and finite methods, boundary elements, etc. must be used.

Calculations of the sound power of the engine were performed using the method developed at the Department of Thermal Science and Automotive Engines of MADI [2], which includes the following stages: engine layout parameters formation, calculation of its working cycle, noise modeling and calculation.

The sound power of the engine is calculated by the formula:

$$P_W(kf_0) = z_S(kf_0) \cdot \rho c \cdot S_{ICE} \overline{v}^2_{e(S)}(kf_0),$$

where $z_S(kf_0)$ – relative coefficient of resistance to radiation normalized by the area of the outer surfaces of the engine; ρ - air density; *c* – sound speed in the air; ρ*c* – wave resistance of the air; S_{ICE} – engine outer surfaces area; $\overline{v}^2_{e(S)}(kf_0)$ – average squared effective vibration velocity on the outer surface.

The engine is modeled by an equivalent cylindric shell on condition of equality of its masses, lengths and surface area. For the shell, the vibration velocity of its outer surface can be calculated by the formula:

$$\overline{v}^2_{e(S)}(kf_0) = \frac{1}{2\pi \cdot M_{ICE} \cdot T^2} \sum_{k=A}^{N} G^2(kf_0) \frac{1}{z_v(kf_0) \cdot \eta(kf_0) \cdot (kf_0)},$$

where: *k* – ordinal number of harmonic; *G*(kf_0) – spectral density of the load factor at the frequency of the kth harmonic (N·s); $T = \frac{30\tau}{n}$ – working cycles sequence period (s); τ – engine cycle factor; $f_0 = \frac{60}{n}$ – crankshaft rotation frequency (Hz); *n* – number of crankshaft revolutions per minute (rpm); *A* – number of the lower harmonic of the load factor (provided that f_0 is the first); *N* – number of the highest harmonic of the load factor (provided that f_0 is the first).

Experimental noise measurements were carried out using the of hardware-software complex LMS Pimento (figure 1) in compliance with the standard methodology (GOST R 53838-2010, GOST R 51401-99).

4. Results of engine structural noise research when it is running on alternative fuels

Currently, the use of natural gas as a motor fuel is a relevant challenge in the engine industry, so the work on conversion of diesel engines to operate on gas fuel has been and continues to be carried out in MADI [3–5].

An 8-cylinder (D = 12 cm, S = 12 cm) diesel engine was converted to spark ignition (SI) natural gas engine. The main results of the noise measurements for the original diesel and converted SI gas engine are shown in figure 2 and figure 3.

Thus, conversion of the diesel engine to SI gas engine enabled to decrease it noise throughout the speed range by 3...6 dBA.

Comparison of spectra of indicator diagrams (figure 3) in the mid- and high-frequency regions showed a significant decrease in the values of spectral components, which led to the decrease in the overall noise level of the engine.

In one more work performed at the Department, a 6-cylinder diesel engine (D = 10.7 cm, S = 12.4 cm) was converted to gas-diesel process, the compression ratio remained unchanged [3, 4]. The results are shown in figures 4–6.

IOP Conf. Series: Materials Science and Engineering **832** (2020) 012004 doi:10.1088/1757-899X/832/1/012004

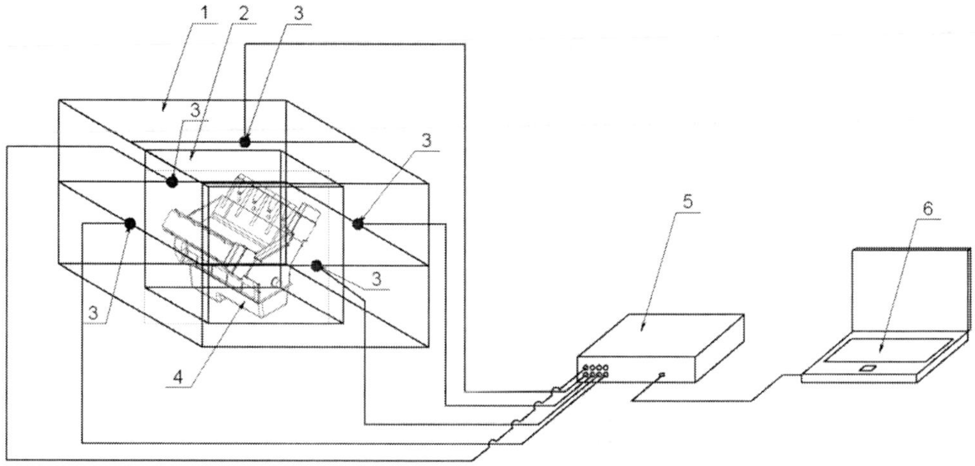

Figure 1. Scheme of the experimental setup:
1 – measuring surface in the shape of a rectangular parallelepiped; 2 – envelope rectangular parallelepiped; 3 – microphone; 4 – engine; 5 – hardware-software complex LMS Pimento; 6 – computer

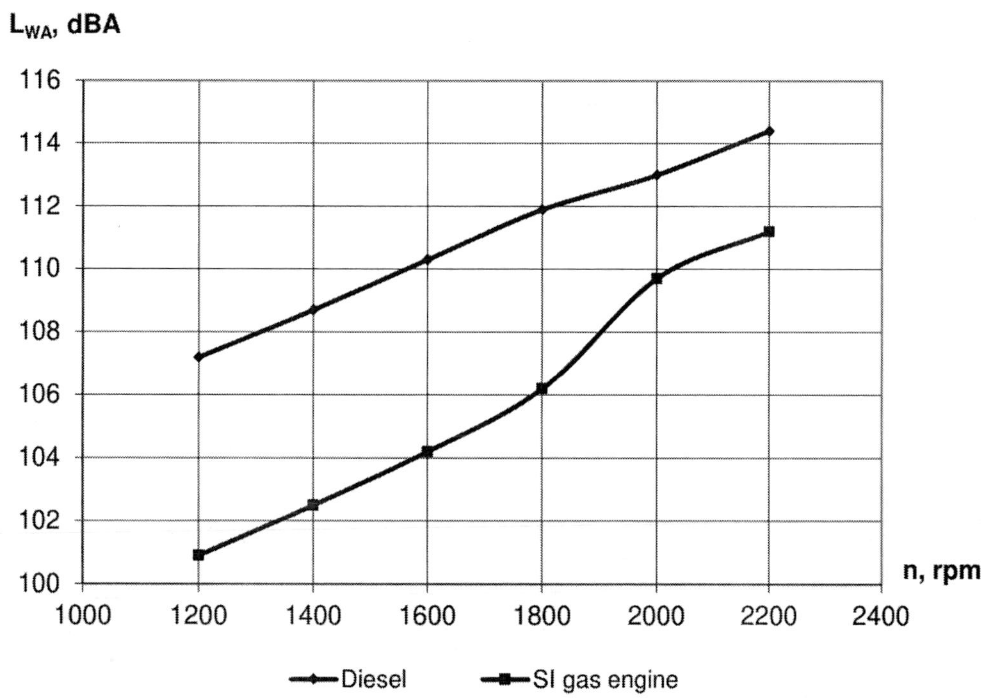

Figure 2. Overall noise levels of working cycles of diesel and SI gas versions of the 8-cylinder engine

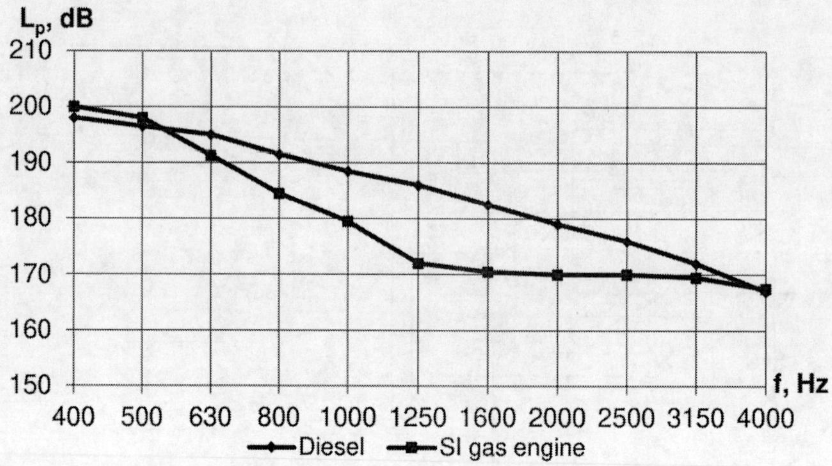

Figure 3. The spectra of indicator diagrams of working cycles of diesel and SI gas versions of the 8-cylinder engine at n = 2200 rpm

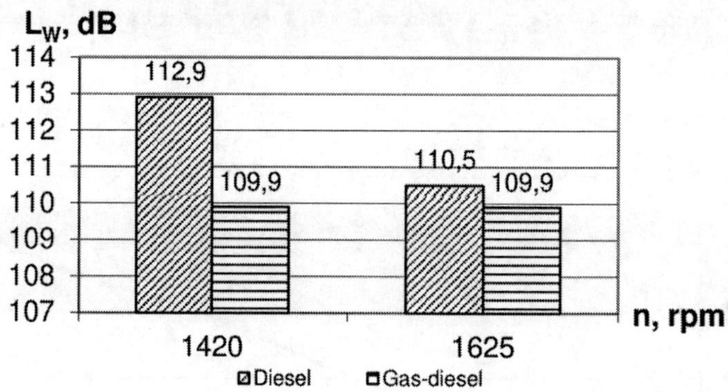

Figure 4. Overall noise levels of diesel and gas-diesel versions of the 6-cylinder engine

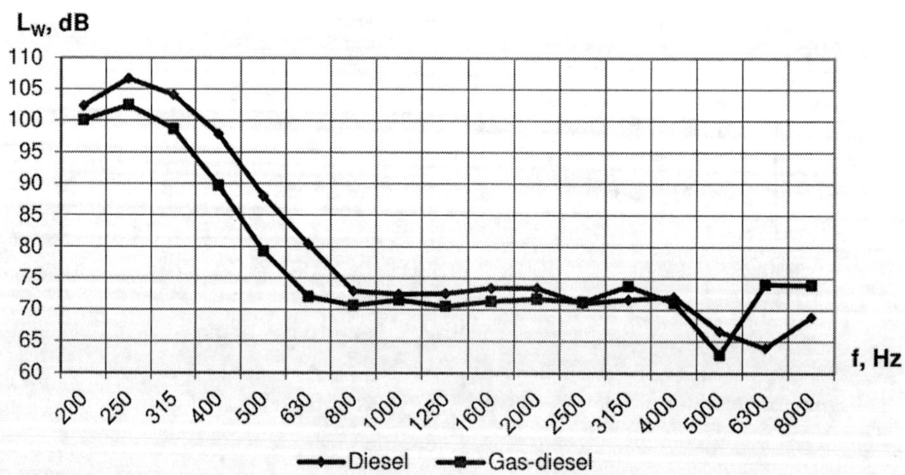

Figure 5. The sound power spectra of diesel and gas-diesel versions of the 6-cylinder engine at n = 1420 rpm

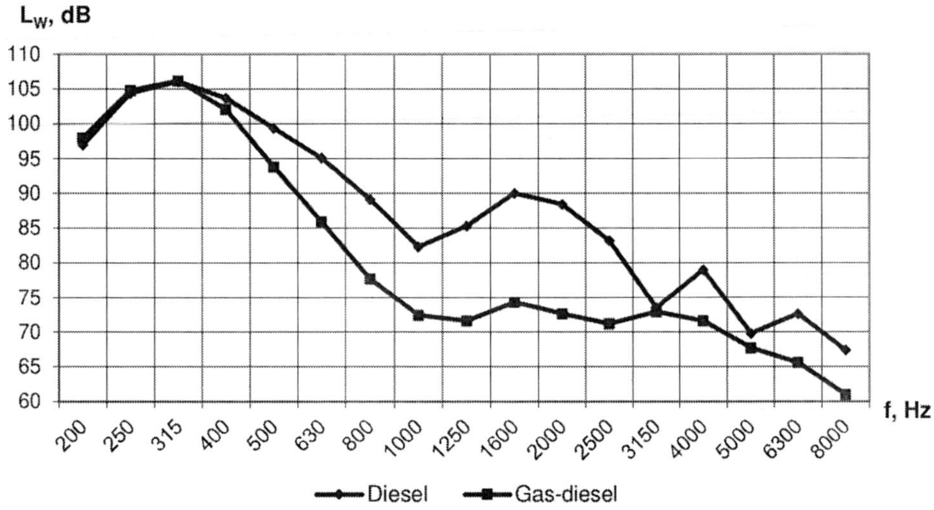

Figure 6. The sound power spectra of the 6-cylinder engine for working cycles of diesel and gas-diesel at n = 1625 rpm

As a result, the noise level of the 6-cylinder engine also decreased by 0.6...2.0 dB, which is significantly less than of the 8-cylinder engine. This is caused by the use of a common-rail fuel system on the original diesel engine, which provides a lower noise level compared to the traditional split-type fuel system, and preservation of the compression ratio, which generally reduces the influence of alternative fuel on the toughness of the working process.

Research carried out at the Department [6–12] showed the possibility of flexible control of fuel injection characteristics when using the CommonRail system. As a result, the harshness of the working process of the diesel can be reduced, which also reduces the diesel noise level.

The Department works on the use of mixed fuels in diesel [13–15], one of which is water-fuel emulsion.

The research of the influence of the water-fuel ratio in the emulsion on engine noise level was performed for a 4-cylinder diesel engine (D = 11 cm, S = 12.5 cm). The results are shown in figures 7–9.

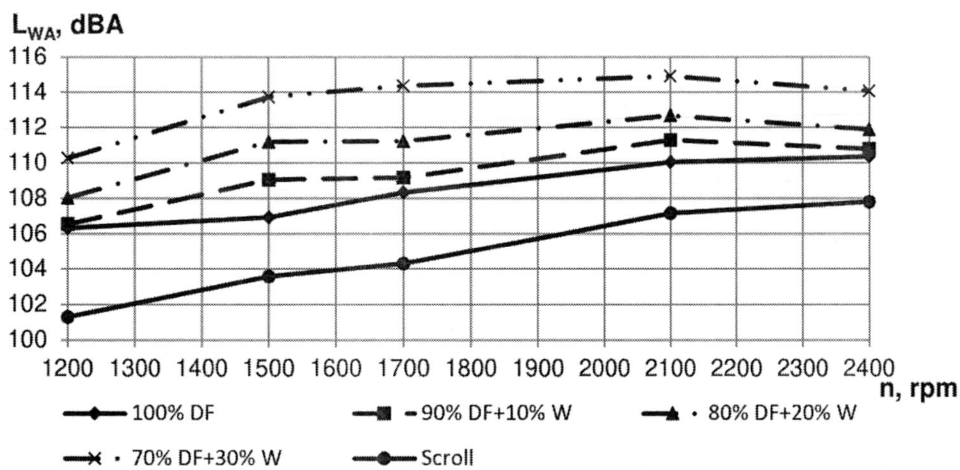

Figure 7. Overall noise levels of the 4-cylinder engine for mixed fuel having different composition

IOP Conf. Series: Materials Science and Engineering **832** (2020) 012004 doi:10.1088/1757-899X/832/1/012004

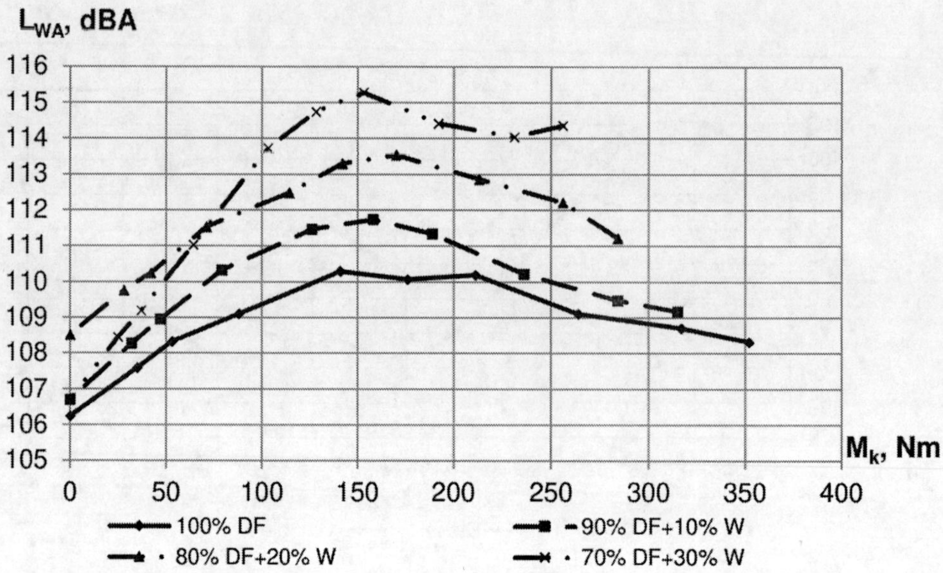

Figure 8. Overall noise levels of the 4-cylinder engine for mixed fuel having different composition at n = 1700 rpm

The overall sound power level for all the emulsion compositions investigated increased when rising of the engine speed. At the same time, the lowest noise corresponds to the pure diesel fuel, and the highest – to the mixture of 70 % of diesel fuel and 30 % of water.

The noise grows when rising the proportion of water in the mixture, which is caused by the increase of the ignition delay period at the same injection advance angle and, as a consequence, by the increase of the combustion process toughness.

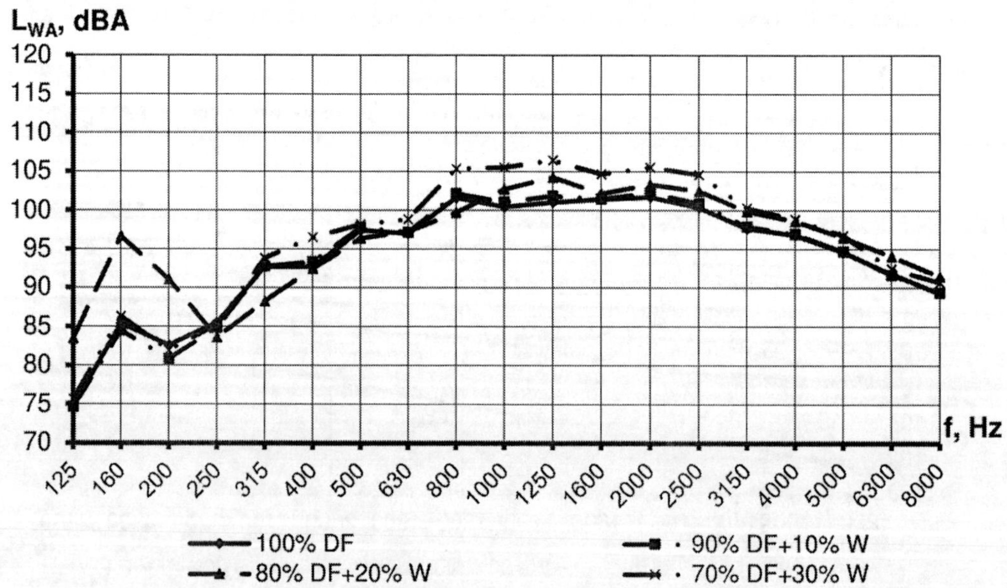

Figure 9. The sound power spectra of the 4-cylinder engine for mixed fuel having different composition at n = 2400 rpm

5. Conclusion

1. The results of conversion of the diesel engine for operation on gas fuel showed that the use of this fuel can significantly reduce the engine noise level.

The noise of the 8-cylinder diesel engine after its conversion to spark ignition gas engine decreased by 3...6 dBA. It is caused by decrease in the toughness of its working process.

After conversion of the 6-cylinder diesel engine to the gas-diesel engine, its noise decreased by 0.6…2 dB. There is little variation in the overall level of noise due to the use of the common-rail fuel system and preservation of the compression ratio of the original diesel engine during conversion.

2. The use of a water-fuel emulsion as a fuel in the 4-cylinder diesel engine maintaining the same value of the injection advance angle, led to the increase in the noise level due to the growth of the ignition delay period. Thus, to reduce the noise when using such a fuel, correction of the injection advance angle is required.

References

[1] Shatrov M G, Yakovenko A L and Krichevskaya T Y 2014 *Noise of automobile internal combustion engines* (Moscow: MADI) 68 p

[2] Shatrov M G and Yakovenko A L 2009 *Vestnik Moskovskogo avtomobil'no-dorozhnogo gosudarstvennogo tekhnicheskogo universiteta (MADI)* **1(16)** 10–8

[3] Shatrov M G, Sinyavski V V, Dunin A Yu, Shishlov I G, Vakulenko A V and Yakovenko A L 2018 Using simulation for development of the systems of automobile gas diesel engine and its operation control Int. J. of Engineer. and Technol. **7(2.28)** 288–95

[4] Shatrov M G, Sinyavski V V, Dunin A Y, Shishlov I G and Vakulenko A V 2017 Method of conversion of high- and middle-speed diesel engines into gas diesel engines. *Facta Univer. Ser. Mechan. Engineer.* **15(3)** 383–95

[5] Shatrov M G, Khachiyan A S, Sinyavskij V V and Shishlov I G 2013 *Transp. na al'ternat. toplive* **4** 29–32

[6] Shatrov M G, Golubkov L N, Dunin A U, Yakovenko A L and Dushkin P V 2015 Influence of high injection pressure on fuel injection performances and diesel engine working process *Thermal Sci.* **19(6)** 2245–53

[7] Shatrov M G, Golubkov L N, Dunin A U, Yakovenko A L and Dushkin P V 2015 Research of the injection pressure 2000 bar and more on diesel engine parameters *Int. J. of Appl. Res.* **10(20)** 41098–102

[8] Shatrov M G, Golubkov L N, Dunin A Yu, Dushkin P V and Yakovenko A L 2017 A method of control of injection rate shape by acting upon electromagnetic control valve of common rail injector *Int. J. of Mechan. Engineer. and Technol.* **8 issue 11** 676–90

[9] Shatrov M G, Golubkov L N, Dunin A U, Dushkin P V and Yakovenko A L 2017 The new generation of common rail fuel injection system for Russian locomotive diesel engines *Pollution Res.* **36(3)** 678–84

[10] Shatrov M G, Malchuk V I, Dunin A U and Yakovenko A L 2016 The influence of location of input edges of injection holes on hydraulic characteristics of injector the diesel fuel system *Int. J. of Appl. Engineer. Res.* **11(20)** 10267–73

[11] Shatrov M G, Malchuk V I, Dunin A Y, Shishlov I G and Sinyavski V V 2018 A control method of fuel distribution by combustion chamber zones and its dependence on injection conditions *Thermal Sci.* **22(5)** 1425–34

[12] Shatrov M G, Golubkov L N, Dunin A U, Yakovenko A L and Dushkin P V 2016 Experimental research of hydrodynamic effects in common rail fuel system in case of multiple injection *Int. J. of Appl. Engineer. Res.* **11(10)** 6949–53

[13] Shatrov M G, Dunin A U, Yakovenko A L and Eziev A A 2013 *Avtomobil' i Elektronika. Sovremennye Tekhnologii* **2** 123

[14] Shatrov M G, Kudryashov B A, Dunin A U, Eziev A A and Livanskij A N 2013 *Izvestiya Volgograd. Gosudarstv. Tekhnich. Univer. Ser. Nazemnye transport. Sist.* **7(21)** 62–6

[15] Mal'chuk V I, Shatrov M G and Dunin A Yu 2007 *Trakt. i sel'skohozyajstv. mashiny* **4** 34–7

Rationale and frequency of the diagnosis of component parts of tractors Belarus in the development of technology maintenance

V I Karagodin and R A Khapugin

Moscow automobile and road state technical University (MADI), 64, Leningradsky Ave., Moscow, 125319, Russia

E-mail: bik250248@yandex.ru

Abstract. The article presents the solution of a new scientific and technical problem of substantiating the frequency of diagnosis and the composition of control and diagnostic operations performed during maintenance of machines based on Belarus-82.1. The methodology of justification of the specified parameters of technological processes of maintenance service takes into account the probability of failures in control periods and provides for increased reliability of cars. It is proved that manufacturers in some cases unreasonably increase the frequency of maintenance for advertising purposes. Experimentally determined parameters of the distribution laws of time between failures of component parts of machinery on the basis of the Belarus-82.1. The new dependence of the probability of the failure of the operating time of machines on existing and recommended strategies ensures their efficiency at all the intervals of achievements. The composition of control and maintenance operations for different types of maintenance is justified not only for warranty period and for the entire service life of the machine. The practical use of the research results presented in the article can significantly improve the reliability of equipment with a slight increase in the cost of maintenance.

1. Introduction

The system of preventive maintenance of machines, originally created as a system of standards that strictly determine the frequency and scope of work in maintenance and repair, with the development of technical diagnosis, has become more flexible and provides for a given frequency of control and diagnostic works and subsequent performance of technical effects on the needs identified as a result of diagnosis.

Issues concerning maintenance intervals are quite deeply explored, but issues concerning frequency and composition of the control and diagnostic operations for maintenance need such a deep study. The reliability of components and machines as a whole depends on the results of solving this problem, which determines the relevance of the tasks, the solution of which is reflected in this article.

2. Collection of information on the reliability of components of machines based on tractors Belarus-82.1

Collection of information on reliability of components of Belarus-82.1 tractors and machines on their basis was carried out at the operational enterprise "Resurs Trans". The information is given in the form of table 1.

IOP Conf. Series: Materials Science and Engineering **832** (2020) 012005 doi:10.1088/1757-899X/832/1/012005

Table 1. Source data (fragment)

Names of components	Time between failures, machine-hour for vehicles with number					
	1	2	3	4	5	6
Clutch	2869	4369	4387	3867	6925	8952
Front tyre		6219		5160	5961	7856
Rear tyre	3566	6219	5129	5160	5961	7856

Based on actual production capabilities, a test plan [N, M, T] was selected. The first letter in the designation of the plan (N) indicates that the test delivered N products, the second (M) that failed during the test machine was restored. The tests shall be terminated at the end of the test time T.

Termination of data collection at the end of the test time T causes a difference in the periods of observations of the reliability of the controlled products [1]. In this case, the resulting samples are considered as censored or repeatedly censored. Censorship is an event that leads to the termination of tests or operational observations of the object before the failure (limit state) of the studied nature. The reasons for censoring are the different times of the beginning and (or) the end of the tests, as well as the need to assess the reliability before the failure of all tested products. In a repeatedly censored sample, the values of developments before censoring are not equal.

Table 1 shows that the front tyres of vehicles 1 and 3 were not replaced during the observation period. They have been modified not to failure and to censorship. Time to the reduction of the first machine is 3566 machine-hours, and the 3rd machine – 5129 machine-hours as it is known that during these developments on the machines rear tires were replaced.

It was hypothesized that the distribution of time to failure of the components of tractors is subject to the distribution of Weibull. This distribution describes the time to failure and failure of non-recoverable and recoverable products, which failure occurs due to fatigue failure. This law is often used in reliability theory to describe uptime and manifests itself in the so-called "weak link" model.

The most important issue in assessing reliability is to determine the volume of tests, which depends on the law of distribution of time to failure. The minimum allowable number of recorded failures in the Weibull distribution for the evaluation of developments for failure with a given limit relative error and confidence probability are presented in [25]. With minimum requirements for accuracy and reliability of the results, the number of recorded failures in the sample should be at least five.

A specially developed program was used to calculate the required parameters. For clutches where the number of failures was 9, the obtained values of the parameters of Weibull distribution a = 7828, b = 2.2, and mean time to failure t_{avg} = 6933, standard deviation MTBF σ = 3336, coefficient of variation v = 0.48 and γ-percent time to failure (at γ=90 %) t_{90} = 2806.

The relatively small number of reported failures affected the accuracy and reliability of the estimates. With a confidence probability of 0.8, the relative error of the obtained parameters is 0.2. For most of the components of tractors, where the number of failures is even smaller, satisfactory accuracy and reliability estimates of reliability could not be obtained.

This led to the need to collect additional information about the reliability that was produced at another enterprise. As a result, other samples of the studied parameters were obtained. To combine them with previously obtained samples, it was necessary to make sure that these samples belong to the same population.

To test the hypothesis that two samples belong to the same population, it is necessary to assess the randomness of the discrepancy between the two sample means and the two sample variances. In the case when the initial data is not enough to construct the distribution law of the studied quantity, the series criterion can be used to verify the belonging of two samples to one General population. Let there be two samples of volumes N_1 and N_2:

$$X_1, X_2, ..., X_{N1};$$
$$Y_1, Y_2, ..., Y_{N2}.$$

Placing all the values of both samples in ascending order, marking the elements from the first sample with a " + " sign and the elements from the second sample with a "–" sign, we get the sequence:

$$X_1 \le X_2 \le Y_1 \le Y_2 \le Y_3 \le X_3 \le X_4 \le Y_4 \le Y_5 \le ... \quad (1)$$

+ + – – – + + – -

1st series 2nd series 3rd series 4th series

Let's call the sequences "+" and "–" series. The number of series is two when the samples are very different from each other. With a slight difference in their number of series will be large. It is known [6] that if both samples X and Y belong to the same population, the probability of obtaining m series in sequence (1) is expressed as a function h(m):

if $$\sum_{m=2}^{d_o} h(m) \le \beta, \quad (2)$$

where β is the significance level, the difference between samples is considered significant. Otherwise, the difference between the samples is insignificant, and they can be considered as belonging to the same population. Usually we take the level of significance β = 0,05.

We tested a hypothesis about belonging of one general set of time between failures of clutches, tractor Belarus-82.1. Variational series of clutch operating time of the first sample:

N_1 = 1438; 2869; 3867; 4369; 4387; 6925; 7029; 7561; 8952.

The statistics of the time between failures of couplings of the second sample:

N_2 = 3025; 3639; 4356; 4573; 4963; 5783; 5835; 5843; 5896; 6254; 6632; 6973; 7253; 7533; 7583; 7647; 7987; 8034; 8422; 8753; 8785; 9457.

Placing all the values of both samples in ascending order, marking the elements from the first sample with a " + " sign and the elements from the second sample with a "–" sign, we get the sequence:

1438; 2869; 3025; 3639; 3867; 4356; 4369; 4387;

\+ – + – +

1st series 2nd series 3rd series 4th series 5th series

4573; 4963; 5783; 5835; 5843; 5896; 6254; 6632; 6925; 6973; 7029;

– + – +

6th series 7th series 8th series 9th series

7253; 7533; 7561; 7583; 7647; 7987; 8034; 8422; 8753; 8785; 8952; 9457

– + – + –

10th series 11th series 12th series 13th series 14th series

The number of received series m = 14. With this number of series, the left part of the condition (2) is 0.298. Since 0.298 > β = 0.05, the differences between the two samples are insignificant, and they can be considered as samples from the same universe. Therefore, the samples can be combined.

In the combined sample for couplings where the number of failures was 31, and the resulting values of the parameters of Weibull distribution a = 7939, b = 3.6, mean time between failure t_{avg} = 7157, the standard deviation of the MTBF σ = 2195, the coefficient of variation v = 0.31, γ-percent time to failure (at γ=90 %) t_{90} = 4267.

The increase in the number of observations has significantly improved the accuracy and reliability of the estimates. With a confidence probability of 0.9, the relative error of the obtained indicators will not exceed 0.1. For most of the components of tractors, for which previously, due to the small number of failures, satisfactory accuracy and reliability estimates of reliability were not obtained at all, the union of samples allowed this to be done.

Failure times of tractor components in machine-hour are shown in table 2.

Relative errors were determined at a confidence probability of 0.8 and 0.9 depending on the number of failures of the component under consideration and on the obtained coefficient of variation of its MTBF.

From the data table 2 it follows that for most of the components obtained acceptable accuracy and reliability of the results that can be practically used. The most accurate and reliable results were obtained for the clutch with the highest number of failures. The relative error in determining the parameters of the Weibull distribution of the clutch failure time is 0.1 at a confidence probability of 0.8 and 0.1 at a confidence probability of 0.9. Close to this accuracy of determining the parameters of the distribution of Weibull's workings on failure is achieved only for the front and rear tires, which also recorded a large number of failures.

The limiting case, when the result still obtained can be considered satisfactory in accuracy, is observed for the generator and the front spar. With a confidence probability of 0.9, the relative error exceeds 0.20, and with a confidence probability of 0.8, it is within 0.20.

Table 2. Results of processing of information on reliability of tractors Belarus-82.1

Name of the failed component	Weibull distribution parameters		Statistical estimates of time-to-failure				Relative error at confidence probability	
	a	b	average MTBF	standard deviation	coefficient of variation	90 % opera-ting time	0.80	0.90
			Elements of governance mechanisms					
Tie rod end	1074	4.18	9762	2633	0.27	6269	0.15	0.2
Steering column	14307	3.09	12792	4530	0.35	6904	0.15	–
Steering booster	14307	3.09	12792	4530	0.35	6904	0.15	–
			Elements of the carrier system					
Front spar	14590	3.17	13062	4518	0.35	7175	0.2	–
Front tyre	8579	3.4	7708	2499	0.32	4432	0.1	0.15
Rear tyre	8598	3.7	7756	2350	0.3	4660	0.1	0.15
			The elements of the transmission					
Clutch	7939	3.6	7157	2195	0.31	4267	0.1	0.1
Transmission	11679	4.2	10619	2839	0.27	6851	0.15	–
Transfer case	12278	4.3	11169	2960	0.27	7239	0.15	–
Outboard bearing	11553	4.09	10485	2881	0.27	6666	0.15	–
			Engine elements and additional equipment					
Starter	12622	2.8	11233	4399	0.39	5585	0.15	0.2
Generator	14966	2.3	13257	6146	0.46	5592	0.2	–
Accumulator	13446	3.1	12031	4206	0.35	6556	0.15	–
Fuel injection pump	12651	3.5	11387	3581	0.31	6681	0.15	–
Radiator cooling system	12121	4.2	11017	2957	0.27	7094	0.15	–
			Elements of hydraulic drive and attachments					
Rubber products hydraulic systems	10811	3.5	9729	3069	0.32	5697	0.15	0.2
High pressure hoses	14458	2.8	12871	5003	0.39	6441	0.15	–
Support arms for attachments	12123	3.9	10968	3169	0.29	6779	0.15	–
Rear suspension arms for attachments	10684	3.6	9628	2968	0.31	5721	0.15	0.2

The values of the mean time between failures of the components of the machine vary from 7157 machine hours for the clutch to 13257 machine hours for the generator, i.e. almost twice. The minimum value of the coefficient of variation is 0.27 for the transmission, the maximum is 0.46 for the generator, i.e. the variation of the operating time for failure of the components differs almost twice. This shows that at close mean time between failures, different components can have a significantly different nature of the increase in the probability of failure with increasing time between failures.

This is illustrated by the 90 percent failure times of the various components. So, a starter with an average time to failure 11233 machine hours has a 90 percent operating only 5585 machine hours, while the radiator of the cooling system with an average time to failure 11017 machine hours (lower than the starter) has a 90 percent operating time 7094 machine hours (nearly one and a half times higher than the starter). Similar conclusions follow, for example, from the comparison of the developments of the generator and the front spar, the tip of the tie rod and the support arms of the attachments. This requires an analysis of the dynamics of failure probabilities of components with increasing operating time. To this end, the failure probabilities were calculated for operating time up to 10 thousand machine hours of individual elements and systems.

3. Study of the dynamics of failure probabilities of components and machines in general

The probability of failure of the system element was determined by the formula:

$$F(t)=1-e^{-(t/a)^b}, \tag{3}$$

where t is operating time, machine-hour;

a and b are the parameters of the Weibull distribution (table 2).

The probability of failure of the k-th system was determined based on the probability of failure of its elements by the formula [34]:

$$P_{fk}(t_i)=1-\prod_{j\in J_k}\left[1-P_{fj}(t_i)\right], \tag{4}$$

where $P_{fk}(t_i)$ – the probability of failure of the k-th system when t_i is running;

j – index of the component part;

J_k – set of components included in the k-th system;

$P_{fj}(t_i)$ – the probability of failure of the j-th component of the operating time t_i.

The probability of failure of the parts of the machine, the previous routine diagnosis in miscontrolling period is defined as the quotient of the number of failures by the total number of cars that passed the routine maintenance and diagnosis to the beginning of the reporting period miscontrolling and worked until the end of this period:

$$P_o(l_i)=\frac{n(l_i)}{A(l_i)}, \tag{5}$$

where $n(l_i)$ – the number of failures of the considered component parts of the machine in miscontrolling period l_i;

$A(l_i)$ – the total number of machines that underwent scheduled maintenance and diagnostics at the beginning of the miscontrolling period l_i and operated until the end of this period.

The probability of failure of the machine as a whole is determined by the probabilities of failure of components that do not pass diagnosis, and the probabilities of failure of diagnosed components. The first probabilities on the i-th operating time interval $P_f(t_i)$ are determined by the formula (2). The probabilities of failure of the diagnosed components are determined by the formula (3), and the probability $P_{fm}(t_i)$ failure of the machine as a whole-by the formula:

$$P_{fm}(t_i)=1-\prod_{j\in H}\left[1-P_{fj}(t_i)\right]\cdot\prod_{j\in D}\left[1-P_{fj}(l_i)\right], \tag{6}$$

where j – the index of the component;

H – a subset of undiagnosed constituents;

D – a subset of the components to be diagnosed.

Schedule routine maintenance of tractors Belarus-82.1 recommended by the manufacturer, provides routine adjustments, a forced substitution of certain elements and fluids, as well as routine checks when servicing the various elements and systems of the tractor, which include the above elements, limiting the reliability and elements of the testing are necessary due to requirements of safety and performance. A fragment of the schedule of planned maintenance, reflecting the control and diagnostic operations to check the technical condition of the elements that limit the reliability of the tractor, is presented in table 3.

Table 3. Fragment of the schedule of installed maintenance

Controlled systems and elements	Operating time, machine-hour						
	125	250	375	500	625	750	875
Steering	1	1	1	1	1	1	1
Carrier system				1			
Clutch	1	1	1	1	1	1	1
Transmission				1			
Engine				1			
Electrics				1			
Hydraulic equipment				1			

Preliminary analysis of the reliability of tractors showed that due to the large inter-control period of diagnosis of the transmission, electrical equipment and hydraulic drive, the probability of failure of equipment, although less than in the absence of diagnosis, is still very high. Therefore, an option was worked out, providing for a reduction in the frequency of monitoring of these systems to 250 machine-hours. A fragment of the recommended schedule of planned maintenance is presented in table 4.

Table 4. Fragment of the recommended maintenance schedule

Controlled systems and elements	Operating time, machine-hour						
	125	250	375	500	625	750	875
Steering	1	1	1	1	1	1	1
Carrier system				1			
Clutch	1	1	1	1	1	1	1
Transmission		1		1		1	
Engine				1			
Electrics		1		1		1	
Hydraulic equipment		1		1		1	

Figure 1 shows the probability of failure of equipment at intervals of operation for the three options under consideration: in the absence of diagnosis, with the frequency of diagnosis established by the manufacturer and with the recommended frequency of diagnosis. From the presented data it can be seen that the timely replacement of the diagnosis of elements with low reliability in the maintenance schedule, recommended by the manufacturer, leads to a significant increase in the reliability of equipment. However, this is not the best option. The transition to the recommended frequency of diagnosis can reduce the probability of failure of equipment by more than one and a half time.

Thus, the customer of technical service is offered a choice of two options: either to carry out diagnostics and maintenance according to the operational documentation of the manufacturer, or to follow our recommendations and, increasing their costs for maintenance by about 5 %, to achieve an increase in the reliability of machines 1.5 times.

4. Conclusion

1. The study experimentally determined parameters of the distribution laws of time between failures of component parts of machines on base tractor Belarus-82.1.

2. The result of simulation based on the new probability of the failure of the achievements of the machines recommended by the manufacturer and we offer the periodicity of the diagnosis and composition of the control and diagnostic operations on all intervals of the developments.

3. Reducing the frequency of diagnosis of some systems reduces the probability of failure of the machine in half, which can be recommended to customers of technical service and operating companies to reduce downtime.

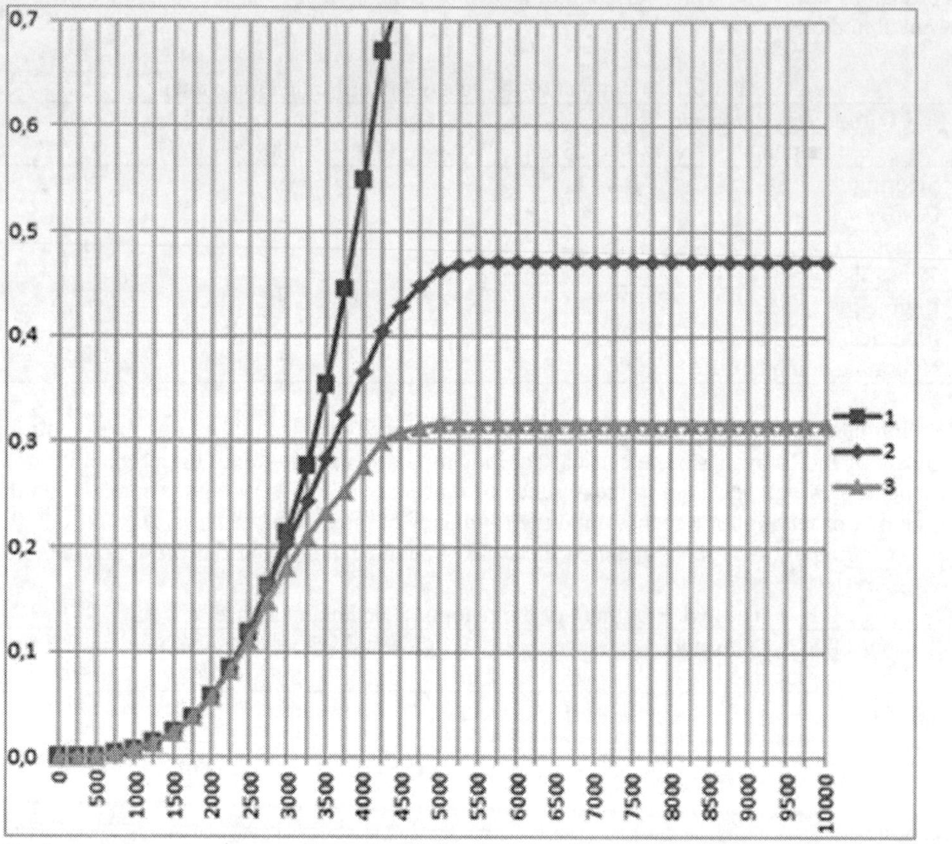

Figure 1. Probability of failure at operating intervals: 1 – in the absence of diagnosis; 2 – at the frequency of diagnosis established by the manufacturer; 3 – at the recommended frequency of diagnosis

References

[1] Khapugin R A and Karagodin V I 2018 Increase of accuracy and reliability of information about reliability of components of a machine based on tractor Belarus-82.1 *Youth and scientific-technical progress: a collection of papers* **1** 93–6

[2] Karagodin V I 2009 *Efficiency of car service system* (Moscow: Tekhpoligraftsentr) p 164

[3] Karagodin V I 2011 *Methods of substantiation of the structure of the repair cycle of cars and their components* (Moscow: Tekhpoligraftsentr) p 157

[4] Khapugin R A and Karagodin V I 2019 Substantiation of periodicity of maintenance of tractors Belarus *Lifting, construction, road, track machines and robotic systems: collection of reports* **1** pp 160–7

[5] Ivchenko G I 2010 *Introduction to mathematical statistics* (Moscow: Statistics) p 310

[6] Karagodin V I and Malyutin V O 2014 On the problem of predicting the complexity of repair mechanisms based on the results of diagnosis *XVIII Moscow international interuniversity*

conference "Lifting, construction, road, track machines and robotic systems" (vol 2) (Moscow: MADI) pp 172–4

[7] Karagodin V I and Pavlov A P 2018 Ensuring the reliability of land transport and technological means by reservation methods *Repair, restoration, modernization* **vol 5** pp 36-40

[8] Karagodin V I 2017 Repair modifications of products *Bulletin of the Moscow automobile and road state technical University (MADI)* **1** 55–9

[9] Felix Ng and Jennifer A Harding 2016 An eco-approach to optimise efficiency and productivity of a hydraulic excavator *Journal of Cleaner Production* **112** 3966–76

[10] W Lee Hansen 2009 A Total Quality Improvement Approach to Student Learning *Educational Innovation In Economics And Business Administration: The Case Of Problem-based Learning* **1** 20–29

IOP Conf. Series: Materials Science and Engineering **832** (2020) 012006 doi:10.1088/1757-899X/832/1/012006

Improvement of tribological characteristics of rubbing surfaces as one of ways to increase energy efficiency of power plants

M Yu Karelina[1], A V Pydrin[2], S M Gaidar[2], A S Vorontsov[3] and L Y Qeblawi[4]

[1]Department of Machinery Parts and Theory of Mechanisms, Moscow Automobile and Road State Technical University (MADI), 64, Leningradsky Ave., Moscow, 125319, Russia
[2]Department of Materials Science and Engineering Technology, Moscow Timiryazev Agricultural Academy, 49, Timiryazevskaya st., Moscow, 127550, Russia
[3]Department of Materials Science and Resource-Saving Technologies, Yanka Kupala State University of Grodno, 22, Ozheshko str, Grodno, 230023, Belarus
[4]Moscow Automobile and Road State Technical University (MADI), 64, Leningradsky Ave., Moscow, 125319, Russia

E-mail: karelinamu@mail.ru

Abstract. In this study, the team of authors solves the problem of increasing the energy efficiency of power plants by reducing losses while lowering the coefficient of friction. This problem is solved with the help of a multifunctional additive to lubricating oils, which are a surface-active substance (surfactant) capable of changing the friction conditions of contacting surfaces in a small amount when introduced into tribological units. The tests were carried out on a friction machine II5018 on a block-roller friction pair (the material of the roller is gray cast iron SCh20, the material of the block is steel 65). During the study, the moment and coefficient of friction were determined. A friction coefficient decrease of up to 50 % was obtained, which proves the effectiveness of the proposed compositions.

1. Introduction

Today, the problem of energy efficiency of power plants is relevant. Power plants are becoming increasingly loaded, and at the same time, they are subject to increased requirements for reliability and resource.

There are several ways to fulfill these requirements:

- Changing the design of the power plant.
- The use of various types of coatings (sprayed, deposited galvanic, etc.)

These methods, although promising, require significant costs both for research and development, and for testing and subsequent implementation of developments in production. In this study, the authors try to solve the problem in another way.

2. Problem Statement

In this work, the team of authors sets the task to solve this problem by using surface-active substances (surfactants). These substances are able to change surface energy, affecting the processes occurring at the phase boundary. In particular, in this study, borates of ethanolamides of carboxylic acids (BECA) were used [1]. The structure of the substance's molecule is shown in Fig. 1. These substances are multifunctional additives to motor and gear oils, have a wide range of applications, for example, they can significantly reduce friction and wear, and increase the corrosion resistance of ferrous and non-ferrous alloys [2, 3]. This article discusses the effectiveness of these compounds as an additive to industrial oil I-20A.

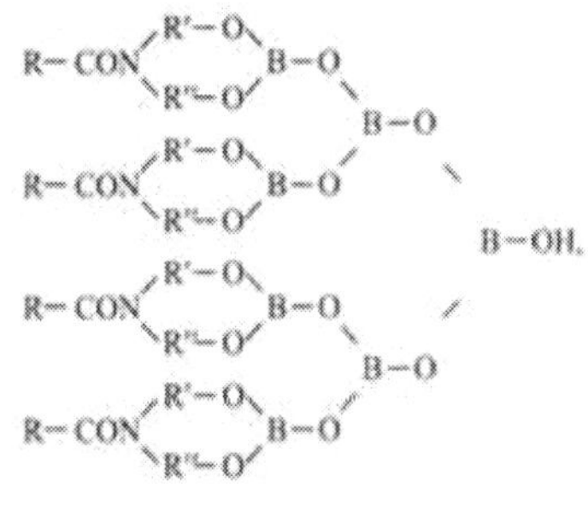

where R– (CH_2– CH_2 – CH_2–)
R',R''– (–CH_2– CH_2–)

Figure 1. The structural formula of the borate of ethanolamides of carboxylic acids.

3. Materials and Methods

For the study, two types of oil were taken:

- Industrial oil I-20A
- Industrial oil I-20A with the addition of 10 % BECA.

The tests were carried out on a friction machine II5018, on a pair of friction roller-block. The block is made of steel 65, the roller is made of cast iron grade SCh20. The appearance of the friction machine, block and roller is shown in Figure 2.

Figure 2. a – Appearance of the roller and test block; b – appearance of the friction machine II5018.

During the tests, the speed of rotation of the roller was constant, the clamping force of the samples changed from 200 to 800 N in increments of 200 N. The moment of friction force was determined, which was later used to determine the coefficient of friction by the formula:

$$k = \frac{M_{fr}}{F_N \times R} \tag{1}$$

where M_{fr} is the frictional moment;

F_N is the force of mutual clamping of samples;

R is the roller radius, equal to 0.025m.

4. Discussion of Results

The data obtained are summarized in Table 1, and the dependences of the friction moment and temperature on the clamping force of the samples for the studied compositions have also been plotted. The indicated dependencies are presented in Figures 3 and 4.

Table 1. The obtained experimental data

Composition	Clamping force of samples F_N , N	Friction moment M_{fr}, N*m	Coefficient of friction k
Oil I-20A	200	1.05	0.21
	400	1.5	0.15
	600	2.25	0.15
	800	2.85	0.143
Industrial oil I-20A with addition of 10 % BECA	200	0.75	0.15
	400	0.9	0.09
	600	1.2	0.08
	800	1.5	0.075

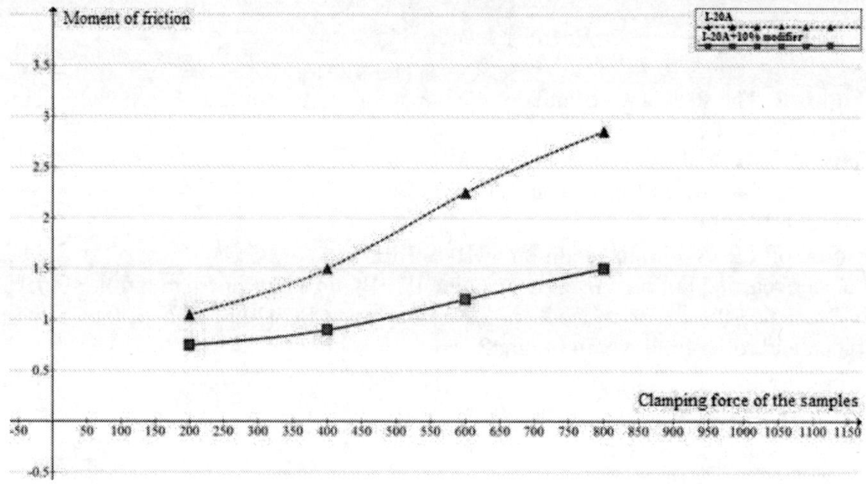

Figure 3. Dependence of the moment of friction force on the clamping force of samples

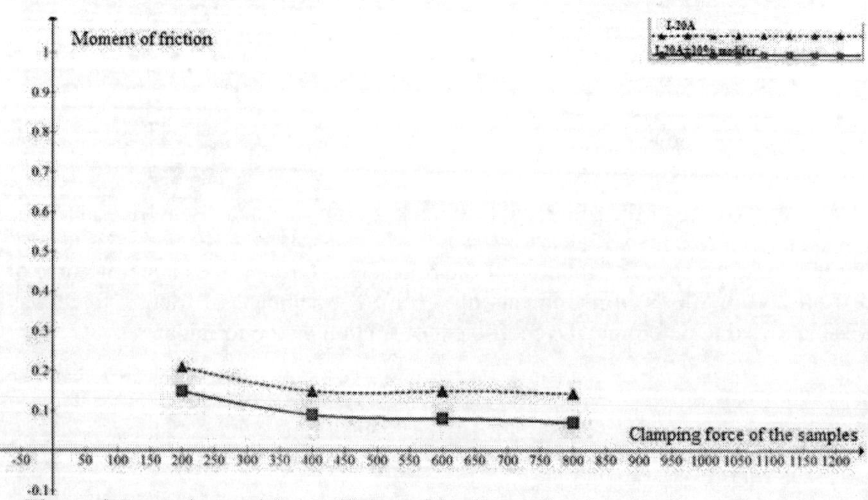

Figure 4. Dependence of the coefficient of friction on the clamping force of samples

IOP Conf. Series: Materials Science and Engineering **832** (2020) 012006 doi:10.1088/1757-899X/832/1/012006

From the above dependencies it can be seen that the use of BECA as an additive to lubricating oils leads to a significant decrease in the friction coefficient, which, accordingly, leads to an increase in the efficiency of the mechanism.

5. Conclusion

In the present work, a method for reducing friction in tribological conjugations using surfactants is clearly shown. It is known that with a decrease in the coefficient of friction, the temperature in a unit, its wear and tear also decrease, and its reliability increases. To quantify these parameters and their dependence on the experimental conditions, an additional series of experiments is required, the results of which will be presented by the authors in future works.

References

[1] Gaidar S M, Pydrin A V and Karelina M Yu 2014 Technology of conservation of automotive diesel engines with a working and conservation composition *Machinery and equipment for the village* **12** 18–23

[2] Gaidar S M, Dmitrievsky A L and Bykov K V 2012 Improving the performance of internal combustion engines using nanomaterials *Gruzovik* **11** 28–35

[3] Gaidar S M 2012 Improving engine performance with nanotechnologies *Works of GOSNITI* **110(1)** 91–5

IOP Conf. Series: Materials Science and Engineering **832** (2020) 012007 doi:10.1088/1757-899X/832/1/012007

Marine propulsion study of planning small-class amphibious vehicles

M Yu Karelina[1], V V Filatov[1,4], A V Klimov[1], D Yu Malakhov[2] and V G Sorokin[3]

[1]Department of Machinery Parts and Theory of Mechanisms, Moscow Automobile and Road State Technical University (MADI), 64, Leningradsky Ave., Moscow, 125319, Russia
[2]Department of Tractors and Amphibious Vehicles, Moscow Automobile and Road State Technical University (MADI), 64, Leningradsky Ave., Moscow, 125319, Russia
[3]Department of Materials Science and Resource-Saving Technologies, Yanka Kupala State University of Grodno, 22, Ozheshko str, Grodno, 230023, Belarus

E-mail: 2vfilatov@gmail.com

Abstract. The authors analyze the propulsion of the planing hulls of the various small-class amphibious vehicles. The magnitude of the resistance forces depending on the movement speed is determined experimentally. For the correct formulation of experiments and their results, the study of various models of the planing small-class amphibious vehicles must comply with the dynamic similarity of the viscous friction and gravity forces. To compare planning effect provided by the design features of various amphibious models a relative measure of the hydrodynamic qualities or its inverse value can be used. In this article, numerical simulation by means of software complex is conducted to determine the hydrodynamic properties of the different planing small-class amphibious models. To verify the adequacy and accuracy of the generated numerical model, a comparison of the computer simulation results with the data of the full-scale experiment is used. Propulsion performance is compared for amphibious models with different hull contours and additional equipment.

1. Introduction

The propulsion of the planing object is largely determined by the shape of its body. Hydrodynamic quality of high-speed planing boats is obviously better than that of planing small-class amphibious vehicles due to the presence of the latter wheel arches and exposed chassis components. To determine the hydrodynamic quality of the amphibious vehicle models with different hull contours and the composition of additional equipment, one can use the basic provisions of the theory of motion of the planning surface object described in [1] and [2].

The condition of steady (uniform) translational motion of a self-propelled surface object is expressed as follows:

$$P_e = R_t, \quad (1)$$

where P_e – effective thrust developed by the propeller, kN,

R_t – the value of the total (towing) resistance, kN.

Therefore, when designing the propellers of a self-propelled surface object, it is necessary to know the force of movement water resistance. It is considered that the total hydrodynamic resistance of amphibious machines R_b, kN, is represented as the sum of three components:

$$R_t = R_{fr} + R_{vortex} + R_{wave}, \tag{2}$$

where R_{fr} – the frictional resistance caused by the viscosity of water, kN,

R_{vortex} – vortex resistance the shape of the hull and suspension of the car, kN,

R_{wave} – wave resistance due to the redistribution of pressure, kN.

A detailed description of the resistance forces components when moving on the water with speed v_w, m/s, is presented in [3]. The value of the total resistance can be represented as follows:

$$R_t = \zeta \frac{\rho v_w^2}{2} \Omega, \tag{3}$$

where $\rho = 1000$ kg/m^3 – water density,

Ω – wetted surface area, m^2,

ζ – the dimensionless ratio of the full water resistance.

To determine analytically the values of the coefficient ζ and the wetted surface area Ω is an extremely time-consuming task due to the complex nature of the hydrostatic and hydrodynamic forces distribution of the amphibian on the water. The magnitude of the resistance forces to the surface objects movement depending on the speed is determined experimentally.

For the correct formulation of experiments and generalization of their results in the study of various amphibian models must comply with the dynamic similarity of the forces of viscous friction and gravity. The Froude number Fr_D characterizes the ratio of inertia forces to gravity.

The condition of transition of the surface object to the planing mode is:

$$v_w > 3(g(\frac{D}{\gamma})^{1/3})^{1/2}, \tag{4}$$

where $g = 9.81$ m/s^2 – gravitational acceleration,

D – weight displacement, N,

$\gamma = 9810$ N/m^3 – specific weight of water.

We can compare the planing effect provided by the design features of the various models by the relative criterion of hydrodynamic quality k_g or by its inverse value ε:

$$k_g = \frac{D}{R_t}, \tag{5}$$

$$\varepsilon = \frac{R_t}{D}. \tag{6}$$

The lower the reverse quality ε, the less power is required to overcome the transition mode.

Thus, having obtained in the course of experimental studies of a series of amphibious models with different configuration towing characteristics $R_t = f(v_w)$ and the change of the trim $\theta = f(v_w)$, we can reconstruct these graphs in relative terms, i.e. $\varepsilon = f(Fr_D)$ and $\theta = f(Fr_D)$. According to the data obtained, we can also determine the dependence of the towing power on the speed $N_t = f(v_w)$ and the specific towing power on the Froude number $N_{tsp} = f(Fr_D)$. The presence of these dependencies for the planing small-class amphibious vehicles with different hull design and additional equipment will allow us to determine the machines characteristics in first approximation at the stage of their design.

2. Problem statement

We can conduct a numerical simulation by means of software complex for the solution of a problem of the hydrodynamic properties determination of the different amphibious models [4, 5]. In conditions of limited material resources, the main advantage of computer modeling is the possibility of experiments on objects, full-scale modeling of which involves the creation of a series of samples and the use of expensive infrastructure. In our case, the simulation of test pools for full-scale models of the bodies of surface objects is made. Computer simulation also allows you to measure any parameters at any point in the computational domain. The main disadvantage of computer modeling is the cost of computing resources and time.

The solver of the software complex is based on the equations of computational fluid dynamics characterizing flow processes: the equations of continuity, state, Navier-Stokes and Euler, as well as the laws of conservation of momentum and energy. Taking into account the peculiarities of the

IOP Conf. Series: Materials Science and Engineering **832** (2020) 012007 doi:10.1088/1757-899X/832/1/012007

problem to be solved, the general equations can be supplemented with partial dependencies, taking into account turbulence, transport of substances, multiphase, etc.

For a specific problem, taking into account its features, a system of nonlinear differential equations is compiled, the solution of which is carried out by means of discretization methods. The general methodology of the numerical experiment is shown in Figure 1.

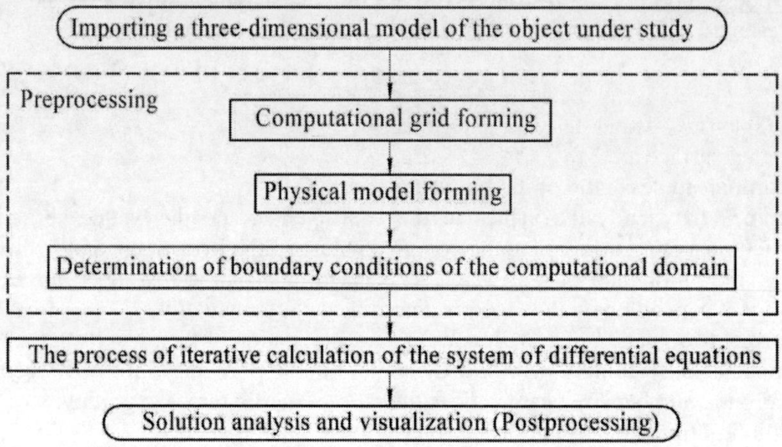

Figure 1. General methodology of the numerical experiment

The geometry of the object under study is given by a three-dimensional solid-state model built in the CAD-system (computer added design), including the surfaces that form the external contours of the object (planing small-class amphibious vehicles), as well as the boundaries of the movable and stationary part of the computational domain. The problem is symmetric with respect to the diametrical plane of the machine; therefore, in order to save computing resources, the motion of half of the object in the computational domain truncated by the corresponding plane is simulated.

Calculation model (preprocessing) is prepared. The formation of the computational grid (Figure 2) occurs in three stages.

At the first stage, the surface of the imported model is transformed into triangular cells (triangulation) and a surface grid is created. Then the volume grid is formed. To discretize the computational domain, a prismatic grid with cell truncation at the boundaries is used. An additional model of prismatic layers is used to improve the quality of reproduction of processes directly on the border of the object with the environment. The final step is to set the relationship between the movable and fixed grids on the model of the superimposed grid.

In the next step, structure of the physical model based on the problem to be solved is set. In our case, the physical model is formed from the following submodels: "three-dimensional", "non-stationary implicit", "Eulerian multiphase", "volume of liquid VOF", "turbulent flow", "k-omega model of turbulence", "waves VOF", "gravity" and some others. Then parametric synthesis of the physical model is carried out: the Eulerian phases are set, the type and velocity of the flow, the position of the free surface relative to the amphibious vehicle model, the freedom degree and inertial characteristics (mass and center of mass, moments of inertia) of the object under study.

Let us proceed to the definition of boundary conditions. To do this, set the types of boundaries for each of the objects of the computational environment: the model of amphibious vehicle ("wall"), the plane of entry ("speed at the entrance"), the plane of symmetry ("plane of symmetry"), as well as mobile ("superimposed grid") and fixed ("pressure at the exit") calculation areas.

The next stage of preprocessing is the choice of the solver settings – the time step and the number of internal iterations. When determining these parameters, it is necessary to ensure the convergence and the necessary accuracy of the solution of the problem.

In conclusion, the formation of the computational model will determine the list of recorded output parameters. As a result of computer simulation of amphibious models propulsion, we obtain dependences on the speed v_w, m/s, the following values: movement resistance R_t, N, and trim θ, °.

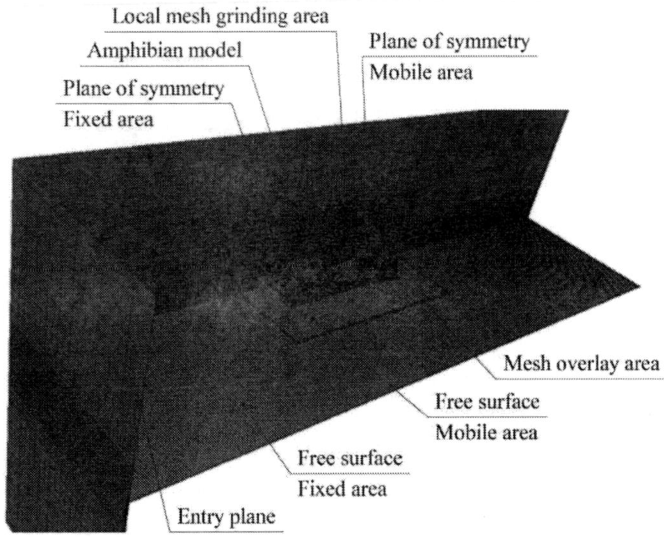

Figure 2. Generated calculation grid

Also the graphical plots for the parameters depending on the speed v_w, m/s, can be obtained: pressure on the surface of the hull p_h, Pa, phase speed of the environment v_{ph}, m/s, free surface.

After preprocessing, initialization of the solution is made and the iterative calculation process is started. At the last stage (postprocessing) it is necessary to process the output. For each model under test, we obtain graphs of the towing power dependence on the speed of $N_t = f(v_w)$. Also make absolute function $R_t = f(v_w)$ and $\theta = f(v_w)$ relative to the mean: $\varepsilon = f(Fr_D)$, $\theta = f(Fr_D)$ and $N_{tsp} = f(Fr_D)$. These indicators for models with different hull contours and additional equipment will be compared, and then conclusions about the impact of the latter on the propulsion of the planing small-class amphibious vehicles will be made.

Each model has a number from "0" to "5" in the order of the experiment. To verify the adequacy and accuracy of the generated numerical model, a comparison of the computer simulation results with the data of the full-scale experiment is made. Model "0" is the standard planing hull of the high-speed small class boat. Theoretical drawings when landing on a flat keel, configuration and weight and size characteristics of all the models are shown in Figure 3.

The drawings show coordinate systems, waterlines, and the positions of the gravity (X_G, Z_G) and magnitude (X_C, Z_C) centers. Graphs of reverse quality and trim depending on the Froude number for the body of the model "0" obtained during the full-scale experiment are shown in paper [6].

The types of used planing hull contours are monohedron (body with constant deadrise angle of the bottom from the transom to amidships 10–17°), low-deadrise hull (the hull with a constant deadrise angle of the bottom from the transom to amidships to 10°) and trimaran (main hull with keel and two side sponsons). A more detailed description of the types of contours of the hulls of high-speed small vessels, as well as their advantages and disadvantages is given in [2].

The effect of additional equipment on the hydrodynamic properties of planing small-class amphibious vehicles is considered, which includes the following elements: bow air tank, bow folding (or retractable) shield, stern air tanks, and transom plates.

The results of computer simulation with the data of full-scale experiment for the model "0" are compared. The combined graphs $\varepsilon = f(Fr_D)$ and $\theta = f(Fr_D)$ obtained during full-scale tests and numerical experiments are shown in Figure 4.

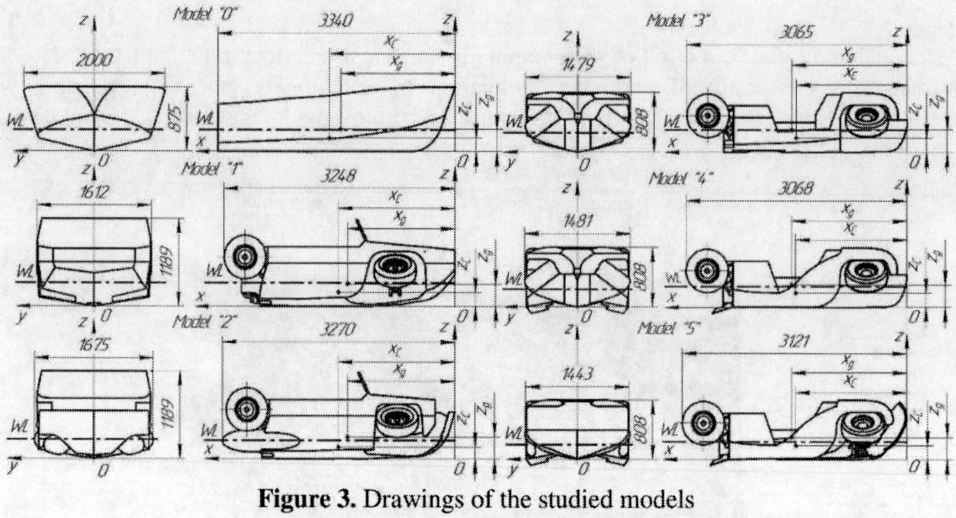

Figure 3. Drawings of the studied models

As can be seen from the graphs, in the Fr_D < 2.2 zone, the maximum divergence of values is 5 %, and the parameters obtained from field tests are more important. In the Fr_D > 2.2 zone, the maximum divergence is 6 %, and the values of the parameters determined during the numerical experiment prevail. Thus, the degree of accuracy of the generated computational model, determined by the quality of the computational grid, is 6 %.

The maximum propulsion values of the studied models are presented in Table 1.

Table 1. Values of the maximum propulsion indicators of the studied models

Parameter	Number of experiment					
	0	1*	2	3**	4	5
ε	0.19	0.74	0.48	0.79	0.47	0.56
θ, °	6.6	–	19.4	22.0	19.0	19.9
N_{tsp}, kW/t	28.4	72.2	47.5	54.2	27.8	32.7

* – trim is constant; ** – does not reach a maximum speed of 15 m/s

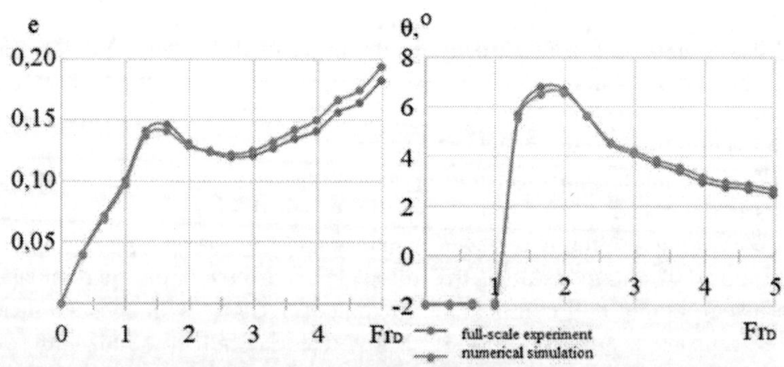

Figure 4. Comparison of field tests and numerical experiment data for model "0"

3. Research Questions

1. The study of hydrodynamic properties of the planing small-class amphibious vehicles with different designs of the hull and additional equipment.

2. Development of constructive measures to improve the propulsion of the planing small-class amphibious vehicles.

4. Purpose of the Study

The purpose of this study is to obtain graphs of the main propulsion indicators depending on the Froude number for the amphibian models series.

5. Methods of the Research

In the article the authors used the analytical, comparative, numerical simulation methods as well as the method of the system approach.

6. Findings

1. Model "0" has the lowest values $\varepsilon = 0.19$, $\theta = 6.6°$, $N_{tsp} = 28.4$ kW/t in the range of Froude numbers $Fr_D < 5$. In addition, this case has a maximum value of static buoyancy reserve 61 % among all the studied models.

2. Model "1" has ε 3.9 times greater and N_{tsp} 2.5 times greater than those of the etalon model "0". Static buoyancy reserve is 56 %. This model of planing small-class amphibious vehicle has the worst hydrodynamic quality of all considered.

3. Model "2" has a 2.5 times greater ε, 2.9 times greater θ and 1.7 times greater N_{tsp} than the standard. The use of such equipment allows reducing ε by a factor of 1.6 and 1.5 times N_{tsp} relatively planing small-class amphibious vehicle without him. In addition, the static buoyancy reserve increases 1.1 times and is 61 %.

4. Model "3" has a 4.2 times larger ε, 3.3 times larger θ and 1.9 times greater N_{tsp} than the model "0". In addition, this case has minimum static buoyancy reserve 44 % among all investigated models. In the dispersal of amphibians to this configuration, there is the effect of high trim, which requires using the high power engine or trim tabs to enter the clean mode of gliding.

5. Model "4" has a 2.5 times greater ε, 2.9 times greater θ than the standard. N_{tsp} is comparable to the standard. Static buoyancy reserve is 48 %. This type of amphibian has the best hydrodynamic quality with respect to all studied, provides a minimum trim on the stern and requires the lowest energy intensity of the amphibian.

6. Model "5" is 2.9 times larger than ε, 3 times larger θ and 1.2 times greater N_{tsp} than the standard. The static buoyancy reserve is equal to 53 %. This type of amphibian is inferior to the previous one in propulsion, but in terms of the set of water-moving properties, it is the best of all studied.

7. Conclusion

The propulsion study of planing small-class amphibious vehicle models with different configurations is made. As a result, it was found that with the help of additional equipment it is possible to provide less energy-consuming transition of the machine from displacement to planing mode and achieve high speeds (at least 15 m/s).

The best propulsion of all the studied models is amphibious vehicle with the contours of the "trimaran", a retractable nose shield, air tank and controlled transom plates ($\varepsilon = 0.47$, $\theta = 19°$, $N_{tsp} = 27.8$ kW/t).

References

[1] Apukhtin A P 1953 *Water resistance to the movement of ships* (Leningrad: Mashgiz)
[2] Egorov I T 1971 *Hydrodynamics of high-speed vessels* (Leningrad: Sudostroenie)
[3] Basin A M 1969 *Propulsion and handling of ships* (Moscow: Transport)
[4] Filatov V V 2017 Hydrodynamic study of perspective high-speed small-class amphibious vehicle *Bull. of civil engineers* **2(61)** 219
[5] Pechenyk A V 2015 Optimization of ship hull form to reduce resistance to movement *Digital Marine Technol.* **2** 10–20
[6] Clement E P and Blaunt D L 1963 Resistance tests of a systematic series of planing hull forms *Society of Naval Architects and Marine Engineers* 491

IOP Conf. Series: Materials Science and Engineering **832** (2020) 012008 doi:10.1088/1757-899X/832/1/012008

Quality research of paintwork of Volkswagen Tuareg car body elements

F A Kipriyanov, R A Shushkov, A S Mikhailov, N I Kuznetsova, S V Gaididei and O I Barinova

Vologda State Dairy Farming Academy by N.V. Vereshchagin, Vologda, Molochnoe 160555, Russia

E-mail: kipriyanovfa@bk.ru

Abstract. Appearance is one of the criteria forming the aesthetic perception of a particular vehicle. However, the appearance of the car is a factor determining the consumers' opinion about the brand of the car. Many vehicles during their operation for many different reasons undergo repair and restoration operations. As a rule, repairing body elements the paintwork is performed, the quality of which and the observance of technologies and standards during work ultimately determine the durability of the paintwork. The quality of the paintwork, the presence of damage, fragments of destruction is one of the essential aspects that interest consumers in the secondary market. The article presents the results of a study of the paintwork quality of the Volkswagen Tuareg car body elements; the presence of defects is identified, the causes of their occurrence are detected.

1. Introduction

The color quality of the car body has become the main aesthetic factor affecting the process of buying a car [1]. Vehicle manufacturers, recognizing the demand of the market, focus on aesthetics, environmental protection, and the quality of the paint applied to the car. The color and quality of the paint largely determine the decision of the buyer [2]. Growing customer expectations regarding the appearance of cars are forcing car paint shops to scrutinize car bodies. Currently, qualified workers visually inspect each car body for the identification and elimination of paint defects. However, people can neither detect nor judge defects objectively and reliably for a more extended period [3].

The complicated and broad requirements that the exterior of the car must comply with prompted BMW to make the technological leap from liquid, transparent coating to powder coating capabilities. A transparent coating creates an unsurpassed gloss effect, and a transparent coating on a powder basis makes the surface of the car even shinier [4].

According to the current classification of coatings for industrial products by their appearance, decorative coatings for cars are class I–II coatings (out of seven regulated). Decorative coatings have two different contact surfaces: one contact surface is with the external environment (usually gaseous or liquid); the other contact medium is with a solid or substrate.

Decorative coatings are films bonded by adhesive forces to a hard surface. The structure of the decorative coating of the car consists of several layers.

Phosphating the metal obtains the first layer. Phosphating is the processing of metal by exclusive means based on phosphate salts, as a result of which a protective film appears on the surface among the options for protecting the metal from corrosion by phosphating, the most known phosphating primer. Hydroabrasive phosphating and chemical treatment of metal are also applied. In addition to protecting the metal, the film provides increased adhesion of the metal to the coating materials.

The second layer is soil, which is the base of the entire coating. The purpose of the second layer:

- provide adhesion of the decorative coating to the substrate;
- protect the substrate from the effects of destructive factors;
- protect the subsequent layers of the coating from the effects of the substrate (for example, alkalis or plasticizer made of plastic).

The third layer is the base coat. The purpose of the third layer:

- strengthen the protective functions of the primer layer;
- ensure the alignment of the soil and the substrate;
- maximize the properties of the coating (light reflection, color, diffusion inhibition);
- increase the protective properties of the coating to mechanical stress.

The last (top) layer is varnish (if necessary). The purpose of the fourth layer:

- basic protection against the effects of the atmosphere (UV radiation, precipitation, air humidity), chemical agents (acids, alkalis, aggressive gases) and solvents (liquid motor fuel);
- solutions to functional problems (decorative, optical, electrical, and other properties).

The decorative coating, initially, laid down specific properties established by the manufacturer. The car operation leads to its aging, the accumulation of damage, and destruction due to interaction with the environment. As a result, defects appear in the product.

According to the requirements of GOST 9.311-87, corrosion lesions are determined with the naked eye with visual acuity of 1.0-0.8, with normal color sensation at a distance of 25 cm from the controlled surface with a sample illumination of at least 300 lux.

There is a possibility to establish the type of corrosion damage by using a magnifying glass or other optical means, for example, a microscope.

2. Research Objectives

1. To identify the presence of a car Volkswagen Tuareg defects in the decorative coating of moldings.
2. If there are defects in the decorative coating of the moldings of the Volkswagen Touareg car, identify the cause of their occurrence.

3. Materials and methods

The object of research is a Volkswagen Tuareg passenger car, 2016 model year.

Research equipment and tools:

- Camera SONY Cyber-shot 14.1 megapixels.
- Magnifier with eight magnification.
- ET-11P feeler gage.
- Digital microscope Digital microscope 500.
- Transparent mesh, area 1 dm 2.
- Ruler.

4. Results and discussions

Surface treatment plays an essential role in the perception of product quality and is the first that a potential buyer sees. Today, the final body repair of cars and trucks must inevitably provide the required surface quality. Defects that occur in the paint shop, such as dust particles, are eliminated by manual grinding/polishing, which leads to other types of defects when the last polishing step is performed incorrectly or is not completed [5]. With the improvement of painting techniques in the automotive industry on car bodies, it is possible to achieve high-quality paints with high gloss, clear images, and deep orange peel. In this case, another defect, called the “metal texture effect,” is becoming increasingly crucial for the appearance of the paint [6]. The automatic detection of small defects (up to 0.2 mm in diameter) on the surfaces of the car body after the painting process is currently one of the most severe problems of quality control in the automotive industry. Although several systems have been developed over the past decade to solve this problem, as far as we know, they were focused exclusively on flat surfaces and were unable to inspect other parts of the surfaces, namely style lines, edges, and corners, as well as deep concavities [7].

The inspection of the vehicle was on the premises of the technic zone in the daytime, with artificial lighting.

In the presence of the car owner, a specialist checked the decorative coating of moldings condition and the paintwork of the car body.

During a visual inspection found:

- the decorative molding on the right front fender has a chrome finish;
- the decorative molding on the right front door has a chrome finish;
- the decorative molding on the right rear door has a chrome finish;
- the decorative trim (top) on the rear bumper has a chrome finish;
- the decorative trim (lower) on the rear bumper has a chrome finish;
- the decorative molding on the left rear door has a chrome finish;
- the decorative molding on the left front door has a chrome finish;
- the decorative molding on the left front wing has a chrome finish;
- the thickness of the car paintwork is in the range of 177 to 348 microns.

Analysis of measurements of the thickness of the paintwork suggests that the tailgate was repainted.

Measurements of the thickness of the tailgate are in the range from 305 to 348 microns. The remaining ferromagnetic elements of the body are in the factory color. The thickness of the paint coating is in the range from 177 to 271 microns. The thickness of the factory paint should be in the range from 140 to 180 microns. The large values of the thickness of the paintwork are explained by the fact that a protective coating (presumably quartz) is additionally applied to the body. The state of the paintwork evidences the fact of applying a protective layer after preliminary washing and approval of the car owner.

An external inspection of the decorative coating (front and rear door and front wing moldings) was performed, including using a Digital microscope 500 electron microscope. As a result of the inspection, it was found that on all the elements mentioned above of the car body there are local areas of coating destruction on which the substance was formed gray to brown, similar in appearance to corrosion products. Following GOST 28246-2006, these areas of destruction are classified as "metal corrosion".

According to the methodology proposed in GOST 9.311-87, the area of the destroyed coating is determined using the following steps: on the surface to be evaluated, experts apply a plate of transparent material with a mesh or wire mesh applied to it (the side of the square can be 10 mm or another, which depends on the size of the samples and areas of destroyed areas) or draw a rectangular shape and calculate its area.

Assessment of the molding of the front left wing led to the determination of the following types of corrosion damage:

A – surface tarnishing (non-removable translucent film) on 20 % of the surface;

G – corrosion of the coating with the formation of oxide-like products (unremovable dark spots and spots) on 3 % of the surface;

D – corrosion of the coating with the formation of salt-like products (gray spots and dots) on 2 % of the surface;

K – corrosion of the base metal (spots and dots) on 1% of the surface.

$$S_{decorative\_coating} = \sum_{i=1}^{N} K_i \times S_i = 0{,}2\times20+0{,}8\times3+1\times2+1\times1=8{,}4\%.$$

Assessment of the molding of the front left door led to the determination of the following types of corrosion damage:

A – surface tarnishing (non-removable translucent film) on 20 % of the surface;

G – corrosion of the coating with the formation of oxide-like products (unremovable dark spots and spots) on 2 % of the surface;

D – corrosion of the coating with the formation of salt-like products (gray spots and dots) on 4 % of the surface;

E – corrosion of the sublayer (spots and dots) on 1 % of the surface.

$$S_{decorative\_coating} = \sum_{i=1}^{N} K_i \times S_i = 0{,}2\times20+0{,}8\times2+1\times4+1\times1=10{,}6\ \%.$$

Assessment of the rear left door molding led to the identification of the following types of corrosion damage:

A – surface tarnishing (non-removable translucent film) on 20 % of the surface;

G – corrosion of the coating with the formation of oxide-like products (unremovable dark spots and spots) on 2 % of the surface;

D – corrosion of the coating with the formation of salt-like products (gray spots and dots) on 2% of the surface;

K – corrosion of the base metal (spots and dots) on 1 % of the surface.

$$S_{decorative\_coating} = \sum_{i=1}^{N} K_i \times S_i = 0{,}2\times20+0{,}8\times2+1\times2+1\times1=8{,}6\ \%.$$

Assessment of the front right wing molding led to the determination of the following types of corrosion damage:

A – surface tarnishing (non-removable translucent film) on 20 % of the surface;

G – corrosion of the coating with the formation of oxide-like products (unremovable dark spots and spots) on 1 % of the surface;

D – corrosion of the coating with the formation of salt-like products (gray spots and dots) on 2 % of the surface;

E – corrosion of the sublayer (spots and dots) on 1 % of the surface.

$$S_{decorative\_coating} = \sum_{i=1}^{N} K_i \times S_i = 0{,}2\times20+0{,}8\times1+1\times2+1\times1=7{,}8\ \%.$$

Assessment of the front right door molding led to the determination of the following types of corrosion damage:

A – surface tarnishing (non-removable translucent film) on 20 % of the surface;

G – corrosion of the coating with the formation of oxide-like products (unremovable dark spots and spots) on 2 % of the surface;

D – corrosion of the coating with the formation of salt-like products (gray spots and dots) on 2 % of the surface;

K – corrosion of the base metal (spots and dots) on 2 % of the surface.

$$S_{decorativ\_coating} = \sum_{i=1}^{N} K_i \times S_i = 0{,}2\times20+0{,}8\times2+1\times2+1\times2=9{,}6\ \%.$$

Assessment of the rear right door molding led to the determination of the following types of corrosion damage:

A – surface tarnishing (non-removable translucent film) on 15 % of the surface;

G – corrosion of the coating with the formation of oxide-like products (unremovable dark spots and spots) on 2 % of the surface;

D – corrosion of the coating with the formation of salt-like products (gray spots and dots) on 2 % of the surface;

K – corrosion of the base metal (spots and dots) on 1 % of the surface.

$$S_{decorative\_coating} = \sum_{i=1}^{N} K_i \times S_i = 0{,}2\times15+0{,}8\times2+1\times2+1\times1=7{,}6\ \%.$$

IOP Conf. Series: Materials Science and Engineering **832** (2020) 012008 doi:10.1088/1757-899X/832/1/012008

Assessment of the upper lining on the rear bumper led to the determination of the following types of lesions:

Peeling (expansion) of the decorative coating under chrome on an area of 3 %.

Evaluating the lower lining on the rear bumper led to the identification of the peeling (expansion) of the decorative coating under chrome on an area of 2 %.

Determine the estimated score:

- Decorative properties of the coating moldings – AD 4.
- Decorative overlays of a back bumper – HELL 5.

By the value of the area of corrosion damage to the base metal (2 %), we determine the estimated score of the protective properties of the coating – AZ 6.

For types of destruction of protective properties, if necessary, it is allowed to put in brackets the designation of the size of a typical defect.

The study found that the decorative body moldings of the Volkswagen Tuareg car have the following groups of damage (defects) of the coating:

- Fading of the coating surface.
- Corrosion of the coating with the formation of oxide-like products.
- Corrosion of the coating to form salt-like products
- Corrosion of the base metal.
- Peeling (swelling) of the coating.

A detailed analysis of damage with magnification up to 500 times (using an electron microscope) allows us to unequivocally state that these defects occurred both with damage to the decorative layer (Fig. 1, 2), and without mechanical damage (Fig. 3).

Figure 1. Image of the enlarged image of the destruction of the molding on the left side

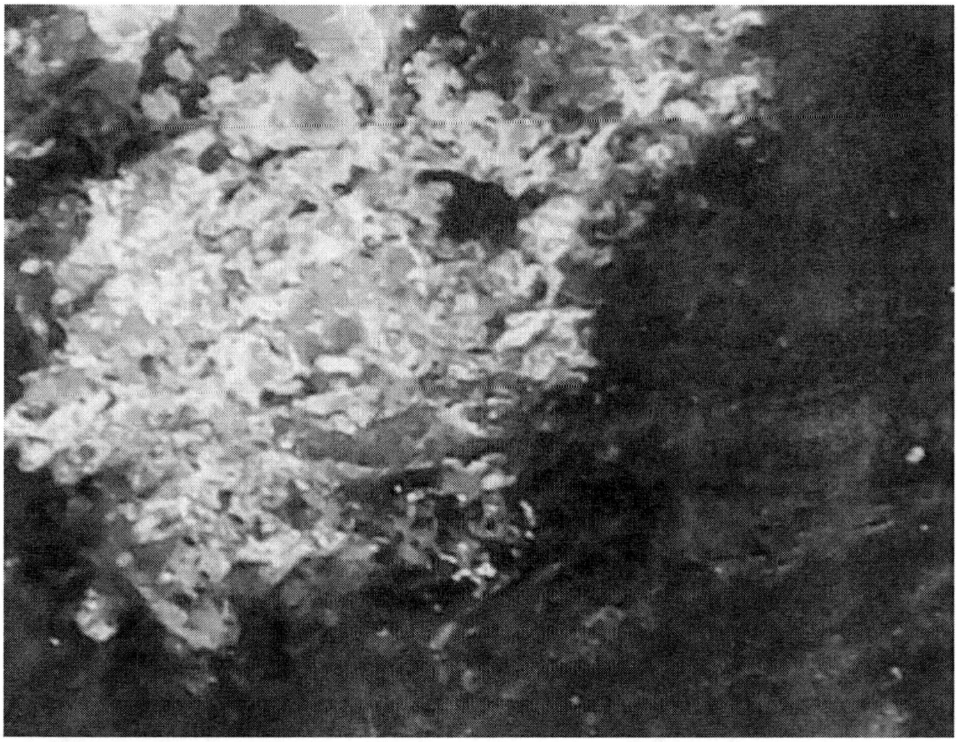

Figure 2. Image of an enlarged image of the destruction of the molding on the right side

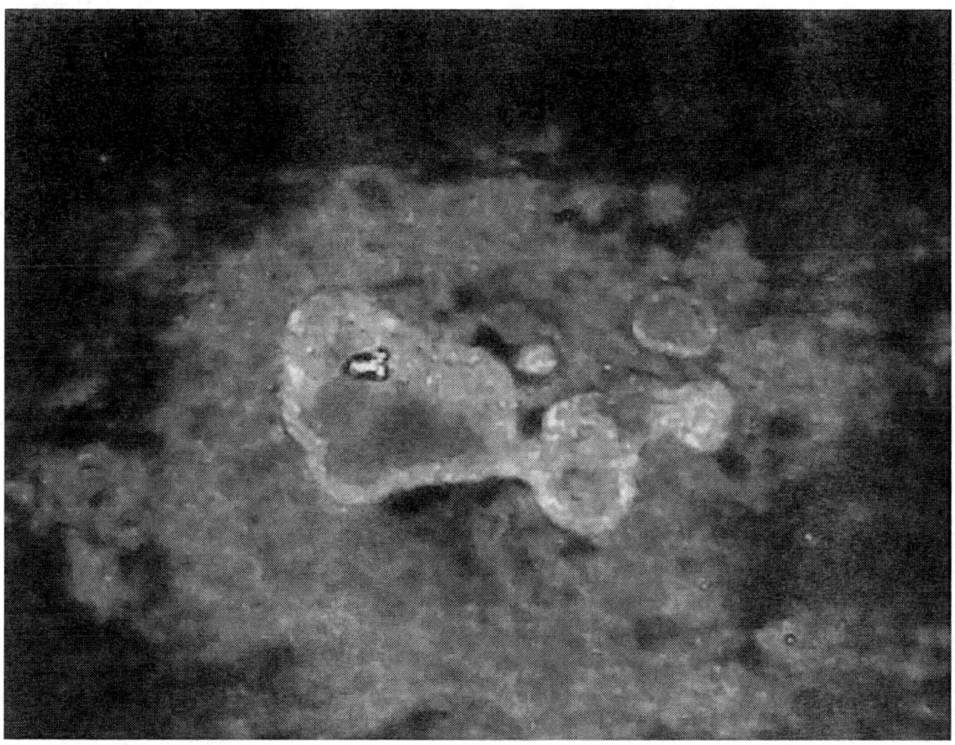

Figure 3. Image of an enlarged image of the destruction of the molding on the right side

IOP Conf. Series: Materials Science and Engineering **832** (2020) 012008 doi:10.1088/1757-899X/832/1/012008

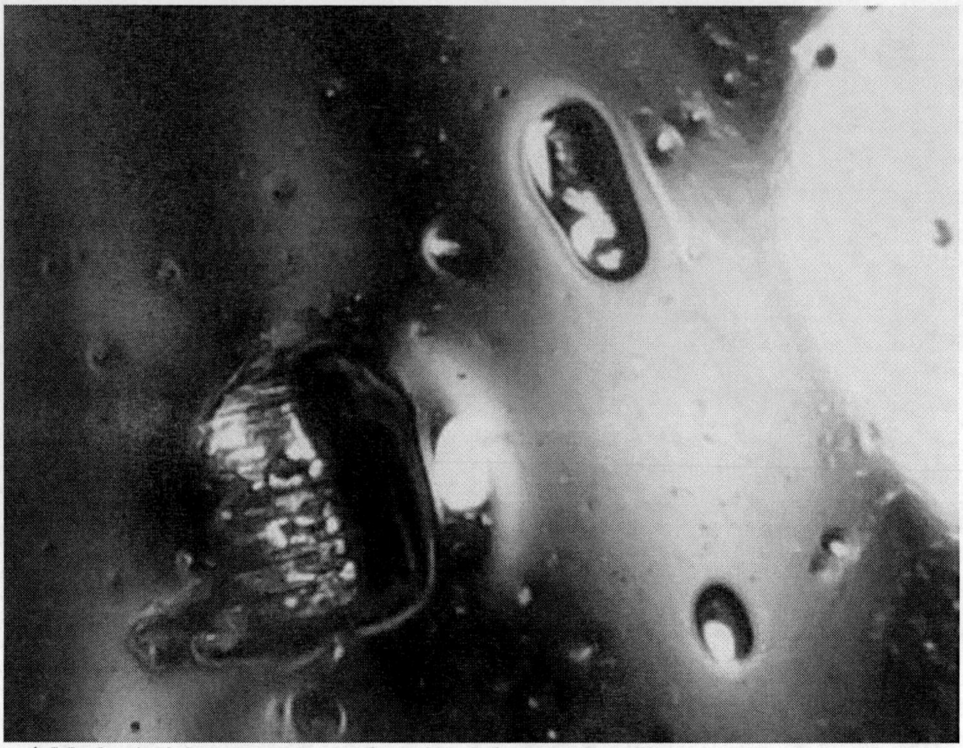

Figure 4. Mechanical damage and manufacturing defects on the grille

The expansion (peeling) of the decorative coating on the upper and lower plates of the rear bumper (Fig. 2, Fig. 3) has no visible damage from mechanical stress, and was formed as a result of insufficient adhesion, which is seen when comparing with how mechanical damage is reflected in the grill radiator (Fig. 4).

It should be noted that the decorative moldings on the door frames do not have similar damage, although they were in the same operating conditions.

The decorative coating has defects, including without external mechanical damage; however, not all compared decorative elements have characteristic defects. Given this fact, we can make an unambiguous conclusion that the defects of moldings (destruction of the decorative layer) on the right front wing, right front door, right rear door, left rear door, left front door, left front wing, on the rear bumper overlays as a result of a violation of the coating technology.

5. Conclusion

The car Volkswagen Touareg, there are defects in the decorative coating of moldings namely; tarnishing of the surface of the coating, corrosion of the coating with the formation of oxide-like and salt-like products, corrosion of the sublayer and base metal, delamination of the coating (expansion).

Manufacturing defects have moldings; front right-wing, right front door, right rear door, left the rear door, left the front door, left front wing, and upper and lower rear bumper trim.

Defects in the decorative coating of moldings and linings for the rear bumper of a Volkswagen Tuareg car were formed as a result of a violation of the coating technology.

References

[1] Takebuchi S, Nakamura T, Asami H and Amasaka K 2012 The automobile exterior color design approach model *J. of Japan Industrial Management Association* **62(6)** 303–10 Retrieved from: www.scopus.com

[2] Thomson J 2005 Innovation in car paints *Polymers Paint Colour J.* **195(4488)** 30–1 Retrieved from: www.scopus.com

[3] Kieselbach K K, Nöthen M and Heuer H (2019 Development of a visual inspection system and the corresponding algorithm for the detection and subsequent classification of paint defects on car bodies in the automotive industry *J. of Coatings Technol. and Res.* **16(4)** 1033–42 doi:10.1007/s11998-018-00178-y

[4] Wimmer W 2000 Powder clear coat – A quantum leap in automotive paint technology *SAE Technical Papers* doi:10.4271/2000-01-1359

[5] Rebeggiani S, Wagner M, Mazal J, Rosén B and Dahlén M 2018 Detection of paint polishing defects *Surface Topography: Metrology and Properties* **6(2)** doi:10.1088/2051-672X/aabfb5

[6] Kang H, Butler C and Yang Q 1999 Evaluation of metal texture effect on paint appearance. *Paper presented at the Proceedings of SPIE – the International Society for Optical Engineering* **3835** 202–9 Retrieved from: www.scopus.com

[7] Molina J, Solanes J E, Arnal L and Tornero J 2017 On the detection of defects on specular car body surfaces *Robotics and Computer-Integrated Manufacturing* **48** 263–78 doi:10.1016/j.rcim.2017.04.009

IOP Conf. Series: Materials Science and Engineering **832** (2020) 012009 doi:10.1088/1757-899X/832/1/012009

The problem of manipulation and angular orientation of gripping devices of construction robots

E V Marsova, S B Benevolenskiy, M Uu Abdulkhanova, V S Ershov and A.G Savelyev

Moscow Automobile and Road State Technical University (MADI), 64, Leningradsky Ave., Moscow, 125319, Russia

E-mail: dm@madi.ru

Abstract. The article describes the method of simple kinematic connections, which allows to organize at the matrix level the solution of problems of spherical trigonometry and angular orientation, which is reduced to the decomposition of the original matrix description of the problem into a system of simple matrices, the sequential solution of which individually or in combinations, allows to produce unambiguous results in the absence of restrictions in the original data. The mathematical apparatus of homogeneous transformations makes it possible to calculate the values of angular reversals in all intermediate joints, from the first to the n-1st, uniquely determining their required spatial orientation, as well as the spatial orientation of the end section of the switchgear.

A number of Russian and foreign research centers are working on the development of crane-manipulators with various types of bonds for the installation of building elements of buildings. Depending on the specific purpose, they use rigid, flexible and combined connections. Some constructions are equipped with a device for temporary support orienting body on the overlap of the working horizon in order to ensure the necessary positioning accuracy. A feature of these designs is the use of special load-gripping devices with several degrees of mobility, providing accurate orientation of the mounted elements. These devices provide correction of the angular positions of the mounted structure in two planes, a turnabout the vertical axis and its vertical implantation.

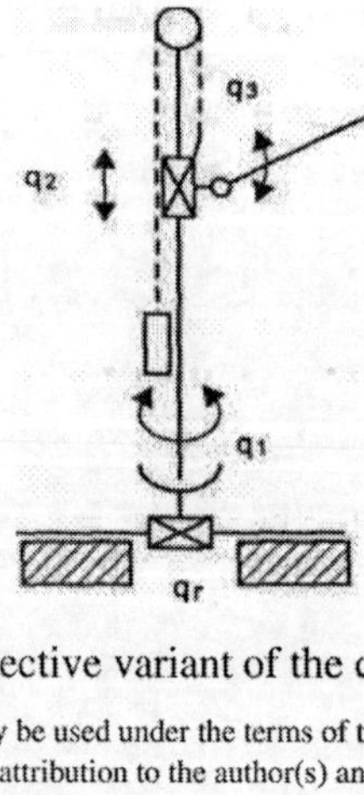

Figure 1. Perspective variant of the crane-manipulator for installation works

Consideration of the kinetic schemes of cranes allows us to assess their greatest potential for use in construction and installation operations, automation and robotization. Successful in this sense is the scheme with rotational kinematic pairs, presented in Figure 1.

The presence of rigid connections of links allows to minimize the positioning errors, freedom of maneuvering while installation of structural building elements.

The crane-manipulator has six local degrees of freedom, which ensures the movement of structural elements into the working area and fixing to a predetermined position. This method of positioning building structures is quite versatile and allows you to use it in the creation of capture devices of construction and installation of robotic systems that perform in addition to the standard functions of capture and stabilization in a given position of the installation elements, their orientation in three dimensions. Therefore, the problem of manipulation and angular orientation is reduced to solving the problems of spatial movement of capture devices. The efficiency of work functions is determined by the modularity of the hardware, software redundancy and the level of intellectual organization of the control system. The implementation of such functions requires the development of an analytical model that implements a universal set of operations for processing digital information provided by the sensor system. The principles of initial information processing incorporated in the model should ensure the independence of its structure and algorithmic support from structural modifications of the kinetic chain of capture devices.

Modeling of processes of trajectory movement and fixation in the space of the gripper device uses the results of the inverse kinematics problem, the solution of which gives a set of variable parameters in the nodes of the joint of the manipulation system in the form of a matrix.

Used in practice scalar analytical and search methods for solving such a kinetic problem are not effective enough because of the loss of accuracy of reproduction of the trajectory of the capture devices when converting the matrix of the initial kinetic relations. The proposed method of simplest kinetic bonds is devoid of this drawback.

Any problem of angular orientation of coordinate systems (CS) with a common origin is described by the matrix relation:

$$\prod_{i=1}^{i=r} R_l(\alpha_i) = I, \tag{1}$$

I – unit diagonal matrix; $R_l(\alpha_i)$ – matrix of angular rotations of CS around one of its axes.

Then the α_i angle of the axes Z Y X respectively will have the form:

$$R_1(\alpha_i) = \begin{vmatrix} \cos\alpha_i & -\sin\alpha_i & 0 \\ \sin\alpha_i & \cos\alpha_i & 0 \\ 1 & 0 & 1 \end{vmatrix}, \tag{2}$$

$R_1(\alpha_i)$, $R_2(\alpha_i)$, $R_3(\alpha_i)$ – matrices of the angular rotations CS:

$$R_2(\alpha_i) = \begin{vmatrix} \cos\alpha_i & 0 & \sin\alpha_i \\ 0 & 1 & 0 \\ -\sin\alpha_i & 0 & \cos\alpha_i \end{vmatrix}, \tag{3}$$

$$R_3(\alpha_i) = \begin{vmatrix} 1 & 0 & 0 \\ 0 & \cos\alpha_i & -\sin\alpha_i \\ 0 & \sin\alpha_i & \cos\alpha_i \end{vmatrix}. \tag{4}$$

The ratios (1) show that there is a sequence of individual rotations of the CS, by which it is possible to realize the angular rotations of an arbitrary state of the Cartesian CS to its initial position.

The ratios (1) capture a kinematic connection in the tasks of the angular orientation and at the smallest number (five) factors you can obtain the simplest kinematic connection in the form of simple matrices angular rotations of the components with the following main properties.

IOP Conf. Series: Materials Science and Engineering **832** (2020) 012009 doi:10.1088/1757-899X/832/1/012009

1. The matrix $R_l(\alpha_i)$ *is* orthogonal, if for any l = 1, 2, 3 the relation is valid:

$$R_l^{-1}(\alpha_i) = R_l^T(\alpha_i), \tag{5}$$

($^{-1}$), (T) are the degrees of inverse and transpose matrices.

2. A series of successive rotations of the CS relative to any of the axes does not complicate the transformation matrix of angular coordinates:

$$\prod_{i=1}^{n} R_l(\alpha_i) = R_l\left(\sum_{i=1}^{n} \alpha_i\right), \tag{6}$$

n is any number.

3. Transposed matrix of elementary angular rotation around one of the axes in the forward direction is adequate to the matrix of the same rotation in the reverse direction:

$$R_l^T(\alpha) = R_l(-\alpha). \tag{7}$$

So if: $\alpha = \pm\pi$,, then:

$$R_l(\pm\pi) = R_l^T(\pm\pi). \tag{8}$$

Rotation around one of the axes of the CS can be adequately replaced by the same magnitude and opposite direction of the rotation angle around the axis, replacing its direction to the opposite.

The formulas of such transformation of coordinates are given in table 1

Table 1. Euler Matrix Transposition Formulas

$R_1^T(\pi)R_2(\alpha)R_1^T(\pi) = R_2^T(\alpha)$	$R_2^T(\pi)R_1(\alpha)R_2^T(\pi) = R_1^T(\alpha)$
$R_2^T(\pi)R_3(\alpha)R_2^T(\pi) = R_3^T(\alpha)$	$R_3^T(\pi)R_2(\alpha)R_3^T(\pi) = R_2^T(\alpha)$
$R_3^T(\pi)R_2(\alpha)R_3^T(\pi) = R_1^T(\alpha)$	$R_1^T(\pi)R_3(\alpha)R_1^T(\pi) = R_3^T(\alpha)$

4. The property of degeneracy matrices of simple angular rotations: matrix of simple angular rotations equal to each other in the absence of rotations around the respective axes of rotation, that is:

$$R_1^T(\alpha)\big|_{\alpha=0} = R_1^T(\beta)\big|_{\beta=0} = R_1^T(\gamma)\big|_{\gamma=0} = I. \tag{9}$$

5. Any angular rotation of the Cartesian coordinate system can be reduced to the replacement of the notation of its axes.

Since there are only two variants of the simplest angular rotations of the original coordinate system associated with the replacement of the designation of its axes, it turns out only six possible modifications of such transformations recorded in table 2, which boil down to two additional turns to the corner π /2.

Table 2. Types of modifications

Direct transformations	Reverse transformations
$R_3(\pi/2)R_1(\alpha)R_3^T(\pi/2) = R_2(\alpha)$	$R_3(\pi/2)R_2(\alpha)R_3^T(\pi/2) = R_1^T(\alpha)$
$R_1(\pi/2)R_2(\alpha)R_1^T(\pi/2) = R_3(\alpha)$	$R_1(\pi/2)R_3(\alpha)R_1^T(\pi/2) = R_2^T(\alpha)$
$R_2(\pi/2)R_3(\alpha)R_2^T(\pi/2) = R_1(\alpha)$	$R_2(\pi/2)R_1(\alpha)R_2^T(\pi/2) = R_3^T(\alpha)$

The simplest kinematic bond (SKB) is a sequence of angular transformations with the number of rotations of the axes equal to five.

There are only three distinct types of SKB, which are shown in Figure. 2. These sequences of axis rotations will be the simplest, displaying expressions with the least number of matrices to be multiplied.

Given the direction of successive transformations, rotations of the axes, we obtain matrix relations for all groups of SKB with rotations:

A. At the angles b, d, e, c, a:

$$R_1(b)R_2(a) = R_2(d)R_3(e)R_1(c); \tag{10}$$

B. At the angles a, m, l, p, n:

$$R_2(a)R_3(m) = R_3(n)R_1(p)R_2(l); \tag{11}$$

C. At the angles f, b, g, k, h:

$$R_3(f)R_1(b) = R_1(n)R_2(k)R_3(g); \tag{12}$$

Further simplifications of at least one of the matrix relations (10 – 12), zeroing any of the angular coordinates, will cause the zeroing of the remaining angles.

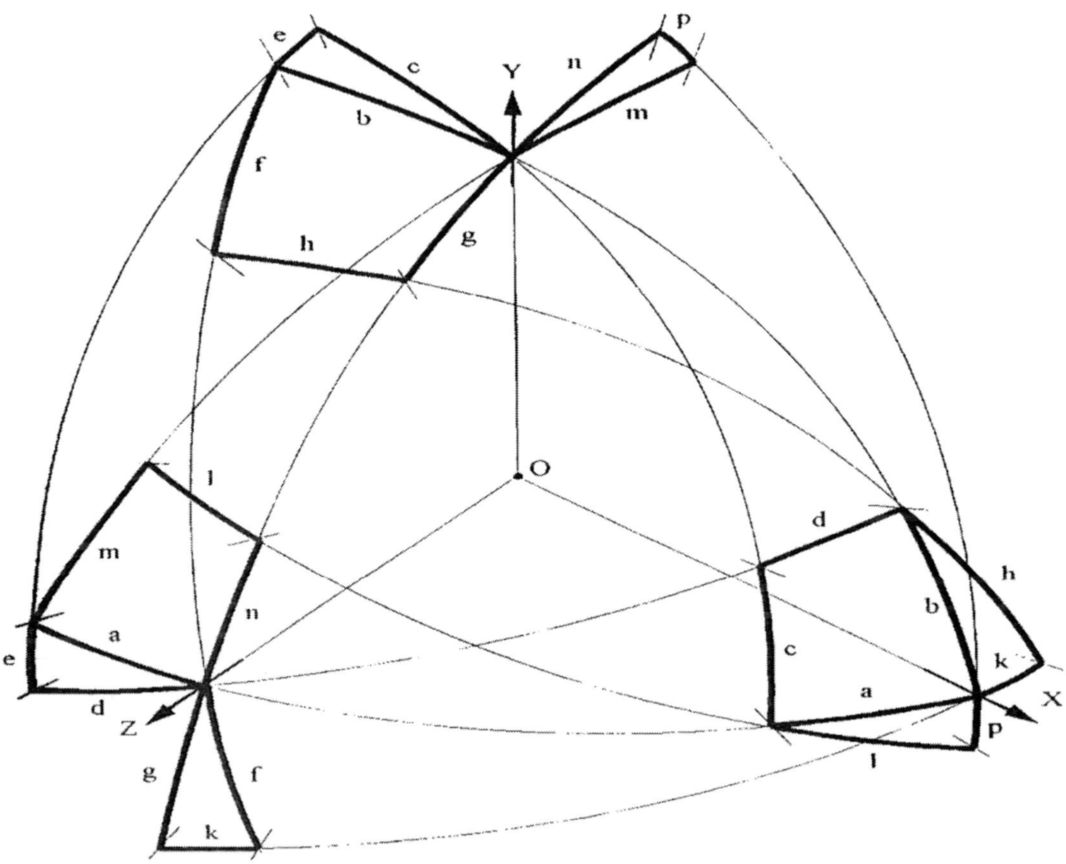

Figure 2. Geometry of the simplest kinematic bonds

It can be stated that the method of the simplest kinematic connections allows to organize at the matrix level the solution of a wide range of problems of spherical trigonometry and angular

IOP Conf. Series: Materials Science and Engineering **832** (2020) 012009 doi:10.1088/1757-899X/832/1/012009

orientation. The basic principle of the solution is the decomposition of the original matrix description of the problem into a system of elementary ones, the sequential solution of which individually or in combinations (depending on convenience), allows to produce unambiguous results in the absence of restrictions in the original data. The mathematical apparatus of homogeneous transformations makes it possible to calculate the values of angular reversals in all intermediate joints, starting from the first and ending with the n-1st, uniquely determining their required spatial orientation, as well as the spatial orientation of the end section of the RS.

References

[1] Bataev S B, Vorobyov V A, Degtyarev V and Mazhibayev O M 1996 *Methods and means of automation of road construction works and machines* (Almaty: Gylym) p 262

[2] Bulgakov A G, Gerner I and Kaden R 1990 *Studies and practical examples of the organization of production and use of robots in the construction industry* (*Machines, mechanisms, equipment and tools* M VNIINTPI vol 1) p 48

[3] Bulgakov A G, Gerner I and Kaden R 1991 *Microprocessors in automation systems of construction equipment* (*Technology of construction and installation works* M VNIINTPI vol.3) p 52

[4] Bulgakov A G and Sukhomlinov A D 1989 *Application of laser information-measuring systems in construction.* (*Technology of construction and installation works* M VNIIESM vol 3) p 53

[5] Bulgakov A G and Schindler I 1994 *Means and systems of automation in construction machinery. Technology and automation of construction.* (M VNIINTPI vol 4) 56 p

IOP Conf. Series: Materials Science and Engineering **832** (2020) 012010 doi:10.1088/1757-899X/832/1/012010

Regulation of liquefied petroleum gas consumption for a passenger car in the mountain conditions of the Republic of Tajikistan

B K Mirov, Y V Panov, V A Zenchenko and P V Bushuev

Moscow Automobile and Road Construction State Technical University (MADI), 64, Leningradsky ave., Moscow, 125319, Russia

E-mail: mirov.ttu@mail.ru, panovyur@mail.ru, zvaeats@yandex.ru, helpsol@yandex.ru

Abstract. In the article the main results of research work to determine the consumption rates of liquefied petroleum gas (LPG) for a car brand KIA Optima are presented. The research was accomplished for high-mountain conditions of operating in the Republic of Tajikistan (RT). The basic factors, influencing fuel and economic indicators of the car in mountain conditions of operation, are considered, and the main analytical dependences, received on the basis of the developed mathematical models, are presented. The practical realization of the developed mathematical models allows for individual rationing of fuel consumption, taking into account the parameters of the mountain road and, accordingly, the assessment of the road complexity. On the basis of the experimental data obtained, graphical dependences of fuel consumption in mountain conditions of vehicle movement are constructed and presented.

1. Introduction

The effective using of automobile transport (AT) in the mountain conditions of the RT (93 % of area is occupied by mountains) is an important task, because mountain roads are characterized by a large number of ascents, descents and sharp turns, which significantly increases the fuel consumption [4–6, 8]. In such operating conditions, it is required to conduct theoretical and experimental research, allowing realizing the effective control of the gas flow rate regulation during vehicles operation [3, 8].

2. Main part

The practical realization of tasks related to the regulation of fuel consumption of vehicles, running on LPG in the mountain conditions, provides for the construction of mathematical models and their realization, enabling the determination of fuel economy of bottled gas driven automobiles by the values of factors, describing the routes. The basis for the construction of mathematical models is theoretical and experimental research of fuel consumption estimates, depending on the complexity of the route in the movement of vehicles, reflected in the works [1–4].

In expression (1) the transformed extended formula of the estimation of route movement complexity with set of the factors, reflecting parameters of mountain conditions of vehicles operation, is presented:

IOP Conf. Series: Materials Science and Engineering **832** (2020) 012010 doi:10.1088/1757-899X/832/1/012010

$$Y_{Ai} = \frac{L_H \cdot \overline{H}_i \cdot K_i \cdot X_{RH}^{Z_{RH}} \cdot X_{\Pi}^{Z_{\Pi}}}{V^2 \cdot X_t^{Z_t} \cdot X_T^{Z_T} \cdot X_P^{Z_P}}, \tag{1}$$

where: L_H – the lower tolerance limit of the normalized sections lengths, accepted in the calculation as a reference point, providing sufficient accuracy and sensitivity of measurements of all researched values, km (accepted for a given probability of 0.8÷0.9 [5]; $\overline{H}$ – average height of the section on the route, km; K – steepness of ascent or descent of the section on the route, α; X_{RH} – relative humidity of the ambient air, (RH) %; X_{Π} – frequency of turns, (П) ea.; V – average speed of the vehicle, km/h; X_t – average movement time of the vehicle, sec., min., hour; X_T – ambient air temperature, (T) °C; X_P – ambient air atmospheric pressure, P; $Z_{RH}, Z_{\Pi}, Z_t, Z_T, Z_P$ – takes 0 if the relevant factor is not taken into account in the complexity formula, and 1 if the factor in question is taken into account.

Given that in the expression (1), the parameters under consideration have different dimensions of the value of the route complexity index, it should be subject to regulation through the standard deviation, i.e.:

$$\sigma(Y_A) = \sqrt{\frac{\sum_{i=1}^{n}(Y_{Ai} - \overline{Y})^2}{n-1}}. \tag{2}$$

In the final form, the normalized value of the route difficulty can be written as:

$$Y_{Bi} = \frac{L_H \cdot \overline{H}_i \cdot K_i \cdot X_{RH}^{Z_{RH}} \cdot X_{\Pi}^{Z_{\Pi}}}{V^2 \cdot X_t^{Z_t} \cdot X_T^{Z_T} \cdot X_P^{Z_P} \cdot \sigma(Y_A)}. \tag{3}$$

Given the fact that in the expression (3) there is a factor of the slope angle of the road (K_i), taking depending on the ascent or descent positive or negative value respectively, the expression (3) of the complexity assessment is converted in order to shift the result to the region of positive values, i.e.:

$$Y_{Ci} = Y_{Bi} - Y_{B\min}, \tag{4}$$

where $Y_{B\min}$ – minimum values Y_{Bi}.

As a target function of fuel efficiency indicators (Q_{LPG}) in l/100km of the vehicle in mountain conditions operation on the route, the model of the form is accepted:

$$Q_{LPG} = f(Y_{Ci}). \tag{5}$$

In the process of constructing models of fuel consumption of the vehicle, the analysis of experimental data obtained by the values of the factors under consideration was made. The analysis included checking for the presence of the interconnection factor with the resulting sign.

Construction of models, describing the researching phenomena, is carried out with the mathematical apparatus of correlation and regression analysis using. Mathematical models of the form can be used as target functionals, linking fuel consumption with the complexity indicator of the route of the vehicle, running on LPG:

$$Q_{LPG} = a + в \cdot Y_{Ci}, \tag{6}$$

$$Q_{LPG} = a + в \cdot \ln \cdot Y_{Ci}, \tag{7}$$

$$Q_{LPG} = a \cdot e^{в \cdot Y_{Ci}}, \quad (8)$$

$$Q_{LPG} = a \cdot Y_{Ci}^{в}, \quad (9)$$

$$Q_{LPG} = a + в \cdot Y_{Ci} + c \cdot Y_{Ci}^{2}. \quad (10)$$

For understanding the nature of the individual factors link, characterizing the driving conditions impact on fuel consumption, a differentiated assessment of the degree of their impact on fuel efficiency was carried out, i.e.:

$$Q_{LPG} = f(X),\ X = (\{V\},\{t\},\{K\},\{\overline{H}\},\{L_y\}). \quad (11)$$

Detection of the dependence of gas fuel consumption on operating conditions was revealed in the process of experimental researches on a mountain highway with a maximal elevation difference on the Dushanbe-Shahristan-Dushanbe route in the RT [3–5]. As the object of research a passenger car of KIA Optima brand was chosen in which gas equipment of the (HBO) 4th generation was installed. The modern ultrasonic fuel level sensor (UFLS) Sigma (hereinafter – the sensor) is used as measuring equipment to determine the gas flow rate on the vehicle. The sensor is connected to the satellite monitoring system of transport for storage and processing of data about the amount of fuel in the gas cylinder [4, 5].

Within the framework of the experimental research, in order to determine the dependence of LPG consumption on traffic conditions, the factors and their influence on the target function of the route complexity [2] were identified, and then their influence on the fuel consumption of the vehicle (Q_{LPG}). As such factors (in the summer test conditions) have been chosen: V_i – the average vehicle speed on the section, km/h; $\overline{H}$ – the average height of land on the route of the vehicle, km; K_i – the steepness of ascent of the sections on the route of the vehicle, α;

On the basis of primary data the analytical dependence of complexity of movement of the motor vehicle on each *i*-m site of a route, depending on set of above noted factors, taking into account normalization on a standard deviation, was revealed $\sigma(Y)$ [4].

Dependences of fuel consumption on the route complexity has regular character – as the complexity of the route increases, fuel consumption increases. The function graph of fuel consumption dependence on the route complexity is close to the quadratic dependence.

Processing of the collected information allowed to obtain mathematical models of fuel consumption (Q_{LPG}) as a function of the route complexity (Y) in the forward (12) and in the reverse direction (13) on the Dushanbe – Shahristan highway. Graphical dependences of fuel consumption on the complexity of the route are shown in figure 1:

$$Q_{LPG} = 1{,}4019 \cdot \tilde{Y}^2 - 0{,}0698 \cdot \tilde{Y} + 3{,}0287, \quad (12)$$

$$Q_{LPG} = 1{,}6722 \cdot \tilde{Y}^2 - 2{,}8195 \cdot \tilde{Y} + 5{,}2367, \quad (13)$$

where $\tilde{Y}$ – takes the form:

$$\tilde{Y}_i = Y_i - Y_{\min}, \quad (14)$$

Y_i is defined from the expression:

$$Y_i = \frac{L_H \cdot \overline{H} \cdot K}{V_i^2 \cdot \sigma(Y^*)}, \quad (15)$$

where $\sigma(Y^*)$ – mean square deviation of the parameter Y_i^* [4]:

IOP Conf. Series: Materials Science and Engineering **832** (2020) 012010 doi:10.1088/1757-899X/832/1/012010

$$\sigma(Y^*) = \sqrt{\sum_{i=1}^{n} \frac{(Y_i^* - \overline{Y}_i^*)^2}{n-1}}. \tag{16}$$

This is Y_i^* reflected in the form of the equation (16):

$$Y_i^* = \frac{L_H \cdot \overline{H} \cdot K_i}{V_i^2}. \tag{17}$$

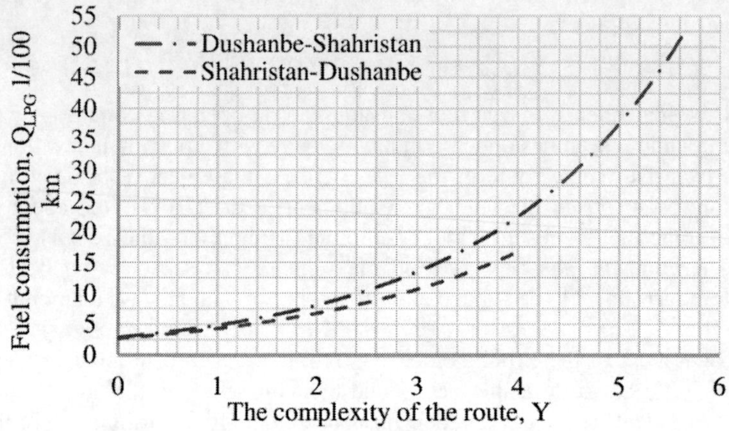

Figure 1. Dependences of LPG consumption on the complexity of the route on the Dushanbe-Shahristan highway

Large values of fuel consumption in the forward direction "Dushanbe-Shahristan" in comparison with the reverse direction of movement on the route "Shahristan-Dushanbe" are associated with different heights of the 1st and 2nd parts of the route. When moving in the forward direction, ascents prevail, and when moving in the opposite direction – descents, which causes a stronger dependence of the consumption on the complexity in the forward direction than in the reverse.

The analysis of the dependence of fuel consumption on the complexity of the route reveals the influence of the average vehicle speed and, accordingly, build the dependence of fuel consumption on the average speed in the considered sections (fig. 2), as well as to obtain a mathematical model of fuel consumption at the average speed of the vehicle in the forward (18) and in the reverse direction (19) on the Dushanbe – Shahristan highway:

$$Q_{LPG} = 113{,}37 \cdot V^{-0{,}596}, \tag{18}$$

$$Q_{LPG} = 25{,}115 \cdot V^{-0{,}306} \tag{19}$$

Figure 2 illustrates a regular decrease in fuel consumption with an increase in the average speed of movement on the site, which is associated with a decrease in the complexity of the movement.

At the same time, on the basis of information about the initial and final heights of the route sections obtained with the help of the satellite transport monitoring system [4, 5], the average height of the Dushanbe – Shahristan highway sections above sea level was determined. Equations were identified and graphical dependences of their influence on the vehicle fuel consumption were constructed (fig. 3). The equations of fuel consumption, depending on the average height of the section on the straight line (20) and in the opposite direction (21), are described by logarithmic dependences and have the form:

$$Q_{LPG} = 5{,}7462 \cdot \ln(\overline{H}) + 9{,}6406, \tag{20}$$

$$Q_{LPG} = 6{,}4057 \cdot \ln(\overline{H}) + 4{,}5158, \tag{21}$$

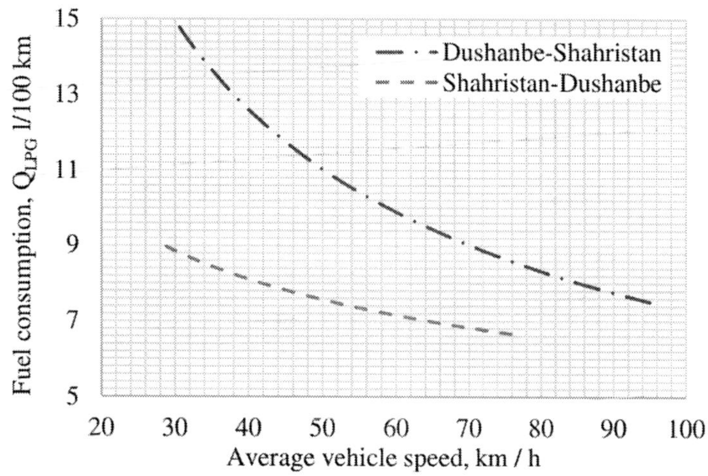

Figure 2. Dependences of fuel consumption on the average vehicle speed on the Dushanbe-Shahristan highway

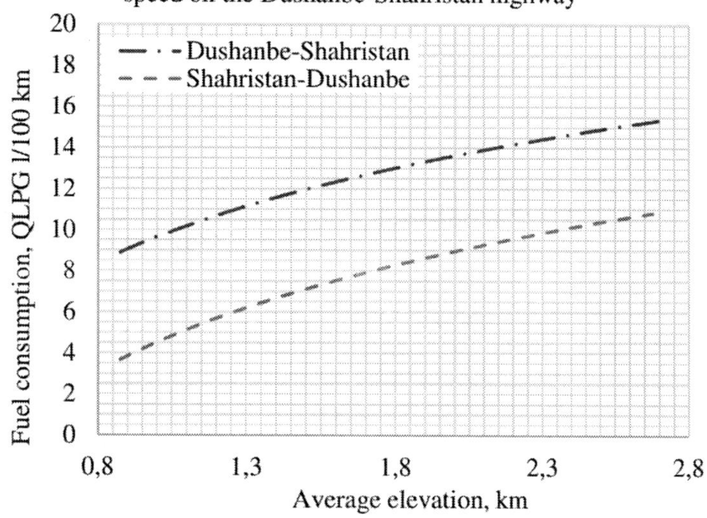

Figure 3. Dependence of LPG consumption on the average height of the section on the Dushanbe-Shahristan highway

Graphical dependencies (fig. 3) show that with the growth of the average height of the section on the route there is an increase in fuel consumption and vice versa.

At the same time, taking into account such factors as the average vehicle speed and the average section height on the Dushanbe-Shahristan highway, it is possible to identify their combined effect on fuel consumption (fig. 4) and obtain the corresponding analytical dependencies (22)–(23):

$$Q_{LPG} = -22{,}7578 + 0{,}266 \cdot \overline{H} + 32{,}4185 \cdot V + 0{,}001 \cdot \overline{H}^2 - 0{,}2776 \cdot \overline{H} \cdot V - 3{,}8156 \cdot V^2; \qquad (22)$$

$$Q_{LPG} = -11{,}6702 + 0{,}5731 \cdot \overline{H} + 3{,}503 \cdot V - 0{,}0025 \cdot \overline{H}^2 - 0{,}2292 \cdot \overline{H} \cdot V + 3{,}5304 \cdot V^2. \qquad (23)$$

From the presented three-dimensional graphs (fig. 4) it follows that fuel consumption depends not only on the speed of the vehicle, but also on the altitude above sea level (a.s.l.). When driving in the city of Dushanbe (0.8–1.0 km a.s.l.) fuel consumption from the speed of the vehicle is in proportion. When driving in the mountains at the level of road tunnels "Istiqlol" and "Shahristan" (2.2–2.8 km a.s.l.) [4, 5] the factors of traffic complexity change the dependence of fuel consumption on speed to the reverse: than less the speed, the more consumption.

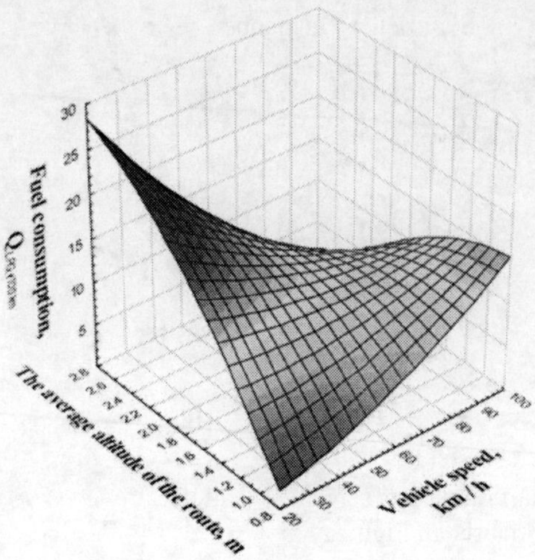

a)

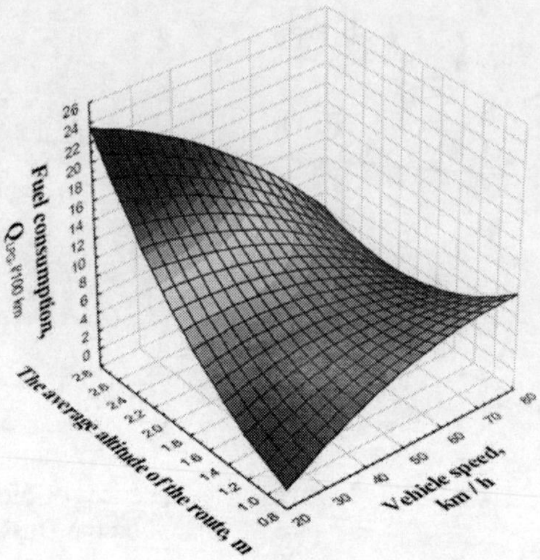

b)

Figure 4. Dependences of fuel consumption on the average height of the section and speed of the vehicle on the Dushanbe – Shahristan (a) and Shahristan – Dushanbe (b) highway

On the basis of primary experimental data, such as the initial and final height of the section, as well as its length, the steepness of ascent of the sections on the route of the vehicle was determined.

The equations of fuel consumption, depending on the steepness of ascent in the forward (24) and in the reverse direction (25), have the form:

$$Q_{LPG} = 9{,}2355 \cdot e^{8{,}5761 \cdot K}, \tag{24}$$

$$Q_{LPG} = 7{,}8271 \cdot e^{7{,}1728 \cdot K}. \tag{25}$$

A graphical representation of the dependence of fuel consumption on the steepness of the route sections is shown in fig. 5.

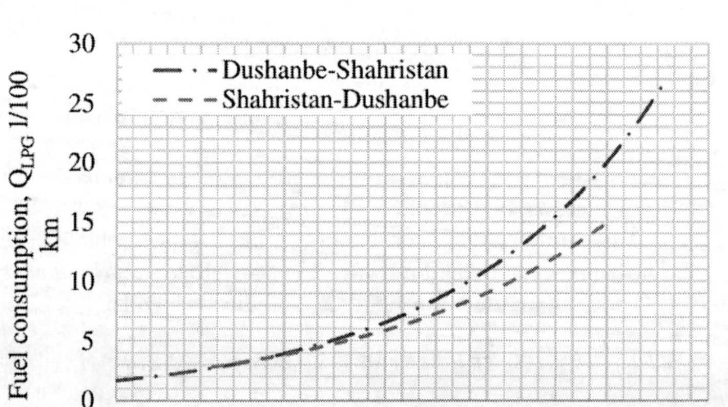

Figure 5. Dependence of fuel consumption on the steepness of ascent on the Dushanbe – Shahristan highway

IOP Conf. Series: Materials Science and Engineering **832** (2020) 012010 doi:10.1088/1757-899X/832/1/012010

It can be seen from the graph (fig. 5) that with an increase in the steepness of ascent, fuel consumption increases. For fig. 6 the estimation of fuel efficiency of Kia Optima car on Dushanbe-Shahristan-Dushanbe sections is graphically reflected, depending on the steepness of ascent and average speed of movement both in forward (26) and reverse (27) directions:

$$Q_{LPG} = 10{,}0679 + 0{,}0169 \cdot K + 199{,}7453 \cdot V - 0{,}0008 \cdot K^2 - 1{,}3463 \cdot K \cdot V + 613{,}256 \cdot V^2; \quad (26)$$

$$Q_{LPG} = 1{,}4294 + 0{,}306 \cdot K + 123{,}131 \cdot V - 0{,}0034 \cdot K^2 - 0{,}8887 \cdot K \cdot V + 432{,}0705 \cdot V^2; \quad (27)$$

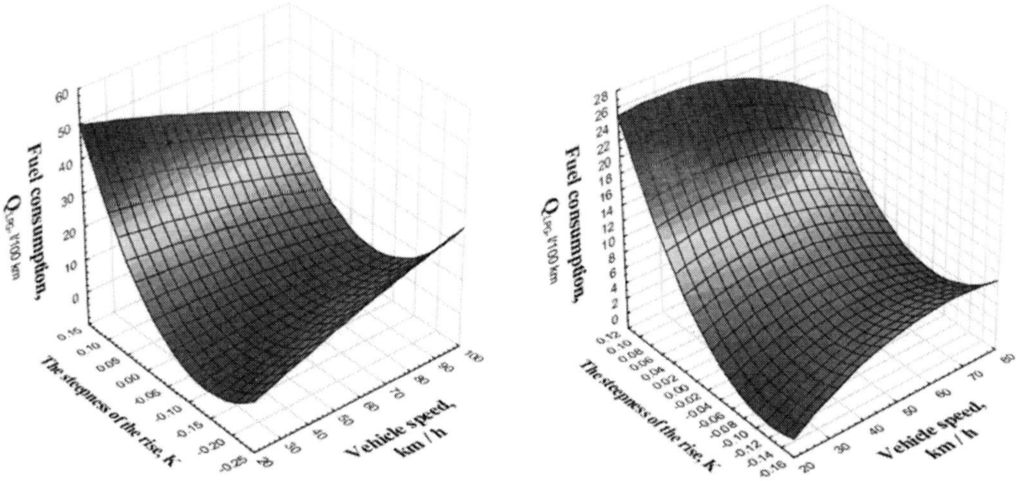

a) b)

Figure 6. Dependences of fuel consumption on the steepness of the ascent and the average speed of the vehicle

At almost all speeds, there is a characteristic increase in fuel consumption with an increase in the steepness of the ascent, which is a natural result.

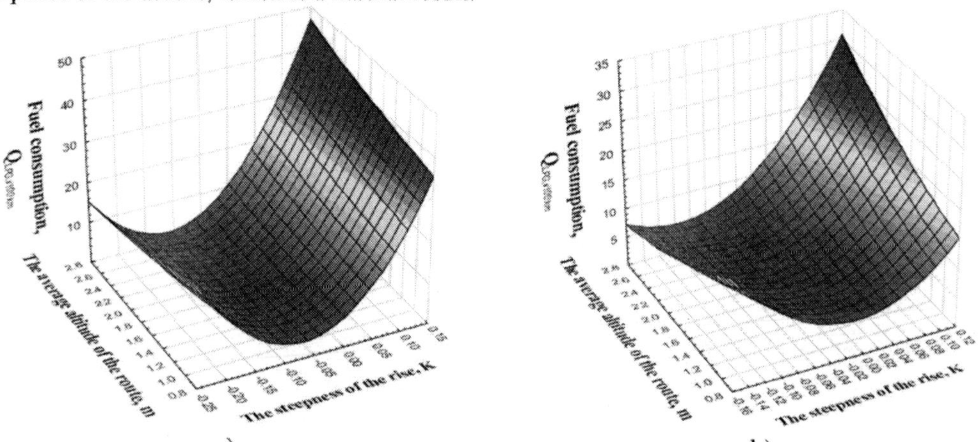

a) b)

Figure 7. Dependences of fuel consumption on the average height of the section and the steepness of ascent of the vehicle on the Dushanbe – Shahristan (a) and Shahristan – Dushanbe (b)

The combined effect of the steepness of the ascent and the average height of the site above sea level on fuel consumption is shown in fig. 7. The equations describing the respective interconnections of the Dushanbe – Shahristan (28) and Shahristan – Dushanbe (29) highway have the form:

$$Q_{LPG} = 7{,}7082 + 69{,}2213 \cdot K + 1{,}2756 \cdot V - 658{,}7635 \cdot K^2 + 27{,}1641 \cdot K \cdot V + 0{,}7993 \cdot V^2; \quad (28)$$

$$Q_{LPG} = 8{,}1683 - 31{,}6496 \cdot K - 3{,}7472 \cdot V + 557{,}1181 \cdot K^2 + 49{,}4571 \cdot K \cdot V + 1{,}5821 \cdot V^2. \quad (29)$$

As follows from the graphical dependencies fig. 7, the highest fuel consumption is observed at maximal ascent at maximal altitude. The minimal fuel consumption is almost independent of the average height and is observed on a moderate (not maximal) descent. At the maximal descent, there is a slight increase in the allowance, which is associated with transient modes of operation of the ICE in such modes.

Using of models (5)–(10) creates conditions for reliable formation of allowances to route norms for extreme mountain conditions of the Republic of Tajikistan. At the same time, the developed model of gas fuel consumption allows adapting the original formula of NIIAT, for passenger vehicles to the level of individual route norms (refer expression 30):

$$Q_{br} = 0{,}01 \cdot Q_H \cdot L_y \cdot (1 + 0{,}01 \cdot D), \quad (30)$$

where Q_{br} – fuel consumption of the vehicle corresponding to the base rate (corresponds to the minimal height above sea level);

Q_H – standard fuel consumption per vehicle mileage;

L_y – the length of the plot having a minimal height above sea level, km;

D – allowance correction factor corresponding to the base rate of consumption (minimal height of the route section above sea level) in percentage.

By converting the original expression (30), it is possible to determine the fuel allowance:

$$D = \left[\frac{Q_E}{0{,}01 \cdot Q_{br} \cdot L_y} - 1 \right] \cdot 100\,\%, \quad (31)$$

where Q_E – the fuel consumption of the vehicle corresponding to experimental data.

Expression (31) allows to determine the value of the allowance on any considered part of the route. This coefficient of allowance, determined on the basis of approximation of experimental data, corresponds to the operating conditions of vehicles in the Republic of Tajikistan.

If it is necessary to estimate the allowance for fuel consumption for high-altitude operating conditions relative to the standard fuel consumption of a car D_T, use an expression of the form:

$$D_T = \left[\frac{Q_P^T}{0{,}01 \cdot Q_H \cdot L_y} - 1 \right] \cdot 100\%, \quad (32)$$

where: Q_P^T – fuel consumption, corresponding to the accepted models (see expressions (5-3), was calculated through the complexity of traffic conditions:

$$Q_P^T = f(Y_{Ci}). \quad (33)$$

The individual route norm is determined by the calculation of the correction factor for the allowance D_T, is consistent with expression (32) for the selected sections of the route and the subsequent summation to assess the allowance for the entire route as a whole.

An example of dependency allowances D_T for specific sections of the route are shown in Figures 8-9 of which shows a high degree of convergence on D_T and Q_T.

IOP Conf. Series: Materials Science and Engineering **832** (2020) 012010 doi:10.1088/1757-899X/832/1/012010

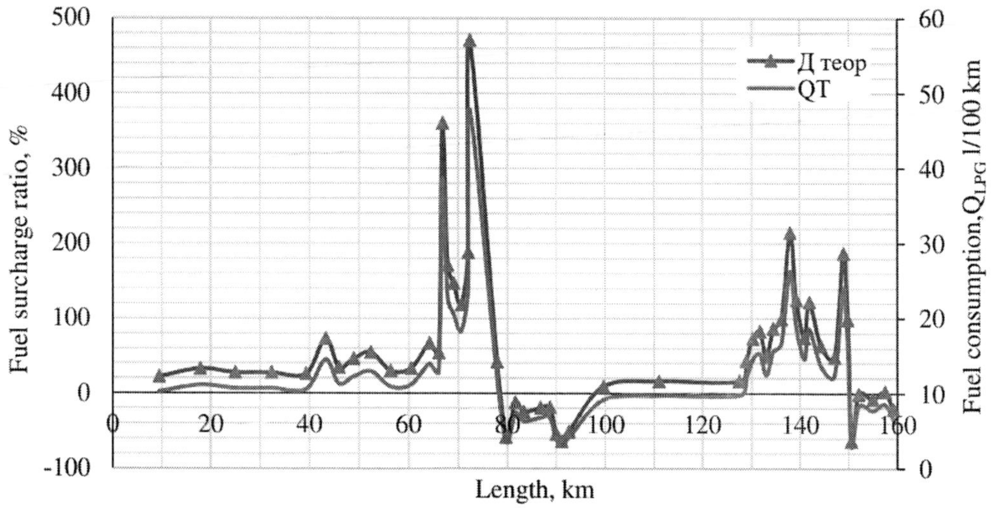

Figure 8. Dependence of the theoretical allowance D_T on the route section and fuel consumption in the corresponding section Q_T Dushanbe – Shahristan

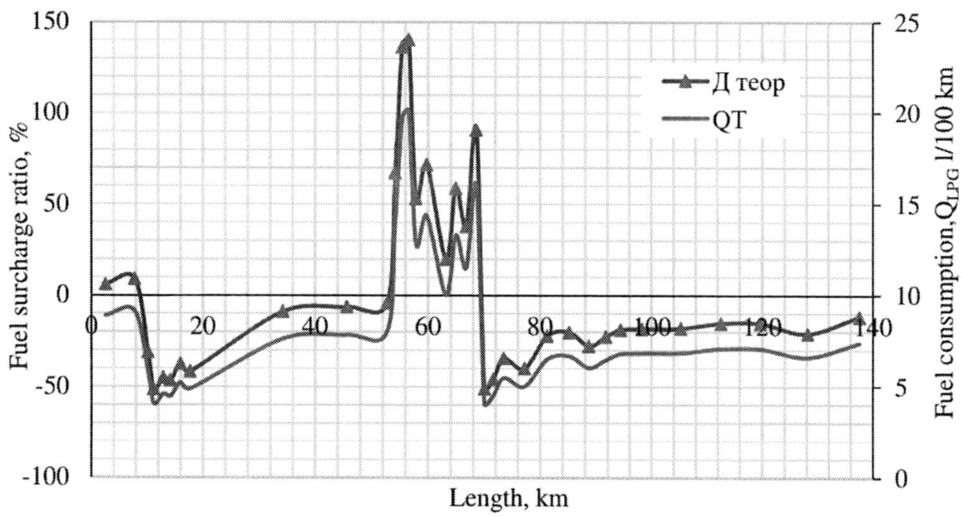

Figure 9. Dependence of the theoretical allowance D_T on the route section and fuel consumption in the corresponding section Q_T Shahristan – Dushanbe

3. Conclusion

Based on the obtained research results, it follows that the proposed mathematical models make it possible to objectively describe the functional dependence of the LPG flow rate and surcharges on it from the generalizing parameter of the route complexity in the mountain conditions of the Republic of Tajikistan. On the basis of the offered models it is possible to receive individual norms of the consumption of gas fuel for any sites and their quantity of the considered road profile. The graphic dependences, constructed on the basis of experimental data, allow to judge about functional dependence of fuel consumption on the generalizing parameter of complexity of a route in concrete conditions. The research results, proposed approaches, principles and models create the conditions for the formation of a regulatory basis by its development and the equation consumption of LPG cars in the mountain conditions of the Republic of Tajikistan.

IOP Conf. Series: Materials Science and Engineering **832** (2020) 012010 doi:10.1088/1757-899X/832/1/012010

References

[1] Zenchenko V A, Rementsov A N, Pavlov A V and Sotskov A V 2012 On the issue of modeling the characteristics of traffic flows and assessing the complexity of traffic *Gruzovik* **8** 34

[2] Zenchenko V A, Rementsov A N, Pavlov A V and Sotskov A V 2012 The rationale for the selection of indicators for assessing traffic traffic *Gruzovik* **5** 38

[3] Mirov B K, Panov Yu V, Bushuev P V and Zenchenko V A 2019 The results of the study of LPG consumption in the mountainous conditions of the Republic of Tajikistan *Problems of technical operation and car service of rolling stock of automobile transport* ed Maksimov V A (Moscow: Tekhpoligraftsentr) pp 117–20

[4] Panov Yu V, Mirov B K, Zenchenko V A and Bushuev P V 2019 Automotive on-board telemetry system and fuel level measuring system for rationing and controlling LPG consumption in the mountains of Tajikistan *Transport on alternative fuel* **2** p 63

[5] Panov Yu V. Mirov B K and Vulman R K 2018 Information technology for monitoring and regulation of LPG consumption for cars. *Problems of technical operation and car service of rolling stock of automobile transport* ed Maksimov V A (Moscow: Tekhpoligraftsentr) pp 166–72

[6] Panov Yu V and Mirov B K 2017 Overview of approaches to adjusting fuel consumption in Tajikistan *Problems of technical operation and car service of rolling stock of automobile transport* ed Maksimov V A (Moscow: Tekhpoligraftsentr) pp 152–5

[7] Panov Yu V 2013 *Automotive gas injection systems. Device, installation, operation* (Moscow: LLC "IDTR")

[8] Mirov B K 2018 Problems of Car transfer to gas fuel under the conditions of Tajikistan *Proceedings of the Third International Conference of European Academy of Science* (Bonn: EAS and Heinemannstraße) pp 31–2

IOP Conf. Series: Materials Science and Engineering **832** (2020) 012011 doi:10.1088/1757-899X/832/1/012011

The results of modeling and evaluating the relative slip coefficients during the interaction of the pneumatic tire with the supporting surface to determine the forces acting in the contact patch

A N Sova[1], G S Mazlumyan[1], O V Egorov[1], S S Shadrin[1] and V A Sova[2]

[1]Moscow Automobile and Road State Technical University (MADI), 64, Leningradsky ave., Moscow, 125319, Russia
[2]Bauman Moscow State Technical University, b. 1, 5, Baumanskaya 2-a str., Moscow, 105005, Russia

E-mail: slsova@mail.ru

Abstract. The article describes the results of modeling and estimates of the relative slip coefficients during the interaction of the pneumatic tire with the supporting surface in order to determine the forces acting in the contact spot. The justified and developed design scheme for the interaction of the wheel propulsion with the supporting surface is presented. The characteristics of the relative slip coefficients, the longitudinal traction characteristics are presented. Lateral traction characteristics, Matlab Simulink friction coefficient calculation procedure are proposed.

When creating a simulation mathematical model of the movement of a wheeled vehicle, the most important task is a reliable description of the physics of the interaction of the pneumatic tire with the supporting surface to determine the forces acting in the contact patch [1].

The results of modeling and estimation of relative slippage coefficients are presented below. In the case when the car is moving without a slip, the slip speed is the difference in wheel peripheral speed $V_{Rij} = \omega_{ij} \cdot r_{stat}$ (equivalent linear speed of the center of the wheel disk, corresponding to the free rolling mode without angular velocity change ω_{ij}) and linear velocity of the center of the wheel disk projected onto the abutment surface (in the case under consideration $V_{Wij} = V_{CoG}$). When introducing the concepts of creep speeds in the longitudinal and transverse directions in the specialized literature, different approaches are encountered. Thus, Burckhardt defines the longitudinal directions of the force acting in the contact patch of the pneumatic tire with the supporting surface and the creep speed as corresponding to the direction of the wheel movement (Fig. 1) [2], while Reimpell as the longitudinal direction selects directly the longitudinal axis of the wheel [3]. We will use the Burkhard approach. Thus, the longitudinal direction of the slip velocity vector is determined in the direction of the linear velocity of the center of the wheel disc projected onto the abutment surface ($\vec{V}_{Wij}$).

Formulas for calculating the relative slippage coefficients in the longitudinal and transverse directions for different driving modes are presented in table 1. Note that the coefficient of relative slippage in the longitudinal direction, depending on the mode of movement, can take values from –1 to +1, inclusively. This is achieved due to the fact that the speed differences (Table 1) are divided by the corresponding large speeds, i.e. for the braking mode of motion, the denominator has a speed value V_W, while for traction mode $V_R \cdot \cos\alpha$.

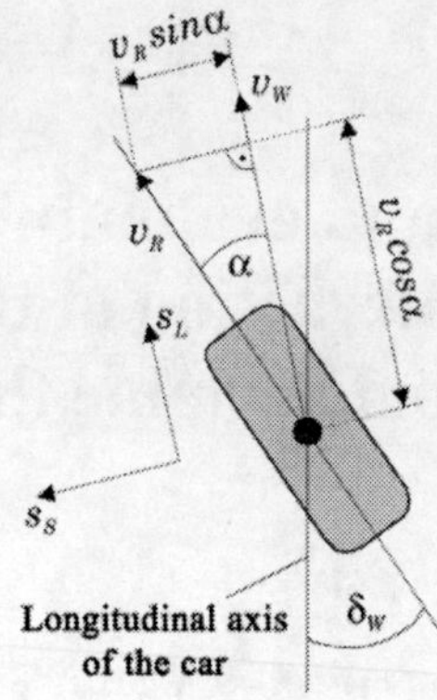

Figure 1. The design scheme of the coefficients of relative slippage

Table 1. Calculation of relative slippage coefficients

Name of coefficient	Brake mode $V_R \cdot \cos\alpha \le V_W$	Traction mode $V_R \cdot \cos\alpha > V_W$
Coefficient of relative slippage in the longitudinal direction – S_L	$S_L = \frac{V_R \cdot \cos\alpha - V_W}{V_W}$	$S_L = \frac{V_R \cdot \cos\alpha - V_W}{V_R \cdot \cos\alpha}$
Coefficient of relative slippage in the transverse direction – S_S	$S_S = \frac{V_R \cdot \sin\alpha}{V_W}$	$S_S = \tan\alpha$

The coefficient of total relative wheel slip takes values from 0 to +1 inclusive and is calculated by the formula:

$$S_{\mathrm{Res}} = \sqrt{S_L^2 + S_S^2} \tag{1}$$

The results of modeling and estimation of the coefficient of adhesion of the pneumatic tire with the supporting surface are given below.

The friction coefficient can be defined as the ratio of the friction force acting in the contact patch of the pneumatic tire with the supporting surface to the vertical load acting from the wheel side on the supporting surface:

$$\mu_{\mathrm{Res}} = \frac{F_{fric}}{F_Z} \tag{2}$$

The calculation of the coefficient of adhesion as a function of the total relative wheel slip is proposed to be performed using the Burkhard method [2]:

$$\mu_{\mathrm{Res}}(S_{\mathrm{Res}}) = c_1 \cdot (1 - \exp(-c_2 \cdot S_{\mathrm{Res}})) - c_3 \cdot S_{\mathrm{Res}} \tag{3}$$

Expression (3) can be supplemented by the coefficient c_4, describing the condition of movement of a car with high speeds, and the coefficient c_5, characterizing increased vertical load on the wheel. Both coefficients can take a maximum value of 1, and lead to a decrease in the coefficient of adhesion. Incorrectly set pressure in pneumatic tires can also lead to a decrease in traction coefficient, however, this effect can be neglected when the pressure fluctuates within 0,3 kgf/cm^2. In view of the above, expression (3) can be represented as follows:

$$\mu_{\mathrm{Res}}(S_{\mathrm{Res}}) = (c_1 \cdot (1 - \exp(-c_2 \cdot S_{\mathrm{Res}})) - c_3 \cdot S_{\mathrm{Res}}) \cdot \exp(-c_4 \cdot S_{\mathrm{Res}} \cdot V_{CoG}) \cdot (1 - c_5 \cdot F_Z^2) \tag{4}$$

Coefficient Values c_1, c_2 and c_3 for various types of coatings of the supporting surface are presented in table 2 [1, 2, 4].

Table 2. Coefficient Values c_i

Type of pavement	c_1	c_2	c_3
dry asphalt	1,2801	23,99	0,52
wet asphalt	0,857	33,822	0,347
dry concrete	1,1973	25,168	0,5373
dry gravel	1,3713	6,4565	0,6691
wet gravel	0,4004	33,708	0,1204
snow	0,1946	94,129	0,0646
ice	0,05	306,39	0

Coefficient c_4 takes values from 0.02 to 0.04 s/m. For example, for a coefficient of total relative slippage of 0.1 (10 %) and a vehicle speed of 20 m / s (72 km/h), the overall coefficient of adhesion decreases by about 8 %. Coefficient $c_5 = 0,0015 \ (1/\kappa H)^2$ and dynamic change of vertical load to 8 kN will lead to a decrease in adhesion coefficient by 9.6 % [1]. A graphical representation of the functions of the dependence of the coefficient of adhesion on the total relative slippage for different types of coatings is shown in Fig. 2 and applies only to quasistatic modes of vehicle motion, i.e. does not describe rapidly changing dynamic processes. An analysis of the time dependence of side reactions in the contact patch of pneumatic tires with the supporting surface on rapidly changing the steering angle is presented in sources [1], [5] and [6].

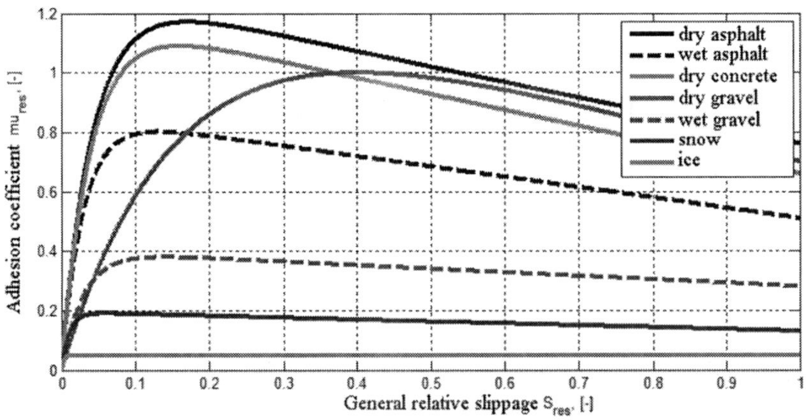

Figure 2. Traction coefficient characteristics

The direction of the vector of general relative slippage coincides with the direction of action of the adhesion force, based on which the concepts of adhesion coefficients in the longitudinal and transverse directions are calculated, calculated by the formulas:

$$\mu_L = \mu_{Res} \cdot \frac{S_L}{S_{Res}} \quad \text{and} \quad \mu_S = \mu_{Res} \cdot \frac{S_S}{S_{Res}} \tag{5}$$

Fig. 3 shows the so-called "Kamm circle", illustrating the conditional distribution of the total adhesion coefficient on the adhesion coefficients in the longitudinal and transverse directions. The contact spot of the pneumatic tire with the supporting surface, the vehicle speed and the load on the wheel determine the maximum coefficient of adhesion $\mu_{Res,MAX}$, which can be conditionally distributed on the coefficients of adhesion in the longitudinal and transverse directions. The signs of the friction coefficients will determine the direction of action of the corresponding forces.

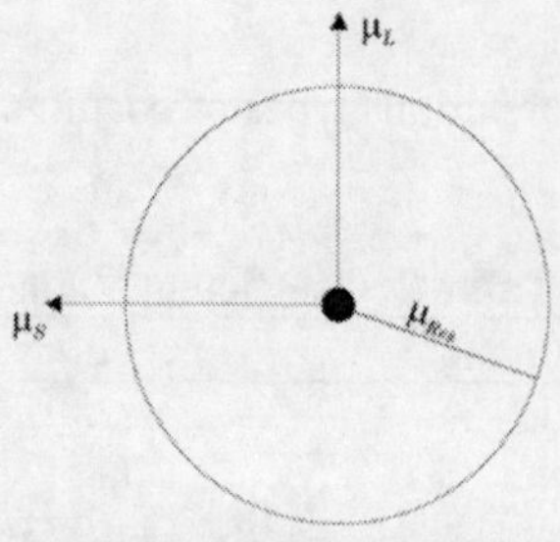

Figure 3. Kamm Circle

We note that if the pneumatic tire has a directional tread, the maximum adhesion coefficient in the longitudinal direction will be greater than the maximum adhesion coefficient in the transverse direction, and therefore, the Kamm circle is transformed into an ellipse. This circumstance can be taken into account in (5) by introducing the coefficient k_S,
then

$$\mu_L = \mu_{\mathrm{Res}} \cdot \frac{S_L}{S_{\mathrm{Res}}} \quad \text{and} \quad \mu_S = k_S \cdot \mu_{\mathrm{Res}} \cdot \frac{S_S}{S_{\mathrm{Res}}}, \tag{6}$$

when $k_S = 1$ formulas (5) and (6) become identical. Typically, for low-profile tires, the coefficient k_S takes values from 0.9 to 0.95 [1]. Coefficients of adhesion can change with intense acceleration and braking, but these effects are not considered in this case. The Kamm circle makes it possible to determine the maximum possible adhesion force in the contact spot, the absolute value of which cannot be exceeded by the forces realized, i.e.:

$$\sqrt{F_{WLij}^2 + F_{WSij}^2} \le \mu_{\mathrm{Res}} \cdot F_{Zij} \tag{7}$$

The following three figures show the traction characteristics of pneumatic tires depending on the driving angles for dry asphalt. We note that in practice, the steering angle is not constant and changes during braking or during turning motion.

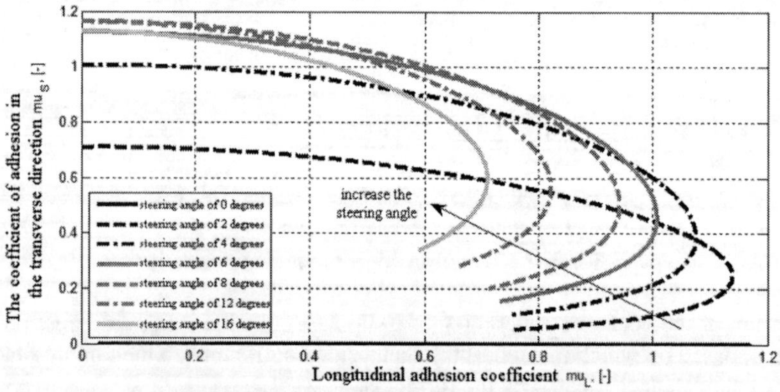

Figure 4. Traction coefficient characteristics

Figure 4 shows the dependence of the coefficient of adhesion of the pneumatic tire in the transverse direction on the coefficient of adhesion in the longitudinal direction for different values of the steering angle. A lateral reaction in the contact patch of the pneumatic tire with the supporting surface does not occur in the absence of a retraction angle, and therefore, in Fig. 4, the curve corresponding to the retraction angle of 0° is located on the abscissa axis. If there is a 2 ° retraction angle, it becomes possible to realize the adhesion coefficient in the transverse direction up to $\mu_S = 0{,}7$, because relative slippage in the transverse direction already reaches $S_S = 3{,}5\%$, which provides high adhesion on dry supporting surfaces. Figure 4 shows that an increase in the angle of inclination leads to a decrease in the longitudinal coefficient of adhesion.

Fig. 5 shows the functions of the longitudinal coefficient of adhesion from the longitudinal relative slip depending on the steering angle change.

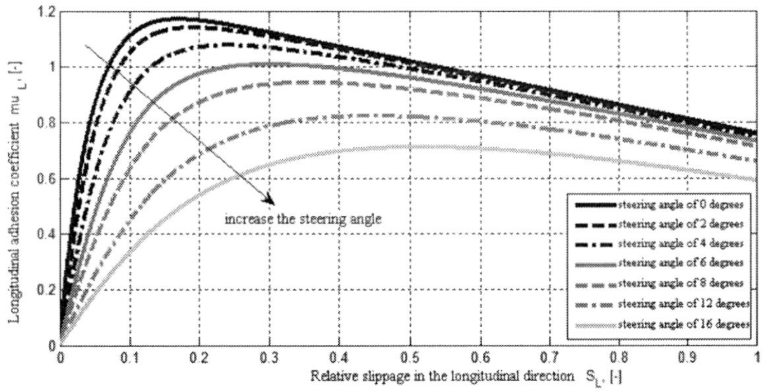

Figure 5. Longitudinal traction characteristics

When braking in a corner, there is a decrease in the longitudinal coefficient of adhesion and an increase in the braking distance, while the steering angle change causes a side reaction in the contact patch and a change in the trajectory of movement. The maximum steering angle of the steered wheels depends on the type of vehicle and is usually approximately 30°. For stable modes of car movement, the steering angle does not exceed 16°, the yaw angle is in the same range.

Fig. 6 shows the functions of the transverse coefficient of adhesion as a function of longitudinal relative slippage depending on the changing steering angle. When turning without braking, it is possible to realize the maximum lateral grip coefficient. With a pull angle of approximately 8° μ_S reaches its maximum value. With increasing longitudinal relative slippage S_L a sharp drop in the values of the transverse coefficient of adhesion can only partially be offset by an increase in the steering angle change.

The graphs shown in Figures 2, 4–6 were constructed using dependencies (1), (2), (3), (5) in the MatLab environment [7]. Fig. 7 shows the implementation of calculating the adhesion coefficients of a pneumatic tire with a supporting surface in the graphic programming language MatLab Simulink.

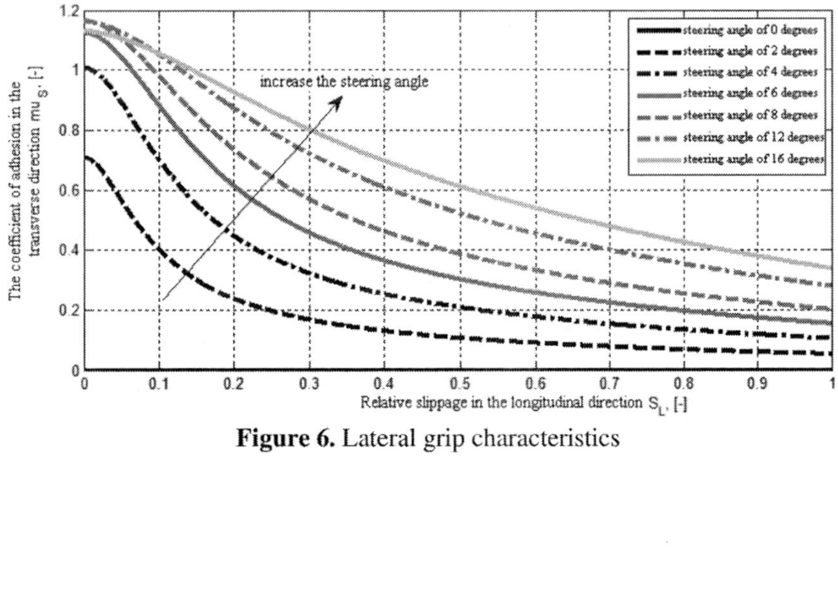

Figure 6. Lateral grip characteristics

IOP Conf. Series: Materials Science and Engineering **832** (2020) 012011 doi:10.1088/1757-899X/832/1/012011

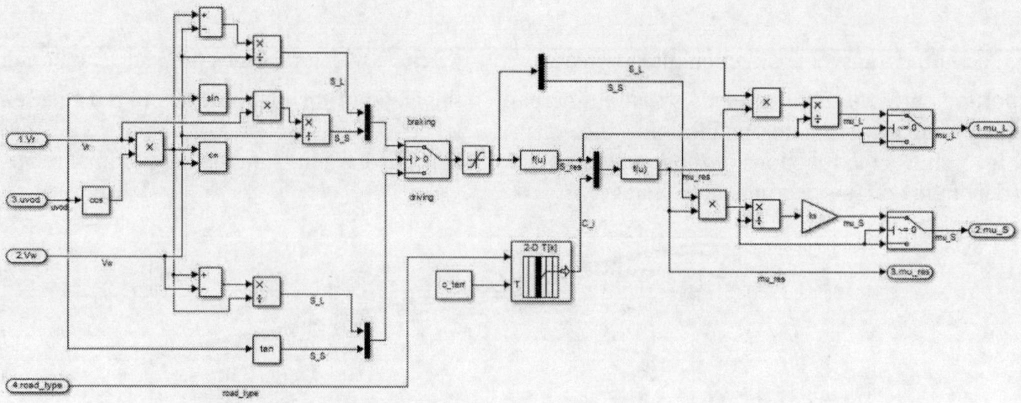

Figure 7. Matlab Simulink adhesion coefficient calculation routine

Thus, the results of simulation modeling and estimation of relative slippage coefficients in the Simulink environment were developed for a more complete simulation of the interaction of a car wheel as part of complex programs for modeling the dynamics of vehicles.

References

[1] Kiencke U and Nielsen L 2005. *Automotive Control Systems – For Engine, Driveline, and Vehicle* 2rd ed (Berlin, Heidelberg: Springer-Verlag) 512 p

[2] Burckhardt M 1993 *Fahrwerktechnik: Radschlupf-Regelsysteme* Vogel Fachbuch (Würzburg)

[3] Reimpell J and Sponagel P 1995 *Fahrwerktechnik: Reifen und Räder* Vogel Fachbuch (Würzburg)

[4] Mitschke M and Wallentowitz H 2004 *Dynamik der Kraftfahrzeuge* 4rd ed (Berlin: Springer Verlag)

[5] Ammon D 1993 *Radlastschwankungen, Seitenführungsvermögen und Fahrsicherheit* vol 1088 of VDI Fortschrittberichte (Düsseldorf: VDI Verlag)

[6] Shadrin S S 2009 *A methodology for calculating the controllability and stability of a vehicle based on the results of field tests* PhD dissertation (Moscow: MADI) 132 p

[7] Shadrin S S, Ivanov A M and Prikhodko V M 2014 Experimental and calculated procedure for determining the adhesion properties of the vehicle pneumatic tires in use *Adv. Environ. Biol.* **8(13)** 294–7

[8] Shadrin S S, Ivanov A M and Prikhodko V M 2015 Usage of Smartphones in the Education Process of MADI during Vehicle Road Tests Conduction *Modern Applied Sci.* **9(1)** 83–8

IOP Conf. Series: Materials Science and Engineering **832** (2020) 012012 doi:10.1088/1757-899X/832/1/012012

The results of modeling the physical processes of interaction of a pneumatic tire with a supporting surface to determine the forces acting in the contact patch

A N Sova[1], G S Mazlumyan[1], O V Egorov[1], S S Shadrin[1] and V A Sova[2]

[1]Moscow Automobile and Road State Technical University (MADI), 64, Leningradsky ave., Moscow, 125319, Russia

[2]Bauman Moscow State Technical University, b. 1, 5, Baumanskaya 2-a street, Moscow, 105005, Russia

E-mail: slsova@mail.ru

Abstract. The article presents the results of modeling the physical processes of the interaction of a pneumatic tire with a supporting surface in order to determine the forces acting in the contact spot. This justifies and develops a diagram of a generalized model of a Simulink car wheel, presents a Simulink circuit block for a mathematical model of a car wheel. Simulink circuit blocks are used for calculating displacements of the reaction application point for calculating the reaction point, the absolute speed and angle of rotation of the wheel, the force, angular and peripheral wheel rotation bones and rolling resistance forces.

An important step in creating a mathematical simulation model of the movement of a wheeled vehicle is the modeling and evaluation of the physical processes of interaction of a car wheel with a supporting surface to determine the forces acting in the contact patch. The initial parameters for calculating a single wheel are: relative slippage coefficients, adhesion coefficients and angle of withdrawal [1].

Figure 1 shows a generalized simulation model of a Simulink automobile wheel for use as part of complex programs for modeling the dynamics of vehicles.

Input, dynamically changing parameters for calculating the block under consideration are [1–8]:

1) the vertical reaction attributable to the wheel in question;

2) the angle of rotation of the wheel;

3) the angular velocity of rotation of the vehicle body relative to the vertical axis;

4) the absolute value of the velocity of the center of mass of the vehicle in a moving coordinate system;

5) the angle of deviation of the velocity vector of the center of mass of the vehicle relative to the central longitudinal axis;

6) the type number of the supporting surface that is currently in the contact patch (dry asphalt, wet asphalt, dry concrete, dry gravel, wet gravel, snow, ice);

7) rolling resistance coefficient corresponding to the type of support coating;

8) traction moment brought to the wheel;

9) the braking torque supplied to the wheel.

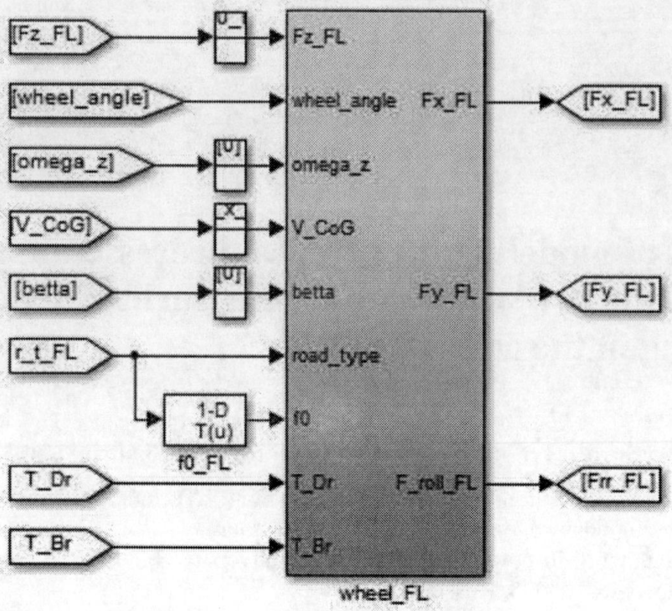

Figure 1. Diagram of a generalized model of a Simulink car wheel

The input constants for calculating the wheel block are accepted:

1) coefficients c_i, specifying the characteristics of μ – S diagrams of different types of coatings;

2) the coefficients for calculating the longitudinal and transverse displacement of the reaction point in the contact spot relative to the geometric center of the contact spot caused by the angle of the longitudinal axis of the pin axis and pressure redistribution in the contact under the influence of lateral force;

3) the ellipse coefficient of the "Kamm circle";

4) coefficients for adjusting the side reaction depending on the changing angle of withdrawal and vertical reaction (dynamic non-linear withdrawal);

5) coefficient taking into account the increase in rolling resistance with increasing speed;

6) dynamic radius of the wheel;

7) moment of inertia of the wheel;

8) a predefined angle of convergence.

The output, dynamically changing parameters when calculating the wheel block are:

1) a longitudinal reaction in the moving coordinate system of the center of mass of the vehicle (i.e. oriented parallel to the longitudinal axis of the vehicle);

2) lateral reaction in the moving coordinate system of the center of mass of the vehicle;

3) rolling resistance force.

Fig. 2 shows the contents of a block of a car wheel (Fig. 1), which, in turn, consists of calculation blocks:

- displacements of the reaction application point in the contact spot (Fig. 3);
- coordinates of reaction application points relative to the center of mass of the vehicle (Fig. 4);
- the absolute speed of the wheel and the angle of withdrawal (Fig. 5);
- relative slippage and adhesion factors;
- forces in the coordinate system associated with the center of mass of the vehicle (Fig. 6);
- angular and peripheral wheel speeds and rolling resistance forces (Fig. 7).

Thus, a generalized simulation model of the Simulink automobile wheel was developed for use as part of complex programs for modeling the dynamics of vehicles.

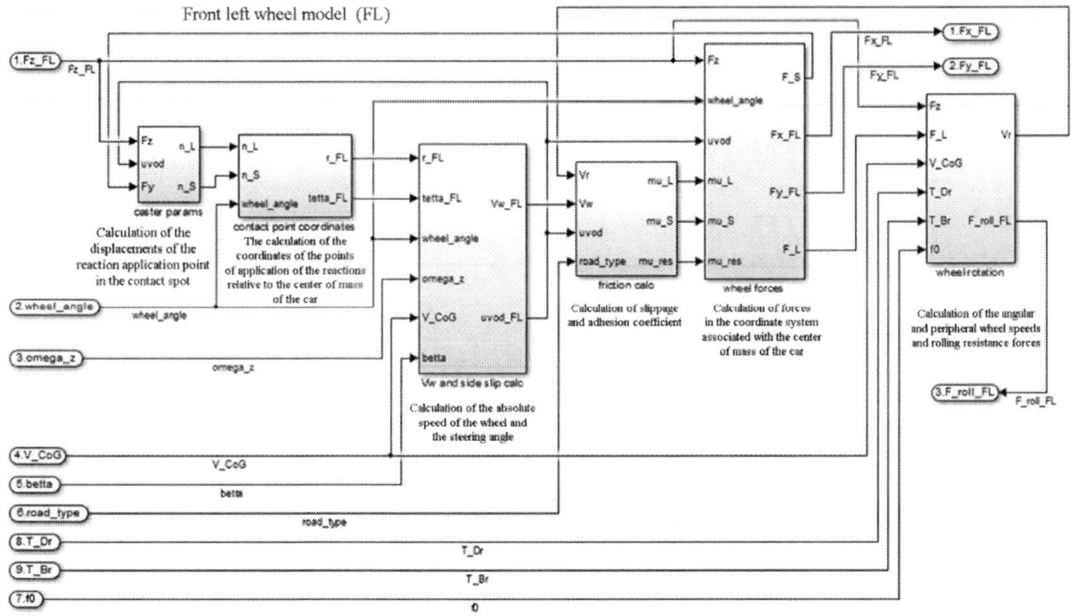

Figure 2. Simulink diagram of a mathematical model of a car wheel

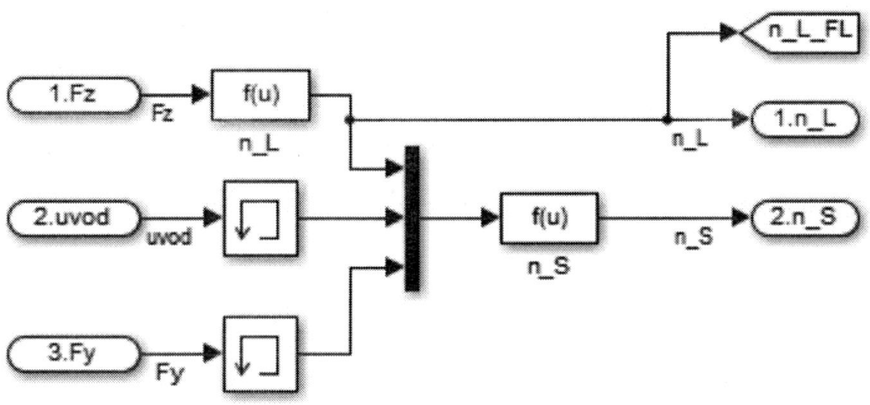

Figure 3. Block for calculating the displacements of the reaction application point

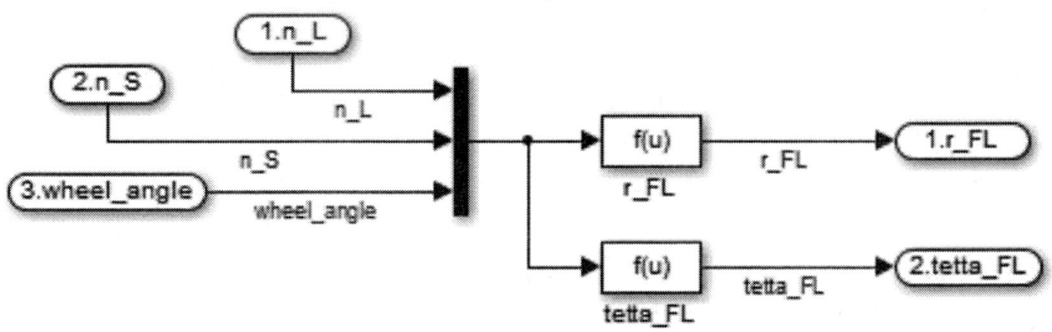

Figure 4. Reaction point calculation unit

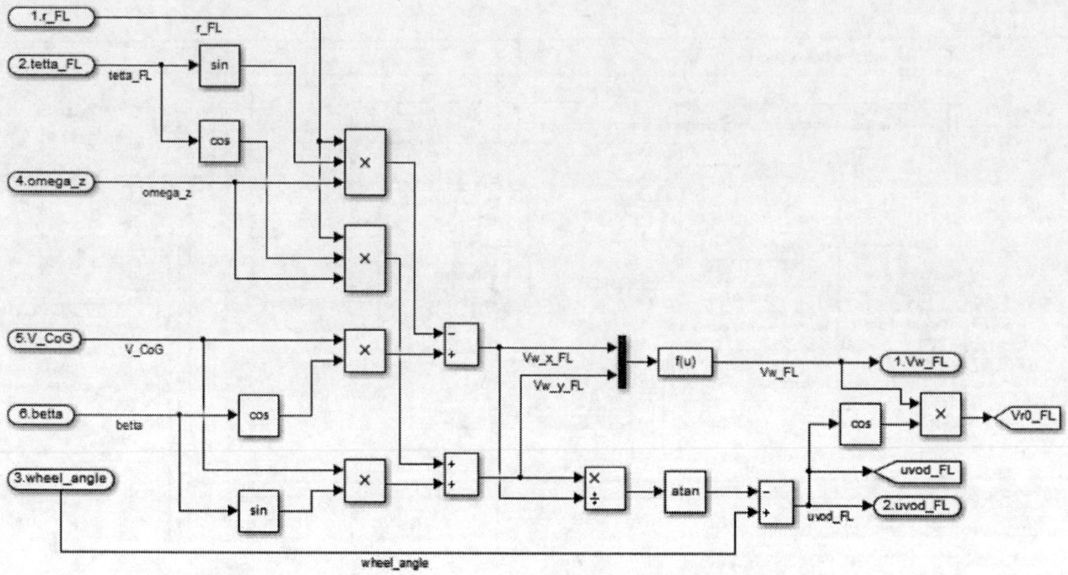

Figure 5. Block for calculating absolute wheel speed and angle

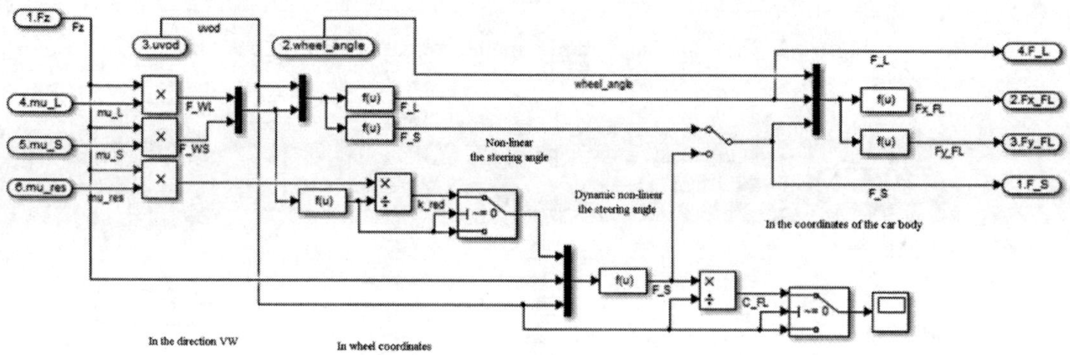

Figure 6. Force calculation unit

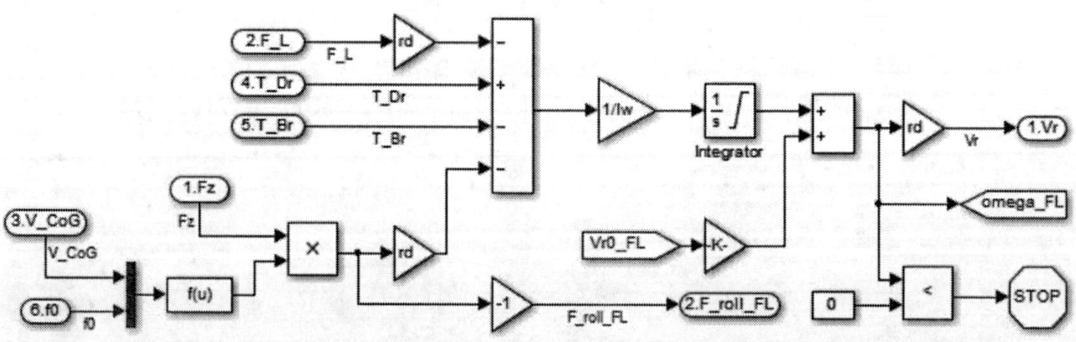

Figure 7. Block for calculating angular and peripheral wheel speeds and rolling resistance forces

References

[1] Kiencke U and Nielsen L 2005 *Automotive Control Systems – For Engine, Driveline, and Vehicle* 2rd ed (Berlin Heidelberg: Springer-Verlag) 512 p
[2] Burckhardt M 1993 *Fahrwerktechnik: Radschlupf-Regelsysteme* Vogel Fachbuch (Würzburg)

[3] Reimpell J and Sponagel P 1995 *Fahrwerktechnik: Reifen und Räder* Vogel Fachbuch (Würzburg)
[4] Mitschke M and Wallentowitz H 2004 *Dynamik der Kraftfahrzeuge* 4rd ed (Berlin: Springer Verlag)
[5] Ammon D 1993 *Radlastschwankungen, Seitenführungsvermögen und Fahrsicherheit* vol 1088 of VDI Fortschrittberichte (Düsseldorf: VDI Verlag)
[6] Shadrin S S 2009 *A methodology for calculating the controllability and stability of a vehicle based on the results of field tests* PhD dissertation (Moscow: MADI) 132 p
[7] Shadrin S S, Ivanov A M and Prikhodko V M 2014 Experimental and calculated procedure for determining the adhesion properties of the vehicle pneumatic tires in use *Adv. Environ. Biol.* **8(13)** 294–7
[8] Shadrin S S, Ivanov A M and Prikhodko V M 2015 Usage of Smartphones in the Education Process of MADI during Vehicle Road Tests Conduction *Modern Applied Sci.* **9(1)** 83–8

The results of modeling and evaluating the dynamics of the movement of a wheeled tractor

A N Sova, G S Mazlumyan, O V Egorov, S A Eruslankin and S S Shadrin

Moscow Automobile and Road State Technical University (MADI), 64, Leningradsky ave., Moscow, 125319, Russia

E-mail: slsova@mail.ru

Abstract. The article describes the results of modeling and evaluating the dynamics of movement of a wheeled tractor. It presents blocks of Simulink schemes for calculating vertical reactions, the gravitational projection of a tractor, the drag force, and the dynamics of a tractor body.

When creating a simulation mathematical model of the movement of a road train with an active trailed link, one of the most important tasks is modeling the dynamics of the tractor [1–4].

The object of research is a wheeled vehicle – a truck tractor based on the MAZ-537 chassis, equipped with a hydromechanical transmission (GMT).

The movement of the tractor, represented by an equivalent circuit, will be described by the following system of equations [2–4]:

$$\begin{cases} m_{CoG}\cdot\begin{bmatrix}\ddot{x}_{In}\\ \ddot{y}_{In}\\ \ddot{z}_{In}\end{bmatrix}=T_{UIn}\cdot\begin{bmatrix}F_{XFL}+F_{XFR}+F_{XRL}+F_{XRR}+F_{windX}+F_{GX}+F_{RR}+F_{Xstr}\\ F_{YFL}+F_{YFR}+F_{YRL}+F_{YRR}+F_{windY}+F_{GY}+F_{Ystr}\\ F_{ZFL}+F_{ZFR}+F_{ZRL}+F_{ZRR}+F_{windZ}+F_{GZ}\end{bmatrix}\\ m_{CoG}=(F_{ZFL}+F_{ZFR}+F_{ZRL}+F_{ZRR})/g=f(\ddot{x}_{In},\ddot{y}_{In},F_{Zstr})\\ T_{UIn}=\begin{bmatrix}\cos\psi\cdot\cos\chi & -\sin\psi\cdot\cos\varphi-\cos\psi\cdot\sin\chi\cdot\sin\varphi & \sin\psi\cdot\sin\varphi-\cos\psi\cdot\sin\chi\cdot\cos\varphi\\ \sin\psi\cdot\cos\chi & \cos\psi\cdot\cos\varphi+\sin\psi\cdot\sin\chi\cdot\sin\varphi & -\cos\psi\cdot\sin\varphi-\sin\psi\cdot\sin\chi\cdot\cos\varphi\\ -\sin\chi & \cos\chi\cdot\sin\varphi & \cos\chi\cdot\cos\varphi\end{bmatrix}\\ \begin{bmatrix}F_{windX}\\ F_{windY}\\ F_{windZ}\end{bmatrix}=\begin{bmatrix}-0.5\cdot c_{airX}\cdot A_L\cdot\rho\cdot(V_{CoGX}-V_{windX}\cdot\cos\psi-V_{windY}\cdot\sin\psi)^2\\ -0.5\cdot c_{airY}\cdot A_S\cdot\rho\cdot(V_{CoGY}+V_{windX}\cdot\sin\psi-V_{windY}\cdot\cos\psi)^2\cdot sign(V_{windX}\cdot\sin\psi-V_{windY}\cdot\cos\psi)\\ 0\end{bmatrix}\\ \begin{bmatrix}F_{GX}\\ F_{GY}\\ F_{GZ}\end{bmatrix}=\begin{bmatrix}\cos(\chi_{road}) & \sin(\chi_{road})\cdot\sin(\varphi_{road}) & \sin(\chi_{road})\cdot\cos(\varphi_{road})\\ 0 & \cos(\varphi_{road}) & -\sin(\varphi_{road})\\ -\sin(\chi_{road}) & \cos(\chi_{road})\cdot\sin(\varphi_{road}) & \cos(\chi_{road})\cdot\cos(\varphi_{road})\end{bmatrix}\cdot\begin{bmatrix}0\\ 0\\ m_{CoG}\cdot g\end{bmatrix}\\ I_Z\cdot\ddot{\psi}=(F_{YFR}+F_{YFL})\cdot l_F-(F_{YRL}+F_{YRR})\cdot l_R+(F_{XFR}-F_{XFL})\cdot 0.5\cdot b_F+(F_{XRR}-F_{XRL})\cdot 0.5\cdot b_R+F_{Ystr}\cdot l_{astr}\\ I_X\cdot\ddot{\varphi}=(F_{ZFL}-F_{ZFR})\cdot 0.5\cdot b_F+(F_{ZRL}-F_{ZRR})\cdot 0.5\cdot b_R-m_{CoG}\cdot\ddot{y}_{In}\cdot h_{CoG}\\ I_Y\cdot\ddot{\chi}=-(F_{ZFL}+F_{ZFR})\cdot l_F+(F_{ZRL}+F_{ZRR})\cdot l_R+m_{CoG}\cdot\ddot{x}_I\cdot h_{CoG}\\ I_{Wi}\cdot\dot{\varpi}_i=T_{Dr,i}-T_{Br,i}-r_d\cdot F_{Li}\end{cases}$$

IOP Conf. Series: Materials Science and Engineering **832** (2020) 012013 doi:10.1088/1757-899X/832/1/012013

where m_{CoG} – the mass of the tractor, when simulated, is a variable due to the changing vertical load on the fifth wheel coupling caused by the redistribution of the vertical reactions of the semi-trailer due to the dynamics of movement;

$\ddot{x}_{In}, \ddot{y}_{In}, \ddot{z}_{In}$ – projections of the acceleration of the center of mass of the tractor in the inertial reference system on the Cartesian axis of the coordinate system;

$T_{UIn} = T_{RotZ} \cdot T_{RotY} \cdot T_{RotX}$ – rotation matrix for transferring the corresponding vectors from the coordinate system associated with the center of mass of the tractor to the inertial reference system; индексы X, Y, Z determine the directionality of the vector relative to the main central axes of the tractor;

F – force; indices FL, FR, RL, RR determine the belonging of the vector to the wheels of the tractor, respectively, front left, front right, rear left and rear right in the direction of travel; index str determines the point of application of the vector (force, speed, acceleration) to the center of the fifth wheel coupling of the tractor; index $wind$ – index of aerodynamic drag or wind speed;

F_{GX}, F_{GY}, F_{GZ} – projections of the tractor gravity vector onto the main central axes of the tractor;

V_{CoG} – speed of the center of mass of the tractor in the moving coordinate system;

ψ – yaw angle;

φ – roll angle (in the transverse direction);

χ – pitch angle (roll angle in the longitudinal direction);

χ_{road} – longitudinal angle of inclination of the profile of the supporting surface (road);

φ_{road} – lateral slope of the road;

c_{airX}, c_{airY} – drag coefficients in the longitudinal / transverse directions;

A_L, A_S – the frontal area of the road train, the lateral area of the tractor;

ρ – air density;

g – acceleration of gravity;

I_X, I_Y, I_Z – the main moments of inertia of the tractor;

l_F, l_R – distance from the center of gravity of the tractor to the axis of the front / rear wheels;

l_{astr} – distance from the center of gravity of the tractor to the fifth wheel coupling;

b_F, b_R – track of front / rear wheels of the tractor;

h_{CoG} – height of the center of mass of the tractor;

I_{Wi} – moments of inertia of the equivalent wheels of the tractor (see figure 2);

$\dot{\omega}_i$ – angular acceleration of the equivalent wheel of the tractor;

$T_{Dr,i}$ – torque supplied to the wheel mover of the tractor;

$T_{Br,i}$ – the braking moment supplied to the wheeled tractor mover;

r_d – dynamic radius of the wheel;

F_{Li} – the force of interaction with the supporting surface in the longitudinal plane of the wheel propulsion.

Fig. 1 shows a block for calculating vertical reactions, Fig. 2 – a block for calculating the projection of the gravity of the tractor on the axis of the moving coordinate system associated with the center of mass of the tractor, and Fig. 3 – a block for calculating the aerodynamic drag forces.

Fig. 4 shows the software implementation of the unit for calculating the dynamics of the movement of the body of the tractor in the MatLab Simulink environment.

Thus, the results of mathematical modeling of a wheeled tractor in the Simulink environment were developed for further simulation of the dynamics of road trains with an active trailer link when driving on roads of various categories and explored terrain.

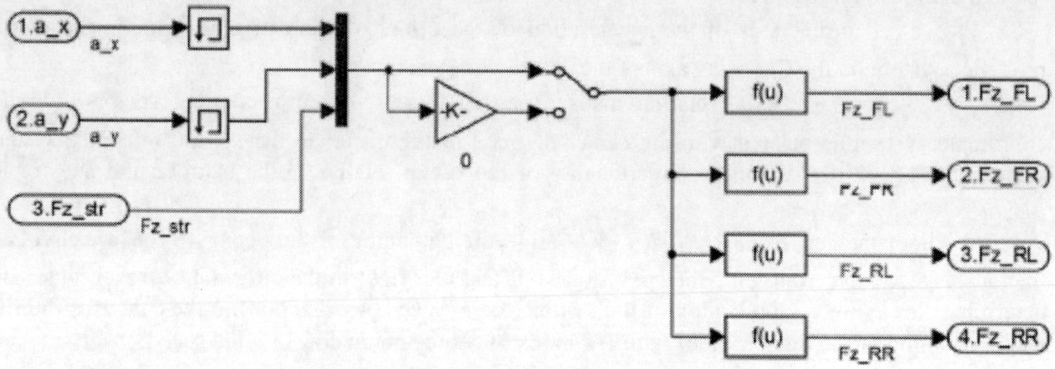

Figure 1. Block for calculating vertical reactions

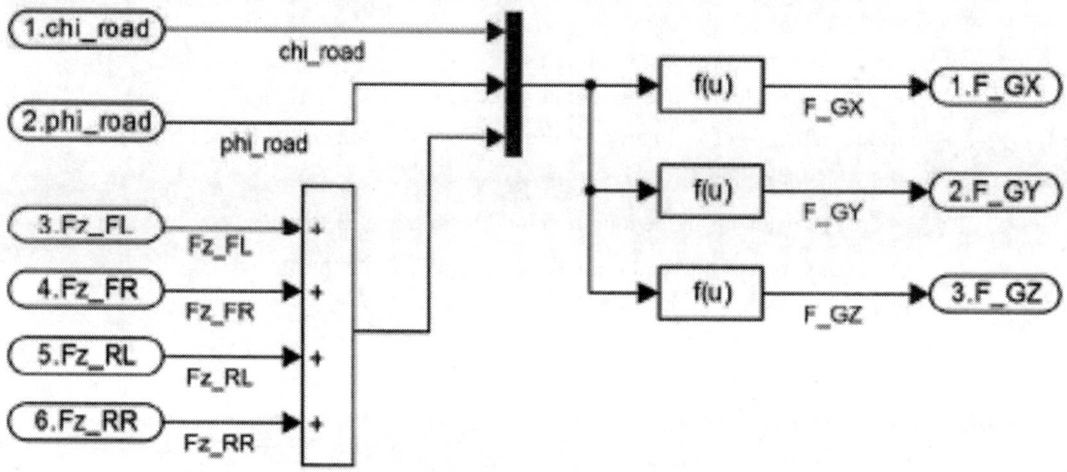

Figure 2. Tractor gravity projection calculation unit

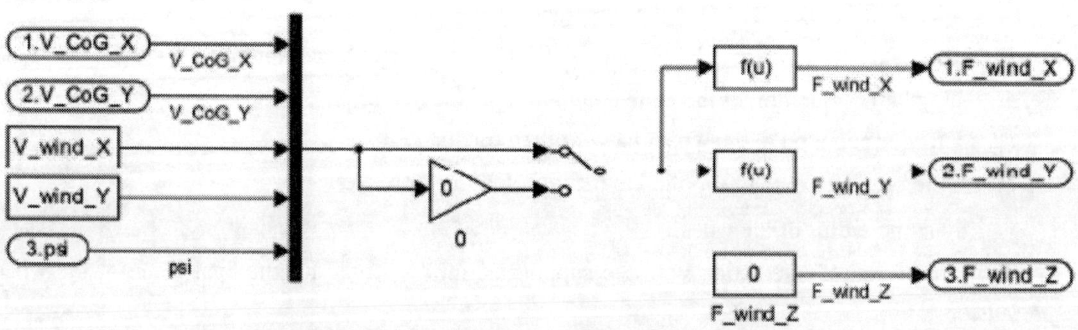

Figure 3. Aerodynamic drag force calculation unit

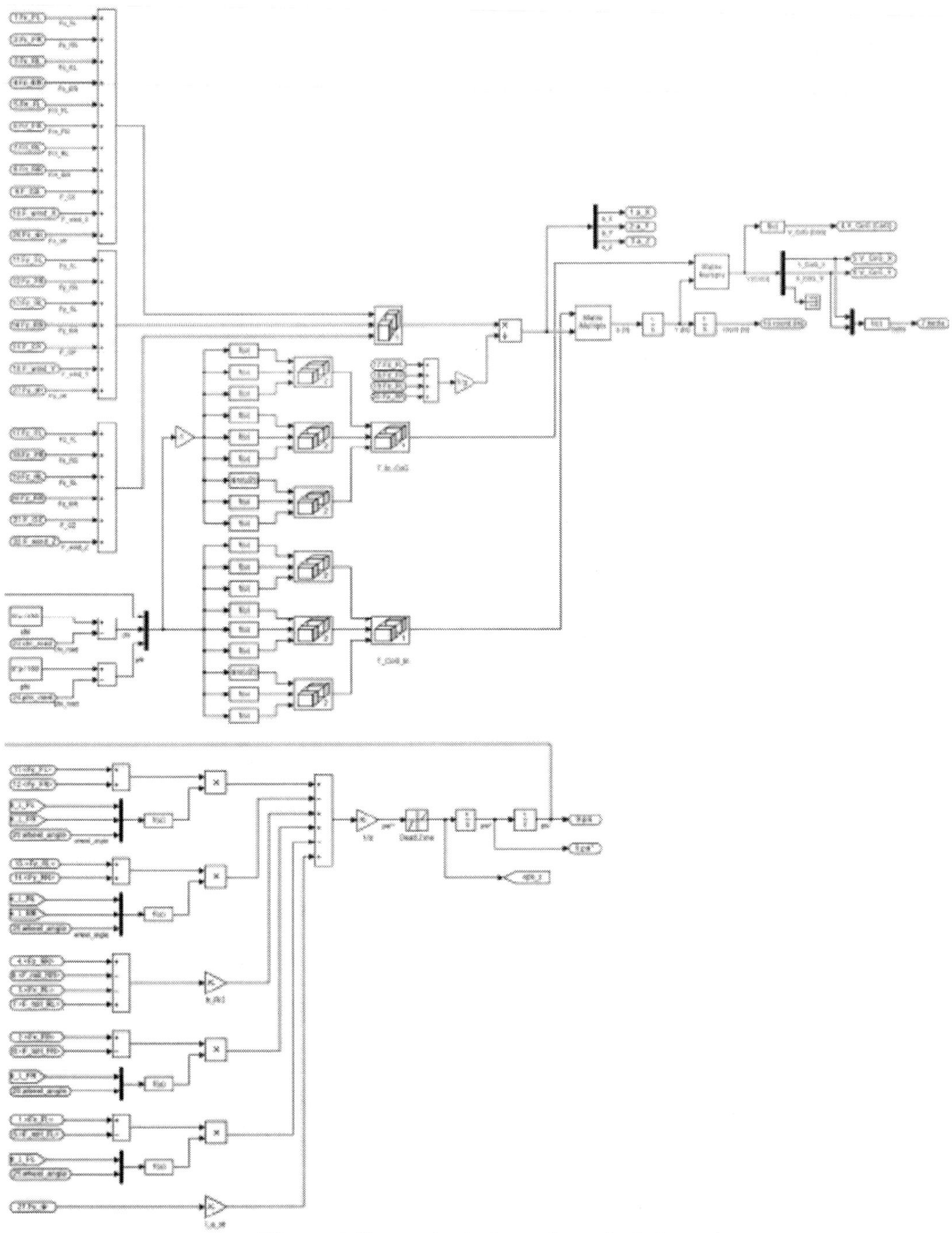

Figure 4. Tractor body dynamics calculation unit

References

[1] Kiencke U and Nielsen L 2005 *Automotive Control Systems – For Engine, Driveline, and Vehicle* 2rd ed (Berlin Heidelberg: Springer-Verlag) 512 p

[2] Shadrin S S 2009 *A methodology for calculating the controllability and stability of a vehicle based on the results of field tests* PhD dissertation (Moscow: MADI) 132 p

IOP Conf. Series: Materials Science and Engineering **832** (2020) 012013 doi:10.1088/1757-899X/832/1/012013

[3] Shadrin S S, Ivanov A M and Prikhodko V M 2014 Experimental and calculated procedure for determining the adhesion properties of the vehicle pneumatic tires in use *Adv. Environ. Biol.* **8(13)** 294–7

[4] Shadrin S S, Ivanov A M and Prikhodko V M 2015 Usage of Smartphones in the Education Process of MADI during Vehicle Road Tests Conduction *Modern Applied Sci.* **9(1)** 83–8

The results of modeling and evaluating the dynamics of the active semitrailer

A N Sova, G S Mazlumyan, O V Egorov, S A Eruslankin and S S Shadrin

Moscow Automobile and Road State Technical University (MADI), 64, Leningradsky ave., Moscow, 125319, Russia

E-mail: slsova@mail.ru

Abstract. The article describes the results of modeling and evaluating the dynamics of an active semi-trailer. It presents the developed blocks for calculating the vertical reactions of the semi-trailer, the projections of the gravity of the semi-trailer, and the aerodynamic drag forces of the semi-trailer, trailer, the angle of rotation of the wheels of the semi-trailer, the dynamics of the body of the semi-trailer.

When creating a simulation mathematical model of the movement of a road train with an active trailed link, the main task is to simulate the dynamics of the active trailed link [1–4].

The movement of the semitrailer will be described by a system of equations identical to the above except that the accelerations of the center of mass of the semitrailer will be calculated from the accelerations of the center of mass of the tractor in the moving coordinate system by transferring them, taking into account the rotational movement of the tractor, from the center of mass of the tractor to the fifth wheel and further, taking into account the rotational motion of the semitrailer and the folding angle, to the center of mass of the semitrailer, which will be the condition for communication.

Thus, the movement of the semi-trailer (equivalent circuit, Fig. 1) will be described by the following system of equations [1–4]:

$$\begin{cases} m_{str}\cdot\begin{bmatrix}\ddot{x}_{In\_str}\\ \ddot{y}_{In\_str}\\ \ddot{z}_{In\_str}\end{bmatrix}=T_{UIn\_str}\cdot\begin{bmatrix}F_{XRL\_str}+F_{XRR\_str}+F_{GX\_str}+F_{RR\_str}+F_{Xstr\_str}\\ F_{YRL\_str}+F_{YRR\_str}+F_{windY\_str}+F_{GY\_str}+F_{Ystr\_str}\\ F_{ZRL\_str}+F_{ZRR\_str}+F_{windZ\_str}+F_{GZ\_str}\end{bmatrix}\\ m_{str}=(F_{ZRL\_str}+F_{ZRR\_str})/g=f(\ddot{x}_{In\_str},\ddot{y}_{In\_str})\\ T_{UIn\_str}=T_{RotZ\_str}\cdot T_{RotY\_str}\cdot T_{RotX\_str}\\ \begin{bmatrix}F_{windX\_str}\\ F_{windY\_str}\\ F_{windZ\_str}\end{bmatrix}=\begin{bmatrix}0\\ \{-0.5\cdot c_{airY}\cdot A_{S\_str}\cdot\rho\cdot(V_{CoGY\_str}+V_{windX}\cdot\sin\psi_{str}-V_{windY}\cdot\cos\psi_{str})^2\cdot\\ \cdot sign(V_{windX}\cdot\sin\psi_{str}-V_{windY}\cdot\cos\psi_{str})\}\\ 0\end{bmatrix}\\ \begin{bmatrix}F_{GX\_str}\\ F_{GY\_str}\\ F_{GZ\_str}\end{bmatrix}=\begin{bmatrix}\cos(\chi_{road\_str}) & \sin(\chi_{road\_str})\cdot\sin(\varphi_{road\_str}) & \sin(\chi_{road\_str})\cdot\cos(\varphi_{road\_str})\\ 0 & \cos(\varphi_{road\_str}) & -\sin(\varphi_{road\_str})\\ -\sin(\chi_{road\_str}) & \cos(\chi_{road\_str})\cdot\sin(\varphi_{road\_str}) & \cos(\chi_{road\_str})\cdot\cos(\varphi_{road\_str})\end{bmatrix}\cdot\begin{bmatrix}0\\ 0\\ m_{str}\cdot g\end{bmatrix}\end{cases}$$

$$\begin{cases} I_{Z\_str} \cdot \ddot{\psi}_{str} = F_{Ystr\_str} \cdot l_{Fstr} - (F_{YRL\_str} + F_{YRR\_str}) \cdot l_{Rstr} + (F_{XRR\_str} - F_{XRL\_str}) \cdot 0.5 \cdot b_{Rstr} \\ I_{X\_str} \cdot \ddot{\varphi}_{str} = (F_{ZRL\_str} - F_{ZRR\_str}) \cdot 0.5 \cdot b_{Rstr} - m_{str} \cdot \ddot{y}_{In\_str} \cdot h_{CoG\_str} \\ I_{Y\_str} \cdot \ddot{\chi}_{str} = -F_{ZF\_str} \cdot l_{Fstr} + (F_{ZRL\_str} + F_{ZRR\_str}) \cdot l_{Rstr} + m_{str} \cdot \ddot{x}_{In\_str} \cdot h_{CoG\_str} \\ I_{Wi} \cdot \dot{\varpi}_i = T_{Dr,i} - T_{Br,i} - r_d \cdot F_{Li} \end{cases}$$

where m_{str} – the mass of the semi-trailer falling on the rear axle;

$\ddot{x}_{In\_str}, \ddot{y}_{In\_str}, \ddot{z}_{In\_str}$ – projections of the acceleration of the center of mass of the semitrailer in the inertial reference frame on the Cartesian axis of the coordinate system;

T_{UIn_str} – a rotation matrix for transferring the corresponding vectors from the moving coordinate system associated with the center of mass of the semi-trailer to the inertial reference frame;

indexes X, Y, Z determine the directionality of the vector relative to the main central axes of the semi-trailer;

the index $\_str$ determines the belonging of the parameter in question to the semi-trailer;

A_{S_str} – lateral area of the semi-trailer;

l_{Fstr}, l_{Rstr} – distance from the center of gravity of the semitrailer to the fifth wheel and the rear axle;

l_{astr} – distance from the center of gravity of the tractor to the fifth wheel coupling;

b_{Rstr} – track of the rear wheels of the semi-trailer;

h_{CoG_str} – height of the center of mass of the semi-trailer.

Communication equations for solving systems of differential equations for the joint movement of the tractor and semitrailer:

$$\begin{bmatrix} \ddot{x}_{str} \\ \ddot{y}_{str} \end{bmatrix} = T_{RotZ}(\lambda) \cdot \begin{bmatrix} \ddot{x} + \dot{\psi}^2 \cdot l_{astr} \\ \ddot{y} - \ddot{\psi} \cdot l_{astr} \end{bmatrix} + \begin{bmatrix} \dot{\psi}_{str}^2 \cdot l_{Fstr} \\ \ddot{\psi}_{str} \cdot l_{Fstr} \end{bmatrix},$$

where $\ddot{x}_{str}, \ddot{y}_{str}$ – projections of the acceleration of the center of mass of the semitrailer in the moving coordinate system (in the axes of the semitrailer);

$T_{RotZ}(\lambda) = \begin{bmatrix} \cos\lambda & -\sin\lambda \\ \sin\lambda & \cos\lambda \end{bmatrix}$ – rotation matrix about a vertical axis;

λ – folding angle of a two-link road train;

$\ddot{x}, \ddot{y}$ – projections of the acceleration of the center of mass of the tractor in the moving coordinate system (in the axes of the tractor);

ψ – yaw angle of the tractor;

ψ_{str} – yaw angle of the semi-trailer.

Fig. 1 shows a block for calculating vertical reactions, Fig. 2 shows a block for calculating the projections of gravity of the semi-trailer on the axis of the moving coordinate system associated with the center of mass of the semitrailer, and Fig. 3 shows a block for calculating the aerodynamic drag forces of the semitrailer.

Fig. 4 shows a block for calculating the steering angle of the steering wheels of a semitrailer depending on the folding angle of a two-link road train.

Fig. 5 shows the software implementation of the block for calculating the dynamics of the body of a semi-trailer in the MatLab Simulink environment.

Thus, a simulation mathematical model of the movement of the active trailed link of a two-link road train was developed. The model has 12 degrees of freedom (6 degrees of freedom of the semitrailer body, rotation of 2 wheels, turns of 2 wheels, individual type of supporting surface under each of 2 wheels when moving). The modularity of the constructed model for the movement of the active trailed link of a two-link road train in the Simulink environment allows us to go from the equivalent circuit of the research object to the actual one for the refinement of VIP control algorithms at the next stage. The developed mathematical model can be used when performing research and developmental technological work, as well as in the educational process in the preparation of specialists in specialized specialties [4].

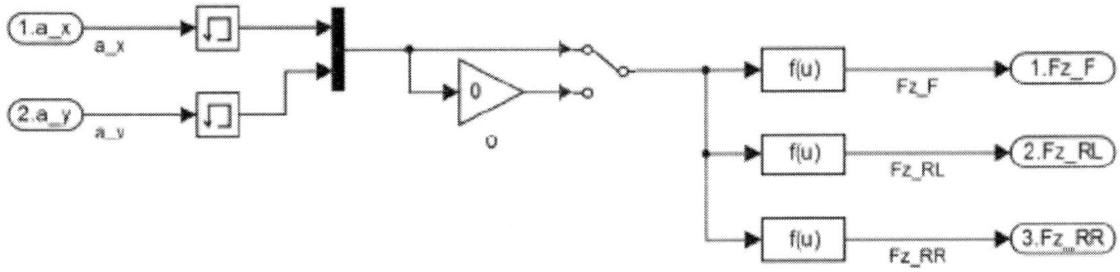

Figure 1. Block for calculating the vertical reactions of the semi-trailer

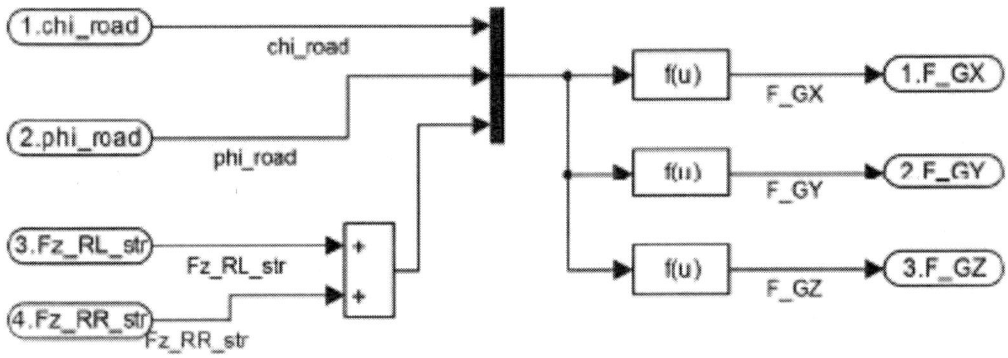

Figure 2. Block for calculating the projections of gravity of the semitrailer

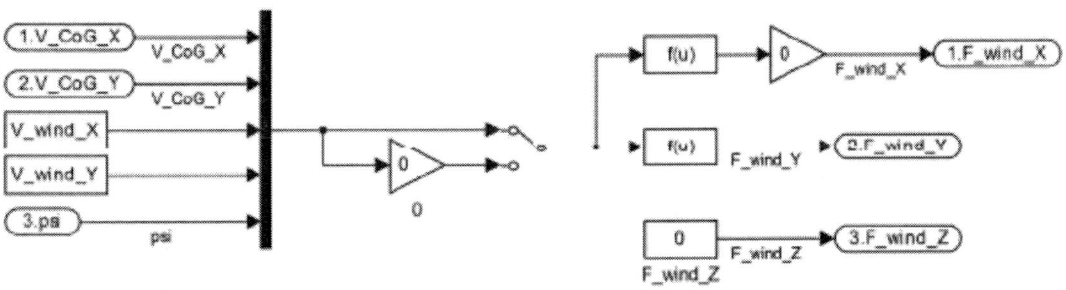

Figure 3. Unit for calculating the aerodynamic drag forces of a semi-trailer

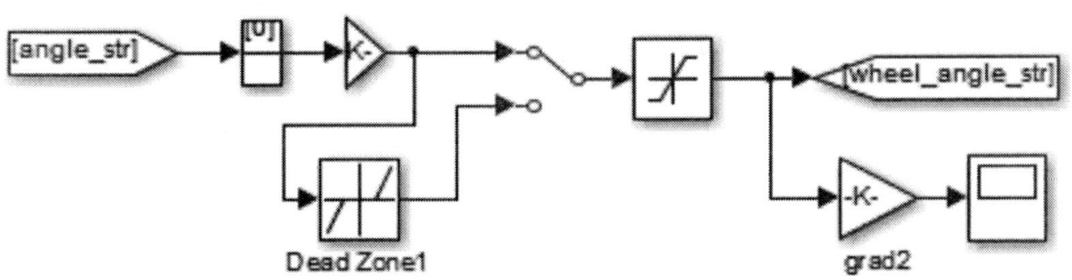

Figure 4. Block for calculating the angle of rotation of the wheels of the semi-trailer

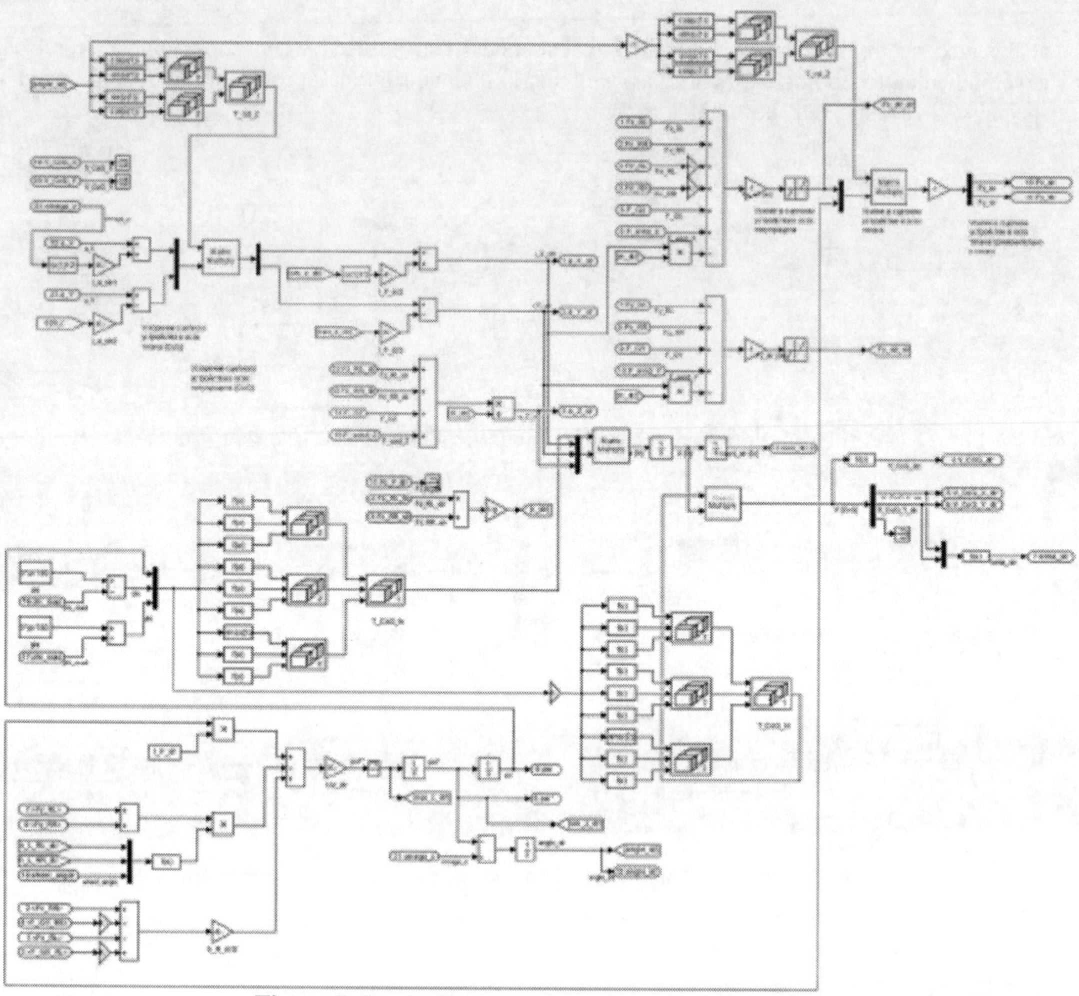

Figure 5. Semitrailer body dynamics calculation unit

References

[1] Kiencke U and Nielsen L 2005. *Automotive Control Systems – For Engine, Driveline, and Vehicle* 2rd ed (Berlin Heidelberg: Springer-Verlag) 512 p

[2] Shadrin S S 2009 *A methodology for calculating the controllability and stability of a vehicle based on the results of field tests* PhD dissertation (Moscow: MADI) 132 p

[3] Shadrin S S, Ivanov A M and Prikhodko V M 2014 Experimental and calculated procedure for determining the adhesion properties of the vehicle pneumatic tires in use *Adv. Environ. Biol.* **8(13)** 294–7

[4] Shadrin S S, Ivanov A M and Prikhodko V M 2015 Usage of Smartphones in the Education Process of MADI during Vehicle Road Tests Conduction *Modern Applied Sci.* **9(1)** 83–8

IOP Conf. Series: Materials Science and Engineering **832** (2020) 012015 doi:10.1088/1757-899X/832/1/012015

The results of modeling and evaluating the dynamics of a road train with an active trailer link based on valve-inductor electric machines

A N Sova[1], G S Mazlumyan[1], O V Egorov[1], S A Eruslankin[1], S S Shadrin[1] and V A Sova[2]

[1]Moscow Automobile and Road State Technical University (MADI), 64, Leningradsky ave., Moscow, 125319, Russia

[2]Bauman Moscow State Technical University, b. 1, 5, Baumanskaya 2-a str., Moscow, 105005, Russia

E-mail: slsova@mail.ru

Abstract. The article describes the results of modeling and evaluating the dynamics of a road train with an active trailer link based on valve-inductor electric machines. It gives a general view and an equivalent road train diagram, shows a generalized simulation model of the movement of a two-link road train developed in the Simulink environment for studying motion in various road conditions. We proposed a reference surface model based on the terrain map, blocks were developed for calculating the road tilt angles separately for the tractor and the semitrailer, taking into account the type of coating under each road train wheel at the current time. The visualization of the movement of the road train is shown, the results of modeling and evaluation of the dynamics of the road train with an active trailer link based on valve-induction electric machines for creating control systems for individual drives of the electric motor-wheel of an active semi-trailer and developing a program and methodology of full-scale experimental research representations.

The object of research is a special vehicle (Fig. 1), which is a two-link road train consisting of a four-axle truck tractor based on a wheeled tractor equipped with a hydromechanical transmission, and a four-axle active semi-trailer with controlled motor wheels equipped with an electromechanical transmission based on valve-induction electric drives. All wheels of a two-link truck train are swivel.

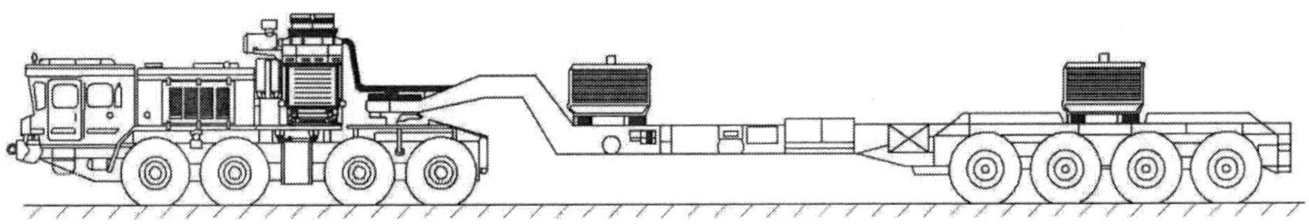

Figure 1. General view of a two-link road train with an active trailed link based on valve-inductor electric machines

When individually controlling the traction forces of the active wheels of the semitrailer, the initial was the calculated external characteristic of the valve-inductor electric drive, reduced to the wheel propulsion, shown in Fig. 2.

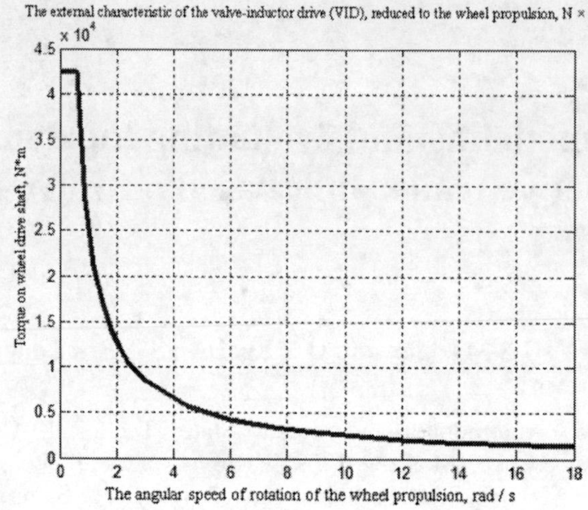

Figure 2. External characteristic of a valve-inductor drive, reduced to a wheel propeller

For calculations, the external characteristic of the valve-inductor electric drive was used, the gear ratio of the gearbox installed between the valve-inductor electric drive and the wheel mover was 34.85 and the gearbox efficiency was 0.944. In order to simplify the mathematical model and reduce the calculation time, we will use the equivalent circuit of the research object presented in Fig. 3.

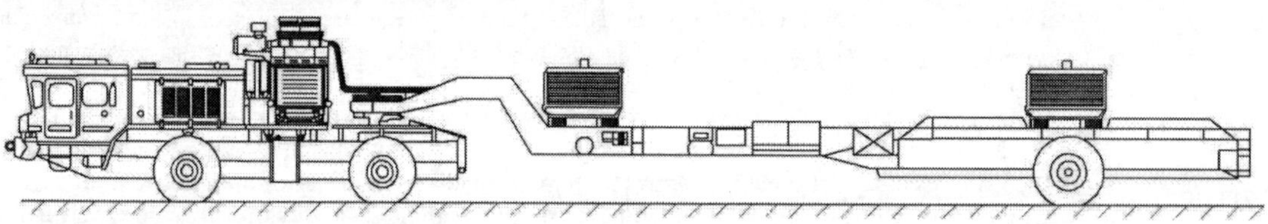

Figure 3. Equivalent circuit of a two-link road train with an active trailed link based on valve-inductor electric machines

The developed mathematical model of the motion of a two-link road train with an active trailer link based on valve-induction electric machines has 30 degrees of freedom (6 degrees of freedom of the tractor body, 6 – of the body of the trailer, rotation of 6 wheels, 6 wheel turns, individual type of bearing surface under each of 6 wheels when driving).

The modularity of the constructed Simulink model allows us to move from the equivalent circuit of the research object to the actual one for the next step to refine the control algorithms of the valve-inductor drive.

The developed mathematical model can be used in the educational process in the preparation of specialists in specialized specialties [8].

Figure 4 shows a generalized simulation model of the movement of a two-link road train of the equivalent circuit developed in the Simulink environment.

To simulate the movement of a two-link road train with an active trailed link based on valve-inductor electric machines in various road conditions, the terrain map was set separately in a mathematical model, an example of which is shown in Fig. 5.

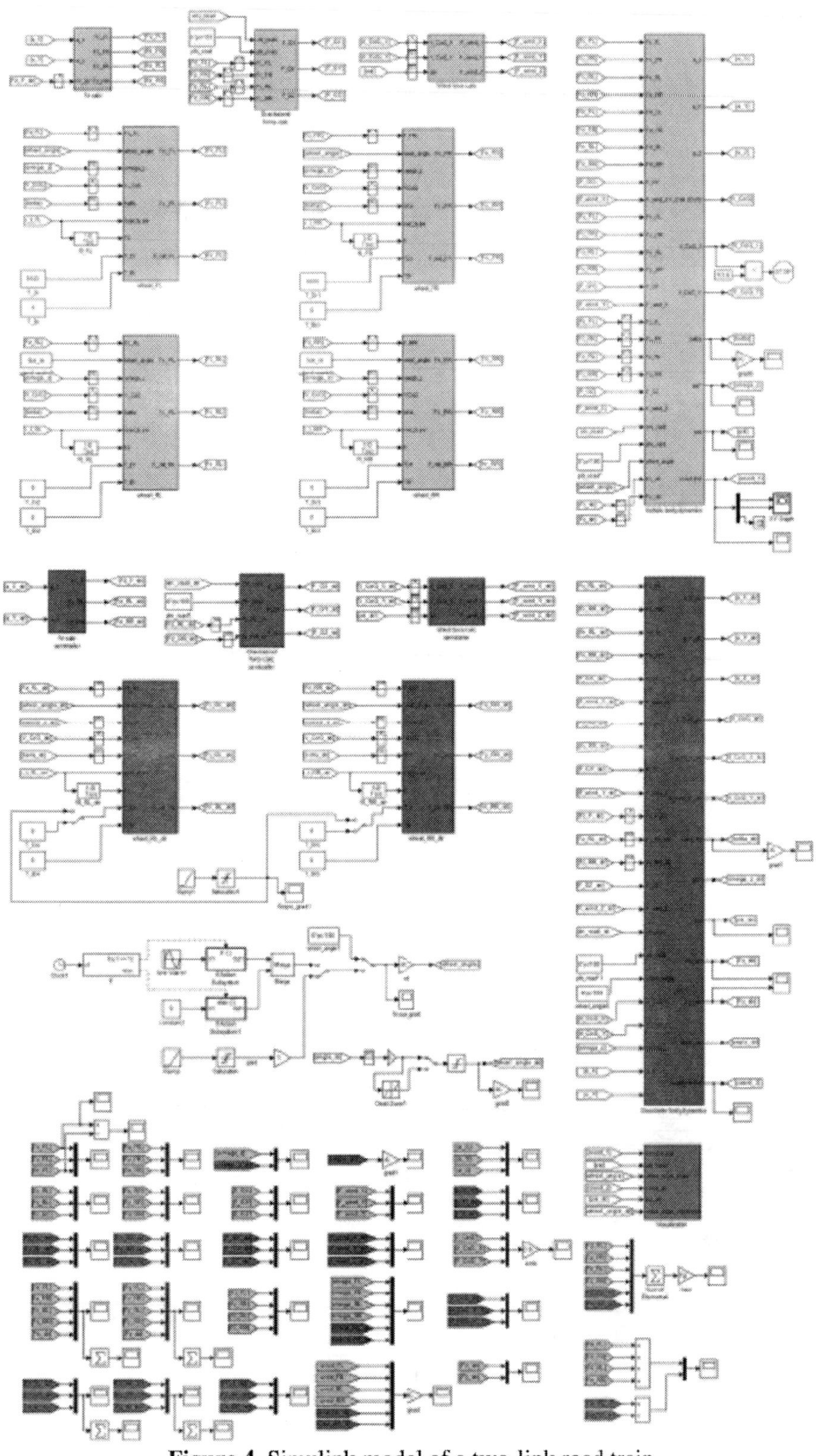

Figure 4. Simulink model of a two-link road train

Thus, the model specified the longitudinal and transverse slopes of the supporting surface, as well as segments of the following types of coatings with individual adhesion characteristics [1]:

- dry asphalt;
- wet asphalt;
- dry concrete;
- dry gravel;
- wet gravel;
- snow;
- ice.

Each of the 7 types of supporting coatings corresponded to individual rolling resistance coefficients of the wheel propulsors [2].

Fig. 6 shows a block that calculates the slope angles of the road separately for the tractor and semi-trailer and determines the type of supporting coating under each wheel of the road train at the current time. In addition, a code was developed for the visualization program for the movement of a road train with an active trailer link based on valve-induction electric machines.

An example of graphical visualization of the movement of a road train (equivalent circuit) along a circular path is shown in Fig. 7.

The results of simulation and assessment of the dynamics of the movement of a road train with an active trailer link based on valve-induction electric machines when applying constant torques of 9500 N·m to the wheels of the front axle of the tractor and sinusoidal action on the steering wheel on a flat horizontal surface with dry asphalt coating are presented in Fig. 8–21.

The developed simulation mathematical model of the movement of a two-link road train with an active trailed link based on valve-inductor electric machines includes:

1) a mathematical model of the movement of a tractor of a two-link road train with an active trailed link in the environment of MatLab Simulink;

2) a mathematical model of the movement of a two-link road train with an active trailed link based on valve-inductor electric machines in a MatLab Simulink environment;

3) a mathematical model of a virtual experimental site in the environment of MatLab Simulink for experimental research and testing of a two-link road train with an active trailed link based on valve-inductor electric machines.

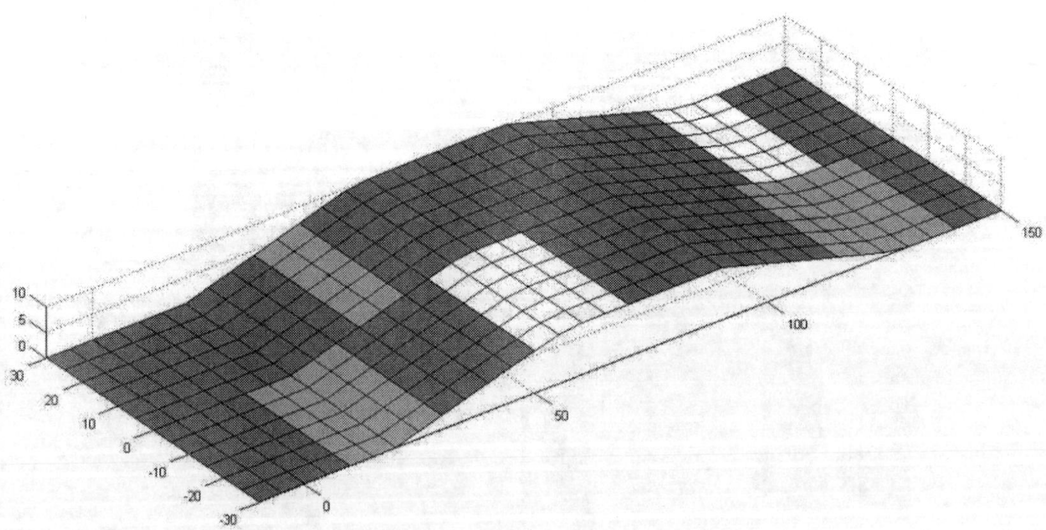

Figure 5. Segmented Terrain Map

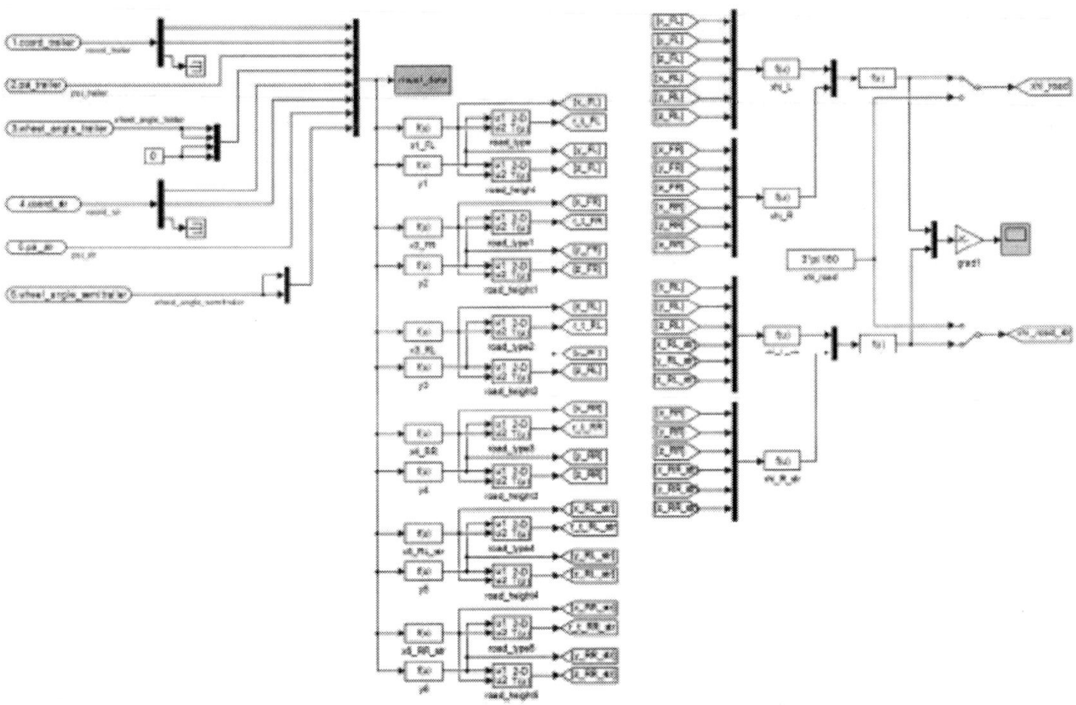

Figure 6. Block for calculating the current slopes of the road and determining the type of supporting coating under each wheel of the road train

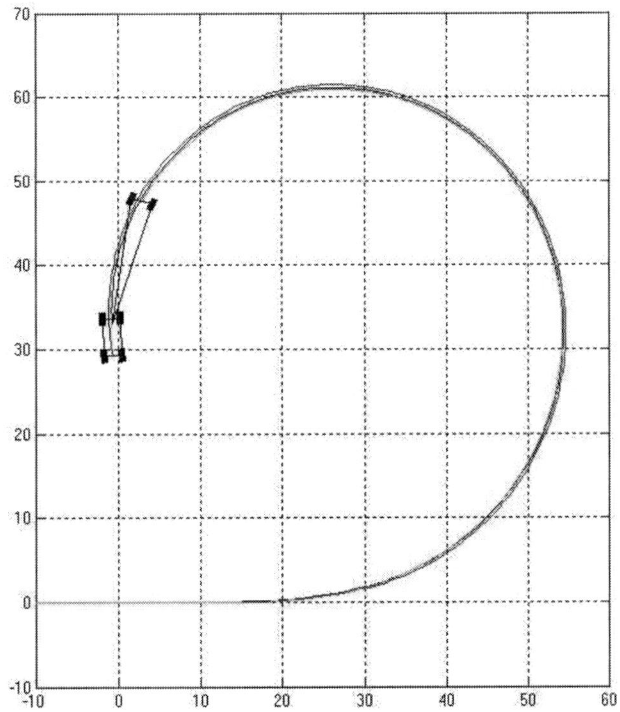

Figure 7. Visualization of the movement of a two-link road train with an active trailed link based on valve-induction electric machines

IOP Conf. Series: Materials Science and Engineering **832** (2020) 012015 doi:10.1088/1757-899X/832/1/012015

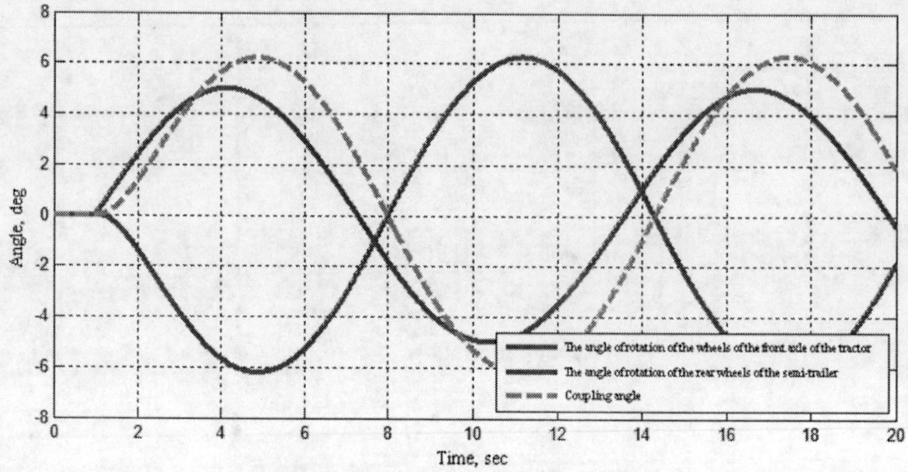

Figure 8. Wheel angle and hitch angle

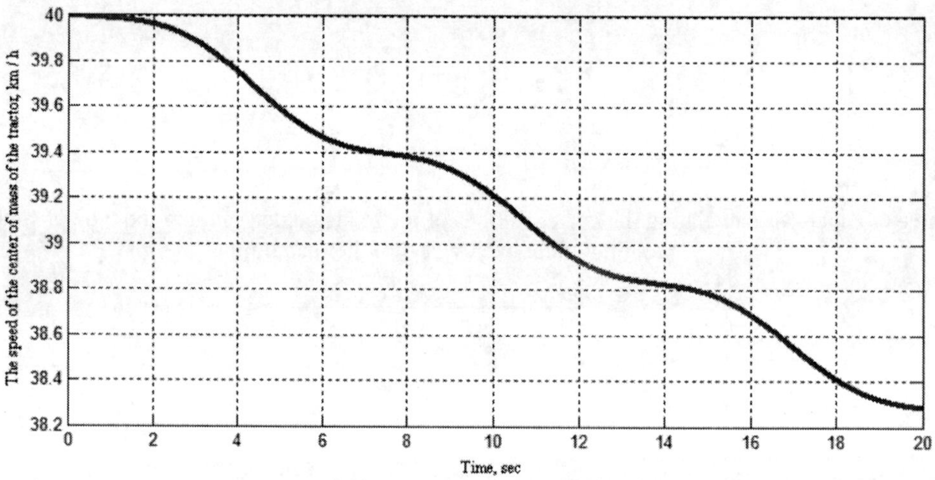

Figure 9. Road train speed

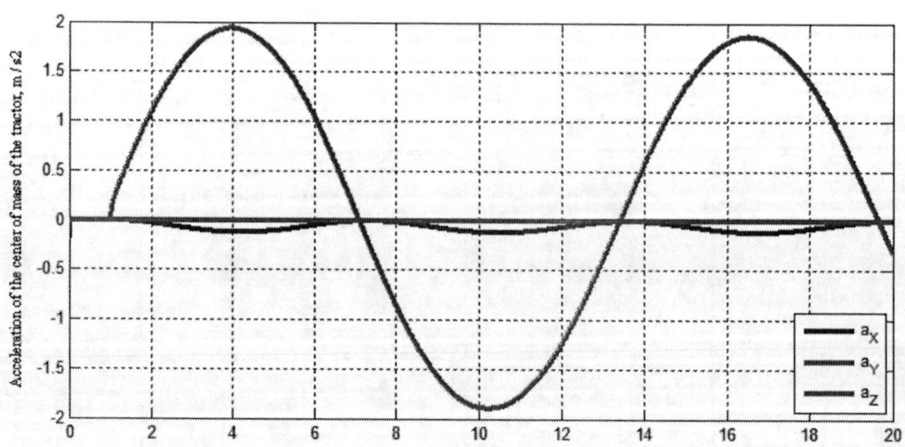

Figure 10. Acceleration of the center of mass of the tractor

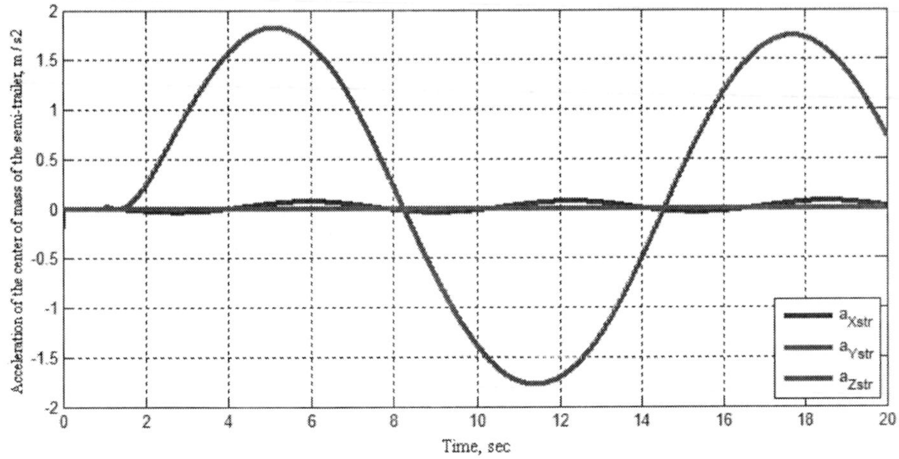

Figure 11. Acceleration of the center of mass of the semi-trailer

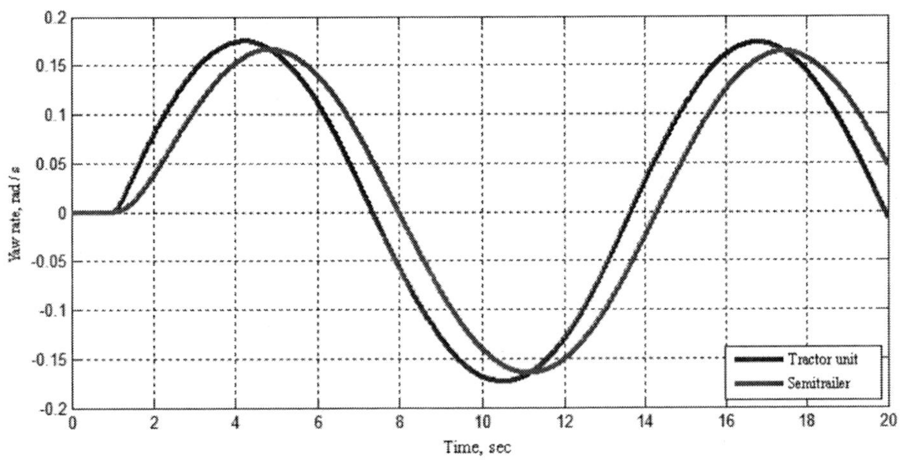

Figure 12. Angular speeds of the tractor and semitrailer

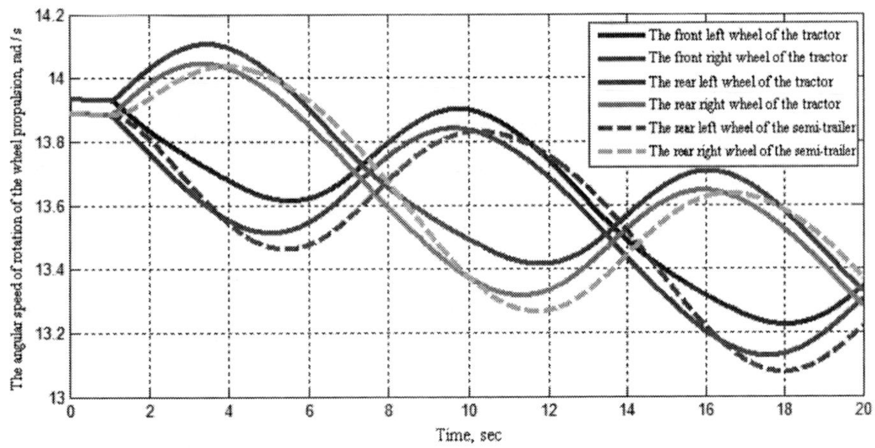

Figure 13. Wheel speeds

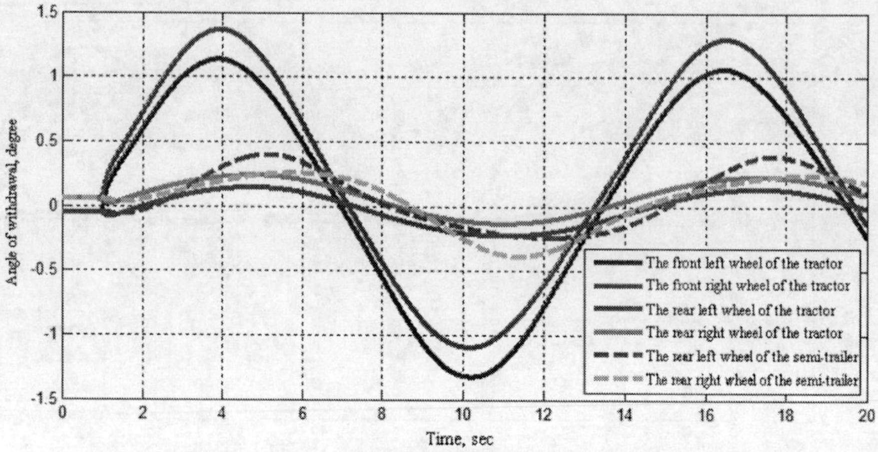

Figure 14. Angles

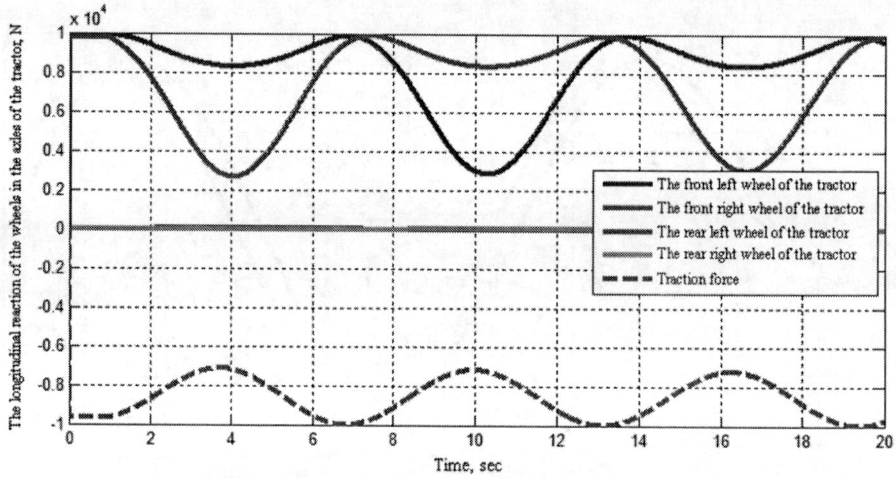

Figure 15. Longitudinal reactions of the tractor

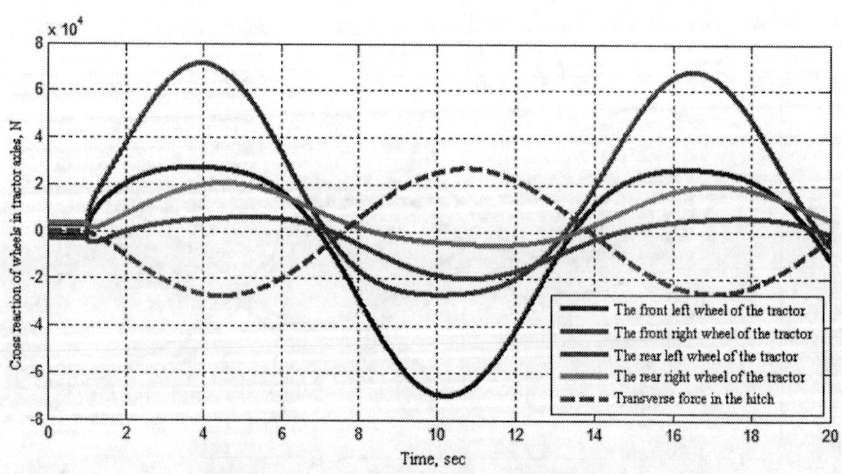

Figure 16. Lateral reactions of the tractor

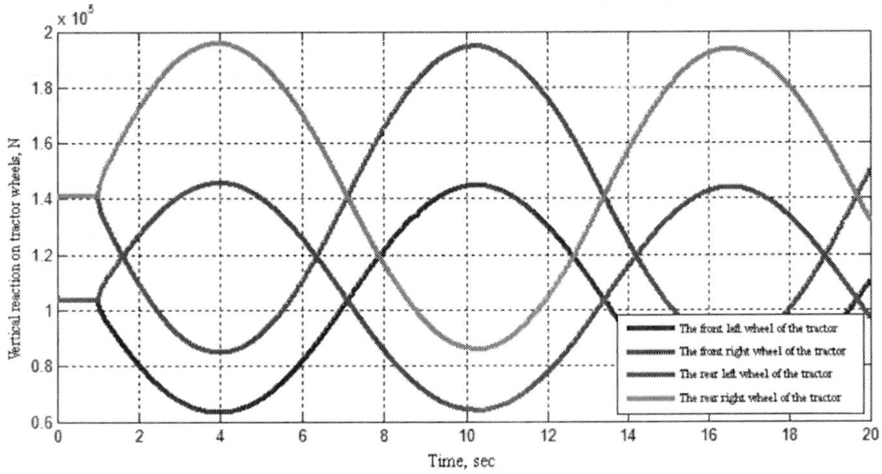

Figure 17. Vertical tractor reactions

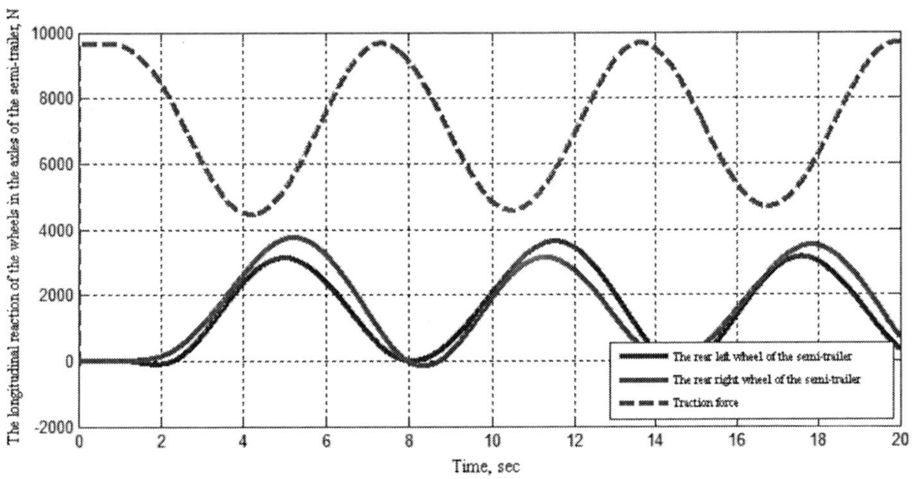

Figure 18. Semi-trailer longitudinal reactions

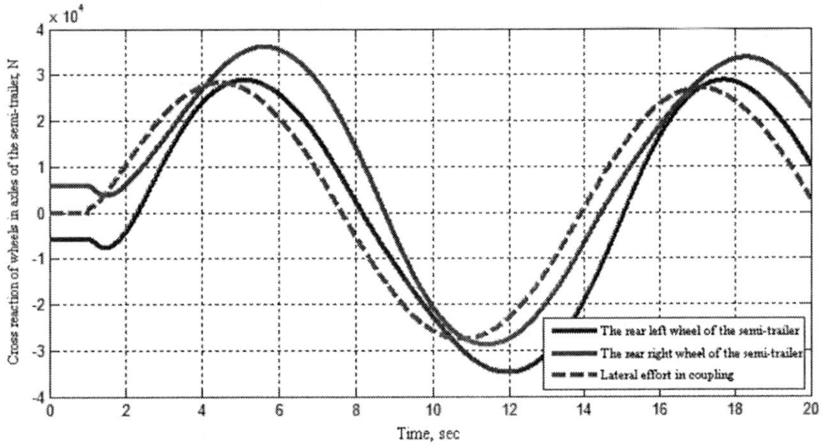

Figure 19. Side reactions of the semi-trailer

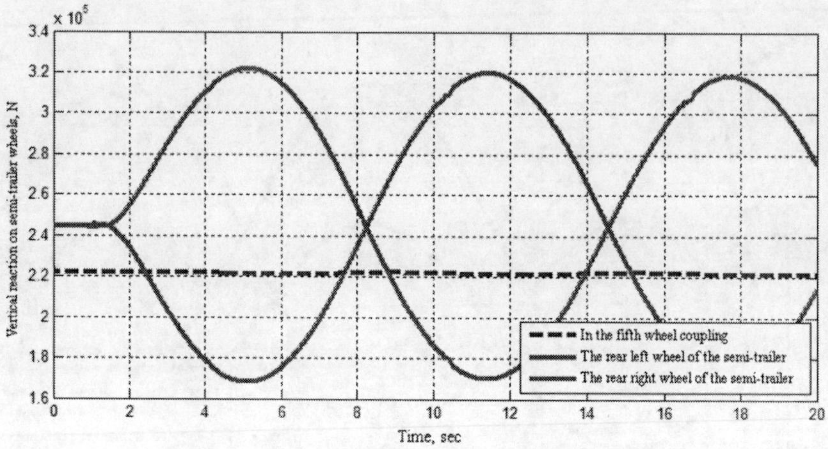

Figure 20. Vertical reactions of the semi-trailer

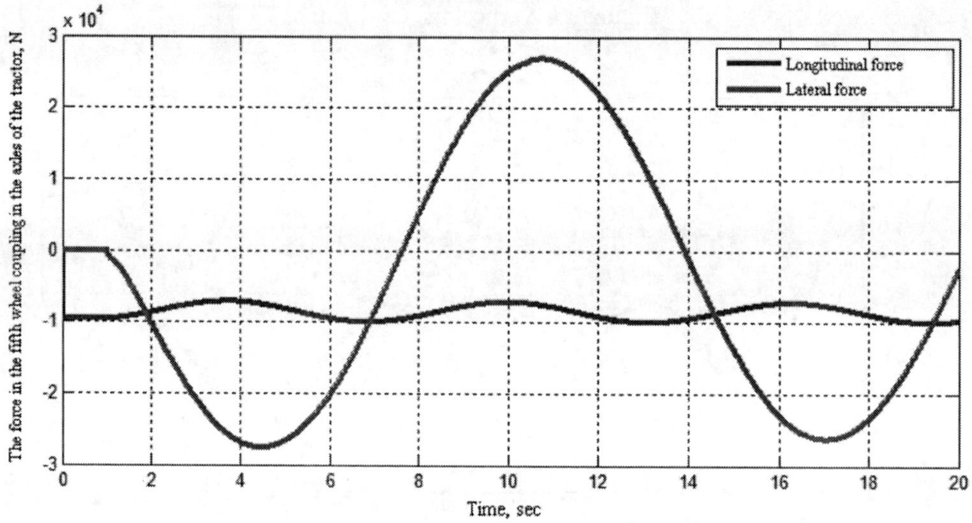

Figure 21. Force in fifth wheel coupling

The performed simulation of the curvilinear motion of a two-link road train with an active trailed link based on valve-inductor electric machines made it possible to justify and develop:

- a control system for an individual drive of an electric motor-wheels of an active semi-trailer;
- the program and methodology of full-scale experimental research.

Field experiments, pilot testing and testing of a two-link road train with an active trailed link based on valve-inductor electric machines made it possible to verify the adequacy of the simulation model and make the necessary refinements, as well as confirm the improvement of the technical and operational properties of the research object using the developed control system.

References

[1] Kiencke U and Nielsen L 2005. Automotive Control Systems – For Engine, Driveline, and Vehicle 2rd ed (Berlin Heidelberg: Springer-Verlag) 512 p

[2] Burckhardt M 1993 *Fahrwerktechnik: Radschlupf-Regelsysteme* Vogel Fachbuch (Würzburg)

[3] Reimpell J and Sponagel P 1995 Fahrwerktechnik: Reifen und Räder Vogel Fachbuch (Würzburg)

IOP Conf. Series: Materials Science and Engineering **832** (2020) 012015 doi:10.1088/1757-899X/832/1/012015

[4] Mitschke M and Wallentowitz H 2004 *Dynamik der Kraftfahrzeuge* 4rd ed (Berlin: Springer Verlag)

[5] Ammon D 1993 *Radlastschwankungen, Seitenführungsvermögen und Fahrsicherheit* vol 1088 of VDI Fortschrittberichte (Düsseldorf: VDI Verlag)

[6] Shadrin S S 2009 *A methodology for calculating the controllability and stability of a vehicle based on the results of field tests* PhD dissertation (Moscow: MADI) 132 p

[7] Shadrin S S, Ivanov A M and Prikhodko V M 2014 Experimental and calculated procedure for determining the adhesion properties of the vehicle pneumatic tires in use *Adv. Environ. Biol.* **8(13)** 294–7

[8] Shadrin S S, Ivanov A M and Prikhodko V M 2015 Usage of Smartphones in the Education Process of MADI during Vehicle Road Tests Conduction *Modern Applied Sci.* 9(1) 83–8

IOP Conf. Series: Materials Science and Engineering **832** (2020) 012016 doi:10.1088/1757-899X/832/1/012016

Improving the efficiency of diagnostics and repair of ICE and road construction machinery in ISTK

D Stepanov, Yu Krasnov, D. Antonov

Moscow Automobile and Road State Technical University (MADI), 64, Leningradsky Ave., Moscow, 125319, Russia

E-mail: Stepanov_d@istk.ru , juri_krasnov@mail.ru

Abstract. This article contains analysis of possible ways to improve the efficiency of diagnostics and repair of internal combustion engines using information systems. Common repair methods for these machines are reviewed here. The analysis revealed disadvantages, which increase, time and cost of repairs. The lack of computerized workplaces reduces the productivity of engine repair department. As a solution to this problem, it was proposed to equip the mechanics' workplace of the ICE service department with a computerized widescreen LCD panel and a tablet with preinstalled software. This system allows to control the work process in real time. The system increases the productivity of the service department, reduces costs and time for diagnostics and repairs. The system also allows to analyze the possibility of reusing spare parts.

Keywords: automation, repair efficiency, diagnostics, internal combustion engines.

1. Introduction

Komatsu Technical Support Service Center in Obukhovo, Moscow Region, is the largest Komatsu service center in the Central Federal District. The center provides quality repair and maintenance services of Komatsu construction machinery. Nowadays the company is able to perform all main types of regular repairs and overhaul of any road construction machinery manufactured by Komatsu. These are bulldozers, excavators, articulated dump trucks, pipe layers, wheel loaders, backhoe loaders.

In order to increase the efficiency of repair tasks, a work automation system was developed in cooperation with the manufacturer to correct existing disadvantages and solve current problems.

The aims of automation are to increase the efficiency of repairs and to make repair process more transparent. The system designed to reduce the number of mistakes and dependence on the human factor, to accelerate service related tasks, and reduce repair costs. This system not only boosts productivity, but also improves control over the repairs process. It also provides online monitoring of the overall repair time, time spent per each process stage (for example, time from the ICE arrives to department to the moment when ICE is assembled). These features helps to control the service duration standards and evaluate the work quality of employees and the entire service department. Only availability of objective information can help to identify current problems and suggest the right solutions.

IOP Conf. Series: Materials Science and Engineering **832** (2020) 012016 doi:10.1088/1757-899X/832/1/012016

2. Problem description

Before the whole system was integrated in service process, we had an examination of the existing service system. Time of work, efficiency and productivity of the service department, repair mechanics surveys were analyzed. As a result, we revealed a number of existing problems.

For example, when repairing engines and transmissions of road-building machinery, the traditional methods was used: the mechanic had only paper instructions with entire step-by-step disassembling algorithm, with tables indicating nominal dimensions of parts, i.e. piston diameter, cartridge diameter, crankshaft position and other data, as well as techniques to determine defects in machine parts. The papers got dirty and quickly became unusable. Moreover, these manuals are usually made in black and white, which makes it difficult for interpretation, and the information becomes outdated in terms of constant upgrades and modifications.

The entire data obtained during the measurement was processed manually. Analysis of the spare parts reuse possibility during troubleshooting was not even included in mechanics tasks. Moreover, the old system was unable to eliminate human-factored faults. Customers also experienced some inconvenience when making a decision. Diagnostic results were provided in tables which needs additional interpretation.

Changes to the existing method will:

1) Improve repair times.
2) Increase the transparency of repair (the system will make it possible to control speed standards, evaluate work quality, and make problem-solving decisions).
3) Make it able to control the amount of data (the system has an easy understandable reports of the results of troubleshooting).
4) Coordinate actions (simplified mechanic-manager-client-warehouse interaction).
5) Improve service effectiveness.
6) Reduce staff training time.
7) Reduce human-factored problems (step-by-step manuals with color pictures and guidelines minimizes the risks of mistakes).

3. Features of the integrated automated engine repair and diagnostics system.

We started to search for new solutions while trying to make these improvements possible. Together with Komatsu CIS, we implemented the control system for engines and transmissions disassembling and troubleshooting in the Obukhovo Technical Center. The system involves the following mechanics work layout:

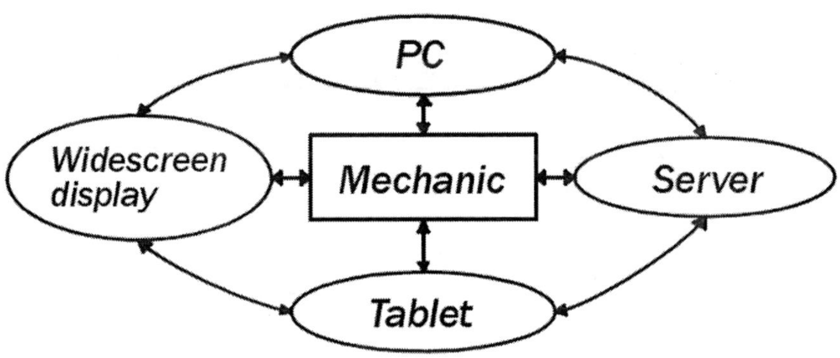

Figure 1. Mechanic-system interaction layout

As part of the program implementation, the service department was upgraded. The mechanic's workplace was computerized and equipped with a widescreen LCD panel, which greatly improves the work with digital materials. A tablet is integrated into the system, with preinstalled software which allows to control the work process in real time by registering all the data in the system.

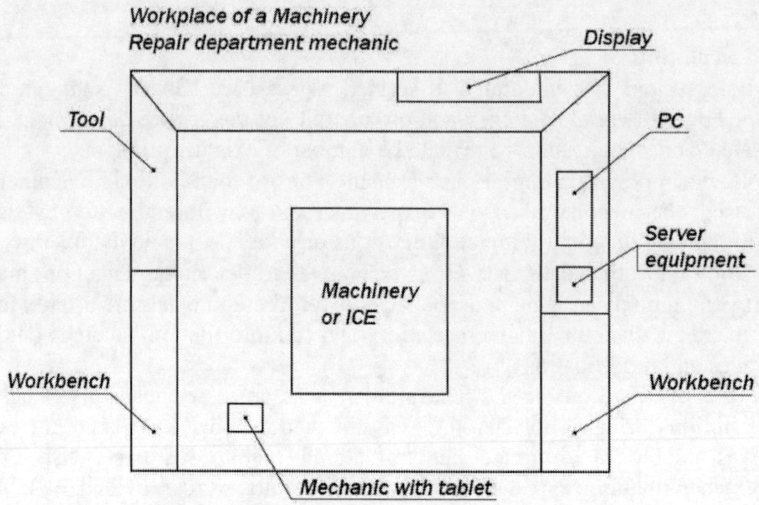

Figure 2. Upgraded mechanic's workplace

Interaction with information system built on technological processes, which can be divided into stages. Disassembly is one of the first stages. The intuitive interface of the program allows the mechanic to get started without longtime training. To start working the system requests the input of basic repair data, the name of the mechanic conducting the repair, the place of the repair, as well as all the necessary information on the unit to be disassembled.

After entering the main data, a menu window pops up. Measurement values needs to be entered in the working window, and this window shows data on:

- Possibility of spare parts reusing.
- Spare parts to be disposed of.
- Spare parts to be replaced.

The software will also indicate the parts that need to be removed to ease the disassembly process. If defective parts are found, the software allows to add a picture on which you can mark additional information.

Modern interface and color indication highlight each category of spare parts with its own color, e.g. white – the spare part may be reused, yellow – the spare part needs mechanical processing, green – the part must be replaced, black – the part must be examined carefully, red – pause or delay the process, blue – detailed disassembling of the unit is needed (e.g., water pump).

Reason for replacement		
Not to be reused	Missing	Worn
Damaged	Disrupted	Upgrade
Age replacement	Fails the specification	--------
--------	--------	--------
--------	--------	--------
--------	--------	--------
--------	--------	--------
--------	--------	--------
--------	--------	--------

Figure 3. System interface menu

After the disassembly and troubleshooting are completed, the software offers to arrange the parts for repair, analyzes location, replacement possibility and timing of the earliest delivery.
There are several key stages is the process of parts arrangement. Firstly, generating of the parts selection plan, then selecting parts (the system provides all necessary information i.e. part name and index, quantity of parts, pictures of parts, and which parts are missing.

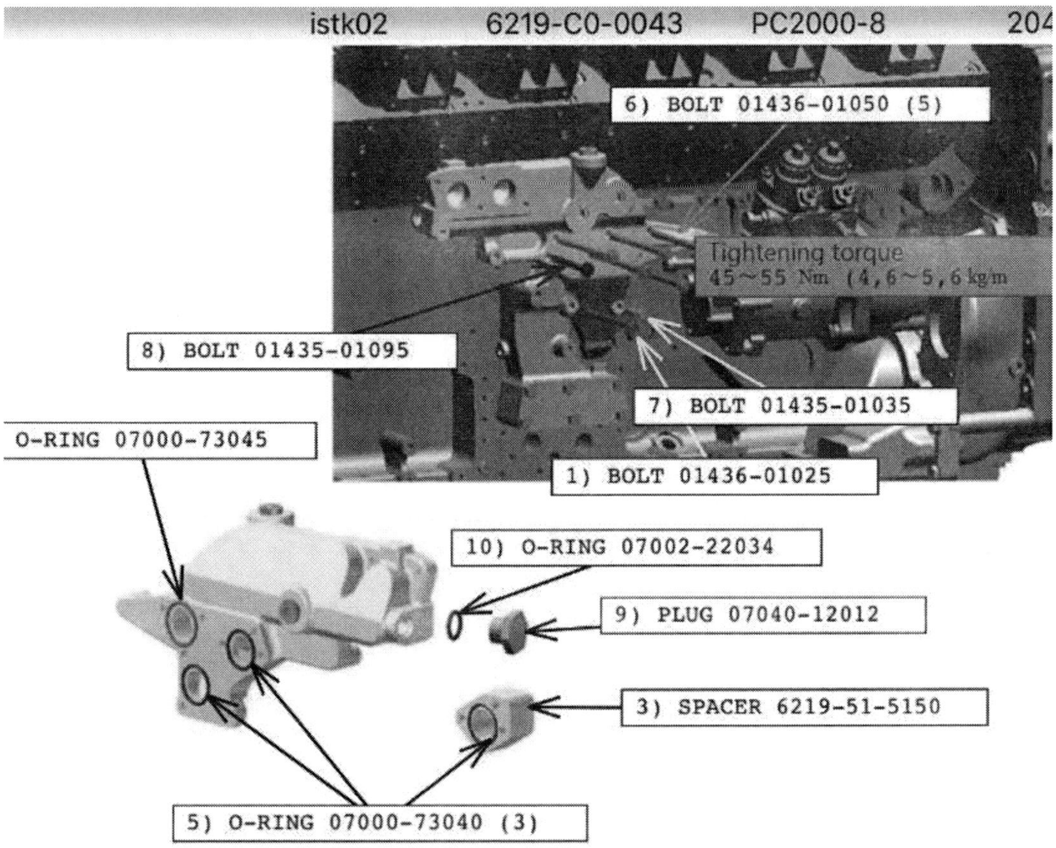

Figure 4. Spare parts catalog of built-in the system interface

After all the spare parts arrived at the warehouse and are received by a mechanic, the software proceeds to the next stage – assembling.

The assembly process starts with generation of a process plan which considers various details i.e. start date, expected repair completion date, model of the unit, unit catalog number, unit serial number, machinery model, type of work, work order number, branch office code/number, and the customer name.

During the assembly, the software indicates not only the installation location, but also the tightening torques of mounting bolts and nuts.

After installing and checking each spare part, the mechanic shall put his electronic signature.

When the unit is assembled, all the necessary data is stored in the system. Data can be stored for an unlimited amount of time, and can be used to analyze information at any time.

Just two months have passed since the implementation of the system, however it has definitely increased labor efficiency and productivity.

For example, when repairing the Komatsu SAA12V140E-3 engine (12 cylinders, 30 l, 976 hp, 4-stroke, water-cooled, V-shaped, with direct injection, turbocharged), the repair time was significantly reduced with the quality maintained.

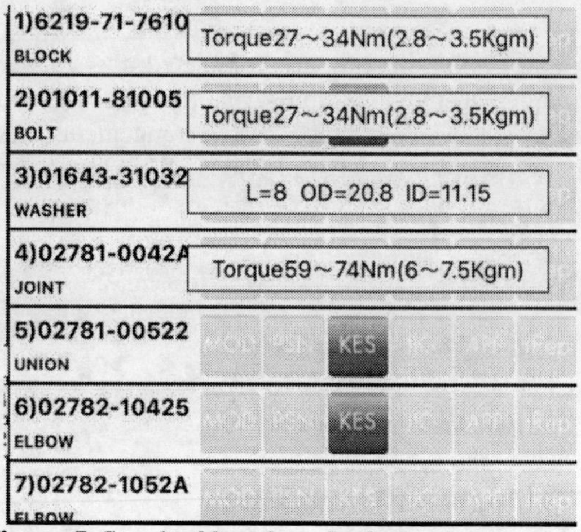

Figure 5. Standard benchmarks for engine assembly

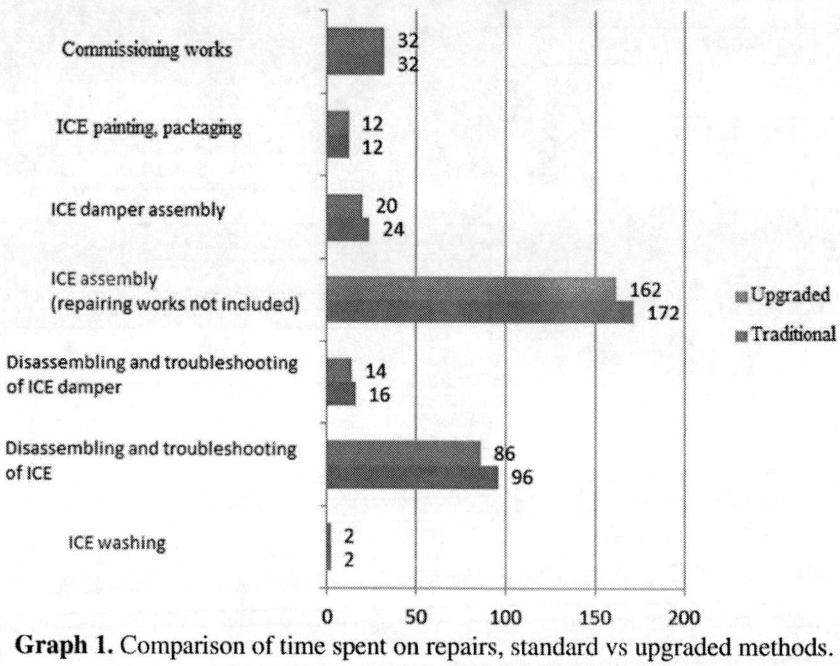

Graph 1. Comparison of time spent on repairs, standard vs upgraded methods.

So, the total hours spent using the standard method is 354 working hours, and using the automated system – 328 working hours, which makes a difference of 26 working hours. This difference reduces repair time, equipment downtime, and also allows to set more competitive repair prices.

We invite you to our Obukhovo Technical Center, where you can find out more about the new system, and its advantages and features.

4. Conclusion

The software is currently under testing and adaptation process. However, even now we may indicate the advantages i.e. operation ease, availability of up-to-date disassembly instructions, spare parts catalog integrated and constantly updated, step-by-step assembly process and repair quality control. The system also provides great opportunities for optimizing the work cycle by analyzing the problems that occurs during the work and suggesting quick solutions.

IOP Conf. Series: Materials Science and Engineering **832** (2020) 012016 doi:10.1088/1757-899X/832/1/012016

5. References

[1] Vasiliev B S, Dolgopolov B P, Dotsenko G N et al 2012 *Repair of road cars, cars and tractors: textbook. allowance* ed. V A Zorina (Moscow: Mastery; Higher School) 496 p

[2] Sinelnikov A F and Konoplin A Yu 2014 *Guidelines for laboratory work on the repair of blocks and cylinder liners by machining* (Moscow: MADI) 40 p

[3] Dovbnya A A and Poyedinshchikov I I 1994 Assessment of the effectiveness of management in the implementation of the goal of the policy in the field of quality *Standards and quality* **3** 12

[4] Osipov B V 1994 Computerization of the enterprise quality system *Certification* **3** 29

[5] Rakhlin K M 1994 Organization of accounting and assessment of enterprise costs for quality *Standards and quality* **3** 34

[6] Komatsu 2009 *Shop manual 114-3 Series Diesel Engine* 36 p

[7] Komatsu 2003 *Shop manual 125-3 Series Diesel Engine* 362 p

[8] Komatsu 2001 *Shop manual component of engine* 648 p

[9] Komatsu 2003 *Specifications & application handbook edition* **24** 889 p

[10] Komatsu 1999 *Standard repair work master & Stanrate program*

[11] Komatsu 2009 *Shop manual 170E-5 Series Diesel Engine* 36 p

IOP Conf. Series: Materials Science and Engineering **832** (2020) 012017 doi:10.1088/1757-899X/832/1/012017

Problems of digital diagnostics for prospect of autonomous vehicles implementation

M V Yashina and Ngoulou-A-Ndzeli

Moscow Automobile and Road State Technical University (MADI), 64, Leningradsky ave., Moscow, 125319, Russia

E-mail: becker20000@yahoo.fr

Abstract. This article outlines the principles and problems associated with the development of intelligent transport systems and, in particular, new trends in the creation and implementation of unmanned vehicles (autonomous vehicles). The alliance between technological evolution and the automotive industry allows us to highlight emerging and future challenges for smart vehicles. In the context of autonomous vehicles, there are a number of issues that need to be addressed. It is assumed that such vehicles must be reliable; therefore, real-time diagnostic issues are relevant. At the beginning of the article, a global vehicle model is considered. Further discusses a management strategy that provides recommendations for the development of diagnostic measures for AV. The results of high-precision joint modeling shown in the article prove the effectiveness of the proposed control technique and new opportunities.

1. Introduction

The automotive sector is constantly evolving. Every day we are getting closer and closer to the revolution of autonomous vehicles, which is one of the most complex systems. This topic still requires solving a large number of problems. Decision-making algorithms should still cover more critical scenarios, sensors should still become more reliable, control strategies should still increase their reliability and adaptability, etc. In the field of land vehicles, the only effector is the bus. Its potential varies with friction and vertical loading. In addition, it must be taken into account that the longitudinal and transverse forces are connected, and their behavior is very nonlinear. The dynamics of the roll can affect the dynamics of the speed of movement, and the lateral speed can inhibit the longitudinal, etc.

In addition, the dynamics at low speed differs from the dynamics at high speed. To prove the safety of autonomous vehicles, it is necessary to ensure dynamics control at the limit of vehicle controllability. Another problem is related to the uncertainty of vehicle parameters. The total weight can be uncertain and can vary, tire stiffness in corners can be uncertain, and it can also get old and so on. The management strategy should be constant regardless of how many passengers are in the vehicle alone or whether the road is dry or wet, whether the tires are new or already worn out, etc. [1, 2].

Diagnostics – this is the determination of the technical condition of the car, its components and components without disassembly and the conclusion about the need for repair, maintenance or serviceability, is an integral part of the technical process of maintenance and repair and is intended to improve the quality of work at minimal cost.

There are two types of technical diagnostics – general and advanced.

IOP Conf. Series: Materials Science and Engineering **832** (2020) 012017 doi:10.1088/1757-899X/832/1/012017

In general diagnostics (D-1), the technical condition is determined according to a common criterion – an object is "suitable" or "unsuitable", can it be operated further without preventive or repair actions [3, 4].

With in-depth diagnostics (D-2), step-by-step diagnostics of the technical condition of units and assemblies is carried out, and places and causes of hidden malfunctions and failures of these nodes are identified. As a result, a diagnostic report is made with an accurate diagnosis and the amount of necessary adjusting and repair actions [5–7].

There are classifications of diagnostic methods according to various criteria.

Diagnostic methods are divided into subjective (organoleptic) and objective (instrumental).

Subjective diagnostic methods:

- visual inspection;
- listening;
- check by touch and smell.

An external inspection determines the condition of the seals, fuel, oil, electrolyte leakage, damage to external parts; listening – knocks, noises and other sounds that differ from normal workers; tapping – threaded, rivet, key and welding joints; touch – places of heating parts, vibration, run out, fluid viscosity; sense of smell – the state of the clutch by a characteristic odor, gasoline leak, etc.

In order to establish quantitative modifications of the parameters of the technical state of the vehicle, an objective diagnosis is established, that is to say with the aid of equipment and special devices. The hardware can be integrated into the machine or connected to it. Built-in sensors include sensors, light signals, operating hours counter, filter clogging indicator, and more. Connected sensors include media, peripherals, etc.

Almost all modern cars are saturated with a variety of electronic systems. This determines the appearance, as well as the demand, of a type of diagnosis such as the computer, able to identify efficiently and quickly all the main problems of a car.

Computer diagnostics of a car is a process in which the codes of possible malfunctions are read on the main nodes, the erasure of these codes and their subsequent correction. For this, dealership scanners and other systems can be used. These include OEM, multifunctional stands, portable readers. Modern diagnostic equipment and software allow you to read and detect the slightest changes in the operation of engine control systems, transmission, instrument panels and others.

All current data is displayed on a single channel multimeter in real time. At the same time, you can plot up to 4 technical parameter calendars, choosing the most convenient form of display. Modern diagnostic systems, which can only be provided by a certified car service, also allow the recoding of parameters. This is done to increase the power characteristics of the automatic chip setting.

For example, the control unit is often reconfigured to optimize it for a given vehicle. It includes idle adjustment or fuel system adjustment. And by downloading additional plug-ins, you can reprogram the car's electronics on the interface of the most recent models in that line and those about to leave the assembly line. The system automatically identifies the differences without the need to manually set the initial and final parameters.

As a general rule, the computer diagnosis of a car is made either when the system itself diagnoses malfunctions on the dashboard (error icons come on), or when the motorist himself observes the incorrect operation from some nodes / systems, you need to make sure (for example, before buying a used vehicle) what technical condition is really the vehicle. You can also perform computer diagnostics, at least once a year, according to the recommendations of the experts (depending on the technical condition of your vehicle).

In fact, computer diagnostics is a very effective and modern way to carry out the most complete verification of the electronic systems of a motor vehicle in order to identify and prevent malfunctions. With this, it is possible to obtain true information on the current state of the control units, parts and components of the vehicle.

The computer diagnosis of the vehicle is performed on embedded systems, via special diagnostic connectors. A rather complicated scanner with serious software is connected, which reads all the codes transmitted by the vehicle.

The codes received are deciphered by specialists, using special programs as well. On the basis of the information received, it can be concluded that there are certain failures. The computer diagnosis itself can be divided into several operations, including suspend diagnostics. This is necessary if the motorist detects uneven rubber wear, shocks or buzzing during sharp turns or when driving at constant speed on rough roads. In addition, if a demolition of the rear or front axles is observed during sharp turns, ABC triggers prematurely or an increase in the free wheel of the steering wheel is noted.

A computerized diagnosis of the engine is carried out if the motorist begins to notice that the engine heats for a long time, that fuel consumption has increased, that the engine is unstable or that it starts badly, that it has lost power, exhaust, there is outside noise, the idle is lowered / increased. During the diagnosis, the following elements are checked: injection system; Power source; measured compression.

A computerized diagnosis of automatic transmissions must be carried out if one of the reports does not start, jerks, noises or slips are visible when shifting gears, fuel consumption is increased, oil leaks are noted. During the diagnosis, the error codes of the automatic transmission control unit are read, the sensor readings for the working fluid temperature and the throttle position are evaluated, as well as the position of the selector lever.

If we compare the computer diagnostics of the car with more traditional diagnostics, the first can most likely be considered the pinnacle of diagnostic technology, because it quickly and effectively detects almost all malfunctions and does not require much time and effort.

There is also another diagnostic method, such as:

- ALDL (assembly line diagnostic system) – a car diagnostic system developed by General Motors and preceding the OBD-I standard. Prior to minor changes, ALDL was called Assembly Line Communications Link or ALCL. These two terms are synonymous. This system was not a clear standard and was therefore accepted as a specification for communication with a vehicle. There are three different ALDL connectors: 5-pin, 10-pin, and 12-pin – the latter is more common on GM vehicles. Earlier versions used 160 bits per second, while later versions used 8192 bits per second and used bidirectional communication with the traction control module (PCM) [8].
- OBD-I (On-Board Diagnostics) – On-board diagnostics that regulate intentions to motivate automakers to develop a reliable emissions control system.
- OBD 1.5 is a partial implementation of the OBD-II, which General Motors used on some cars in 1994 and 1995 (General Motors did not use the term OBD 1.5 in the documentation of these cars; called OBD and OBD-II sections). in the user manual).
- OBD-II (Embedded Diagnostics) – Embedded diagnostics, a standard developed in the mid-1990s, allow complete engine control. Allows you to monitor additional body parts and peripherals and diagnose a car control network. In this standard, manufacturers use various vehicle connection protocols.

ISO 9141-2
ISO 14230 Keyword Protocol 2000
SAE J1850 VPW
SAE J1850 PWM
ISO 15765-4 CAN (Controller Area Network)
ODB-II Socket contacts. Max. 0.6m from the steering wheel.

The OBD-II specification provides a standardized hardware interface and is a SAE J1962 compliant diagnostic link connector (DLC) with 16 contacts (2x8) for connecting diagnostic equipment to the vehicle in the form of a trapezoid. Unlike the OBD-I connector, which is sometimes under the hood of a car, the OBD-II connector must be near the steering wheel or within reach of the driver. SAE J1962 determines the location of the pins on the connector:

In the context of autonomous vehicles, a number of issues need to be addressed. It is assumed that these vehicles will perform multiple maneuvers and may be overloaded.

At the beginning of the article, a global vehicle model is considered. What follows is a management strategy that, according to the authors, should be used in the context of sufficiently activated vehicles.

The results of the high precision assembly modeling presented in the article prove the effectiveness of the proposed control technique and new features [9, 10].

The article is structured as follows: Development of a global vehicle model that can help control engineers learn more about dynamic vehicle couplings. The traffic control is synthesized using a very congested vehicle. A discussion of the outstanding issues in terms of reliability and quality objectives is presented. Findings and future work are described.

The authors believe that car manufacturers and equipment manufacturers will join forces to develop and standardize the proposed control architecture for future passenger cars.

2. Analysis

It is important to study: maneuverability by following the speed of movement and lateral stability by minimizing the lateral slip angle. The four-wheel vehicle model was subsequently examined in the case of a low-level direction for ESP. The clutches were not controlled to a high level of control since two different vehicle models were considered. In addition, no lateral speed adjustment was provided, whereas a vehicle equipped with active front and rear steering systems (AFS and ARS) can allow lateral transient movement, for example to avoid an obstacle.

The current state shows that we are using complex and reliable controllers based on unrelated simplified vehicle models, or that we are separating a complex vehicle model to use simplified controllers in each direction. The authors do not give priority to either the first or the second approach. The new approach has been sufficiently studied, in which a relatively sophisticated and robust high level controller is used, based on a relatively complex model of four-wheeled vehicle with an optimal coordination strategy. The objective is to evaluate dynamic coupling at the vehicle level to justify the structure of the high-level controller.

A twin longitudinal transverse control requires a paired longitudinal transverse vehicle model.

Future autonomous vehicles (AV) must simultaneously drive longitudinal, lateral and, ultimately, vertical controls. These speakers are connected because they are connected to the same system. For global chassis control (GCC), the vehicle's internal dynamic clutches must be considered. These connections may be more restrictive or, conversely, more relaxed depending on the chassis systems integrated in the same vehicle. For example, with two-wheel steering (2WS), a vehicle can avoid an obstacle only by changing its course (angle of movement). Unlike the vehicle 2WS, vehicle 4WS, four-wheel steering in one direction, you can bypass an obstacle without changing its direction. Since it is impossible to know the future hardware design of cars, and since the design of cars may differ from different manufacturers, the authors of the article are developing a new detailed global car simulation.

In order to replace the driver and make cars autonomous, it is necessary to introduce additional embedded systems. In addition, in order to distinguish a car brand from competitors, various subsystems can be implemented by various manufacturers. One thing is certain, future vehicles can be heavily overloaded.

A layered architecture is a good option for future automotive control tasks. This architecture provides, in particular, the following criteria: adaptability, fault tolerance, dynamic reconfiguration, extensibility and modularity.

The car is equipped with an Active Rear Steering (ARS) system, a vehicle-based braking dynamics control system (VDC) and two rear wheel electric motors for vectorizing rear torque (RTV). The generalized efforts required to move the vehicle can be optimally distributed using optimization strategies based on optimization (CA) on four buses. Then, such bus forces can be converted into actuator commands and activate the corresponding system, avoiding any internal conflicts.

IOP Conf. Series: Materials Science and Engineering **832** (2020) 012017 doi:10.1088/1757-899X/832/1/012017

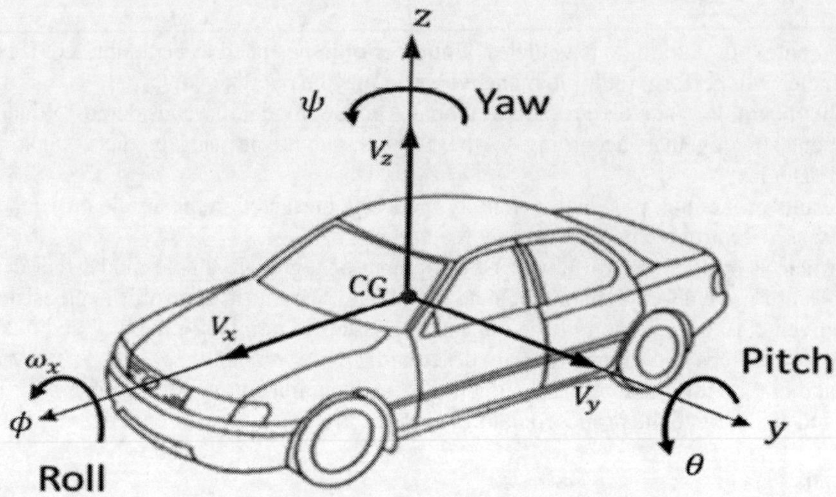

Figure 1. Vehicle axis system ISO 8855-2011

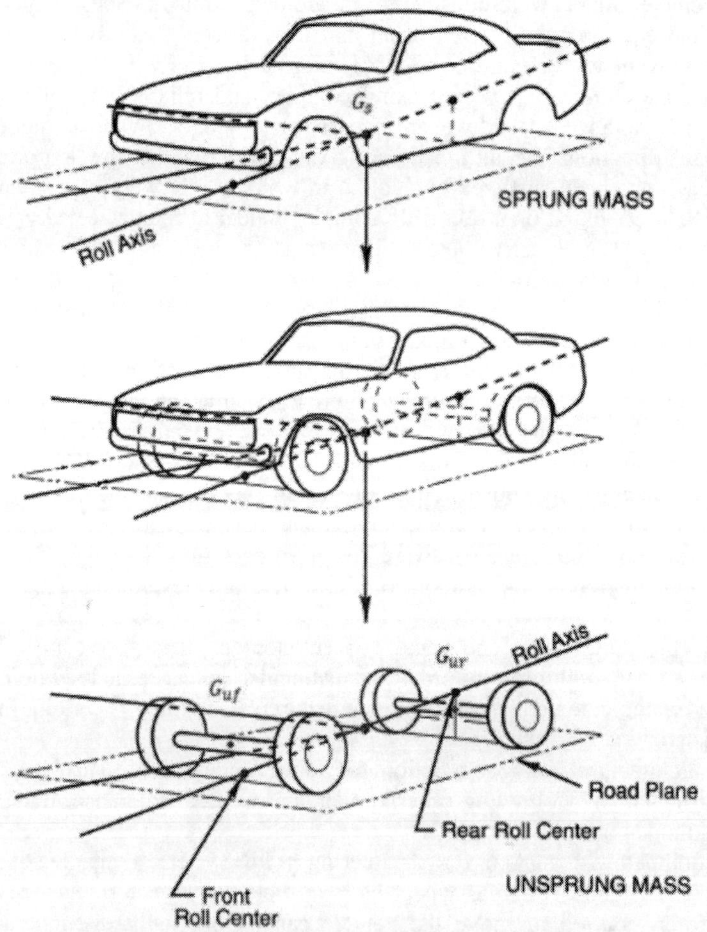

Figure 2. The decay of sprung and unsprung masses.

There are three types of control:

- High level control

The goal here is to calculate the required forces on the vehicle propeller to track the desired speeds. Dynamics at this point is characterized by inertial parameters such as mass and moment of inertia. These parameters are usually uncertain, which requires a certain degree of reliability. A multi-input multi-output (MIMO) controller is required to account for various connections. Since the vehicle is equipped with ARS, VDC and RTV and access to active suspensions is not permitted, only a flat vehicle model can be considered at a high level. However, when evaluating tire potential, the importance of vertical dynamics should be considered.

$$\sum = Ss + Suf + Sur$$

Σ: total body mass M and center of gravity (CoG) G, Ss: spring-loaded mass Ms and CoG Gs, Suf: front unsprung mass Muf and CoG Guf, Sur: rear unsprung mass Mur and CoG Gur.

- Medium level control [8].

This intermediate level aims at coordinating the wheel systems in order to avoid differences and to generate high level general forces. To do this, all the forces must be distributed optimally on the four tires in order to activate the system with the desired level of effort. Since the number of forces on the tires exceeds the total forces that must be applied to the propeller of the vehicle, the system is over-activated, which leads us to the CA problem.

3. Conclusion

Diagnostic issues for future autonomous vehicles are considered. These vehicles need several built-in systems to work efficiently and therefore need an optimal intersystem coordination strategy. In addition, as the driver becomes a passenger, motion sickness should be avoided, creating new driving behavior to provide comfort and confidence in the senses.

References

[1] Agrar. Bull. of the Urals 2012 **10-1(102)** 38–40

[2] Novichev A V and Berdyshev I V 2016 Development of a stand for testing brakes *Youth and Sci.* **5** 133

[3] Trompet G M, Alexandrov V A 2007 The technology of diagnostics of mechanical treatment with control devices *Technologies for repair, restoration and hardening of machine parts, mechanisms, equipment, tools and technological equipment: Coll. of mater. 9th Int. prakt. conf.* pp 357–359 (St. Petersburg: Publishing house of the Polytechnic (University))

[4] Trompet G M, Alexandrov V A and Volynkin V V *Optimization of methods for separating processing errors during tuning means of control control*

[5] Grintsevich V I, Boys S V and Kozlov G G 2012 *Technological processes of diagnosis and car maintenance* Laboratory workshop (Krasnoyarsk: Sib. Feder. un-t) 204 p

[6] Borts A D, Zakin Ya Kh and Ivanov Yu V 1979 *Diagnostics of the technical condition of the car* (Moscow: Transport) 160 p

[7] Gazaryan A A 2000 *Technical operation, maintenance and repair of motor vehicles* Practical recommendations and regulatory framework (Moscow: Third Rome) 263 p

[8] *Classification of diagnostic tools* Retrieved from: http://avtomeh.ucoz.net/index/klassifikacija_sredstv_diagnostirovanija_avtomobile

[9] Kissai M, Monsuez B, Mouton X, Martinez D and Tapus A 2019 *Adaptive Robust Vehicle Motion Control for Future Over-Actuated Vehicles*

[10] Napolsky G M and Yakubovich I A 2010 *Basics of technological design of stations maintenance of cars* Textbook allowance (Magadan: Publishing House of the North-Western State University) 87 p

[11] Byshov N V et al 2012 *Design of technological processes THAT, repair and diagnostics of cars on motor vehicles enterprises and service stations* (Ryazan: Publishing house of RSTU) 162 p

[12] Kissai M, Monsuez B, Mouton X, Martinez D and Tapus A 2019 Adaptive robust vehicle motion control for future over-actuated vehicles *Machines* **7(2)** 26

Remote identification of damages of freight by means of transport telematics

R Zaikin, S Filatov, D Komkova and D Efimenko

Moscow Automobile and Road Construction State Technical University (MADI), 64, Leningradskiy ave., Moscow, 125319, Russia

E-mail: Romio1406@inbox.ru, tdmadi@mail.ru, daryia22@mail.ru, ed2002@mail.ru

Abstract. The article informs about the possibilities of means of transport telematics, which are mentioned in the confirmation or refutation of the road accident and the assessment of possible damages during the road traffic accident. In addition, the article examines aspects of the practical application of information on the dynamic parameters of a vehicle obtained by means of transport telematics.

Keywords: means of transport telematics, detection of cargo damage.

1. Introduction

Globalization presupposes a high rate of mutual adaptation of national economies and open markets for trade operations. Cargo transportation is of fundamental importance in the implementation of international economic activity. To meet the needs of consumers of transport companies' services, cargo transportation must be safe, timely and accessible to a wide range of users, but it is especially important that the goods are delivered to their destination safely and securely.

According the information from the Ministry of Transport of the Russian Federation, from January to September 2016, the freight turnover of road transport amounted to 169.8 billion ton-kilometers. According to the traffic police to data, more than 6 million trucks have been registered in Russia, and the number of violations of traffic rules and the use of vehicles by truck drivers amounted to 2,903,701 cases. This statistics allows drawing a conclusion that the losses connected with accidents or careless driving can be considerable that increases importance of the research connected with quality of freight automobile transportation.

The problem of the quality of transportation of goods truck involves the formulation a number of tasks aimed at identifying the causes of costs, developing methods for estimating losses, developing recommendations for the formation of a risk system and developing requirements for the safety of the transportation process in order to improve the quality and competitiveness of freight road transport.

It should be noted that the analysis of information carried out in this direction showed the lack of practice in the use of means of transport telematics for the operational control of the quality of the transportation process, consisting in the detection of accidents and careless driving. Moreover, no methodology has been found for the remote evaluation of cargo damages in the performance of dangerous maneuvers or accidents. This circumstance narrows the range of potential applications of transport telematics and creates prerequisites for carrying out research in the field of improving the quality of freight motor transport, taking into account the use of information obtained by means of transport telematics.

2. Transport telematics

The term "telematics" is derived from the words "telecommunications" and "informatics". Accordingly, the concept of "transport telematics" covers the area of using the capabilities of telecommunication technologies and informatics in solving technological problems in transport. Telematics means a set of interrelated automated systems that solve the tasks of traffic management, monitoring and managing the operation of all modes of transport (individual, public, freight), informing citizens and enterprises about the organization of transport services on the territory of the region [1].

The main technologies used in transport telematics systems in road transport and in the road sector are:

- coordinate-time and navigation technologies;
- geoinformation technologies;
- telecommunication technologies, including mobile communication and navigation technologies;
- technologies for collecting, storing and processing information on a computer [2].

3. Detection of cargo damage

In theory, overloading refers to the ratio of body weight to gravity acting on this body or increasing the body weight when additional acceleration acts on it. Overload is expressed by the term g, in physics means – acceleration of gravity. The term g is used in astronautics, aviation, motor sport. Overloads occur when the vehicle speed changes (braking, accelerating or changing the trajectory). At large (more than 5g) overload values, negative consequences can be observed: the movement of loose cargo, the destruction of structural elements of the vehicle and cargo, the deterioration of the well-being or health of people subjected to the forces that caused the overload.

To detect damage to the cargo, it is necessary to know the values of the overloads recorded by the monitoring system in order to calculate the force that acts on the cargo during the accident. Perhaps this is due to the built-in accelerometer, which measures the projections of accelerations in three planes (longitudinal, transverse, vertical) and in the event of an accident, the values of overloads increase, thereby signaling a dangerous road accident.

As a simplified example of the detection of cargo damage, which is a concrete construction structure weighing 150 kg, it is possible to describe a situation in which the monitoring system fixes an overload in the longitudinal direction equal to 0.05 g. It is known, that the concrete design has the module of elasticity of 1,1 MPas (for cellular autoclave curing of the B1 brand) which is equal to the applied force in 11,22 kilograms. At speedup 0,05g force operating on freight makes 7,5 N or 0,075 MPas, or 0,764 kilograms of force. This approach has drawbacks, which will be mentioned at the end of the article.

4. Experiment

For confirmation of a hypothesis of a possibility of application of means of transport telematics for the purpose of identification of damages of freight during transportation the experiment regarding a possibility of fixation and drive in real time of information on the arising overloads in the database of the operator of monitoring system was made.

The components of the experiment were as follows.

Test stand-laboratory equipment that allows simulating an accident without the use of a vehicle.

The monitoring system is a system installed on a wheeled vehicle and designed to determine the location and traffic parameters of a vehicle according to GLONASS or GLONASS signals in conjunction with other global navigation satellite systems, the transmission of a message and other information about the vehicle in a traffic accident.

The operator – the organization which installed monitoring system collected and stores the data obtained from the monitoring system in the database.

The software is the software of the Operator.

The experiment consisted in simulating an accident with a test stand, on which the monitoring system was given an acceleration in the range from 1 to 10 g with an exposure time of 25 to 100 ms. The device was fixed to the platform, the platform accelerated, then struck against the obstacle. If the accelerometer fixes an acceleration greater than 1 g for 25 ms along the x and y axes and / or 2 g on the vertical z axis, the monitoring system records data into the random access memory for subsequent transmission to the operator's server. Accelerometer readings were recorded for each of the three coordinate axes (x, y, z) 3 seconds before and 3 seconds after exceeding the specified acceleration thresholds with a frequency of at least 100 records per second.

As a result of the experiment, the following information was received in the database of the monitoring system operator:

- the date and time of the event;
- coordinates of the vehicle;
- vehicle speed;
- identification number (VIN);
- identification number of the body;
- state registration plate;
- the values of the overloads that occurred;
- confirmation of the device at the time of a dangerous event.

1. State registration plate of the vehicle: **T47 ****
2. Vehicle VIN: **K17 ****
3. Date and time of the accident (date and time of operation of the accelerometer): **29-JAN-17 12.45.06.0000 PM + 00: 00**
4. Coordinates of the accident **55.80309906479152 37.529457807540894**
5. The speed of the vehicle at the time of the accident: **0 km / h**
6. The values of the accelerations along the three coordinate axes: **X = 2.375, Y = 1.85. Z = 1.002**
7. Data efficiency of the navigation information unit: **YES**

Figure 1. Information entered into the database

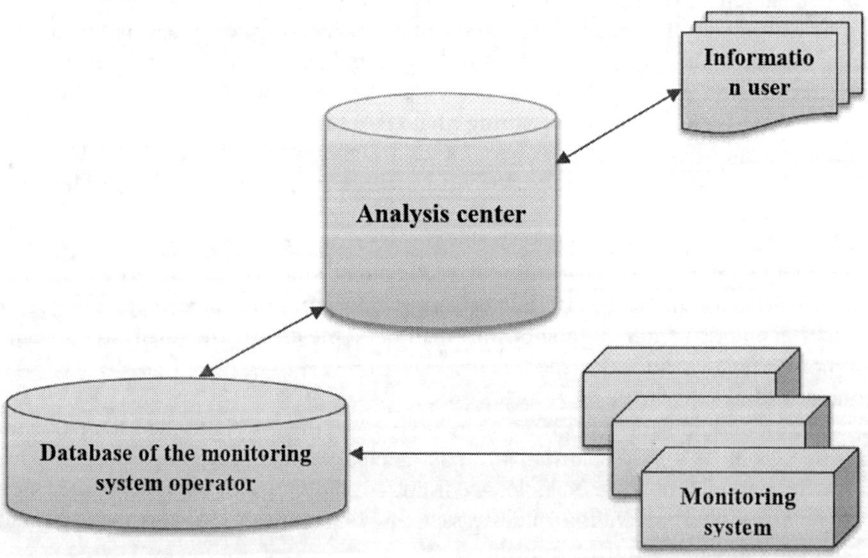

Figure 2. Scheme of information interaction

The experiment confirmed the possibility of using means of transport telematics for remote detection of cargo damages arising from road accidents or careless driving.

5. Method of calculation of the loadings influencing a transport package

All loadings operating on a transport package can be divided into static loadings which are shown at rest or uniform motion without speedup and dynamic which are shown in the form of action on a package of inertial forces in the course of transportation with change by speed, and also at blows and pushes. As dangerous road incidents are characterized by dynamic, but not uniform motion with identical speed, for calculation the corresponding method will be used. The dynamic loadings operating on a transport package include vertical, longitudinal and cross forces.

Vertical force is determined by a formula:

$$Fv = vsM\left(qvr - 1\right), \tag{1}$$

vs = vertical speedup in shares of g, m/s^2;
qvr = quantity of vertical rows;
M = mass of freight in a container, in a transport package, kg.

Longitudinal force is determined by a formula:

$$Fl = ls\left(culd - 1\right)M, \tag{2}$$

ls = longitudinal speedup in shares of g, m/s^2;
$culd$ = number of cargo units in the longitudinal direction,
M = mass of freight in a container, in a transport package, kg.

Cross force is determined by a formula:

$$Fc = cs\left(cucd - 1\right)M, \tag{3}$$

cs = cross speedup in shares of g, m/s^2;
$cucd$ = number of cargo units in the cross direction;
M = mass of freight in a container, in a transport package, kg.

Examples of calculations:

Calculating the loads acting on the transport package made up of boxes with household appliances, we will use the values given in Table 1 as the values of the accelerations.

Table 1. The size of impact of blows on a container and freight under various conditions of transportation by the motor transport

Conditions of transportation	Load value g, m/s^2
On asphalt surface	1.5
On a dirt road	3.5

Table 2. Calculation data

Material, from which freight is made	Module of elasticity of material (E)
Iron	200 HPa

On asphalt surface:

$$Fv = 1.5 \times 9.81 \times 6 \times 3 = 265\ H\ (2.65 HPa) \tag{4}$$

$$Fl = 1.5 \times 9.81 \times 6 \times 1 = 88\ H\ (0.88 HPa) \tag{5}$$

$$Fc = 1.5 \times 9.81 \times 6 \times 2 = 177 H (1.77 HPa) \tag{6}$$

On a dirt road:

$$Fv = 3.5 \times 9.81 \times 6 \times 3 = 618\ H\ (6.18 HPa) \tag{7}$$

$$Fl = 3.5 \times 9.81 \times 6 \times 1 = 206\ H\ (2.06 HPa) \tag{8}$$

$$Fc = 3.5 \times 9.81 \times 6 \times 2 = 412\ H\ (4.12 HPa) \tag{9}$$

Calculating the loads acting on the transport package made up of boxes with glass products, we will use the values given in Table 3 as the acceleration values.

Table 3. The size of impact of blows on a container and freight under various conditions of transportation by the motor transport

Conditions of transportation	Load value g, m/s²
On asphalt surface	1.5
On a dirt road	3.5

Table 4. Calculation data

Material, from which freight is made	Module of elasticity of material (E)
Glass	56 HPa

On asphalt surface:

$$Fv = 1.5 \times 9.81 \times 13 \times 1 \;=\; 191.3\,H\;(1.91 HPa) \tag{10}$$

$$Fl = \;1.5 \times 9.81 \times 13 \times 3 \;=\; 574\,H\;(5.74 HPa) \tag{11}$$

$$Fc = 1.5 \times 9.81 \times 13 \times 2 \;=\; 382.6 H\left(3.83\;HPa\right) \tag{12}$$

On a dirt road:

$$Fv = \;3.5 \times 9.81 \times 13 \times 1 \;=\; 446\,H\;(4.46 HPa) \tag{13}$$

$$Fl = 3.5 \times 9.81 \times 13 \times 4 \;=\; 1339\,H\;(13.39 HPa) \tag{14}$$

$$Fc = 3.5 \times 9.81 \times 13 \times 2 \;=\; 893\,H\;(8.93 HPa) \tag{15}$$

The calculations made allow us to assume that in all cases there is no destruction, but if the conditions have been increased by increasing the weight of the cargo and the number of boxes, the force acting on the cargo will increase.

$$Fl = 3.5 \times 9.81 \times 40 \times 5 = \;6867 H\left(68.67\;HPa\right) \tag{16}$$

6. The next stage of scientific work

This method has drawbacks associated with the abstract representation of the application of the force acting on the load, as well as the area of this impact, which greatly affects the accuracy of the calculation.

In addition to these formulas to clarify the results you need to know:

- class, load capacity of the semitrailer/container;
- the number of overpack located inside the semitrailer/container;
- scheme for placing transport packages inside the semitrailer/container;
- type of pallets used;
- layout of the boxes with cargo on the pallet;
- characteristic of the cargo;
- Characteristics of cargo packing.

When these factors are taken into account, opportunities are opened in the field of developing a risk system and finalizing the requirements for the safety of the transportation process, which in this case must take into account the data obtained by means of transport telematics.

7. Conclusion

Thus, it is theoretically possible to use means of transport telematics for the purpose of remote analysis of the transport process for the detection of traffic accidents, detection and evaluation of cargo damages resulting from accidents or careless driving. The information transmitted by the monitoring system, as well as the developed methodology, may be of interest to the following users:

- transport companies will be able to monitor the safety of the transportation process in order to maintain competitiveness, which is to reduce the risk of damage or loss of cargo, in addition, this information may be proof of the absence of the fault of the carrier for the damage caused to the goods in case the cargo was damaged not during transport.

- the insurance companies will quickly get information about the occurrence of an accident and information about damage to the cargo as a result of this event in order to settle the losses, which will minimize the risk of financial losses associated with fraud and speed up the process of registration of insured events;
- the customs, in order to optimize operational activities, will be able to obtain information about an accident that occurred while the vehicle was traveling to perform customs operations;
- consumers of transport services need to have an idea of how well the transport companies provide services in order to further select the most suitable and safe, based on the statistics of emergency situations and violations of the mode of transport of goods;
- The Russian Association of International Road Carriers will be able to collect statistical data on the quality of transport services for compiling the rating of the best carriers and further taking measures to improve the efficiency of the transport process on the territory of the Eurasian Economic Union.
- The application of means of transport telematics in the implementation of international road transport facilitates:
- creation of the practice of using indications of navigation and information devices when verifying the fact of an accident;
- increase of evidence, objectivity and accuracy of detecting accidents;
- creation of additional mechanisms to resolve differences between insurers and Insurants;
- creation of a system for assessing the quality of goods road transport;
- optimization of the operational activities of customs authorities in the performance of customs operations.

References

[1] Efimenko D B, Vasilenko R V, Filatov S A and Zaikin R N 2015 Topical issues of customs regulations of the activities of international road carriers *J. Motor transport enterprise* **10** 5–9
[2] Vlasov V M 2013 *Telematics in the Transport road sector: proc. the allowance* (Moscow: MADI)
[3] Melnikov T E, Melnikov E S and Efimenko D B 2015 Freight motor transport: issues of institutionalization of the subjects of transportation activities *J. Motor transport enterprise* **6** 3–6
[4] Efimenko D B, Krutova T A and Filatov S A 2013 The Development of the technical equipment of the navigation system of dispatching management of international road haulage *J. Motor transport enterprise* **6** 16–29
[5] Efimenko D B, Filatov S A, Sergeev S V and Vasilenko R V 2015 Improvement of information support of processes of freight transport management road transport *J. In the world of scientific discoveries* **6(66)** 261–9
[6] Mel'nikova T E and Obukhov N I 2019 Legal aspects of improving traffic safety when using smart transport systems *J. Science and technology in the road industry*
[7] Efimenko D B, Zaikin R N and Filatov S A 2018 Development of logistics monitoring systems of international goods movement and digital infrastructure of cargo traffic promotion Proc. of the Int. sci. and pract. Conf. "*Progress of vehicles and systems*" ed. by I A Kalyaev, F L Chernousko, V M Prikhodko (Volgograd, 09–11 October 2018)
[8] Ledovskiy A A and Filatov S A (2018) Assessment of the effect of the introduction of automated control systems of freight transport *J. of Transport*

Use of self-recovering slowly-hardening concrete to ensure the longevity of highways

B A Asmatulayev, R B Asmatulayev and N B Asmatulayev
LLP Kazakh Research and Design Institute «DORTRANS», Kazakhstan

E-mail: boris-aisa@mail.ru

Abstract. At the beginning of this century, the new requirements for the durability of highways are made according to the modern concepts of the US and EU. The article describes the main results of theoretical, experimental studies and long-term monitoring of highways' sections, built using the self-recovering road concretes that meet the new requirements. The durability of road concretes substantiates the results of physical and chemical studies, physical and mechanical tests of various compositions of belite cements, concrete and concrete cores. It was shown that the strength of the concrete continues to increase for more than 39 years of roads' maintenance. It is assumed that the path to durable «Roman concrete» is found.

Keywords: concept, durability, highways, road concrete, belite cement, research, physical-chemical properties, physical and mechanical tests, strength.

1. Introduction

Currently, due to changes in traffic and increase of the load capacity of vehicles up to 12–13 tons per axle, the service life of road pavements of the highways declined sharply: the durability of asphalt concrete coatings in Russia and Kazakhstan decreased to 5–6 years [1–5], and of the cement concrete, including abroad – to 20–25 years [6–11]. The current critical situation, in reducing the service life of road pavements, is caused by mismatch of the technical state of the road structures and the actual composition, intensity and dynamic traffic flow. The effectiveness of the highways' construction is justified at the service life of at least 50 years [6–11]. At the beginning of this century, this was the impetus for the development of new concepts: 1) concept of the «road pavements with a long life» on the European terminology; 2) concept of the «eternal road pavements» in accordance with the terminology adopted in the USA [6, 11]. In connection with these concepts, the Technical Committee C 4.3 «Road pavements» of the World Road Association (PIARC-AIPCR) has developed a Program and Methods of examination of the road pavements, created an Expert Committee on the questionnaire survey of the durability of the roads sections. Experts from 10 countries (for 100 road sections) established the basic principles and conditions to ensure the longevity/durability of roads based on concepts of the US and EU in order of their importance:

- ensuring the comfort and safety of traffic on the roads;
- stability of roads under the influence of transportations and climate changes;
- increasing bearing capacity of pavement layers from bottom to top;
- increased strength of the coatings' materials for compression to perceive heavy loads from the top.

Under the terms of the roads' durability, on the modern concepts of the US and the EU, to extend the life of roads up to 50 years or more it is necessary to ensure that the bearing capacity of the pavement layers increases from bottom to top. This condition is completely contrary to the methods of designing the road pavements, adopted in Russia and Kazakhstan, on which the strength of the layers, conversely, decreases downwards. The main conditions to achieve the durability of the highways on the new concepts are increasing the bearing capacity of the lower layers of road pavements, subject to compliance with economic and technical efficiency of road construction. The idea of creating a more solid foundation, than the actual covering, is not new [14]. For example, during construction of a long-lasting road structure on the road with very heavy traffic (USA, California), the bottom layer contained more binding agent than the other layers. The same idea was expressed in the 1970–80^s in the works of A.O. Salle, A.E. Merzlikin and B.S. Radovsky and was realized in Russia by A.E. Merzlikin on the Ring Road's section in the city of Kazan during construction of the layered foundation of the cement-modified soil [6]. The implementation of such structural changes in the design of road pavements, using traditional materials and techniques, certainly, results in a significant increase in the cost of road construction.

The authors of this work have conducted a many years research and have developed new alternative efficient self-recovering road concretes, which allow significantly enhance the durability of the highways' bases with the disposal of industrial technological wastes of Kazakhstan [15–19]. Durability of the proposed concretes is ensured by a radical change in the structure of the road concrete, by moving from a crystalline structure to a thixotropic one of the cement stone. For the scientific justification of the concrete durability, the authors carried out studies of the properties of new belite cements and road concretes on their basis in a year-round operation of highways in the sharply continental climatic conditions of Kazakhstan [24–31].

2. Methods of studying the physical and mechanical properties of belite cements based on phosphoric granulated slag.

Preparation of belite cement was carried out by joint grinding of granulated phosphorus slag with cement dust in the laboratory ball mills (MBL type) to the fineness, characterized by the residue on a sieve of 0.08 mm (not more than 15 %) or to a specific surface of at least 3000 cm^2/g.

Before grinding, granulated phosphorus slag had been dried to a constant weight at a temperature not exceeding 200 °C. Samples preparation for the determination of the strength properties of belite cement was performed by compression of binding agents in the optimum moisture. This, in our opinion, most realistically reflects the conditions of formation and hardening at their use in road concrete.

The study of strength and deformation characteristics of belite cement was carried out on the following samples: test beams with a size of 40x40x160 mm and the cylinders with diameter and height of 50 mm. Forming of the samples was performed by standard methods – by compacting on a press in metal molds with bilateral liners under a load of 15 MPa and within 3 minutes.

Strength, deformation properties and frost resistance of the road concrete on the basis of slowly-hardening belite cements was studied on the samples – test beams with a size of 100x100x400 mm and the cylinders with diameter and height of 100 mm. The concrete mix was compacted in a hydraulic press under a load of 20 MPa for 3 minutes. The number of samples was at least three based on repeatability when measuring that ensures the reliability of the experience equal to 0.95 with a relative error of 3σ.

Storage of test samples was carried out in the water-sealed baths; in the future, these conditions of hardening will be called as normal. Currently, the existing normative documents allow construction of basements of soils reinforced with cement at lower temperatures (from +5 to –10 °C), and at strengthening with slag binding agents – only at positive air temperatures. Therefore, on the first phase of the study, we have studied the impact of low positive and small negative temperatures (up to –10 °C, typical for the winter periods of fifth road climatic zone) on the properties of the slag binding agent during its hardening.

Some of the samples were kept under normal conditions (series 2), the others – at constant low positive and negative temperatures at a given mode of hardening in cold rooms. Some samples were placed in cold storage immediately after preparation, others – after the preliminary storage at positive temperatures. Moreover, one series of samples was kept in a periodic variation of temperature within time (series 13), i.e. with a gradual decrease from +5 to –10 °C, and then with an increase from –10 to +5 °C, and after the end of this cycle, the samples were placed in the water-sealed baths. Duration of holding samples of a series 13 (under certain temperatures) was set at the averaged data of the onset and duration of multiyear average daily air temperatures, typical for the regions of Kazakhstan, located in the fifth road-climatic zone. The actual length of low temperatures (in the year) from 0 to –10 °C is 140 days; for laboratory tests and ease of comparison with control samples, hardened under normal conditions, we used 150 days with intervals of no less than 15 or 30 days. For storing the samples in refrigeration chambers, they were wrapped in a polyethylene film. Before the test, the samples were thawed for at least 4 hours in moist sand. Initially, the samples were tested for bending tension, and the halves of samples – for compression. Then, with loads of 0.2–0.4 of tensile strength in bending, on 3 beams, we set the amount of elastic deflection and measured the elastic modulus of the material.

3. Methods of physical and chemical research of slag binding agent

To study the physical and chemical processes occurring at hardening of slag binding agent, after durability test, from the middle of the samples we selected the samples for petrographic, X-ray, thermographic and electron-microscopic analysis. Immediately after the sampling, samples have been crushed in an agate mortar and treated with ethanol and ether to remove free water from the system.

Differential thermal analysis (DTA) has been performed on a Hungarian derivatograph of the system "R. Paulik, I. Paulik, D. Erdei" at a heating rate of 10° per minute, at the sensitivity of the weight of 500 mg and with the sample weight of 0.5–1.0 g. Analysis was performed in platinum crucibles. As the reference sample we used calcined aluminum oxide (Al_2O_3). Some thermographic investigations have been carried out on the thermographic device TU-IM at a heating rate of 100° per minute and with the sample weight of 0.01g. Analysis was performed in ceramic crucibles.

The mineralogical composition was studied at X-ray diffraction data, performed on the X-ray machine URS-50 IM with an attachment PK-2 and applying a copper anode, at the anode current of 10uA and a tube voltage of 35 kV. Recording of radiographs was performed on a standard chart tape at 2 degrees per minute. The sample was placed in a flat cell of 1.5 mm thickness. To decrypt thermal- and radiographs, we used the published data from thermography and radiography.

Mineralogical and petrographic studies were performed on the microscope (brand MIN-8). Study of hardening products has been performed in immersion preparations and transparent sections. Because of the strong hygroscopic properties of the material, from powders of the binding agents we were preparing a liquid paste in admixture with an alcohol, which was applied by a thin layer on a glass slide, and then dried. Further preparation and review of the immersion preparations was carried out by the usual method. When preparing transparent thin sections, kerosene was used. Counting of components was performed in thin sections under the microscope by a point method. The essence of the method is that in the polished sections we evenly distributed a large number of observation points under the microscope on a square grid. At the same time, we counted the number of points got on each mineral. According to the theory of probability, these amounts are proportional to the volumes, occupied by the relevant minerals in thin section. When counting, method of fields has been applied. For this purpose we used the net eyepiece, wherein the total number of strand intersections (nodes) was 400. In each field, we counted the number of points (nodes), attributable to certain mineral. View of 10 fields in thin section ensured sufficient reliability of quantifying. Within one field of view, counting was performed 3 times. We have determined the average mineral content for the field. The total number of fields for each of the section, where we determined the volume content of the glass phase, was 10. Thus, the average percentage of the mineral was determined by 30 estimates, that increases the accuracy of the result for average values.

Electron microscopic studies have been performed on the microscope UEMB-100.

4. Main part (Results and Discussion)

In Kazakhstan to confirm the idea of building strong and durable roads, a number of experienced road sections with monolithic bases of slowly-hardening high-strength road concrete, coated with hot and cold asphalt concrete, were built in 80–90s [15]. Calculation of road constructions was carried out according to the traditional method of nonrigid road clothes; in calculation the indices of slowly-hardening reinforced materials of 90-days samples with a compressive strength of not more than M60 (RoK ST 973) were taken into account. Here further strengthening of concrete for more than twice (during the years of operation) was not taken into account. By the way, method of calculation of road pavements/cloths with long-term strengthening concretes, called below as «self-recovering road concrete», is still not developed. The longevity of these road constructions is confirmed by the works [15], and below we will give justification for this effect.

To improve the durability of roads, the Ministry of Transport and Communications of Kazakhstan in 2006 made a decision on designing and construction of roads of international importance (under the traffic load of not less than 13 tons per axle) with cement -concrete and asphalt stone mastic coatings.

Construction of cement concrete pavements on new technologies required the study of foreign experience in Germany and the US for the joint development of Kazakhstan's regulatory and technical documents [16,17].

Currently, in the Republic of Kazakhstan there are more than 600 kilometers of roads with cement concrete pavements. International highway "Western Europe – Western China" (2900 km) as well as the highway "Astana-Karaganda" (more than 220 km) is under construction. During the construction of cement concrete pavements, there used a technology of laying rigid concrete mixtures with sliding formwork, using high energy compaction of high-performance road concreting machines. The cost of roads' construction has increased significantly, since the regulations [16, 17] recommended to use expensive high-strength Portland cement and high-quality stone aggregates, that requires substantial financial and material costs. This is further amplified by the annual increase in the deficit of traditional road-construction materials (bitumen, cement, road mastic, rubble, mineral powder) and an increase in the cost of land acquisition.

Professor V.A. Zolotarev [18] confirms that «Today in the world road construction there are three main priorities. The first priority is durability of roads, which is generalized by the concept of "eternal roads" (USA) or «roads with a high life expectancy» (EC) with renewal of the upper layer every 6–7 years. The second and third priorities are interrelated: energy saving, environmentally friendly technologies, with re-use of old asphalt concrete, which is 100 % processed material [18].

Resource Saving with re-use of all materials of the reconstructed roads is not less important for high material-intensive road construction [19–22]. Besides that, Kazakhstan accumulated in the dumps huge amounts of technogenic mineral waste (hereinafter – TMW), suitable for road construction. Ministry of Environment Protection (RoK MEP) accounted 775 facilities of TMW, which have accumulated about 34 billion of tons; at the same time there is a trend of their annual growth for 700–800 mln tons. For these priority directions, as a result of years of research, the authors developed and tested long-life monolithic bases, using the secondary road-building materials and industrial technogenic mineral wastes (TMW) [15–17, 19]. Evidences of the effectiveness of using TMW and monolithic bases of road clothes (in terms of their durability) are Roman roads, some parts of which are preserved in the state suitable for traffic nowadays [23]. During construction of roads in Italy, the volcanic ash from the town of Pozzuoli (near the city of Naples) [23]) and volcanic tuff from the banks of the Rhine have been used as the binding agents. Subsequently, all these additives have been named "pozzolanic", as they had the same property – ability to solidify in the presence of lime and water. Coryphaeus of road science – Prof. V.F. Babkov – gave examples of road structures on the Roman roads, which had monolithic bases and preserved until today. It was found [15] that Kazakh TMW, in particular, blast furnace / phosphorous slags, bauxite sludges, fly ash, phosphogypsum etc., being heat-treated, similar to volcanic ash and tuff, have the same durable cemented properties. To build long-lasting monolithic bases, Kazakhstan first developed compositions of slowly-hardening belite cement [24] and mastered the use of long-lasting «self-recovering road concretes» [25], including, by using the subgrade «Zhertas», for which there is a patent in the Republic of Kazakhstan [18].

IOP Conf. Series: Materials Science and Engineering **832** (2020) 012019 doi:10.1088/1757-899X/832/1/012019

Kazakh scientists [19–21] repeatedly noted the prospects of using TMW in combination with cement. Many foreign studies have found [22–25] that there are potential opportunities to increase the durability of cement concrete pavements when using such mixtures. In 2002 for the construction of airport taxiways (United States, Texas), cement (50 %) in combination with ground granulated slag (25 %) and fly ash (25 %) have been used. Concrete was tested in 2008 and it was approved that the service life of cement concrete pavement was at least 120 years [32].

The results of years of research and monitoring on the experimental sections of roads in Kazakhstan, being in operation for the past 31–39 years, confirm the accuracy of the performed theoretical and experimental studies. Below are only the main results of research and testing of the roads confirming the durability and self-recovery of road concrete, used in road construction in Kazakhstan since 1976. The use of new slowly-hardening road concrete allows significantly improving the durability of concrete roads. This effect is substantiated by a radical change in the macro- and microstructure of road concrete, by moving from a crystalline to a thixotropic structure of the cement stone.

It is well known [39] that technological modes of construction of concrete roads are caused by the setting time of Portland cements and the hydration rate of four main clinker minerals: **C_3S; C_2S; C_3A** and **C_4AF** (see Table 1). The prevailing content in Portland cement clinker of the following materials: highly basic tricalcium silicate (**C_3S – Alite),** tricalcium aluminate (**C_3A**) and tetracalcium alyumoferrita (**C_4AF),** the degree of hydration of which is 100 % in 28 days, causes rapid setting and hardening of road Portland cement and concrete with formation of a cement stone, basically, of a crystal structure. In the same time, the hydration degree of tricalcium silicate (**C_2S – Belite)** only after 180 days barely reaches 50 %, with amorphous formations of a thixotropic structure. Therefore, to provide slow solidification of the new cements, the content of slowly-hardening low-basic dicalcium silicate (**C_2S – Belite**) is increased in their compositions up to 60–85 %; and the number of fast-setting materials (highly-basic silicates **C_3S,** aluminates **C_3A** and alumina ferrite **C_4AF)** is minimized. Table 1 compares the chemical and mineralogical compositions of traditional Portland cements (hereinafter – alite cements) and slowly hardening cements (hereinafter – belite cements).

Table 1. Chemical and mineralogical compositions of alite cements and slowly-hardening belite cements

Cement type	**Chemical composition, mass. %**			
	CaO	**SiO_2**	**Al_2O_3**	**Fe_2O_3**
Alite*	60–67	17–25	3–8	0.2–6
Belite*	33–46	39–61	3–10	3–5
	Mineralogical composition, mass. %			
	C_3S (alite)	**C_2S (belite)**	**C_3A**	**C_4AF**
Alite*	40–75	5–25	2–15	5–20
Belite*	10–35	60–85	3–5	2–7

Notes* – code names of cements according to the predominant content of minerals: C_3S – alite, C_2S– belite.

Phosphorus and blast furnace slags, bauxite sludges and ash of TPP (thermal power plant) are the main components of belite cements [24] and self-recovering road concretes [3].

Physico-chemical studies [24, 25] confirmed that according to the mineralogical composition, belite cement stone is mainly represented by the slow-hardening low-basic silicate C_2S (Belite). Formation of the structure of a slowly hardening cement stone, during its hardening within 8 years, is shown in the photos of fractures of cement beams, tested for tensile bending (Fig. 1).

Thus, the nature of the slow decomposition of the cement grains and the occurrence of formations can be clearly traced in Fig. 1, a–d. Fig. 1, a shows undecomposed cement grains; then, over time, there are grains with formed peripheral shell, which gradually grows (Fig. 1, b) and then amorphous formations with a gelly-like structure appeared (Fig. 1, c).

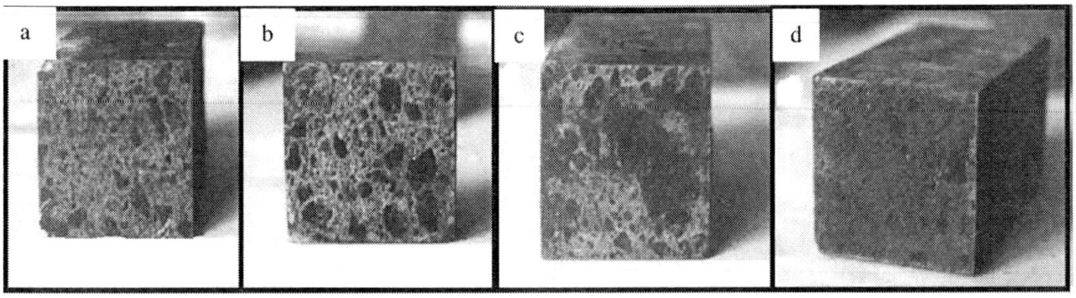

Figure 1. Photos of fractures of the samples, hardened under normal conditions and tested after 1 year (a), 3 years (b), 6 years (c) and 8 years (d)

Amorphism of these formations is caused by fuzziness and diffuseness of their edges and chaotic growth in all directions. Along with amorfites, there are single crystals of C-S-H (Fig. 1«b»). In immersion gel is presented as a colorless isotropic mass with an index of refraction 1.330–1.567.

The number of gelly formations in samples raises with increasing age of the samples. But even after 8 years of hardening under normal conditions, belite cement samples have unhydrated grains, indicating the potential possibility of the cement for further hardening.

Radiographs of belite cement stone, shown in Figure 2, confirm the data obtained. Radiographs of the cement stone samples, previously stored under normal conditions and then hardened at the temperatures of +5, 0, –5 and –10 °C within three months (Fig. 2, a), as well as the samples, hardened without preincubation at normal conditions (Fig. 2, b), are almost similar.

Diffraction lines, typical of hydrosilicates of tobermorite group, are combined with the lines of calcite, that somewhat complicates their decryption. So, interplanar distances of 1.875; 1.913; 2.095; 2.283 Ao correspond to interplanar distances of calcite. Lines 1.814; 2.88; 3.03 Ao match interplanar distances of silicate C-S-H. Radiographs of all samples contain lines with a maximum of 3.34 Ao and the lines close to the values of 2.82; 1.980; 1.817 Ao, relevant to interplanar characteristics of the quartz.

Thus, regardless of the temperature conditions of hardening of the belite cement stone (within experiment) and timing of incubation, the phase composition of formations has not changed significantly. Retardation of hydration processes for belite cement (with decreasing temperature of hardening) is confirmed by the decrease in diffraction line, corresponding to the interplanar distance 3.027 Ao, compared with the line 3.039 Ao, obtained in the process of hardening at positive temperatures.

A number of studies noted that construction of roads with slowly hardening road mixtures should be carried out in early summer or at least two months before the first frost, in order to give them the opportunity to harden before winter. In order to test this hypothesis, we examined the effect of hardening temperature for further growth of the strength of the belite cement stone samples, previously stored (within 30 days) under normal conditions (Fig. 3, a, b).

Figure 3 shows the results of tested samples in the series 5–8, which after the preliminary storage for 1 month (30 days) under normal conditions, have been hardened for 5 months at the temperatures of +5, 0, –5, –10 °C, and then – under normal conditions. The test results of the samples (series 5–8) were compared with the samples of series 2, being permanently hardened under normal conditions.

The data obtained indicate that low positive and negative temperatures slow down the process of hardening of belite cement stone, previously stored under normal conditions. Moreover, the higher the temperature of hardening, the slower the curing and hardening.

So the strength of the samples of months of age (at normal hardening), after three months of their storage at a temperature of + 5 °C (series 5) increased in compression by 33 %, when bending – by 38 %, and at 0 °C (series 6) – by 22 % (in compression) and 20 % (when bending) respectively. At the temperatures of –5 °C (series 7) and –10 °C (series 8), strength growth is practically not observed (its decrease is even marked). Upon further storage in normal conditions, strength of all samples of series 5–8 raises (compression – up to 10 %, tensile bending – up to 18 %), compared with the strength of samples (series 2), which were constantly kept at normal conditions. Increased bending strength indicates an increase in dispersion of formations, which leads to improved deformation properties of the cement stone structure.

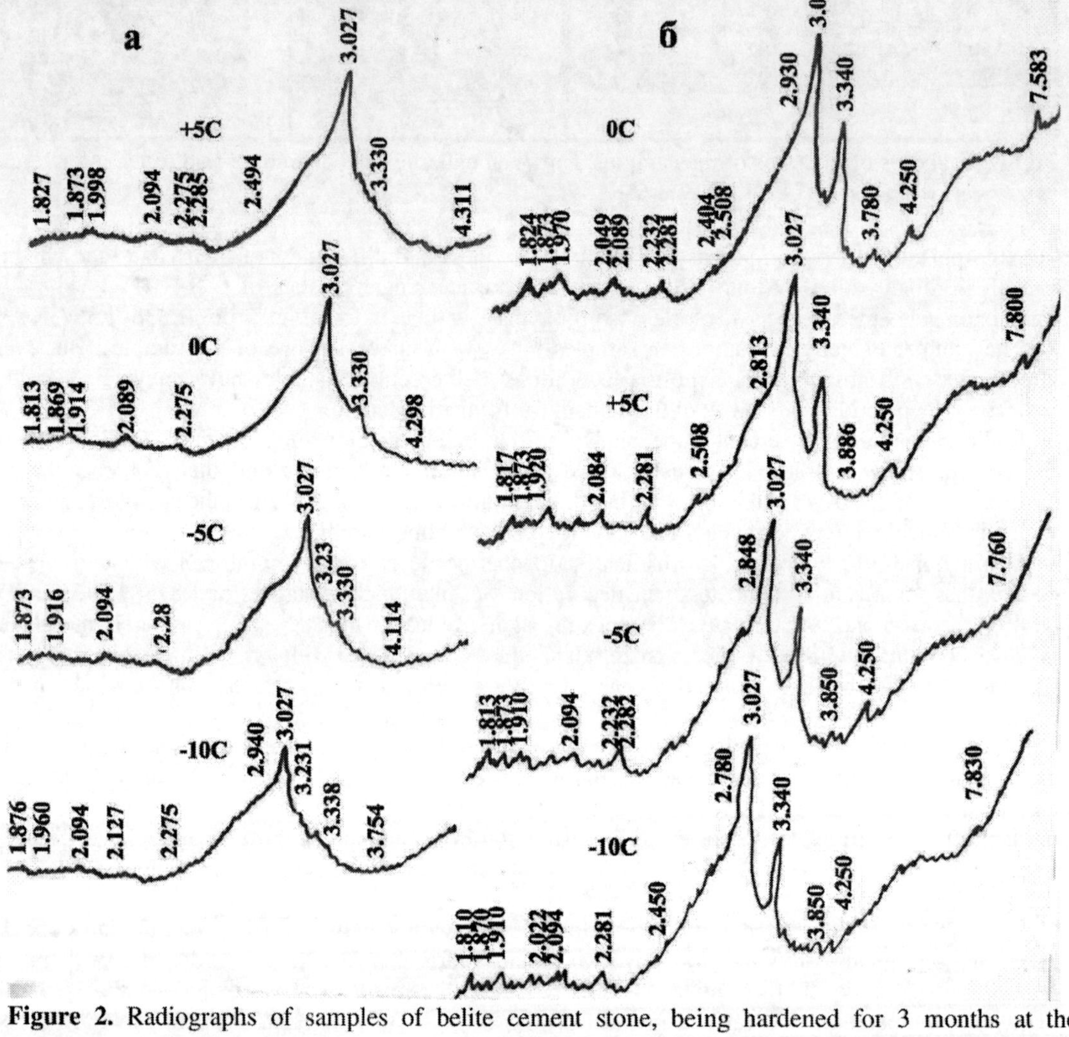

Figure 2. Radiographs of samples of belite cement stone, being hardened for 3 months at the temperatures of +5, 0, –5, –10 °C: a – after 1 month of storage under normal conditions; b – without aging under normal conditions

When keeping/storing the samples at positive temperatures in the water-sealed baths or under normal conditions, cement stone is gaining strength within 2–3 years and further strength is stabilized. The compressive strength reaches 130 MPa and tensile bending – up to 10.7–12 MPa, indicating high-strength and deformation properties of belite cement.

Test of cores of self-recovering road concrete on the basis of belite cements, drilled from the monolithic bases of experimental road sections has shown that in a sharply-continental climatic conditions of Kazakhstan, increase in the strength of concrete occurs during the period of long-term operation of roads.

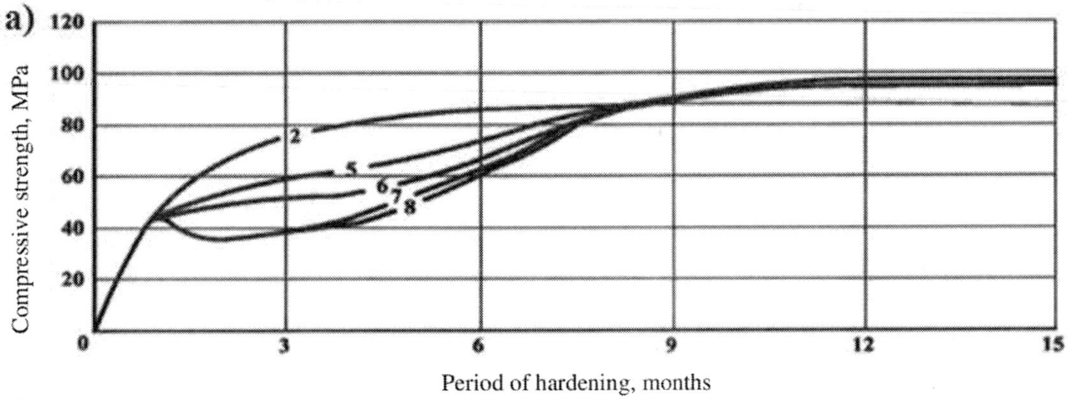

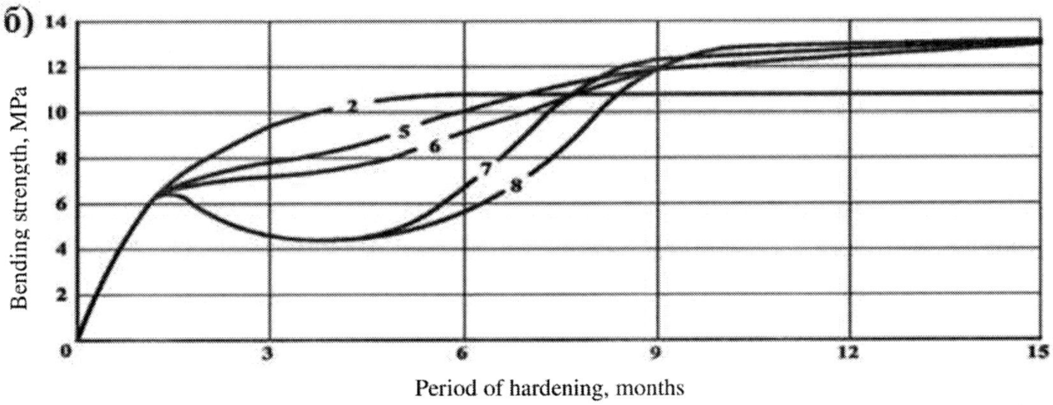

Figure 3. The kinetics of change in strength of samples of belite cement stone over time (series 2, series 5,6,7,8), preliminary stored for 1 month under normal conditions: a – compressive strength; б – tensile strength in bending

Compositions of concrete mixes for experienced road sections were selected from the rubble-sand and gravel-sand aggregates: aggregates – 80–90 %; belite cement (C_2S up to 80 %) – 10–15 %; (C_2S up to 70 %) – 10–18 %; (C_2S from 60 %) – 10–20 %. The water-cement ratio is chosen within 0.35–0.45.

Compressive strength of concrete, tested in vitro, was: at the age of 90 days – 10–20 MPa, after 2 years – 20–30 MPa.

Table 2 shows the results of the test of cores, drilled from the concrete base of the road «Fogolevka-Zhdanovo» PK 54-74, built in December 1977, and PK 74-90, built in November 1978.

Table 2. The test results of the cores from the concrete base of the road «Fogolevka-Zhdanovo»

Name of the indicator measured	Test of cores in 1989, MPa (Concrete age – 12 years)	Test of cores in 2005, MPa (Concrete age – 28 years)
Compressive strength	36.4, 36.7, 36.5 Average 36.5 (M 350)	48.6, 49.0, 48.8 Average 48.8 (M 450)
Bending strength	–	8.6, 8.3

At design marks of road concrete, chosen for construction of experimental roads, their strength at the age of 90 days is M100–M200; the actual strength of concrete at the age of 28–39 years of operation reached M250–M450, i.e. increased more than 2 times (Fig. 4).

This indicates that belite road concretes have the property of self-recovering and prolonged hardening in conditions of constant dynamic vibration transport and climatic loads. To explain the effect, the authors have conducted detailed physico-chemical studies. The results of petrographic, X-ray, differential-thermal analysis and monitoring with a scanning electron microscope have shown that at hardening of belite cements, basic structure-forming formations in concrete are gelled low-basic hydrosilicates of calcium (type C-S-H.

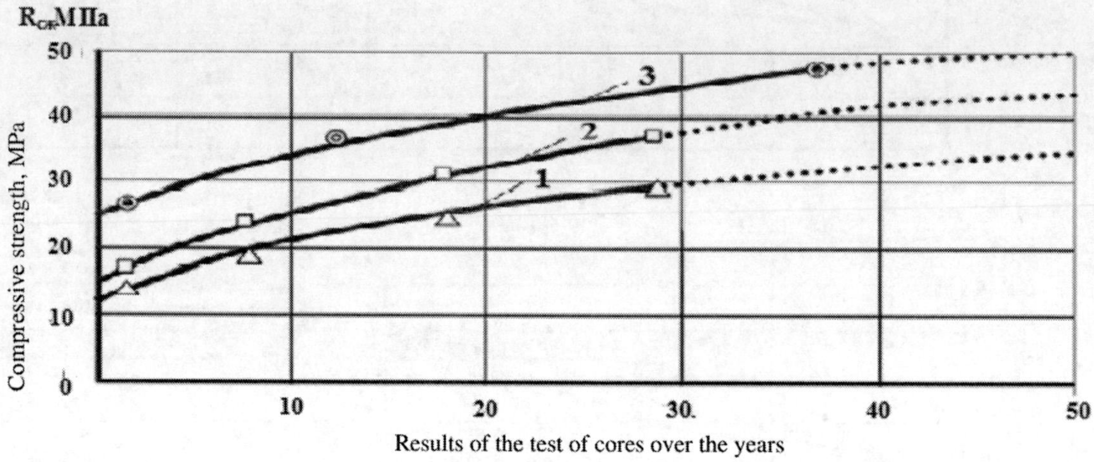

Figure 4. Kinetics of concrete strength of self-recovering concrete road bases of experimental roads, built in 1976–1984. The content of belite (C_2S) in belite cement: 1 – 60; 2 – 70; 3 – 80 %

The obtained experimental results suggest that from all theories of hardening of mineral binding agents, it is possible to highlight a colloid-chemical theory of V. Michaelis, which obviously will be more reasonable to explain the process of hardening of belite cements.

At normal temperature hydrosilicates C-S-H are formed as plate submicro-crystals, the average length of which is close to 10000 A° (1 micron), and the width and the thickness are 360–560 A° and 20–30 A°. Due to the very small sizes of hydrosilicates, as well as their ability to adsorb water, hydrosilicates have properties of colloids. These calcium hydrosilicates have bonds, different from the rigid crystallization bonds. The structure, formed by amorphous products, takes precedence over crystallization bonds, because it does not cause significant crystallization pressure and contacts are thermodynamically stable. Hydrosilicates of the type C-S-H have similarities with layered minerals of swelling clays. This similarity is shown in the ability to give some of the water (between the layers of the crystal lattice) reversibly. Loss or saturation of water is accompanied by a change of the distance between the layers of the crystal lattice of hydrosilicate C-S-H, which leads to changes in the material strength. Further maintaining of the material under wet conditions provides adsorption of moisture with gel, replenishment of bonding water films between the layers of the hydrosilicate lattice and restoring the strength of the material (Fig. 3–7). Therefore belite road concretes have the ability of self-recovering depending on the temperature and climatic changes and dynamic traffic loads. This is also confirmed by the change in the number of firmly bound water in the cement stone (Fig. 5).

The kinetics of change in strength over time (Fig. 3) and the kinetics of changing the number of firmly bound water (Fig. 5) in belite cement stone, depending on the temperature of samples, are similar, that confirms the validity of theoretical assumptions about the self-recovering properties of the belite cement.

When keeping the cement stone at low temperatures (samples of series 5–8), decrease in strength is accompanied by the displacement of firmly bound water from the fibrous formations in an amount of 10–30 % by weight of moisture, available in their capillaries. When storing samples (series 5–8) under normal conditions, a number of firmly bound water and the strength of the cement stone are restored within a month. Further keeping samples (series 5–8) under normal conditions within three months leads to increased strength and amount of tightly bound water unlike samples (series 2), hardened under normal conditions. This indicates a deepening of the hydration processes of the cement grains and increasing dispersion of formations at low storing temperatures, that also increases the strength of the cement stone (Fig. 3, 4).

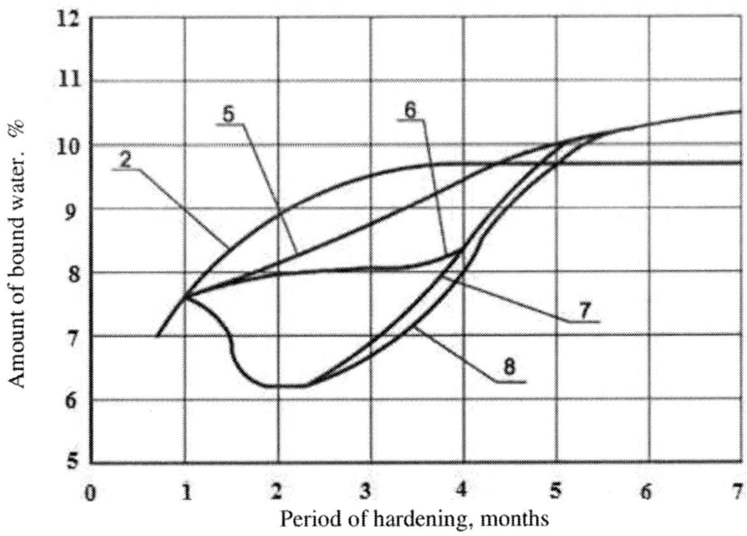

Figure 5. Kinetics of changes in the number of firmly bound water in the belite cement stone during hardening at different temperatures

When testing concrete samples at the age of 90 days to frost resistance, there were performed up to 200 cycles of freezing and thawing. As it can be seen from Fig.6, there is a slight decrease in strength due to squeezing of moisture out of the capillaries and reducing of its amount. With further exposure of the samples under standard conditions, concrete strength is fully restored and even exceeds the strength of 90-days concrete samples, shown in Fig. 3.

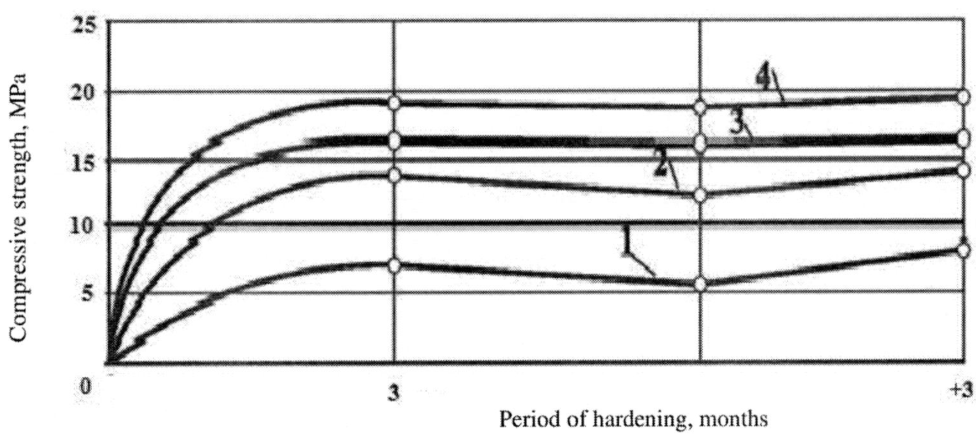

Figure 6. Self-recovering of strength of the road concrete, tested for frost resistance (MP3-200), depending on the amount of belite cement: 1 – 15; 2 – 20; 3 – 25; 4 – 30; mas. % of the cement with the content of C_2S 60–65 %

IOP Conf. Series: Materials Science and Engineering **832** (2020) 012019 doi:10.1088/1757-899X/832/1/012019

Road belite concretes have a slow hardening compared to traditional alite cements, but the strength characteristics of concretes at the age of 180 days are almost equal; the deformative indices of belite concrete even exceed the indexes of alite one. Moreover the tensile strength in bending is 31 % higher, and elastic modulus is below for 5000 MPa (Table 3).

Table 3. Comparison of indices of road cement concrete on the basis of alite cement and slowly-hardening concrete on the belite cement.

Composition of the road concrete, mas. %				Tensile strength at the age of 180 days, MPa (average value)			Elastic modulus
Fractions of rubble, mm 5–10	Fractions of rubble, mm 10–20	Sand M_{cr}=2.5	Cement, %	R_{compr}	$R_{bend.}$	R_{bend}/R_{compr}	E_y
15	34	29	Belite cement, 15 %	30.9	5.9	0.19	30000
15	34	29	Alite cement, M40 0.15 %	30.0	4.5	0.15	35000

High deformation properties of slowly hardening concrete evidence of a high dispersion and tensile strength (cohesive bond) of the new-formations of belite cements stone.

Presented X-ray diffraction, thermographic and electron microscopic studies confirm that in the slowly hardening road concretes, technological and operational advantages are provided by gelly structure of hardening of belite cements. Road concretes preserve their «self-recovering» properties due to water permeability of fibrous formations and continuing increase of hydration of the cement stone with an increase in strength during the years of roads' operation. In the structure of the traditional alite Portland-stone, contrary, low content of gelly dicalcium hydrosilicates fills only the free space inside the main frame, formed due to accretion of large crystalline hydrates. The basis of the frame is highly basic hydrosilicates of calcium, hydroaluminates, hydroaluminaferrites, hydrosulfoaluminates and calcium hydroxides. These crystal hydrates are hard leakproof, that prevents the penetration of moisture to the non-hydrated cement grains (microconcrete Jung is more than 40 %), and therefore does not have the property of self-recovery after fracture. The service life of concrete covers is not more than 25–30 years at their operation in the sharply continental conditions.

It was found that formation of mostly rigid crystallization bonds predetermines the limited terms of existence of thixotropic properties in the cement stone of the traditional Portland cements. This, in turn, reduces the productivity of complex road concreting machines, because it requires strict adherence to all technological operations and conditions in the linear-in-line road building.

5. Conclusions

The presented results of years of research and practical experience of monitoring the behavior of different road structures, built on the basis of the proposed innovative materials and technologies, as well as requirements to ensure the longevity of roads in accordance with modern concepts of the US and EU, allow us to make the following conclusions.

1) To improve the durability of roads up to 50 years or more, in accordance with modern concepts of the US and EU, it is advisable to use «self-recovering road concretes» on the basis of belite cements for the construction of the «support road bases with a high life expectancy». At this, updating the upper layer of the coating (made of various high-strength asphalt concretes) should be carried out every 5–7 years.

2) Do not exclude ways to improve the durability of cement concrete pavements by adjusting the mineralogical compositions of traditional Portland cements and increasing the number of hydrosilicates $\mathbf{C_2S}$, which will require changes to the requirements for road Portland cements and concretes on their basis.

3) The use of road constructions with increasing strength «upwards», in accordance with the new concepts, will require changes in design techniques and calculation of road pavements (non-rigid and rigid types).

4) The widespread use in the construction and reconstruction of roads of hydraulically active

technogenic wastes (fly ash from TPPs, phosphorus and blast furnace slags, bauxite sludges, phosphogypsum) and re-use of recycled road construction materials and soils will improve the quality of construction, reduce costs, improve the quality of the environment and ensure the environmental safety of road construction.

References

[1] Kochetkov A V, Kokodeeva N V, Rapoport P B et al, 2011 Calculation and design of road pavements *Highways* **12** 86–94
[2] Patkin I A, Poshekhonova T A and Rogachev P N 2011 *On the issue of new methods of assessment of cement concrete performance for road and bridge constructions.* Collection of works: Roads and bridges iss. 26/2 (Moscow: Rosavtodor) pp 295–310
[3] Asmatulayev B A, Tursumuratov M T and Asmatulayev R B 2012 Experience and prospects of the use of innovative technologies for soil stabilization during the construction of international highways in Kazakhstan *Transport and Roads of Kazakhstan* **1–2** 21–6
[4] Asmatulayev B A, Asmatulayev R B and Shestakov V N 2012 Theory and practice of innovative technologies in road construction in Kazakhstan *Mater. of the 66th Int. Sci. and Pract. Conf. B* **1** 33–38 (Omsk: Federal State Educational Institution of Higher Professional Education "SIBADI")
[5] Khristolyubov I N and Makhrov E Y 2012 Assessment of performance of road pavements with the bases from rubble-sand mixtures, fortified with mineral binders in conditions of Altai territory *Mater. of the 66th Int. Sci. and Pract. Conf. B* **1** 118–122 (Omsk: Federal State Educational Institution of Higher Professional Education "SIBADI")
[6] Radovskiy B S 2011 *The concept of the eternal road pavements* Catalog Road Construction Equipment pp 120–132
[7] WSDOT Pavement Policy 2011 (Washington: State Department of Transportation) 104 p
[8] Hall K and Dawood D 2007 *Long Life Concrete Pavements in Europe and Canada* Report no FHWA- PL-07-027 pp 1–84
[9] *Standard Specifications for Road, Bridge, and Municipal Construction* 2014 (Washington: DOT) 971 p
[10] Jackson N 2014 *Using Existing Pavement in Place and Achieving Long Life* (TRB, Washington D.C. SHRP 2) 133 p
[11] David E, Richard W and David H 2009 *Perpetual Asphalt Pavements, Asphalt Pavement Alliance*
[12] Ushakov V V 2002 Repair of cement concrete pavements of the highways Survey information *Information center of highways* **iss. 6** 28
[13] Nosov V P 2008 Causes of the formation of ledges on the cement concrete pavements of highways *Science and technology in the road sector* **iss. 3** 23–6
[14] Radovskiy B.S. 2006 *The problem of increasing the durability of road pavements and methods of its solutions in the US* Catalog Directory: Road Construction Equipment
[15] Asmatulayev B A, Bekbulatov Sh H, Asmatulayev R B and Tursumuratov M T 2013 Self-recovering road concretes. Prospects of use *Roads of CIS* **5(34)** 92–6
[16] Asmatulayev B A 2008 *RoK 218-66-2008. Recommendations for the construction of rigid road pavements with high-performance road machines* (Almaty: MTK KAD) 41 p
[17] Asmatulayev B A 2007 *RoK 218-61-2007. Recommendations regarding the selection of mixtures of road concretes for laying by formless technology* (Almaty: MTK KAD) 40 p
[18] Zolotarev V A 2012 Fifth Congress "Evroasfalt-Evrobitum" *Science and technology in the road sector* **4** 40–1
[19] Asmatulayev B A 1999 Construction of road pavements with a re-use of materials of reconstructed highways (Almaty: LLP "Evero") 212 p
[20] BS EN 13263-1 2005 *Silica fume for concrete* Part 1 *Definitions, requirements and conformity criteria* Part 2 *Confomiity evaluation*

[21] Standard specification for silica fume foruse as a mineral admixture in hydraulic – cement concrete, mortar, and grout 2014 *ASTM* 1240
[22] Lothenbach B K and Scrivener R D 2011 Supplementary cementitious materials Hooton Review Article *Cement and Concr. Res.* **41(1)**
[23] Krechetov V A 1991 *Roman concrete* (From the history of the building and construction techniques of ancient Rome) (Moscow: Stroyizdat) 111 p
[24] Asmatulayev B A., Asmatulayev R B., Tursumuratov M T et al 2009 *Cement clinker and the cement clinker production process* Patent of the Republic of Kazakhstan no 21157, published in 15.05.2009, Bul. no 5
[25] Asmatulayev B A, Nadirov N K, Asmatulayev R B et al 2015 *Self-recovering road concrete* Patent of the Republic of Kazakhstan no 29852, published in 15.05.2015, Bul. no 5
[26] Asmatulayev B A, Asmatulayev R B, Tursumuratov M T et al 2011 *The complex composition of the soil bases for road and airfield construction "Zhertas"* Patent of the Republic of Kazakhstan no 24251, published in 15.07.2011, Bul. no 7
[27] Asmatulayev B A, Nadirov N K, Asmatulayev R B et al 2007 *A method of construction of highways* Patent of the Republic of Kazakhstan no 18277, published in 15.02.2007, Bul. no 2
[28] Asmatulayev B A., Asmatulayev R B., Tursumuratov M T et al 2009 *Road concrete for coverings and bases of highways* Patent of the Republic of Kazakhstan no 20697, published in 15.01.2009, Bul. no 1
[29] Pshembaev M K 2006 *Rationale for the technical and technological parameters of the reconstruction of asphalt concrete pavements by using high-performance machinery* PhD dissertation thesis 25 p
[30] Asmatulayev B A, Sadykov Zh O and Pshembaev M K 2006 *RoK 218-56-2006. Recommendations for the use of cold recycling technology in the construction of road pavements, using active technogenic wastes and by-products* (Astana) 41 p
[31] Asmatulayev B A 2009 *The use of road concretes from slowly hardening binders in the construction of road pavements with improved surfaces in Kazakhstan* Collection (Moscow: MADI) pp 29–33
[32] Radovskiy B S 2015 Cement concrete pavements in the US: construction *Highways* **4(100)** 56–62
[33] Fuller W B 1992 The laws of proportioning concrete / Fuller, W.B. and S.E. Thompson *Trans. Am. Soc. Civ. Eng.* **33** 222–98
[34] Mehta P K and Aitcin P C 1990 Principles Underlying the Production of High- Performance Concrete *Cement, Concrete and Aggregates J.* **12(2)** 70–8
[35] Mehta P K 2004 High performance, high volume fly ash concrete for sustainable development *Proc. of the Int. Workshop on Sustainable Development and Concrete Technol.* (Berkeley, USA: University of California) p 12
[36] Richard P and Cheyrezy M 1995 Composition of reactive powder concretes *Cement and Concr. Res.* **25(7)** 1501–11
[37] Wille K, Naaman A E and Parra-Montesinos G J 2011 Ultra-High Performance Concrete with Compressive Strength Exceeding 150 MPa (22ksi) *A Simpler Way, ACI Mater. J.* **108(1)** 46–54
[38] Li Mo and Li V C 2009 Influence of material ductility on performance of concrete repair *ACI Mater. J.* September-October 419–28
[39] ACI. 212.3R-10 *Report on Chemical Admixtures for Concrete* 2010 61 p

Use of ceramic dust waste in the composition of road plates

T S Himich, S A Matveev, V A Utkin and G M Kadisov

Siberian State Automobile and Highway University, 5, Mira ave., Omsk, 644080, Russia

E-mail: dfsibadi@mail.ru

Abstract. The use of industrial waste – ceramic dust in the production of cement concrete mixtures is considered. A comparison of the characteristics of ceramic dust and building sand is done. The analysis of the physicomechanical characteristics of the cement-concrete mixture with a ceramic dust content of 10 and 50 % of the amount of building sand in the mixture was performed. Samples from this mixture were tested in a certified construction laboratory. It is shown that the use of ceramic dust in the composition of the cement-concrete mixture as a filler increases the average concrete strength up to 44 %, without significantly changing neither its structure, nor other physicomechanical properties. The possibility of using cement-concrete in construction practice, in which 50 % of sand is replaced by waste, has been proved.

1. Introduction

Waste management implies their complete or partial elimination, minimization or, where possible, reuse of materials that could otherwise become waste. Waste recycling or reuse is an essential element of sustainable resource management [1, 2].

Reuse of industrial waste in construction is widespread in the form of various additives and substitutes in building materials [3]. Thus, in the manufacture of concrete mixtures, granular blast furnace slag [4], spent polymers [5], and ceramic waste [6–9] are used. Wastes generated in the process of coal mining in the form of coarse-grained or finely dispersed fractions (coal dust) are widely used in the production of road-building materials. Along with these wastes, a by-product of foundry-quartz sand is actively used as a fine aggregate in concrete mixtures [10]. Full or partial replacement of natural aggregates in a self-compacting concrete mixture with industrial waste showed that replacing 50 or 100 % of natural aggregates with recycled concrete aggregates gives concrete with properties comparable to the reference sample [11].

The quality of the aggregate material has a significant impact on the physicomechanical properties of concrete, since aggregate occupies almost 70–80 percent of the total concrete volume. The influence of small aggregates on the properties of concrete is studied in articles [12–13]. The calculation method for determining the sand content in heavy concrete is presented in [14], where the calculation equations for determining the sand content in a mixture of concrete aggregates are substantiated. This ensures the lowest viscosity of the concrete mix and prevents water separation. The properties of concrete with various types of aggregates are studied in [15–19]. The mechanism of the effect of the particle size distribution of aggregate on the strength of concrete is analyzed in [20].

2. Materials and methods

One of the production wastes associated with the machining of metal products is ceramic dust or abrasive powder having a dusty state of aggregation. The starting material for it is a ceramic fraction.

IOP Conf. Series: Materials Science and Engineering **832** (2020) 012020 doi:10.1088/1757-899X/832/1/012020

Ceramic dust has a fourth class of environmental hazard and does not have hazardous properties. The waste is black and irregularly shaped particles. Its component composition is presented in table 1.

Table 1. The composition of ceramic dust

Component name	% by weight
Silica	44.06
Calcium oxide	19.57
Iron oxide	18.98
Aluminium oxide	8.66
Magnesium oxide	7.40
Titanium oxide	0.60
Nickel	0.1773
Manganese	0.4524
Zinc	0.0837
Copper	0.0102
Lead	0.0064

From the analysis of the table it follows that ceramic dust has a significant similarity with sand since it consists of 44 % silicon oxide.

The question arises about the possibility of using ceramic dust waste as a filler in the manufacture of road slabs, which are used not only in road construction, but also in the arrangement of any construction site.

A study of the physical and technical properties of the waste was carried out in the laboratory of Building Materials SibADI. A comparison was also made of the indicators with the requirements of GOST 8735-88 "Sand for construction work. Test methods" to sand class 2. The grain composition of ceramic dust is presented in table 2.

Table 2. Grain composition of ceramic dust

Name the residues	Residues,% by weight, on sieves 2.5	1.25	0.63	0.315	0.16	Pass through a sieve with a mesh No. 0.16, % by weight
Private	0.5	7.7	24.9	34.3	19.8	12.8
Full	0.5	8.2	33.1	67.4	87.2	100

From table 2 it is seen that the largest percentage of the content of particles of ceramic dust accounts for 0.315 sieve and is 34%. Ceramic dust modulus is $M_к$=1.96.

Comparative characteristics of ceramic dust and building sand are presented in table 3.

Table 3. Comparative characteristics of ceramic dust and building sand

Indicator	Ceramic dust value	GOST 8735-88 requirements for sand of the II class (small)
True density, kg / m^3	3100	not standardized
Bulk density, kg / m^3	1655	not standardized
Intergranular voidness	46.7	not standardized
Fineness modulus	1.96	1.5-2.0
Grain content (total residue on sieve No. 063 is finer than 0.16 mm)	33.1	10-30
The content of clay and dust particles	0.5	no more than 5

From the analysis of the comparative characteristics presented in table 3, it follows that sand of class 2 (fine) and ceramic dust have similar physical and mechanical properties and satisfy the

IOP Conf. Series: Materials Science and Engineering **832** (2020) 012020 doi:10.1088/1757-899X/832/1/012020

requirements of GOST 8735-88 "Sand for construction work. Test methods", which allows the use of ceramic dust as a fine aggregate for concrete mixtures, replacing it with fully or partially sand.

For testing, three concrete mixes were selected, from which samples were made of cubes of the reference (clean) concrete mix and samples containing 10 and 50 % ceramic dust instead of sand. The compositions of concrete mixtures are presented in table 4.

Table 4. The composition of concrete mixtures

Compositions	Pure mixture, kg	Mix with 10 % ceramic dust, kg	Mix with 50 % ceramic dust, kg
Cement	4,8	4,8	4,8
Sand	8,4	7,56	4,2
Ceramic dust	–	0,84	4,2
Crushed stone	21,24	21,24	21,24
Water	2,2	2,2	2,3

Testing of the samples was carried out in accordance with GOST 10181 - 2014 "Concrete mixtures. Test Methods". At the same time, the requirements of the following standards were taken into account: GOST 26633-2015 "Concrete heavy and fine-grained. Specifications", GOST 13015-2012 "Concrete and reinforced concrete products for construction. General technical requirements. Rules for acceptance, marking, transportation and storage", GOST 10060 – 2012 "Concretes. Methods for determining frost resistance".

3. The discussion of the results

The test results of samples of concrete mix are presented in table 5.

Table 5. Test results of samples made from concrete mix

Name indicator	Clean mixture	Mixture with 10 % ceramic dust	Mixture with 50 % ceramic dust	GOST 13015-2012 requirements
Cone draft, cm	4	4.3	4.1	not standardized
The average density of the concrete mixture, kg/m^3	2500	2520	2530	not standardized
The average strength of concrete after steaming, MPa	26.6	31.9	28.8	Not less than 70 % of concrete class
The average strength of concrete at the age of 28 days, MPa	29.5	40.7	42.5	Not lower than 26.19±4.5
Average abrasion of concrete mix, g/cm^2	0.17	0.125	0.1	No more than 0.7 g/cm^2
Concrete frost resistance	0.20	0.25	0.21	0.18–0.25 (F 200)

From the analysis of table 5 it follows that the sediment cone and the average density of the concrete mixture containing ceramic dust, slightly differ from similar parameters of the clean mixture, not exceeding the values of 7.5 and 3.6 %, respectively. The average strength of concrete increases with an increase in the percentage of ceramic dust in its composition. So the average concrete strength at the age of 28 days increases to 44 % when replacing 50 % of sand with ceramic dust. The rate of abrasion of a concrete mixture containing ceramic dust is reduced compared to the rate of a clean mixture, but still corresponds to the grade of abrasion G1 – (0.7 g/cm), established by the standard for structures operating in conditions of increased traffic intensity (road plates, sidewalk plates, etc.). The frost resistance of a concrete mixture with ceramic dust slightly exceeds the frost resistance of a clean mixture.

All the indicators shown comply with the requirements of GOST 13015-2012; therefore, all the mixtures presented are suitable for the manufacture of products for structures operating in conditions of increased traffic.

In accordance with GOST 33148-2014 "General automobile roads. intensity: sidewalk plates and road plates".

For the production of road slabs with non-tensioning reinforcement, the strength class of concrete must be at least B30 "Reinforced concrete road plates. Technical requirements". Concrete mixtures where part of the sand is replaced with ceramic dust, have just such a strength class, therefore they are suitable for the production of road plates.

4. Conclusions

1. The study of the physical and technical properties of ceramic dust showed that it can be used as a fine aggregate for cement concrete mixtures.

2. The developed composition of cement concrete with 50 % ceramic dust content as a fine aggregate has a strength class of B30, abrasion resistance of 0.1 g/cm^2 and in accordance with GOST 33148-2014 can be used for the manufacture of road plates.

3. The frost resistance indicator makes it possible to use road plates for arranging construction sites and laying roads in winter.

References

[1] Napier T 2016 *Construction Waste Management Retrieved* from: https://wbdg.org/recources/construction-waste-management

[2] Sokolov E M and Moskvichev Y A 2006 *Recycling of production and consumption waste.* (Yaroslavl: Yaroslavl State Technical University Publishing House) 388 p.

[3] Dvorkin L, Dvorkin O and Ribakov Y 2016 *Construction Materials Based on Industrial Waste Products* (New York: Nova Science Publishers) 242 p.

[4] Kononova O V, Anisimov S N, Smirnov A O and Leshkanov A Y 2016 The effectiveness of granulated blast furnace slag in concrete with an additive based on polycarboxylate ether *Modern high technolog.* **6-2** 259–63

[5] Galvão J C A, Portella K F, Joukoski A, Mendes R and Ferreira E S 2010 *Use of Waste Polymers in Concrete for Repair of Dam Hydraulic Surfaces* (Elsevier) pp 1049–54

[6] Juan A, Medina C, Ignacio Guerra M, Morán J M, Aguado P J, Sánchez de Rojas M I and Frías Mand Rodríguez O 2012 *Re-Use of Ceramic Wastes in Construction Ceramic Materials* (Rijeka, Croatia: Sciyo) pp 197–211

[7] Higashiyama H, Yagishita F, Sano M and Takahashi O 2012 Compressive Strength and Resistance to Chloride Penetration of Mortars Using Ceramic Waste as Fine Aggregate *Construct. and Building Mater.* **26** 96–101

[8] Suzuki M, Meddah M S and Sato R 2009 Use of Porous Ceramic Waste Aggregates for Internal Curing of High Performance Concrete *Cement and Concrete Res.* **39** 373–81

[9] Puertas F, Garcia-Diaz I, Palacios M, Gazulla M F, Gomez M P and Orduna M 2010 Clinkers and Cements Obtained from Raw Mix Containing Ceramic Waste as a Raw Material: Characterization, Hydration and Leaching Studies *Cement and concrete composites* (Elsevier) 175–86

[10] Mishra B and Mishra R S 2015 A Stud on Use of Industrial Wastes in Rural Road Construction *Int J. of Innovat. Res. in Sci., Engineer. and Technol.* **4(11)** 10387–98

[11] Kenai S, Menadi B, Debbih A and Kadri E H 2014 Effect of Recycled Concrete Aggregates and Natural Pozzolana on Rheology of Self-Compacting Concrete *Key Engineer. Mater.* **600** 256–63

[12] Kronlof A 1994 Effect of very fine aggregate on concrete strength *Mater. and Structures* **27(1)** 15–25

[13] Muhit I B, Haque S and Alam M R 2013 *American J. of Civil Engineer. and Architect.* **1(5)** 103–6 DOI: 10.12691/ajcea-1-5-3

[14] Dvorkin L I 2018 *Magazine of Civil Engineer.* **7** 186–97

[15] Kalashnikov V I, Tarakanov O V, Kuzneczov Y S, Volodin V M and Belyakova E A 2012

Magazine of Civil Engineer. **8** 47–53
[16] Sivakumar N, Muthukumar S, Sivakumar V, Gowtham D and Muthuraj V 2014 *Int. J. of Engineer. and Sci.* **4(1)** 27–36
[17] Abdullahi M 2012 Effect of aggregate type on Compressive strength of concrete *Int. J. of Civil. and Structur. Engineer.* **2(3)** 791–800
[18] Liu J, Wu K, Wang Y and Yang Y 2017 Effects of fly ash diatomite admixture with variable particle sizes on the mechanical properties and porosity of concrete *J. of Wuhan Univer. of Technol.-Mater.* **32**(5) 1072–9
[19] Zimbili O, Salim W and Ndambuki M 2014 A Review on the Usage of Ceramic Wastes in Concrete Production *Int. J. of Civil. and Environmental Engineer.* **8(1)** 91–5
[20] Wu J, Feng M, Meo X and Xu J 2018 *J. Construct. and Building Mater.* **193** 295–311

Application of the speed prediction model (V_{85}) (V85) in the design of freeway interchanges

R M Jafarov and V P Zaluga

Moscow Automobile and Road State Technical University (MADI), 64, Leningradsky ave., Moscow, 125319, Russia

E-mail: dzhafaroff89@gmail.com

Abstract. The design of freeway interchanges is performed on the basis of technical and economic comparison of options for a large number of indicators. Design decisions depend on a large number of factors: the class of intersecting streets and roads, the level of service, traffic intensity, the estimated speed at the ramps, terrain, urban development, hydrogeological conditions of construction, etc. An integral indicator of the quality of design decisions is the maximum safe speed. In most countries of the world, the speed of single passenger cars is 85 % of the security (V_{85}), which may differ from the calculated speed at the maximum safe speed. In the article, on the basis of experimental data, empirical models for calculating speeds (V_{85}) are obtained at freeway interchanges with directional ramps. Experimental studies were performed at freeway interchanges in the city of Moscow using a car laboratory according to the method of following the leader, as well as by processing video from a quadrocopter. The results of the study make it possible to predict the speed of movement (V_{85}) at the ramps of freeway interchange and to compare the freeway interchanges schemes according to the criterion of speed, loss of time, relative safety and traffic capacity.

1. Introduction

The effectiveness of design solutions for freeway interchange depends on a large number of factors; the key ones are throughput, security, time loss, and capital costs. Mathematically, this position can be written in the form of the following expression:

$$\sum_1^n(Vn; Pn; Kb) \to \max, \text{ with } \sum_{n=1}^i(Tn; E) \to \min \qquad (1)$$

where Vn is the speed at the exit section, km / h; Pn – throughput of the exit section, vehicles/h; Kb – accident rate at the ramp section; Tn is the ramp time, s; E – capital expenditures for the construction of the ramp section, million rubles; n is the number of ramp sections, pieces.

Let us give a brief description of each variable presented in formula (1).

Travel time (T_n) – the travel time on the ramp depends on the path parameters (S_n) and speed (V_n).

The path (S_n) is the length of the ramp section, which depends on the layout of freeway interchange and the dimensions of the geometric elements in the plan and longitudinal profile, which are assigned depending on the category of intersecting streets and roads (read the estimated speed – V_r), and also local accommodation conditions. The path parameters (S_n) also determine capital investments (K_t).

Speed (V_n) – is a comprehensive indicator of the quality of the transport process when driving along ramps of a freeway interchange. The combination of geometric elements in plan and profile, the width of the carriageway, as well as the road situation (signs, markings, fences, etc.), form the initial data for the choice of speed by the participants in the movement.

Capital investments (K_t) – the cost of construction, which also depends on the path length (Sn), local accommodation conditions (land acquisition area, demolition of buildings, reconstruction of utilities, deforestation, etc.), type and size of artificial structures, additional costs for the organization of construction, etc.

Relative traffic safety (K_b) – characterizes the degree of constancy in the behavior of the driver when passing adjacent sections at ramps. This parameter can be estimated by the method of traffic safety factors proposed by Professor V.F. Babkov [1]. The methodology is based on an assessment of the degree of reduction of the maximum safe speed of 85 % security in the study area ($V^{account}{}_{85}$) relative to the maximum safe speed on the approach to this site (Vex $_{85}$).

The traffic freeway interchange capacity (P) is determined by the capacity of its individual ramps. The main reason for the decrease in the traffic capacity and the formation of congestion at the ramps is the presence of sections of a sharp decrease in speed (this issue is not considered in detail in this article).

From the foregoing, it can be concluded that all key indicators (Pn, Kb, Tn) can be estimated by knowing the maximum safe speed ($\mathbf{V_{85}}$) in the appropriate road conditions.

2. Statement of the problem

It was found that drivers choose the high-speed mode in free conditions based on the balance between the minimum time and safety. In world practice, the speed of movement that characterizes a transport structure from the point of view of maintaining a balance between minimum time and maximum traffic safety is taken as the speed of movement of single passenger cars in free conditions of 85% of security (V_{85}), i.e. a speed that exceeds only 15 % of observed cars.

The choice of speed at the exits of freeway interchanges in free conditions is affected by:

- the value of geometric elements (the radius of the curve in the plan is r, the angle of rotation is α; the longitudinal slope is i, the distance of visibility is L_{view}, the width of the carriageway is B_{floor}, etc.);
- arrangement of elements of arrangement (road signs, marking, fencing and much more);
- experience and psychophysiological characteristics of the driver; natural and climatic factors;
- the operational condition of the carriageway.

However, there are key factors that are laid at the stage of development of the project and which have the greatest impact on the value of the speed chosen by the driver in free conditions:

- the radius of the curve in the plan (r);
- longitudinal slope (i);
- the angle of rotation (α).

At the moment, there are a large number of empirical dependencies for calculating the speed of movement V_{85} [2–8]. Most of the dependencies were obtained between 1980 and 2008 in the conditions of country two-way roads.

For traffic conditions at the ramp sections of freeway interchange, the work of scientists from China is known today [9], who in 2014 proposed empirical models for calculating the speed V_{85} depending on the parameters of the curvature of the route (rad / km) and the amount of slope of the turn (%) for sections of deceleration, uniform movement and acceleration at the ramp. However, it is not clear from the work to which type of ramp sections these dependencies (flyover or tunnel) are applicable and how to determine the speed in a straight section in a plan with zero curvature.

The analysis shows that all existing models for calculating the speed of movement have little convergence with the results obtained in Russia. The reason is the difference in road conditions, the difference in approaches to the designation of the geometric parameters of the ramps, the composition of the traffic flow, climatic conditions and driving style.

3. Materials and Methods

To obtain reliable empirical models for calculating the velocities V_{85} in the period of 2015-2019, field studies were carried out for 10 freeway interchanges, including 18 ramps with a different combination of geometric elements in Moscow and the Moscow region.

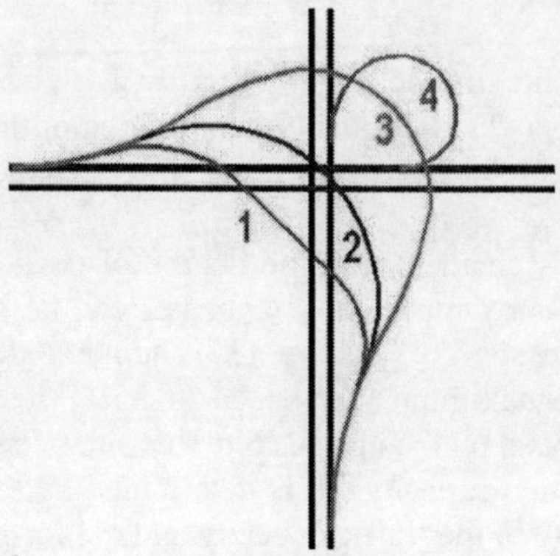

Figure 1. Planning schemes for freeway interchanges' ramps: 1–3 – semi-directional ramp; 4 – loop ramp

All the ramps were of a flyover type, with 2 lanes, the total width of the carriageway: 10.4 m $\leq$ b $\leq$ 11.2 m, the radius of the curve in the plan: 30 m $\leq$ r $\leq$ 270 m, longitudinal slope: – 55 $\leq$ I $\leq$ + 50 ‰, the angle of turn of the track $40^0 \leq \alpha \leq 180^0$, the transverse slope of the turn: 20 ‰.

Velocity measurements were performed using a laboratory car, which was equipped with a high-frequency (10 Hz/s) GPS logger with 2 cameras of the English company Racelogic. This equipment made it possible to continuously record diagrams of speeds, accelerations (longitudinal and transverse) and trajectories of movement of single cars in free conditions along the entire route along the ramp. Using specialized software, Circuit Tools and Vbox Tools, velocity plots and other motion characteristics were analyzed (Fig. 2).

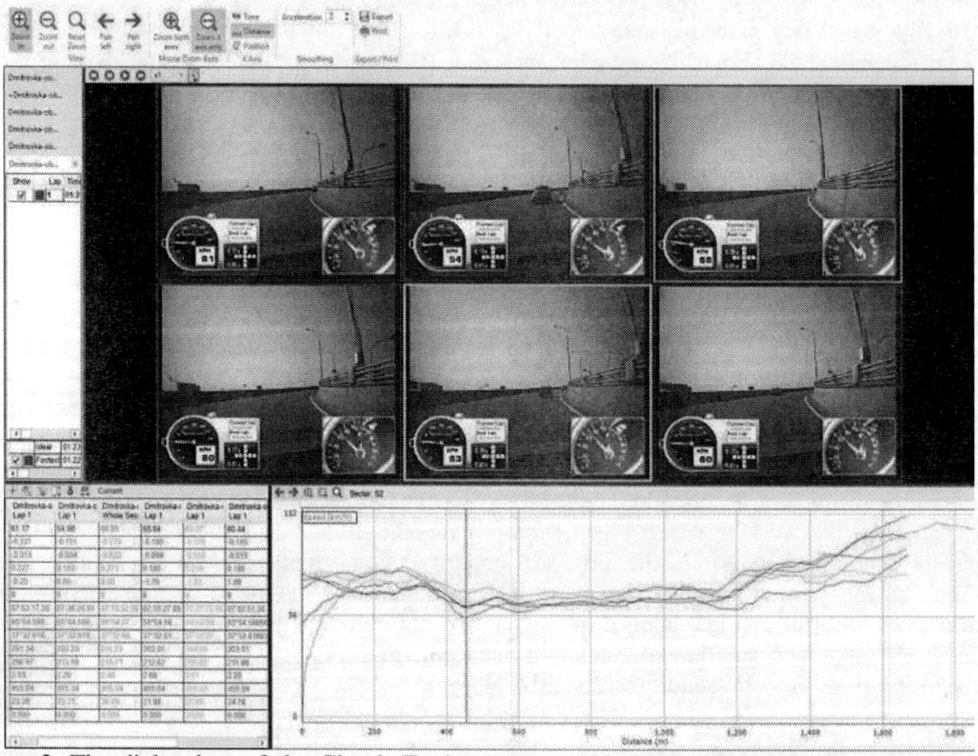

Figure 2. The dialog box of the Circuit Tools program with the results of 6 rides at the freeway interchange at the intersection of Dmitrovskoye Highway and Moscow Ring Road in Moscow

From the diagrams of motion velocities (Fig. 1), characteristic points of variation of the motion velocities (extrema in the velocity diagram) were revealed, including the midpoints of the curves in the plan (minimum speeds of motion), the points of the beginning of braking in front of the curves in the plane, and the points of the beginning of acceleration at the exit from curves, as well as sections of maximum speed on straight sections of ramps.

To increase the objectivity of the experimental data, additional studies using a quadrocopter were also conducted. The video recording from the quadrocopter was processed using the Sony Vegas Pro 13 program, which allowed frame-by-frame scrolling of the video (25–30 frames per second) and with high accuracy to determine the travel time between the characteristic points recorded on the video.

By regression analysis of the experimental data, cumulative curves of the distribution of movement speeds at the characteristic points of the ramps were obtained (Fig. 3).

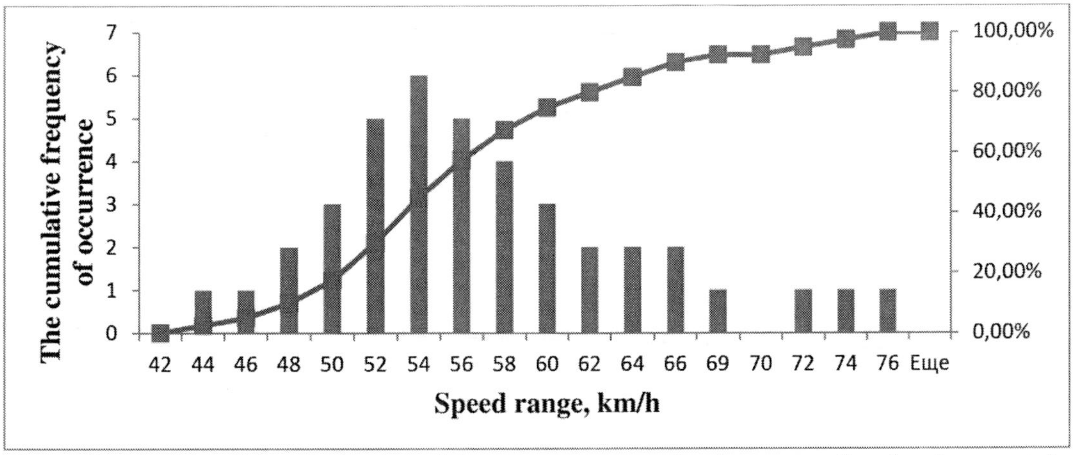

Figure 3. Schedule of the distribution of free movement speeds of cars at a directional ramp at the intersection of the Moscow Ring Road and Leninsky Prospekt in the city of Moscow at a characteristic ramp point with geometric parameters: 1 angle, α1 = 44 °, R1 = 100 m, i = +30 ‰.

Based on the results of processing the experimental data, graphs of the dependence of the velocity V_{85} were constructed for various curve radii in the (r) plane (Fig. 4).

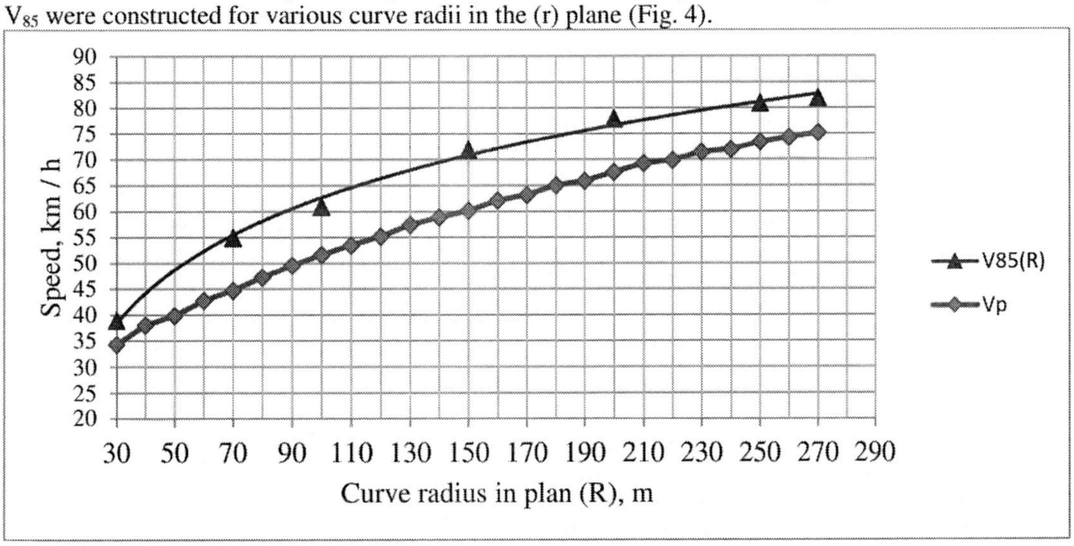

Figure 4. The graph of the dependence of the speed of movement (V85) on the radius of the curve in terms of the radius (r), V_p – calculated speed

4. Discussion of the results

The data shown on the graph (Fig. 3) can be described in the form of mathematical formulas:

$$V_{85} = 20.162\ln(r) - 30.163 \quad R^2 = 0.995 \tag{2}$$

$$V_p = (127r(\mu \pm i_{in}))^{1\backslash 2}$$

where r is the radius of the curve in the plan; i_{in} – a bias of a turn, ‰; μ is the shear force coefficient adopted according to table 1; R^2 – the accuracy of the approximation, characterizing the convergence of the proposed model with experimental data.

Table 1. The values of the shear force coefficient (μ) in the Russian Federation for roads outside settlements, depending on the estimated speed

Estimated speed (V_r), km/h	150	120	100	80	60	50	40	30
Shear coefficient (μ)	0.08	0.09	0.12	0.14	0.17	0.19	0.23	0.28

The graphs show that the speed (V85) at the ramps is 9–20 % higher than the calculated speed (VR). Formula (2) is true for the parts of the curves in terms of radius 30m < r < 270 m with transitions normative long, with an angle 40° < α < 60°, the longitudinal slope of 30 ‰ < I < +30 ‰ for two-lane roadway cross-slope (super elevation) 20 ‰ < IB < 30 ‰.

The influence of the longitudinal slope (i) can be taken into account by introducing the following correction factors:

- when the angle of rotation α < 60° and the slope on the approach to the curve (r) on the rise up to +30 ‰ < I ≤ +50 ‰ and the curve position on a slope in the range of +30 ‰ < I ≤ +40 ‰ coefficient equal $K_{prod} = 1.03$;
- when the rotation angle is α < 60 ° and the slopes after the curve (r) on the descent are within –50 ‰ ≤ i < –30 ‰ and the curve is located on the slope within –40 ‰ ≤ i < –30 ‰, the coefficient is $K_{prod} = 0.97$.

The influence of the rotation angle (α) in combination with the longitudinal slope (i) can be taken into account by introducing the following correction factors:

- when the rotation angle α > 60 ° in the middle of a circular curve located on longitudinal slopes –30 ‰ ≤ I ≤ +30 ‰, the coefficient is $K_{\alpha i} = 0.97$;
- at an angle of rotation α> 60 ° at the beginning of a circular curve in the lifting section with a slope of +30 ‰ < I < +55 ‰ the coefficient is $K_{ai} = 0.94$;
- when the rotation angle α > 60 ° at the end of the circular curve in the descent section with a slope of –30 ‰ < i < –50 ‰, the coefficient is $K_{\alpha i} = 1.06$.

Due to the fact that the plan of the freeway interchange ramps consists not only of curves of different radius (r), but also sections of lines, the empirical equations for determining the speed (V_{85}) depending on the value of the longitudinal slope (i) and the radius of the curve in plan (r) at the end of the line were also obtained (3–5).

With a longitudinal slope of rise +40 ‰ < I < + 55 ‰ the speed on the straight section is:

$$V_{85} = 0.0786r + 72.571 \quad R^2 = 0.9912 \tag{3}$$

With a longitudinal slope of rise + 40 ‰ < I < + 55‰ the speed on the straight section is:

$$V_{85} = 0.11r + 64.167 \quad R^2 = 0.9973 \tag{4}$$

With a longitudinal slope of rise +40 ‰ < I <+55 ‰ the speed on the straight section is:

$$V_{85} = 0.0688r + 79.533 \quad R^2 = 0.9888 \tag{5}$$

To determine the position of the extremum point (maximum speed) and the section with constant speed at the ramp, it is necessary to know the magnitude of the longitudinal accelerations (a_{ij}, m/s^2), which depends on the degree of change of speed ($\Delta V_{ij} = V_j - V_i$) and the values of the longitudinal slope.

Deceleration or acceleration on a straight section with a longitudinal slope of +30‰<i<+55‰ after leaving the main road, depending on the speed difference ($\Delta V_{i\text{-}j}$) is equal to:

$$a_{ij} = 0.0697\,\Delta V_{ij} + 0.0696 \text{ at } V_i < V_j \quad R^2 = 0.6524 \tag{6}$$

$$a_{ij} = -0.079\ \Delta V_{ij} - 0.0208 \text{ at } V_i > V_j \quad R^2 = 0.8591 \tag{7}$$

The deceleration in a straight section with a longitudinal slope of +30‰ < I <+ 55 ‰ before the curve of any radius (r) depending on the speed difference (ΔV_{i-j}) s equal to:

$$a_{ij} = -0.0342\ \Delta V_{ij} + 0.1049 \qquad R^2 = 0.7849 \tag{8}$$

The acceleration in a straight section with a longitudinal slope of –20 ‰ < i < +20 ‰ after a curve of any radius (r) depending on the speed difference (ΔV_{ij}) is equal to:

$$a_{ij} = 0.0231\ \Delta V_{ij} + 0.1776 \qquad R^2 = 0.6589 \tag{9}$$

The deceleration in a straight section with a longitudinal slope of +30‰<i<+55‰ before the curve of any radius (r) depending on the speed difference (ΔV_{i-j}) s equal to:

$$a_{ij} = -0.0315\ \Delta V_{ij} - 0.1096 \qquad R^2 = 0.6068 \tag{10}$$

The acceleration in a straight section with a longitudinal slope of –20 ‰ <i <+20 ‰ after a curve of any radius (r) depending on the speed difference (ΔV_{ij}) is equal to:

$$a_{ij} = 0.0315\ \Delta V_{ij} + 0.1664 \qquad R^2 = 0.7482 \tag{11}$$

To determine the length of the section of uniform movement with constant speed, it is necessary to calculate the linear position of the point at the ramp at which the extreme (maximum) speed value $V_{85}$is reached:

$$S_{ij} = (V_j^{\ 2} - V_i^{\ 2})\,(26a_{ij})^{-1}. \tag{12}$$

Let us give an example of plotting the motion velocities V_{85} for a directed ramp at the intersection of Leninsky Prospekt and the Moscow Ring Road in Moscow The design scheme is shown in Fig. 5.

Figure 5. The design scheme for plotting traffic speeds at a directional ramp at the intersection of Leninsky Prospekt and the Moscow Ring Road in Moscow

The methodology includes the following steps:

- Step 1. Determine the characteristic points of change in speed at a directional ramp.
- Step 2. Calculate the V85 speeds at characteristic points at the ramp using formulas (2–5).

- Step 3. Calculate the lengths of the characteristic sections (S) using formula (12) using formulas (6–11).
- Step 4. Plotting in MS Excel, as shown in Fig. 6.

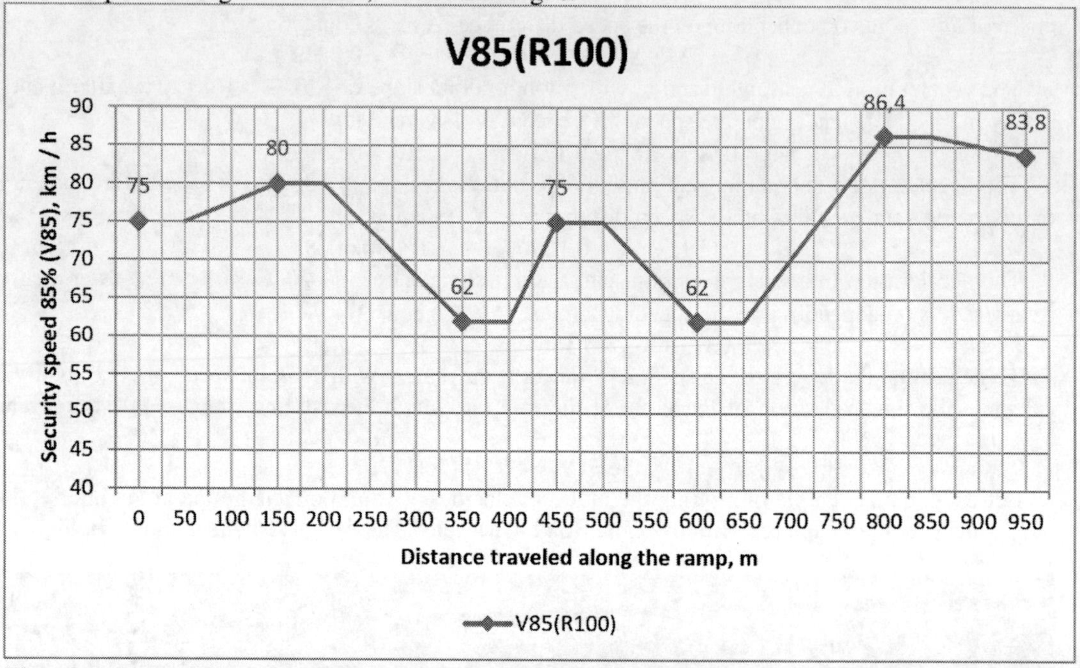

Figure 6. Let us give an example of plotting the motion velocities V_{85} for a directed ramp at the intersection of Leninsky Prospekt and the Moscow Ring Road in Moscow

5. Conclusion

It has been established that an integral indicator of the quality of the transport process is the maximum safe speed equal to the speed of 85 % of security (V_{85}). The choice of speed of movement is influenced by many factors, but the greatest influence during free movement is exerted by the radius of the curve in the plan (r), the longitudinal slope (i), the angle of rotation (α) and, as a result, the distance of visibility in the plan (L_{view}).

It is proved that the developed methodology for predicting the speed of movement (V_{85}) is effective in the techno-economic comparison of planning schemes of freeway interchanges by the criterion of travel time, relative traffic safety and relative traffic capacity (this issue was not considered in this article), as well as in solving a wide range of tasks of transport planning and traffic management.

References

[1] Babkov V F 1993 *Road conditions and traffic safety* (Moscow: Transport)
[2] Liapis ED, Psarianos B and Kasapi E 2001 Speed Behavior Analysis at Curved Ramp Sections of Minor Interchanges *Transportation Research Record* **1751** 35–43
[3] Abdul-Mawjoud Ayman A and Sofia Gandhi G 2008 Development of models for predicting speed on horizontal curves for two-lane rural highways *The Arabian J. for Sci. and Engineer.* **33(2B)**
[4] Cantisania G and Di Vitoa Michele 2012 CCV: A New Model for S85 *SIIV – 5th Int. Congr. – Sustainability of Road Infrastructures* (Rome, Italy: University of Roma La Sapienza Department of Civil & Environmental Engineering)
[5] Castro M 2010 Speed models for highway consistency analysis *A Colombian case study. 4th Int. Symp. on Highway Geometric Design* (Valencia: Universidad Politécnica de Madrid)
[6] Collins K M and Krammes R A 1996 Preliminary Validation of a Speed-Profile Model for

Design Consistency Evaluation *Transportat. Res. Record* **1523** 11–21

[7] Fitzpatrick K and Collins J 2000 Speed profile model for two lane rural highways *Transportat. Res. Record* **1737** 42–9

[8] Syed Khairi Syed Abbas, Muhammad Akram Adnana and Intan Rohani Endut 2011 Exploration of 85th Percentile Operating Speed Model on Horizontal Curve: A Case Study for Two-Lane Rural Highways *Procedia Social and Behavioral Sci.* **16** 352–63

[9] Zhi-yong Zhang, Xiao-yun Hao and Wen-bin Wu 2014 Research on the Running Speed Prediction Model of Interchange *Procedia Social and Behavioral Sci.* **138** 340–9

Prospects for estimating the pavements structural capacity based on rolling wheel deflection

B M Khroustalev[1,4], U A Veranko[1,4], V V Zankavich[1], A U Busel[1,5], B Shang[2], J Shi[3] and Yu Liu[2]

[1]Henan center for outstanding overseas scientists, Henan Gaoyuan highway maintenance technology Co., Ltd, № 6 Gaoyuan road, Xinxiang, 453003, China
[2]National engineering laboratory for highway maintenance equipment, № 6 Gaoyuan road, Xinxiang, 453003, China
[3]Henan provincial key laboratory of highway detection and maintenance, № 6 Gaoyuan road, Xinxiang, 453003, China
[4]Science and technology park of Belarusian national technical university «Polytechnic», 24 Ya. Kolasa str., Minsk, 220013, Belarus
[5]State enterprise «Belarusian road scientific and research institute «BeldorNII», 60 4th Zagorodnyi ln., Minsk, 220073, Belarus

E-mail: vitali.zankavich@chngaoyuan.com

Abstract. This article contains the results of the study of elastic deflection of semi-rigid pavement defined by the method of a rolling wheel (RWD) under different axial loads. For the investigations, an exclusive mobile complex was developed, allowing measurements to be made in a motion for various axial loads from 100 to 250 kN. It has been observed that the dynamic method of testing a semi-rigid pavement with a rolling wheel with different load levels allows determining non-linear characteristics of the deformation curve and deformation features of the package of asphalt concrete layers with appropriate loads, that cannot be done with the standard method of estimating the elastic deflection under a load of 100 kN. The results of the study can be used to develop criteria for assessing the strength, reliability, and deformability of semi-rigid pavement. It is possible to determine the likelihood of a joint failure-free operation of a rigid base and non-rigid layer with a resource forecast and the feasibility of its cold recycling.

Keywords: recycling, elastic deflection, reliability, residual service life, rolling wheel deflection method, semi-rigid pavement

1. Introduction

Nowadays, more and more attention regarding the practice of repair and reconstruction of roads is being paid to the reuse of materials of the structural layers of old pavements, in connection with which both technological and material research methods of increasing the reliability and efficiency of their regeneration processes are being developed.

Technological solutions for the regeneration of exploitation ability in terms of the life cycle of pavements (layers) have their advantages and disadvantages in terms of their implementation and achievement of maximum technical and economic efficiency indicators. Recently, cold pavement regeneration technologies, both through the central asphalt concrete plant and in-situ, have become very popular [1, 2].

Most often, bitumen emulsions or foamed bitumen are used as a regenerating substance or a binder [3–6], in the proportions determined at the design stage of the composition of cold mixes. Portland cement, fly ash or quicklime are added to increase the stability of cold mixes and achieve higher structural characteristics of concretes prepared on their basis in road pavements.

In most cases, to increase the reliability of the newly constructed in this way pavements, the reinforcement layers are made from hot and warm asphalt concrete mixes, or protective layers (from floated emulsion-mineral mixes, in the form of surface dressing, impregnations, etc.).

The most important stages of the cold pavement recycling system are its diagnostics with the development of optimal methods for regenerating the properties of the materials of the old structural layers and making a new design [7]:

- collection of initial data (historical data, service life, traffic volume, types of traffic loads, etc.);
- preliminary studies and selection of pavement sections with an identical condition;
- detailed studies of the pavement condition on the equal sections;
- the primary design of new pavement based on the predicted properties of the regenerated materials of the structural layers;
- laboratory design of the compositions of the regenerated materials for designing structural layers of new pavement;
- final design and calculation of the new pavement;
- technical and economic analysis of options for cold recycling of an existing pavement.

The primary indicator that currently determines the structural strength of existing pavement in the process of diagnostics is elastic deflection. This is due to the simplicity of the experiment and great difficulties in the experimental evaluation of stress fields and deformations of a real road structure.

As a result of the combined effect of traffic loads and weather and climatic factors, stress and strain occur in the structural layers of the pavement. The durability of the whole structure depends on the magnitude of these stresses and the reaction of payment layer materials to them. The ability to accurately estimate stresses and deformations during the entire service life of pavement at the design stage is the crucial point of most modern methods of its design and calculation.

The stresses and deformations that occur in the structural layers of the pavement, due to the complexity of the properties of the materials from which they are constructed, depending on many factors, including the load time and temperature. Calculating them analytically requires solving many equations. It can be reasonably stated that at present, it is theoretically impossible, with a high level of convergence, to perform calculations of the pavement resource during its life cycle.

Due to the described difficulties of accurately determining stresses and deformations by calculation over decades, most countries have used empirical pavement design techniques that were based on experimental results (such as AASHTO road tests in the USA in 1962, 1965), or the long-term observations for the behavior of various pavements during the whole period of their service. Many pavement calculation schemes adopted abroad are based on methodologies centered around these studies, and the lack of theoretical ideas about the stress-strain state is compensated for by a massive amount of experimental data. For example, even in the USA, only in 2002, pavement calculation methodologies that did not directly consider research data were developed [8].

Empirical dependencies in the form of equations, graphs, nomograms, link some measured indicators of the pavement (most often the value of the elastic deflection on the surface of the road) with the service life. Generally, these dependencies are poorly correlated with the results beyond the limits of the data used for developing.

In many countries, so far, the design criterion assumes that the road structure meets the requirements of strength and reliability in terms of the elastic deflection [9-11]. The value of the minimum total required module of elasticity of the structure is calculated by empirical dependencies on the traffic volume at the end of the service life, or on the total number of load applications over the entire service life. Only elastic deflection can be tested experimentally in a simple and accessible way.

The value of deflection on the road surface is taken as its vertical deformation under the applied static or dynamic load. The most sophisticated measuring instruments record vertical deformations at many points, which makes it possible to obtain the outlines of a "deflection bowl". The "deflection bowl" is the area of the layer that is deformed when a load is applied.

All tools used to determine the deflection on the layer can be divided into two categories:

- Static;
- Dynamic.

The primary tool for the static determination of the elastic deflection is the lever-type deflection indicator, or "Benkelman beam". Among the dynamic methods, there are tools operating on the principle of the vibration load application and the principle of a falling load or rolling wheel.

2. Studies of the pavement elastic deflection

Despite quite complicated methods of bringing the measured values of elastic deflection to standard conditions, very often they are not enough for proper application of design solutions for major repairs and reconstruction, since it is difficult to calculate the residual service life of materials of structural layers, their resistance to deformations under the influence of modern traffic loads. Fig. 1a shows an example of a typical (phenomenological) curve (model) of changes in deflection during the pavement life cycle when the development of defects may not occur predictably, but rather incrementally while ensuring the required strength indicators, determined based on the estimated value of the elastic deflection.

In this regard, there is also a difficulty with a reasonable choice of areas that are most suitable for cold recycling of pavements. The situation is even more complicated when the bases are composed of materials containing mineral binders (concrete of varying strength, reinforced materials), for so-called semi-rigid road structures with non-rigid (usually asphalt concrete) layers and with rigid (reinforced by mineral binders) base layers.

This is primarily due to the accumulation of intrastructural damageability in the asphalt concrete layers and layers reinforced with mineral binders. Damageability is the accumulation of defects in the structure of the material of the pavement structural layer in the course of its operation that leads, when there are too many defects, to unacceptable deformation or destruction. The development of damageability takes place in three stages and leads to a decrease in the strength and deformation characteristics of the materials of the road construction layers:

1. The destruction at the micro-level and sub-micro level, which cannot be fixed based on existing methods and tools.
2. The formation of micro-defects and the development of intrastructural damageability, which can be set with special tools and pursuant to observations for several years.
3. The development of macro-defects which can be recorded visually (i.e., defects that are visible on the layer).

By area of coverage, all three stages are uneven. Therefore, it is important to have appropriate monitoring in order to identify priority areas for prevention and proper repair.

Thus, the existing method of assessing the strength of pavement by measuring deflections and determining the module of elasticity is quite simple, however, it is often ineffective in terms of taking into account the actual service life reserve, and does not allow for simple and adequate dependencies linking the module of elasticity of the pavement with indicators of operational condition pavement such as cracking (Fig. 1b), roughness (Fig. 1c), rutting (Fig. 1d), etc. [12].

Diagnostic methods based on visual fixation of defects and measurement of elastic deflection only state the actual condition of the pavement but do not allow determining the reason for the defects that have appeared and the forecast of their development over time, which makes it difficult to assign the type and term of repair correctly. Errors may be even higher in the study of semi-rigid pavement. Another disadvantage of existing methods for estimating deflections, especially when assessing the strength of pavements with increased reliability (including semi-rigid), is often a small value of deflections from a standard load of 100 kN (within the experimental error).

IOP Conf. Series: Materials Science and Engineering **832** (2020) 012022 doi:10.1088/1757-899X/832/1/012022

It is known that one of the most effective ways to determine the elastic deflection of pavement is to dynamically measure from a rolling wheel (RWD) [13–17], as the vehicle's effect on the pavement is simulated to the maximum. This method allows obtaining much more information about the characteristics of the road surface. We have developed a mobile complex that allows measurements to be made in a motion for various axial loads from 10 to 25 tons (Fig. 2) based on a laser Doppler measurement system.

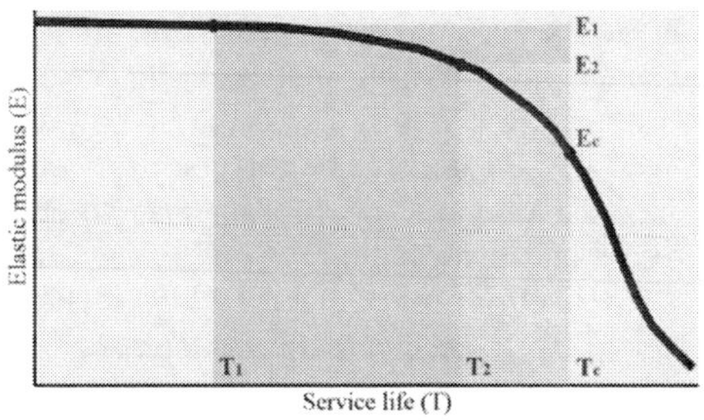

a) Elasticity modulus over service life

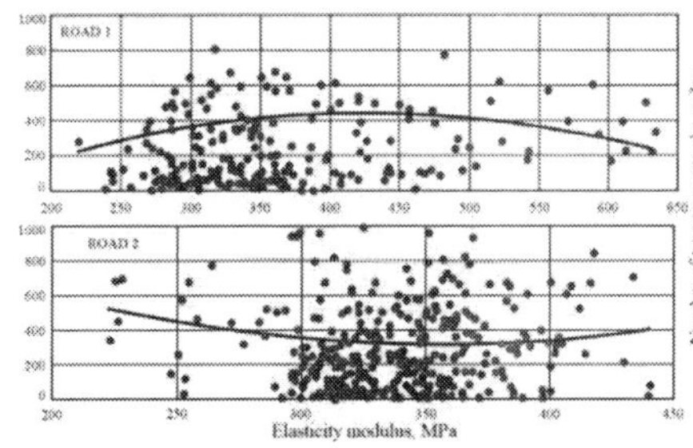

b) Elasticity modulus vs Cracks

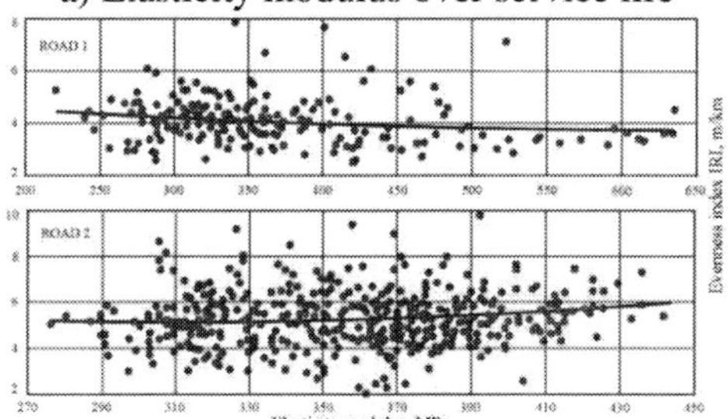

c) Elasticity modulus vs IRI

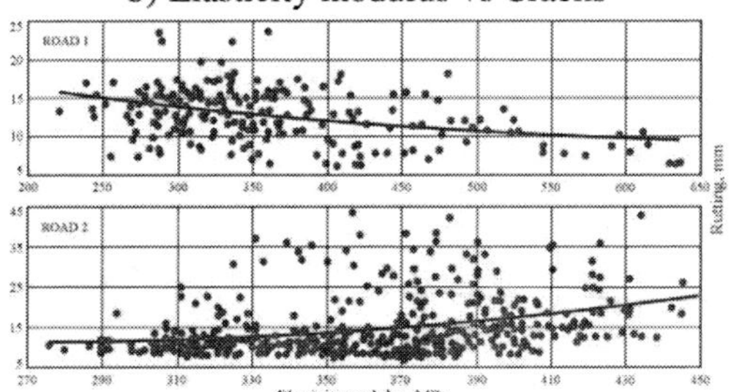

d) Elasticity modulus vs Rutting

Figure 1. Features and results of the elastic deflection study

Figure 2. Mobile Measuring Complex (RWD)

However, even this method of direct measurement of the deflection is insufficient for assessing the actual reliability of the pavement and determining the optimal terms of maintenance and repair which is the most crucial element of the pavement management system. Corresponding post-processing of the obtained data of evaluating vertical deformations is required. This is since the decrease in the module of elasticity of structural layers from rigid (elastic) and non-rigid (viscoelastic) materials, for example, due to an increase in intrastructural defects, affects the durability of pavement, including elastic deflection, to varying degrees.

Fig. 3 shows the dependence obtained by carrying out calculations from which it can be seen that if the module of elasticity of the thick rigid foundation due to the accumulation of intrastructural damage decreases by 2–2.5 times, this will lead to an increase in the amount of pavement elastic deflection in 1.6–2.3 times, whereas, and if the module of the upper non-rigid layers is also decreased by the same value, compared to the base, thickness will increase by 1.2–1.25 times. The calculations were made according to the dependencies developed by Korsunsky based on the fundamental solutions of Burmister, Kogan and others for the evaluation of stresses and deformations in multilayer structures. In this case, if you operate only with the value of elastic deflection from a standard load of 100 kN, it is impossible to identify the cause of the defects and, especially, to predict their development.

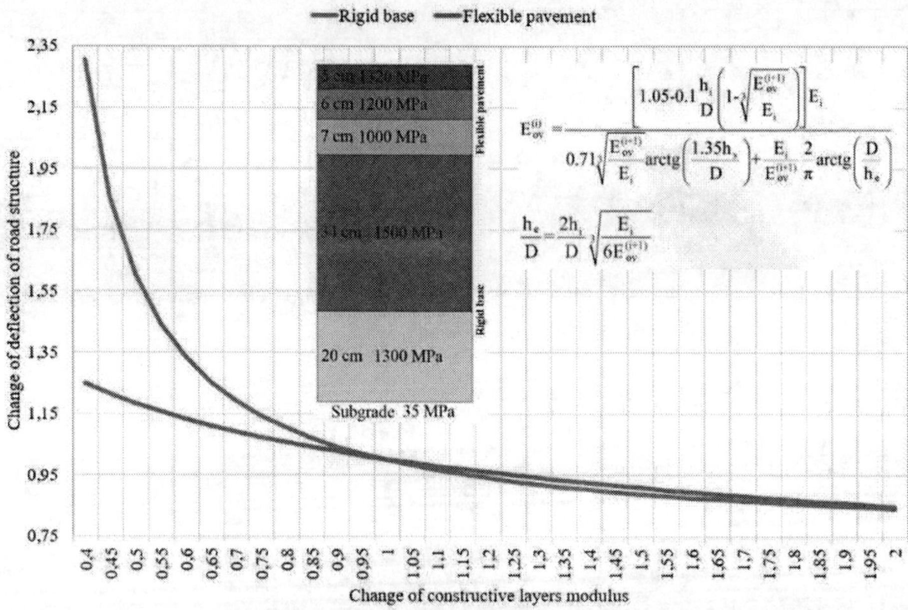

$$E_{ov}^{(i)} = \frac{\left[1.05 - 0.1\frac{h_i}{D}\left(1 - \sqrt[3]{\frac{E_{ov}^{(i+1)}}{E_i}}\right)\right] E_i}{0.71\sqrt[3]{\frac{E_{ov}^{(i+1)}}{E_i}} \operatorname{arctg}\left(\frac{1.35h_i}{D}\right) + \frac{E_i}{E_{ov}^{(i+1)}}\frac{2}{\pi}\operatorname{arctg}\left(\frac{D}{h_e}\right)}$$

$$\frac{h_e}{D} = \frac{2h_i}{D}\sqrt[3]{\frac{E_i}{6E_{ov}^{(i+1)}}}$$

Figure 3. Change of the elastic deflection of the semi-rigid pavement

With a higher increase in damage and decrease in the module of rigid layers, the deflection can rise avalanche-like. If the pavement structure has deviations from the quality of materials or the quality of its structure, then the effect may be even higher.

The process of assessing the reliability of such pavements is hampered due to their high structural strength, when the value of elastic characteristics, for example, elastic deflection, is insignificant when tested with standard loads (e.g., 100 kN).

Currently, the calculations of road structures are mostly focused on the fact that the pavement is represented as a half-space (and a multi-layer one) of a certain rigidity (elasticity) [18–23]. At the same time, the disadvantage of the elastic half-space model is that it does not limit, for example, the load-carrying capacity of the compressible zone at the base of the road structure. In actual conditions of interaction between the pavement and the base, the base and the soil of the subgrade under the influence of dynamic traffic loads, the load-carrying capacity of the structure is limited, and this feature significantly influences the nature of contact stress distribution and, consequently, the accuracy and efficiency of models for assessment reliability and durability of the pavement.

With an unlimited load-carrying capacity of a multi-layered rigid (elastic) half-space, the increase in the calculated value of the deflection occurs almost linearly, where the deflection value is directly proportional to the value of the actual load. This does not consider the features of semi-rigid pavement combining rigid bases and non-rigid upper layers. Because of this, there are often failures in the design and calculation of pavements, when, regardless of the material type of the structural layer, the strength of the pavement increases only based on the value of the module of the elasticity and thickness of the layers. This is fundamentally wrong for real road structures and especially for semi-rigid pavements when calculating their reliability and durability, establishing their residual life.

In order to increase the efficiency and confidence of studies of the strength and reliability of semi-rigid pavements, rolling wheel tests (RWD) were carried out with fixing elastic deflection for three design axial loads: 13, 17 and 21 tons (Fig. 4) in 2018 on the non-stop traffic highway (the year of pavement is 2007) on an area of 5 km at a design temperature of asphalt concrete pavement about 20 °C.

Fig. 5 presents the data of the results of studies of elastic deflection determined by the rolling wheel method, for individual sections (1–6) in the form of flow diagrams in the coordinates of "load-deflection".

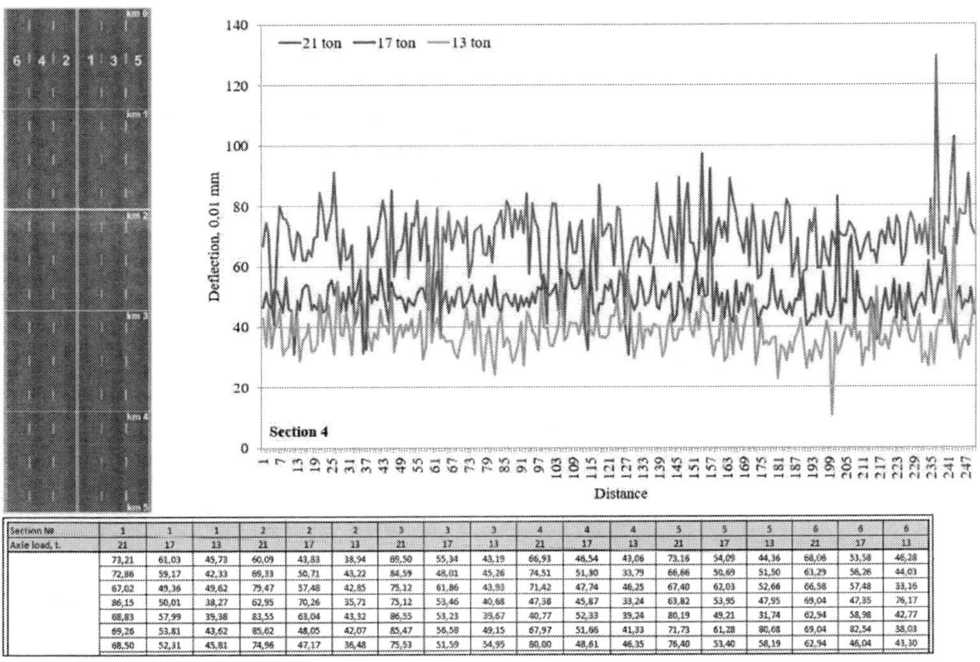

Section №	1	1	1	2	2	2	3	3	3	4	4	4	5	5	5	6	6	6
Axle load, t.	21	17	13	21	17	13	21	17	13	21	17	13	21	17	13	21	17	13
	73,21	61,03	45,73	60,09	43,83	38,94	69,50	55,34	43,19	66,93	46,54	43,06	73,16	54,09	44,36	68,06	53,58	46,28
	72,86	59,17	42,33	69,33	50,71	43,22	84,59	48,01	45,26	74,51	51,30	33,79	66,66	50,69	51,50	63,29	56,26	44,03
	67,02	49,36	49,62	79,47	57,48	42,85	75,12	61,86	43,93	71,42	47,74	46,25	67,40	62,03	52,66	66,58	57,48	33,16
	86,15	50,01	38,27	62,95	70,26	35,71	75,12	53,46	40,68	47,38	45,87	33,24	63,82	53,95	47,95	69,04	47,35	76,17
	68,83	57,99	39,38	83,55	63,04	43,32	86,55	53,23	39,67	40,77	52,33	39,24	80,19	49,21	31,74	62,94	58,98	42,77
	69,26	53,81	43,62	85,62	48,05	42,07	85,47	56,58	49,15	67,97	51,66	41,33	71,73	61,28	80,68	69,04	82,54	38,03
	68,50	52,31	45,81	74,96	47,17	36,48	75,53	51,59	54,95	80,00	48,61	46,35	76,40	53,40	58,19	62,94	46,04	43,30

Figure 4. Example of the results of testing the elastic deflection by the method of a rolling wheel

As shown by the results of statistical processing of experimental data with a large proportion of correlation ($R_2 \geq 0.95$), the value of the elastic deflection of a semi-rigid pavement (l_i) depends on the value of the axial load (P_i) with the subordination of the X reciprocal linear regression model (reciprocal X model) of the form:

$$P_i = a - \frac{b}{l_i}, \quad (1)$$

where a and b are the coefficients of the model, in case of $l_i \to \infty$, $P_i \to P_{max} = a$.

IOP Conf. Series: Materials Science and Engineering **832** (2020) 012022 doi:10.1088/1757-899X/832/1/012022

Thus, the development of dynamic deflection (from the real vehicle wheel, including the loads above the standard, equal to 100 kN) does not occur in the dependence close to linear, but has pronounced (characteristic) levels that can be captured when building correlation dependencies of the form (1) for various loads (the parameters of correlation dependencies for average values of samples of elastic deflection values for the studied areas are presented in Table 1):

1. The level of growth of linear elastic deformation (instantaneous elastic) which may occur, for example, due to: the presence of serious inconsistencies in the strength of the base (low resistance) and the thickness of the layer (small thickness); the presence of defects at the boundary between the separation of a rigid base and a non-rigid upper layers; compressing a package of non-rigid layers; experimental error.
2. The level of axial load (dynamic elastic deflection) at which the deformation curve transitions to a non-linear stage that is caused by the base reaction and non-rigid layer properties.
3. The level of non-linear growth of vertical deformation considering the reaction of a semi-rigid pavement to a disturbance from the traffic load of various value.
4. The level of maximum axial load (P_{max}) above which the increase in deformation can occur unpredictably and can lead to disruption of the pavement continuity, even with a small number of impacts of traffic loads of this level.

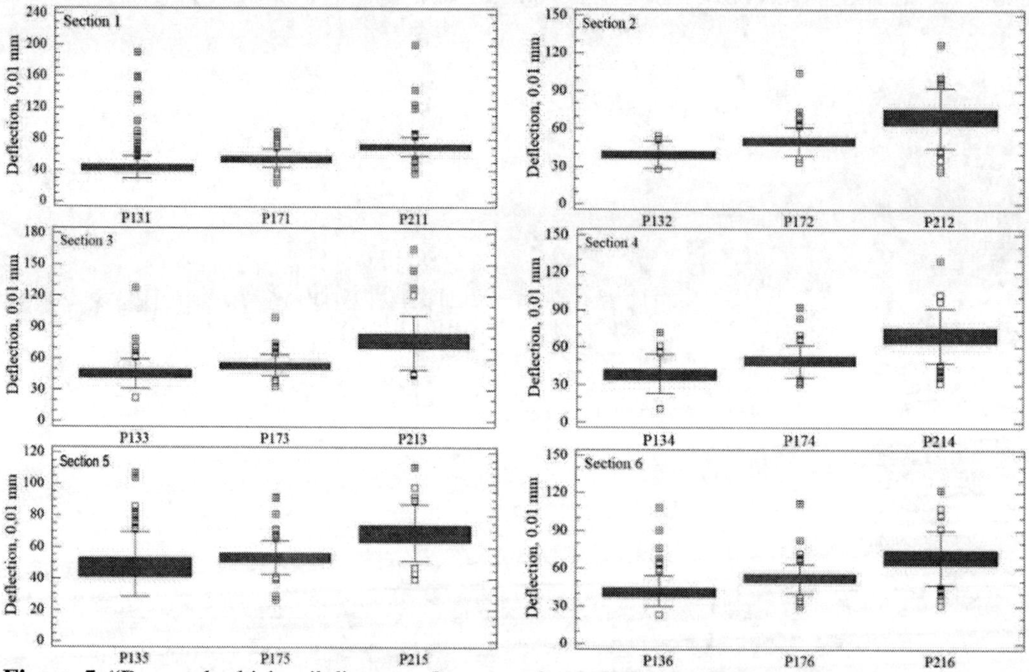

Figure 5. "Box-and-whiskers" diagrams for research results by sections

Table 1. Results of the elastic deflection study

Section No.	Average sampling elastic deflection value (0.01 mm) for loads, t			Parameters of correlation dependence (1)	
	13	17	21	a	b
1	46.29	54.45	71.41	35.80	1045.10
2	39.58	50.48	69.65	31.53	733.42
3	46.40	53.14	76.85	32.98	898.09
4	38.27	49.51	69.18	30.88	685.16
5	48.79	53.62	68.95	39.76	1272.46
6	42.20	52.97	68.41	33.78	880.06

Taking into account the fact that the measurements were made on the highway with the same design structure, the variation of values is quite high: the calculated minimum value (P_{max}) is 30.88 tons, and the maximum value is 39.76 tons, the difference between which is almost equal to the standard axial load of 10 tons. From the presented data, for semi-rigid pavements (pavements of increased deformability), elastic deflection under a load of 10 tons is not very meaningful for assessing reliability and durability.

This is since under a standard load of 10 tons, the development of deformation occurs in the linear elastic stage when the reaction of the rigid base practically does not affect its connection with the subgrade soil. Deformations of the package of asphalt concrete layers insignificantly affect the elastic deformation value with spreading of the deflection bowl over a larger area. As shown in Fig. 3, semi-rigid pavement is more sensitive to an increase in damage (as a result, a module reduction) in the concrete base layers than in asphalt concrete pavements, in terms of evaluation, for example, of the load-carrying capacity of such road structures. In this connection, in order to assess the suitability of rigid bases for further use in the process of reconstruction and repair of roads, which is also essential when performing pre-design diagnostics of pavements while developing solutions for cold recycling of structural layers, tests should be performed at loads that exceed standard loads of 100 kN in 1.5–2 times. Only in this case, the entire pavement together with the soil starts to resist, and therefore, the non-linearity of the deformation process manifests itself, which is associated primarily with a non-linear change of the pavement reaction module (to a higher degree of a rigid base on the unpaved half-space).

Thus, the value of the maximum load (P_{max}) can be used to assess the ability (reliability) of semi-rigid pavement as a whole, and to a higher degree of the rigid base, to the accommodation of heavy traffic loads, including their cyclic effects.

Pavements in many countries are mostly designed based on existing methodologies focused on calculations of the model of a layered elastic system, which allow calculating stresses and deformations at any point of the structure. With such an analysis, it is assumed that the materials of the pavement layers are uniform, isotropic, and linearly elastic. Mathematically, this model is based on the theory of elastic layers developed by Boussinesq in the late 19th century [24]. This model is grounded on very simple mathematical dependencies; however, it requires accepting many significant assumptions, which negatively affects the accuracy of the calculations. The main ones are the following:

- layers are continuing in the longitudinal direction;
- the bottom layer (subgrade soil) has an unlimited thickness;
- layer materials are active only in the elastic stage.

Under such conditions, the increase in the pavement reaction module (R_p) to the disturbance by the traffic load is continuously growing.

However, a completely different situation is observed with an actual pattern of deformation of semi-rigid pavement. Fig. 6 shows the dependencies connecting the pavement reaction module (N/m) with the value of axial load and elastic deflection concerning the correlation dependences indicated in Table 1. As is evident from these dependencies, when it is built in the indicated axes, there is some extremum of the values of the pavement reaction module. This maximum unambiguously suggests the transition of the road structure to a non-linear state with an appropriate deformation (stress).

Based on the maximum values of the pavement reaction, it is possible to determine the value of axial (wheel) load (P_e) and vertical deformation (d_e) corresponding to this extremum (Table 2). The load value corresponds to the moment when the deformation of the layer ceases to be elastic-linear and subsequently develops non-linearly.

Thus, we have one of the essential points of the pavement deformation curve, which shows the moment (according to the load (deformation)) when it ceases to be deformed in the elastic stage, i.e. deformations cease to be recurrent, and damage starts to accumulate in the structure of structural layers and their durability decreases.

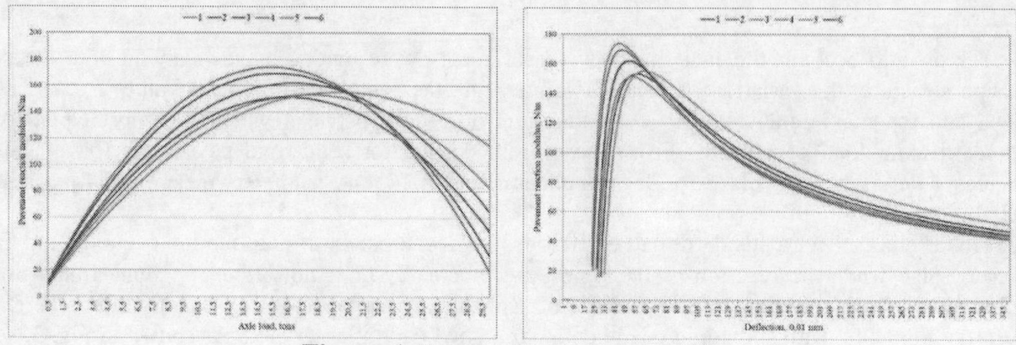

Figure 6. Pavement reaction module

Table 2. Results of the maximum axial load calculation (P_e)

Parameter	Value of the parameter for section No.					
	1	2	3	4	5	6
Axial load (P_e), kN	18.0	16.0	16.5	15.5	20.5	17.0
Maximum deformation (d_e), mm	1.53	1.69	1.51	1.73	1.55	1.62

The above dependencies indicate that the deformation of the semi-rigid pavement under the traffic load is mainly non-linear. With an increase in the traffic load to a level of about 15-20 tons per axis (for the sections considered), pronounced deformations appear, primarily due to the properties of a non-rigid asphalt concrete pavement and the subgrade soil, which are transited from the actual load to a non-linear stage when corresponding strengths achieved. In this case, considering the accumulation of residual deformations in the subgrade soil, there is, as mentioned above, the accumulation of damage in the asphalt concrete pavement layers and concrete base.

Thus, the dynamic method of pavement testing with a rolling wheel with different load levels allows determining non-linear characteristics of the deformation curve and deformation features of the package of asphalt concrete pavement layers with appropriate loads, including their cyclic strength under the conditions of actual traffic loads [25]. Such studies with only a large proportion of approximations can be performed using the standard method for determining the elastic deflection under a load of 100 kN.

3. Conclusion

Nowadays, more and more widely old materials of pavement constructive layers are reused for their reconstruction and repair, both with their processing and regeneration of properties and in the form of undisturbed structures and as separate elements of new road structures. In this regard, to assess the technical and economic characteristics of such projects and increase their efficiency, much attention is paid to the assessment of the residual resource of old pavements.

In many countries, the design criterion, which determines the strength and reliability of the pavement, implies compliance with the requirements for the elastic deflection value. The value of deflection on the layer surface is taken as its vertical deformation under the applied static or dynamic load.

However, despite quite complicated methods of bringing the measured values of elastic deflection to standard conditions, very often they are not enough for proper application of design solutions for major repairs and reconstruction, since it is difficult to calculate the residual life of materials of structural layers, their resistance to deformations under the influence of modern traffic loads.

In order to increase the efficiency and reliability of studies of the strength and reliability of semi-rigid pavements (structures of increased strength), rolling wheel tests (RWD) were carried out with fixing elastic deflection for three design axial loads: 13, 17 and 21 tons.

According to the results of the study, it was found that when testing semi-rigid pavements with elastic deflection fixation with at least 2–3 different axial loads in the range over 10 to 20 tons, it becomes possible to establish the non-linear characteristics of the deformation curve and deformation features of the package of asphalt concrete layers with the corresponding loads that cannot be done with the standard method for determining the elastic deflection under a load of 10 tons.

Thus, when developing criteria for assessing the strength, reliability, and deformability of semi-rigid pavement, it becomes possible to estimate the likelihood of a joint failure-free operation of a rigid base and non-rigid layer with a resource forecast and the feasibility of its cold recycling, which is the direction for future research.

Acknowledgements

This study was supported by Henan center for outstanding overseas scientists: [grant number GZS2018006]; Intergovernmental international cooperation on science and technology key innovation project from the ministry of science and technology of the people republic of China: [grant number 2016YFE0111000]; Transportation science and technology project of Henan province [grant number 2017J5]; Major science and technology project of Xinxiang [grant number ZD19007].

References

[1] Vaitkus A, Gražulyt J, Jukneviciut-Žilinskien L and Andrejevas V 2017 Review of Lithuanian Experience in Asphalt Pavements Cold Recycling *Proceeding of the 10th Int. Conf. "Environmental Engineer."* (Vilnius: VGTU Press) Article ID: enviro.2017.153. https://doi.org/10.3846/enviro.2017.153

[2] Turk J, Mauko Pranijć A, Mladenovič A, Cotič Z and Jurjavčič P 2016 Environmental comparison of two alternative road pavement rehabilitation techniques: cold-in-place-recycling versus traditional reconstruction *J. of Cleaner Production* **121** 45–55 Retrieved from: https://doi.org/10.1016/ j.jclepro.2016.02.040

[3] Cox C B and Howard L I 2016 Cold In-Place Recycling Characterization for Single-Component or Multiple-Component Binder Systems *J. of Mater. in Civil Engineer.* **28** 04016118 DOI: 10.1061/(ASCE)MT.1943-5533.0001621

[4] Dal Ben M and Jenkins K J 2014 Performance of cold recycling materials with foamed bitumen and increasing percentage of reclaimed asphalt pavement *Road Mater. and Pavement Design* **15(2)** 348–71 Retrieved from: https://doi.org/10.1080/14680629.2013.872051

[5] Guatimosim F V, Vasconcelos K L, Bernucci L L B and Jenkins K J 2018 Laboratory and field evaluation of cold recycling mixture with foamed asphalt *Road Mater. and Pavement Design* **19(2)** 385–99 Retrieved from: https://doi.org/10.1080/14680629.2016.1261726

[6] Kuna K, Airey G and Thom N 2017 Mix design considerations of foamed bitumen mixtures with reclaimed asphalt pavement material *Int. J. of Pavement Engineer.* **18(10)** 902–15 Retrieved from: https://doi.org/10.1080/10298436.2015.1126271

[7] CR201 2016 *Recommended Mix Design Guidelines for Cold Recycling Using Emulsified Asphalt Recycling Agent* Guidelines (Annapolis, USA: Asphalt Recycling & Reclaiming Association) Retrieved from: https://www.arra.org/page/CR201Guideline?& hhsearchterms=%22cr201%22

[8] Li Q, Xiao D, Wang K, D Hall K and Qiu Y 2013 Mechanistic-empirical pavement design guide (MEPDG): A bird's-eye view *J. of Modern Transport.* **19** 114–33 Retrieved from: https://doi.org/10.1007/BF03325749

[9] Sarker P, Tutumluer E and Lackey S 2016 Nondestructive Deflection Testing based Mechanistic-Empirical Overlay Thickness Design Approach for Low Volume Roads: Case Studies *Procedia engineer.* **143** 945–53 Retrieved from: https://doi.org/10.1016/j.proeng.2016.06.080

[10] Horak E, Hefer A and Maina J W 2015 Determination of pavement number for flexible pavements using FWD deflection bowl information *Proc. of the 34th Annual Southern African Transport Conf.* (Pretoria: University of Pretoria) Retrieved from: https://repository.up.ac.za/handle /2263/57730

[11] Roodi G H, Phillips J R and Zornberg J G 2016 Evaluation of Vertical Deflections in Geosynthetic Reinforced Pavements Constructed on Expansive Subgrades *Proc. of the 3rd Pan-American Conf. on Geosynthetics* (Miami: GeoAmericas 2016) pp 1970–91 Retrieved from: http://www.caee.utexas.edu/prof/zornberg/pdfs/CP/Roodi_Phillips_Zorn-berg_2016.pdf
[12] Veranko V A, Zankavich V V and Ladyshau A A 2015 *Durable asphalt concrete pavements of roads, bridges and streets* (Minsk: ArtDizayn)
[13] Hausman J J and Steele D A 2011 Using the Rolling Wheel Deflectometer (RWD) to Improve Pavement Management Decisions *Proc. of the 8th Int. Conf. on Managing Pavement Assets "Fulfilling the social, economic, and environmental responsibility for sustainable, well managed, better roads"* (Washington: Transportation Research Board of the National Academies) Retrieved from: http://worldcat.org/isbn/9789561412309
[14] Flintsch G, Katicha S, Bryce J, Ferne B, Nell S and Diefenderfer B 2013 Assessment of continuous pavement deflection measuring technologies *Transport. Res. Board 2013: no. SHRP 2 Report S2-R06F*-RW-1 (Washington: Transportation Research Board) Retrieved from: https://trid.trb.org/view/ 1244232
[15] Elseifi M A, Abdel-Khalek A M, Gaspard K, Zhang Z and Ismail S 2011 Evaluation of continuous deflection testing using the rolling wheel deflectometer in Louisiana *J. of Transportat. Engineer.* **138(4)** 414–22 Retrieved from: https://ascelibrary.org/doi/abs/10.1061/(ASCE)TE.1943-5436.0000349
[16] Zhang Z, Gaspard K and Elseifi M A 2016 Evaluating pavement management treatment selection utilizing continuous deflection measurements in flexible pavements *Int. J. of Pavement Engineer.* **17(5)** 414–22 DOI: 10.1080/10298436.2014.993198
[17] Nasimifar M, Siddharthan R V, Gonzalo R, Rada G R and Nazarian S 2017 Dynamic analyses of traffic speed deflection devices *Int. J. of Pavement Engineer.* **18(5)** 381–90 DOI: 10.1080/10298436.2015.1088152
[18] Judycki J, Jaskuła P, Pszczoła M, Ryś D, Jaczewski M, Alenowicz J, Dołżycki B and Stienss M 2016 New polish catalogue of typical flexible and semi-rigid pavements *Proc. of XI Int. Road Safety Seminar GAMBIT 2016* **122** 8 (MATEC Web of Conf.) Retrieved from: https://doi.org/10.1051/matecconf/201712204002
[19] Little D N, Allen D H and Bhasin A 2018 *Modeling and Design of Flexible Pavements and Materials* (Berlin: Springer) DOI: 10.1007/978-3-319-58443-0
[20] Narzary B K and Ahamad K U 2018 Estimating elastic modulus of California bearing ratio test sample using finite element model *Construct. and Build. Mater.* **175** 601–9 Retrieved from: https://doi.org/10.1016/j.conbuildmat.2018.04.228
[21] Bodin D, Chupin O and Denneman E 2017 Viscoelastic asphalt pavement simulations and simplified elastic pavement models based on an "equivalent asphalt modulus" concept *J. of Testing and Evaluation* **45(6)** 1887–95 Retrieved from: https://doi.org/10.1520/JTE20160652
[22] Mikolaj J, Remek L and Macula M 2017 Asphalt Concrete Overlay Optimization Based on Pavement Performance Models *Advances in Materials Science and Engineering* (Article ID 6063508**:** 10 p.) Retrieved from: https://doi.org/10.1155/2017/6063508
[23] Loganathan K, Isied M, Coca A M, Souliman M, Romanoschi S and Dessouky S 2019 Development of Deflection Parameters to Evaluate the Structural Capacity of Flexible Pavements at the Network Level: Case Study for the State of Texas *In MATEC Web of Conferences 2019* **271** 5 p Retrieved from: https://doi.org/10.1051/matecconf/201927108003
[24] Graczyk M, Rafa J and Zofka A 2016 The selected problems of multi-layer pavements–influence of composite impacts vehicles and climatic factors on the behavior of roads pavements *Transportation Res. Proc.* **14** 2487–96 Retrieved from: https://doi.org/10.1016/j.trpro.2016.05.316
[25] Li Z, Liu T, Shi J, Veranko U and Zankavich V 2017 Fatigue resistance of asphalt concrete pavements. Peculiarity and assessments of potentials *The Baltic J. of Road and Bridge Engineer.* **XII(4)** 270–5 Retrieved from: http://dx.doi.org/10.3846/bjrbe.2017.34

Determination of the capacity of the two-lane acceleration lanes with a central island

A V Kostsov

Moscow Automobile and Road Construction State Technical University (MADI), 64, Leningradsky ave., Moscow, 125319, Russia

E-mail: kostsov_msfs@bk.ru

Abstract. Interconnection sections of exit ramps at different levels with motorways are one of the main sources of traffic hindrance, so special attention should be paid to ensuring the capacity of such areas. Dual-lane motorway connections are often made using two-lane acceleration lanes, with safety islands as a method of increasing their capacity. In order to develop recommendations for determining the capacity of such acceleration lanes, the author carried out experimental studies of the patterns of DLC maneuvers within the highways on the sections bordering the entrances. The studies were carried out by recording the trajectories of vehicles that perform a maneuver of exit from the entrance to the highway. On the basis of the carried out researches an improvement of a method of calculation of capacity of two-lane acceleration lanes is offered. The results of the studies are recommended for use in the design of highways and intersections at different levels of them, as well as for the reconstruction of the highway lanes in order to increase their capacity.

1. Introduction

Interconnection sections of exit ramps at different levels with motorways are one of the main sources of traffic hindrance [1], so special attention should be paid to ensuring the capacity of such areas. The significant amount of cars traffic performing the exit to the main direction of traffic leads to fluctuations in the speed of the traffic flow, i.e. the formation of road waves [2], which causes a drop in capacity and increases the risk of accidents [3]. One of the ways to reduce the negative impact of these processes when connecting exits by two-lane transition-speed lanes is the device of security islands (Fig. 1), which makes it possible to increase the distance between the areas of convergence of traffic flows.

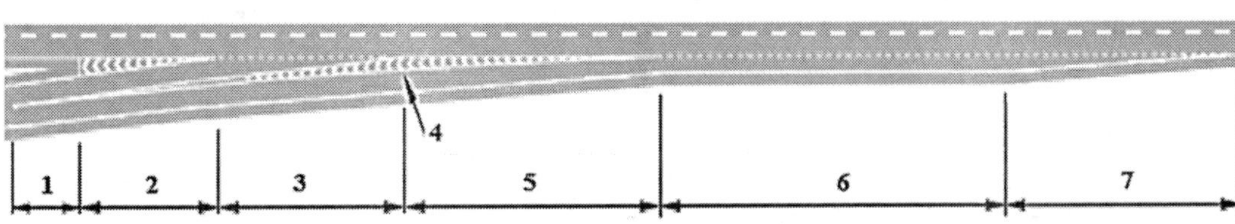

Figure 1. Schematic diagram of the interface area with the use of a two-lane acceleration lane and a central separating island: 1,2 – exit, 3,6 – traffic flow merger section, 4 – traffic island, 5 – maneuvering section (L_m^d), 7 – driving off section.

In order to assess the capacity of two-lane acceleration lanes containing traffic islands, it is possible to apply the boundary interval theory, according to which the capacity of the merging area is determined by road conditions and traffic intensity of the conflicting traffic flow. In this case, to determine the capacity of the second (in the course of traffic) section of the interface, it is necessary to have the value of traffic intensity at least on the right lane of the motorway before the second section of the interface (N_{a2}), which can be found from (1):

$$N_{a2} = N_{a1} + Q_r - N_{12} \quad (1)$$

where

N_{a1} – traffic intensity at the rightmost lane of the motorway before the first section of the interface, cars/h;

Q_l – traffic intensity on the left lane of the exit, cars/h;

N_{12} – traffic intensity of vehicles that have completed the transposition manoeuvre from the extreme right to the central motorway lane, cars/h.

The analysis of the dependence (1) shows that in order to determine the capacity of the second (in the course of traffic) merging section, it is necessary to assess the traffic intensity of vehicles that have made a transposition manoeuvre from the extreme right to the central lanes of the motorway. The study of the works performed on this subject has shown that up to now the characteristics of lane changes and their influencing factors on motorways have been studied mainly in terms of driver behavior [4] and road conditions [5, 6]. The most famous work is that of P. Gipps, who first proposed a traffic flow model that takes into account the lane-change of cars on a multi-lane road [7]. This model assumes that the main drivers' motives for changing lanes are:

- maintaining the desired speed;
- the need to be in a given lane to perform exit manoeuvres from the road or change direction.

The model developed by P. Gipps was further developed in works [8, 9] where its applicability to highway conditions was proved. K. I. Ahmed [10] applied the model in conditions of high levels of vehicle density (Z>0.8). Further studies of the process of lane change are based on risk-oriented models [11] and also use intellectual algorithms such as neural networks [12] and methods of fuzzy interference [13].

There is practically no empirical research into patterns of lane changes. Insufficient attention is paid to the search for patterns of execution of DLC maneuvers in the movement of vehicles from the exits to the highway. Despite the fact that this type of maneuvers has a significant impact on the capacity of the highway, in most of the existing movement models the patterns of DLC maneuvers at the exit are ignored or solved primitively. This is also noted in [14, 15]. This can be explained by the lack of technical capacity to carry out this type of research until recently, as it involves the need to observe the movement of vehicles over long distances, which prevents the use of traditional types and methods of research.

2. Research method

For experimental studying of regularities of the transport flows interaction following on the main lanes of the highway and making the maneuver of reorganisations from the rightmost to the central lanes of the highway, the author carried out researches of movement regularities of the cars moving from exit to the main transport flow. The traffic service level was identified as C as the one taken for calculations in the course of road design and complying with road economically feasible operation. Simultaneously with the registration of the load level in the measurement area, the laboratory car was used to study the trajectories of cars entering the main traffic flow from the exit. In the course of the measurements, the trajectories of the vehicles were studied by measuring the distance covered by the vehicle under study within the right lane of the multi-lane road at the exit of the vehicle from the motorway exit (L1, Fig. 2). The beginning of the report was the moment of entry to the motorway from the acceleration lane (the passage of the car's axis of the roadway marking line). Distances covered were measured by means of Garmin Mobile PC (V5.00.60g) software package which determines the distance covered by a car using the data of NAVSTAR and GLONASS global positioning systems. The error of mobile laboratory longitudinal and latitudinal location coordinates determination in the course of above mentioned measurements did not exceed 1.5 to 3.0 m. The covered distance data was registered by means of a studied car movement videotape and indicators of the covered distance and further processing of the videotape in office conditions.

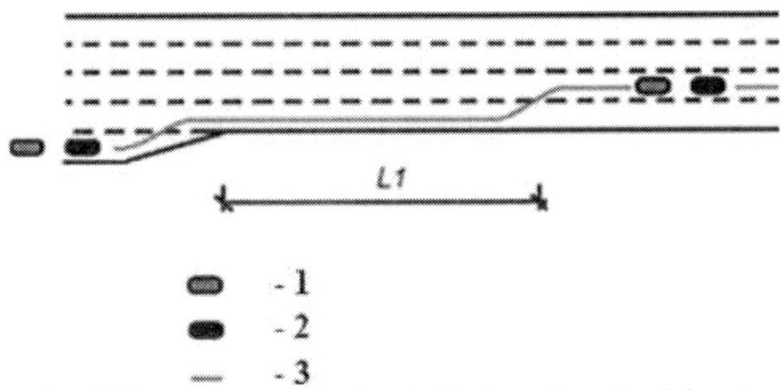

Figure 2. Trajectory of vehicle movement in the studied section: 1 – laboratory car; 2 – leader car; 3 – trajectory of movement

3. The Results of the Study

On the basis of experimental data the curve of distribution of maneuvers of rearrangements of cars between lanes of movement in functional dependence on distances before the beginning of exit which is presented in fig. 3 is received.

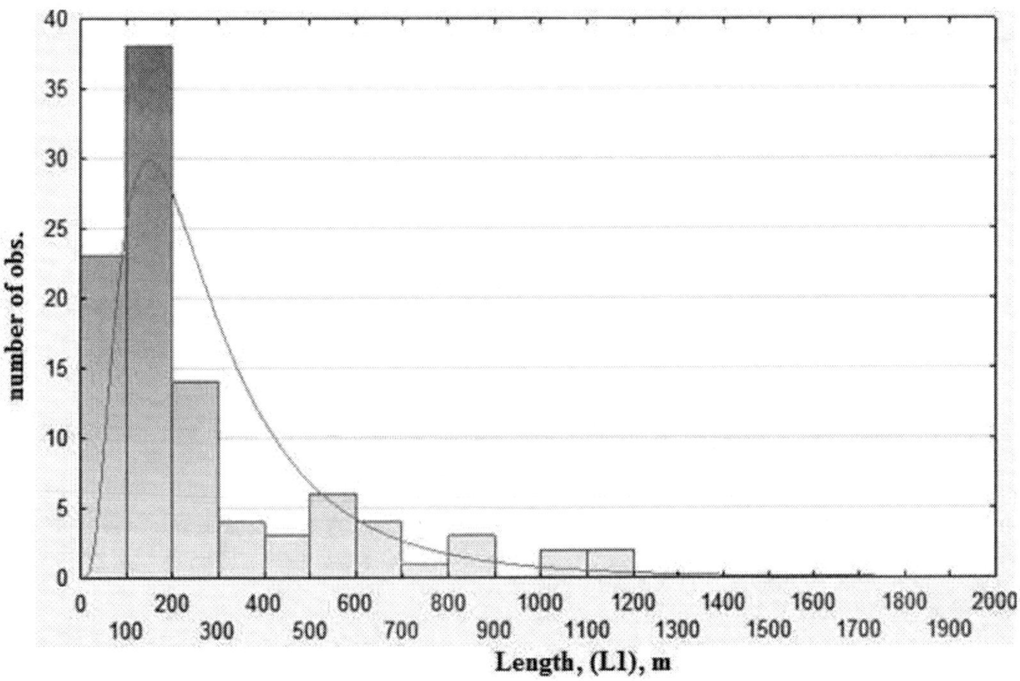

Figure 3. Histogram and distribution curve of rearrangement maneuvers between lanes in functional dependence on the distances before the start of the exit

4. Design considerations for two-way acceleration lanes

Integrating the distribution curve obtained in the course of the studies, we obtain (2):

$$N_{12} = \int_0^{L_m^d} f(L_m^d)dL. \quad (2)$$

Substituting (2) in (1) we get:

$$N_{a2} = N_{a1} + Q_r - \int_0^{L_m^d} f(L_m^d)dL. \quad (3)$$

Using the data in Fig. 3, the expression (3) allows carrying out a quantitative assessment of the traffic intensity at the rightmost lane of the motorway in front of the second section of the interface, and therefore of the entire interface section. The author's analysis of the capacity of the interface node shown in Fig. 1 with account of (3) has shown that the capacity of such a node at the introduction of central traffic islands of length from 100 to 700 m leads to an increase in its capacity by 4–18 % (Table 1) in the presence of a reserve capacity of the highway. Thus, the arrangement of central traffic islands on two-lane acceleration lanes can be considered an effective measure to increase the capacity of the merging areas of intensive traffic flows, and the results of the studies are recommended for introduction into the design practice.

References

[1] Mauch M and Cassidy M J 2002 Freeway traffic oscillations: observations and predictions Proc. of the 15th Int. symp. *Transportation and traffic theory* pp 653–74
[2] Sultan B, Brackstone M, Waterson B and Boer E 2002 Modeling the dynamic cut-ins situation Transport *Res. Record: J. Transport. Res. Board* **1803(1)** 45–51 DOI: 10.3141/1803-07
[3] Coifman B, Krishnamurthy S and Wang X (2005) Lane-change manoeuvres consuming freeway capacity *Traffic Granular Flow* 3–14
[4] Sun D. and Elefteriadou L 2011 Lane-changing behavior on urban streets: a focus group-based study Appl. Ergonom. **42(5)** 682–91
[5] Tang T-Q, S C Wong, H-J Huang, and P Zhang 2009 Macroscopic modeling of lane-changing for two-lane traffic flow *J. of Advan. Transportat.* **43(3)** 245–73
[6] Toledo T, Choudhury C F and Ben-Akiva M E 2005 Lane-changing model with explicit target lane choice *Transportat. Res. Record*, **1934** 157–165
[7] Gipps P G 1986 A Model for the Structure of Lane Changing Decisions Transportat. Res. **20B** 403–14
[8] Yang Q and Koutsopoulos H N 1996 A Microscopic Traffic Simulator for Evaluation of Dynamic Traffic Management Systems *Transportat. Res.* **4C** 113–29
[9] Hidas P and Behbahanizadeh K 1999 Microscopic Simulation of Lane Changing under Incident Conditions, in Proceedings of the 14th International Symposium *Theory of Traffic Flow and Transportation* pp 53–69
[10] Ahmed K I 1999 *Modeling Drivers Acceleration and Lane Changing Behavior* (Massachusetts Instit. of Technol.)
[11] Hamdar S H 2009 *Modeling Driver Behavior as a Stochastic Hazard-based Risk-Taking Process* PhD thesis (Evanston, USA: Northwestern University)
[12] Wei C-H 2001 Developing freeway lane-changing support systems using artificial neural networks J. of Advan. Transportat. **35(1)** 47–65
[13] Mai T, Jiang R. and Chung E A (2016) Cooperative Intelligent Transport Systems (C-ITS)-based lane-changing advisory for weaving sections J. of Advan. Transportat. **50(5)** 752–68
[14] NCHRP Report 687: *Guidelines for Ramp and Interchange Spacing*. TRB 2011 Retrieved from: http://www.cmfclearinghouse.org/studydocs/nchrp_rpt_687.pdf.
[15] Hu X and Sun J 2019 Trajectory optimization of connected and autonomous vehicles at a multilane freeway merging area *Transportat. Res.* Part C: *Emerg. Technol.* **101** 111–25

IOP Conf. Series: Materials Science and Engineering **832** (2020) 012024 doi:10.1088/1757-899X/832/1/012024

Comparative assessment of the technological properties of mastic and sulfur-mastic asphalt concrete mixtures

E V Kotlyarski, Y E Vasilev and V I Kochnev

Moscow Automobile and Road State Technical University (MADI), 64, Leningradsky ave., Moscow, 125319, Russia

E-mail: eco46@mail.ru

Abstract. The technological properties of mastic and sulfur-mastic asphalt concrete mixtures have been evaluated. The hypothesis has been confirmed that the compositions of mastic asphalt concrete mixtures with crystallization-coagulation structure have increased fracture resistance. Optimal technological modes of producing and paving are recommended.

1. Introduction

Mastic asphalt is road-building material, a type of asphalt concrete formed after cooling in the road surface of the mastic asphalt concrete mixture.

According to GOST R 54401-2011 "Hot road mastic asphalt concrete. Technical requirements "mixture of mastic asphalt concrete is" a moving mixture, with minimal residual porosity, consisting of the mineral part (crush, sand and mineral powder) and viscous oil road bitumen (with polymer or other additives (or without them) as an astringent. The mastic mix is laid at a mixture temperature of at least 190 degrees Celsius without being sealed. Depending on the temperature, size and time of the load application, the mastic asphalt concrete at operating temperatures shows elastic-elastic and viscous-plastic properties [1].

Mastic asphalt concrete differs from traditional asphalt concretes by the increased to 7.5–10 % (by mass) content of bitumen and increased to 20–30 % shares of mineral powder. Thus, the content of asphalt astringent (AA), consisting of mineral powder and bitumen, increases to 28 % and above. The content of the rubble (grains larger than 5 mm) is from 0 to 50 % of the mass, which at this concentration causes the formation of a half-framed and frameless structure of asphalt concrete. The main feature of mastic asphalt concretes is that the strength and deformational properties of the material are determined by the rheological properties of the asphalt astringent substance or its microstructure [2, 3].

2. Setting a problem

In traditional compacted asphalt concretes, the so-called macrostructure (the content and shape of rubble and sand) [2] plays a much larger role in the formation of the physical and mechanical properties of materials. Increased content of AA causes the fluidity of mastic mixtures, which, with the right selection of rational composition, allows you to stack them without sealing.

This explains the main transport and operational properties of the mastic asphalt concrete – waterproofness, high grip with the underlying structural scraps of road clothes (bridge canvas), fatigue fracture resistance and the ability to extinguish (damp) oscillations at sign-shifting transport loads in a wide range of frequencies and amplitude of oscillations on road artificial structures throughout the life cycle of the construction layer [4]. The material shows high resistance to corrosion, antibacterial resistance, resistance to anti-ice reagents, is environmentally friendly.

Compared to traditional compacted, the main advantages of mastic asphalt concretes are high water resistance and fatigue durability.

The disadvantages of mastic asphalt concrete include poor resistance to plastic rut formation, the need to use special equipment and equipment in the delivery and laying of the material, low crack resistance in winter when used in it composition of road bitumen of increased viscosity (with a penetration of less than 50). Technological difficulties of laying the mixture if necessary to control the design longitudinal and cross slopes. All this significantly limits the scope of the use of mastic blends.

Wider use of mastic asphalt concretes found in road and bridge construction, on city streets in the tramway zone, on sidewalks.

Mastic mixtures are actively used in pit repair especially in winter, as they do not require a seal during which traditional hot mixtures are quickly cooled.

During the years of the use of mastic asphalt concrete in our country and abroad, numerous attempts have been made to optimize the compositions and properties of the material, however, at the end of the 20th century, principled materials, technological and constructive solutions no one was found [5, 6].

3. Materials and methods

Asphalt concrete and asphalt concrete mixtures, including mastic, are polymineral highly concentrated dispersal systems [7, 8], allowing the use of modern ideas about structural formation using basic positions of physical-chemical mechanics and contact interaction theory in highly concentrated dispersal systems and theoretical prerequisites for the management of these processes in asphalt concretes

In such systems, in the process of technological reworking, spatial structures arise, the properties of which are largely determined by superficial phenomena at the boundaries of the phase section. The type of these structures is represented in the rice according to [9, 10] in figure 1 and is determined by the type of contact between the particles of the dispersal phases:

- reversible coagulation contacts formed through the thickness-equal layer of liquid dispersion medium;
- irreversibly collapsing true phase contacts or interweaving contacts;
- strong true contacts that are formed as a result of chemical and phase transformations in the process, and after the completion of reactions polymerization of liquid astringent, crystallization, cooling of melts, plastic deformation of dispersed particles, interweaving fibers and bonding them, and other.

In various real-world building composites there are also mixed types of contact structures, for example, coagulation-crystallization. The properties and strength of such dispersal systems are determined by the ratio of these types of contact interactions. Depending on the task at which the technological operations are carried out, as noted above, it is its mobility, which reduces the energy costs when stirring the mastic asphalt concrete mixture, and does not require sealing mechanisms during its styling and sealing. This is achieved by heating up to high technological temperatures and basal macrostructure of a mixture with a high content of asphalt astringent substance (AV).

However, the operational requirements for shifting resistance of asphalt concrete coatings force the use of high viscous road bitumen when designing mastic mixtures, which, as noted, significantly reduce crack resistance in winter. It is not always possible to achieve an optimal balance of these contradictory requirements.

The task of materials has been successfully solved by the addition of a modified sulfur mixture in the composition of the asphalt concrete mixture [11, 12], which has a phase transition temperature in the presence of bitumen in the range of 113–120 °C in the amount of up to 30 %.

As a working hypothesis of the study it is accepted that in the mastic compositions with the addition of sulfur this possibility is realized by the creation of dispersal systems with a mixed micro-contact structure. At the technological stage at high temperatures (above 120 degrees), only coagulation contacts are formed in the mixture, which ensures its mobility. After the completion of all technological operations in the process of structural formation when cooling, part of the contacts due to the phase transition of the sulfur component forms a more complex coagulation-crystallization structure. In the resulting dispersal system, crystallizing contacts at positive summer temperatures provide a shift resistance of the material in a constructive layer, and coagulation – crack resistance in winter.

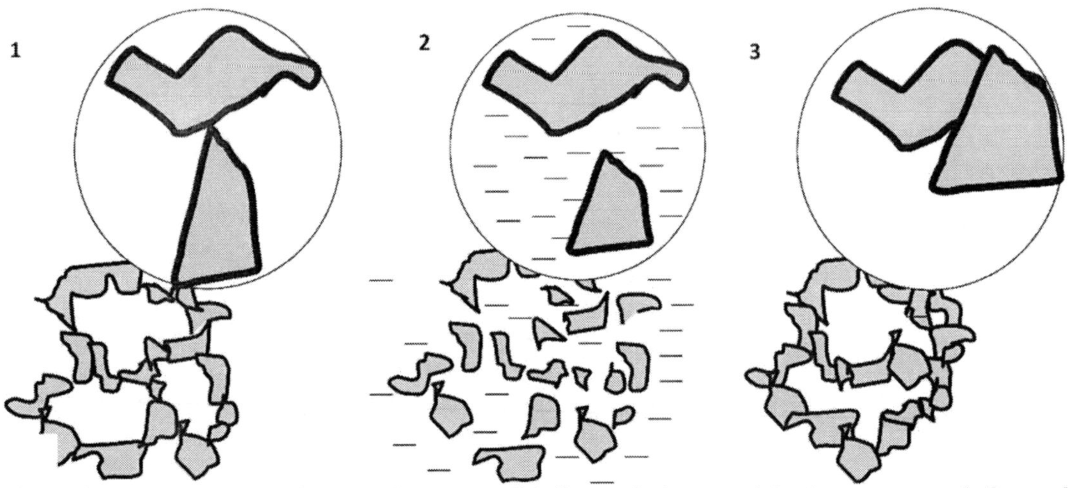

Figure 1. The main types of contact between solid dispersal phase particles in concentrated dispersal systems (by P.A. Rebinder [9, 13]): 1 – atomic contact; 2 – coagulation contact; 3 – phase contact.

Academician P.A. Rebinder [9, 13] considered the processes of structural formation in dispersal systems. Given there is additiveness of the strength of elementary contacts:

$$P_m = k_1 \cdot F_c \cdot n^{2/3} = k_2 \cdot F_c \cdot \varphi \cdot S_{sp}^2 = k_3 \cdot F_c \cdot \varphi \cdot d_i^{-2} \quad (1)$$

where: P_m – limit the voltage of the shift; F_c – average grip strength in contact between particles; n – average number of contacts; φ – relative density of structure; S_{sp}, specific surface of the particles of the dispersal phase of the highly concentrated dispersal system; d_i – the characteristic size of these particles.

Comparative assessment of the quality of the initial effects on the asphalt concrete mixture and the elementary contacts formed in the asphalt concrete were made by the magnitude of the average grip force in contact between Fc particles. Professor E.D. Yakhnin proposed to define it by known experimentally defined values of the limit voltage shift P_m and the geometric parameters of the researched dispersal system [14]:

$$F_c = P_m \cdot d \cdot \rho \cdot \frac{(V_S - V_m) \cdot (V_{s0} - V_m)}{V_{S0} - V_S} \quad (2)$$

where P_m – limit the voltage of the shift; d – characteristic size of dispersal phase particles; ρ – dispersal phase density; V_s – the specific volume, i.e. the volume attributed to the unit of the mass of the structure; V_m – the specific volume, i.e. the volume attributed to the unit of the mass of the structure; – critical specific volume of the system, above which the structure is not formed and P_m=0 when $V_s = V_{s_0}$, but and P_m=0 when $V_s < V_{s_0}$.

Processes of formation of bitumen film on mineral materials began to be studied in the 60s at the Department of Road-Building Materials HADI under the direction of Professor M.I. Volkov. Thus, I.M. Borsch with the help of chromatographic and fluorescent analyses found that as a result of interaction on the surface of mineral particles formed adsorption-salt layers of associated bitumen.

On the basis of these theoretical provisions in his doctoral work and subsequent publications, Professor I.V. Korolev used the concept of "bitumo-intensiveness" that allowed the author to propose a calculated methodology for determining the optimal content of bitumen in bitumomineral materials.

The strength of the dispersal system depends on the average number of contacts between particles (***n***).

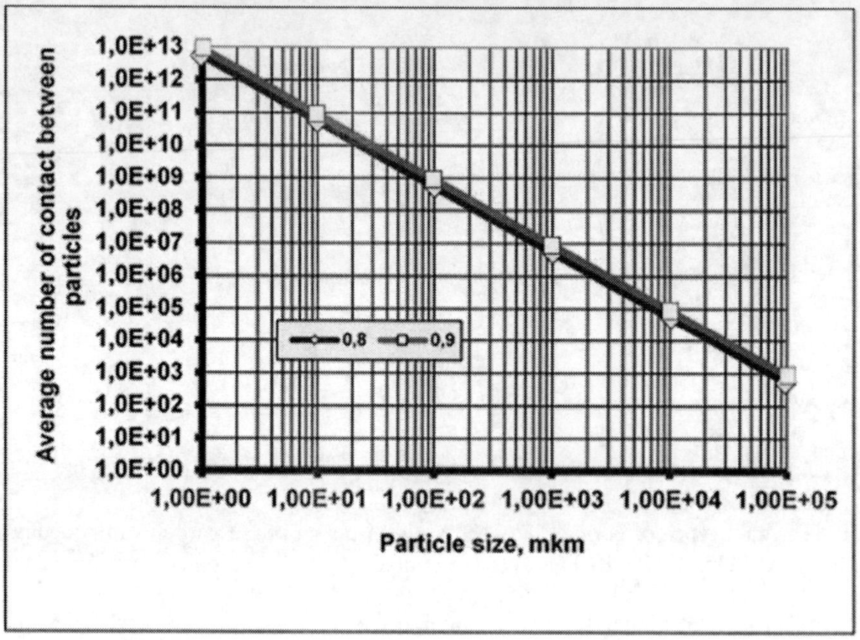

Figure 2. Change in the number of contacts between globular monodispersive particles depending on their size

The average number of contacts between the particles of the highly concentrated dispersal phase, which make up the asphalt concrete mixture and asphalt concrete with a known approximation can be determined by the nomogram [15] or by empirical dependence (3):

$$n = 6{,}7373 \cdot e^{5{,}44 \cdot \varphi} \cdot d_{eq}^{\ -2} \qquad (3)$$

where: φ – relative density, $\mathbf{d_{eq}}$ – characteristic particle size.

The value of the characteristic particle size (***di***) is statistical in nature and depends on the quantitative ratio of particles by fractions of each material, shape and size ratios.Его можно определить по формуле (4):

$$d_i = \sum_{d_{\min}}^{d_{\max}} \frac{\left(k_f \cdot k_d \cdot a_i\right)}{100} \qquad (4)$$

where $\boldsymbol{a_i}$ – private remnants of **i**-faction particles, k_f – particle shape ratio, k_d – particle density ratio, $\boldsymbol{d_i}$ – fractional particle size.

To quantify the nature of bitumen interaction with the mineral surface, a medium-strength index of the grip in contact between particles can be used. According to the dependence of E.D. Yahnin (2) on experimentally defined values of the limit voltage of the shift P_m and the geometric parameters of the researched dispersal system (characteristic size of particles of the dispersal phase d, density of the dispersal phase ρ, specific volume V_s, volume adjustment V_m and critical unit of the dispersal system V_{s_0}) [14].

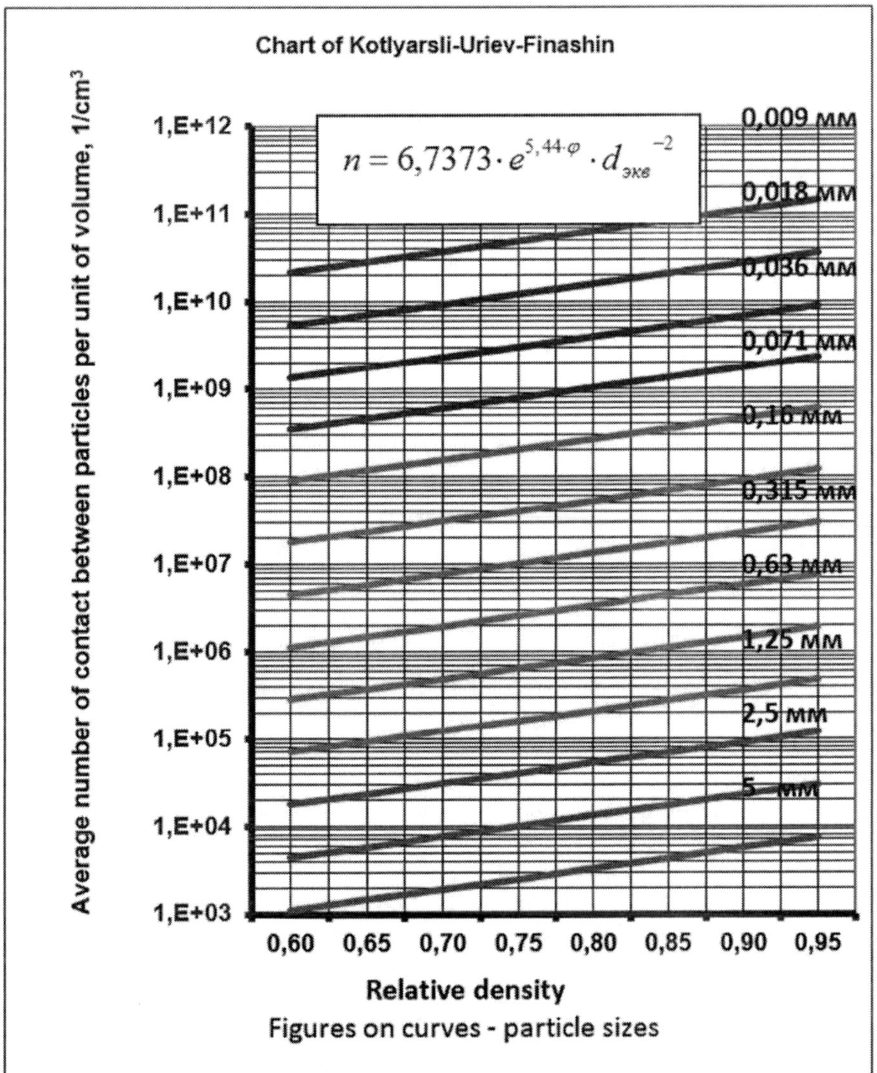

Figure 3. Nomogram for determining the average number of contacts between particles [15]

After the transformations, the formula of E.D. Yahnin [15] takes the form (5). In order to assess the structure and properties of highly concentrated dispersal systems, it is advisable to use the criterion of its relative density in some cases, as it is done in the expression (1.8) proposed by P.A. Rebinder, N.V. Mikhailov and S.Y. Shalyt [17]:

At the same time $\varphi = \dfrac{\rho_i}{\rho}$ and $\varphi = \dfrac{\rho_p}{\rho}$, then

$$F_c = P_m \cdot d^2 \cdot \frac{\rho'}{\rho_n} \cdot \frac{(1-\varphi_i)\cdot(1-\varphi_p)}{(\varphi_i - \varphi_p)} \tag{5}$$

where φ_p – relative density of loose asphalt concrete mixture; φ_i – relative density of compacted asphalt concrete mixture, P_m, d, ρ' - same as expressions (1, 2). At the same time $\varphi_p \langle \varphi_i \langle 1{,}0$.

At the Department of Road-Building Materials MADI developed a method of calculating the optimal content of bitumen, which allowed to determine the bitumointensity of particles measuring 1 – 71 microns for various mineral materials based on results obtained experimentally [6].

The initial information used the values of bitumomy, experimentally obtained by I.V. Korolev and A.M. Gridchin by using statistical methods of mathematical forecasting.

The physical and chemical processes at the border of the phase section of the high-concentrated dispersal system will depend on the area of contact interaction.

The total surface of all the elementary contacts of the highly concentrated dispersal system after the transformations was according to the formula of P.A. Rebinder, N.V. Mikhailov and S.Y. Shalyt (1) [17]:

$$\sum_0^m S_i = \frac{1}{k_1} = \sum_{d_{\min}}^{d_{\max}} \frac{1}{k_1} \tag{6}$$

where **n** – average number of contact between particles per unit of volume, $\mathbf{k_1}$ – size factor.

$$k_1 = \left(\frac{P_m}{F_c}\right) \cdot n^{2/3} \tag{7}$$

After that, the average area of a single (elementary) contact was defined as:

$$S_i = \sum_{d_{\min}}^{d_{\max}} \frac{S_i}{n} = \frac{1}{k_i} \cdot \frac{1}{n} = \left(\frac{F_c}{P_m}\right) \cdot n^{-1/3} \tag{8}$$

Assuming that single contacts are shaped like a circle, you can estimate their average size:

$$\delta_i = \sqrt{\frac{4 \cdot S_i}{\pi}} = 2 \cdot \sqrt{\frac{S_i}{\pi}} \tag{9}$$

An equally important characteristic of the structure of asphalt concrete is the magnitude of the average strength P_i of a single contact. It can be defined by the formula (10):

$$P_i = \frac{F_c}{S_i} = \frac{P_m}{n^{1/3}} \tag{10}$$

4. Discussion of the results

The structural and mechanical characteristics of asphalt-concrete mixtures and asphalt concrete, considered above, allowed to develop a calculated-experimental method of assessing the structural and mechanical properties of the compositional material and its structure on the different technological stages of its production and in the course of operation, including physically sound quantitative criteria of the structure and properties of the mastic asphalt concrete mixture and mastic asphalt concrete as polydisperpolysper polymineral compositional material. This makes it possible to regulate the technological parameters of mixtures in a direction, as well as to more reliably predict the behavior of the material at real operational impacts.

In the laboratory of MADI we tested compositions of mastic asphalt concrete mixtures with the addition of different amounts of sulfur.

The composition of the mineral part of the asphalt concrete mixtures fully complies with the requirements of GOST 544401. Four series of samples with modified sulphur have been tested (table 1).

Table 1. The mastic of asphalt concrete mixture type I

Series No	Rubble	Sand	Mineral powder	Bitumen	Modified sulfur
1	40.0	35.0	25.0	7.79	0
2	40.0	35.0	25.0	7.79	10
3	40.0	35.0	25.0	7.79	20
4	40.0	35.0	25.0	7.79	30

In table 2 the results of the definition of rationed indicators of physical and mechanical properties for GOST 54401 series 1 and 3 are given.

Table 2. Effect of temperature on standard mechanical properties mastic mixes series 1 and 3

A view of mastic asphalt concrete	The strength limit when compressed at temperature, °C			The strength limit when shifting 50 °C
	0	20	50	
Traditional	7,39	2,74	0,71	0,95
Sulfur asphalt concrete	9,30	4,30	3,75	3,45
	Coefficients of heat resistance, %			
Traditional	2.70	1.00	0.26	
Sulfur asphalt concrete	2.16	1.00	0.92	

The results show that the mastic asphalt concrete mixture of series 3 with 20 % sulphur content has significantly higher rates and noticeably lower temperature sensitivity compared to the composition of series 1 on pure bitumen BND 60/90.

It should be noted that at a temperature of 50 degrees Celsius, this composition has 3.63 times higher values of the maximum voltage of the shift.

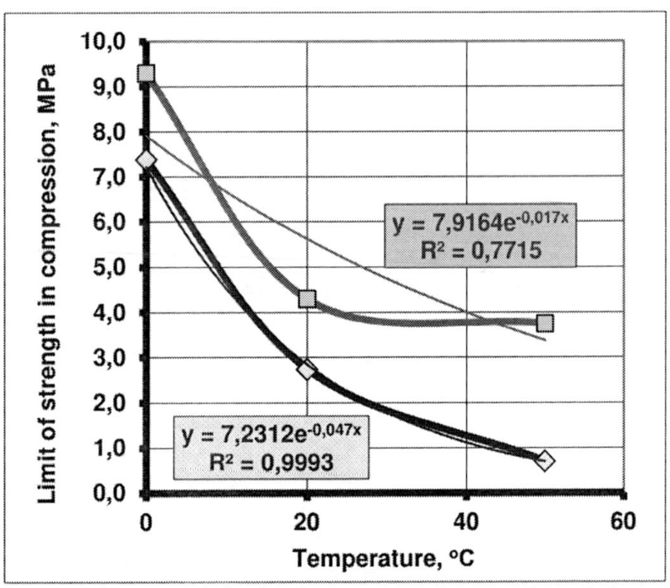

Figure 4. Dependence of the strength limit when compressing mastic asphalt concrete mixtures on temperature

IOP Conf. Series: Materials Science and Engineering **832** (2020) 012024 doi:10.1088/1757-899X/832/1/012024

Using correlation allotment analysis, the correlations shown in Figure 4 of the correlations of the strength limit change at temperature compression were determined, on the basis of which a mathematical prediction of the change of this indicator as an area was made, low operating and high technological temperatures.

According to the method [6], they were identified and presented in table 3 values of the maximum voltage shift in a wide range of operational and technological temperatures.

Table 3. Change in the estimated values of the maximum voltage shift depending on temperature, MP

Temperature		Extreme shear tension, MPA Bitumen/sulfur ratios		Technological criteria
°C	**°K**	**100/0**	**90/10**	
-20	253	22.18	10.23	
0	273	8.71	7.27	
20	293	3.42	5.16	
50	323	0.84	3.09	
80	353	0.21	1.85	
100	373	0.0817	1.31	
119	392	0.0336	0.95	
121	394	0.0306	0.92	Sulphur crystallization
140	413	0.0126	0.66	
160	433	0.0050	0.47	Producing the serous mixture
180	453	0.0019	0.33	
200	473	0.0008	0.24	Pawing a traditional mastic mix

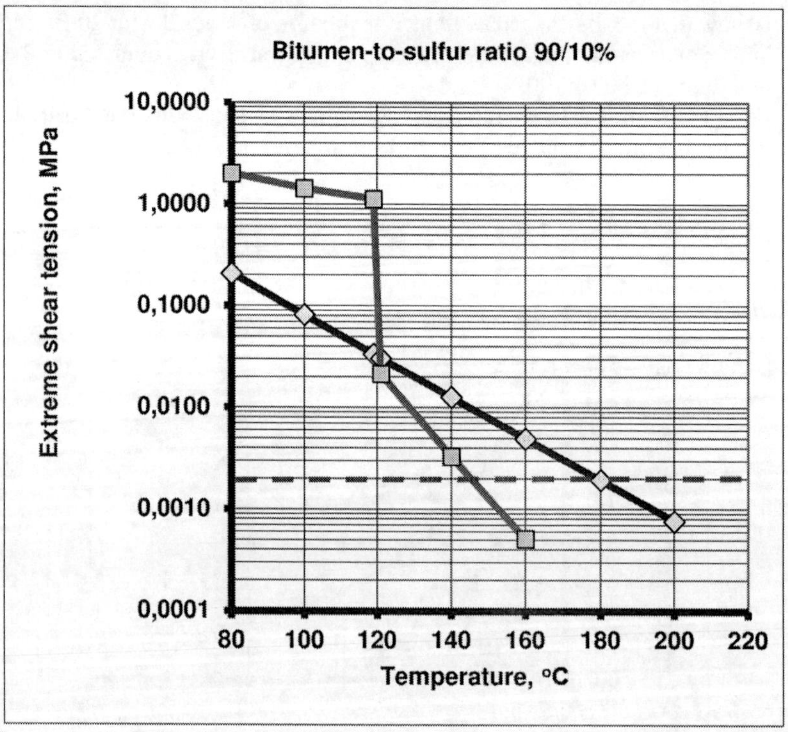

Figure 5. Changing the technological properties of mastic blends when stirring

Table 4. Change in the estimated values of the maximum voltage shift depending on temperature, MPa

Temperature		Extreme shear tension, MPA Bitumen/sulfur ratios		Technological criteria
°C	°K	**100/0**	**80/20**	
-20	253	22.18	10.23	
0	273	8.71	7.27	
20	293	3.42	5.16	
50	323	0.84	3.09	
80	353	0.21	1.85	
100	373	0.0817	1.31	
115	388	0.0405	1.02	Sulphur crystallization
117	390	0.0369	0.98	
140	413	0.0126	0.66	Producing the serous mixture
160	433	0.0050	0.47	
180	453	0.0019	0.33	Pawing a traditional mastic mix
200	473	0.0008	0.24	

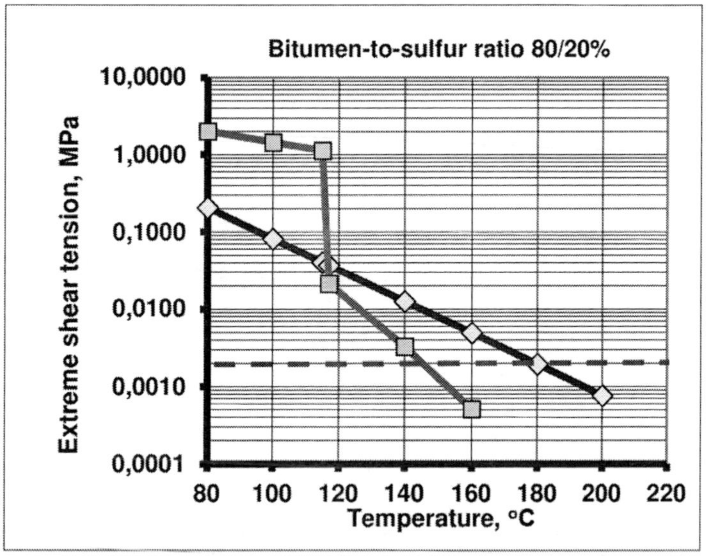

Figure 6. Changing the technological properties of mastic blends when stirring

Table 5. Change in the estimated values of the maximum voltage shift depending on temperature, MPa

Temperature		Extreme shear tension, MPa Bitumen/sulfur ratios		Technological criteria
°C	°K	**100/0**	**70/30**	
-20	253	22.18	10.23	
0	273	8.71	7.27	
20	293	3.42	5.16	
50	323	0.84	3.09	
80	353	0.21	1.85	
100	373	0.0817	1.31	

115	388	0.0405	1.02	Sulphur crystallization
117	390	0.0369	0.98	
140	413	0.0126	0.66	Producing the serous mixture
160	433	0.0050	0.47	
180	453	0.0019	0.33	Pawing a traditional mastic mix
200	473	0.0008	0.24	

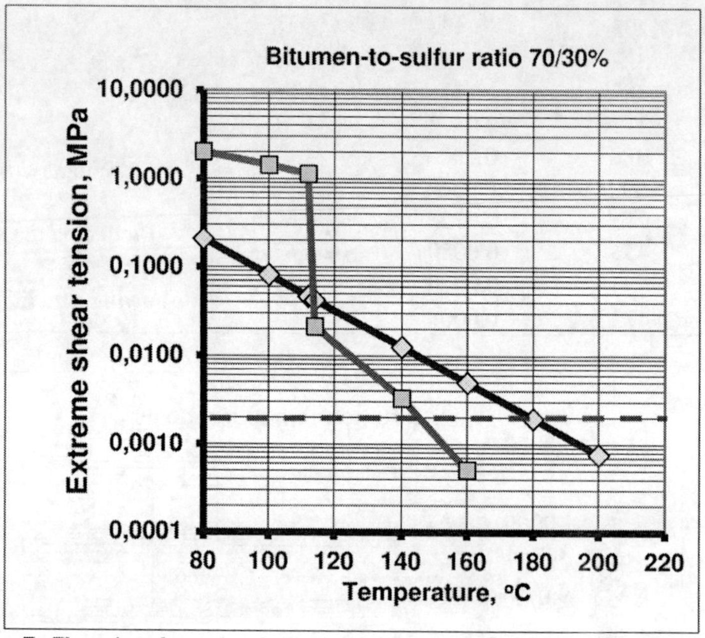

Figure 7. Changing the technological properties of mastic blends when stirring

5. Conclusion

Analysis of the presented results allows us to formulate the following conclusions:

- The hypothesis has been confirmed that the compositions of mastic asphalt concrete mixtures with crystallization-coagulation structure have increased resintorence and fracture resistance, which allows them to compete successfully with traditional asphalt concrete compacted mixtures with significant technological advantages.
- The mixing and laying of serolytic asphalt concrete mixtures can be made at much lower (about 40–50 degrees) technological temperatures, while sneering lye do not lose their ability to do.
- When designing mastic asphalt concretes, it is necessary to take into account the basic provisions of physical and chemical mechanics, which allows to regulate their structure and properties in order to obtain the required indicators of technological and exploitation characteristics.

References

[1] Shaffer R Kand Mellott D B 1972 Mastic asphalt concrete *Dept. of Transportation. Bureau of Materials Testing and Research* (Pennsylvania) 28 p
[2] Ivanchenko S N, Yarmolinskaya N I and Parfenov A A 2006 Ensuring the quality of asphalt concrete, taking into account the features of the components and the technology of sealing (Khabarovsk: Izd-in The Pacific. State. un-ta) pp 82–83. ISBN 5-7389-0450-8
[3] Rybyev I A 969 Asphalt concretes *Vyss. Sc.* (Moscow) 399 p
[4] Illiopolov S K and Uglova E V 2007 The longevity of asphalt concrete coatings in the face of

increasing dynamic impact of vehicles *Informavtodor* **4**

[5] Melik-Bagdasarov M S et al 1998 *Guide to the use of mastic asphalt concrete mixtures in the construction and repair of urban and highway roads* (Moscow)

[6] Kochnev V I and Kotlyarsky E V 2017 Normation of the technological properties of asphalt concrete mixtures based on analysis of their structural and mechanical properties *The association of articles of the Association of Asphalt Concrete Researchers*

[7] Alexeyev S S, Vasilyev J E, Kotlyarsky E V et al 2002 *Russian patent* 2002109768/03, 16.04.2002

[8] Kotlyarsky E V, Finashin V N and Uryev N B 1987 Formation of the structure of highly concentrated dispersal materials taking into account contact interactions in the process of compaction (example of asphalt concrete) *Colloidal Journal* **1**

[9] Rebinder P A 1966 Physics and Chemical Mechanics of Dispersal Structures *A collection of articles by the SOVIET Union* (Moscow: Science) 3–16 p

[10] Uriev N B 1980 High-concentrated dispersal systems *Chemistry* (Moscow) 319 p

[11] Vasilyev J E et al 2001 *A mixture of sulfur asphalt concrete mastic and mastic sulfur asphalt concrete. Specifications* TY 5718-002-53737504-01 (Moscow: VNIIGAS LLC, ARSP)

[12] Prikhodko V M, Vasilyev J E, Yumashev V M et al 2010 *Functional complex for the preparation of compositional seroasphaltbeton (options)* Patent RU 96 506 U1 no 2010106700/22, 26.02.2010/

[13] Rebinder P A 1979 Selected Works. Surface phenomena in dispersal systems *Physics and chemical mechanics* (Moscow: Science) p 381

[14] Yakhnin E D 1968 On the connection of the strength of the dispersal structure with the forces of interaction between elements. *Reporting by AN USSR* **178(1)** 152

[15] Kotlyarsky E V, Finashin V N and Uryev N B 1987 Contact interactions in the formation of asphalt concrete mixtures in the process of sealing *Improving the quality of construction of asphalt concrete black coatings* (Moscow: Proceedings of the Union) p 65

[16] Kotlyarsky E V and Finashin V N 1987 Stages of the formation of asphalt concrete mixtures when they are compacted *Ways to improve technology and improve the quality of road-building materials* (Moscow: MADI)

[17] Shalyt S Y, Mikhailov N V and Rebinder P A 1957 Effect of active filler and solvent on the properties of bitumen *Colloidal Journal* 19(2) 244–51

IOP Conf. Series: Materials Science and Engineering **832** (2020) 012025 doi:10.1088/1757-899X/832/1/012025

Modeling of live load influence in analisis of bridge structures endurance

N V Kozak

Saint Petersburg State University of Architecture and Civil Engineering, 4, Vtoraja Krasnoarmejskaja Str., St. Petersburg, 190005, Russia

E-mail: kozak.spbgasu@gmail.com

Abstract. This article presents the results of the research phase of existing methods for calculating the endurance of bridge structures. The presented stage of the analysis of the endurance of bridge structures is devoted to the creation of an algorithm for modeling the impact of moving loads on the controlled elements created. The process of obtaining the necessary initial data is described, including finite element modeling of the span and its subsequent structural part to obtain matrices and analyze surface effects. Available initial data and load characteristics were the basis for the development of a particular program for the analysis of the effects of loads on the bridge. It is shown that this algorithm can be used to compare the effects of design loads both among themselves and with the effects of possible actual loads.

1. Introduction

Calculation of the endurance of elements under conditions of the multi-cyclic circulating load is one of the keys to bridging structures [1]. Design of structures, taking into account the fatigue of the elements for the required service life, provides more excellent reliability and operational safety relative to the design, taking into account only short-term loads [2]. At the same time, it should be noted that at present, the general approach to assessing the endurance and cyclic fatigue of elements is not fully formed. The algorithms cited in various regulatory documents are based on significantly different hypotheses. The most common in these algorithms is the use, as one of the parameters of the functions, of a value characterizing the stress-strain state of the point under study, and the value is taken not at a specific point in time, but taking into account the duration and cyclic load [3]. To analyze the ways and methods for determining this parameter, the author, as part of a study of existing methods for calculating the endurance of elements of bridge structures, develops an algorithm for modeling the effects of moving loads on a controlled element.

2. Problem statement

In the analysis of existing methods for calculating the endurance of elements carried out at the previous stage of the study, a significant difference was noted in the methods used in the Russian Federation and the methods standard in several other countries [4]. According to the current standards of SP 35.13330.2011 [5], the indicator of the magnitude of the stress is the maximum absolute value, and the general structure can be expressed by the formula (1):

$$\frac{\sigma_{max}}{C} \leq K \cdot R_y \qquad (1)$$

In the formula (1), K is some composite coefficient, which includes coefficients for reliability, working conditions, and others; C is a coefficient characterizing the properties of the material and structure during cyclic operation; σ_{max} is the maximum absolute voltage in the element; R_y is the elastic limit of the material.

At the same time, as in the national norms of Ukraine [6] and the Eurocode [7], stress swings are used to assess fatigue effects, the general structure corresponds to (2).

$$\Delta\sigma \leq K \cdot \Delta\sigma_C \quad (2)$$

In the formula (2), K is a particular composite coefficient, which includes coefficients for reliability, working conditions, and others; $\Delta\sigma$ is the calculated range of stresses in the structural element; $\Delta\sigma_c$ is the relative threshold of possible stresses for the part category.

Modeling of the processes of the loads' impact on the bridge structures' elements in dynamics over time is associated with the need to compare the previous approaches, as well as to assess their correlation with possible actual loads. The task of this study is to develop a universal algorithm that allows evaluating the effect on the loads' elements that change over time.

3. Methods

The simulation algorithm was developed in the following particular initial conditions:

- Span structure – typical, steel-reinforced concrete, split, with an estimated span of 42.0 m, performed according to the standard project 3.503.9-119.93;
- Controlled parameter – normal (axial) stresses;
- Parameter control points – the lower flange of the wall of the main beam No. 2 in the middle of the span.

The above parameters were chosen for reasons of convenience of monitoring the performance and debugging of the developed algorithm.

A finite span model was created for obtaining information on the magnitude of the loading of the element under consideration. Modeling and further analysis were performed in SOFiSTiK (Fig. 1).

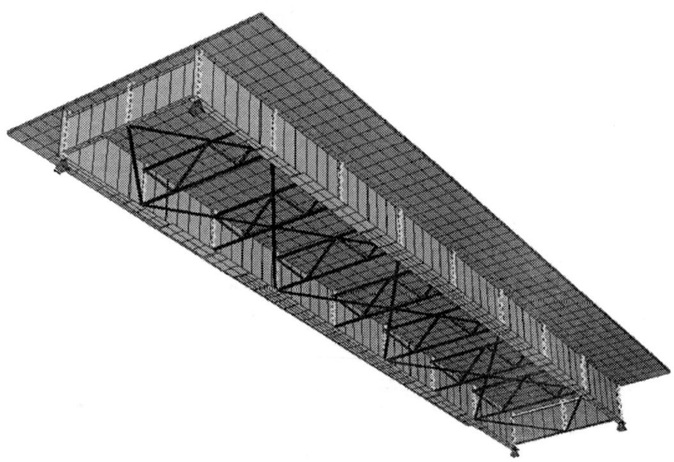

Figure 1. FEA model of bridge substructure

The calculation and analysis of the surfaces of influence on the standard stress value at the control point are based on the created two-dimensional grid of loads with unit loads on the surface of the spanning plate. The grid pitch along the X-axis (along the bridge) was 0.852 m, along the Y-axis (across the bridge) – 0.730 m. Based on the calculation results, a matrix of stress values was generated at the control point of size 21x51. A graphical interpretation of the matrix in the form of an influence surface is presented in Fig. 2.

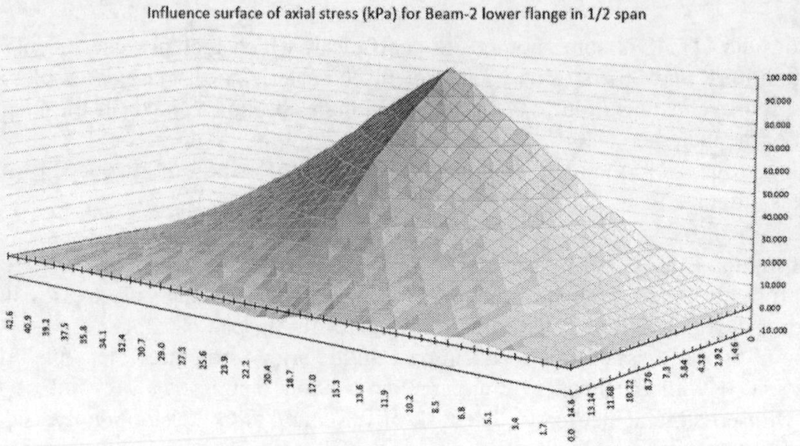

Figure 2. Influence surface of axial stress (kPa) for Beam-2 lower flange in 1/2 span

Obtaining data on the dynamics of changes in stresses in a controlled element depending on time is based on a developed program that runs a given load over the influence surface with a record at the time point of the summing effect from all load components. The main parameters for modeling a load run on a surface are:

- Type (distributed or mobile), scheme and values of the considered loads;
- Functions of movement of given loads of the format x = (t);
- Load arrangement in the transverse direction;
- Time discretization parameter (frequency of reading) and metric parameters of the influence surface matrix (grid step).

Point movable loads (car type) were programmed for movement along the entire length of the influence surface, taking into account the stages of the incomplete location of the load on the surface (entering the surface and leaving the surface). The function of the load movement was taken into account to study the influence of the combined effects of various loads when moving at different speeds.

Distributed loads (line type) specified in the simulation program were taken into account unambiguously or with a restriction on the sign of the arising forces (only tensile or only compressive). To calculate the effect of the distributed load, we used the integration of the surface function over a given line of load action.

The result of the algorithm is a sequence of values of the total control values (for this case, the voltage in the lower zone of the main beam) corresponding to a sequence of time points with a particular frequency.

4. Testing and results

The developed algorithm was used to compare the impact on the control element of the following calculated loads (Fig. 3). The movement of these loads was modeled uniformly, with a unit speed of 1 m/s:

- A14 according to SP 35.13330.2011 (Russia);
- A15 (endurance calculation scheme) according to DBN B.2.3-26: 2010 (Ukraine);
- Model 3 according to EN 1993-2 (Eurocode).

Besides, the following combinations of possible actual loads were considered taking into account possible differences in speed (Fig. 4):

- A truck with a total mass of 30 tons with a unit speed of 1 m/s;
- Passenger car with a conditional total weight of 4 tons and increased to 2 m/s speed;
- A combination of one convoy of trucks moving in the flow in one direction (4 cars, with a

distance of 20 m between each), in a neighboring convoy – cars (10 cars, with a distance of 15 m between each).

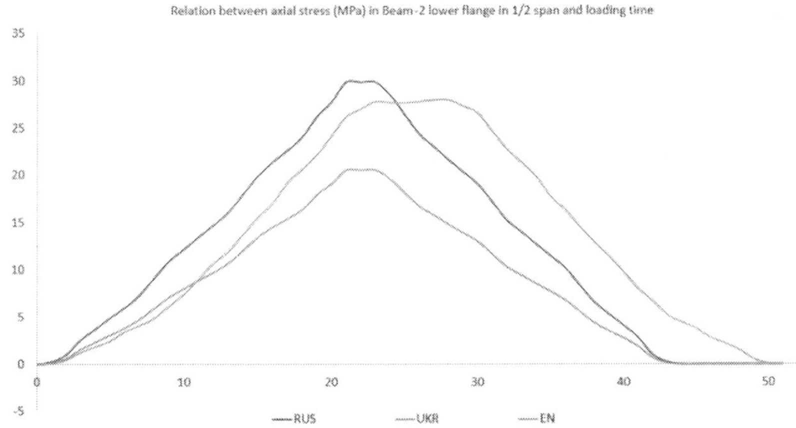

Figure 3. Relation between axial stress (MPa) in Beam-2 lower flange in 1/2 span and loading time for several national design loads (Russian, Ukrainian and Eurocodes)

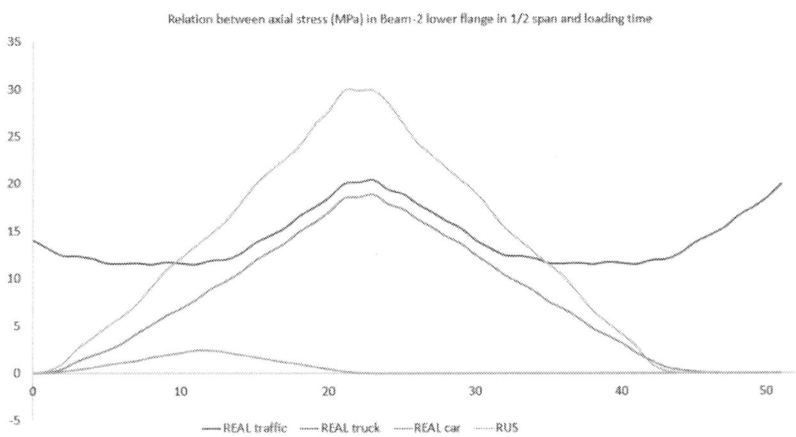

Figure 4. Relation between axial stress (MPa) in Beam-2 lower flange in 1/2 span and loading time for several proposed "real" loads and Russian design loads

As can be seen from Figure 3, the graphs obtained to compare the effects of different types of calculated loads and evaluate both the magnitude of the stresses and the absolute values. The application of the obtained algorithm to compare the possible actually and calculated loads (Fig. 4) helps to analyze the correlation of effects by controlled parameters.

5. Conclusion

Testing of the developed algorithm showed its operability, which allows it to be further used to model the dynamics of actions in more complex structures and under more complex conditions. In continuation of the present study, it is planned to use the described algorithm for modeling the operation of continuous structures in regimes with a frequent change of stress signs, statistical modeling of loads from actual traffic flows, as well as an active study of stresses in nodes along the upper belt of metal structures, at the junction with reinforced concrete slabs.

IOP Conf. Series: Materials Science and Engineering **832** (2020) 012025 doi:10.1088/1757-899X/832/1/012025

Acknowledgments
I express gratitude for the ideological inspiration and scientific advice to the professor of SPbGASU, Ph.D. Bystrov V.A. For help in the implementation of the algorithm, I express gratitude to Korolev E.A.

References
[1] Kuhlmann U, Bove S, Breunig S and Drebenstedt K 2019 Fatigue design of selected details in steel bridges *Structures* Available online 3 June
[2] Bystrov V and Kozak N 2018 Issues and concepts of road transport structures' development and provision of traffic and pedestrian safety *Transportat. Res. Proc.* **36** 103–07
[3] Korneev M M 2015 *Steel-reinforced concrete bridges: theoretical and practical design manual* (St. Petersburg: FSBEI HPE PGUPS)
[4] Kozak N 2019 Comparative analysis of domestic and foreign design standards in the field of endurance element calculations. Innovation and durability of transport infrastructure facilities (materials, structures, and technologies) ed. Klekovkina M P (St. Petersburg: SPbGASU) pp 92–6
[5] SP 35.13330.2011 *Bridges and pipes* 2011 The updated edition of SNiP 2.05.03-84* (Moscow: FSUE TsPP)
[6] EN 1993-2: 2006 Eurocode 3: *Design of steel structures* Part 2 *Steel bridges* (CC0 Universa)
[7] DBN B.2.3-26: 2010. *Bridges and pipes. Steel construction. Rules for the design of a bath* 2011 (Kiev: Ministry of Ukraine)

IOP Conf. Series: Materials Science and Engineering **832** (2020) 012026 doi:10.1088/1757-899X/832/1/012026

Improving the regulatory framework for the design of rigid pavements

V P Nosov, V V Ushakov, A A Fotiadi and A P Stepush

Moscow Automobile and Road State Technical University (MADI), 64, Leningradsky ave., Moscow, 125319, Russia

E-mail: nvpmadi@yandex.ru

Abstract. The study analyses local and foreign experience and trends in the field of improving the design of rigid structures of road pavements with the aim of a significant increase in their reliability and durability. It is shown that while ensuring the necessary quality of construction and timely completion of the full list of works on the current maintenance and repair of the construction of rigid pavements, they can serve without major repairs for 50 years or more. Based on the analysis, it was found that structures with a cement-concrete coating thickness of more than 30 cm on highly stressed highways have a resource of more than 100 million standard axial impacts. One of the factors hindering the practical implementation of new solutions is the lag in the development of the theory of calculation and the difficulty in determining the parameters necessary for calculation that characterize the properties of materials, taking into account their heterogeneity and variability during the designed service life. A catalog of rigid pavement designs is proposed based on the predicted number of standard axial impacts for the design life of the structure before overhaul.

1. Introduction

A significant increase in the reliability and durability of pavement structures requires a detailed analysis of the accumulated experience, taking into account the full life cycle, including the processes of calculating and designing pavements, construction technology and organization, construction quality control, activities carried out in the process of repair and maintenance of roads.

The inconsistency of the situation lies in the fact that the design life of pavements with asphalt coatings for a long time was defined to be in the range of 10–18 years, and that of pavements with cement concrete coatings was 25–30 years. These terms were used by specialists in the process of calculating and designing pavements, when determining the total number of impacts of wheel loads, by checking the coating materials for fatigue strength, for shear in the subgrade, and for the stability of the structure against frost heaving.

Based on these terms, the requirements of the standards for building materials used in the construction of pavements were established. These terms served as the guidelines in the selection of criteria for assessing and monitoring the quality of construction of the subgrade and layers of pavement. Funds for the repair and maintenance of road surfaces were mostly planned based on the indicated terms as well.

However, many examples can be given of the fact that long before the expiration of the design service life, cracks, ruts, potholes, ledges and other damages appear on the surface of the carriageway. As a result, the need for repair occurs before the expiration of the standard terms and the necessary repair work is postponed to a later period.

IOP Conf. Series: Materials Science and Engineering **832** (2020) 012026 doi:10.1088/1757-899X/832/1/012026

As local and international experience shows, the most important feature of roads is their heterogeneity along the length of the road. This leads to the fact that at almost any object, individual sections of road surfaces are destroyed before the design service life; however, others endure much longer. According to Figure 1, 10 % of asphalt concrete pavements require repair after 7 years, while 10 % of cement concrete pavements require repair after 20 years.

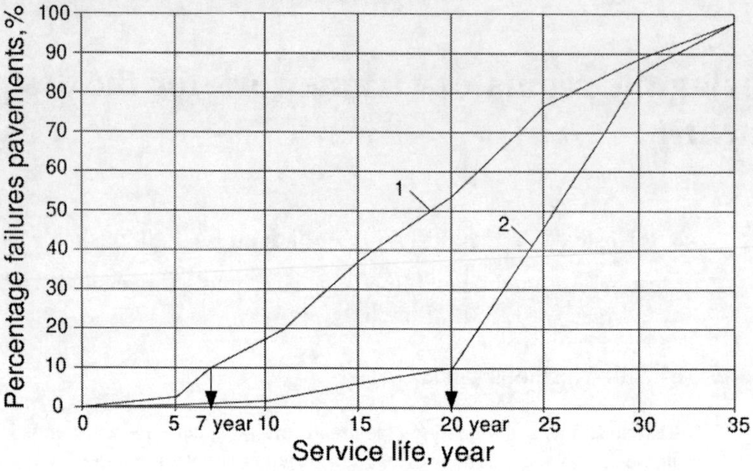

Figure 1. Service life of asphalt concrete (1) and cement concrete (2) pavements on German federal highways

An analysis of the reasons for the difference in design and actual pavement life [1, 2] showed that the totality of all reasons can be divided into the following four groups:

- actual deviations of the values of the intensity and composition of the movement during operation from the predicted values adopted in the design process;
- the actual deviations of the parameters of the constructed structures from those provided for by the project, as well as the heterogeneity of the subgrade and materials for road pavement, often significantly exceeding the guideline values;
- deviations of the actual parameters of atmospheric processes, mainly changes in temperature and the amount of precipitation, during operation from long-term average values accepted in the calculation and design of pavement;
- deviations in the quality and terms of the actually carried out repair and maintenance measures from those stipulated by current standards and rules.

Considering the first group of reasons, one should pay attention to the technological progress and the efficiency of road transport being the most often associated with an increase in axial loads, because it is the factor that allows the most appreciably and at the lowest cost increasing the carrying capacity of vehicles and their economic efficiency. But even under limited axial loads, increasing the number of axles and developing multi-axis trucks and road trains allows raising the carrying capacity to 50 tons and more. Freight and passenger considerably change over 25–30 years, which leads to large deviations in the actual intensity and composition of traffic in comparison with those values adopted in the calculation of pavements.

The second group of reasons includes possible deviations in the design of the subgrade and pavement in comparison with the parameters adopted in the project. This applies primarily to deviations in the thickness of individual structural layers and the significant heterogeneity of the applied subgrades and building materials. It is important to emphasize that most of these deviations are completely natural and are determined by the natural features of the starting materials, the designs and capabilities of road-building machines and equipment for the production of materials and construction technology.

As a result, there are regulated tolerances for deviations of the actual structural parameters and physico-mechanical properties of the materials embedded in them. It is also impossible to exclude unregulated and unacceptable deviations that are the result of violations of technology, the use of low-quality materials and the limited information received in the process of quality control of construction.

As a result, the parameters of the implemented facility differ from those provided by the project and, accordingly, we should expect a decrease in the service life in comparison with the design values.

The third group of reasons includes a set of atmospheric effects that cause changes in temperature and humidity of the subgrade and materials in the layers of the pavement structure. The temperature and shrinkage stresses in the layers with cement depend on how temperature and humidity change, as well as the stiffness and strength of the layers with bitumen. Abnormally high temperatures over a long period can cause large-scale damage on a large part of roads with asphalt pavement due to a significant decrease in the strength of asphalt concrete layers.

The decrease in the strength of materials as a result of alternate freezing and thawing depends on how the meteorological conditions are formed in a particular year or season. The change in the deforming properties and strength of the soil base depends on the duration of the period of autumn moisture accumulation, the intensity of freezing of the soil, the number of thaws in the winter period and the temperature behavior during thawing. These processes are characterized by their own characteristics, depending on the meteorological conditions prevailing in each specific year.

A significant dependence of the strength of materials on their moisture content should be noted as well. An objective assessment of the health of the material should be based on an analysis of the relative duration of its work in the structure in a dry, moderately moist and water-saturated state.

The fourth group of reasons includes deviations from routine maintenance of the road during operation, most often due to insufficient or untimely financing. The regulations envisage the implementation of a sufficiently large number of measures, including the cleaning of the roadway from dust and snow, the uniform distribution of deicing agents on the surface of the roadway, the timely restoration of roughness, the filling of joints and cracks with mastic or bitumen, the sealing of individual irregularities and potholes, and others. The delay in these measures significantly complicates the working conditions of the structure, increases the level of stresses and increases the intensity of destruction. The presence of irregularities increases the load on the structure due to vertical vibrations of vehicles.

The delay with closing cracks and joints in the coatings contributes to their filling with mineral particles and water, which causes the formation of potholes. Untimely patched potholes a significant part of the time are filled with water and, accordingly, create a spot of excess moisture. All this as a result activates the process of accumulation of damage and reduces the service life.

From the above it follows that there is a set of objective reasons that are underconsidered in the road pavement design, which leads to a significant difference between the design and actual service life. It is important to emphasize that the degree of responsibility for premature destruction at each stage of the investment cycle “design, construction, operation” is uncertain, which cannot be considered satisfactory, since excludes targeted elimination of the destruction causes.

A detailed analysis of the results of long-term monitoring of the condition of road surfaces on roads in our country and abroad also showed that at many objects there are separate sections of roads on which the surfaces last much longer than the standard service life. This circumstance served as the basis for the formulation of a wide range of studies in the direction of a significant increase in the durability and service life of pavement.

2. Problem statement

Taking into account the general tendency to increase the service life of pavement, in 2000 the Alliance of Asphalt Concrete Pavements of the USA proposed the concept of Perpetual pavements [3]. Perpetual pavements were defined as asphalt concrete pavements designed and constructed in such a way that they could serve more than 50 years without major repairs, requiring only periodic removal and replacement of the worn part of the top layer of asphalt concrete. The following advantages of such coatings were noted:

1. Low cost of the full life cycle costs due to the reduction of capital repairs;
2. Low loss of users due to a decrease in the total duration of time for repair works;
3. Reducing the environmental impact due to reduced material consumption and the reuse of the material from scraped off top layer.

The expert community working in the field of improving the design and construction of cement concrete pavements is also interested in increasing the overhaul life. A set of scientific studies is being carried out aimed at increasing the service life of up to 65 years.

As a result, the Long-lasting pavement direction was formed aimed at a significant increase in the durability of pavement designs proposed in the USA. Figure 2 shows an algorithm diagram for designing such structures.

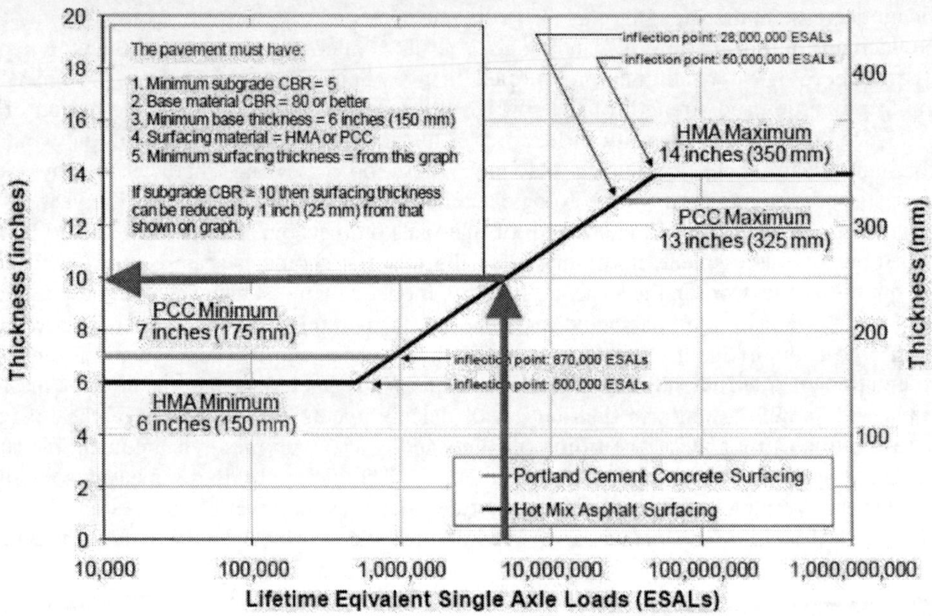

Figure 2. Dependence for determining the thickness of asphalt concrete and cement concrete pavement in the USA on the number of equivalent single axial loads

One of the main new requirements is a significant increase in the thickness of pavement layers for roads with high traffic intensity. An increase in the total stiffness of the coating reduces the load on the underlying layers of the pavement and the soil base. This allows inhibiting the accumulation of damage in these layers and helps to increase the overhaul life. Evidently from the plot shown in Figure 2, the thickness of the coatings is proportional to the number of standard axial loads. At maximum axial loads exceeding 100 million, the required thickness for cement concrete pavements is 32.5 cm and 35 cm for asphalt concrete pavements.

The principles of designing pavements with an extended service life lie in a significant increase in the thickness of the coating layers for roads with high traffic intensity. An increase in the total stiffness of the coating reduces the load on the underlying layers of the pavement and the soil base. This allows inhibiting the accumulation of damage in these layers and helps to increase the overhaul life. Obviously from the graph, the thickness of the coatings is proportional to the number of standard axial loads. At maximum axial loads exceeding 100 million, the required thickness for cement concrete pavements is 32.5 cm and for asphalt concrete 35 cm.

Studies in this direction have spread in European countries [4]. In the early 2000s, a working group was created from representatives of research institutes of interested European countries (Long-Life

Pavement Group), which prepared a review report that included information on experience indicating the possibility of structures serving longer periods without repair than design lifetimes. It was decided to consider durable types of pavements as those types of pavements that will not develop significant damage in the working layer of the subgrade and base layers that require overhaul of the pavement structure. At the same time, measures to replace the top layer are considered acceptable.

It is characteristic that at present the total thickness of the layers of cement concrete pavements assigned during the design of pavements in different countries ranges from 22 cm to 31 cm (see Figure 3).

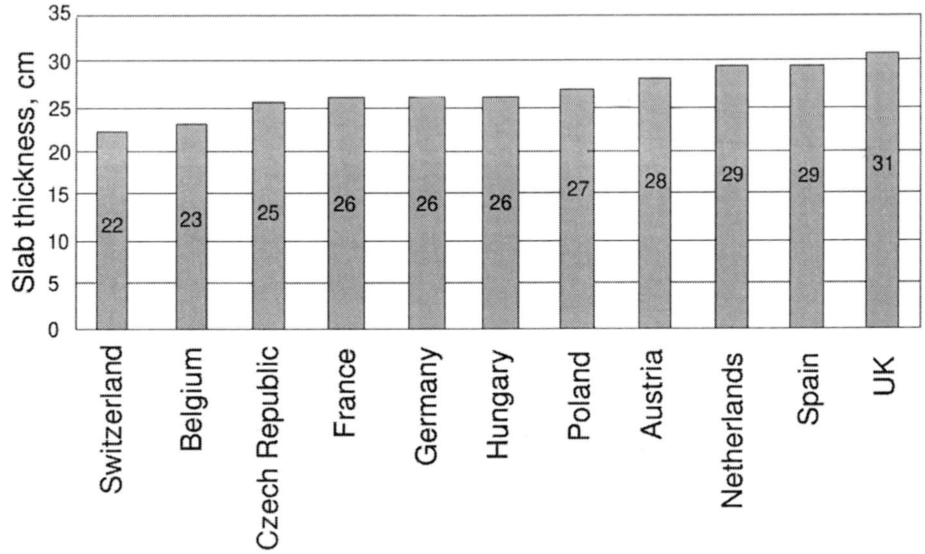

Figure 3. Thickness of concrete coating in different countries of Europe

Road users are making ever greater demands on the operational condition of roads and, as a consequence, on improving methods for designing pavements. Today, road pavements work in difficult conditions of an ever-increasing intensity of vehicle traffic, in a mode where residual deformations quickly accumulate, intensive wear of road surfaces occurs, and various types of damage appear. This leads to a decrease in the service life of the pavements. The use of structures and materials capable of increasing the service life of pavements and coatings is an urgent task.

In the Russian Federation, road design pays unsufficient attention to pavement construction. The factual experience of operating a variety of road structures, with the identification of the most effective solutions, is not fully taken into account. Therefore, the choice in favor of a particular pavement design in many cases is carried out without taking into account the accumulated experience of using well-established standard designs.

Thus, there was a need to improve the existing practice of designing road structures to bring them in line with the international trend of increasing the durability and service life of road pavements.

3. Basics of proposed design methodology for rigid pavement construction

In world practice, there are two fundamental approaches to the design of pavement designs [5]. The first approach is based on calculating the parameters of the stress-strain state of the structure and comparing these parameters with the maximum permissible values. The second approach involves the appointment of parameters of typical designs based on a limited number of the most significant factors. At the same time, great importance is attached to structures tested at testing grounds or in real operating conditions. Typical pavement designs are widely used in the USA, Austria, Germany, Italy, France and other countries.

The basis for the practical account of cement concrete fatigue in structures is the test results of samples under the influence of repeated loads with a constant amplitude and frequency of loading (see Figure 4).

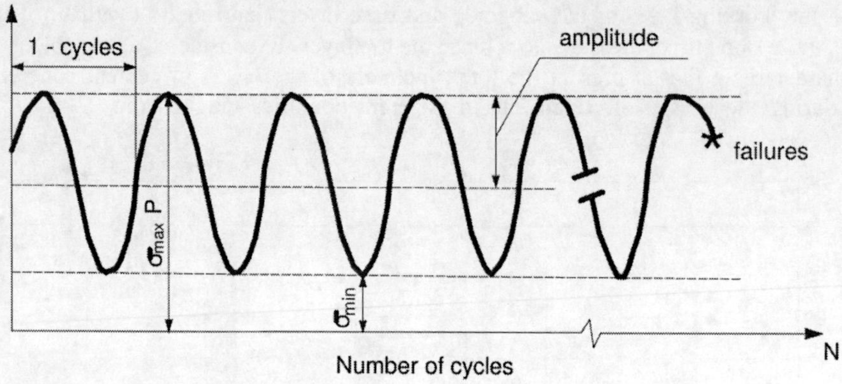

Figure 4. Changes in stress during fatigue tests of concrete samples

According to the results of such tests, the number of cycles necessary for the destruction of the sample at voltages of different levels is established. The stress state level in a cycle is the ratio of the maximum stress to the ultimate strength under a single loading.

On the basis of experiments, it was found that the number of cycles before the destruction of sample N can be represented as function F of the ratio of the maximum stress to the ultimate strength of concrete with single loading R:

$$N = F \cdot \left(\frac{\sigma_{max}}{R} \right) \tag{1}$$

In addition, it was also established that the number of loading cycles before fracture of the sample depends on the characteristics of the loading cycle. The characteristic of the loading cycle is called the ratio of the minimum stress in the cycle to the maximum value.

As a result of numerous experiments to determine the parameters of fatigue curves, a large number of models have been proposed that also take into account a number of other parameters, such as the frequency of loading, the moisture content of cement concrete, the time intervals between cycles, the rate of rise of the load, the type of aggregate in concrete, the type of cement, etc. However, characteristic of all tests was the constancy of the level of stress within one implementation. The dependencies obtained in this way served as a means for predicting the operability and designing various structures in which the parameters of the loading cycles vary significantly during their operation.

In relation to the calculation of cement-concrete coatings of roads and airfields, the empirical dependences presented in Figure 5 are most widely used. It is characteristic that a number of dependences gave fatigue coefficients greater than 1, which would seem to contradict the physical nature of the fatigue phenomenon. However, in one case this is due to the phenomenon of dynamic hardening due to short-term application of the load, (curve 3), and in the other cases (curves 1 and 2), with the design features of pavements, which are plates on an elastic base, statically indeterminable systems with a large number of extra connections.

With regard to road coatings, it is well known that the stresses caused by the passage of cars vary significantly due to differences in the carrying capacity of vehicles, uneven loading, position with respect to the edge of the roadway, condition of the subgrade, temperature regime of the coating and other factors. Figure 6 schematically shows how tensile stress can change in an arbitrary section of a cement concrete pavement under the influence of passing cars. The magnitude of the voltage at an arbitrarily fixed point in time is a random variable, and the function of the voltage change in time can be assigned to the class of random functions.

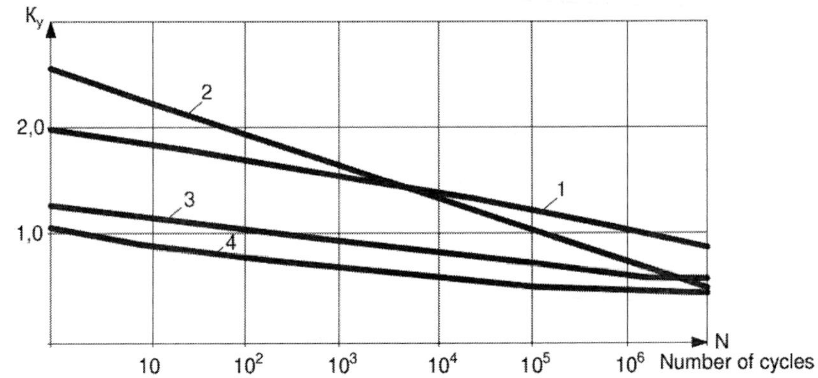

Figure 5. Dependence of cement concrete fatigue coefficients on the number of load application cycles: 1) according to SNiP 11-47-80; 2) according to VSN 197-93; 3) from eq. $K_y = 1.3 - \frac{\lg N}{7}\left(1.3 - \frac{0.5}{1-0.616\cdot\rho}\right), \rho = 0$; 4) as $K_y = 1.08\cdot N^{-0.063}$

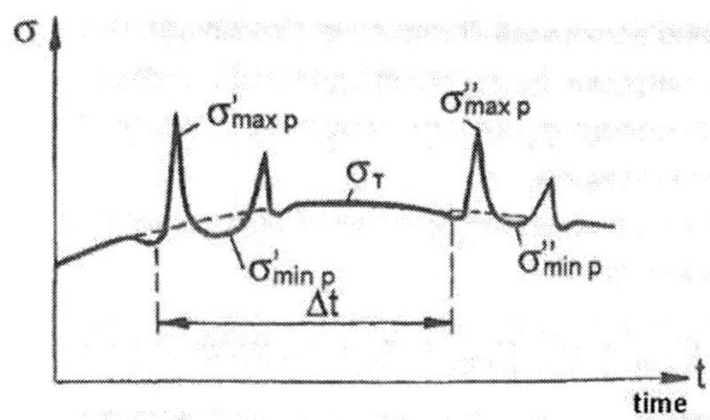

Figure 6. Change in stress in cement concrete coatings under a moving car

To estimate the maximum number of load cycles for a structure undergoing a stress state characterized by an arbitrary random function, the Miner Rule hypothesis (linear summation of damage) has been widely used.

The hypothesis is based on the assumption that each load application cycle can be characterized by a unit damage, which can be defined as:

$$\alpha_i = \frac{1}{N_i}, \tag{2}$$

where N_i is the number of cycles with a given level of stress and a cycle characteristic that causes the destruction of the cement concrete sample.

The process of changing the measure of damage is a function of time, and the value at each fixed point in time can be a characteristic of the current state of the pavement. The closer this value is to 1, the sooner the destruction of the pavement will occur and the less residual durability the structure has at this point in time.

The concept of a measure of damage is also introduced, for which the sum of single damage is taken for the period of interest or after a certain number of cycles of load exposure.

$$D = \sum_{i=1}^{C} \alpha_i, \tag{3}$$

where C is the number of stress repetitions for the time period of interest.

From the definition of a unit damage, it follows that in the process of structure material operation without destruction, the following inequality must be fulfilled:

$$D < 1. \tag{4}$$

D reaching 1 means that the number of cycles of exposure to the load has reached the limit value. The specified ratio allows predicting the performance and calculating the resource structures of rigid pavement.

Therefore, the essence of the proposed methodology is, based on an analysis of the parameters of the stress-strain state of cement concrete layers in the considered pavement structure under a flow of conventional vehicles with standard axial loads, to calculate the design life in the form of the number of standard axial impacts, for example – P11.5.

4. Main results

Based on the calculations in accordance with the above methodology and accounting national experience in the design, construction, operation and diagnostics of rigid road pavements, as well as advanced foreign experience, MADI specialists developed standard GOST R "General automobile roads. Designing rigid pavement. Typical designs".

The standard governs the parameters of rigid pavement with monolithic cement concrete pavement. The standard allows normalizing the most important parameters of the pavement design: the thickness of pavement, the type and parameters of the base, the thickness of additional layer of the base, the distance between compression joints, the diameter of pins and the distance between them. The value of the predicted number of passes of standard axial impacts for the design service life of the pavement structure is taken as the main factor determining the design parameters. As a calculated load, A-11.5 was adopted.

The fundamental principle in choosing a standard design is to predict the number of impacts of standard axial loads of the most loaded road lane for the design period of pavement. At the same time, the predicted flow, including the entire range of cars from light trucks to multi-axle road trains, is replaced by the amount of standard axial loads A-11.5 equivalent in the destructive effect.

The basis for predicting the intensity and composition of the traffic flow is the results of a feasibility study on the need for the construction or reconstruction of a highway.

The predicted composition of the traffic flow should correspond to the categories and types of vehicles in accordance with GOST 32965.

The modulus of elasticity of the soil of the working layer of the subgrade should be at least 40 MPa. For road sections with a lower modulus of elasticity of the working layer of the subgrade, it should be increased through special measures to regulate the water-thermal regime, strengthening or replacing the soil.

The assigned design service life of rigid pavement is at least 30 years. The class of concrete for tensile bending for the standard structures with cement concrete coating presented in this standard is assumed to be Btb4.4 MPa as per GOST 26633.

The material of the base layer is assigned depending on the availability of road building material in the construction area and the number of equivalent impacts of standard axial loads for the design service life of the pavement structure.

The use of typical designs of rigid road pavements will improve the quality of design, increase the reliability road pavements, increase their overhaul life and significantly reduce the cost of road users.

5. Conclusions

For an objective assessment of the advantages and possible disadvantages of the proposed structures, it is extremely important to carry out detailed quality control of construction works and regularly conduct instrumental monitoring of the state of the constructed carriageways with registration of such important characteristics as longitudinal evenness, rut depth, the number and location of cracks and other damage.

Table 1. Parameters of typical designs of rigid pavement with cement concrete coating, depending on the resource P11.5

Parameter name		Types of pavement design					
		R-70	R-50	R-20	R-5	R-1	R-0.3
Number of equivalent impacts of standard axial loads [mln]		≥50	20–50	5–20	1–5	1–0.3	≤0.3
Thickness of the cement concrete coating [cm]		28	27	25	23	20	18
Type and thickness of the base [cm]	Hard rolled concrete	20 –	18 –	16 –	–	–	–
	Cement treated material	– 24	– 22	– 20	–	–	–
	Mineral material not treated with binder	– –	– –	– –	20	18	18
Transverse seams	Distance between seams [m]	6	6	5	5	5	4.5
	Pin diameter [mm]	30	30	25	25	25	25
	Pin length [cm]	50	50	50	50	50	50
	Pin spacing [cm]	25	25	25	25	25	25
Transverse expansion seams	Pin diameter [mm]	–	–	–	25	25	25
	Pin length [cm]	–	–	–	50	50	50
	Pin spacing [cm]	–	–	–	25	25	25
Longitudinal seams	Pin diameter [mm]	20	20	20	20	20	20
	Pin length [cm]	80	80	80	80	80	80
	Pin spacing [cm]	100	100	100	100	100	100

References

[1] Nosov V P 2013 Cement-concrete coatings of highways. Damage prediction based on mathematical modeling: monograph (Moscow: MADI) 228 p

[2] Nosov V P, Dobrov E M, Chistyakov I V, Borisiuk N V and Fotiadi A A 2017 Mathematical Modelling of Cracking Process in Concrete Pavement Highways Int. J. of Applied Engineer. Res. **12(23)** 13158–64

[3] Pavement Technologies Greenroads™ Manual v1.5PT-1 Long-Life Pavement Retrieved from: http://www.transport-links.org/transport_links/filearea/publications/1_764_PA3736_2001.pdf

[4] The full report (Report 250) can be found at: http://www.trl.co.uk/online_store/reports_publications/trl_reports/cat_highway_engineering/report_design_of_long-life_flexible_pavements_for_heavy_traffic.htm

[5] Long Life Pavements and Success Stories ISBN: 2-84060-226-1 PIARC Technical Committee 4.3 Road Pavements. International Standard Book Number 2-84060-226-1

[6] Transportation Research Circular E-C118 Pavement Lessons Learned from the AASHO Road Test and Performance of the Interstate Highway System Transportation Research Board Pavement Management Section July 2007

Development of the formulas for estimating the travel time of rainwater slope runoff and channel flow for the anthropogenic territories of small catchment basins

Paudyal Soorya Prakash, K Y Sokolova and I V Chistyakov

Moscow Automobile and Road State Technical University (MADI), 64, Leningradsky Ave., Moscow, 125319, Russia

E-mail: suprapa@mail.ru

Abstract. The article presents developed formulas for estimating the travel time of rainwater slope runoff and channel flow based on a unified schematization of mixed catchment basins with wide anthropogenic territories together with diversion channels. The estimation of common travel time for the rainwater overland flow on the adjacent watershed and further along the diversion channel is derived by modeling the catchment basins preserving their natural configurations. The necessary hydrological and hydraulic equations for describing the unsteady rainwater slope runoff and channel flow formation are chosen and proposed taking into consideration the proper factors affecting the flow. When developing a general formula for calculating the preliminary total travel time of the slope and channel flow have been taken into consideration the proper patterns of rainwater runoff on the inclined surfaces and stepped sectors applicable for estimation of the unsteady slope runoff and channel discharge (flow rate) in case of using the differential equations of one-dimensional hydraulics. The developed formula can be used for determining the travel time in predicting rainwater runoff for designing the system of road surface drainage and culverts facilities in mixed catchment basins with terraced paddy fields and other slope areas. Authors have proposed some recommendations and drawn the relevant conclusions.

Key words: terraced catchment basins, the principle of "extreme intensities", rainwater slope runoff and channel flow, the travel time of the slope runoff and channel flow, the total estimated rainfall duration.

1. Introduction

Much of the agricultural land of tropics and subtropics of Asia, Africa, Europe and America is used for growing rice. For this purpose, the natural inclined surfaces of the slopes are transformed into a system of terraces (steps) with zero gradient of their bottom and small stable earth dykes (levees or bunds) at their edges to preserve the water for a long period of time. A huge part of the paddy fields is located in the most rain-prone regions, forming a kind of extensive catchment basins with significant areas of anthropogenic territories. Very often, terraced basins are located in priority areas of road transport.

One of the causes of damaging and destroying the small culverts and other drainage facilities on the roads in these regions, often causing significant economic losses, is insufficient reliability of the prediction of rainwater runoff for selecting rational structures and organizing the optimal surface drainage system. Such a situation requires an adequate modeling of the catchment areas using more accurate hydrological and hydraulic equations for estimating rainwater slope runoff and channel flow rate.

The present article discusses the schematization of the surface of mixed catchment basins with differentiated consideration of inclined and stepped slope surface segments in their uninterrupted connection with the diversion channel. The main factors and characteristics of such mixed catchment basins causing the formation of rainwater surface runoff for determining the more accurate travel time of the slope and diversion channel flow are taken into account in traditional assumptions. The necessary transitional approaches have been taken for estimating rainwater runoff when designing drainage facilities by using the differential equations of unsteady slope runoff and channel flow. For describing the unsteady process of forming the rainwater slope runoff and channel discharge, changing in course of time and length the of the catchment area and diversion channel through the mixed surfaces of the catchment basin, it is necessary to determine the possibly utmost accurate time duration for the concentration of rainwater runoff. When using the differential equations of one-dimensional hydraulics, the maximum rainwater runoff at the given point can be determined depending on the estimated time duration that is sufficiently accurate for establishing the extreme discharge, variable along the channel. The proposed formulas for estimating travel time of the mixed slope and diversion channel are substantiated by the logically correct schematization of the catchment basin and the diversion channel, are proposed the hydraulic relations verified in practice.

Applied in estimating rainwater runoff in small catchment basins *the principle of "extreme intensities" represents itself the travel time of a portion of water through the slope and channel from a hydraulically most distant point of watershed boundary to the given point* (closing cross section). When determining the total travel time for catchment basin with anthropogenic terraced areas, it is important to compute differentially the travel time of separate sectors of the inclined and terraced slopes and diversion channel along the catchment area with sequent conjugation by using the relevant hydraulic laws and principles of rainwater runoff and schematization of the catchment basin.

2. Modeling a mixed catchment basin and equations for describing the rainwater slope runoff

According to the method developed by Professor O.V. Andreev and A.F. Shahidov [1] we get too short estimated time durations of rainfall for determining the flow rate in small drainage basins. Such short durations cannot be used for determining the extreme discharges when using the differential equations of the unsteady flow of rainwater proposed below in a continuous process of slope runoff and channel flow in small catchment basins.

The methods proposed in [2] and [3] reflect more reasonably the physical processes of the formation and flow of rainwater. However, these standards include a variety of approximate nomograms, tabular coefficients and various simplified multipliers. For example, in [2] the fundamental parameters are based mainly on the results of calculations with a single rainfall intensity of 0.5 mm/min only for the inclined slope surfaces and later correlated to other cases. In [3] the travel time of the slope runoff is determined taking into account the natural zones depending on the grouped hydro-morphological characteristics of the slopes. This norm is also not acceptable when calculating the travel time of slope runoff for terraced areas.

Along the slopes of the catchment basins in the tropics and subtropics, we meet mixed surfaces, alternating the areas with ordinary inclined surfaces (with different vegetation cover and gradients) *and terraced areas of paddy fields with zero gradients at the bottom of the steps* (Figure 1). The terraces along their edges have stable small earth dykes (levees or bunds) that retain water for a long time. The runoff passes over them like a weir and the flow is broken after passing them, which requires the use of weir formulas.

The total travel time of slope in such mixed catchment areas must be determined by successive consideration of each alternating inclined (Figure 2) and stepped (Figure 3) sectors of the slope strip (since the inclined sector is always followed by a stepped one and vice versa) separately, taking into account the fundamentally different formulas (1) and (2) for the formation and flow of rainwater runoff on them. For the convenience of schematizing such different sectors, it is better to introduce a unified numbering system for these successive sectors along the modeled elementary strip stretching from the beginning of watershed to the diversion channel, taking the beginning sector as inclined (let it be even with just 1m length), followed by a stepped one, and so forth.

The process of the formation and flow of unsteady rainwater runoff with shallow depth along an elementary strip (1m wide) of the inclined slope of catchment surface can be described with sufficient accuracy by the following differential equation of one-dimensional hydraulics [1, 5–10]:

$$\frac{\partial q_{sl}}{\partial x}+\frac{\partial h_{sl}}{\partial t}=a-b;\quad q_{sl}=Ch_{sl}^{z};\quad C=\frac{i_{slav}^{1/2}}{n_{sl}};\quad z=1.5+y_{sl},\tag{1}$$

where q_{sl} – elementary flow rate on the elementary modeling slope strip of 1m width; x – current coordinate, directed from the beginning point of the watershed down the slope; h_{sl} – depth of the slope runoff on the strip; t- time; a, b – intensities of rainfall and runoff losses; C – Chezy coefficient; i_{slav}, n_{slav} – weighted average values of the gradient and roughness coefficient of the considered slope strip; z – an exponent; y_{sl} – an exponent depending on the flow depth and roughness coefficient. For shallow depths, it is recommended $y_{sl}=1.5(n_{slav})^{1/2}...1.7(n_{slav})^{1/2}$ [4]. However, this issue requires serious study and correction.

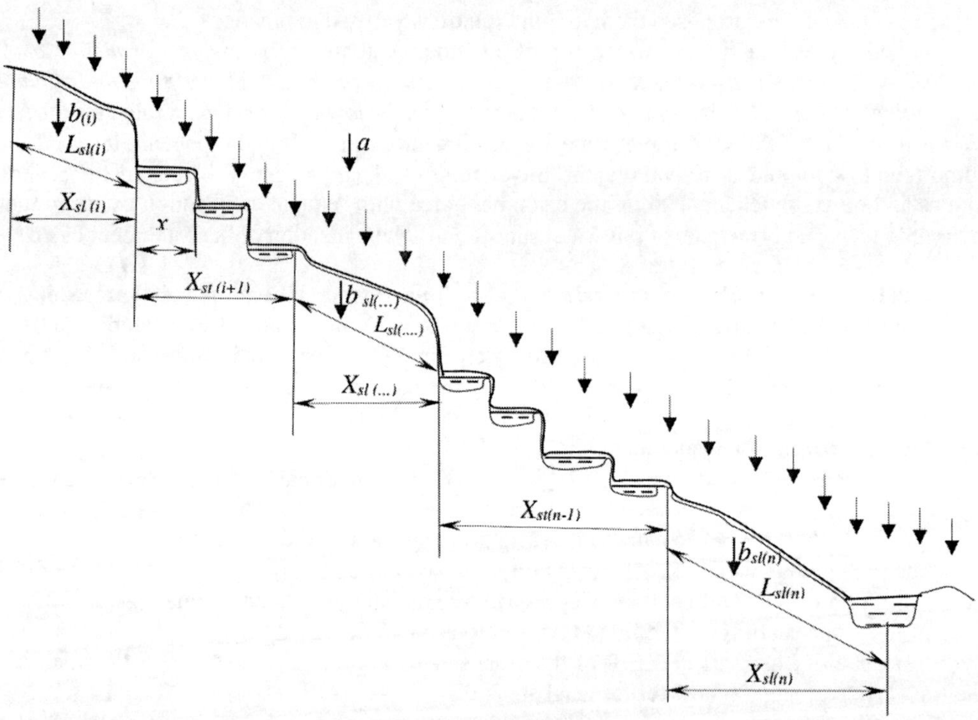

Figure 1. Schematization of the formation of rainwater runoff on mixed slopes: $L_{sl(i)}$ – length of the *i-th* inclined section of the slope strip; $X_{sl(i)}$ - horizontal projection of the inclined length $L_{sl(i)}$; $X_{st\ (i+1)}$ – length of the $(i+1)$ *-th* stepped sector of the slope strip; a, b – intensities of rainfall and runoff losses

The process of the formation and flow of rainwater runoff on terraced slopes with a zero gradient at the bottom of the steps (terraces) with a shallow pond differs from the above-said case and can be described (for a selected model strip of 1m width) by the differential equation of rainwater flow and a scheme of "continuous layer" taking into account the waterfall over the small earth dykes (levees or bunds) by the following equations:

$$\frac{\partial q}{\partial x}+\frac{\partial h}{\partial t}=a;$$
$$q=m\sqrt{2g}\;h^{3/2}, \tag{2}$$

where q – the linear flow rate along the model strip; x – the current coordinate, directed from the beginning point of the watershed down the terraces, equals to the width of the step; t – time; a – rainfall intensity without losses (steps are saturated with water); m – weir coefficient; h – the flow depth over the mud dyke of the considered terrace; g – gravitational acceleration.

When solving the equation (1) numerically to obtain a flow rate with permissible accuracy, for example in finite differences, it is necessary to consider the process in course of time with the optimal time intervals Δt and length steps Δx until a full flow rate is established at the end of a given inclined sector. For this, it is necessary to use a previously calculated in some way the time duration, approximately equaling to the travel time. An approximate travel time of the runoff can be determined, based on the proper schematization of catchment basins and the simulation of slope runoff formation, followed by the use of acceptable hydrological and hydraulic formulas.

The instantaneous average velocity $v_{slav(i)}$ on the considered *i-th* segment of the inclined slope surface (Figure 2) can be determined by the Chezy-Manning formula [7, 5, 8]:

$$v_{slav}=\frac{1}{n_{slav(i)}}i_{slav(i)}^{1/2}\,h_{slav(i)}^{2/3}, \tag{3}$$

where $n_{slav(i)}$, $i_{slav(i)}$ – weighted average values of roughness coefficient and surface gradient of the *i-th* sector of inclined strip; $h_{slav(i)}$ – average depth of the runoff lair on the *i-th* section of the slope strip lengthening $L_{sl(i)}$.

The elementary discharge at the entrance of the *i-th* inclined sector is equal to the elementary discharge at the end of the preceding stepped sector, because the inclined and stepped sectors alternate each other, starting from the inclined one, i.e.:

$$q_{slen(i)}=q_{st(i-1)}. \tag{4}$$

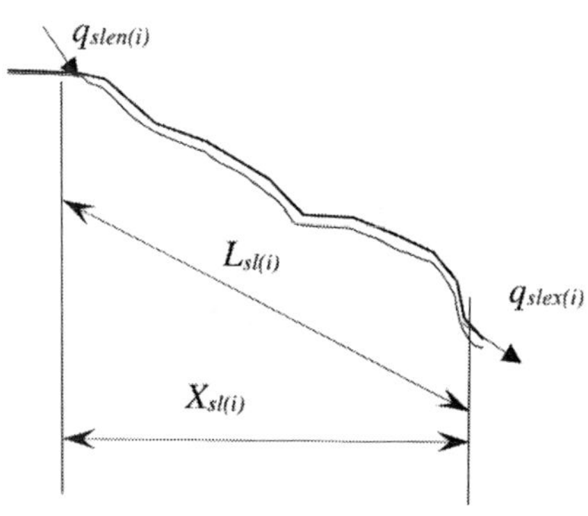

Figure 2. Schematization of runoff formation on the inclined surface: $q_{slen(i)}$, $q_{lex(i)}$ – runoff rates at the entrance and exit

When schematizing the trajectory of the rainwater runoff along an isolated inclined strip from the extent watershed point down to the diversion channel, we can adopt the mean values of the velocity $v_{slav(i)}$, depth of the runoff layer $h_{slav(i)}$ and the unit discharge rate $q_{slav(i)}$ within the length of the declined sector $L_{sl(i)}$ for determining the slope runoff travel time $t_{sl(i)}$ [4, 6, 10–12]. Since, at the beginning point of the model strip with 1m width (at the boundary line of the watershed) the discharge is zero and reaches the maximum at the end of the strip with a length X_{sl} equaling to $\boldsymbol{a}X_{sl}$, formed due to the rainfall intensity $\boldsymbol{a}$. The designed intensity of the rainfall duration should be defined as the exceeding intensity, remained after subtracting the losses and transformed from the hourly intensity to the estimated one multiplying by the transform factor, depended on the travel time. Based on this, it can be taken that along the *i-th* sector of the inclined slope an average runoff rate equaling to $q_{slav(i)} \approx 0.5\boldsymbol{a}X_{sl(i)} = v_{slav(i)} \cdot h_{slav(i)}$ is formed due to the estimated intensity $\boldsymbol{a}$. Taking into account the discharge at the beginning point (at the entrance) of the *i-th* slope sector $q_{slen(i)}$ from the preceding one and an additionally formed average discharge due to the intensity $\boldsymbol{a}$, equaling to $0.5\boldsymbol{a}X_{sl(i)}$ on the very sector and using the expression (3) and (4) for the modeled strip we get the following formulas:

$$h_{slav(i)} = \left(\frac{q_{slav(i)} n_{slav(i)}}{i^{1/2}_{slav(i)}} \right)^{3/5};$$

$$q_{slav(i)} = q_{slen(i)} + 0.5aX_{sl(i)},$$

$$v_{slav(i)} = \frac{q_{slav(i)}}{h_{slav(i)}} = \frac{\left(q_{slav(i)} + 0.5aX_{sl(i)} \right)^{2/5} i^{3/10}_{slav(i)}}{n^{3/5}_{slav(i)}}; \quad (5)$$

where $q_{slav(i)}$ – the average linear discharge on the sector $L_{sl(i)}$; $q_{slen(i)}$ – discharge at the entrance of the sector, equaling to the discharge $q_{stex(i-1)}$ at the end of the $(i-1)$*-th* sector.

Using the formula (5), we get the travel time $t_{sl(i)}$ for the inclined section of slope $L_{sl(i)}$:

$$t_{sl(i)} = \frac{L_{sl(i)}}{v_{slav(i)}} = \frac{L_{sl(i)} n^{3/5}_{sl(i)}}{i^{3/10}_{slav(i)} \left(q_{slen(i)} + 0.5aX_{sl(i)} \right)^{2/5}}. \quad (6)$$

Similarly, for the stepped section of the slope $X_{st(i+1)}$ (Figure 3) the travel time can be determined by taking into account the average speed of the waterfall over the dykes (levees or bunds) at the edges of the steps, neglecting the free waterfall time along almost vertical walls. By such a schematization of the steps (see Figure 3) with a waterfall threshold width of 1m, the traveling time $t_{st(i+1)}$ of the stepped section, average speed $v_{stav(i+1)}$, average depth $h_{stav(i+1)}$ of the stream over the dykes (levees or bunds) and average linear discharge $q_{stav(i+1)}$ on the stepped section $X_{st(i+1)}$ can be described [7, 13] by formulas:

$$t_{st(i+1)} = \frac{X_{st(i+1)}}{v_{stav(i+1)}}; \quad v_{stav(i+1)} = m\sqrt{2g}\, h^{1/2}_{stav(i+1)};$$

$$h_{stav(i+1)} = \left(\frac{q_{stav(i+1)}}{m\sqrt{2g}} \right)^{2/3};$$

$$q_{stav(i+1)} = q_{sten(i+1)} + 0.5aX_{st(i+1)},$$

where m – discharge coefficient of the weir (for weirs of a practical profile with a small depth of head flow, it is usually adopted m = 0.32 ... 0.45); g – gravitational acceleration; $q_{sten(i+1)} = q_{slex(i)}$ – unit discharge at the beginning of the $(i+1)$*-th* stepped section $X_{st(i+1)}$ (i.e., at the end of the *i-th* inclined section $L_{sl(i)}$ of the strip under consideration).

Taking the mean value of m = 0.38 and g = 9.81 m/c^2 in the above formulas, we obtain the travel time $t_{st(i+1)}$ for the stepped section $X_{st(i+1)}$:

$$t_{st(i+1)} = \frac{X_{st(i+1)}}{v_{stav(i+1)}} = \frac{0.71X_{st(i+1)}}{\left(q_{sten(i+1)} + 0.5\,a\,X_{st(i+1)}\right)^{1/3}}. \quad (7)$$

Taking in view the above said, the total travel time T_{slt} of the considered slope strip will be:

$$T_{slt} = \sum_{i}^{n} t_{sl(i)} + \sum_{i+1}^{n+1} t_{st(i+1)}, \quad (8)$$

where $i = 1,3,5,7,\ldots n$.

In the formula (8), the travel time $t_{sl(i)}$ of every inclined section and $t_{st(i+1)}$ of the stepped one in each strip under consideration should be determined in strict sequence by formulas (6) and (7).

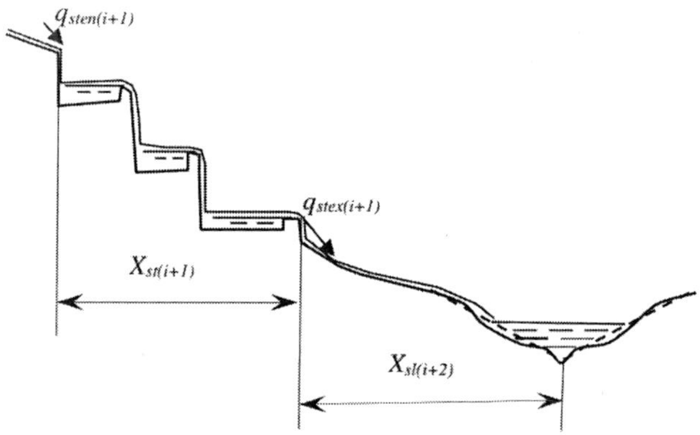

Figure 3. Schematization of the formation of runoff on the stepped secttion: ($q_{sten(i+1)} = q_{slex(i)}$) – linear discharge at the entrance of the *(i+1)-th* stepped secttion of the strip; $q_{stex(i+1)}$ – the same at the exit of the *(i+1)-th* stepped, secttion; $q_{slex(i)}$ – the same at the end of the *i-th* inclined sector of the strip

The total travel time T_{tt} for passing the trajectory from the hydraulically most distant point of watershed boundary line of mixed catchment basins to the closing cross section of the diversion channel consists of two components:

$$T_{tt} = T_{slt} + T_{ch}, \quad (9)$$

where T_{ch} – duration of the flow time from the end of the considered slope strip to the closing cross section of the channel resulting the expected longest travel time.

After determining the travel time for each probable slope strip, the longest of them could have been taken as the longest duration of the downpour, since by that time the full runoff flow rate had already been established at the ends of all other strips. But the total estimated time to reach the closing cross section of the channel consists of two components: the travel time of the catchment basin slope and the time to reach the closing cross section of the channel from the last point of the assumed hydraulically longest slope strip along the single trajectory to the end of the channel. The maximal duration of travel time formed by the slope and channel components of the trajectory together may not necessarily include the strip of the slope with the longest running time.

According to the above schematization of mixed slopes of small catchment basins, the runoff with a continuous layer varying its thickness in time and length from the watershed boundary line downward (starting from zero at the beginning point) reaches its local maximum at its end point along the axis of the diversion channel. Consequently, a continuous lateral inflow with different rates is formed along the sides of the diverting channel at each point, varying in time and length.

If the diversion channel passes across a slope, like a mountain ditch, the inflow is formed only from its upper side. But in case of lowland with a ditch or similar diversion channel between two opposite slopes like the road upper side ditches it occurs from both sides. This continuous lateral inflow merges into the channel stream and flows along the single open channel according to other hydraulic laws and characteristics forming the channel discharge.

For describing the unsteady flow of rainwater in open channels with continuous lateral inflow can be used a system of differential equations of mathematical physics [4, 5, 10, 12]:
the equation of continuity

$$\frac{\partial Q}{\partial l}+\frac{\partial \omega}{\partial t}=q \tag{10}$$

and the equation of dynamic equilibrium

$$i_{ch}=\frac{\partial h}{\partial l}+\frac{\alpha_0 q}{g\omega}(\theta-v)=\frac{\alpha}{2g}\frac{\partial v^2}{\partial l}+\frac{\alpha_0 \partial v}{\partial t}+\frac{Q^2}{K^2}, \tag{11}$$

where Q – discharge; l – length along the channel axis; ω – area of the live cross section; t – time; q – lateral continuous inflow along the channel per unit of time; i_{ch} – gradient of the channel bed; h – depth of the flow in the channel; g – gravitational acceleration; θ – projection of the lateral inflow velocity on the direction of the flow velocity in the channel; v – the channel flow velocity; α_0, α – the Coriolis and Boussinesq coefficients; K– discharge characteristic.

The equations (10) and (11) can be solved more easily in finite differences. To ensure the accuracy of the channel flow discharge computation varying in time and length, it is necessary to determine the optimal intervals (finite elements) for the time Δt_j and length Δl_m taking in consideration the continuous process until the full discharge is established at the closing cross section of the channel (Fig. 4).This situation occurs only by the expiration of the total travel time.

3. Modelling a diversion channel in conjunction with the adjacent mixed catchment basin. Equations for the description of rainwater channel flow

Based on the foregoing, it can be stated that *the estimated duration of the rainfall of a given probability of exceedance for mixed catchment basins with terraced paddy fields represents the total travel time of a slope and channel flow of a portion of rainwater along a concrete trajectory, beginning from the hydraulically most distant point of the watershed to the closing cross section of the diversion channel.*

When determining the travel time of channel flow T_{ch}, should be taken into account only the part $L_{ch(k)}$ of the channel, extending from its closing cross section to the mouth of the considered *k-th* model slope strip that stretches from the extreme boundary point of the watershed to the axis of the channel, assuming that the flow of a portion of water along this strip of catchment to the channel axis and further to its closing cross section creates the trajectory for forming the longest travel time. Within the length $L_{ch(k)}$, there can be many characteristic model slope strips located either on one side (on the left or right of it as in Figure 3) of the diversion channel (for example, in case of upland ditch) or on the both sides – as along the side ditch (Figure 4). Thus, a continuous lateral inflow exists along the diversion channel forming a variable flow rate at every point of it in course of time and length, establishing a maximum flow rate at the closing cross section by the end of the longest travel time.

In the equations (10) and (11) an average lateral inflow $q_{(mj)av}$ along the *m-th* interval of the diverting channel at the *j-th* interval of time from the left ($q^l_{(mj)av}$) and right ($q^r_{(mj)av}$) of it (for example, in case of a road side ditch) can be determined by the formula (12):

$$q_{(mj)av}=q^l_{(mj)av}+q^r_{(mj)av}. \tag{12}$$

The maximum flow rate at the closing cross section of the channel Q_{max} can be determined, based on either of the following main hypothetical assumptions: 1) through the maximum inflows at the mouths of the model slope strips q_{max}, (see Figure 4); 2) through the mean values of the continuous lateral inflow rates q_{av}. This question should be analyzed by detailed simulations depended on the solution of the equations (2), (10) and (11). The analysis of various schemes of modeling and formation of the discharge along the diversion channel is also necessary. Here we can take either 1) an average area of the live cross section of the part $L_{ch}(i)$ taking the parabolicity of the curvature of free surface flow [4, 6], or 2) the average flow rate along the entire length of the channel. The later can be formed by the continuous lateral inflow (equaling to half of the maximum flow rate at the closing cross section), by the constant (boundary) flow rate at the beginning of the channel Q_{en} (sometimes it may take place) and by other concentrated discharges along the channel (if there exist some springs or other concentrated tributaries). The second assumption may slightly increase the travel time. But at the same time this increases the reliability of the runoff computation when using the partial differential equations (2), (10) and (11), because the computed flow rates may also be underestimated due to an underestimated travel time. When the travel time exceeds its real value it causes no more influence on the maximal hydraulic outputs of the flow after the full runoff has been established, whatever may be the time duration and the rainfall intensity [4, 5, 10].

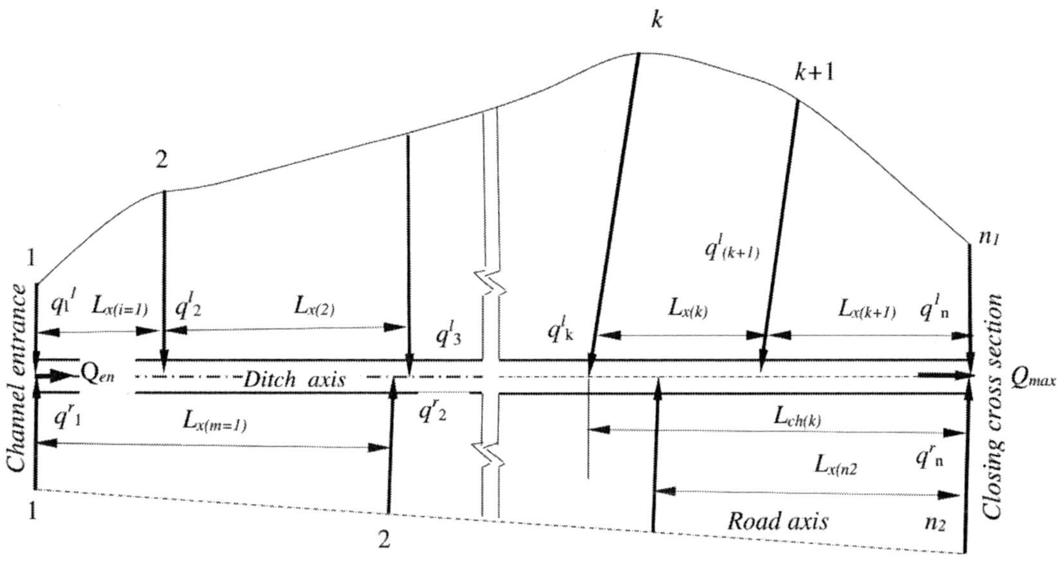

Figure 4. Unified schematization of the slopes and diversion channel with lateral inflow (a road side ditch is presented for an example)

4. Development of a formula for estimating the travel time of rainwater channel flow in mixed catchment basins with terraced surface plots

When the travel time along the model slope strip under consideration beginning from the hydraulically most remote point of the watershed boundary to its mouth at the axis of the diversion channel is reached, the full inflow rate is already established at the mouths of all other strips. Therefore, simplifying the task, for a maximum discharge Q_{max} at the closing cross section of the channel can be used the sum of the average products of lateral inflow rates of each adjacent pair of model slope strips and the distances between them $L_{x(i)}$ at their mouths (Figure 4) taking into account the input discharge Q_{en} at the beginning of the channel and the springs or other concentrated tributaries Q_{trib} along it (if they exist) as shown in the formula (13):

$$Q_{max} = Q_{en} + Q_{trib} + \frac{1}{2}\sum_{i=1}^{i=n_1}\left(q_{(i)}^{l} + q_{(i+1)}^{l}\right)L_{x(i)} + \frac{1}{2}\sum_{m=1}^{m=n_2}\left(q_{(m)}^{r} + q_{(m+1)}^{r}\right)L_{x(m)}, \tag{13}$$

where $i = 1,2,3 \ldots n_1$ – numbers of the strips on the left slope; m = $1,2,3 \ldots n_2$ – the same on the right side; Q_{en} – constant inflow rate at the inlet of the diversion channel (if it exits), m^3/c; Q_{trib} – discharge from the springs or other concentrated tributaries; $L_{x(i)}$, – length of the *i-th* section of the channel between adjacent strips of the left slope, m; $L_{x(m)}$ – the same of the *m-th* section of right slope, m; q_i, $q_{(i+1)}$ – inflow rates at the mouths of adjacent elementary strips on the $L_{x(i)}$ sector from the left slope, m^2/s; q_m, $q_{(m+1)}$ – the same from the right-hand slope $L_{x(m)}$, m^2/s.

Taking for the average flow rate Q_{chav} over the entire length of the diversion channel as half of the maximal flow rate Q_{max} (14) at its closing cross section formed by the lateral inflows that can be obtained by the formula (13) plus the flow rate at the inlet of the channel and other concentrated tributaries, we get:

$$Q_{chav} = Q_{en} + Q_{trib} + \frac{1}{4}\sum_{i=1}^{i=n_1}\left(q_{(i)}^{l} + q_{(i+1)}^{l}\right)L_{x(i)} + \frac{1}{4}\sum_{m=1}^{m=n_2}\left(q_{(m)}^{r} + q_{(m+1)}^{r}\right)L_{x(m)}. \tag{14}$$

In a diversion channel the average instantaneous flow velocity can be determined by the formula:

$$v_{chav} = \frac{1}{n_{chav}} I_{chav}^{1/2} R_{chav}^{2/3},$$

where n_{chav}, I_{chav} – weighted average values of the roughness coefficient and bed gradient; R_{chav} - hydraulic radius of live cross section of the channel.

Simplifying the shape of the diversion channel as triangular with weighted average side gradients m_{1chav} and m_{2chav}, the average values of the area of channel cross section ω_{chav}, the flow depth h_{chav} and the wetted perimeter χ_{chav} will be:

$$\omega_{chav} = \frac{1}{2}\left(m_{1chav} + m_{2chav}\right)h_{chav}^2 = m_{chav}h_{chav}^2,$$

$$h_{chav} = \sqrt{\frac{\omega_{chav}}{m_{chav}}}; \qquad \chi_{chav} = h_{chav}\left(\sqrt{1+m_{1chav}^2} + \sqrt{1+m_{2chav}^2}\right).$$

Based on this, the average instantaneous flow velocity will be:

$$v_{chav} = \frac{\left(\frac{\omega_{chav}}{\chi_{chav}}\right)^{2/3} I_{chav}^{1/2}}{n_{chav}} = \frac{\left(m_{chav}h_{chav}\right)^{2/3} I_{chav}^{1/2}}{n_{chav}\left(\sqrt{1+m_{1chav}^2} + \sqrt{1+m_{2chav}^2}\right)^{2/3}}. \tag{15}$$

Taking into account the above said and $\upsilon_{chav} * \omega_{chav} = Q_{chav}$, the equation for the average depth of the flow becomes:

$$h_{chav} = \sqrt{\frac{Q_{chav}}{m_{chav}v_{chav}}} = \sqrt{\frac{n_{chav}Q_{chav}\left(\sqrt{1+m_{1chav}^2} + \sqrt{1+m_{2chav}^2}\right)^{2/3}}{m_{chav}^{5/3} I_{chav}^{1/2} h_{chav}^{2/3}}}$$

or
$$h_{chav} = \frac{\left(n_{chav}Q_{chav}\right)^{3/8}\left(\sqrt{1+m_{1chav}^2} + \sqrt{1+m_{2chav}^2}\right)^{1/4}}{m_{chav}^{5/8} I_{chav}^{3/16}}, \tag{16}$$

where Q_{chav} – average discharge computed by using the formula (14).

The duration of the travel time of channel flow T_{ch} is the time that takes for the water to reach from mouth of the the considered k-th slope strip to the closing cross section of the channel with a length $L_{ch(k)}$ and an average velocity υ_{chav}, computed by the formula (15). The average depth of the flow is calculated by formula (16). Having the distance $L_{ch(k)}$ and the flow velocity υ_{chav}, we can easily determine T_{ch}:

$$T_{ch} = \frac{L_{ch(k)}}{v_{chav}} = \frac{n_{chav}^{3/4} L_{ch(k)} \left(\sqrt{1+m_{1chav}^2} + \sqrt{1+m_{2chav}^2} \right)^{1/2}}{I_{chav}^{3/8} \left(m_{chav} Q_{chav} \right)^{1/4}}. \qquad (17)$$

Taking in view the above said, the formula (9) is used to compute the total travel time T_{tt} for rainwater runoff from small catchments with a mixed surface and the T_{slt} and T_{ch} are computed by formulas (6), (7), (8) and (17).

5. Some recommendations

1. When determining the travel time by the proposed method for a considered part of the slope, should be used the weighted average gradients and surface roughness coefficients of the model strips taking into account their differences in different segments of the given distance.
2. As the small catchment basins in rain-prone regions with a long rainy season are taken in account, it is permissible to neglect the various kinds of losses due to the constant saturation of soils with moisture, wetted vegetations and filled up microrelief cavities because of the frequent precipitation.
3. When calculating total travel time T_{slt} for the slope, it is necessary to convert the intensity of a one-hour-long shower to the intensity of the estimated duration by using a transformation factor $K_t = (60/T_{tt})^{2/3}$, where the T_{tt} is taken in minutes. This requires determining the travel time T_{slt} on the finally adopted design slope strip based on the most probable shower intensities and the remained part $L_{ch(k)}$ of the diversion channel.
4. The main goal of the present method is to determine the approximate travel time of the slope and channel flow and use it when working out and testing a simulation model for estimating the rainwater slope runoff and channel flow based on the differential equations of one-dimensional hydraulics. Therefore, the estimation formulas of the slope runoff and the channel flow as well as the cross section of the diversion channel are taken permissibly simplified.
5. Having obtained the travel time by this method, it is necessary to conduct numerical simulation studies of the real catchments and forms of the diversion channels with their own characteristics based on the above proposed hydrological and hydraulic equations until establishing the full flow on the slopes and channels separately and clarify the derived formulas.
6. It is necessary to simultaneously conduct researches aimed to working out formulas for calculating optimal intervals of time and length to solve the proposed differential equations in finite differences in order to obtain the hydrological and hydraulic characteristics of the rainwater slope runoff and channel flow with a given accuracy.

6. Main conclusions

1. A new method is proposed for unified schematization of a mixed catchment basin with natural and terraced rice field sectors of the slopes in conjunction with a diversion channel, taking into account the peculiarities of small drainage areas with anthropogenic plots.
2. The equations of one-dimensional hydraulics based on the laws of conservation of matter and energy are proposed for describing the rainfall slope runoff as a sequent and uninterrupted process.
3. The main characteristics of the slope surfaces of mixed catchment basins and chosen diversion channels influencing the formation of rainwater runoff and travel time of the slope and diversion channel are taken into account.

4. A unified formula has been developed for determining the total travel time of the slope runoff and channel flow on mixed drainage basins with terraced areas, which can be used in calculating slope runoff and channel discharges based on the differential equations of unsteady flow of storm water.

5. The developed formula based on the methods of mathematical modeling, gives a more accurate travel time of slope and channel flow than the methods recommended by existing standards, and is a novelty for the case of anthropogenic territories.

References

[1] Andreev O V and Shahidov A F 1977 *Metodicheskie razrabotki po raschyotu livnevogo stoka s malyh vodosborov* (Moscow: MADI) 8 p

[2] VSN 63-76 1976 *Instrukciya po raschetu livnevogo stoka vody s malyh bassejnov* (Moscow: Orgstroj Mintransstroya SSSR) 101 p

[3] SP 33-101-2003 2004 *Opredelenie osnovnyh raschetnyh gidrologicheskih harakteristik* (Moscow: Gosstroya Rossii, FGUP CPP) 36 p

[4] Bolshakov V A and Kurganovich A A 1983 *Gidrologicheskie i gidravlicheskie raschyoty malyh iskusstvennyh sooruzhenij: uchebnoe posobie dlya vuzov* (Kiev: Golovnoe izd-vo izdatel'skogo ob"edineniya "Vysha shkola") 280 p

[5] Paudyal S P 1990 *Proektirovanie optimal'nogo poverhnostnogo dorozhnogo vodootvoda* Cand. dissertation thesis (Moscow: MADI) 20 p

[6] Paudyal' S P 2005 Opredelenie raschyotnoj prodolzhitel'nosti livnevogo stoka s malyh vodosborov dlya proektirovaniya optimal'nyh konstrukcij sistemy poverhnostnogo dorozhnogo vodootvoda *Proektirovanie avtomobil'nyh dorog* (Moscow: MADI (GTU)) pp 117–26

[7] Paudyal S P 2011 Osobennosti formirovaniya raskhodov livnevogo sklonovogo stoka v terrasirovannyh risovyh dolinah Nepala s zemlyanymi valikovymi regulyacionnymi ustrojstvami *Proektirovanie avtomobil'nyh dorog* (Moscow: MADI (GTU)) pp 43–8

[8] Fedotov G A 1986 *Avtomatizirovannoe proektirovanie avtomobil'nyh dorog* (Moscow: Transport) 317 p

[9] CHebotaryov A I, Serpik B I 1973 O raschyote parametrov gidrologicheskogo rezhima *Meteorolog. i gidrolog.* **1** 55–67

[10] Chistyakov I V 1988 *Metodika detal'nogo raschyota livnevogo stoka s malyh vodosborov pri proektirovanii malyh mostov i vodopropusknyh trub na avtomobil'nyh dorogah* Cand. dissertation thesis (Moscow: MADI) 18 p

[11] Paudyal S P 2014 Issledovanie vliyaniya harakteristik poverhnosti smeshannyh vodosbornyh bassejnov s terrasirovannymi uchastkami na vremya sklonovogo dobeganiya livnevogo stoka v terrasirovannyh risovyh dolinah Nepala s zemlyanymi valikovymi regulyacionnymi ustrojstvami *Vestn. Moskovsk. avtomobil'no-dorozhn. Gosudarstv. Tekhnichesk. Univer. (MADI)* **1(36)** 92–9

[12] Gavrilenko S M 1980 *Issledovaniya neustanovivshegosya dvizheniya livnevyh vod v otkrytyh ruslah s bokovoj pritochnost'yu* Cand. dissertation thesis (Kiev) 21 p

[13] Chugaev R R 1982 *Gidravlika* Uchebnik (Leningrad: Energoizdat) 672 p

Ensuring the processability of construction disposal during design

A P Pavlov and N N Mitrohin

Moscow Automobile and Road State Technical University (MADI), 64, Leningradsky ave., Moscow, 125319, Russia

E-mail: 89037628407@mail.ru

Abstract. The level of organization of recycling and recycling processes in Russia has been assessed. The questions of the level of consumption of resources in production are presented, which gives an idea of the potential possibilities of unification not only of design solutions, but also of the nomenclature of used materials. It has been established that technological efficiency of recycling provides not only reduction of total costs for the recycling process itself, but also allows to more fully use the potential of working capacity of parts laid down during their production. The possibilities of ensuring the processability of the product design during its disposal at the design stage are considered. One of the most effective methods of ensuring the processability of the structure during recycling is to use the system of repair kits during design, which allows simultaneously increasing the processability during production, operation and repair. The proposed approach makes it possible to significantly reduce not only the time spent in repair of ground-based transport and technological means (GTTM), but also to significantly realize the possibility of providing technological effectiveness in recycling GTTM themselves and their units and parts. The use of repair kits formed at the design stage allows increasing efficiency not only during maintenance and during repair of GTTM, but also during their utilization and recycling.

1. Introduction

The formation of Russia as a technologically developed power and its integration into the global economy will be impossible without the introduction and improvement of methods of efficient and environmentally sound use of natural and technogenic resources. This problem can be solved only by creating closed low-waste cycles, as well as by stimulating and providing technical support for the integrated use of secondary resources generated in the production and consumption processes. The less garbage in the country, the cleaner and safer it is for human habitation, the richer and stronger on the world stage.

Extensive growth in resource consumption has already led to a certain raw material dependence in the economies of many countries, including Russia, and to the aggravation of environmental problems associated with the environmental impact of the extraction of raw materials, production and the need to solve the problem of final disposal of production and consumption waste.

2. Problem statement

Continuous development and improvement of the vehicles design, which include: cars and trucks, buses, special communal vehicles on an automobile and original chassis, construction and road vehicles, vehicles for servicing aircraft and maintaining airports, etc. leads to saturation of the consumer market, and in some areas to glut.

This is directly related to the increasing consumption of natural resources, as can be seen from the data presented in Fig. 1.

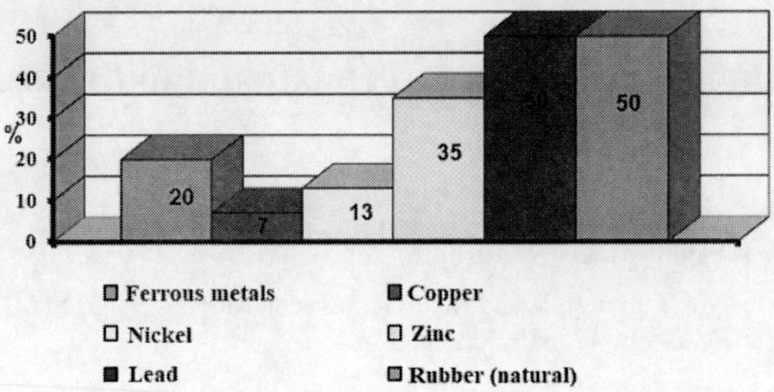

Figure 1. Consumption of natural resources in production

Moreover, the efficiency of the use of consumed resources is close to 0.

For example, for the manufacture of 1 ton of parts and assembly units used in a car, 150 tons of natural substances are processed, i.e. of each 1 ton of the natural substances in the car remains – 0.7 %. The remaining 99.3 % is wasted.

This situation poses a challenge for designers to increase the efficiency of the use of minerals. The service life of these machines varies from 5 to 30 years, but in any case, the question arises that the operation ends sooner or later and there is an acute question of the disposal of units and materials from which they are made.

The only thing that is not logical in this process is the late-arising question – what to do with the remains of machines.

For example, at the beginning of 2016, about 1,500 million tons of toxic production and consumption waste were accumulated at enterprises in various industries. About 90 million tons of toxic industrial waste (IW) are generated annually at enterprises of the Russian Federation, of which 87 million tons are classified as hazard classes III and IV. The amount of consumer waste, or municipal solid waste (MSW), is increasing annually in Russia by 30 million tons.

In 2016, enterprises used in their own production about 40 million tons (40 %) and completely neutralized 9 million tons (10 %) of the total amount of waste generated over the year. The remaining waste was transferred to landfills for disposal.

An analysis of the nomenclature of the materials used (see Table 1) in the manufacture of machines for leading machine manufacturers showed all their diversity and the difficulty of solving the problem of ensuring their disposal, taking into account the need to comply the efficiency and environmental friendliness of these processes.

Data analysis of table 1 makes it possible to conclude that the disposal of machinery is not only a reduction in environmental pollution, but also a reduction in the consumption of natural resources and energy. The environmental and economic effect of the utilization of vehicles consists of:

- the economic component, including cost reduction in the secondary use of materials and the associated decrease in the production of materials from fossil natural resources, reduction in energy consumption;
- environmental component, including reduction of pollution of soil, water, atmospheric air from the effects of abandoned and not recycled cars, reduction of environmental pollution when used in the production of recycled materials.

Table 1. Materials used in the manufacture of automobiles

Name of material	Material content in a typical automobile, % of weight		
	USA	Japan	Europe
Steel and iron	67	72,2	65
Plasties	8	10,1	12
Glass	2,8	2,8	2,5
Rubber	4,2	3,1	6
Liquids and oils	6	3,4	2,5
Nonferrous metal	6	6,2	8
Other materials (paint, insulation, wiring)	4	2,2	4
Weight, kg	1438	1270	1185

The product life cycle ends with the disposal phase. For cars and tractors (especially cars), the relevance of safe disposal is increasing every year, as the number of used cars grows, and existing car dumps are overcrowded and poison the atmosphere. Therefore, the international community establishes uniform principles to address this problem. Thus, the European Union Directive 2000/53 / EU "Decommissioned vehicles" and the Commission Directive dated March 30, 2011/37 / EU amending Appendix II to Directive 2000/53 / EC establish:

- participating countries shall take the necessary measures to organize in their territory a sufficient number of points of reception for used cars and create a system for recycling and disposal of old cars and worn-out components;
- no later than January 1, 2006, all vehicles that are out of service must be disposed of at least 85 % of the total mass, and the level of reuse of parts and recycling of materials must be at least 80 % of the mass of the car (without burning).

Therefore, at the stage of design preparation for the production of a new product (automobile, road vehicle and their components), it is necessary not only to evaluate, but also to constructively ensure the processability at the design stage.

The processability of disposal is determined by the following key indicators:

- the time of disassembling the product into its component parts;
- the possibility of reuse of parts in order to ensure the development of the full resource inherent in them during production;
- nomenclature of the used connections, unification of threaded connections;
- the number of tools required for disassembling;
- the time necessary for drainage (i.e., freeing the vehicle of fuel, grease, brake and coolant, etc.);
- the time of sorting by groups of materials for the application of various recycling processes and the suitability of parts for reuse.

In various works, as one of the criteria, it is proposed to use, for example, specific indicators of dismantling manufacturability. This is the ratio of the mass of the dismantled unit, part, set to the time spent on dismantling.

However, the use of this indicator has some difficulties, since a car, for example, combines, as mentioned above, a large number of materials of various nature, the specific weight indicators of which are also varied, and the complexity of dismantling, for example, steel parts does not differ much from the complexity of dismantling parts from polymers, which is quite clearly seen in Fig. 2.

Due to the characteristics of polymeric materials, which have significantly worse recycling characteristics, the operations of dismantling plastic, rubber, and electrical products (including electronic) from mixed materials are separately considered and analyzed.

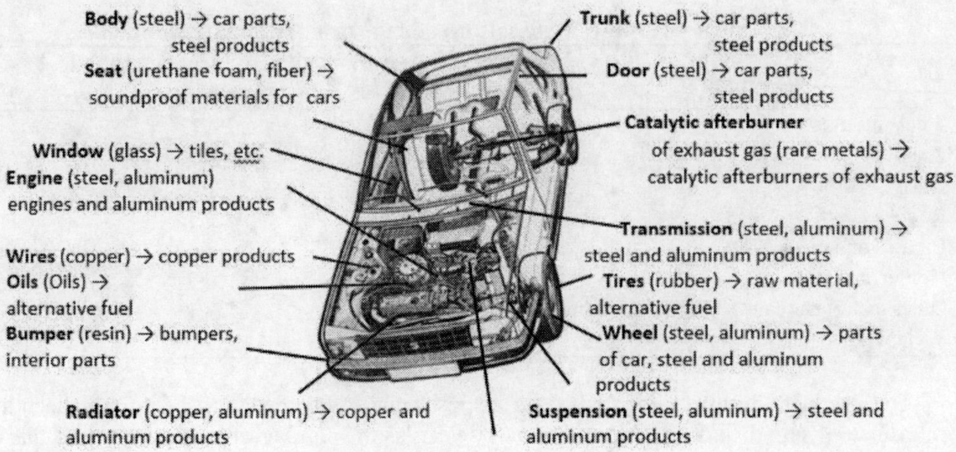

Figure 2. Reuse of waste products obtained during car recycling as raw materials for industry

Also, some authors consider the hierarchy of application of fasteners optimal from the point of view of optimization of disposal, which is given in table. 2.

Priority is for the use of fasteners for automotive components.

Table 2. Types of Recommended Compounds

Priority Sequence Number	Connection type	Name of fasteners
1	Quick couplings	Magnetic, Velcro, latch, clip, ¼ or ½ turn fasteners, button, clasp, clips, pin
2	Detachable connections	Screw, bolt, nut with convenient heads of the same size for the tool and accessibility for power tools. Adhesive joints (gluing), allowing to quickly separate parts without the use of tools. Screw, bolt, nut with heads of several sizes, requiring replacement of the tool with accessibility for electric pneumatic tools. Screw, bolt, nut with heads of several sizes with difficult accessibility for power tools
3	One-piece connections	Rivet, soldering, welding, gluing, foaming, thermal shrinkage

Thus, simplifying the disassembly of the product into its components, reducing the dismantling time, reducing the drain time of all liquids and oils used in the operation, using environmentally friendly materials, etc., the designer can increase the processability of the car during disposal at the design stage.

3. Problem solving methods

Ensuring the processability of the product is a multifaceted work to improve the design of the machine as a whole, its individual components and parts with the goal of the most rational production, maintenance, repair and disposal.

Ensuring the processability of the design is laid at the very early stages of design. The following tasks are solved:

- technological control of design documentation;
- assessment of the level of processability of the design;

- development of the design for processability;
- design changes.

Each of these tasks needs to be addressed, both in relation to the product as a whole, and to individual assembly units, parts and blanks.

Therefore, it is extremely important to carry out the design development for processability at the early stages of design, when the huge cost of funds for the technical preparation of production has not yet been made. Otherwise, the design will have to be "adjusted" for the already purchased and installed equipment, which can significantly reduce the service (functional, consumer) quality of the product, in particular a car, road-building machine, tractor or their individual units and parts, or purchase new equipment.

The processability of the product design during disposal is manifested in operations such as preparation for decommissioning, disassembly of the product, transportation, storage and disposal. In preparation for decommissioning and disassembly, the main attention is paid to the environmental characteristics of the product (radioactivity, the separation of materials by brand, weight and size) of the components of the product.

Therefore, the product designed by the constructor must meet the requirements not only of its functional purpose, but also of the processes of its manufacture (assembly, preparation of blanks, manufacturing of parts), maintenance, repair and disposal.

Ensuring the processability of the product design is carried out at the level of parts, assembly units and the product as a whole. The most general requirements for processability of product design:

- rationality of the division, layout of products and their components;
- widespread use of the principles of constructive and technological continuity, unification, standardization and simplification;
- rational limitation of the number of grades and assortments of materials used;
- widespread use of non-deficient materials and materials, the processing of which does not cause difficulties;
- rational designation of tolerances and roughness parameters of the machined surfaces;
- ensuring the convenience of basing parts in the product and workpieces during their processing and, if possible, achieving sufficient structural rigidity;
- compliance with the conditions ensuring the simplification of assembly work and the possibility of their mechanization and automation;
- creation of parts of such structural forms that allow the use of more productive processing methods and the use of high-performance equipment;
- reducing the variety of types of machined surfaces and geometric dimensions of the same type of structural elements of the part;
- the maximum possible simplification of the design of assembly units and parts;
- the possibility of using advanced technological processes, high-performance equipment and more advanced methods of labor organization, which requires the presence of certain structural elements in parts or assembly units;
- ease of maintenance and repair during operation, which requires the introduction of certain elements into the design.

Based on the requirements formulated above for the processability of the product's design from the standpoint of ensuring its more efficient disposal, it is necessary to consider some problems that can only be effectively solved at the design stage of the machine.

To reduce the costs of various types of dismantling during disposal, it is necessary to design the principle of forming repair kits in the design of machines, the use of which will not only optimize the time required for disassembly, but also ensure a more complete use of the resource of each part of the kit and the machine as a whole.

In accordance with [1], a product repair kit is a combination of its assembled parts, united by functional unity, forming a closed dimensional group with full technological completeness. The closure of the group is realized, for example, in the fact that when it is installed in the product, it forms a closed dimensional chain with the accuracy parameters of the Assembly size within the standard.

From the definition of a repair kit, it follows:

- replacement of the repair kit with the corresponding new (repaired) repair kit should not require additional technological operations to completely restore its functional properties and reliability of the product;
- in contrast to a detailed replacement, a complete replacement does not destroy (does not worsen) due to repair actions the functional work and reliability indicators of the unit as a system.

In accordance with the definition introduced, if the product has more than one repair kit, then the product's repair kits must be independent in aggregate, i.e. the operability of any of the repair kits of the product does not depend on whether or not any combination of its other kits has failed.

Thus, the proposed system of repair kits provides not only compliance with the principles of processability of product design during design and production, but also provides optimal conditions for the operation, maintenance, repair and disposal of machines at the end of their term of use [2].

Another no less important area for improving the processability of product design is the wider use of the potential applications of polymer materials, not only for gluing machine parts, but for the fundamental replacement of the conditions for their connection with each other, for example, the use of shaft-sleeve retainers, threaded retainers, and liquid gaskets.

The most promising direction of ensuring the processability of the design of products in this direction is the use of shaft-sleeve retainers, which not only interconnect mating parts, but also allow you to fundamentally change the conditions of such connections [3].

It is known that today, constructors widely use one-piece connections (interference fit) or key and splined connections to transmit torque.

The first type of joints is characterized by the creation of additional internal stresses in the details, which leads to an increase in their stress-strain state and, as a consequence, a decrease in their service life and the inability to continue to use them at all, or necessitates additional technological operations to restore their operability. And this is an additional time and money cost [4].

In the second case, it is necessary to carry out additional machining of the surfaces to be joined under the key or slots and to ensure the production of additional parts – keys.

Such joints are based on landings with a gap (or transitional ones), and this, in turn, leads to accelerated wear of such joints and serious problems in restoring their operability, that is, causes problems of further efficient use of their resource.

4. Conclusion

The use of shaft-sleeve retainers to replace such joints allows not only to reduce costs during their manufacture and assembly, but also during disassembly and makes it possible to more deeply use their working capacity, while unconditionally ensuring the quality of the assembly joint [4].

Consideration of the possibilities of applying modern and promising materials and methods in the design of products will certainly provide an increase in the processability of the design of products not only for operation, but also for the disposal of machines after the end of their service life.

In our opinion, the most promising direction for improving the processability of the design of products is the replacement of metal parts with parts made of polymeric materials. This replacement is especially relevant for parts of friction pairs, for example, camshaft support bushings or plain bearing shells for crankshafts.

Today, the physicomechanical properties of composite materials are no worse than the properties of metals, and often, in terms of the main parameters characterizing the performance of friction pairs, they exceed them substantially.

References

[1] Apsin V P, Dehterinsky L V and Norkin S B 1996 *Modeling of machine recovery processes* (Moscow: Transport) 253 p

[2] Pavlov A P et al 2013 *The theory of the potential for efficiency and repair backup reliability of aging technical systems* textbook allowance (Moscow: MADI) 104 p
[3] Baurova N I and Zorin V A 2016 *The use of polymer composite materials in the manufacture and repair of machines* a textbook (Moscow: MADI) 264 p
[4] Pavlov A P and Trefilov R V 2013 Improving the durability of machines by the backup of their compounds using polymer materials *Coll. of sci. papers "Problems of creating information technology"* **22** 38–44 (Moscow: MADI)

Using foam polyurethane sealers for strengthening of soils of a road bed of transport constructions

E S Polyntsev and A V Kvitko

Saint Petersburg State University of Architecture and Civil Engineering, Vtoraya Krasnoarmeiskaya, St. Petersburg, 190005, Russia

E-mail: www.esp@mail.ru

Abstract. The article presents an overview of the methods of producing alternative building materials by strengthening soils with organic binders based on synthetic resins. Their use in road construction would meet the needs of construction organizations in high-quality building materials and ensure reliable operation of soil masses of subgrade road transport structures throughout estimated service life. The authors proposed a new, alternative method for soil stabilization using urethane foam sealants, which have proven themselves in civil engineering in the form of mounting foams. The paper describes a method for strengthening soil with sealants, its labor production technology, properties and physical and mechanical characteristics. The results of the trial experience of soil strengthening with polyurethane foam sealant are described and conclusions are made based on the work.

Keywords: road, subgrade, soil strengthening, pavement, organic binder, polyurethane foam sealant, mounting foam.

1. Introduction

During the construction of roads, in most cases, cohesive soils act as soils of natural bases. This is due to the fact that the resource of local building materials suitable for the construction of the subgrade and the working layer often cannot fully meet the needs of builders. In order to reduce the cost of construction, the subgrade is built from local silty and clay soils.

During the life cycle of the transport structure, the subgrade is exposed to external climatic influences. The factors affecting the subgrade are diverse and depend on the time of year. The main factors affecting the efficiency of the pavement are waterlogging and freezing of the soil of the subgrade. The experience of operating transport facilities shows that the service life of pavement on roads passing through waterlogged areas is much less, and the main destruction occurs at the stages of excessive waterlogging of the subgrade.

Waterlogging of cohesive soils of the subgrade leads to a decrease in their strength more than twice, thereby reducing the overall strength of pavement by more than 25 %. The decrease in soil strength with increasing humidity is clearly visible on the graph in Fig. 1 [1]. In regions with persistently low temperature in winter, the maximum decrease in the strength of the pavement occurs in spring, when the thawing of frozen soils goes beyond the additional base layer into the working layer of the subgrade [2].

With this circumstance, the question of how to build a subgrade that can withstand several overhaul cycles and at the same time preserve the strength characteristics sufficient for the safe operation of the structure is even more acute.

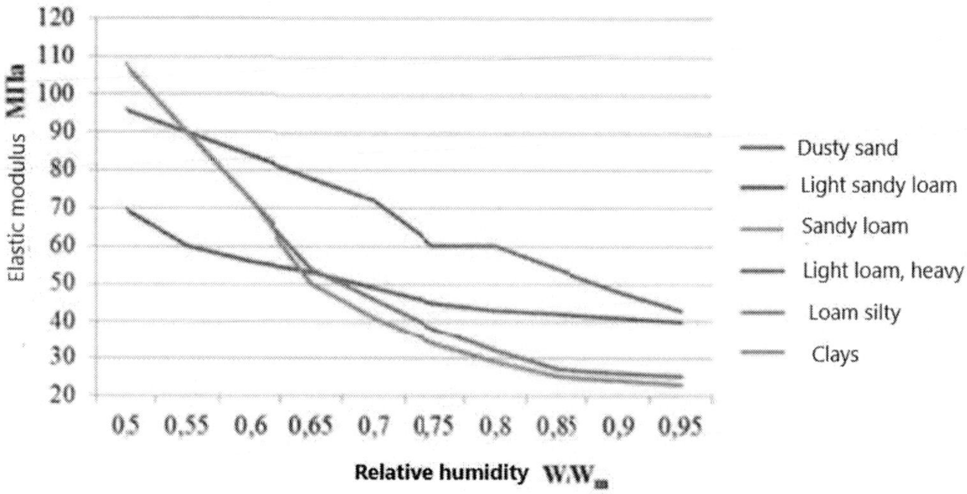

Figure 1. Dynamics of soil strength

The solution to the above problem should be sought in achieving such a condition of the subgrade that its humidity during the year does not exceed the optimum humidity. Since the strength properties of soils, in this state, almost reach a maximum [4].

Currently, in order to solve this problem, reinforcement of soil subgrade is carried out by mechanical, physical or chemical methods. However, based on the results of previous studies [5], we can say that the known methods have a large number of limitations and disadvantages. In particular, it lies in the properties of soils that are not suitable for strengthening. In the Russian Federation regulatory documents [6–8] limit the suitability of the soil for reinforcement in the following indicators: the number of plasticity, grain composition, content of humus particles, the degree of salinity, humidity, etc. Therefore, an even more urgent question arises about the search for scientific solutions to extend the life cycle of the subgrade under any soil conditions.

The method of strengthening the soil of the subgrade with polyurethane foam sealants is based on the use of polyurethane foam that has positively proven in civil engineering (hereinafter – Sealant). Polyurethane foam sealant is a thermosetting polymer belonging to the condensation type, consisting of isocyanate, polyol and auxiliary substances (catalysts, blowing agents, etc.).

Mounting foam has a number of irreplaceable properties in construction, due to which it can provide the necessary quality of soil reinforcement of any types. These properties include:

- speed of strengthening process (full hardening of foam maximum 24 hours, standard from 4 to 8 hours);
- high adhesion of sealant with soil particles (the adhesion of foam higher to wet surfaces, which in the conditions of strengthening of water-saturated soils is undoubtedly a plus);
- increase in the volume of the introduced sealant (up to 20 times). Due to a significant increase in volume, the foam penetrates into the micropores of the soil, absorbing the necessary for the formation of its own structure and displacing excess moisture;
- drainage of weak water-saturated soils (the cross-linking mechanism of one-component polyurethane sealants is carried out due to the interaction of isocyanate groups with moisture) [9];
- formation of a layer with a low coefficient of thermal conductivity (thermal conductivity of the foam 0.003 W/mK).

A method of strengthening the soil Sealants can be used on the contact systems «pavement – weak foundation» (Fig. 2, a), «pavement – subgrade» (Fig. 2, b), and in the middle part of the arranged subgrade (Fig. 2). Due to its properties, the Sealant can not only increase the water/frost resistance and strength characteristics of the soil, but also provide the formation of a capillary and vapor

IOP Conf. Series: Materials Science and Engineering **832** (2020) 012029 doi:10.1088/1757-899X/832/1/012029

impermeable insulating layer, which is especially important in the second road-climatic zone, when dealing with the formation of the gleyed soil subgrade [5].

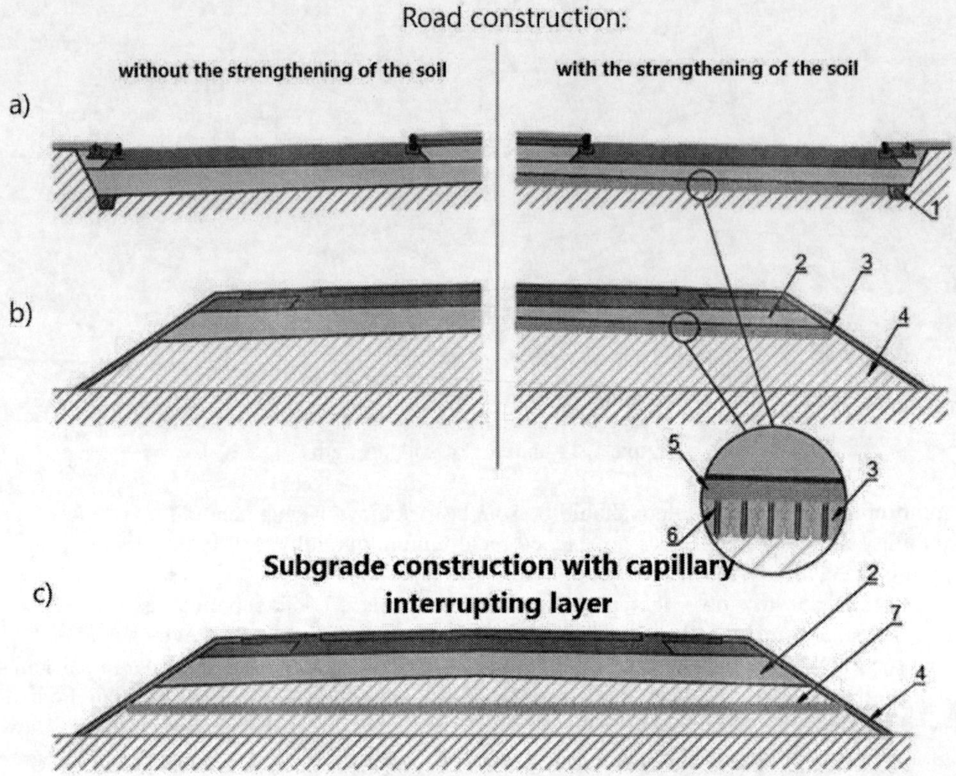

Figure 2. The device of the pavement of the road with reinforced soil according to the system: a) «Pavement – weak foundation»; b) «Pavement – subgrade»; c) In the middle part of the arranged subgrade; (1) – the accompanying drainage removed from the soil reinforcement zone; (2) – additional (frost protection) layer; (3) – the working layer of the subgrade, reinforced with a sealant; (4) – subgrade of the local silty or clay soils; (5) – leveling sand layer; (6) – the spike of the working body for the supply of Sealant; b(7) – Capillary-breaking layer reinforced with mounting foam.

In Russia and a number of foreign countries studies are conducted to strengthen the soil bases and coatings of roads with cold-cured synthetic resins [10, 11]. As binders, additives are used: urea-formaldehyde resins of various compositions, furfuraniline resins; resorcinol-formaldehyde resins; acrylic compounds; lignosulfate and lignoprotenic compounds; organosilicon and other resins [12].

In the course of the research, based on the results obtained, the following conclusions were made: additives of synthetic resins provide the formation of a strong and hydrophobic structure, water and frost resistance of the strengthened soil; samples obtained by strengthening with synthetic resins from loamy soils withstood up to 50 cycles of freezing and thawing, from sandy soil – 115 cycles; light loamy and clay soils with moisture equal to the lower yield stress, acquire water resistance and considerable strength; condensation polymers are less sensitive to soil impurities and air oxygen than polymerization polymers, which contributes to more stable curing [12].

Also, experiments on soil strengthening with resorcinol-formaldehyde resins showed the ability of the binder to bind water in an amount exceeding the mass of gummers 2–2.5 times. This allowed researchers to conclude that it is advisable to use it to strengthen waterlogged soils [12].

Currently, the research conducted in the field of application of polyurethane binders in road construction is mainly aimed at strengthening slopes, recesses, bulk structures, cones of bridges and overpasses, the surface of which is strengthened by rubble or gravel, in order to protect them from external environmental factors [13–15].

In preparation for the experimental part of the research, on the basis of the Department of «Roads, bridges and tunnels» of the St. Petersburg state University of Architecture and Civil Engineering, a trial experiment was prepared and conducted to strengthen the cohesive soil with a sealant. As a binder was chosen single-component, all-season, polyurethane foam. The experiment showed that when a binder is introduced into the soil (to a depth of 10–15 cm), there is an active interaction (sticking), soil and sealant. In the process of primary and secondary expansion of the binder, there is a wedging of foam into the soil, the soil is «absorbed», thus forming a single system of «soil – binder». The expansion of the sealant passed evenly in all directions, and amounted to about 10 cm from the puncture. At the same time, a cylindrical layer of strengthened soil was formed with irregularly shaped edges with a diameter of about 20 cm, which is a monolith of solidified foam interspersed with soil. It was also observed partial protrusion on the surface, with the formation of a «cap» of frozen foam. The sample section is shown in Figure 3.

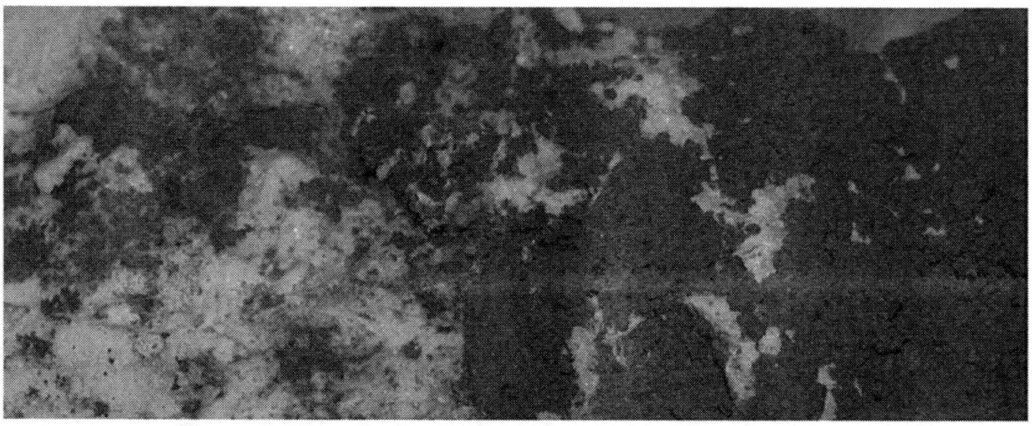

Figure 3. Slice of soil sample reinforced with polyurethane foam sealant

With additional mechanical mixing of the soil and the injected foam, immediately after injection into the soil, a layer of reinforced soil was formed with more evenly distributed soil particles.

Technology of production for strengthening the soil of the subgrade transportation construction with polyurethane foam sealants includes three main operations:

1) Preparation of the ground base of the subgrade.

1.1. The device of pavement in the recess includes excavation to design level (10 cm below the sole of the pavement), followed by planning.

1.2. The device of the pavement mound include: raising the body of the mound up to design level (10 cm below the sole of the pavement), followed by planning.

2) Putting a binder.

The binder is introduced into the ground by injection, using a special frame with spikes mounted on construction equipment (see Fig. 4). The spike is immersed in the soil to a depth of 10–15 cm, after which the sealant is injected.

The spikes are fixed on the frame of the working body of construction equipment. Sealant is supplied from the cylinder to the working body occurs through special tube-channels.

3) The device of the protective-leveling sand layer.

Distribution of a protective-leveling sand layer with a thickness of 10 cm, according to the technology «Push», with compaction and layout. The compaction of the layer should begin not earlier than 8 hours after the introduction of the binder. Then we arranged a pavement.

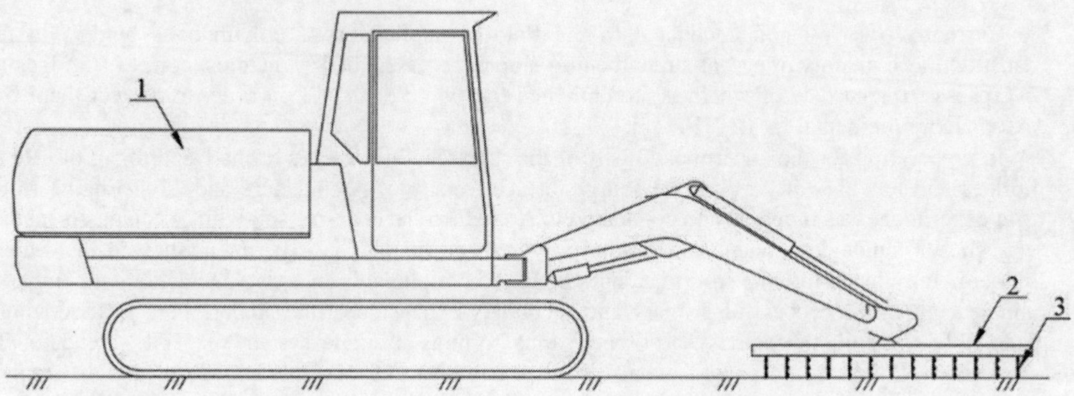

Figure 4. Tractor equipped with a frame with spikes for putting sealant: (1) – capacity for storage and transportation of Sealant; (2) – frame with spikes (3) – spike

2. Summary

1. According to the results of this work, it can be concluded that the method of strengthening the soil with polyurethane foam sealant is relevant, promising and more effective way, compared with the applied ones. Due to its simplicity, extensive scope, fast and high-quality final result, the method will be in demand in the field of road and civil engineering.

2. Polyurethane foam sealant refers to the condensation type, which means it is less sensitive to impurities in soils and oxygen, which significantly expands its scope.

3. The resulting building material is characterized by a low coefficient of thermal conductivity and allows:

- significantly reducing the thickness of the additional layer of pavement;
- providing a stable water-heating regime of the subgrade;
- maintaining the high strength characteristics of the subgrade at the level of strength of the soil components at optimal humidity;
- preventing frost heave of the subgrade.

4. The conducted trial experience of strengthening of the soil with a sealant showed the effectiveness and efficiency of the proposed method. Ease of entry into the ground and the speed of strength can ensure continuity of construction and installation work on the sections of the route with specific soils.

5. The technology of work on strengthening of the soil of the subgrade of transport facilities with polyurethane foam sealants is easy to use, allows getting the desired result in a short time and does not require special expensive equipment.

References

[1] Kamenchukov A V and Bogdanovskaya K I 2015 Performance assessment of pavement *International collection of scientific papers Far East. Highways and traffic safety* (Khabarovsk: Pacific State University) pp 59–62

[2] Kamenchukov A V, Yarmolinsky V A and Lopashuk V V 2016 Trends in the Strength of Pavements *International collection of scientific papers Far East. Highways and traffic safety* **16** 91–6

[3] *On the standards of financial costs and the Rules for calculating the size of the budget allocations of the federal budget for the overhaul, repair and maintenance of federal roads* Decree of the Government of the Russian Federation dated May 30, 2017 No 658

[4] Tulaev A Ya, Avseenko A A and Malitsky L S 1987 *Construction of streets and city roads at 2 p. Part 1 Construction of a subgrade* (Moscow: Stroyizdat) 480 p

[5] Polyntsev E S 2018 Strengthening of soil massifs of the bases of highways with polyurethane foam sealants *Materials of the 71st All-Russian Scientific and Practical Conference of Students, Graduate Students and Young Scientists "Actual Problems of Modern Construction"* (St. Petersburg: St. Petersburg State University of Architecture and Civil Engineering) pp 21–7

[6] *Guidelines for soils and materials reinforced with organic binders* 2003 ROSAVTODOR Approved March 15, 2003 (Moscow) 68 pp.

[7] *Mixtures of crushed stone-gravel-sand and soils treated with inorganic binders for road and airfield construction* 2005 GOST 23558-94: edition (August 2005) with Izm. No 1.2. Gosstroy of Russia Introduction. 01/01/1995 (Moscow: Standartinform) 2005 16 p

[8] *Organic-mineral mixtures and soils fortified with organic binders for road and airfield construction* 1997 GOST 30491-97 Gosstroy of Russia Introduction. 09/01/1997 (Moscow: GUP TsPP) 26 p

[9] Timakova K A, Loginova S E and Panova Yu T 2013 One and two-component polyurethane sealants *Modern high technology* **8-1** 107–8

[10] Pershin M N, Platonov A P and Glinskaya K P 1968 Theory and methods of strengthening waterlogged soils with cold cured resins *Materials for the VIth All. conference on consolidation and compaction of soils ("Theory and methods of art. soil improvement of various petrographic types")* (Moscow: Publishing House of Moscow State University) pp 174–8

[11] Platonov A P 1966 *Resins of cold curing and their use in soil strengthening* (Leningrad) 141 p

[12] Bezruk V M, Guryachkov I L, Lukanina T M and Agapova R A 1982 *Reinforced soils (Properties and applications in road construction)* (Moscow: Transport) 231 p

[13] Kochetkov A V and Yankovsky L V 2017 The use of polyurethane binders for the repair of slopes, cones of bridges and overpasses *Transport. Transport facilities. Ecology* **3** 106–20

[14] Zadiraka A A and Gagarin Yu A 2017 The use of polyurethane composite compositions for the construction of bases and/or coatings of transport structures *Bulletin of BSTU named after V.G. Shukhov* **4** 72–5

[15] *Guidelines for the use of polyurethane binder to strengthen slopes, excavations, bulk structures, cones of bridges and overpasses* 2017 ODM 218.3.093-2017. Rosavtodor Raspor. from 06.28.2017 (Moscow) 84 p

IOP Conf. Series: Materials Science and Engineering **832** (2020) 012030 doi:10.1088/1757-899X/832/1/012030

Study of coarse-grained soils design characteristics

P I Pospelov and T A Rasulov

Moscow Automobile and Road State Technical University (MADI), 64, Leningradsky Ave., Moscow, 125319, Russia

E-mail: tojras80@email.ru

Abstract. The article presents the results of deformation characteristic studies for soils containing various amounts of rock clastic material during its use as a working layer of the subgrade and the foundations of pavements. The studies were performed on coarse soil models using the geotechnical software package Plaxis, which avoided the cumbersome and laborious laboratory tests of soil mixtures. Also, to compare the obtained results of numerical modeling, an analytical analysis of coarse soil deformation conditions was performed. They presented the graphical representation of the relative strain and the deformation modulus of soils belonging to large-clastic varieties, as well as the pattern of their change depending on the percentage of fragments larger than 2 mm. Thus, it was first established that the dependence of the relative vertical deformations of soils on the quantitative content of the coarse fractions in them is strictly linear. However, as for its deformation modulus, the same dependence is no longer linear, but parabolic. The obtained patterns are confirmed by analytical calculations and processing of previously performed studies on physical models. They proposed the methods to refine the calculated characteristics of coarse soil with a varying amount of rock fragments for pavement calculation.

1. Introduction

Coarse-grained soils are widely used both in the construction of hydroelectric station dams, and in the erection of subgrade and construction of foundations of various types of engineering as well as transport structures. These soils have the advantages of good deformation and strength characteristics as compared with cohesive soils.

Depending on the percentage of large debris, the soil massif may be characterized by the deformation-strength index of one or another type of constituent component (fine aggregate or skeletal fraction). The influence of a large debris on the deformation characteristics of soils begins to take place when its soil content is more than 10 % by mass. With large fragment content increase up to 60–70 %, the value of vertical deformation of the soil decreases, and a further increase of skeletal fractions does not significantly affect the change in the deformation characteristics of coarse soil [1].

As they already noted, the main influence on the deformation modulus of heterogeneous soils is the percentage of large fragments. The content of large fragments of about 30 % increases the deformation modulus by 13 % [2]. This is due to the massif rigidity increase, mainly by large fragments.

The classification of soils has no specific quantitative and qualitative characteristics of fine aggregate that regulate the classification of coarse material to one or another group of soil. However, the calculated characteristics of these types of soils are not found in regulatory documents on the design of pavement [3].

2. Research methods

The main methods which determine the deformation characteristics of coarse soil are laboratory test methods and an analytical method. In this work, it was planned to conduct experiments in the PLAXIS finite element software package and determine the deformation characteristics of coarse soil by the analytical method.

2.1. Numerical simulation via PLAXIS program

The geometric model in PLAXIS 2D consists of points, lines and clusters, by which soil massifs should be separated by types and separate layers, objects by structures and stages of construction, as well as by load effects.

Suppose that coarse soil has a continuous particle size distribution and is located in a discrete medium. For convenience, we divide all fractions into two groups: the particles up to 2 mm and more than 2 mm – up to 200 mm. Considering that all fractions smaller than 2 mm have a negligible effect on the stress-strain state of inclusion volume, we model them as a continuous medium. After entering the general settings, we create a geometric model with a block size of 1.0 × 1.0 m. Inside the model, we select large fragments in the form of polyhedral particles. We reproduce the filling of large fragments taking into account the percentage (from 10 to 80 % by volume).

The boundary conditions were set as follows: full fixation (embedment) at the base of the model, and sliding embedment on vertical faces (U_x = 0; U_y = free). A uniformly distributed load of 0.5 MPa acts on the block. The calculated parameters of the elements are shown in table 1.

Table 1. Main characteristics of materials

Item №	Indicator name	Designation	Un. of meas.	Amount	
				skelet.	Stone-free
1	Soil model	Model	–	LE	MC
2	Soil behavior type	Type	–	Drained	Drained
3	Soil specific gravity	γ_{unsat}	kN/m^3	20.5	16.5
4	Saturated soil specific gravity	γ_{sat}	kN/m^3	21.0	18.5
5	Elasticity modulus	E_{ref}	kN/m^2	350000	25000
6	Poisson's ratio	υ	–	0.27	0.35
7	Clutch	C_{ref}	kN/m^2	-	10.0
8	Internal soil angle	φ	°	-	21

Based on the simulation results, a sediment block of 1.0 × 1.0 m was determined under the influence of a static load with different contents of fine earth aggregate (sandy loam, loam, clay) and rock fragments of one size or another (Figure 1).

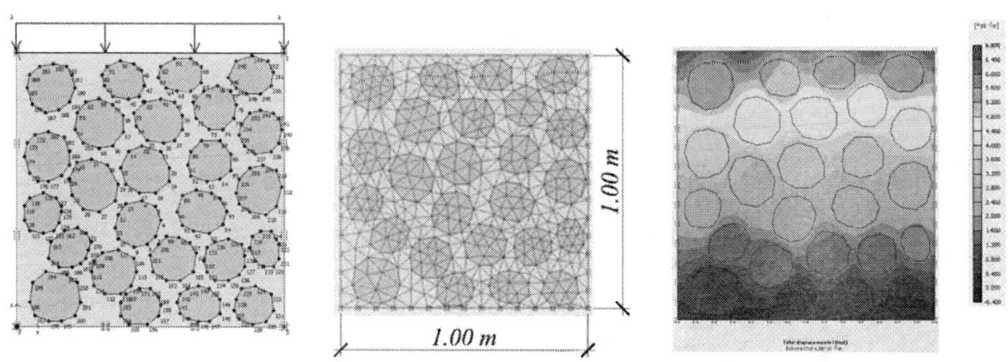

a b c)

Figure 1. Coarse soil model: a) soil model with 50 % rock fragments under a uniformly distributed load of 0.5 MPa; b) finite element mesh; c) isofields of vertical displacements.

It was determined that the dependence of soil sediment block under the influence of a static load on the percentage of large fragments is linear, and the dependence of the elastic modulus on this factor is non-linear (Figure 2).

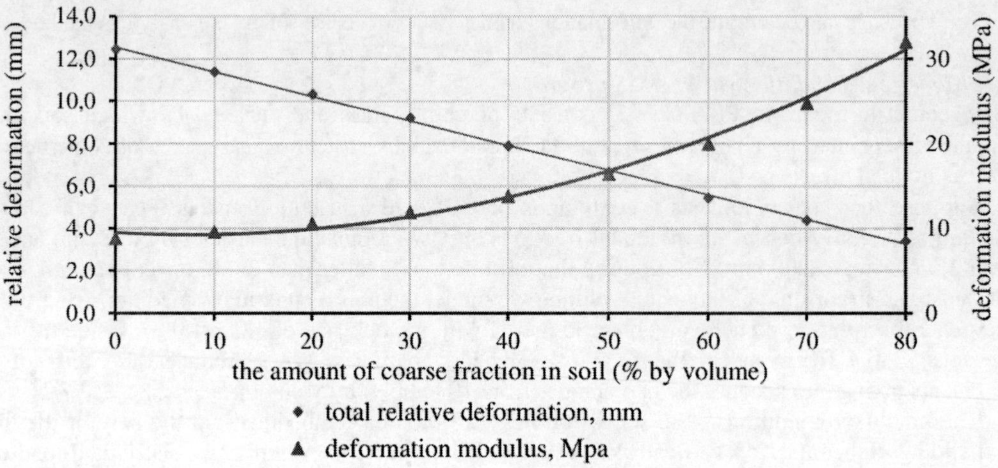

Figure 2. Dependence of the relative strain and the deformation modulus of the soil block on the content of large rock fragments

The increase of rock fragment amount at the same aggregate moisture (sandy loam, loam and clay) leads to the increase of soil elasticity modulus. When the content of large particles is more than 30% by volume, the formation of contacts between particles occurs mainly due to "skeletal" particles. If the content of small particles constituting the aggregate of the soil mixture exceeds 30% by volume, then the gradual development of contact stresses between adjacent small particles begins, as well as contact stresses between smaller and "skeletal" fractions [4]. In this case, the magnitude of the deformation modulus decreases and tends to the characteristic of the aggregate itself.

2.2. Analytical method for coarse soil deformation characteristic study

Similar results were obtained during the theoretical study of coarse soil block deformation (Figure 3), conditionally consisting of two layers. The upper layer with the thickness equal to the relative total volume of the coarse fraction (skeleton) in soil, and the other, with the thickness equal to the relative total volume of the fine earth aggregate [5].

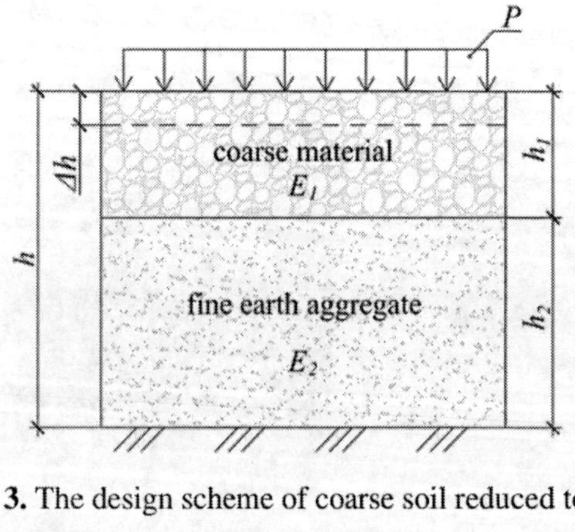

h_1 and h_2 – variables

Figure 3. The design scheme of coarse soil reduced to a two-layer structure.

The module of the total soil deformation is calculated by the following formula:

$$E_o = \frac{p}{\Delta h} h \,. \tag{1}$$

The total deformation of two-layer structure Δh is equal to the sum of the deformations of the upper and lower layers:

$$\Delta h = \Delta h_1 + \Delta h_2 \,. \tag{2}$$

The total deformation of the considered two-layer structure depends on the elastic modulus and thickness of each layer (skeleton and aggregate). This allows the equation (2) to be represented as follows:

$$\Delta h = \frac{p}{E_1} h_1 + \frac{p}{E_2} h_2 = p\left(\frac{h_1}{E_1} + \frac{h_2}{E_2}\right) = p\left(\frac{h_1 E_2 + h_2 E_1}{E_1 E_2}\right). \tag{3}$$

To find the thickness value of each layer, we take $h_1 = nh$, then $h_2 = (1-n)h$, where $n = 0{,}1; 0{,}2; \ldots 1{,}0$. Based on this, the deformation of the structure will be the following:

$$\Delta h = p\left[\frac{nE_2 + (1-n)E_1}{E_1 E_2}\right] h = p\left[\frac{n}{E_1} + \frac{(1-n)}{E_2}\right] h. \tag{4}$$

The formula for the deformation modulus calculation of two-layer structure material taking into account the calculated total deformation (formula 4) can be represented as follows:

$$E_o = \frac{E_1 E_2}{E_1(1-n) + E_2 n} \,. \tag{5}$$

Thus, the result of mathematical modeling and theoretical analysis indicates the presence of a linear dependence of the relative subsidence of the soil mass on the number of rock fragments and a nonlinear dependence of the deformation (elasticity) modulus on the ratio of coarse soil and fine earth aggregate, which is in good agreement with similar field measurements [1].

3. The purpose of elasticity modulus of coarse-grained soil

The results were used as the basis for required soil design characteristics calculation for the working layer of the subgrade. To obtain the estimated characteristics of coarse soil, it was necessary to have data both on the calculated characteristics of fine soil, and the values of coarse soil deformation and strength indicators containing more than 70 % of rock material. The standard values of the elastic modulus of gravel mixtures C5 with the largest particle size of 40 mm were adopted as the calculated values of the deformation characteristics of rocky soils.

Figure 4 shows one of the graphs for calculated elastic moduli determination for some types of soils.

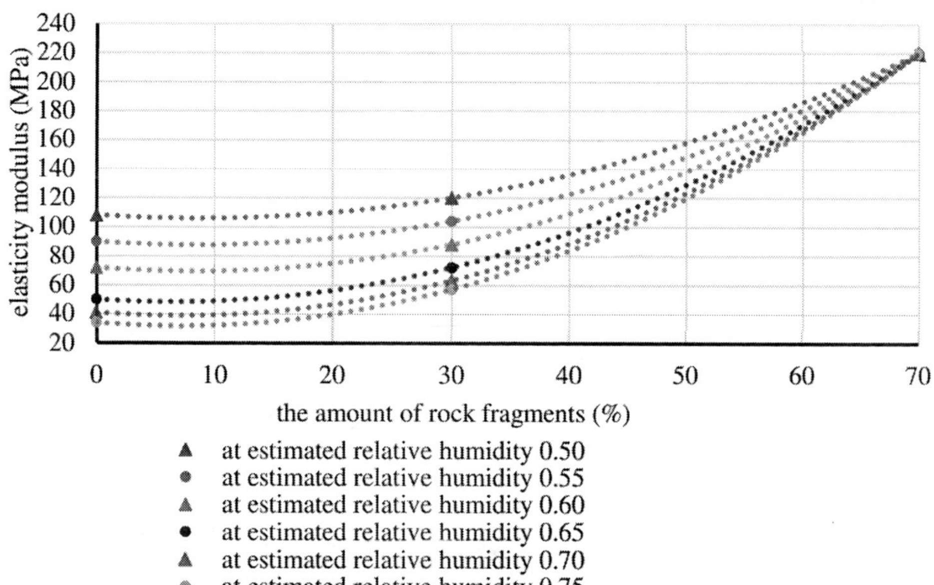

Figure 4. The dependence of the elastic modulus of light loam, heavy loam and clay on the percentage of large fragments at different values of relative humidity.

Based on the obtained correlation equations, the elastic moduli were calculated for the soil containing a given amount of rock material, which allowed us to propose the calculated soil characteristics shown in Table 2.

Table 2. Recommended calculated values of elasticity modulus of soils with the content of rock material (with the degree of solubility in water $q_{sr} \leq 0{,}01$).

Road climatic zone	The scheme of the working layer moistening	Number of large debris in soil, %	The average value of the relative humidity of the filler soil $\overline{W}_{tabl}$ (the fractions of W_T) / soil elasticity modulus (MPa)				
			light sandy loam	dusty sand	light, heavy and heavy silty clay loam	clay	sandy loam, heavy silty sandy loam, light silty clay loam
IV	1	**< 15**	0,53 / 64	0,55 / 90	0,57 / 83	0,57 / 83	0,60 / 72
		15 – 30	0,53 / 70	0,55 / 95	0,57 / 88	0,57 / 88	0,60 / 77
		30 – 50	0,53 / 103	0,55 / 123	0,57 / 118	0,57 / 118	0,60 / 109
		50 – 70	0,53 / 173	0,55 / 181	0,57 / 178	0,57 / 178	0,60 / 175
		> 70	220	220	220	220	220
V	1	**< 15**	0,52 / 66	0,53 / 92	0,54 / 94	0,54 / 94	0,57 / 83
		15 – 30	0,52 / 72	0,53 / 98	0,54 / 99	0,54 / 99	0,57 / 88
		30 – 50	0,52 / 105	0,53 / 124	0,54 / 125	0,54 / 125	0,57 / 118
		50 – 70	0,52 / 173	0,53 / 181	0,54 / 182	0,54 / 182	0,57 / 178
		> 70	220	220	220	220	220

4. Summary

1. Mathematical modeling was performed using the PLAXIS software package and the theoretical analysis of coarse soil deformability depending on the percentage of skeletal material.

2. They determined the regularity of the change concerning the magnitude of the vertical deformation of coarse soil depending on the percentage of large debris in soil.

3. They determined the regularity of elastic modulus of coarse soil change depending on the percentage of large fragments. With their insignificant content (up to 30%), the load is mainly perceived by the aggregate, since large fragments do not contact each other, but rather "float". With a debris content of more than 30%, a gradual increase of coarse soil elastic modulus begins.

4. They determined the estimated values of the deformation and shear characteristics of some varieties of coarse soil for the design of pavement.

References

[1] Dobrov E M and Lyubchenko V A 1981 *Coarse soil in road construction* (Moscow: Transport)
[2] Mirny A Yu 2013 *Mechanical properties of heterogeneous soils as bases and material for earthworks* PhD dissertation (Moscow) 175 p
[3] ODN 218.046-01 Non-rigid pavement design (instead of VSN 46-83) *State Road Service (Rosavtodor)* 2001 (Moscow: Informavtodor) 146 p
[4] Korolev VA and Zhang Shengrong 2018 Modeling of physical-mechanical properties of sand-gravel soil mixtures by PFC-software *Engineer. Geology* **XIII(1–2)** 6–20 DOI 10.25296/1993-5056-2018-13-1-2-6-20.
[5] Pospelov P I and Rasulov T A 2019 Estimated soil characteristics of the subgrade working layer *Sci. and technol. in the road industry* **3(89)** 13–4

IOP Conf. Series: Materials Science and Engineering **832** (2020) 012031 doi:10.1088/1757-899X/832/1/012031

Aging of bitumen on the surface of crushed granite at high temperatures

M G Salikhov, E V Veyukov, R Y Kalugin and A O Yendeletov

Volga region state University of technology, 17, Panfilova Str., Yoshkar-Ola, 424000, Russia

E-mail: VeukovEV@mail.ru

Abstract. When exposed to high temperatures, bituminous materials including mixtures of crushed stone and bitumen (black gravel) quickly change their physical and mechanical properties. In the process of heating up, the structure and composition of the bitumen binding agent go through complex transformations due to the processes of polymerization, polymerization of carbons, evaporation of volatile components, etc. These processes depend on the thickness of bitumen films on the surfaces of mineral particles, depending on the type and size of mineral particles, exposure to air, magnitude and intensity of the process, high temperatures and other factors. Since bitunomineral mixtures contain mineral particles that can vary in size hundreds of times, films of different thickness and directivity are formed on their surfaces in the process of mixing with bitumen. This raises the need for a differential study of their behavior at high temperatures and changes in their adhesive properties in particular. At this, due to the lack of a standard methodology for analysis of these processes, a special estimating experimental method has been developed (see the Patent of the Russian Federation for the invention No. 2654954 [22]). The aging of the bitumen binding agent on the surfaces of mineral materials was analysed using a dimensionless aging factor for local indicators of physical and mechanical properties. According to the research based on this method, the most sensitive indicator is the compressive strenght limit at 50 degrees Celsius for standard cylindrical samples, formed from bituminous mixtures pre-heated at high temperatures. The aging of crushed granite mixtures of fr. 5...10 mm with petroleum road bitumen 90/130 was calculated at various ratios – 1:25; 1:30; 1:35; 1:40, pre-aged at + 150 °C. The aging dynamics was assesed by the values of aging intensity according to the aging coefficient. Due to the experiments, the optimum ratio of bitumen and mineral part (crushed granite) with the best resistance to aging at high temperatures and the dynamics of changing the aging intensity at different time intervals was determined. The results also include mathematical models describing the aging of mixtures under study, analysis of the experimental results and conclusions.

Key words: crushed granite, petroleum road bitumen, bitumen film, heating, high temperature, aging coefficient and intensity, compressive strength limit, mathematical model.

1. Introduction

It is known that at the stage of composing asphalt concretes and other bituminous materials, large fillers are subject to strict requirements: they must ensure good compatibility with a binding agent and meet the required values for strength, durability, frost resistance and other indicators. Crushed acidic rock (granite) with negative surfaces of its grains is often used for that purpose.

As road bitumens are considered acidic environments, the interaction of their surfaces is mostly physical due to the Van der Waals forces [1–4]. As a result, in the process of making bituminous mixtures, surfaces of crushed granite get covered with poorly oriented bitumen films with a low degree of stratification by the group composition [2, 4].

It is also known that the specific surface of crushed stone fillers in the asphalt concretes of the optimal structure is 2.8...3.0 m^2/kg, for mineral powder – 300...500 m^2/kg [2]. As a result, the share of oriented bitumen on the crushed stone surfaces is relatively small. Since crushed stone is the main framework (skeleton) of bituminous mixtures, combining it into a single whole with an adhesive substance – bitumen – is of great importance for ensuring the carrying capacity of this material in general.

Ensuring that the crushed stone frame, consisting of separate stone pieces, is functioning well altogether under the influence of external forces in a constructive layer of road clothing is one of the key requirements for successful functioning of bituminous materials. Thus, resistance to mechanical stresses arising in the body of bitumomineral materials is provided by the mechanical strength of grains of stone materials (crushed stone), the cohesion strength of bitumen films and adhesive strength (adhesion) of the binding agent to the crushed stone surfaces.

The required values of the above indicators are to be provided starting from the stage of composing them. However, it is known that due to natural processes and under the influence of water, high changing positive and negative temperatures, solar radiation, dynamic and cyclical loads when being used individually or jointly during the operation, the properties of all components change as well, usually for the worst.

This process is especially evident in organic components. The result is a change in operational properties of structural layers of road surfaces, such as cracks, potholes, etc. These processes often cause deep irreversible structural changes in materials resulting in weakening of intergranular bonds of stones, less deformability, less corrosion resistance, crack resistance, wear resistance, frost resistance of the coating material, etc. Such processes are called aging.

2. Setting the problem and research objective

The speed and intensity of aging of bitunomineral materials over time seems to depend on the thickness and structure of bitumen films, size and type of grains of mineral materials, the ratio of bitumen and filling agent, etc. [2]. It is evident that aging of bitumen also depends on exposing to oxygen of contact areas, contamination and electrochemical state of mineral material surfaces and the magnitude of current temperatures.

With these considerations in mind, one should expect that the aggressive environment affects the structure of dispersed mineral materials formed on the surfaces of bitumen films (particles of mineral powder, sand and crushed stone) in different wways There is little information in the literature on the study of this problem [1, 2, 4–16]. While it is of high practical importance and theoretical interest.

The objective of this research study is justification of the optimal ratio of bitumen and crushed stone in bitunomineral mixtures, in terms of the highest resistance to their aging at high temperatures. For experiments, a mixture of crushed granite M 1200 fr. 5...10 mm and petrolium road bitumen 90/130 in the ratios of 1:25, 1:30, 1:35 and 1:40 were used, pre-warmed up at 150 degrees Celsius for 0, 1, 3, 5 and 7 hours.

3. Justification of selected methods of the experimental research.

To date, the aging of bitumen is studied according to the PNST-2012 (Preliminary National Standard) [18] and GOST 33140 (State Standard) [19] methods, which assess the aging of bitumen in terms of changing the mass loss of the bitumen sample under study after heating it at the temperature of +163 °C. The drawback of this method is neglecting the influence of the lining surface on aging. Moreover, the structure and thickness of bitumen films are different from their values on the surface of stone materials.

There is also a method [1], that studies changes in the properties of bitumen extracted from their bituminous mixture after heating it with solvents. This method is complex and does not take into account possible changes in properties during the dissolution and extraction of bitumen from the bituminous mixture. As for the study of the aging of bituminous mixtures in general, there are no standard methods so far.

In this regard, a few methods were proposed based on the study of changes in the physical and mechanical and rheological properties of the samples of bituminous mixtures after the exposure to ultrasound, successive exposure to the elevated temperature, freezing-thawing, etc. [12, 18–20].

The above methods are generally complex and require special expensive equipment or long-term and not fully simulating the aging processes of bitumen films within bituminous mixtures. Considering the above mentioned shortcomings of the methods available at the Department of Building Technologies and Roads of the Volga region state University of technology has developed and proposed a new method [22]. It allows studying the aging processes of both multi-component bituminous mixtures, as well as two-component ones – of bitumen and powder, bitumen and sand, bitumen and crushed stone, etc. At this, testing does not take long and requires standard equipment available in construction laboratories, while aging is assessed using a dimensionless, independent of the scale parameter – the aging intensity factor.

The proposed method is as follows:

- Selecting samples of crushed stone, drying them to the constant weight at 150...160 °C.
- Making samples of bitumen by dewatering and heating up to the operating temperature.
- Making mixtures of crushed stone and bitumen (CS:B) in various ratios by thorough mixing.
- Placing the resulting mixtures evenly on separate metal trays and then into a preheated to the estimated temperature (in this case + 150 degrees Celsius) ventilated oven. Then, heating them up during the planned period of time (0, 1, 3, 5, 7, etc. hours).
- After the time expiries, taking trays with mixtures out of the oven and molding standard cylindrical samples (of 71.4 mm diameter and height) from the heated mixtures at pressure of 40 MPa for 3 minutes. After 1 day of exposure to air at room temperature, the samples are tested according to standard GOST 12801-98 [17] methods in order to determine values of their physical and mechanical properties.
- According to experimental results, calculating the aging factor values based on the determined value of the physical and mechanical parameter $K_{Ag}^{Ind_i}$ and aging intensity $I_{Ag}^{K_{Ag}^{Ind_i}}$.
- Determining mathematical models and evaluating aging processes according to parameters $K_{Ag}^{Ind_i}$ and $I_{Ag}^{K_{Ag}^{Ind_i}}$.
- Developing proposals for determining the optimal ratio of bitumen and crushed stone (B:CS) in terms of the highest resistance to aging at high temperatures and selecting measures for reducing them.

4. Results of the experiments and their analysis

This research study used samples of crushed granite M 1200 fr. 5... 10 mm and viscous petrolium road bitumen 90/130 in the bitumen to crushed stone (B:CS) ratios of 1:25, 1:30, 1:35 and 1:40, which were carefully mixed at 140... 160 °C, then placed on separate trays and heated at the temperature of 150 °C for 0, 1, 3, 5 and 7 hours.

After expiration of the estimated period of time, standard cylindrical samples were made of them and then tested. In particular, the compressive strength limit at 50 °C was determined. Some of the results of these experiments are presented in Table 1.

Mathematical models with a high degree of approximation obtained in CurveExpert [23] are as follows:

a) for $K_{Ag}^{R_{compr}^{+50\,°C}}$:

$$K_{Ag}^{R_{compr}^{+50\,°C}} = 9{,}8999 \cdot 10^{-002e^{-1{,}008 \cdot t_{warm}}}$$

$$S = 0{,}0639 \text{ (standard error)}; r = 0{,}9909 \text{ (correlation coefficient)};$$

б) for $I_{Ag}^{K_{Ag}^{R_{compr}^{+50\,°C}}}$:

$$I_{Ag}^{K_{Ag}^{R_{compr}^{+50\,°C}}} = \frac{1}{167{,}1559 - 168{,}63 \cdot t_{warm}^{\;0{,}066}}$$
$$S = 0{,}0333;\; r = 0{,}9967.$$

Table 1. Results of the experiments

№ subpar.	Bitumen:Crushed stone ratio	t_{warm}, hour	δ_{cp}, gr/cm³	$R_{compr}^{+50\,°C}$ MPa	$K_{CT}^{R_{compr}^{+50\,°C}}$	$I_{Ag}^{K_{Ag}^{R_{compr}^{+50\,°C}}}$
1		0	2,33	0,73	1	
2		1	2,26	0,23	0,32	–0,68
3	1:25	3	2,24	0,11	0,15	–0,17
4		5	2,19	0,07	0,01	–0,14
5		7	2,27	0,08	0,01	0,00
6		0	2,24	0,26	1	
7		1	2,19	0,12	0,46	–0,54
8	1:30	3	2,25	0,12	0,46	0,00
9		5	2,18	0,11	0,42	–0,02
10		7	2,17	0,14	0,54	0,06
11		0	2,19	0,13	1	
12		1	2,23	0,11	0,85	–0,15
13	1:35	3	2,23	0,21	1,38	0,09
14		5	2,21	0,15	1,15	0,05
15		7	2,19	0,09	0,69	–0,54
16		0	2,21	0,12	1	
17		1	2,2	0,11	0,92	–0,08
18	1:40	3	2,15	0,09	0,75	–0,17
19		5	2,22	0,2	1,67	0,46
20		7	2,24	0,08	0,67	–0,34

According to the table data analysis, maximum values of medium density and tensile strength of bituminous mixtures are at the ratio of bitumen and crushed stone (B:CS) of 1:25 (see lines 4 and 5).

Further analysis of experimental findings indicates that mixtures of bitumen and crushed stone of all ratios are aged within first 3–5 hours of heating. It means that such mixtures reach their limit in terms of aging, and structural layers of road clothing in this state should be expected to have the least resistance to destruction at external influences. It is evident that polymerization processes in bitumen films end in 3...5 hours.

To explain the experimental data pattern, let's have a look at conditions for formation of bitumen films on surfaces of crushed stone grains. At the bitumen:crushed stone ratios of 1:25, 1:30, 1:35, 1:40, the actual content of bitumen in bituminous mixtures is 3.99; 3,34; 2.86 and 2.50 % respectively. According to I.V. Korolev [2], the specific surface of crushed granite fr. 5...10 mm is 0.71 m^2/kg.

At the bitumen:crushed stone ratios of 1:25, 1:30, 1:35 and 1:40, the proportion of blackened surfaces at the average thickness of the bitumen film of $6.4 \cdot 10^{-4}$ cm is 88.5; 73,7; 63,2; and 63.0 % respectively. This implies that the highest adherence of bitumen films is in the mixtures with the most blackened surface of crushed stone grains: the mixture of bitumen and crushed stone with the ratio of 1:25, that was confirmed by the tests results as well.

Analyzing the samples aging dynamics during the heating, we can see that samples with a high bitumen content have higher speed and intensity of aging (columns 6 and 7 of Table 1). The maximum aging values in all samples are during the first hour of heating, then this process is reduced.

Therefore, in order to reduce the aging process in early periods of life of bituminous mixtures, we should search for technological and other methods. According to studies by Bahrah G.S.[6], Korolev I.V. [2], Gezentswey L.B. [1], Uglovaya E.V. [9], Skripkin A.R., Starkov G.B., Kolesnik D.A. [12], Bratchun V.I. and Stukalov A.A. [8], Saenko S. S. [11], American, Chinese and Japanese scientists [15, 16], scientists from Belgorod State University of Technology [24] and others [20], to achieve this, mixtures should be made at the lowest technological temperature, in the environment of inert gases, with adding of special anti-aging and surface active agents, using binding agents of polymer and bitumen, rubber and bitumen, sulfur and bitumen, etc.

5. Conclusions

1. Experimental findings let us conclude that bituminous mixtures with the 1:25 ratio of bitumen and crushed stone have the highest resistance to the aging of bitumen at high temperatures. This corresponds to bitumen content in the mixture of about 4.0% of the crushed stone mass.

2. At high temperatures (150 °C), bitumen in the mixtures of bitumen and crushed stone ages intensely during the first hour of heating, then this process stabilizes.

3. Resistance of samples of the bitumen and crushed stone mixtures compressed at 50°C is sharply reduced due to aging. In order to ensure the resistance of structures using mixtures of bitumen and crushed stone, methods for regulating this process at all stages of their life cycle should be developed.

References

[1] Gezentsway L B, Gorelyshev N V, Boguslavsky A M and Korolev I V 1985 *Road asphalt concrete* (Moscow: Transport)
[2] Korolev I V 1986 *Ways of saving bitumen in road construction* (Moscow: Transport)
[3] Kuchma M I 1980 *Surfactants in road construction* (Moscow: Transport)
[4] Kovalev Y N 1990 *Activation-technological mechanics of road asphalt concrete* (Moscow: Transport)
[5] Rybiev I A 1969 *Asphalt concretes* (Moscow: Vysshaya shkola)
[6] Bahrah G S 1973 Considering the aging process at planning the composition of bituminous mixtures *Avtomobilnye dorogi* **9** 8–9
[7] Gohman L M and Amosova N V 1988 Study of the effect of bitumen quality on the processes of its aging in thin films *Oil chemistry and refining* **2** 6–8
[8] Bratchun V I, Gulyak D V and Bespalov V L 2008 On certain patterns of the aging of concrete mixtures and concretes on organic binders using tar concrets as an example *Vestnik Kharkovskogo Natsionalnogo Avtomobilnogo dorozhnogo univer.* (Kharkiv: HADI)
[9] Iliopolov S K, Mardirosova B V and Uglova E V 1994 Development of bitumen aging processes in asphalt concrete pavements of motor roads *Izvestia vuzov. Series: Construction* **3** 48–52

[10] Uglova E V 1993 *Increasing the resistance to the aging of bitumen in asphalt concrete pavements in the South of Russia environment* Master of technical sciences' thesis
[11] Saenko S S 2008 *Methods for minimizing the aging of bitumen in a working boiler when making hot asphalt mixtures* PhD dissertation thesis
[12] Skripkin A D, Starkov G B and Kolesnik D A 2010 Aging of bitumen in the technological process of its preparation for production of asphalt mixtures *Collection of articles and reports of the annual scientific session of the Associat. of researchers of asphalt concrete* (Moscow: MADI) pp 46–53
[13] Stukalov A A 2015 *Aging of asphalt concrete mixtures and ways to increase their thermal oxidation stability* PhD dissertation
[14] Salikhov M G, Ilivanov V Yu, Malyanova L I and Veyukov E V 2018 Studying the temperature aging of modified crushed stone and mastic asphalt concrete with limestone crushing waste *Transport facilities. Ecology* **1** 70–4 (Perm: PNIIPU)
[15] Siddiqui M M 2002 Use of X-ray diffraction in assessing the aging pattern of asphalt fractions *Fuel* **1** 51–8
[16] Zhao Zhaohui A 2003 Study on aging kinetics of asphalt based on softening point *Petroleum Sci. and Technol.* **10** 1575–82
[17] GOST 12801-98 1999 *Materials based on organic binders for road and airfield construction* Experimental methods Enforced by Resolution of the Gosstroy of Russia
[18] PNST 8-2012 (Preliminary national standard) 2014 *Public roads. Viscous petroleum bitumen. Method for determining bitumen resistance to aging under the influence of high temperature and air* (Moscow: Standartinform)
[19] GOST 33140-2014 2015 *Public motor roads. Viscous petroleum road bitumens. The method of determining bitumen resistance to aging under the influence of high temperature and air (RTFOT method)*
[20] Kalgin Y I, Strokin A S and Tyukov E B 2014 *Promising technologies for the construction and repair of road surfaces* (Voronezh: Voronezh state univer. of architect. and construct.)
[21] Tarashchansky E G and Wilmsen I I 1975 *Study of the asphalt concrete aging using impulse ultrasonic method Improving the efficiency of cement and asphalt concrete in Siberia* (Omsk: SibADI) pp 40–61
[22] Salikhov M G, Veyukov E V and Sabirov L R and Malyanova L I 2018 *Method for determining the speed and intensity of aging of asphalt concretes* Patent of the Russian Federation for the invention no. 2654954
[23] Mazurkin P M add Filonov A S 2006 *Mathematical Modeling. Identification of one-factor statistical patterns* (Yoshkar-Ola: MarGTU)
[24] ODM 218.3.021-2011 2012 Methodical recommendations on the selection of asphalt concrete mixtures, taking into account the effect of adhesive additives on the aging of organic binder components in bituminous mixtures *Federal Road Agency* (Moscow: Rosavtodor)
[25] ODM 218.3.020-2012 *Methodical guidelines for ensuring the bitumen sustainability against aging in the technological processes of manufacture and application of asphalt concrete mixtures* (Moscow: Rosavtodor)
[26] Anasuke M 1971 *Method for preventing embrittlement and destruction of asphalt concrete coatings* The patent of Japan no. 51-41766

Development of road transport infrastructure in the regions of the Russian Federation using digital technologies

T A Shpilkina and M A Zhidkova

Moscow Automobile and Road State Technical University (MADI), 64, Leningradsky ave., Moscow, 125319, Russia

E-mail: luk-72@mail.ru, zhidkova_m_a@mail.ru

Abstract. The article studies the problems of road transport infrastructure in the regions of the Russian Federation on the basis of the use of digital technologies. The parameters of the national project "Safe and high-quality roads" (BKAD) are analyzed and the prospects of its implementation by 2024 in the regions of Russia, in which the largest amount of road repair work is planned, are determined. Such regions are Nenets Autonomous Okrug, Kalmykia, Kirov and Kostroma regions, Altai Republic and Tomsk region. By 2024 roads in the Arkhangelsk and Saratov regions will be repaired least of all. Identified regions in which the repair of roads will not change during the period of 2017–2024 are Krasnoyarsk Krai, Voronezh oblast, Chukotka, etc. The data are presented Rosavtodor to equip highways and introduce intelligent transport systems using modern digital technology in 2021 and 2024; the program of measures on the national project BCAD is intended for 2019–2024. The implementation of the presented measures will contribute to the safety of road conditions, the system of assistance to victims of road accidents, and reduce negative emotional stress.

1. Introduction

A key problem for many regions of Russia is the problem of roads, which consists both in their lack and in their rather poor condition. In this regard, the national project "Safe and quality roads" was developed in 2018. It is aimed at solving the problems of road transport infrastructure on the basis of the introduction of modern digital technologies, which, in turn, should improve both the condition of roads and lead eventually to a decrease in road accidents, which is important in modern realities.

Every year on the roads of the country killed so many people that can be equated to the population of a small town. In addition, the implementation of this national project will link the regions with new roads and increase the speed of delivery of goods and passengers throughout the country.

2. Problem statement

In the regions there is a high need to update the territorial socio-economic infrastructure» [1]. In the last 15 years, those regions where there are prospects for development for the population have been actively built up. However, the construction of new residential neighborhoods and office buildings was often chaotic, due to the crisis of the current system of spatial planning. Over the past 30 years, the role and importance of the master plan decreased, as a result, it lost the status of a strategic document [2]. «Some large cities are trying to make high-quality master plans today, but the average plans have a number of drawbacks. They have no goal-setting and vision of the future, they are not based on reliable data, which are often absent, or on the results of research» [2].

As a result – attractive settlements have huge building without competently thought-out construction of roads. This leads to the congestion of the existing road infrastructure, and this, in turn, causes discontent of the population and motorists, increases the process of delivery of goods and transportation of passengers, causes irritation in the population, and adversely affects the health and well-being, while worsening the efficiency of people's work. We can say that «The existing problems in the modern road sector of the country are one of the limiting factors of economic progress» [3].

3. Materials and Methods

The passport of the national project «Safe and quality roads» was developed by the Ministry of transport pursuant to the Decree of the President of the Russian Federation, May 7, 2018 No. 204. «On the national goals and strategic objectives development of the Russian Federation for the period up to 2024». It includes 4 of the Federal projects: «Road network», «System-wide measures for the development of road facilities», «Road Safety» and «Roads of the Ministry of defense of Russia». These projects should be fully completed by 2024. This year, the first stage of the launch of the national project began and already in the regions of the Russian Federation there were questions about the availability of funds for the implementation of these projects and their real implementation. The project «Safe and quality roads» is largely based on regional funds.

In 2019, the share of regions in its financial support is 74 %, and by 2024 it will grow to 95 % [4]. Excise taxes should become a source of funds for the implementation of the national project, that is, from the sale of excisable fuels and lubricants, more and more funds will flow to regional budgets, and by 2024, 100 % of excise taxes on fuel and lubricants will go there. Nevertheless, the amount of expenses for this national project is significant, and, therefore, there is a risk of default [4].

4. Results and Discussion

To understand the implementation of the national project «Safe and quality roads» (hereinafter BCAD), we turn to the targets and indicators of spending on it and other national projects. The target indicators of the national project of BCAD are: increase in the share of regional roads in the standard state, from 43 to 50.9 %; the share of roads of urban agglomerations in the standard state, from 42 to 85 %, that is, almost 2 times. The construction of new roads is not included in this national project [5].

With regard to the expenditure of funds for this national project, the situation is as follows. According to experts of the Analytical credit rating Agency (ACRA), most of the costs in the regions are accounted for by roads – 4.1 trillion rubles, or 84.5 %.

For comparison: «Health» accounts for 265 billion rubles or 5.4 % of the amount funded by the regions (Fig. 1).

As for BCAD, to a greater extent it is based on regional means. The total amount of financing of the national project is 4.77 trillion rubles, 93 % of which will go to road works, while 4.1 trillion rubles will be provided by the budgets of the subjects of the Federation. 299 billion rubles during the first 3 years will come from the Federal budget. Funds from the Federal budget to the regions will give the rate of 12.7 million rubles for 1 km of road requiring repair.

The volumes of expected repair work in the regions are shown in Fig. 2; data for regions where road conditions are almost unchanged are presented in Fig. 3 [5].

As seen in Fig. 2 by 2024 the roads in the Nenets Autonomous Okrug should be repaired – 63.3, 50 % of repaired roads are expected in Kalmykia, Kirov region, Kostroma region, Altai Republic and Tomsk region, 42,7 % in the Ryazan region. By 2024 the roads in the Arkhangelsk region will be repaired least of all – 29.1 % and in the Saratov region – only 24.8 %.

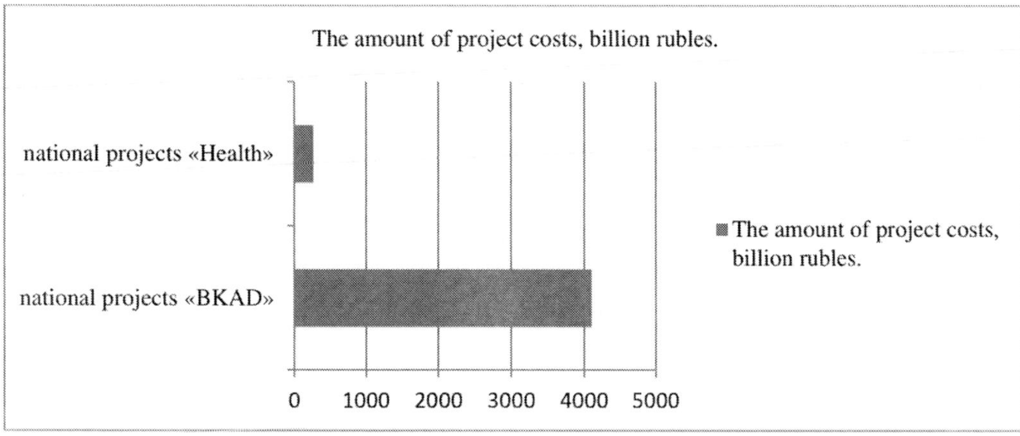

Figure 1. Comparative assessment of expenditures on national projects «BKAD» and «Health»

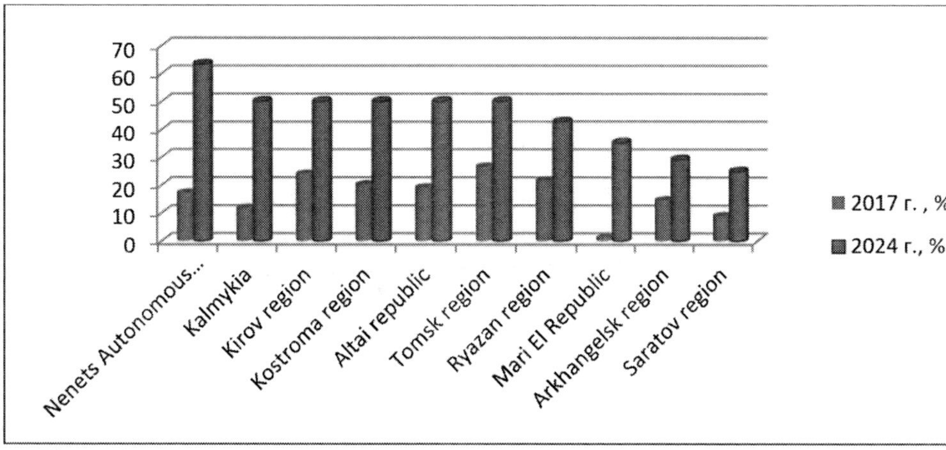

Figure 2. Data on the regions of Russia, where the most roads will be repaired by 2024

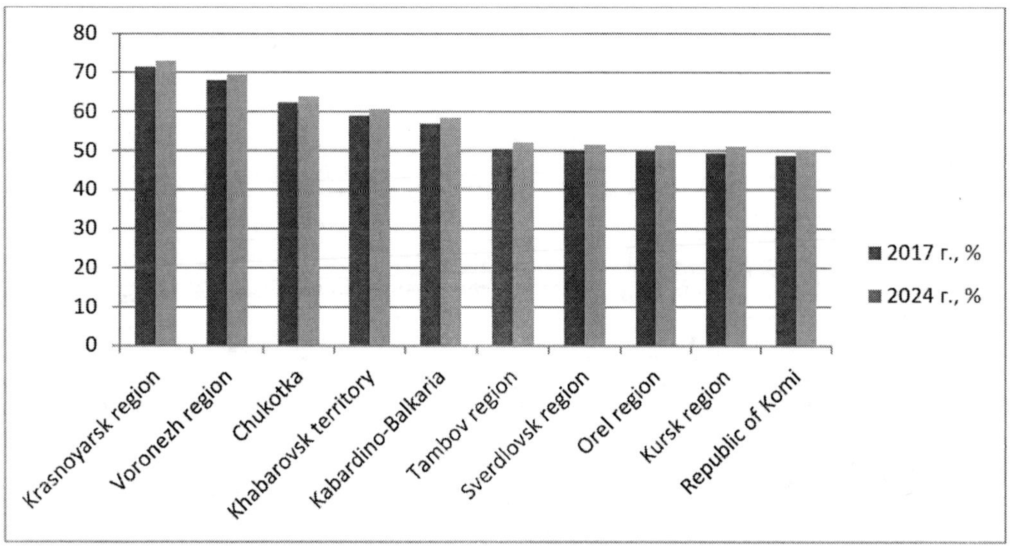

Figure 3. Data for regions where road conditions will be almost unchanged by 2024

IOP Conf. Series: Materials Science and Engineering **832** (2020) 012032 doi:10.1088/1757-899X/832/1/012032

Based on Fig. 3 it follows that in the presented regions for 8 years (from 2017 to 2024) the volume of work on road repair will not change, that is, the values of 2017 and 2024 differ slightly. However, the largest volumes by 2024 will be made in Krasnoyarsk region – it is expected that 73 % quality of roads in the Voronezh region – 69,6 %, on Chukotka – 63,9 %, in Kursk region – 51 % and in the Republic of Komi, 50.3 per cent.

According to experts, the stated values may not be fully realized, and this is due to the fact that «in the medium term, the financial performance of the regions may deteriorate due to the declining growth of tax and non-tax revenues and growing pressure on the expenditure side of the budget» [4]. Of course, the situation over these 5 years can change for the better; a lot depends on the market situation and the appropriate attention of both Federal and regional authorities to the implementation of this national project.

With the active introduction of digital technologies, the question arises as to how this will affect the transport infrastructure. According to Rosavtodor, the situation on the equipment of roads, the introduction of intelligent transport systems, including those focused on the automation of traffic control processes is shown in Table 1 [6].

Table 1. Data of Rosavtodor on equipment of highways and introduction of intelligent transport systems in 2021, 2024

No.	Name	2021	2024
1	Equipping road sections and artificial structures of regional importance with elements of intelligent transport systems focused on automation of traffic management processes (number of sections), %	20	60
2	Implementation of intelligent transport systems, providing automation of traffic management processes in urban agglomerations with cities with a population of over 300 thousand people, %	28	64
3	Equipping sections of roads and artificial structures of Federal importance with elements of intelligent transport systems focused on automation of traffic control processes (number of sections), %	40	120

As can be seen from table 1 equipping road sections and artificial structures of regional importance with elements of intelligent transport systems aimed at automating traffic management processes in the number of sections in 2024 will increase 3 times from 20% to 60%. Implementation of intelligent transport systems that provide for automation of traffic management in urban areas with cities with the population over 300 thousand persons, growth is projected 2.3 times, i.e., from 28% to 64%. A significant increase is expected in the equipment of sections of roads and artificial structures of Federal importance with elements of intelligent transport systems, namely from 40% to 120%, that is, 3 times.

Most regions of Russia are also interested in that other activities will be carried out within the framework of the national project «Safe and quality roads». The programme of activities is presented below.

Table 2. The program of activities of the national project BCAD in 2019–2024

No.	Name	Period
1	It is planned to create and implement a Register of new and best technologies, materials and technological solutions for reuse.	End 2019
2	Adoption of regulatory legal acts aimed at strengthening the responsibility for some of the most dangerous offenses in the field of traffic.	End 2020
3	Adoption of legal acts aimed at improving the system of professional training of drivers.	End 2021
4	It is planned to introduce and ensure the functioning of the information system for monitoring the formation and use of road funds at all levels (SKDF) and the annual updating of the data included in it.	End 2024

In addition, the system of traffic management of vehicles and pedestrians will be developed, the safety of road conditions will be increased; the system of assistance to victims of road accidents will be developed [7].

5. Conclusion

The implementation of these measures will solve many problems that are taking place today. High-quality and safe roads will reduce accidents and the number of accidents that will save lives, accelerate the delivery of goods and passengers, both in the regions of the Russian Federation and between them. «To address existing shortcomings in the system of road funds and increasing the efficiency of funds necessary to create an effective system of control over the quality of work for the expenditure of funds in the road sector, to devise and implement a system of personal responsibility for the proper and efficient use of funds road funds» [3].

This problem is one of the most important, and the implementation of the national project BCAD will have a positive impact on the road transport infrastructure of the country, increase the efficiency of enterprises and reduce the emotional stress of the population, all of this will improve the lives of Russians.

References

[1] Frolova V, Dolina O and Shpilkina T 2018 Mining Enterprises Development as a Financial Source of Environmental Activities in Mining Regions *E3S Web of Conf.* **41(04027)**
[2] Shchukin A 2019 Master plans prevented builders *Expert* **15(1115)** 18–21
[3] Dolina O N, Zhidkova M A and Shpilkina T A 2019 Road funds in Russia: problems of effective use *Bulletin of the Moscow Automobile and road State Technical University (MADI)* **3(58)** 48–54
[4] Mavrina L 2019 Roads under threat *Expert* **21(1120)** 36–8
[5] Buranov I, Hvostik E, Rozhdestvenskaya Ya and Sarhanyants K *Very expensive roads. Will 4.7 trillion rubles help to overcome off-road and accidents.* Retrieved from: http:// www.commersant.ru
[6] *Official website of Rosavtodor.* Retrieved from: www.rosavtodor.ru
[7] *Official website of the Russian Government.* Retrieved from: www.govervment

IOP Conf. Series: Materials Science and Engineering **832** (2020) 012033 doi:10.1088/1757-899X/832/1/012033

Field evaluation of rutting in concrete pavements

V Ushakov[1], D Jelagin[2], G Dyakov[1], S Dmitriev[1] and M Goryachev[1]

[1]Moscow Automobile and Road State Technical University (MADI), 64, Leningradsky ave., Moscow, 125319, Russia

[2]Royal Institute of Technology, Department of Civil and Architectural Engineering, Division of Building Materials, Stockholm, Sweden

E-mail: madi-ushakov@mail.ru, jelagin@kth.se

Abstract. Surface wear by studded tires is a primary cause for rutting in portland cement concrete (PCC) pavements, and may, for the pavement sections with high traffic intensities, become a primary factor controlling the length of pavements maintenance intervals. In the present study, PCC pavement field performance with respect to surface wear is examined based on the transverse surface profile measurements performed yearly during first five years of pavements service life. Continuous transverse surface profile measurements have been performed yearly on a 20 km long rigid pavement section in Moscow region, Russia. Based on the measurements, quantitative parameters describing rut depth evolution are obtained and evaluated; the obtained results are discussed in context of previous research findings reported in the literature. The effect of pavement surface wear on the further rutting accumulation rate is examined. The particular attention is also given to the spatial variability of the measured pavement surface wear characteristics, i.e. rut depth and accumulation rate. The implications of the obtained results on the length of rigid pavement maintenance intervals are discussed.

1. Introduction

Rutting is one of the primary modes of pavements distress, controlling to a great extent the length of road service life. As discussed in several previous studies, a primary cause for rutting in portland cement concrete (PCC) pavements is surface wear by studded tires [5, 6]. In fact, for either PCC or asphalt concrete pavement sections with high traffic intensities, the studded tire damage may become a primary factor controlling the length of pavements maintenance intervals [6, 16]. Furthermore, pavement surface wear by studded tires generates significant amounts of dust and represents thus a significant source of air pollution [20], and may create polishing effects which may affect friction properties. Moreover, ruts are also creating serious safety risks when water cannot flow off or when vehicles suffer fishtailing when changing the traffic lanes. Accordingly, understanding the material, traffic and environment parameters controlling the rate of pavement surface wear is of profound importance for identifying pavement construction practices that are optimal from life cycle cost perspective as well as for quantifying full socio-economic effects of using studded tires. The present study aims at providing further insight into the issue through examining PCC pavement field performance with respect to surface wear.

The factors controlling the accumulation rate of studded tire wear have received considerable attention in the literature [14]. Damage of aggregates due to stud impacts with subsequent removal of material fragments by the scraping action of the stud may contribute significantly to the pavement surface wear [15]. Accordingly, ensuring adequate resistance of the aggregates to contact induced damage is crucial for ensuring adequate pavement performance with respect to surface wear [7, 12, 16].

The factors affecting the surface wear rate may be categorized as pavement, environment, traffic, vehicle and tire [13]. Regarding the pavement the most important factors were found to be the section geometry, pavement surface material and surface condition (i.e. it's temperature and the presence of water, snow or ice). Generally, PCC pavements are reported to have a slower rutting rate as compared to the asphalt ones [5, 20]. It has however to be pointed out that for the case of asphalt pavements, isolating load related rutting from surface wear is a difficult task [5]. Furthermore, the rate of surface wear in PCC pavements varies significantly with time. The PCC pavements wear faster during their first 5 years of service, which may be attributed to the fact that wear is dominated by the removal of cement paste from pavements surface finishing [6]. The wear rate is reduced when the damage to underlying aggregates becomes a controlling factor.

Traffic and vehicle characteristics affect pavement surface wear rates profoundly; obviously, high share of studded tire traffic on the road and long periods of winter road conditions promote surface wear. Studded tire characteristics, in particular the number of studs, their weight and geometry affect the pavement surface damage rate profoundly [17]. The pavement surface damage induced by studded tires depends also on vehicle's driving regimes and traffic speed. In areas of acceleration and braking, the pavement wear is generally found to be higher as compared to the sections with uniform traffic speeds [14]. In fact, the influence of traffic speed on the pavement wear rate may be profound; for instance, according to the data presented by Unhola [17], changing the uniform traffic speed from 80 to 120 km/h results in approximately 3.5-time increase in pavement wear.

It may be concluded from the above that pavement surface wear by studded tires is a complex multi-factor problem. Accordingly, wear rates reported in different studies vary considerably [6, 14]. Correspondingly, the influence of studded tire traffic on service life of PCC pavements has to be examined at representative traffic and environmental conditions. Several PCC pavements constructed recently in Russia have exhibited severe rutting during few years after construction [18, 19]. As argued in their study, damage from studded tires may be a primary cause for the excessive rutting observed. The pavement sections examined by Ushakov and Diakov [18] belong to major highways with a speed limit of 130 km/h. Presently, there is, however, only a very limited amount of the quantitative data on surface damage accumulation in high speed PCC pavement sections.

In the present study the implications of studded tire traffic on service life of rigid pavements are evaluated for traffic and environmental loading conditions representative for the Moscow region in Russia. In particular, transverse surface profile measurements have been performed yearly on a 20 km long rigid pavement section in Moscow region. Based on the measurements obtained during the first five years of service life, quantitative parameters describing rut depth evolution are presented. The effect of pavement surface state and traffic characteristics on the accumulation rate of the rut depth is examined. The results are discussed in context of previous research findings from literature. Particular attention is given to the spatial variability of the measured characteristics of pavement surface wear, i.e. rut depth and rutting accumulation rate. The implications of the obtained results on the length maintenance intervals for rigid pavements are discussed.

2. Field Performance Study

The examined road section is located at federal highway M4 "Don" between kilometer marks of km 52+000 and km 71+000. The reconstruction of the road section, finished in 2011, included both road widening and re-paving. During reconstruction, the width of the median strip has been reduced and an additional lane in each direction has been constructed. The plan of the reconstructed road section is shown in Figure 1 along with the pertinent dimensions (in meters) as well as lane and wheel path notation used in this report. The inner to outer lanes are designated from 1 to 3. The north- and south-bound lanes are designated with "+" and "–" signs respectively. In order to designate the measurements performed in the left and right wheel paths letters "R" and "L" are added to the lane designation.

The structural design of the road section is uniform for the whole section. As part of the reconstruction, the existing PCC pavement course has been removed and the section has been repaved. The resulting pavement cross-section is depicted in Figure 2. The section is exposed to quite intensive

IOP Conf. Series: Materials Science and Engineering **832** (2020) 012033 doi:10.1088/1757-899X/832/1/012033

traffic, with average daily vehicle count exceeding 55 thousands. During winter driving conditions, the share of studded tires on passenger cars is close to 100%. The maximum allowable speed limit along the section is set to 130 km/h for the vehicles with the gross weight below 3.5 tons. As a result, the heavy vehicle traffic is basically fully confined to the outmost road lanes (i.e. lanes ±3).

In order to investigate the rate of rutting accumulation in the road section described above, continuous transverse profile measurements have been performed yearly between 2012 and 2016. As the pavement reconstruction has been finished in 2011, the obtained results correspond to the first 5 years of pavements life. Profile measurements have been performed with the road surface testing vehicle developed at MADI University, Russia. The surface testing vehicle, illustrated in Figure 3, combines video recording of the road surface with continuous longitudinal and transverse profile measurements. The laser profile measurement system incorporates accelerometers to compensate for vehicle movements and allows for the road transverse profile measurement with an approximately 1.2 meters step. The transverse profile measurement results are also illustrated in Figure 3. In order to calculated the rut depth, for a given measured transverse profile, a straight line, with a slope corresponding to the sections superelevation at this point, is fitted to the profile. The rut depth is obtained then as a vertical distance from the fitted line to lowest points in the left and right wheel paths.

Based on yearly transverse profile measurements, the pavement surface deterioration is examined as follows. The measurements illustrated in Figure 3 are used to calculate average rut depths at 25 meter increments for all lanes and wheel paths. Rutting accumulation rate is obtained then as difference between rut depths measured in two consecutive years. In order to examine the spatial distribution of pavement surface wear, cumulative distributions of the rut depths are calculated for all lanes examined and the results are compared with average values per lane.

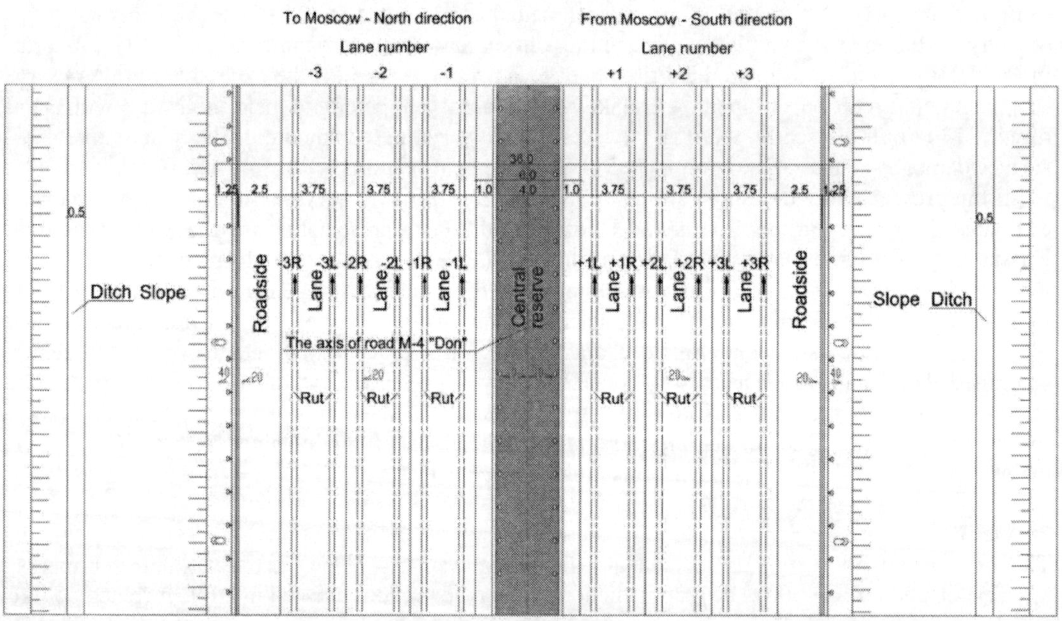

Figure 1. Plan of the reconstructed road section

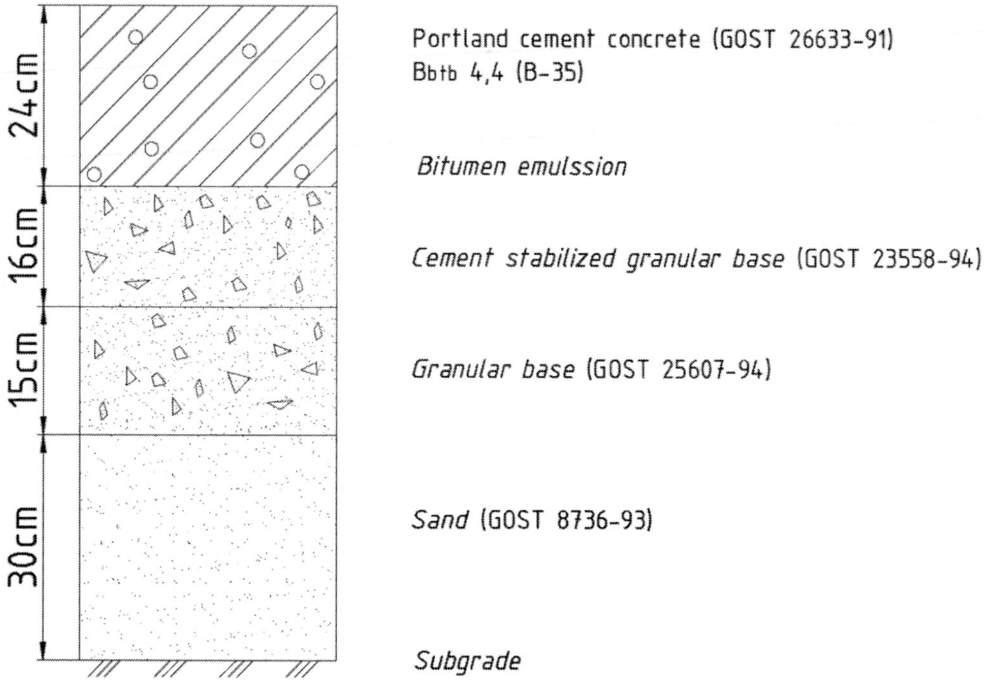

Figure 2. Structure of the rigid pavement section after reconstruction.

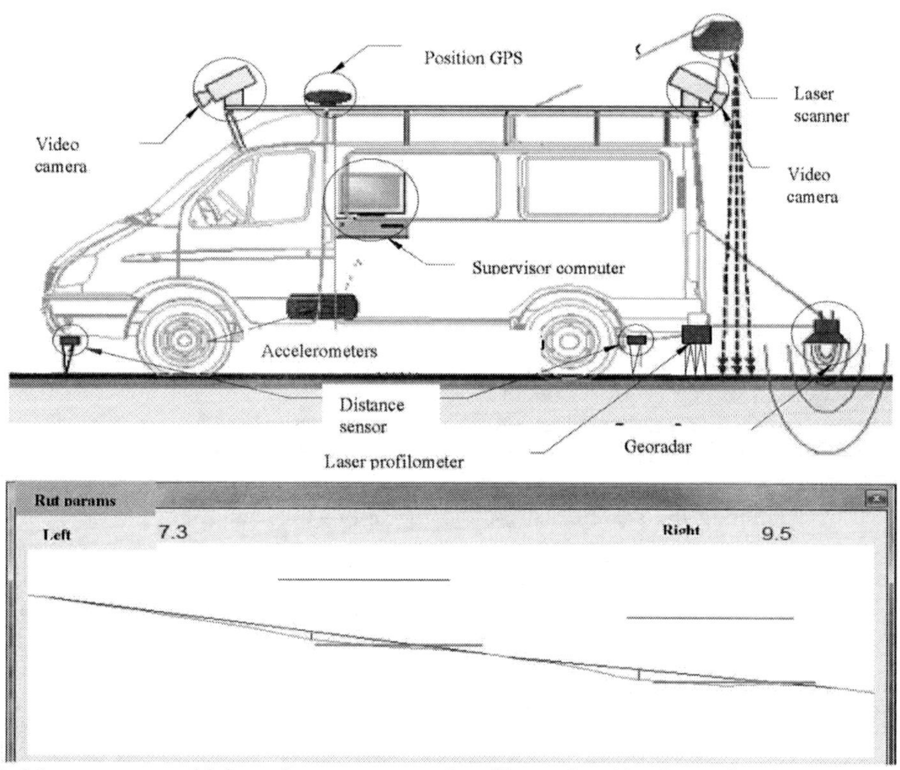

Figure 3. Measurement vehicle and measurement data illustration (the data point shown corresponds to lane –1, km 66+500)

3. Results and Discussion

3.1. Rutting evolution with time

Attention was focused first on the evolution of the transverse pavement profiles; results are presented using lane and wheel path notations of Figure 1. Figures 4 and 5 show the evolution of average rut depth in each wheel path over time as obtained from transverse profile measurements for North and South bound lanes. As may be seen in Figures 4-5, the measurements for the left and right wheel paths are fairly consistent for all the lanes. In most cases the difference in rut depth measurements is less than 1 mm and the maximum difference between the measurements obtained in the left and right wheel path is approximately 2 mm. Obviously, the observed surface wear is highly dependent on the lane. According to Figures 4–5, the inner lanes accumulate rutting at least twice as fast as than the outmost lanes. This was expected, as the outmost lanes are primarily used by heavy traffic and accordingly both traffic speed and proportion of studded tire traffic is higher for the lanes 1 and 2.

As it may also be noticed in Figures 4–5, the rut depth accumulation rate varies significantly with time. A particularly high rutting rate is observed during the first year of service for all the lanes examined. This observation may be associated with the faster wear of cement paste from the pavements surface finishing [6]. The average rutting accumulation rates per lane are also reported in Table 1. As may be seen in Table 1, after the first service year, the rutting accumulation rate drops significantly and starts to increase again during service years 3–5.

It has to be pointed out that for all the cases examined the observed rutting rates are quite high, being above 0.5 mm/year for virtually all the cases examined. At average wear rates above 0.5 mm/year the surface wear will become a primary factor controlling service life of rigid pavements [6]. The average wear rates measured during first five years of pavements service are above 2mm/year for lanes 1 and 2, cf. Table 1. Also, in contrast to observations reported by Cotter and Muench [6], rutting rate does not reduce with years of service. Rutting rate reduction was attributed in their study to change of rutting mechanism from wear of the cement paste to aggregate damage. This qualitative discrepancy between the results in Table 1 and the ones reported by Cotter and Muench [6] may be explained by the differences in traffic speed, material properties and winter maintenance approaches. In order to evaluate whether material related factors are responsible for the excessive wear observed, the intention is, as part of the future studies, to conduct an experimental study for quantifying the resistance of both cement concrete and aggregates to abrasive wear.

3.2. Spatial distribution of rutting

In order to investigate spatial distribution of surface wear, maximum rut depth values for each 25 meter interval of the pavement section have been calculated based on the transverse profiles measured after 5th years of the roads service life. The rut depth distribution obtained is presented in Figure 6 for the North- and South-bound lanes 1, having their average rut depth values of 10.5 and 12 mm respectively. As may be seen, there is considerable spatial variation in the pavements surface wear, with rut depths varying between approximately 5 and 20 mm for both lanes examined. Regions with extremely high rutting (with rut depth of approximately 20 mm) are not concentrated to any given location, but rather more or less uniformly distributed along the section. These regions, representing severe driving safety concern with respect to hydroplaning, are not captured adequately with the average rut depth measures. Accordingly, the results presented in Figure 6 emphasize the importance of examining the spatial distribution of rutting for supporting pavement maintenance decisions.

Based on the high gradients of rutting presented in Figure 6 (rut depth often changes more than 5 mm over 250 meters intervals), it may be argued that the observed variability is due to traffic characteristics, i.e. it's speed, density and driving conditions. This conclusion is indirectly supported by the consistently lower rut depths observed at the ends of the monitored section, as the toll stations located there results in both lower traffic speed and more uniform distribution of traffic over the lanes. However, a detailed investigation of the factors affecting variability of rut depth lies beyond the framework of the present study.

The spatial variability of rutting measured in lanes 1 after 5 years of service is further examined in Figure 7, where cumulative distribution of rut depth along the lanes is presented. As discussed above, the average rut depths for the north- and south-bound lanes 1 differ by approximately 15 %, and, as seen in Figure 7, cumulative distribution of rut depths in the North-bound lane is somewhat lower than for the South-bound lane. At the same time, the tails of the cumulative distributions presented basically coincide, with basically 100 % of the section having a rut depth of at least 5 mm and 10 % of the section stretches having a rut of at least 17 mm. Accordingly, results presented in Figure 7, indicate that the average rut depth does not provide a fully adequate measure of pavements surface wear. In particular, the difference in the length of the repairs for the north- and south-bound lanes 1 may vary between 0 and 30 %, depending on the threshold rut depth (i.e. rut depth requiring maintenance). This again emphasizes the importance of incorporating spatial characteristics of rutting into pavement management decisions.

In order to examine the influence of rut depth on accumulation rate of rutting, average yearly rutting rates are presented in Figure 8 as function of the rut depth measured in the previous year. The average rutting rates reported in Figure 8 for the south-bound lane 1, were obtained as differences between the profiles measured in two consecutive years. Based on the rutting rates calculated for each 25-meter stretch of the lane, average rutting rates have been determined for the rut depth intervals presented in Figure 8. As may be noticed in Figure 8, rutting rates are significantly higher at the points where accumulated rut depth exceeds 6 mm. Primary factors responsible for this increase in the pavement surface degradation rate may be speculated to be as follows. As discussed, presence of water at the pavement surface increases the pavement wear from studded tires [14, 20]. Obviously, presence of significant amount of rut depth will hinder water removal from the road surface and thus increase further rutting accumulation rate. Deep ruts may furthermore increase dynamic tire road interaction forces as well as reduce lateral traffic wander, thus concentrating wear of pavement surface to narrower regions.

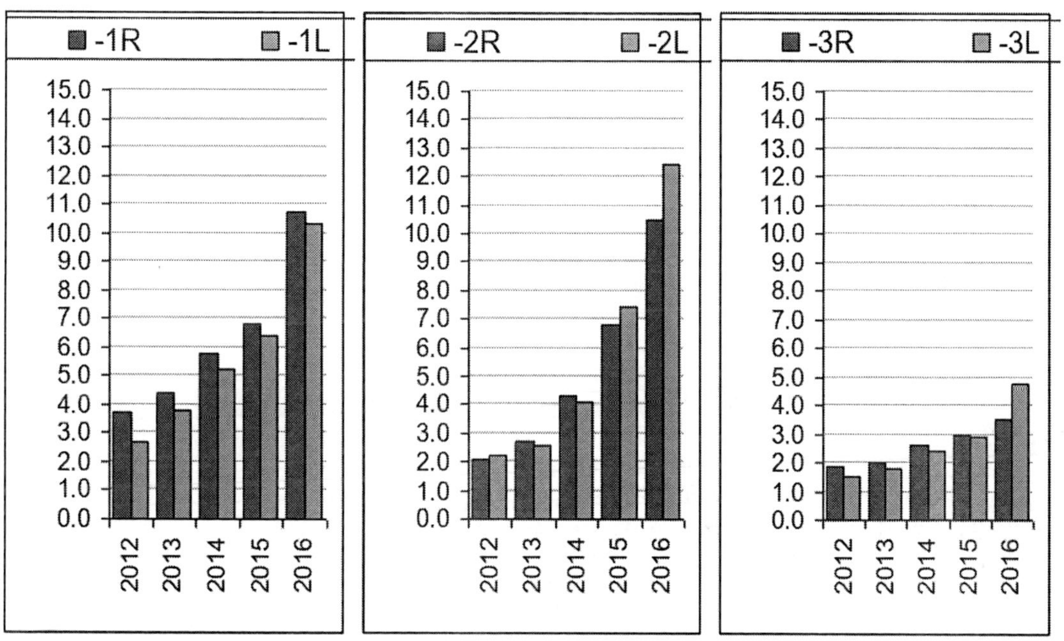

Figure 4. Rut depth accumulation in North-bound lanes

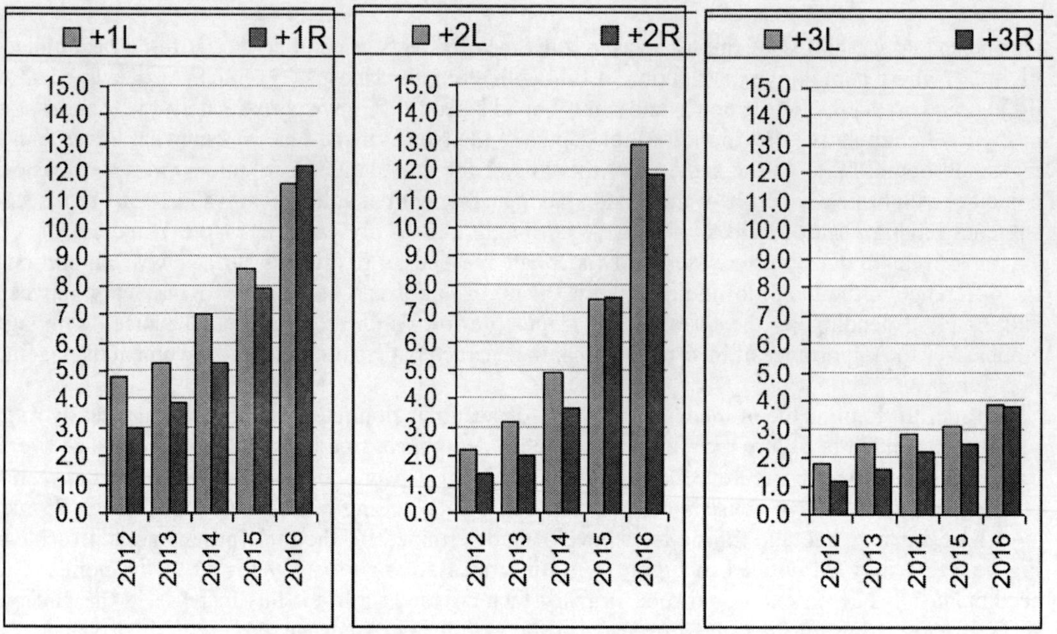

Figure 5. Rut depth accumulation in South-bound lanes.

Table 1. Average rutting rate (in mm).

Years of service	Lane					
	−3	−2	−1	+1	+2	+3
1	1.7	2.2	3.2	3.9	1.8	1.5
2	0.2	0.5	0.9	0.7	0.8	0.6
3	0.6	1.6	1.4	1.6	1.7	0.5
4	0.5	2.9	1.1	2.1	3.3	0.3
5	0.8	4.4	3.9	3.7	4.9	1.1
Average	*0.8*	*2.3*	*2.1*	*2.4*	*2.5*	*0.8*

3.3. Discussion

From the results presented in Figures 4-8 and in Table 1, it may be concluded that rutting controls the service life of the rigid pavement section examined. The rut formation is fully attributed to the surface wear by the studded tire traffic as underlined by significantly lower rutting rate in the outer lanes where traffic is predominantly composed of heavy vehicles equipped with stud-less winter tires.

Major corrective maintenance techniques for restoring smoothness of rigid pavements include surface grinding and construction of either asphalt or PCC wearing courses. by However, while diamond grinding is the most economical solution with respect to restoring pavement smoothness it does not address the underlying structural and/or material defects of the pavement [6]. Accordingly, given the severity of rutting observed within the first 5 years of pavement life, construction of wearing course overlay is a most feasible maintenance method.

Hard and coarse aggregates tend to slow (but not prevent) studded tire wear for both asphalt and cement concrete overlays [6, 16]. Accordingly, it is crucial that aggregates with the best performance in durability and soundness tests such as Los Angeles (LA), micro-Deval abrasion and aggregate soundness tests [1–3] are used in the wearing courses. As pointed out above in connection to Table 1, rutting rates measured in this study are not decreasing with years of service, in contrast to the results reported in Cotter and Muench [6]. In fact, as illustrated in Figure 8, rutting rate rather tends to increase with the accumulated rut depth. These observations provide an indication of inadequate abrasion resistance of the aggregates. With respect to cement concrete wearing courses, several concrete additives with the potential to improve concrete's resistance to abrasive wear have been proposed and evaluated [4, 19]. The evidence with respect to the effect of those additives on the material wear performance in the field is still lacking and needs to be investigated further [4].

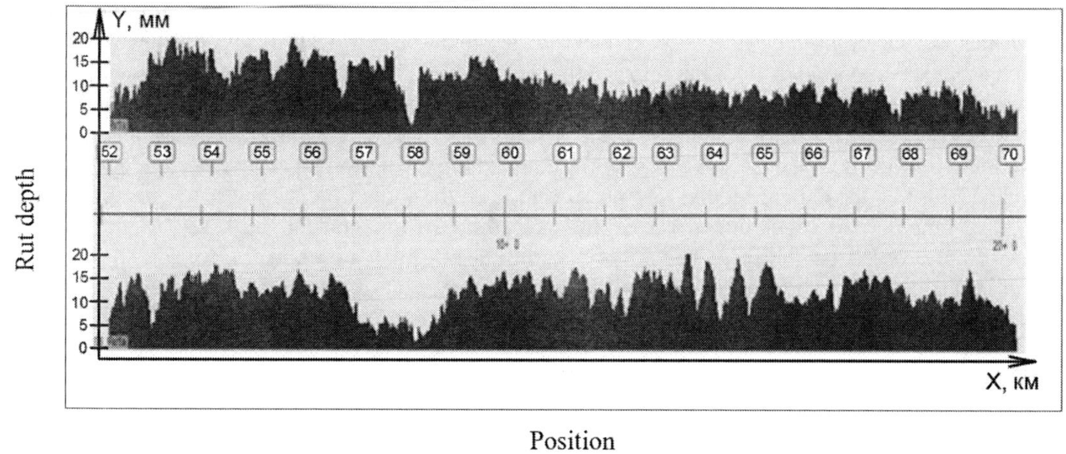

Figure 6. Spatial variation of the measured rut depth for lanes 1; a) North bound; b) South bound.

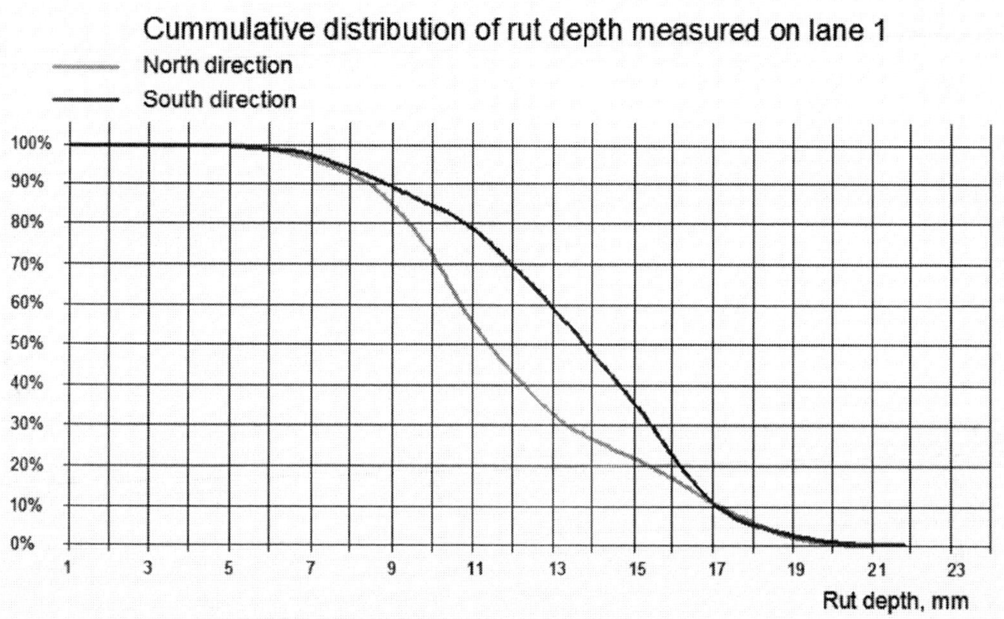

Figure 7. Cumulative distribution of rut depth measured on lanes 1, north and south bound after 5 years of service life.

In addition to material related factors, extremely high rutting rate observed in this study may be attributed to traffic parameters, and in particular, traffic speed may have a profound influence on the pavement surface wear [17]. The section examined belongs to the primary highway network with the maximum allowable speed limit of 130 km/h, being above the speed ranges commonly accounted for in the literature. It is thus essential to investigate the resistance of wearing courses to studded tires at speeds

IOP Conf. Series: Materials Science and Engineering **832** (2020) 012033 doi:10.1088/1757-899X/832/1/012033

representative for the intended application. In this context, accelerated pavement testing under representative vehicle-road interaction conditions (i.e. speed, tire type, temperature and humidity), may provide a useful link between material properties and expected performance in the field. Hence, it is the intention of future studies, to identify the optimal wearing courses by combining extensive material characterization, with particular focus on aggregates durability and soundness, and the evaluation of road surface wear, based on the road simulator tests at representative climate conditions and at speeds ≥ 130 km/h.

It has also to be pointed out, that the dependency of the rutting rate from accumulated rut depth emphasizes the importance of taking into account spatial variation in the pavement surface wear as well as variation of pavement surface wear rate evolution with time for accurate estimation of the remaining pavement service life. In fact, taking a rut depth of 20 mm as a maximum allowable rutting, the remaining service life of a PCC pavement under studded tire traffic can be estimated to be approximately 4 years, based on the data in Figures 4–5 and Table 1. At the same time, as seen in Figure 7, more than 70 % of lane 1 have accumulated rut depths exceeding 10 mm during the first 5 years of service. Accordingly, based on the results in Figure 8, rutting rate may be expected to exceed 5 mm/year, resulting in less than 2 years of remaining life . Furthermore, according to Figure 8, sections with rutting above 14 mm (25 to 50 % of lanes 1, as illustrated in Figure 7), may, be expected to accumulate a maximum allowable rut depth of 20 mm within a year. It may be concluded thus that yearly transverse profile measurements improve estimation of the remaining service life and are therefore very useful tools in pavement maintenance and rehabilitation planning.

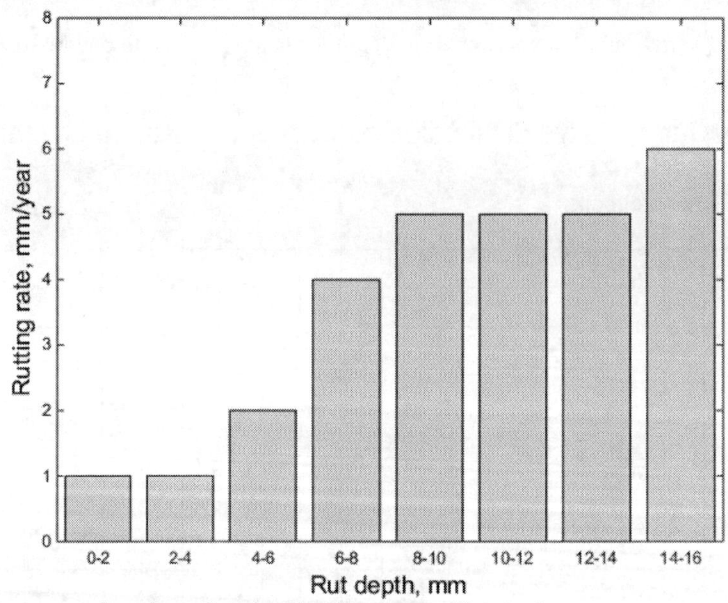

Figure 8. Rutting rate as function of rutting depth measured in previous year.

4. Conclusions

A field study of surface deterioration of a rigid pavement has been performed in order to evaluate the influence of surface wear on the service life at traffic and climate conditions representative for the highway network in the Moscow region, Russia. For the road section examined, yearly transverse profile measurements have been performed for the first 5 years of pavements service. Based on the profile measurements, average rut depths and yearly rutting accumulation rates have been calculated at 25 meter increments for all pavement lanes. The main findings of the study may be summarized as follows:

- The surface wear of the road is significantly higher for the inner lanes as compared to the outmost lanes. In particular, rutting accumulated during first five years of pavements service life has been found to be at least 2 times higher for the inner lanes as compared to the outmost lanes. Furthermore, rutting accumulation rates measured at 5th service year were approximately 4 times higher for the inner lanes. These differences are attributed to traffic composition, as the outmost lanes are primarily used by slow heavy traffic with stud-less tires whereas the inner lanes 1 and 2 was were subject to high speed traffic with a high share of studded tire vehicles

- The rut depths and rates measured for lanes 1 and 2 indicate that pavement wear by studded tire traffic is quite significant. Indeed, assuming 20 mm maximum allowable rut depth shows that wear by studded tires becomes a dominant mode of failure controlling the length of maintenance intervals. Supported by results from literature, severe rutting of the section may be attributed to high traffic speed and inadequate aggregate durability. The detailed experimental investigation of the influence of material parameters on pavement wearing rate was beyond the scope of this study and will be performed as a part of future work.
- Measured rut depths and rutting accumulation rates were found to exhibit significant spatial variability, where regions with excessive rutting were in general uniformly distributed throughout the section. Furthermore, rutting rates were found to be dependent on the accumulated rut depths. In particular the sections with rutting exceeding 6 mm were found to have at least twice as high rutting rate as compared to those with less rutting. This may be attributed to the fact that presence of significant amount of rutting may hinder water removal from the road surface, increase dynamic forces applied to the pavement surface and reduce lateral wander of the traffic.
- Based on observed variability of measured rut depths and rates it was shown that yearly continuous transverse profile measurements allow more accurate predictions of the remaining pavement service life as compared to those based on the average rut depths and rates. Accordingly, taking into account spatial variability of rutting and rutting rate evolution with time may improve pavement maintenance and rehabilitation planning.

References

[1] AASHTOT-327 2004 *Standard test method for resistance of coarse aggregate to degradation by abrasion in the micro-Deval apparatus, standard specifications for transportation materials and methods for sampling and testing* (Washington, D.C.)
[2] AASHTOT-96 2004 *Standard method of test for resistance to degradation of small-size coarse aggregate by abrasion and impact in the Los Angeles machine, standard specifications for transportation materials, and methods for sampling, and testing* (Washington, D.C.)
[3] AASHTOT-104 2004 *Standard method of test for soundness of aggregate by use of sodium sulfate or magnesium sulfate, standard specifications for transportation materials, and methods for sampling and testing*. Part-2A: Tests. (Washington, D.C.)
[4] Anderson K W, Uhlmeyer J and Pierce L M 2007 *Studded Tire Wear Resistance of PCC Pavements* (WA, Olympia: Washington State Department of Transportation)
[5] Brunette B E. and Lundy J R 1996 *Use and effects of studded tires on Oregon pavements Transportation Research Record* **1536** 64–72
[6] Cotter A, Muench T S 2010 *Studded Tire Wear on Portland Cement Concrete Pavement in the Washington State Department of Transportation Route Network*, University of Washington, Washington State Transportation Center, Washington State Department of Transportation
[7] Celma C C, Jelagin D, Partl M and Larsson P L 2017 Contact-induced deformation and damage of rocks used in pavement materials *J. of Mater. and Design* **133(5)**
[8] GOST 26633-91 *Heavy-weight and sand Portland cements concretes* Specifications
[9] GOST 23558-94 *Crushed stone-gravel-sandy mixtures, and soils treated by inorganic binders for road and airfield construction* Specifications
[10] GOST 25607-94 *Crushed stone-gravel-sandy mixtures for road and airfield surfacings and bases* Specifications
[11] GOST 8736-93 *Sand for construction works* Specifications
[12] Jacobson T 1995 *Study of the wear resistance of bituminous mixes to studded tyres* Tests with

slabs of bituminous mixes inserted in roads and in the VTI's road simulator, VTI särtryck, Nr 245, VTI Väg-och transportforskningsinstitutet.

[13] Keyser J H 1970 *Effect of Studded Tires on the Durability of Road Surfacing* (Montreal, Canada: Ecole Polytechnique, and Highway Research Board, Highway Research Record) January no 331 38 p

[14] Lundy J R, Hicks R G, Scholz T V and Esch D C 1992 Wheel Track Rutting Due to Studded Tires *Transportation Research Record* **1348**.

[15] Niemi A 1978 *Technical Raid Studies Related to Studded Tires* Proc., *Road Paving Research* (Helsinki, Finland, Helsinki University of Technology) pp 25–33

[16] Snilsberg B, Saba R G and Uthus N 2016 Asphalt pavement wear by studded tires – Effects of aggregate grading and amount of coarse aggregate *Proc. of 6th Eurasphalt and Eurobitumine Congress*.

[17] Unhola T 1997 Studded tires the Finnish way Proc. of the 5th Int. Symp. on Cold Region Developm. pp 609–612 (Anchorage, AK)

[18] Ushakov V and Dyakov G 2014a Investigation of abrasive wear of concrete pavements *J. Sci. and Engineer. for Highways* 67

[19] Ushakov V and Dyakov G 2014b Concrete pavements with polymer-modified concrete surface layers *J. Vestnik MADI* 37

[20] Zubeck H; Aleshire L, Harvey S, Porhola S and Larson E 2004 *Socio-Economic Effects of Studded Tire Use in Alaska* (Anchorage, AK: University of Alaska Anchorage) http://www.engr.uaa.alaska.edu/research/upload/StuddedTiresInAlaska.pdf.

Optimization of complex frost-resisting additives in line with operational requirements towards freezing temperature in road dressing constructions

E A Vdovin and V F Stroganov

Kazan State University of Architecture and Civil Engineering, 1, Zelenay str., Kazan, 420043, Russia

E-mail: vdovin007@mail.ru

Abstract. The subject of the study is to optimize the compositions containing frost-resisting complex organic and inorganic additives (FRA) that provide the required level towards freezing temperature when constructing cement-mineral layers (CML) of road dressing (pavement). As a prerequisite to carry out these studies is the necessity to preserve the liquid phase in CML during asphalt laying at low (subzero) air temperatures, as well as ensuring a given level stated to the performance criteria and durability of road dressing layers. In our studies, we applied the comprehensive approach to the experiment, concluding in a choice of FRA components and in the use of experiment planning. The influence of functional additives and their quantitative composition on the freezing temperature (T_z) during the experiment optimization in line with the D-optimal second-order plan has been considered. A mathematical model of the process has been obtained. We have found that the most significant mixture is the content of sodium formate (SF), compared to calcium chloride ($CaCl_2$) and a superplasticizer based on the polycondensation product of naphthaleisulfonic acid and formaldehyde (C-3), the amount of which varies in the series 3>1>0.3. The obtained results on optimization allow to choose FRA compositions that provide the ability to perform construction works in conditions of subzero temperatures up to –15.6 °C.

1. Introduction

The importance of the transport system development is much disscussed in [1–4] and one of the limiting factors for increasing a number of constructions is the necessity to carry out works in adverse conditions – at low air temperatures. The other way in development of the country's transport strategy is the use of CML modified with functional additives (stone materials and soils reinforced with cement) for road dressing [5–27]. Construction works conducted at subzero air temperatures (up to –15 °C) allows to use CML for road dressing, chloride compounds, sodium and calcium nitrite, potash and other additives that ensure the presence of a liquid phase in mixtures and the process of cement hydration [28–36]. However, these substances do not increase frost resistance of cement-reinforced materials and impair the mixture workability. Frost-resisting additives, in addition to lowering the ice point of hardening water and accelerating the hardening processes and CML plasticization, should improve the material performance and their durability. It is known that the most widespread compounds are sodium formate (SF) – the sodium salt of formic acid, as frost-resisting additive to cement concrete and CML [37–40]. However, the high efficiency of FRA is provided by the use of additives, having polyfunctional characteristics. We have developed the complex FRA based on SF using a hardening accelerator — calcium chloride ($CaCl_2$) and C-3 superplasticizer — the product of the polycondensation of naphthalenesulfonic acid and formaldehyde. It should be noted that the superplasticizer included in the complex composition is intended to reduce humidity, improve technological properties and increase CML density. Calcium chloride, in addition to accelerating the process of hardening, enhances the anti-frost effect of SF, and its reduced dosage prevents some destruction development at hardening processes of cement-mineral material. Strength gain at CML hardening in the condition of subzero temperatures, providing the operational requirements, significantly depends on T_z water solutions of frost-resisting additives. Thus, the aim of our work is to optimize the compositions of frost-resisting complex additives, taking into account the operational requirements for freezing temperature at road dressing constructions.

2. Materials and methods

2.1. Materials

Components of complex frost-resisting additives:

- sodium formate – sodium salt of formic acid (HCOONa), a water-soluble product under "ChP" (chemically pure) brand;
- crystalline calcium chloride ($CaCl_2$), a water-soluble product under 'ChP' (chemically pure) brand;
- superplasticizer C-3 – sodium salt of the condensation product β-naphthalenesulfonic acid and formaldehyde. Used as 2.5 % water solution with pH in a range of 7–9.
- hardening water – water that complies with the requirements of EN 1008: 2002.

2.2. Methods

- optimization towards the composition of complex FRA was carried out by the method of experiment planning according to the second order D-optimal plan. The levels of variation and conventional concrete signs are presented in table 1. According to the results of digital processing, a regression dependence of T_z on the components dose in FRA was obtained. The significance of the coefficients to be calculated for regression model equations was estimated with the critical values by the Duncan's test, and the adequacy of the models was tested on the basis of Fisher's variance ratio. Then, according to the obtained regression equations, there was built the dependences of T_z on the components dose in FRA with isolines of the equal values of the water-ice transition temperature, which was in water solutions of the complex additives (WSCA).
- preparation of WSCA: the components of complex additives (SF, $CaCl_2$ and C-3) were sequentially added into the required amount of hardening water (the figures are introduced in Table 2), mixing it for 2–5 minutes at the temperature of 20 ± 2 °C;
- determination of the freezing temperature (T_z): prepared WSCAs were placed in the TFPT apparatus (manufactured by PM Tamson Instruments) with Tamson TLC80-14 creamer, allowing to do measurements up to –70 °C according to ASTM D1177 – ASTM D852 – ASTM D1015 – ASTM D1493 – ASTM D6875 standards. For each WSCA with a different components dose, the values of T_z have been determined.

3. Results and Discussion

To achieve the goal we applied the integrated approach to conduct the experiment, which consisted in a choice of FRA components and in the use of experiment planning. The influence of functional additives and their quantitative composition on the freezing temperature was examined with the digital technologies. It enabled to optimize the experiment according to D-optimal second-order plan and design a mathematical model of the process.

As factors (X_1, X_2, X_3), when planning the experiment, we took the quantitative value of each FRA component ($CaCl_2$, SF, C-3) in % from the cement mass in CML (Table 1). $CaCl_2$ varied within 1.0–3.0 %, SF 5.0–9.0 %, C-3 0.5–2.0 %. The levels of variation and conventional concrete signs are presented in Table 1.

Based on the data presented in table 1, we designed the matrix on the experiment planning with coded and natural values of the factors and determined T_z for each complex FRA. The results are shown in Table 2.

Table 1. The levels of variation and conventional concrete signs

Factors	Name of factors	Level of variation		
		low (−1)	medium (0)	upper (+1)
X_1	$CaCl_2$ consumption in % from the cement mass	1.0	2.0	3.0
X 2	SF consumption in % from the cement mass	5.0	7.0	9.0
X 3	C-3 consumption in % from the cement mass	0.5	1.25	2.0

Table 2. Ice point of WSCA solutions

Additives, % from the cement mass			T_z WSCA, °C
SF	$CaCl_2$	C-3	
5.0	1.0	0.5	–8.0
5.0	1.0	1.25	–8.1
5.0	1.0	2.0	–8.5
7.0	1.0	0.5	–12.0
7.0	1.0	1.25	–12.1
7.0	1.0	2.0	–12.5
9.0	1.0	0.5	–13.6
9.0	1.0	1.25	–13.7
9.0	1.0	2.0	–14.0
5.0	2.0	0.5	–8.8
5.0	2.0	1.25	–8.9
5.0	2.0	2.0	–9.2
7.0	2.0	0.5	–12.8
7.0	2.0	1.25	–12.9
7.0	2.0	2.0	–13.2
9.0	2.0	0.5	–14.4
9.0	2.0	1.25	–14.5
9.0	2.0	2.0	–14.8
5.0	3.0	0.5	–9.6
5.0	3.0	1.25	–9.7
5.0	3.0	2.0	–10.0
7.0	3.0	0.5	–13.6
7.0	3.0	1.25	–13.7
7.0	3.0	2.0	–14.1
9.0	3.0	0.5	–15.1
9.0	3.0	1.25	–15.2
9.0	3.0	2.0	–15.6

The results have been processed with some statistical methods and due to this processing the mathematical model of the process has been designed. It has been described by the regression equation depicting the dependence between the freezing temperature T_z WSCA and the component dose in FRA:

$$T_z = -12.9 - 0.78X_1 - 2.78X_2 - 0.22X_3 - 1.25X_2^2 - 0.15X_3^3. \quad (1)$$

The significance of the coefficients to be calculated for regression models equations was estimated with the critical values by the Duncan's test, and the adequacy of the models was tested on the basis of Fisher's variance ratio.

Table 3. Calculated values of coefficients of regression equation

Type of dependency	Calculated values of coefficients of regression equation									
	B_0	B_1	B_2	B_3	B_{11}	B_{22}	B_{33}	B_{12}	B_{13}	B_{23}
$T_z = f(X_i)$	–12.9	–0.78	2.78	–0.22	0.00	1.25	–0.15	0.00	0.00	0.00

Table 4. Statistical evaluations of regression coefficients

Type of dependency	Statistical evaluation									
	S_y^2	$S^2 B_0$	$S^2 B_i$	$S^2 B_{ii}$	$S^2 B_{ij}$	ΔB_0	ΔB_{ij}	ΔB_i	ΔB_{ii}	$F_{calculation}$	F_{tab}
$T_z = f(X_i)$	0.023	0.004	0.002	0.009	0.003	0.131	0.097	0.187	0.108	2.10	2.7

We have found that the most significant element in FRA mixture is the content of sodium formate (SF), compared to calcium chloride and a superplasticizer C-3, the amount of which varies in the series 3>1>0.

With the regression equation (1), the experimental results were digitally optimized and the dependence between the freezing temperature of WSCA from FRA components in the form of isolines in the range from –9 °C to –15 °C in increments of 1 °C has been depicted (Fig. 1). The obtained results on optimization allow to choose FRA compositions that provide the ability to perform construction works in conditions of subzero temperatures up to –15.6 °C.

4. Conclusion

1. In line with the purpose stated for this work we conducted optimization on the compositions of complex frost-resisting additives based on sodium formate (SF), calcium chloride and C-3 superplasticizer. The operational requirements for freezing temperature when constructing road dressing with CML were taken into account. The obtained results on optimization allow to choose FRA compositions that provide the ability to perform construction works in conditions of subzero temperatures up to –15.6 °C.

2. The influence of functional additives of sodium formate, calcium chloride and C-3 superplasticizer and their quantitative composition on the ice point of WSCA has been considered. The results have been processed with some statistical methods and due to this processing we gained the math model of the process, which is described by the regression equation depicting the dependence between the freezing temperature T_z WSCA and the component dose in FRA. We have found that the most significant element in FRA mixture is the content of sodium formate (SF), compared to calcium chloride and a superplasticizer C-3 the amount of which varies in the series 3>1>0.3.

3. With the regression equation (1), the experimental results were digitally optimized and the dependence between the freezing temperature of WSCA from FRA components in the form of isolines in the range from –9 °C to –15 °C in increments of 1 °C has been depicted (Fig. 1).

IOP Conf. Series: Materials Science and Engineering **832** (2020) 012034 doi:10.1088/1757-899X/832/1/012034

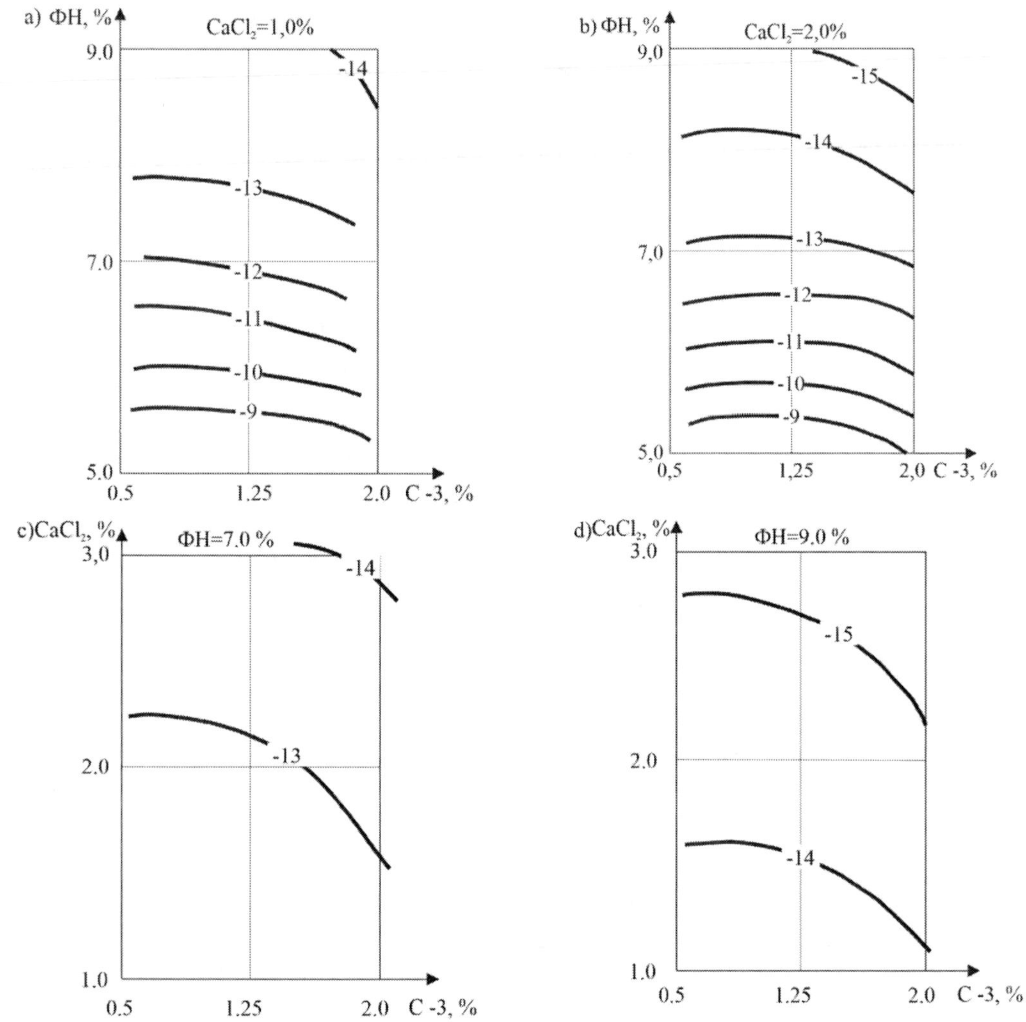

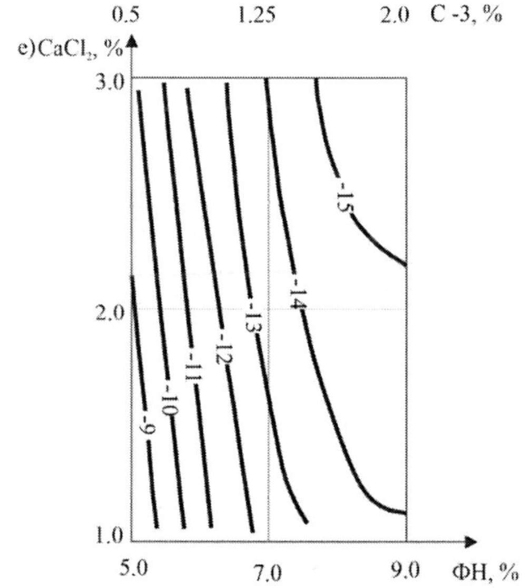

Figure 1. Dependences of T_z WSCA on components dose in FRA:
a) with $CaCl_2$= 1.0 %;
b) with $CaCl_2$= 2.0 %;
c) with SF= 7.0 %;
d) with SF = 9.0 %;
e) with C-3 = 2.0 %

References

[1] Ludlow D, Sakhrani V and Wu C 2019 *Managing State Transportation Research Programs* (TRB, Wachington, DC) pp 164

[2] Kim A, Sadatsafavi H, Anderson S D and Bishop P 2017 Preparing for the Future of Transportation Construction: Strategies for State Transportation Agencies *J. of Management in Engineering* **33(3)**

[3] Kim A, Sadatsafavi H, Anderson S D and Bishop P 2017 Using Scenario Planning for Identifying Major Future Trends and Their Implications for State Transportation Agencies *Int. Conf. on Sustainable Infrastructure* pp 237–49

[4] Kermanshachi S, Anderson S D and Goodrum P T 2017 Project Scoping Process Model Development to Achieve On-Time and On-Budget Delivery of Highway Projects *Transportation Res. Record: J. of the Transportation Res. Board* **2630** 147–55

[5] Saxena S K, Kumarb M and Singh N B 2018 Effect of Alccofine powder on the properties of Pond fly ash based Geopolymer mortar under different conditions *Environmental Technol. and Innovation* **9** 232–42

[6] Kavyateja B V, Jawahar J G and Sashidhar C 2019 Investigation on ternary blended self compacting concrete using fly ash and alccofine *Int. J. of Recent Technol. and Engineer.* **7** 447–51

[7] Hu L, Hao J and Wang L 2014 Laboratory evaluation of cement treated aggregate containing crushed clay brick *J. of Traffic and Transportation Engineer. (English Edition)* **1(5)** 371–82

[8] Visser A T 2017 Potential of South African road technology for application in China *J. of Traffic and Transportation Engineer. (English Edition)* **4(2)** 113–7

[9] Iwański M, Buczyński P and Mazurek G 2016 Optimization of the road binder used in the base layer in the road construction *Construction and Building Materials* **125** 1044–54

[10] Öz H Ö 2018 Properties of pervious concretes partially incorporating acidic pumice as coarse aggregate *Construction and Building Mater.* **166** 601–9

[11] Mymrin V, Aibuldinov E K, Alekseev K, Petukhov V, Avanci M A, Kholodov A, Taskin A, Catai R E and Iarozinski A 2019 Efficient road base material from Kazakhstan's natural loam strengthened by ground cooled ferrous slag activated by lime production waste *J. of Cleaner Production* **231** 1428–36

[12] Rana A, Kalla P, Verma H K and Mohnot J K 2016 Recycling of dimensional stone waste in concrete: A review *J. of Cleaner Production* **135** 312–31

[13] Poulikakos L D et al 2017 Harvesting the unexplored potential of European waste materials for road construction. *Resources, Conservation and Recycling* **116** 32–44

[14] Vishwakarma V and Ramachandran D 2018 Green Concrete mix using solid waste and nanoparticles as alternatives – A review *Construction and Building Mater.* **162** 96–103

[15] Autelitano F and Giuliani F 2016 Electric arc furnace slags in cement-treated materials for road construction: Mechanical and durability properties *Construction and Building Mater.* **113** 280–9

[16] Mengue E M, Mroueh H, Lancelot L and MedjoEko R 2017 Physicochemical and consolidation properties of compacted lateritic soil treated with cement *Soils and Foundations* **57(1)** 60–79

[17] Melese E, Baaj H, Tighe S, Zupko S and Smith T 2019 Characterization of full-depth reclaimed pavement materials treated with hydraulic road binders *Construction and Building Mater.* **(226)** 778–92

[18] Oluwatuyi O E., Adeola B O, Alhassan E A, Nnochiri E S, Modupe A E, Elemile O O and Obayanjua T A 2018 Ameliorating effect of milled eggshell on cement stabilized lateritic soil for highway construction *Case Studies in Construction Mate.* **9** e00191

[19] Dołżycki B and Jaskuła P 2019 Review and evaluation of cold recycling with bitumen emulsion and cement for rehabilitation of old pavements *J. of Traffic and Transportation Engineer.(English Edition)* **6(4)** 311–23

[20] Ribeiro D, Néri R and Cardoso R 2016 Influence of Water Content in the UCS of Soil-Cement

Mixtures for Different Cement Dosages *Procedia Engineer.* **143** 59–66
[21] Mola-Abasi H and Shooshpasha I 2016 Influence of zeolite and cement additions on mechanical behavior of sandy soil *J. of Rock Mechanics and Geotechnical Engineer.* **8(5)** 746–52
[22] Hassan H F, Taha R, Rawas A A, Shandoudi B A, Gheithi Kh A and Barami A A 2005 Potential uses of petroleum-contaminated soil in highway construction *Construction and Building Mater.* **19(8)** 646–52
[23] Anggraini V, Asadi A, Syamsir A and Huat B B K 2017 Three point bending flexural strength of cement treated tropical marine soil reinforced by lime treated natural fiber *Measurement* **111** 158–66
[24] Nguyen L, Fatahi B and Khabbaz H 2016 Predicting the Behavior of Fiber Reinforced Cement Treated Clay *Procedia Engineer.* **143** 153–60
[25] Rezaeian M, VazFerreira P M and Ekinci A 2019 Mechanical behavior of a compacted well-graded granular material with and without cement *Soils and Foundations* **59(3)** 687–98
[26] Lenoir Th, Preteseille M and Ricordel S 2016 Contribution of the fiber reinforcement on the fatigue behavior of two cement-modified soils *Int. J. of Fatigue* **93(1)** 71–81
[27] Ilyina O H and Konovalov N V 2013 Road-building material based on local mineral materials treated with a complex binder *Izvestiya KGASU* **2(24)** 295–300
[28] Karagöl F, Demirboğa R, Kaygusuz M A, Yadollahi M M and Polat R 2013 The influence of calcium nitrate as antifreeze admixture on the compressive strength of concrete exposed to low temperatures *Cold Regions Sci. and Technol.* **89** 30–5
[29] Karagöl F, Demirboğa R and Waleed H Kh 2015 Behavior of fresh and hardened concretes with antifreeze admixtures in deep-freeze low temperatures and exterior winter conditions *Construction and Building Mater.* **76** 388–95
[30] Aïtcin P C and Eberhardt A B 2016 Historical background of the development of concrete admixtures *Sci. and Technol. of Concrete Admixtures* 666
[31] Polat R 2016 The effect of antifreeze additives on fresh concrete finally to freezing and thawing cycles *Cold Regions Science and Technology* **127** 10–7
[32] Çullua M and Arslan M 2013 The effects of antifreeze use on physical and mechanical properties of concrete produced in cold weather *Composites Part B: Engineering* **50** 202–9
[33] Demirboğa R, Karagöl F, Polat R and Kaygusuz M A 2014 The effects of urea on strength gaining of fresh concrete under the cold weather conditions *Construction and Building Mater.* **64** 114–20
[34] Çullu M and Arslan M 2014 The effects of chemical attacks on physical and mechanical properties of concrete produced under cold weather conditions *Construction and Building Mater.* **57** 53–60
[35] Xie N, Dang Yu and Shi X 2019 New insights into how $MgCl_2$ deteriorates Portland cement concrete *Cement and Concrete Res.* **120** 244–55
[36] Hartmann A, Khakhutov M and Buhl J-Ch 2014 Hydrothermal synthesis of CSH-phases (tobermorite) under influence of Ca-formate *Mater. Res. Bull.* **51** 389–96
[37] Tene O 2011 *Mixture of substances, applicable, in particular, as an additive for concrete mix* US Pat. 2433096 Rus. Federation no 2009134179/03; published 10.11.11
[38] Milovidova O V, Sychev V M, Rozhkov E V, Pliner S J and Shmotiev S F 2018 *Composition for stabilization of natural and industrial soils* US Pat. 2670468 Rus. Federation no 0002670468, published 10.29.18
[39] Gushchin S V and Drozd A A 2014 Effect of sodium formate on hardening of concrete at low temperatures *Problems of modern concrete and reinforced concrete* **6**
[40] Lanovetsky S V 2015 Investigation of the stability of supersaturated solutions of sodium formate *Engineer. Bull. of the Don* **2(2)**

The complex method of estimation of highway maintenance quality taking into account the International Roughness Index

E V Zhustareva and V I Bochkarev

Moscow Automobile and Road State Technical University (MADI), 64, Leningradsky ave., Moscow, 125319, Russia

E-mail: joustareva@mail.ru

Abstract. The work deals with improving the method of a complex estimation of highway maintenance quality. The methods of estimation of highway maintenance quality were analyzed. The most complete and reasonable estimation is reached using a complex method. This method takes into account the influence of different transport and operating characteristics on providing secured speed of cars. The car speed impacts on the efficiency of motor traffic. One of the indices influencing the car speed is a road roughness. Today, in the Russian Federation the road roughness is estimated using the International Roughness Index (IRI). But the complex method doesn't take into account the state-of-art measuring means as well as the International Roughness Index (IRI). The theoretical and experimental studies were carried out to establish the dependence between the IRI index and a partial coefficient of providing the design speed of cars. The model to define the partial coefficient of providing a design speed based on the International Roughness Index (IRI) was offered. The table for calculation of the partial coefficient of providing a design speed was offered as well.

1. Introduction

The problem of rational use of financial resources granted to the programs of development of the road/traffic complex of the Russian Federation is of great importance under modern economic conditions. One of these programs is the National project "Safe and high quality of highways" developed by the Ministry of Transport of Russia in order to implement the provisions of the President Order. The National project involves almost all the country – 83 regions of Russia (except Moscow and St. Petersburg) and 104 urban agglomerations.

To realize such large-scale national projects and to define the priority of works it is necessary to estimate highway maintenance quality (TES AD).

2. Statement of problem

The complex method of estimation of highway maintenance quality developed by Prof. Vasilyev A.P. [1] allows evaluating in full measure and reasonably the road maintenance, taking into account the influence of different operating characteristics on their consumer properties.

The advantages of the complex method of TES AD estimation are as follows:

- using a common criterion (secured design speed) for estimation of highway functionality;
- using a relative index (coefficient of securing design speed) showing the influence of road different transport and operating characteristics on the secured speed that allows comparing the degree of influence of such characteristics on the estimation criterion chosen;

• possibility to define the preferable type of repairing works taking into account a profitability of works offered.

It should be noted that some organizations involved in road activity act according to the regulations developed on the basis of the complex method of TES AD estimation.

The complex method of TES AD estimation assumes the calculation of a partial coefficient of design speed securing (K_{ds}) that is the ratio of the real maximal speed of a single autocar (V_{max}), secured by road safety, to the basic design speed (V_b) [1]:

$$K_{ds} = \frac{V_{max}}{V_b}. \tag{1}$$

The reference speed of 120 km/h is accepted as a basic design one [1].

According to the complex method of TES AD estimation ten partial coefficients are defined, each of them shows the influence of a concrete road operating characteristic on the secured speed of movement.

One of the important characteristics is a pavement longitudinal roughness. The pavement longitudinal roughness impacts on speed, traffic safety, fuel rate, comfort, ecology, tear and wear of car suspension. Partial coefficient number six (K_{ds6}) takes into account the influence of the longitudinal roughness on the speed.

The standard documents acting in the Russian Federation [2, 3] specify the measurement of the longitudinal roughness using profilometric devices, in this case the IRI international index is a result of measurement.

According to the complex method of TES AD estimation K_{ds6} coefficient is calculated using the data of the actual longitudinal roughness measured by PKRS devices (control device of the roughness and the pavement adhesion of the wheel) and THK (bumpometer) (Table 1). The method doesn't consider the measurement of the longitudinal roughness using profilometric devices. Therefore, the necessity came to a head to update the method of complex estimation.

Table 1. K_{ds6} of partial coefficient values

Roughness according to THK, cm/km	K_{ds6}	Roughness according to PKRS, cm/km	K_{ds6}
less 60	1.25	до 300	1.25
70	1.15	350	1.20
80	1.07	400	1.12
90	0.96	500	0.98
100	0.92	600	0.84
120	0.75	700	0.72
140	0.67	800	0.65
160	0.63	900	0.59
200	0.57	1000	0.55
250	0.50	1100	0.51
300	0.43	1200	0.43
350	0.37	1400	0.33
400	0.31	1600	0.28
450	0.25	1800	0.24
more 500	0.20	2000	0.20

3. Solution

For setting dependence $K_{ds6} = f(IRI)$ one can use the known functional dependences between the results of measurements of the longitudinal roughness obtained using different measurement instrumentation (Table 2). According to these data graphic dependencies were plotted (Figure 1).

Table 2. Dependences between the indices of estimation of longitudinal roughness obtained by different measurement instrumentation and the international index *IRI*

Author, source	Dependence formula	Equation №
Prof. V.P. Zhigarev, on the basis of theoretical modeling of the process of vehicle oscillating movement [4]	$\frac{PKRS}{IRI} = 128$	E1
Empirical dependence [5]	$IRI = 0.0073 \cdot PKRS - 0.33$	E2
Empirical dependence [5]	$IRI = 0.032 \cdot THK + 0.4$	E3
Dependence obtained by Prof. O.A. Krasikov [6]	$IRI = 0.0241 \cdot THK + 0.8253$	E4

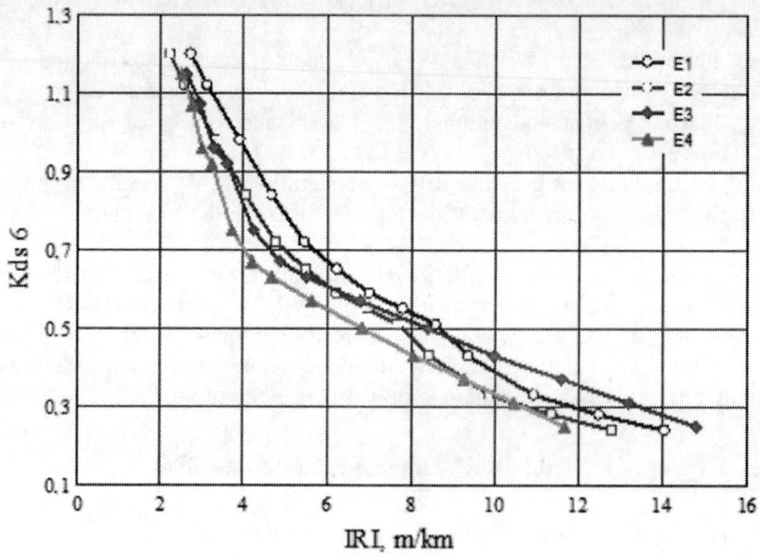

Figure 1. K_{ds6} dependence on IRI: figures on curves designate the type of the initial dependence in Table 2

The analysis of data showed that the type power function is most suitable for describing the dependence:

$$K_{ds6} = A \cdot IRI^{-B}, \qquad (2)$$

A and B are the coefficients needed more precise.

The coefficients of equation (2) were defined and the next formula was obtained:

$$K_{ds6} = 2.469 \cdot IRI^{-0.8} \qquad (3)$$

with correlation coefficient R=0.976.

To prove analytical dependence (3) the experimental studies were carried out. The autocar speeds and the IRI longitudinal roughness were measured on experimental objects that were sectors of highways. The first object was "Odintsovo northern roundabout way", the second one –"R-22 Kaspy". Four sectors of each object were examined. A total volume of measurements of speed was 47.894 and 51.211 respectively.

The roughness measurements of the traffic area of these sectors were carried out by the DVS-4-IK mobile laboratory (MADI). The laboratory made a video film and measurements of the microprofile. The registration of height marks was carried out at pitch of 125 mm along the movement direction.

To define and fix the speed of the car movement the KRIS-P mobile photoradars were used. A maximal speed of the autocar movement was defined by cumulative curves at 85 % providing on every experimental sectors. K_{ds6} was obtained by formula (1).

The functional dependence was defined using the data of speed and pavement roughness:

$$K_{ds6} = 0.972 \cdot IRI^{-0.453}. \qquad (4)$$

The value of the correlation coefficient R=0.969 shows a high degree of the statistic interdependence between the indices studied.

Figure 2 presents analytical and empirical dependences. It is evident that these dependences are similar.

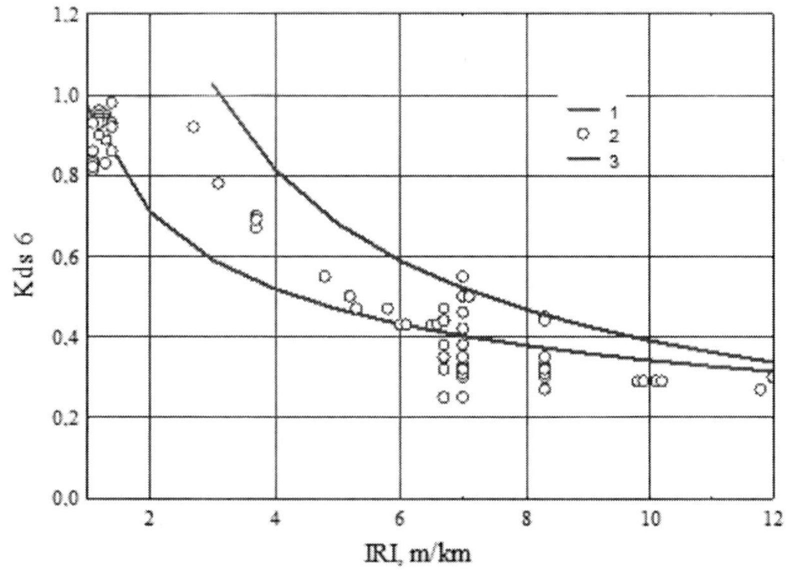

Figure 2. Comparison of analytical and empirical dependences $K_{ds6} = f(IRI)$: 1 – line corresponding to formula (4), 2 – line corresponding to formula (3), 3 – experimental points

The values for K_{ds6} designation at longitudinal roughness measurement by profilometric devices were tabled on the basis of studies carried out (Table 3).

Table 3. Correspondence of K_{ds6} to *IRI*

Roughness by profilometer (*IRI*), m/km	K_{ds6}
less 0.6	1.23
0.8	1.08
1.0	0.97
1.2	0.89
1.5	0.81
2.0	0.71
3.0	0.59
5.0	0.47
7.0	0.40
10.0	0.34
14.0	0.29
more 20.0	0.25

4. Conclusions

The studies resulted in the following conclusions:

- It is proved that the power function suits for describing dependence $K_{ds6} = f(IRI)$ and is written as $K_{ds6} = 0.972 \cdot IRI^{-0.453}$.
- To get K_{ds6} when carrying out the complex method of estimation TES AD the use of Table 3 can be recommended.

Nomenclature

TES AD	highway maintenance quality	V_b	basic design speed
K_{ds}	partial coefficient of design speed securing	*PKRS*	control device of the roughness and the pavement adhesion of the wheel
THK	bumpometer		

References

[1] ODN 218.0.006-2002. Rules for the diagnosis and assessment of road conditions. 2002. Federal road agency (Rosavtodor). Retrieved from: http://libnorm.ru/Files1/10/10874/index.htm.

[2] GOST 33101-2014. Automobile roads of general use. Road pavement. Roughness measurement methods. 2016. Federal agency on technical regulation and metrology (Rosstandart). Retrieved from: http://libnorm.ru/Directory2/1/4293758/4293758083.htm.

[3] GOST R 56925-2016. Automobile roads and airfields. Roughness measurement methods for base courses and pavements. 2016. Standartinform. Retrieved from: http://libnorm.ru/Directory2/1/4293754/4293754705.htm.

[4] Borisevich V B and Zhigarev V P 1999 Estimation of road smoothness using the international index of roughness and the index of PKRS device *Design and experimental studies of vibrating load and strength of structural elements of transport vehicle* ed V Zhigarev (Moscow: MADI-TU) pp 4–18

[5] Recommendations on adaptation of initial data in model HDM-4 to Russian conditions. 2004. Federal road agency (Rosavtodor). Retrieved from: http://rosavtodor.ru/eye/page/81/14644 (accessed October 18, 2019)

[6] ODM 218.11.001-2015 Methodological recommendations on record keeping of the increase of dynamic load impact as unevenness accumulates and on definition of the coefficient of dynamics depending on the evenness index. 2015. Federal road agency (Rosavtodor). Retrieved from: http://libnorm.ru/lis2/64332-0.htm

IOP Conf. Series: Materials Science and Engineering **832** (2020) 012036 doi:10.1088/1757-899X/832/1/012036

Analysis of thermophysical properties of asphalt mixtures

V A Zorin, N I Baurova and E A Kosenko

Moscow Automobile and Road State Technical University (MADI), 64, Leningradsky ave., Moscow, 125319, Russia

E-mail: madi-dm@list.ru, nbaurova@mail.ru

Abstract: The article describes the main stages of experimental design of thermophysical properties of asphalt mixtures. Based on the results experimental studies, we have constructed graphs showing the dependence of the cooling time of the asphalt concrete mixture on environmental conditions and the method of thermal insulation of the asphalt concrete mixture during its transportation. Thermograms of experimental samples clearly show the effectiveness of various methods of thermal insulation of asphalt concrete mixes. The choice of a regression model characterizing the dependence of the cooling time of an asphalt concrete mix on variable factors is substantiated.

Keywords: analysis, asphalt mix, experiment, quality indicators, thermal properties.

1. Introduction

Ensuring the quality of asphalt concrete pavements of roads while maintaining high rates of their construction and repair is impossible without improving the technology of installation and compliance with technological regimes. One of the main tasks facing the producers of asphalt-laying works is to maintain the temperature of the asphalt concrete mixture at the stages of its transportation, laying and compaction. For different types and brands of asphalt mixes, this value may be different and is set by the regulatory documents governing the implementation of this type of work.

Currently, the problem of ensuring the temperature regimes for laying asphalt concrete pavement is successfully solved by introducing asphalt concrete loaders into the technological process, which allow one to achieve temperature and fractional homogeneity of the asphalt concrete mixture and increase the rate of work production. With a significant removal of the object of construction or repair of pavements, the possibility of placing mobile asphalt concrete plants near the place of laying asphalt concrete pavement is being considered.

Temperatures can also be provided by arranging transportation of the asphalt concrete mixture using thermal insulating materials and setting the maximum transportation distance taking into account environmental conditions (temperature, humidity, wind speed, etc.), as well as the type of the mixture being transported.

The intensity of cooling of asphalt mixes largely depends on the ambient temperature and is characterized by the coefficient of thermal diffusivity, which is expressed by the formula:

$$a = \frac{\lambda}{c\rho} \tag{1}$$

where a is the coefficient of thermal diffusivity, m^2/s; λ is coefficient of thermal conductivity coefficient, W / (m°C); ρ is the density of the material, kg/m^3; and c is the specific heat capacity, J/(kg °C).

IOP Conf. Series: Materials Science and Engineering **832** (2020) 012036 doi:10.1088/1757-899X/832/1/012036

It is known [1] that the specific heat capacity of the asphalt-concrete mixture is higher, the lower the percentage of crushed stone in its composition. Thus, in accordance with expression (1), it can be concluded that asphalt-concrete mixtures of different granulometric compositions will cool at different speeds, ceteris paribus.

Based on the theoretical analysis of the thermophysical properties of asphalt mixes, it is possible to identify the main factors affecting the intensity of their cooling: the particle size distribution of the mixture, the ambient temperature, the method and means of thermal insulation. The processes occurring in the volume of building materials when constructing pavements differ from the processes that are studied in laboratory conditions; therefore, when processing the results of studies conducted on specially prepared samples (on a scale of 1: 4000), it is necessary to use the mathematical apparatus of similarity theory and physical modeling [2].

2. Experimental program

2.1. Materials

To conduct laboratory research to determine patterns of cooling of asphalt-concrete mixtures and assess the level of significance of determining factors, laboratory samples of asphalt-concrete mixtures of type A, brand I, type B, brand I, type D brand II were prepared.

Tables 1–3 present the composition of the analyzed asphalt-concrete mixtures.

Table 1. Composition of fine-grained asphalt concrete mix, type A, brand I

The name of the material	The composition of the mineral part of asphalt-concrete mixture (bitumen over 100 %)	The composition of the mineral part of asphalt-concrete mixture (bitumen in 100 %)	Dosage of materials for batch 1000 kg in kg
Crushed stone (5–20 mm)	43,0	41,0	410
Crushed stone (5–10 mm gabbro diabase)	11,0	10,5	105
Natural sand	19,0	18,1	181
Sand from the screenings of crushing 0–5 mm	20,0	19,1	191
Mineral powder	7,0	6,7	67
Bitumen brand BND 60/90	4,8	4,6	46

Table 2. Composition of fine-grained asphalt concrete mix, type B, brand I

The name of the material	The composition of the mineral part of asphalt-concrete mixture (bitumen over 100 %)	The composition of the mineral part of asphalt-concrete mixture (bitumen in 100 %)	Dosage of materials for batch 1000 kg in kg
Crushed stone (5–20 mm)	39,0	37,1	371
Crushed stone (5–10 mm gabbro diabase)	8,0	7,6	76
Natural sand	42,5	40,5	405
Mineral powder	10,5	10,0	100
Bitumen brand BND 60/90	5,0	4,8	48

Table 3. Composition of fine-grained asphalt concrete mix type D, brand II

The name of the material	The composition of the mineral part of asphalt-concrete mixture (bitumen over 100 %)	The composition of the mineral part of asphalt-concrete mixture (bitumen in 100 %)	Dosage of materials for batch 1000 kg in kg
Natural sand	82,5	76,7	167
Mineral powder	17,5	16,3	163
Bitumen brand BND 60/90	7,5	7,0	70

2.2. Experimental methods

During the experiment, the change in the rate of cooling of asphalt concrete mixes (Y) depending on the ambient temperature (X_1), the method of thermal insulation (X_2) and the percentage of crushed stone in the composition of asphalt concrete mixtures (X_3) is investigated.

Table 4. Values of the main, upper and lower levels of each factor during the experiment

Name	Factor values		
	X_1	X_2	X_3
Main level	10	Using a tarp cover	45
Upper level	20	Using bunker thermal insulation method	52
Lower level	0	Without the use of thermal insulation	0

To obtain the most accurate research data, an experiment is conducted for three levels for each factor. Then, we suppose that the dependence of the criterion — the cooling time of the transported mixture – on the indicated factors will be expressed by means of the regression equation of the response surface.

Figure 1 shows the measurement scheme of samples of asphalt concrete mixture by five points. Prepared samples of asphalt mixes weighing 5 kg were heated to a temperature of 160 °C and subjected to cooling to 120 °C at different ambient temperatures: at temperature t = 0 °C, t = 10 °C and t = 20 °C.

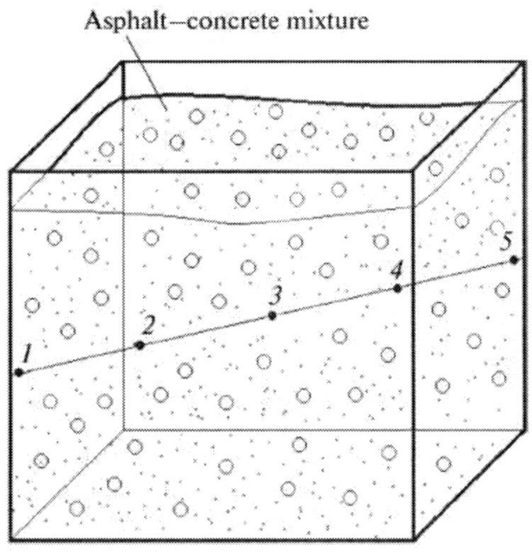

Figure 1. Sample for studying the cooling rate of asphalt–concrete mixtures: (1–5) measurement points

During transportation, the most intensive cooling of the asphalt concrete mixture occurs through the open surface of the mixture in the vehicle body; therefore, it is possible to evaluate the effectiveness of various methods of thermal insulation by measuring the temperature of the asphalt concrete mixture using a thermal imager. For research we used thermal imager Testo 875-1i.

3. Results and discussions

According to the results of the studies we obtained graphs. Figure 2 presents the graphs showing the dependence of the average value of the cooling rate of asphalt concrete mix type A, brand I on the ambient temperature, taking into account the method of thermal insulation.

Graphs of changes in the temperature of asphalt-concrete mix type A, brand I, depending on the method of thermal insulation, taking into account the ambient temperature, are presented in Figure 3.

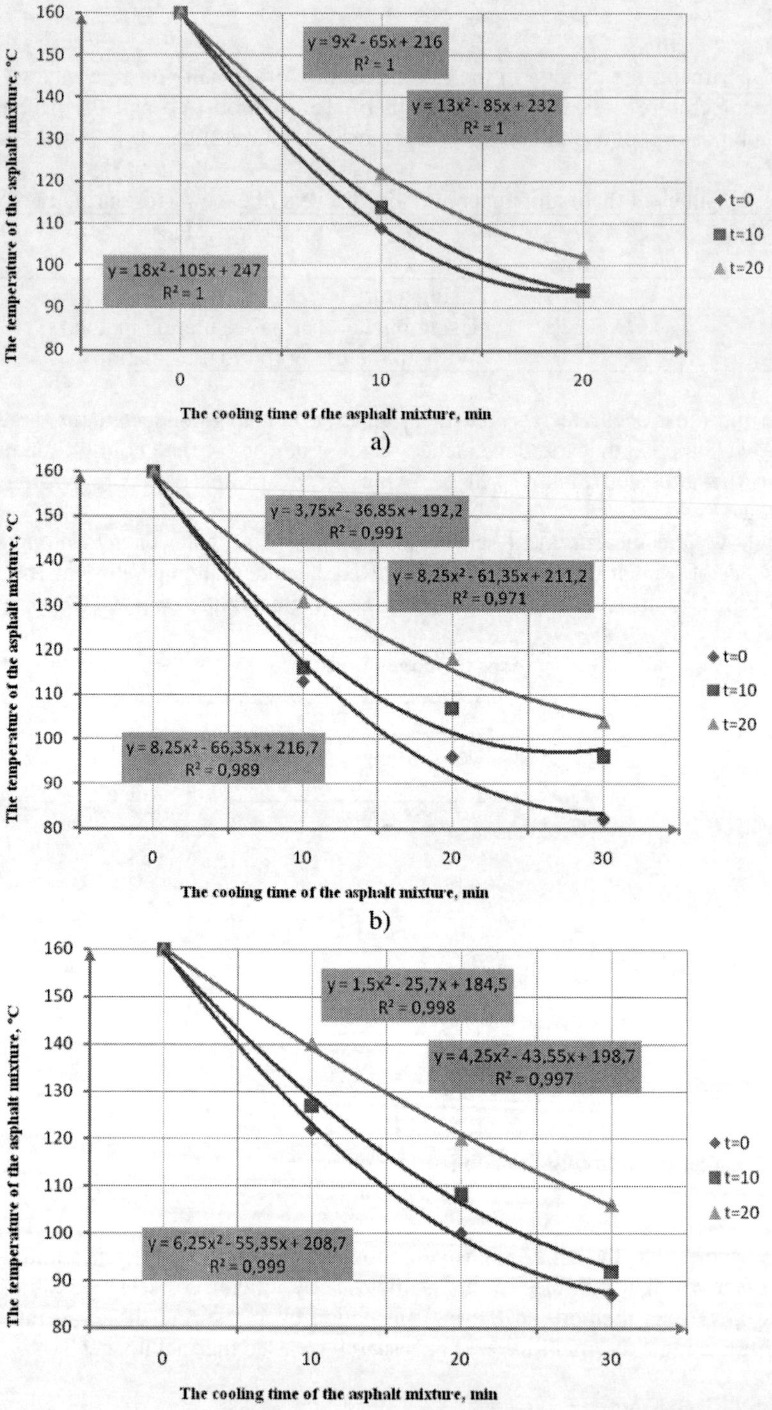

Figure 2. Graphs of temperature change of asphalt-concrete mix type A brand I depending on the ambient temperature, taking into account the method of thermal insulation of the vehicle body: a) without the use of thermal insulation means; b) using cover with a tarpaulin; c) using a bunker thermal insulation method

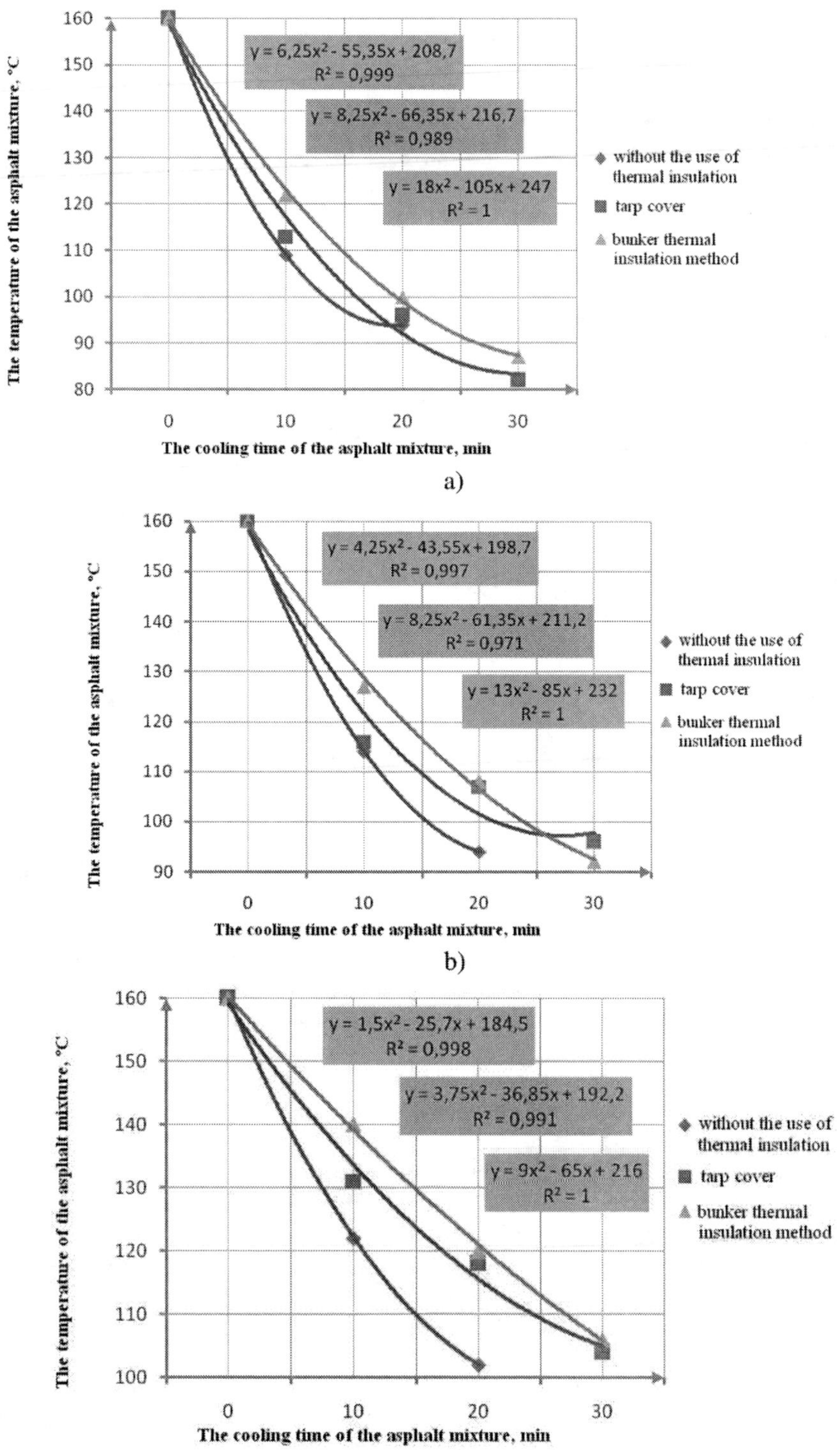

Figure 3. Graphs of changes in the temperature of asphalt-concrete mix type A brand I, depending on the method of thermal insulation, taking into account the ambient temperature: a) at t = 0 °C; b) at t = 10 °C; c) at t = 20 °C

As a result of the research, thermograms of samples were obtained before and at the end of the studies (Figure 4).

The measurement results showed that the lowest rate of cooling of asphalt mixes provides a bunker method of thermal insulation.

Fig. 4. Thermograms of the sample in the study of the process of cooling the asphalt mixture: a) at the beginning of measurements b) at the end of transportation without means of thermal insulation; c) at the end of transportation using a cover with a tarpaulin; d) at the end of transportation using the bunker method of thermal insulation

Analysis of the results of the experiment is the most important and crucial stage of the study, since it is at this stage that the main regularities of the studied phenomena are established. However, any pattern can only reflect a finite number of basic bonds in an infinite set of unexplored bonds, which leads to inaccurate implementation of patterns with some deviations. To account for random deviations, it becomes necessary to apply the methods of probability theory and statistical methods [3].

Statistical methods for assessing the accuracy and mathematical description of the technological process make it possible to build a mathematical model that serves as a control object when carrying out the operation of transporting asphalt concrete mixture in the technological process of paving [4].

As a result of the application of statistical methods for studying the influence of independent variables on the dependent variable, a regression model is defined that reflects the dependence of criterion (Y) on predictors (X) [5].

Most often, the following regression models are used to display the dependence of the criterion on predictors: simple regression, multiple regression, factor regression, polynomial regression, regression of the response surface, regression of the mixture surface [4].

An analysis of the use of regression models is presented in Table 5.

Table 5. Analysis of the use of regression models

No.	Nameofregression	Regression equation	Description
1	Simple regression plan	$Y=b_0+b_1X_1$ $Y=b_0+b_1X_1^n$	Only one predictor is used. If a simple regression plan uses effects of a higher order X, then the values of the plan matrix will be raised to a power corresponding to the nth order level
2	Multiple regression plan	$Y=b_0+b_1X_1+b_2X_2+b_3X_3$	It is a simple regression for 2 or more continuous predictors
3	Full factorial regression plan	$Y=b_0+b_1X_1+b_2X_2+b_3X_1X_2$ $Y=b_0+b_1X_1+b_2X_2+b_3X_3+...$ $...+b_4X_1X_2+b_5X_1X_3+b_6X_2X_3$	Defined as a plan in which all possible observations of continuous predictors are presented. For example, the full factorial regression plan for two continuous predictors X_1 and X_2 will contain the main effects X_1 and X_2 and the effect of their second X_1 and X_2 interaction, which is the product of the values X_1 and X_2, for each observation. The factorial regression plan can also be fractional, and the effects of a higher order can be removed from the plan
4	Polynomial regression plan	$Y=b_0+b_1X_1+b_2X_1^2+...$ $..b_3X_2+b_4X_2^2+b_5X_3+b_6X_3^2$	It contains both the main effects and the effects of higher orders of continuous variables, but it does not include the interaction of predictors. The maximum degree of the effects of polynomial regression can be different for different predictors
5	Second order response surface regression plan	$Y=b_0+b_1X_1+b_2X_1^2+b_3X_2+...$ $...+b_4X_2^2+b_5X_3+b_6X_3^2+...$ $+b_7X_1X_2+b_8X_1X_3+b_9X_2X_3$	It is a hybrid of the polynomial regression plan and the fractional factorial regression plan. This type of plan is used in applied research (for example, in industrial planning and analysis of experiments)
6	Regression surface of the mixture	$Y=b_1X_1+b_2X_1^2+b_3X_2+...$ $...+b_4X_1X_2+b_5X_1X_3+b_6X_2X_3$	It is an analogue of second order factorial regression without a free term

The mathematical model of a full-factor experiment for three factors with three levels for each factor is:

$$Y=b_0+b_1X_1+b_2X_2+b_3X_3+b_{11}X_1X_1+b_{22}X_2X_2+b_{33}X_3X_3+...$$
$$...+b_{12}X_1X_2+b_{13}X_1X_3+b_{23}X_2X_3 \quad (2)$$

where b_0, b_1, b_2, b_3, b_{11}, b_{22}, b_{33}, b_{12}, b_{13}, b_{23} – regression coefficients.

Statistical analysis is performed in accordance with the generally accepted method of processing and analysis of tests [4, 5].

4. Conclusion

The results of experimental studies have shown that the accepted norms for cooling asphalt concrete mixes [6, 7] are averaged and too high at an ambient temperature of 20 °C and underestimated at an ambient temperature of 10 °C. Unjustified overestimation of the cooling time of the asphalt concrete mixture may lead to the rejection of the asphalt concrete mixture due to non-observance of temperature conditions. The use of a smaller period of time for cooling the asphalt concrete mixture during the execution will lead to a decrease in the efficiency of the pavement construction process.

IOP Conf. Series: Materials Science and Engineering **832** (2020) 012036 doi:10.1088/1757-899X/832/1/012036

Thus, it can be concluded that in order to avoid economic losses as a result of a violation of the temperature regimes of the technological process of laying asphalt concrete mixture, it is important to determine the cooling time of the asphalt concrete mixture for each composition and method used for transportation to the pavement construction site.

References

[1] Zubkov A F and Odnol'ko V G 2009 *Technology of construction of asphalt concrete pavements of highways* (Moscow: Mashinostroenie) 224 p

[2] Zorin V A, Baurova, N I and Kosenko E A 2018 Features of research on the thermophysical properties of road-building materials *Polymer Science Series D.* **11(1)** 72–6

[3] Gmurman V E 2008 *Probability theory and mathematical statistics* (Moscow: Higher education) 479 p

[4] Shakalis V V 1973 *Modeling of technological processes* (Moscow: Mashinostroenie) 136 p

[5] Daniel K 1979 *Application of statistics in an industrial experiment* (Moscow: Mir) 299 p

[6] GOST 31015-2002 *Asphalt concrete mixes and crushed stone-mastic asphalt concrete. Technical conditions*

[7] Melik-Bagdasov M S, Gioev K A and Melik-Bagdasov N A 2007 *Construction and repair of road asphalt pavements* (Belgorod: Constant) 157 p

[8] Zorin V A, Baurova N I and Kosenko E A 2017 Detection of defects in components made of dispersion-filled polymeric materials by the method of infrared thermography *Polymer Science Series D.* **10(3)** 241–3

[9] Zorin V A, Baurova N I and Kosenko E A 2018 Analysis of the influence of quantum-mechanical processes on the possibilities of determining the low degree of curing a binder when molding products from polymer composite materials *Polymer Science Series D.* **11(3)** 334–8

IOP Conf. Series: Materials Science and Engineering **832** (2020) 012037 doi:10.1088/1757-899X/832/1/012037

Refined assessment of strength of concrete pavement probabilistic method

A P Stepushin, I V Chistyakov, M G Goryachev, Y S Sadovnikova, A K Kyaing

Moscow Automobile and Road Construction State Technical University (MADI), 125319, Leningradsky prospekt, 64, Moscow, Russian Federation

Email: fumo@madi.ru

Abstract. The object of research is the method of calculating the strength of cement concrete airfield pavement. According to the current regulatory documents, the calculation of the strength of the coating is carried out according to the first limit state, taking into account the internal forces arising in the slab from the impact of the wheel load and temperature. The repeated impact of loads from the wheels of the main supports of various types of air is replaced by the influence of the main support of the heaviest aircraft, causing the greatest bending moment in the design section of the plate. However, the calculations do not take into account the stochastic nature of such factors as the take-off weight of aircraft, the plate thickness, the bed ratio of the elastic foundation, the probability of the combined effect of operating loads from the main bearing wheels and temperature in the calculated section of the plate, as well as the probability distribution of loads from the aircraft spectrum across the width of the airfield elements. The article proposes a method for taking into account the probabilistic-statistical variability of operating loads, physical and mechanical properties of materials and the geometric characteristics of the plate, recommended when developing a new regulatory framework for the design of rigid airfield pavements. The method of statistical linearization of random functions of several random arguments was chosen as the research method. Using the mathematical apparatus of the theory of random functions, the statistical coefficient of the working conditions of the plate is analytically determined, and a numerical analysis of the obtained analytical solution is carried out. It has been established that with the same unfavorable combination of variability of basic arguments, the calculated absolute value of the statistical coefficient of the working conditions of the slab is more influenced by the coefficient of variation of the slab thickness than the coefficient of variation of the strength of cement concrete in bending. The condition for calculating the required thickness of the hard airfield slab of monolithic cement concrete is determined.

1. Introduction

The article describes the analysis of current practice used in Russian Federation and in the CIS countries for new concrete pavements design and evaluation. According to the specification [1] concrete pavements are designed to serve a predicted number of load applications for twenty years at the lowest possible initial and maintenance costs. The main criteria for design of concrete slabs thickness is cracking initiated by aircraft wheel gears and thermal stresses. Repeated loading from the traffic mixture is converted into a single design aircraft. The design aircraft is determined as the most damaging heavy airplane. All annual departures of the traffic mixture are converted to equivalent number of annual departures of the design aircraft. Thickness design specification [1,2,3] reflects only a realistic degree of variation in flexural strength of concrete applying 95% confidence level and thermal stress about 0.05…0.07 MPa considering its different signs during daily and night periods and decreases due to subbase restraint stress. Existing procedure does ignore stochastic nature of other main variables: aircraft takeoff weight, airplane traffic wander, in-situ slab thickness, elastic modulus of concrete, modulus of subgrade reaction and climatic inputs causing thermal bending moments in the concrete and probability of its combined effect with bending moment in slab under traffic loadings expected from specific volume of mixed traffic and its distribution pattern.

The analysis of concrete pavements performance life shows that real serviceability age at civilian airports of Russian Federation and in the CIS countries differ from 3 to 30 years [4-8]. One of the reasons of such deviation in the concrete performance life may be ignoring of the stochastic nature of the design inputs in the previous deterministic design guides. Consequently, it is necessary to develop the current methodology for calculating rigid airfield pavements based on mathematical methods of the reliability theory, as was done in [9–12]. The purpose of this article is to summarize the methodology for calculating cement concrete pavement during its design and assessment of the carrying capacity.

2. Methods

Deterministic design procedure includes comparison of bending moments caused by aircraft loads at the outside edge of a concrete pavement and ultimate bending moment for given slab thickness using the following equation:

$$d_d \leq d_1 \cdot z, \tag{1}$$

where z – factor to convert bending moment increase at the outside edge; d_l – bending moment in concrete slab caused by design aircraft main gear wheels at the interior of the slab; d_d – bending moment in concrete slab caused by the design aircraft main gear arrangement at the outside edge of slab:

$$d_d \leq d_2 \cdot z \tag{2}$$

where m_2 – ultimate bending moment for given slab thickness, flexural strength of concrete at design age and number of load repetitions:

$$d_l = \gamma_2 \cdot R_{btb} \cdot z \cdot \frac{h^2}{6} \tag{3}$$

where γ_2 – factor to convert concrete strength increase with age and warping stresses which was not changed since 1960 year till nowadays (table 1), R_{btb} – design modulus of rupture; h – slab thickness.

The results of flexural fatigue research under continued repetitions of loads failure of concrete beams occur at stresses ratios of less than unity. Flexural fatigue of concrete is reflected in specification [1] by selection of load repetition factor z_l based on the number of the heaviest aircraft undercarriage wheel passes expected during the pavement design life which is supposed to be equal to 20 years:

$$z_1 = 2 - \frac{\log L}{6}, \tag{4}$$

where L – expected number of the heaviest aircraft main gear wheel load applications for design age of the pavement.

It should be noted that equation (4) corresponds to PCA (Portland Cement Association) fatigue model results published in 1973 [13-18] and may be expressed as follows:

$$Z_1 = \frac{\gamma(N)}{\gamma(N = 1 \cdot 10^6)} = \frac{\gamma(N)}{0.5} \tag{5}$$

where Z_l – load repetition factor to reflect fatigue effect in the concrete under repeated loading; $\gamma(N)$ – stress repetition ratio for N_i.

$$\gamma(N) = \frac{\sigma_{N_i}}{R} = 1 - \frac{\lg(N)}{12} \tag{6}$$

where σ_{N_i} – repetitive stress; R – mean flexural strength of concrete; N – number of repetitions to cause concrete flexural fatigue failure; $\gamma(N = 1 \cdot 10^6) = 0.5$ – stress ratio permitting $1 \cdot 10^6$ load repetitions without loss of fatigue resistance of plain cement concrete in flexure test.

From equation (6) one can calculate the number of repetitions N_i to cause flexural rupture under stress σ_{N_i}.

$$N_{\sigma_{N_i}} = 10^{12 \cdot (1-\gamma(N_i))}. \quad (7)$$

Under real pavement performance conditions, stress ratio $\gamma(N_i)$ of concrete depends on stochastic nature of design variables which effect on the bending moments in the slab.

Suggestions proposed here where developed from a study and correlation of the existing design procedures [1], statistical analysis of aircraft centerline deviations on runway and taxiway that were collected in reports [19-36], plate theory [13-19], Miner's fatigue law and theory of stochastic functions [36]

The equation (1) may be rewritten in the form:

$$(\overline{d_F} + \overline{d_t}) \cdot P_{Ft} \le z_1 \cdot \overline{d_1}, \quad (8)$$

where $\overline{d_F}$ – average bending moment under main gear wheels loading calculated by formula (2); $\overline{d_t}$ – average bending moment caused temperature calculated by the following equation [18]:

$$\overline{d_t} = 0.0437\alpha \cdot h^2 \cdot E_b \cdot Q, \quad (9)$$

where α – coefficient of thermal expansion of concrete: $\alpha = 0.00001$; h – slab thickness; E_b – modulus of elasticity of concrete, Pa; Q – average difference in temperature between the top and bottom surfaces of the slab during design period (spring or autumn); P_{Ft} – probability to reflect the combined effect of the functions $\overline{d_F}$ and $\overline{d_t}$ on the summary bending moment in the slab under consideration.

Table 1. Values of γ_2 according to specifications

Pavement Sections	Values of γ_2 according to Specifications			
	Construction standards 120-60	Construction standards 120-70	Construction Norms and Regulations11-47-80	Construction Norms and Regulations 2.05.08-85 Set of rules 121.133.2012
Taxiways	0.8	0.8	0.8	0.8
The ends of runways L= 150 m	0.9	0.8	0.8	0.8
Subdivisions of runway pavements $L_{RUNWAY}/4$ – 150 m, connecting to the runway ends	0.9	0.9	0.9	0.9
Runway middle pavement section	0.9	0.9	0.9	0.9
Edge parts of the runway middle pavement section	1.1	1.1	1.1	1.1
Exits, aprons and other	0.9	0.9	0.9	0.9

special sections

$$P_{Ft} = P(\overline{d_F} + \overline{d_t}) - P(\overline{d_F}) \cdot P(\overline{d_t}), \tag{10}$$

where $P(\overline{d_F})$ – probability of the mean bending moment under main gear wheels; $P(\overline{d_t})$ – probability of the mean thermal bending moment;

$$P(\overline{d_F}) = P(\overline{d_t}) = 0.5, \text{ so by (9) } P_{Ft} = (0.5+0.5) - 0.5 \cdot 0.5 = 0.75,$$

where $\overline{d_1}$ – mean ultimate bending moment for slab thickness, flexural strength of concrete at design age and number of load repetitions calculated by equation (3).

Z_2 – parameter to adjust stochastic nature of the design inputs and design reliability of the concrete pavement [18];

$$Z_2 = \frac{1-\sqrt{1-(1-z^2{}_2 \cdot V^2{}_{dd+dt})(1-z^2{}_2 \cdot V^2{}_{d1})}}{1-z_2{}^2 \cdot V^2_{d1}} \tag{11}$$

where Z_2 – statistical coefficient to account cumulative effects of variability of the takeoff weights of the airplanes, number of the full-load applications, flexural strength and modulus of elasticity of concrete, modulus of subgrade reaction, in-situ pavement slabs thicknesses, surface pavement temperature amplitude and design reliability; z_2 – standardized normal variant for P level of pavement reliability; D_{D_F} – variance of bending moment caused by designed gear load; D_{dt} – variance of bending moment due to temperature gradient between top and the bottom surfaces of slab

$$D_{D_F} = \left(\frac{\partial D_F}{\partial F_d}\right)^2 \cdot S^2{}_{Fd} + \left(\frac{\partial D_F}{\partial E_b}\right)^2 \cdot S^2{}_{Eb} + \left(\frac{\partial D_F}{\partial h}\right)^2 \cdot S^2{}_h + \left(\frac{\partial D_F}{\partial Z_S}\right)^2 \cdot S^2{}_{Zs} \tag{12}$$

$$D_{dt} = \left(\frac{\partial d_t}{\partial t}\right)^2 \cdot S^2{}_t + \left(\frac{\partial d_t}{\partial E_b}\right)^2 \cdot S^2{}_{Eb} + \left(\frac{\partial d_t}{\partial Q}\right)^2 \cdot S^2{}_Q \tag{13}$$

$$V_{d1} = \frac{\sqrt{D_{d1}}}{\overline{d_1}} \tag{14}$$

$$D_{D_1} = \left(\frac{\partial D_1}{\partial R}\right)^2 \cdot S^2{}_R + \left(\frac{\partial D_1}{\partial h}\right)^2 \cdot S^2{}_h + \left(\frac{\partial D_1}{\partial L}\right)^2 \cdot S^2{}_1 \tag{15}$$

where $\overline{D_f}$ – average bending moment in concrete stab caused by the design aircraft main gear wheels; S_{Fd}, S_{Eb}, S_h, S_{Zs}, S_Q, S_R, S_1 – standard deviations of wheel load, modulus elasticity of concrete, slab thickness, modulus of subgrade reaction, amplitude of temperature, flexural strength of concrete and number of loading; $\left(\frac{\partial D}{\partial}\right)$ – first partial derivative of stochastic function of bending moments with respect to the means of random variables: wheel load F_d modulus of elasticity of concrete E_b thickness of slab h, modulus of subgrade Z_s, amplitude of pavement surface temperature Q, flexural strength of concrete R and number of load applications L:

$$\frac{\partial D_F}{\partial F_d} = 0.018306\sqrt{P_Q / \overline{F}^3{}_d \sqrt[4]{\overline{h}^{-3}\overline{E}_b / \overline{Z}_S}} \tag{16}$$

$$\frac{\partial D_F}{\partial E_b} = 0.001749\sqrt{\overline{h}^{-3} / \overline{Z}_S \cdot \overline{E}^{-3}{}_b} - 0.013745\sqrt[4]{\overline{h}^{-3}} / \overline{Z}_S \cdot \overline{E}_b{}^7 \tag{17}$$

$$\frac{\partial D_F}{\partial h} = 0.0113745\sqrt{P_Q \overline{F}_d \sqrt[4]{\overline{E}_b \overline{h}^{-5}} / \overline{Z}_S} + 0.005248\, P_Q \sqrt[3]{\overline{E}_b / \overline{h}\overline{Z}_S} \tag{18}$$

$$\frac{\partial D_F}{\partial Z_s} = 0.022908\sqrt{P_Q \overline{F}_d \sqrt[4]{\overline{E}_b t^3 / \overline{Z}^9{}_s}} - 0.005247\, P_Q \sqrt[2]{\overline{E}_b \overline{h}^3 / \overline{Z}^5{}_s} \tag{19}$$

$$\frac{\partial D_t}{\partial h} = 0.0874\, E_b \alpha \overline{Q}\overline{h} \tag{20}$$

$$\frac{\partial D_t}{\partial E_b} = 0.0437\alpha \overline{Q}\overline{h}^2 \tag{21}$$

$$\frac{\partial D_t}{\partial Q} = 0.0437\alpha_1 E_b \overline{h}^2 \tag{22}$$

$$\frac{\partial D_1}{\partial R} = \alpha_1 \overline{h}^2 (2 - \lg \overline{L}/6) \tag{23}$$

$$\frac{\partial D_1}{\partial h} = 2\alpha_1 \overline{R}\,\overline{h}(2 - \lg \overline{L}/6) \tag{24}$$

$$\frac{\partial D_1}{\partial L} = -0.0724\alpha_1 \overline{R}\overline{h}^2 / \overline{L} \tag{25}$$

where $\overline{D_t}$ – average bending moment due to temperature gradient between top and the bottom surfaces of slab; $\overline{D_1}$ – average ultimate bending moment computed for given slab thickness and design modulus of rupture of the concrete; α_1 – parameter to convert concrete flexural strength increase with age.

3. Results and Discussion

In order to investigate the influence of statistical variability of design parameters and given probability level on the concrete pavement functional life numerical analysis was performed under following conditions: concrete slab thickness $\overline{h}$=0.3 m; coefficient of variation of slab thickness V_h=0.10; average flexural strength of concrete $\overline{R}$=5.13 MPa; coefficient of variation of concrete flexural strength V_R=0.135; average modulus elasticity of concrete $\overline{E}_b = 3.24 \cdot 10^4$ MPa; coefficient of variation of modulus of elasticity of concrete $V_{E_b} = 0.135$; average coefficient of subgrade reaction $\overline{Z}_s = 51.6$ MH/m^3; coefficient of subgrade reaction $V_{Z_s} = 0.30$; average amplitude of concrete surface pavement temperature $\overline{Q}$=7.55 ^{0}C; coefficient of variation of amplitude V_Q=0.33; aircraft average take-off mass $\overline{D} = 206$ t; coefficient of take-off mass variation V_D=0.55; average daily departures $\overline{U_1}$=60; coefficient of daily take off number variation V_1=0.10; number of tandem axes – 2; probability of main wheels passes P_i (x)=0.83.

Values of parameter K_2 calculated by formulas (11) – (25) are given in the table 2.

Table 2. Results of numerical analysis

Parameter	Values				
Design reliability level	0.5	0.6	0.7	0.8	0.9
Parameter Z_2	1.0	0.95	0.90	0.84	0.77

To account the stochastic nature of the design inputs and considering (8) the equation (7) may rewritten as follows

$$N_{\sigma_{N_i}} = 10^{12(1-P_{Ft})\frac{(\overline{D}_f + \overline{D}_t)}{Z_2 D_1})} \tag{26}$$

where D_f – mean of bending moment in concrete stab caused by the wheels of main gear arrangement; D_t – mean of bending moment due to temperature gradient between the top and the bottom surfaces of slab; D_l – mean of ultimate bending moment computed for given slab thickness and design modulus of rupture of concrete; Z_2 – statistical coefficient to account combined effect of variability of wheel loads, number of load repetitions, flexural strength and modulus of elasticity of concrete, modulus of subgrade reaction, thickness of slab and surface pavement temperature amplitude (11).

Detrimental effect D caused by load applications N_i may be calculated by formula

$$D = \frac{1}{N_{\sigma_{N_i}}} = \frac{1}{10^{12(1-P_{Ft}(\frac{(\overline{D}_f+\overline{D}_t)}{Z_2 D_l})}} \qquad (27)$$

In common summary detrimental effect D under repetition cycles of special aircraft traffic mix pattern is expressed by Miner's law [31]

$$\sum_{i=1}^{k} \frac{N_{\sigma_i}}{N_{\sigma_{N_i}}} = 1 \qquad (28)$$

where N_{σ_i} – number of stress repetitions at outside edge of slab for i-th aircraft main undercarriage wheels for the design life may be calculated as follow

$$N_{\sigma_i} = n(365 - T_f)\sum z_{n_i} \cdot n_1 \cdot U_i \cdot P_i(x) \qquad (29)$$

where n – design life of the concrete pavement; T_f – number of days in the year when subgrade is in frozen condition; z_{ni} – coefficient to convert the effect of particular aircraft to the effect of undercarriage assembly of the design aircraft; n_l – number of tandem gears in undercarriage assembly of the particular aircraft; U_i – number of daily departures of particular aircraft; $P_i(x)$ – probability to account lateral movement (airplane wander) and particular airplanes main gear and configurations on the transverse distribution of load applications on the pavement at the anticipated facility;

$$P_i(x) = P(x_1 \cdot x_2 \cdot x_3) = \sum_{x_1}^{x_2} f(x, \overline{x}_i, S_{xi}) \qquad (30)$$

$$f(x, \overline{x}_i, S_{xi}) = \frac{1}{S_{xi}\sqrt{2n}} e^{-(x-\overline{x}_i)^2 / 2S^2_{xi})} \qquad (31)$$

$$x_1 = x - \frac{b}{2}$$
$$x_2 = x + \frac{b}{2}, \qquad (32)$$

where $f(x, \overline{x}_i, S_{xi})$ – normal distribution function; x_i and S_{xi} – mean and standard deviation of aircraft wheel paths from pavement centerline or guideline marking; x – distance from longitudinal axis (centerline or guideline marking) of pavement to design section of slab, where load repetitions are determined; b – design traffic width

$$b = d + 2R_i \qquad (33)$$

where d – distance between centers of contact areas of dual wheels; R_i – radius of main undercarriage wheel contact area for specific aircraft

$$R_i = \sqrt{\frac{F_{d_i}}{\pi \cdot P_Q}} \qquad (34)$$

$$F_{d_i} = \frac{\overline{D}_i \cdot g \cdot z'}{n' \cdot n_w} z_d \gamma_f \qquad (35)$$

where F_{di} – wheel load, N; g – acceleration of gravity; z' – percent of the maximum anticipated takeoff weight of the airplane on the main landing gears; z_d – parameter to account dynamic impact of

IOP Conf. Series: Materials Science and Engineering **832** (2020) 012037 doi:10.1088/1757-899X/832/1/012037

the airplane main gears wheels on the pavement; γ_f – parameter to reflect wings lift; n' – number of main undercarriage assemblies; n_w – number of wheels in undercarriage assembly.

Substitution formula (29) to (26) gives following equation to determine anticipated life of concrete pavement

$$n = \frac{10^{12(1-P_{Ft}(\frac{(\bar{D}_f+\bar{D}_t)}{Z_2 D_1})}}{(365-T_f)} \sum_{i=1}^{k} z_{n_i} \cdot n_{1_i} \cdot u_i \cdot P_i(x) \qquad (36)$$

Proposed equation allows more precisely evaluate number of load applications expected during pavement life from specific volume of mixed traffic, its distribution pattern, variability of flexural strength and elastic modulus of concrete, modulus of subgrade reaction, slab thickness, mean amplitude of surface pavement temperature, number of stress repetitions and design reliability level of pavement structure.

In order to reduce the complexity of calculating the statistical coefficient of K_2 working conditions using formula (11), a program for calculating ''STK'' was developed.

The program ''STK'' provides for printing the statistical coefficient of working conditions at a given level of reliability P = 0.5 ... 0.95 and the initial calculation data. The results of the calculation with the following initial data are given in table 3.

Table 3. The results of the calculation

Parameter	Value	Unit
Road-climatic zone	2	
Type of hydrogeological conditions	2	
Soil of natural base	sandy loam	
The average value of the bedding ratio of the soil base	50	MN/m3
The coefficient of variation of the bedding coefficient of the soil base	0.30	
The average value of the load on the wheel	170	кN
The coefficient of variation of the load on the wheel	0.05	
Internal tire pressure	1.0	MPa
The average strength of cement concrete in tensile bending	5.12	MPa
Coefficient of variation cement concrete strength under bending tension	0.135	
The average value of the modulus of elasticity of cement concrete	32400 MPa	
The coefficient of variation of the modulus of elasticity of cement concrete	0.135	
The average number of applications loads	20000	
The coefficient of variation of the number of applications loads	0.20	
The average value of the amplitude of temperature fluctuations on the surface of the cement concrete	10	°C
The coefficient of variation of the amplitude of temperature fluctuations	0.30	

IOP Conf. Series: Materials Science and Engineering **832** (2020) 012037 doi:10.1088/1757-899X/832/1/012037

The average value of the plate thickness	0.16…0.30 m
The coefficient of variation of the thickness of the plate	0.05
The value of the estimated coefficients: zd=1.20,	

It is shown that functional life of concrete pavements depends on statistical variability of mechanical properties of materials, aircraft traffic mix loads and environmental condition. That is why to predict pavement service life till cracking of concrete due to fatigue consumption will take place can be implemented only with certain probability [37-39]. Procedure proposed in this paper may be considered as first step towards statistical approach in that direction. Similar results in the calculation of building structures can be noted in [3,12,16,22]. However, in the data, the work does not take into account the specifics of the functioning of aerodrome coatings using coefficients characterizing stochastic operating loads. The results of that approach also underline the significance of quality control and statistical evaluations of test data of construction materials used for a particular airport pavement's project in situ. The use of statistical approach provides more realistic data to the cost estimator for a new particular project or pavement overlay design [31, 35, 36]. Probability distribution pattern for different traffic mix and statistical coefficient values, calculated for various regions of Russian Federation and in the CIS countries are proposed for adjustment to aerodrome's construction rules and to modify existing deterministic method [1].

It is important to note that functional pavement life define the moment of initial process of cracking caused by stress repetitions in a critical traffic width where the probability of design aircraft main leg wheels passes is maximum. That is why procedure proposed here requires modification to reflect cracking propagation in concrete slabs after initial cracks have developed. In addition it is of interest to compare numerical value of functional life received by statistical approach to real serviceability age of concrete pavements designed by conventional deterministic methodology.

4. Conclusion

1. The article defines the statistical coefficient of the working conditions of the plate taking into account the operational loads, physical and mechanical properties of materials and the geometric characteristics of the plate.
2. The statistical coefficient of the working conditions of the plate at a given level of reliability for different values of the plate thickness was calculated.
3. Proposed to use the condition (2) for calculating the required thickness of the hard airfield slab coating of monolithic cement concrete.
4. The comparison of the calculated values of the statistical coefficient of the conditions of hard airfield pavements with those adopted in Set of rules 121.133302 shows that the standard coefficient of working conditions γ_2= 0.80 for groups of sections of group A of cement concrete pavements of aerodromes located to the north of 50° north latitude corresponds to the specified level of reliability P = 0.80 for plates with a thickness of 0.16…0.22 m and P <0.80 for plates with a thickness of 0.24…0.30 m.
5. Implemented statistical analysis showed that mean serviceability age of concrete pavements in Russian Federation airports is 11 years. That result has a good compromise with statistical approach [37-39] using suggested form (33) at desired probability level P=0.95.

Reference

[1] SP 121.13330.2012 2012 Aerodromy. Aktualizirovannaya redaktsiya SNiP 32-03-96. [Airfields. Updated edition of Construction Rules and Regulations 32-03-96]. (Moscow: Analitik) p. 97.

[2] Klekovkina M P 2016 Sostoyaniye i nedostatki konstruirovaniya i rascheta dorozhnykh odezhd s tsementobetonnym pokrytiyem [The state and shortcomings of the design and calculation

of pavement with cement concrete pavement] *Vestnik grazhdanskikh inzhenerov* **1**(54) 136-140.

[3] Shvabyuk V I, Rotko S V, Matkova A V 2016 K probleme sovershenstvovannogo rascheta izgotovleniya plit na zhestkom i elasticheskom osnovaniyakh [To the problem of the improved calculation of the manufacture of plates on rigid and elastic bases] *Naukoví notatki* **54** 376-380.

[4] Carpiuc-Prisacari A, Poncelet M, Hild F, Leclerc H, Kazymyrenko K 2017 Comparison between experimental and numerical results of mixed-mode crack propagation in concrete: influence of boundary conditions choice *Cement and Concrete Research* **100** 329-340.

[5] García V J, Zúñiga-Suárez A R, Zuñiga-Torres B C, Villalta-Granda L J, Márquez C O 2017 Brazilian test of concrete specimens subjected to different loading geometries: review and new insights *International Journal of Concrete Structures and Materials* **11**(2) 343-363.

[6] Corrado M, Molinari J-F 2016 Effects of residual stresses on the tensile fatigue behavior of concrete *Cement and Concrete Research* **89** 206-219.

[7] Wei Wu, Jianwu Wang, Chao Zhang, Qingkun Yu 2015 Research on method of the heavy traffic airport cement concrete pavement thickness design, In *Proceedings of International Conference on Materials, Environmental and Biological Engineering* (Paris: Atlantis Press) pp. 521-424.

[8] Bažant Z P 2002 Concrete fracture models: testing and practice *Engineering Fracture Mechanics* **69**(2) 165-205.

[9] Chen X, Chen C, Xu L, Shao Y 2017 Dynamic flexural strength of concrete under high strain rates *Magazine of Concrete Research* **69**(3) 109-119.

[10] Gameliak I, Vyrozhemskyi V, Voloshyna I, Dmitriev M 2016 Research of cement concrete pavement using thermal imaging method. functional pavement design, In *Proceedings of the 4th Chinese-European Workshop on Functional Pavement Design* (Netherlands: CRC Press) pp. 1489-1498.

[11] Tang S W, Li Z J, Yao Y, Andrade C 2015 Recent durability studies on concrete structure *Cement and Concrete Research* **78** 143-154.

[12] Jeong M G, El-Basyouny M 2010 Statistical applications and stochastic analysis for performance-related specification of asphalt quality assurance *Transportation Research Record* **2151** 84-92.

[13] Stepushin A P 1985 Primeneniye metodov teorii sluchaynykh funktsiy k opredeleniyu nesushchey sposobnosti plity armobetononogo pokrytiya [Application of the methods of the theory of random functions to the determination of the bearing capacity of the slab of reinforced concrete pavement] *Sbornik nauchnykh trudov MADI* **1985** 20-33.

[14] Stepushin A P 2016 Naturnyye ispytaniya tsementobetonogo pokrytiya [Field tests of cement concrete pavement] *Vestnik MADI* **4**(47) 98-102.

[15] Stepushin A P, Saburenkova V A 2017 Nadezhnost tsementobetonnykh pokrytiy aerodromov [Reliability of cement concrete coatings of airfields] *Vestnik MADI* **1**(48) 84-89.

[16] Ushakov V V, Yarmolinsky V A, Dobrov E M, Goryachev M G, Lugov S V 2017 Revision of descriptions and calculated properties of road bed soils and asphalt concrete materials upon designing of highway pavements in terms of criteria of residual deformations and fatigue cracking of asphalt concretes *International Journal of Civil Engineering and Technology* **8**(9) 1074-1083.

[17] Van-Hieu Nguyen, Duy-Dong Nguyen, Tatarinov V 2018 Methods of spectral density estimation for airfield pavements *MATEC Web of Conference* **251** 9.

[18] Guide for design of aerodromes 2005 Part 3. Pavements. ICAO Doc. 9157 – AN/ 901. (Montreal: ICAO) p. 349.

[19] Nosov V P, Dobrov E M, Chistyakov I, Borisiuk N V and Fotiadi A 2017 Mathematical modelling of cracking process in concrete pavement highways *International Journal of Applied Research* **12**(23) 13158-13164.

[20] Ushakov V V, Nosov V P, Yarmolinsky V A, Goryachev M G, Lugov S V 2016 Setting frequency of works on the arrangement of wear and protective layers of road surfaces. *International Journal of Applied Engineering Research* **11**(23) 11207-11214.
[21] Kiselev V P, Filimonov V S, Bugaenko M B, Kemenev N V, Ivanova L A 2013 Improving the quality of polymer-modified asphalt coating binder by settling resin of pyrolysis *SibFU Journal. Engineering and Technologies* **6**(8) pp. 895-902.
[22] Paccard R G 1973 *Design of concrete airport pavement* (Illinois: Skokie, Portland Cement Association) p. 61.
[23] Ashford N, Wright P H 1984 *Airport Engineering* (Wiley-Interscience Publication: New York) p. 433.
[24] Zhang J, Weng X, Qu B, Liu J, Yang B, Li Y 2018 Failure modes and mechanisms of pavements in saline foundations *Proceedings of the Institution of Civil Engineers: Transport* **171**(3) 174-182.
[25] Sassani A, Ceylan H, Kim S, Arabzadeh A, Taylor P C, Gopalakrishnan K 2018 Development of carbon fiber-modified electrically conductive concrete for implementation in des moines international airport *Case Studies in Construction Materials* 8 277-291.
[26] Chen X, Chen C, Xu L, Shao Y 2017 Dynamic flexural strength of concrete under high strain rates *Magazine of Concrete Research* **69**(3) 109-119.
[27] Qingkun Yu, Liangcai Cai, Jianwu Wang 2018 Experimental and fem research on airport cement concrete direct-thickening double-deck pavement slabs under aircraft single-wheel dynamic loads *Advances in Materials Science and Engineering* **2018** 16.
[28] Bieliatynskyi A, Krayushkina E, Skrypchenko A 2016 Modern technologies and materials for cement concrete pavement's repair, *In Proceedings of the 9th International Scientific Conference Transbaltica* (Elsevier) pp. 344-347.
[29] Shvabyuk V I, Rotko S V, Matkova A V 2016 To the problem of an improved calculation of bending of plates on rigid and elastic foundations *Naukovi Notatki* **54** 376-380.
[30] Ho Sang V 1975 *Field Survey and Analysis of Aircraft Distribution on Airport Pavements.* Report FAA RD-74-36, p. 286.
[31] Ying-Haur Lee 1997 Zero-maintenance considerations for concrete pavement design. *Pavements Session of the 13th IRF World Meeting* Toronto, Ontario, Canada, p. 12.
[32] Simon M J 2003 Concrete mixture optimization using statistical methods. Final Report FHWA-RD-03-60 (Gaithersburg: National Institute of Standards and Technology) p. 167.
[33] FAA Airport Pavement Design and Evaluation 2009 AC 150/5320-6E (U.S. Department of Transportation) p. 124.
[34] Roesler J, Evangelista F, Domingues M 2007 Effect of gear positions on airfield rigid pavement critical stress locations, In *materials of FAA worldwide airport technology transfer conference*, Atlantic city, New Jersey, USA, p. 13.
[35] AC No 150/5380-6E 2007 Guidelines and procedures for maintenance of airport pavement. Federal Aviation Administration, Advisory Circular.
[36] Simonyan A M, Arutyunyan A S, Akopyan E A 2014 K voprosu o summirovanii ustalostnykh napryazheniy materialov [On the issue of summation of fatigue stresses of materials] *Izvestiya Natsional'noy akademii nauk Armenii* **67**(4) 65-70.
[37] Bin C, Balbo J 2014 Comparing results of airport pavement concrete slab design using damage models of FAARFIELD to MEPDG concrete fatigue model, in *materials of FAA Worldwide Airport Technology Transfer Conference,* Atlantic City, USA.
[38] Davis J 2012 Airport Pavement Design and Evaluation, In *materials FAA Runway Safety and Pavement Maintenance Seminar for Africa*, Retrieved from: http://www.airtech.tc.faa.gov/ATT2014/Papers/P10049%20-%20Bin%20&%20Balbo.pdf.
[39] Rodchenko O V 2013 Computer technologies of finite element modeling of airfield rigid pavement, in *Proceedings of the 16th Conference for Junior Researchers "Science – Future of Lithuania"* Vilnius, Lithuania, pp. 65–70.

IOP Conf. Series: Materials Science and Engineering **832** (2020) 012038 doi:10.1088/1757-899X/832/1/012038

Set of organizational, technical and reconstructive measures aimed at improvement of section performance indicators based on the study of systemic relations and regularities of functioning of railway transport system

V A Olentsevich, Yu I Belogolov, G N Kramynina

Irkutsk State Transport University, 15 Chernyshevsky Street, Irkutsk, 664074, Russia

E-mail: olencevich_va@irgups.ru

Abstract. The organization of operational work should correspond to the corporate interests of the railway transport system: increasing profitability, minimizing transportation costs, increasing the interest of structural units in improving financial and economic results, focusing on high-quality transport customer service. Measures to reduce costs do not give the expected effect, resulting in the decrease in the reliability of technical means and transportation quality, may entail additional costs for the restoration of technical means and lower profitability. A constant increase in cargo turnover necessitates an increase in the throughput capacity of railway lines and sections. One of the main transport enterprises in Eastern Siberia is the East Siberian Railway. At the Bolshoi Lug - Slyudyanka railway section, the maximum ascent and descent magnitude is 18 ‰, the minimum horizontal curve radii are less than 300 meters. In this regard, the traffic operation on the mountain pass section is organized by way of pushing trains. For the estimated year 2023, the target indicators of the comprehensive project of the polygon require the throughput capacity of the section under consideration to increase to 137 pairs of trains per day. The article analyzes the factors affecting the magnitude of the section performance indicators and identifies the "bottlenecks". It proposes the complex of organizational, technical and reconstruction activities. The value of one-time and current costs is determined. The economic effect of changing the performance indicators of structural units due to an increase in the volume of work, as well as the payback period of each event, are calculated.

1. Introduction

The organization of operational work should correspond to the corporate interests of the railway transport system: increasing transportation profitability, minimizing cargo transportation expenses, increasing the interest of all structural units in improving financial and economic results of the work, focusing on high-quality transport customer service [1-5]. The constant increase in cargo turnover necessitates the increase of the throughput of railway lines and sections in the railway transport system [6-7]. The increase, defined as the difference between the required and available throughput capacity, can be achieved by implementing organizational, technical and reconstruction measures. In order to successfully develop railway infrastructure, a set of activities is necessary [8-12]. Firstly, to eliminate the "bottlenecks" of the railway transport system by building the second, as well as third and fourth tracks in the main directions of train traffic volumes, which will increase the throughput and processing capacity of the infrastructure. Secondly, to expand the railway network through the construction of

IOP Conf. Series: Materials Science and Engineering **832** (2020) 012038 doi:10.1088/1757-899X/832/1/012038

strategic, socially significant, high-speed, cargo-forming and technological sites in the railway transport system [13, 14]. Thirdly, to regularly update obsolete infrastructure elements, reconstructing worn-out railway lines, modernizing buildings and facilities, improving the used rolling stock, increasing comfort and speed of transportation in order to ensure their competitiveness in comparison with other modes of transport, etc., and methods of randomly forming solutions [15-18].

2. Results of a study of systemic relationships and patterns of functioning of the railway transport system

Effective operational activities of the railway transport system integrates and implements the activities of elements and subsystems of railway transport, the precise organization of which ensures full satisfaction of the country's needs for transportation at the lowest cost, increases the competitiveness of railway transport, economic stability of the industry, and implements a customer focus program for the industry. The implementation of these areas is the most important criterion of the industry operation in a competitive environment with other modes of transport [19-26].

One of the main transport enterprises in Eastern Siberia, which plays a significant role in the development of regions in accordance with its geographical position, is the East Siberian Railway. The railway consists of four road station operation hubs, divided into railway sections, with their own characteristic technical and operational features.

The length of the Bolshoi Lug - Slyudyanka railway section is 83 kilometers. In the freight-hauling direction, the route profile is characterized by an excessive gradient from the Bolshoi Lug intermediate station to the Podkamennaya intermediate station. With that, the difference in elevations is 500 meters, followed by an equally long descent to Lake Baikal. The maximum ascent and descent magnitude is 18 %, the minimum horizontal curve radii are less than 300 meters. In this regard, the traffic operation on the mountain pass section is organized by way of pushing trains. In the section Bolshoi Lug - Slyudyanka - I, trains are driven by pusher locomotives. Uncoupling pusher locomotives from down trains at Slyudyanka - I station, attaching to up trains (the necessity for attaching, depending on the head locomotive series and train weight) at Slyudyanka - II station. Automatic blocking with tonal rail circuits is used as a system of interval regulation of train traffic on the section. The section includes seven stations with a certain class depending on the volume and complexity of the works: Bolshoi Lug, Podkamennaya, Glubokaya, Andrianovskaya, Angasolka, Slyudyanka-I, Slyudyanka-II.

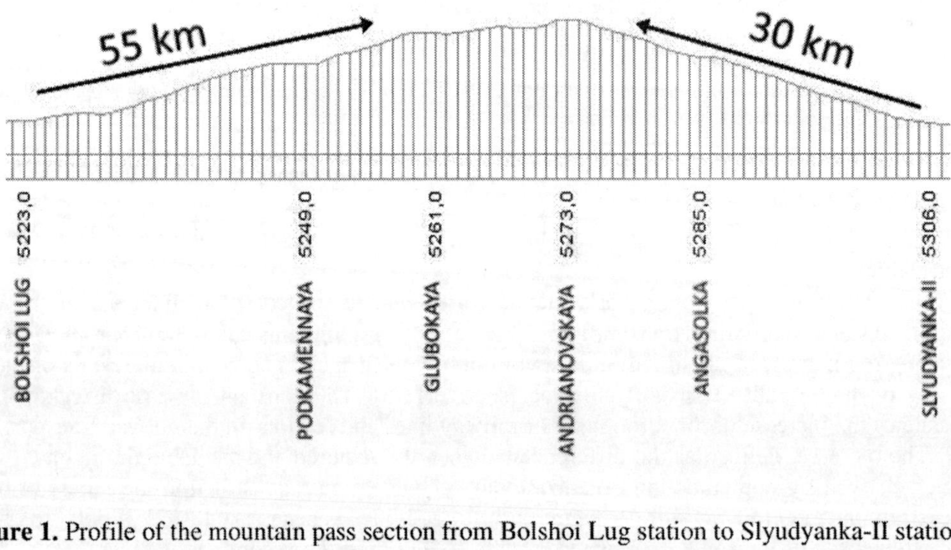

Figure 1. Profile of the mountain pass section from Bolshoi Lug station to Slyudyanka-II station.

One of the main problems in organizing the movement of trains at the mountain pass section under consideration is related to the need to push freight trains within the boundaries of the Bolshoi Lug - Slyudyanka-I stations. The pushing is required for the main part of down freight trains directly from the Bolshoi Lug intermediate station, as well as some up freight trains from the local station Slyudyanka-I.

The current regulatory schedule for the Bolshoi Lug - Slyudyanka-I section provides for the handling of up to 96 pairs of freight trains. The requirements for pusher locomotives make 35 units. Some pusher locomotives are delivered for up freight trains. The remaining train locomotives are returned to the station Bolshoi Lug station as a "reserve". According to established requirements, pusher locomotives are returned fastened together by three units [27-29].

The problems in organizing the pushing of trains are caused by a number of circumstances, the main of which is the insufficient development of the railway infrastructure of the Bolshoi Lug intermediate station. The existing track arrangement of this station provides for the possibility of using only one dead end track for supplying pusher locomotives for down freight trains. The minimum required time for delivery and coupling of a pusher locomotive to a freight train is 16 minutesfrom the "end" of the train, 24 minutes from the "head" of the train. According to the data presented, the feature of the existing amount of freight train traffic for the Bolshoi Lug station is excess downtime of freight trains and, on some days, "congestion" in the operation of the running line adjoining from the west. These station work conditions are further complicated by operations for uncoupling pusher locomotives from up freight trains. During the day, the pusher locomotives are uncoupled from 10-15 up trains. The time required for uncoupling a pusher locomotive, according to the existing technology of operation, is: 10 minutes from the "end" and 14 minutes from the "head".

In addition the "bottleneck" in the work of the considered section is the down yard neck of the Slyudyanka-I station and the Slyudyanka-I - Slyudyanka-II running line. The existing technology for the operation of the section stipulates that a large volume of various kinds of operations takes place at the Slyudyanka-I precinct station: reception and departure of up-and-down passenger and freight trains, uncoupling pushing locomotives from even freight trains, attaching pushing locomotives to up freight trains, forming a fastenings out of pusher locomotives and their departure to the Bolshoi Lug station, departure and reception of motor vehicles during the track possessions, etc. [1, 5, 30].

The main operations are performed at the station due to the need of train or shunting movements in its down yard neck. Due to the existing down yard neck construction, many of the routes envisaged by technological operations are mutually antagonistic. This imposes significant restrictions not only on the operation of the station itself, but also on the possibility of handling trains along the Slyudyanka-I - Slyudyanka-II running line adjacent to its down yard neck.

Analyzing the existing technology of the site and its constituent stations, the designed and constructed schedules for the passage of trains, their performance should also be noted the presence of "reverse removal" of freight trains by suburban ones (where removal means the inability to handle the lines of freight trains along the running line due to the passage of passenger trains). In relation to the section under consideration, the need to remove the paths of the freight train schedule due to the suburban trains is implied due to the fact that the station-to-station travel time of the latter ones is longer than that for freight trains. The fact that commuter trains need more time to overcome the running lines is due to the presence of stops along the route with almost identical restrictions on the maximum speeds of freight and suburban trains.

It is important to note that the issue of lack of throughput capacity is complex, while its causes and solutions must be sought not only within the boundaries of the Bolshoi Lug - Slyudyanka-II section, but also beyond them. Due to the high traffic density and lack of reserves for train handling, any malfunctions that occurred in one place automatically lead to the emergence and growth of unproductive downtime of trains throughout the entire mainline.

For the estimated year 2023, the target indicators of the comprehensive project of the Eastern polygon provide for the need to increase the throughput of the section under consideration to 137 pairs of trains per day, including 107 pairs of freight trains per day and 19 fastenings of down pusher locomotives. The track turnover of the section is going to increase to 116.7 million / ton per year. With the specified

IOP Conf. Series: Materials Science and Engineering **832** (2020) 012038 doi:10.1088/1757-899X/832/1/012038

parameters of the prospective amount of transportation, the available throughput capacity shortfall of the Irkutsk - Slyudyanka section should be 14 pairs of trains by 2023. In freight traffic, subject to the use of all reserved paths provided for in the regulatory schedule, the throughput capacity shortfall will be 11 pairs of trains of various categories.

3. Reconstructive measures aimed at improving the performance of the section, through the construction of additional reception and departure tracks of the station

The measures taken in recent years to reconstruct the railway transport, especially to introduce new types and methods of traction, present new requirements for the operation of these structural units. When solving issues of technical equipment and improving the technology of work, it is necessary to ensure such a relationship between the volume of work performed and the existing fixed assets of the station and its technical means, that the total operating costs are minimal. The optimal solution to this problem can only be achieved by taking into account the real operating conditions of the railway station, i.e. during the non-uniform loading of station devices.

The analysis of the technology of operation of the Bolshoi Lug intermediate station and analysis of the developed daily schedule of work and its indicators made it possible to determine the need for the construction of additional receiving and departure tracks at the station – to restore for operation receiving and departure tracks No. 4 and 15 of the intermediate Bolshoi Lug station.

Table 1. Performance indicators of the intermediate station Bolshoi Lug taking into account the reconstruction activities.

Indicator	Unit of measurement	Indicator value	
		with existing work technology	taking into account the reconstruction activities
Train-hours of detention	t/h	1212,8	1136
Locomotive-hours of detention	l/h	3154,8	3936
The number of handled down trains	trains	98	119
The number of handled up trains	trains	95	106
Number of pusher locomotives	locomotive	28	30

The implementation of the proposed reconstruction activities will improve the technical work of the station by reducing the detention time of trains, and will increase the throughput and operating capacity. Train-hours of detention at the Bolshoi Lug station will be reduced by 1.13 hours, and the throughput capacity will increase by 21 down trains and by 11 up trains.

Carrying out this reconstruction at the Bolshoi Lug station will require a one-time investment of 114,693.83 thousand rubles. In connection with the restoration of additional receiving and departure tracks for the station, the annual operating costs for the maintenance of new fixed assets will change by 19004.27 thousand rubles per year. The introduction of this event gives an economic effect in the amount of 262,764.64 thousand rubles per year. The payback period is 1 year.

4. Set of measures to increase the throughput and operating capabilities of railway stations and adjacent running lines

According to the existing work technology, the pusher locomotives are returned from the Slyudyanka station to the Bolshoi Lug station fastened together by three units each. The authors examined the work of the station Slyudyanka-II and Slyudyanka-I during the commissioning of the third main track between the stations and a change in the operation technology of pushing locomotives. This, in turn, will increase the throughput and operating capacity of the railway station itself and increase the performance of adjacent running lines. The construction of the third main track between the stations will allow accumulating and returning locomotives fastened together by five units each.

Table 2. Indicators of the daily schedule of work of the station Slyudyanka-II with the existing and modified technology of work.

Indicator	Unit of measurement	Indicator value	
		with existing work technology	according to the proposed technology of the section Slyudyanka-II and Slyudyanka-I
Transit car detention without yard operation	hr	$t_{tr}^{w/y} = 0.89$	$t_{tr}^{w/y} = 0.58$
Medium detention of the local car	hr	$t_l = 13.2$	$t_l = 12.4$
The ratio of dual operations	-	$K_{dua} = 1.4$	$K_{dua} = 1.4$
Average detention of the local car under one freight operation	hr	$t_{fr} = 9.43$	$t_{fr} = 8.72$
The working fleet of cars	cars	$n_w = 132$	$n_w = 132$
Station car turnover	Cars	$n_{tu} = 5746$	$n_{tu} = 5993$
The ratio of using check station crews		$\varphi_1 = 0.39$ $\varphi_2 = 0.48$	$\varphi_1 = 0.57$ $\varphi_2 = 0.43$

Thus, it is obvious that a set of measures to increase the effective values of the throughput and operating capabilities of the Slyudyanka-II station makes it possible to reduce the car detention time and improve overall performance. The transit car detention without yard operation was reduced by 0.31 hours, the average detention of the local car was reduced by 0.8 hours, and the average detention of a local car under one freight operation was reduced by 0.72 hours. The value of car turnover will improve by 247 cars due to the reduction in the car detention time. This, in turn, will lead to an increase in the volume of departed cars per day and, as a result, an increase in throughput and processing capacities.

A set of measures to increase the effective values of the throughput and operating capabilities of the Slyudyanka-II station makes it possible to reduce the car detention time and improve overall performance. The transit car detention without yard operation was reduced by 0.29 hours, and the transit car detention with yard operation was reduced by 0.2 hours. The value of car turnover will improve by 3834 cars due to the reduction in the car detention time. This, in turn, will lead to an increase in the volume of departed cars per day and, as a result, an increase in throughput and processing capacities.

Carrying out a set of measures on the Slyudyanka-II and Slyudyanka-I sections will require capital investments in the amount of 883740.249 thousand rubles. Due to the reconstruction, the value of the annual operating costs will change. The increase will amount to 286301.499 thousand rubles per year.

IOP Conf. Series: Materials Science and Engineering **832** (2020) 012038 doi:10.1088/1757-899X/832/1/012038

The economic effect, caused by the increase in the throughput capacity of the stations and the improvement of the section performance, will amount to 348908.391 thousand rubles. The payback period of the project is 15 years.

Table 3. Indicators of the daily schedule of work of the station Slyudyanka-II with the existing and modified technology of work.

Indicator	Unit of measurement	Indicator value	
		with existing work technology	according to the proposed technology of the section Slyudyanka-II and Slyudyanka-I
Transit car detention without yard operation	hr	$t_{tr}^{w/y} = 0.98$	$t_{tr}^{w/y} = 0.69$
Transit car detention without yard operation	hr	$t_{tr}^{w/y} = 1.9$	$t_{tr}^{wi/y} = 1.7$
The working fleet of cars	cars	$n_w = 168$	$n_w = 226$
Station car turnover	cars	$n_{tu} = 8120$	$n_{tu} = 11954$
The ratio of using check station crews		$\varphi_1 = 0.57$ $\varphi_1 = 0.54$	$\varphi_1 = 0.64$ $\varphi_1 = 0.6$

5. Evaluation of measures to increase the voltage level at the limiting running lines of the section

The main purpose of all calculations of the energy supply system is to establish such parameters of this system that would ensure reliable and cost-effective working conditions of the electric railway. The considered section is serviced by five traction substations. The results of traction calculations for the up-and-down train traffic in the section under consideration at different weight standards showed that, according to the existing scheme with a given inter-train interval of 8 minutes, in the section Bolshoy Lug – Slyudyanka-I there are four limiting running lines: Bolshoi Lug – Podkamennaya, Glubokaya – Andrianovskaya, Andrianovskaya – Angasolka, Angasolka – Slyudyanka-II. The inter-train interval was 12, 14, 16, 13 minutes, respectively. Considering the obtained calculation results, we can conclude that to ensure the passage of trains of a given amount of traffic with a given inter-train interval of 8 minutes, measures are required to increase the voltage level at the limiting running lines.

The event will require capital investments in the amount of 65,660,200 thousand rubles. The increase in annual costs will amount to 107,315.738 thousand rubles. The payback period of this proposal is 4 years.

The results of the study of systemic relationships and work patterns of the Bolshoi Lug – Slyudyanka-I section based on a set of organizational, technical and reconstruction measures are listed in Table 4.

Table 4. Performance indicators of the section Bolshoi Lug - Slyudyanka-1.

A section performance indicator	Units of measurement	Section performance indicators			
		with existing work technology	while strengthening the track arrangement of the Bolshoi Lug station	during the construction of an additional main track and a change in the operation technology of pusher locomotives	while strengthening the power supply system
Train-hours	t-hour	514.55	677.88	647.91	725.5
The number of trains of even direction	train				
- freight		79	100	105	114
- passenger		10	10	10	10
- suburban		8	8	8	8
- goods collecting		1	1	1	1
Number of trains of odd direction	train				
- freight		91	95	102	113
- passenger		9	9	9	9
- suburban		8	8	8	8
- goods collecting		1	1	1	1
The number of pusher locomotives	loc.	28	30	21	37
Civil speed	km/h	45.8	46	47.6	39.3
Service speed	km/h	34.8	33.7	37.1	36.7
Service speed ratio	-	0.76	0.73	0.78	0.93
Weight of train	t	4163	4163	4163	4163
The average train formation	car	65	65	65	65
Locomotive journey	hr	6.52	7.7	6.52	6.88
The required number of train locomotives	loc.	23	33	29	31
The average daily mileage of a locomotive	km	728	681	767	718
Locomotive performance	ton-km gross	2497	2835	3193	2990

6. Conclusion

A significant drawback in the operation of the railway transport system, compared to other modes of transport while implementing the customer focus policy, is a rather high level of tariffs for both freight and passenger transportation. In connection with these factors, much attention has always been paid and is being paid to the search for areas of activity to attract additional traffic volumes. A special emphasis in this matter has been placed to directions for reducing tariff levels, but in most cases a solution is possible only in case of the reduction of the operational costs of the railway transport system associated with transportation. The accumulated experience in this field shows that measures to reduce the level of expenses in most cases do not give the expected result and more often lead to a decrease in the reliability of technical means and quality of transportation. In the future, this may entail not only additional costs for restoration of technical means, but also a decrease in profitability due to the loss of traffic. Thus, the main areas of activity in the fight for the client between carriers are: the use of modern digital technologies, customer focus, increasing the throughput and operating capabilities of the infrastructure of the railway transport system that can determine the needs of the client, reduce its transportation costs, simplify access to the railway service, thereby increasing the attractiveness of the industry.

References

[1] Kopylova T A, Mikhailov A Yu 2016 Evaluation of the functioning of intermodal hubs of urban passenger transport *In the collection: In the collection: Transport systems of Siberia. Development of the transport system as a catalyst for the growth of the state economy. International scientific and practical conference* pp 528-532

[2] Kolosov A D, Gozbenko V E, Shtayger M G, Kargapoltsev S K, Balanovskiy A E, Karlina A I, Sivtsov A V, Nebogin S A 2019 Comparative evaluation of austenite grain in high-strength rail steel during welding, thermal processing and plasma surface hardening *IOP Conference Series: Materials Science and Engineering* **560** 012185

[3] Konstantinova M V, Balanovskiy A E, Gozbenko V E, Kargapoltsev S K, Karlina A I, Shtayger M G, Guseva E A, Kuznetsov B O 2019 Application of plasma surface quenching to reduce rail side wear *IOP Conference Series: Materials Science and Engineering* **560** 012146

[4] Guseva E A, Kargapoltsev S K, Balanovskiy A E, Karlina A I, Shtayger M G, Gozbenko V E, Konstantinova M V, Sivtsov A V 2019 Comparative evaluation of corrosion resistance of wheel and rail steels in various media *IOP Conference Series: Materials Science and Engineering* **560** 012181

[5] Grechneva M V, Balanovskiy A E, Gozbenko V E, Kargapoltsev S K, Karlina A I, Shtayger M G, Karlina Yu I, Govorkov A S 2019 Quality and reliability improvement of "tube-tube plate" welded joints during welding by pulse pressure *IOP Conference Series: Materials Science and Engineering* **560** 012142

[6] Grigoryeva N N, Lazareva V I 2016 Innovation as a tool to improve efficiency and competitiveness *Transport infrastructure of the Siberian region: proceedings of the Ninth International Scientific and Practical Conference. Volume 2* (Irkutsk: IrGUPS Publ.) pp 89-92

[7] Yolkin K S, Yolkin D K, Kolosov A D, Ivanov N A, Shtayger M G 2018 Technologies, which allow to reduce an impact of metal silicon production on the environment IOP Conference Series: Materials Science and Engineering 411 012028

[8] Ivankova L N, Kuznetsova T G, Ivankov A N 2016 Optimization of the ground location of station yards using genetic algorithms. *In the collection: Modern approaches to transport and logistics management. Collection of materials of the All-Russian scientific and practical conference* pp 56-62

[9] Vdovin K N, Feoktistov N A, Sinitskii E V, Gorlenko D A, Durov N A 2015 Production of high-manganese steel in arc furnaces. Part 1 *Steel in Translation* Vol. 45 **10** pp 729-732

[10] Vdovin K N, Gorlenko D A, Feoktistov N A 2016 Characteristics of excess phase in cast high-manganese steel *Steel in Translation* Vol. 46 **7** pp 484-488

[11] Vdovin K N, Feoktistov N A, Gorlenko D A 2016 The effect of the cast high-manganese steel primary structure on its properties *Materials Science Forum* Vol. 870 pp 339-344

[12] Shtayger M G, Balanovskiy A E, Kargapoltsev S K, Gozbenko V E, Karlina A I, Karlina Yu I, Govorkov A S, Kuznetsov B O 2019 Investigation of macro and micro structures of compounds of high-strength rails implemented by contact butt welding using burning-off *IOP Conference Series: Materials Science and Engineering* **560** 012190

[13] Ivankova L N, Darmansky S I 2012 Planning an experiment in the study of the operation of

specialized freight stations *Science and technology of transport* **1** pp 63-67
[14] Balanovskiy A E, Shtayger M G, Karlina A I, Kargapoltsev S K, Gozbenko V E, Karlina Yu I, Govorkov A S, Kuznetsov B O 2019 Surface hardening of structural steel by cathode spot of welding arc *IOP Conference Series: Materials Science and Engineering* **560** 012138
[15] Grigoryeva N N 2018 Evaluation of the feasibility of implementing an innovative project for an automated locomotive traffic safety management system at the Irkutsk locomotive service depot *Materials of the 6th International Symposium on Innovation and Sustainability of the Modern Railway ISMR*
[16] Gozbenko V E, Khomenko A P, Kargapoltsev S K, Minaev N V, Karlina A I 2017 Creating of the alternative lubricants and practice of their use *International Journal of Applied Engineering Research* Vol. 12 **22** pp 12369-12372
[17] Khomenko A P, Gozbenko V E, Kargapoltsev S K, Minaev N V, Karlina A I 2017 Comparative analysis of simulation results and test of the dynamics of the wheelset *International Journal of Applied Engineering Research* Vol. 12 **23** pp 13773-13778
[18] Kargapoltsev S K, Khomenko A P, Gozbenko V E, Minaev N V, Karlina A I 2017 Development of new lubricants for reducing the wear of the elements of the path and running parts of rolling stock *International Journal of Applied Engineering Research* Vol. 12 **22** pp 12362-12368
[19] Sivtsov A V, Sheshukov O Y, Tsymbalist M M, Nekrasov I V, Makhnutin A V, Egiazar'yan D K, Orlov P P 2019 Steel Semiproduct Melting Intensification in Electric Arc Furnaces Using Coordinated Control of Electric and Gas Conditions: II. On-Line Control of the State of the Charge and Melt Zones in Electric Arc Furnaces *Russian Metallurgy (Metally)* **6** pp 565-569
[20] Sivtsov A V, Sheshukov O Y, Tsymbalist M M, Nekrasov I V, Makhnutin A V, Egiazar'yan D K, Orlov P P 2018 Steel Semiproduct Melting Intensification in Electric Arc Furnaces Using Coordinated Control of Electric and Gas Conditions: I. Heat Exchange and Structure of the Electric Arc Furnace Laboratory *Russian Metallurgy (Metally)* **12** pp 1108-1113
[21] Nekrasov I V, Sheshukov O Y, Metelkin A A, Sivtsov A V, Tsymbalist M M 2016 Slag conditions in electrosmelting: A review *Steel in Translation* Vol. 46 **6** pp 435-442
[22] Sivtsov A V, Sheshukov O Y, Tsymbalist M M, Nekrasov I V, Egiazar'yan D K 2015 The Valve Effect of an Electric ARC and Problems in Controlling Electric-ARC Furnaces *Metallurgist* Vol. 59 **5-6** pp 380-385
[23] Nekrasov I V, Sheshukov O Y, Metelkin A A, Sivtsov A V, Tsymbalist M M, Egiazar'yan D K 2015 Ensuring Consistent Foamability in Electric-Furnace Slags *Metallurgist* Vol. 59 **3-4** pp 300-304
[24] Sheshukov O Y, Nekrasov I V, Sivtsov A V, Tsymbalist M M, Egiazar'yan D K, Metelkin A A 2014 Dynamic monitoring of slag oxidation and thickness in the ladle-furnace unit *Steel in Translation* Vol. 44 **1** pp 43-46
[25] Poltava Yu O 2018 Modern survey methods of the road network using comparative analysis *The Bulletin of the Angarsk State Technical University* **12** pp 196-198
[26] Kuzmin O V, Tyurneva T G 1997 The Schroeder numbers, their generalizations and applications. In the collection: Asymptotic and enumerative problems of combinatorial analysis. The collection of scientific papers (Irkutsk: Irkutsk State University) pp 117-125
[27] Lombardi P, Styczynski Z, Sokolnikova T, Suslov K 2014 Use of energy storage in Isolated Micro Grids *Power Systems Computation Conference, PSCC* 7038361
[28] Suslov K V, Solonina N N, Smirnov A S 2011 Smart Grid: A new way of receiving primary information on electric power system state *IEEE PES Innovative Smart Grid Technologies Conference Europe* 6162654
[29] Lombardi P, Sokolnikova T, Suslov K V, Styczynski Z 2012 Optimal storage capacity within an autonomous micro grid with a high penetration of Renewable Energy Sources *IEEE PES Innovative Smart Grid Technologies Conference Europe* 6465706
[30] Kuzmin O V, Orkina K P 2006 Building error-correcting codes using a Pascal-type triangle *The Bulletin of the Buryat State University* **13** pp 32-39

IOP Conf. Series: Materials Science and Engineering **832** (2020) 012039 doi:10.1088/1757-899X/832/1/012039

Evaluation of heuristic procedures formation results to select and justify composition of defectiveness parameters of road pavement affecting reliability indicators

A P Bazhanov

Penza state university of architecture and construction engineering, 28 str. Titova, Penza, 440028, Russia
e-mail bajan_p@mail.ru

Abstract. The research presents effective solution to the problem of evaluation of Heuristic procedures formation results, obtained by the Delphi approach, during the selection and justification of the composition of defectiveness determining parameters of road pavement that affect the indicators of their reliability. The article presents the parameters of the defects of road surfaces of the Penza agglomeration, obtained by diagnostics in order to identify the most important of these parameters in the questionnaire of experts and a single matrix of expert assessments. The author developed the algorithm processing the results of the selection of a subgroup of experts with a close opinion, including the need to conduct expert evaluation of the rank type. According to the variance analysis (presented by the author), a statistical assessment of the significance of differences in the distribution of defectiveness parameters of road pavements affecting some generalized indicator of their reliability, the structure of their influence is determined, while the parameters of the distribution of differences of significant and insignificant differences are assigned to different groups by their degree of influence. The structure of their influence is determined, at the same time parameters that have essential and non-essential difference in the distributions according to the degree of their influence are assigned to different groups. This approach allowed leveling the significance of differences in the assessment of the distribution of defectiveness parameters of road pavements that affect some generalized indicator of their reliability, with the assessment of differences in the average values of the ranks. The comparison of the evaluation of parameters that have an insignificant difference in the average values of ranks assigned to one group, with an estimate for the parameters for the average degree of influence attributed to another group, showing their practical unambiguity. The final structure of groups of defectiveness parameters of road pavements is established, taking into account the priority of their influence on the generalized indicator of reliability of these roads. This heuristic analysis made allowed identifying the most essential parameters affecting the quality of road pavements and their reliability.

1. Introduction

One of the solutions to the task of conducting expert studies [1] on the selection and justification [2] defining defectiveness parameters of road pavements that [3] affect some generalized indicator of their reliability [4], involving heuristic procedures [5], was considered in [6].

The above mentioned works are devoted to a detailed review of the content of the main steps for the heuristic procedure implemented at Delphi's approach: the heuristic preparation procedure, the survey of experts [7], and statistical processing of results of expert survey [8].

Therefore, in this article, the content of the quantitative assessment of the stage of implementation of the heuristic procedure for the formation of its results is considered.

2. Materials and Methods

The task is to identify the most important parameters from the questionnaire of experts (Table 1).

Table 1. Questionnaire of experts on the assessment of defectiveness parameters of road pavements of the Penza agglomeration, conducted by the results of their diagnostics [9]

No	Defectiveness parameters	Description of coating defectivenisses parameters
x_1	Pollution bands at the edges of the pavement	The presence of contamination of the pavement from dirt and debris at the edges
x_2	Destroyed and deformed seams not filled with seal on cement concrete pavement	Damage to seams and pavements near them, presence of debris and inclusions of elements of reinforcement in damaged seams
x_3	The grid of cracks along the bands of rolling	The destruction of the pavement without significant distortion of the profile and accompanied by small longitudinal and transverse cracks of arbitrary shape, characteristic of insufficient pavement strength
x_4	Temperature cracks	Cracks resulting from temperature fluctuations and stress concentrations in the pavement layers
x_5	Cracks	Violation of the integrity of the pavement without removing the material with the formation of narrow slits. Cracks of arbitrary outline and location with a width of opening more than 3 mm on the roadway
x_6	Untreated sections of the roadway with sweating of binder	Surplus of binder on the surface of the pavement with a change in its texture and color area of more than 1 m
x_7	Breaks	Destruction of pavement for its entire thickness with a sharp distortion of the transverse profile, accompanied by a grid of cracks
x_8	Potholes	The destruction of the pavement in the form of depressions of various shapes with sharply defined edges (more than 3 cm deep and an area of more than 200 cm)

To this end, there was the survey of 12 experts about 8 - and most important defectiveness parameters of road pavements, most significantly affecting their reliability indicators, the answers of which are summarized in Table 2.

According to Table 2, we check the equality condition for the sums of all its individual columns $\sum_{j=1}^{m} x_{ij}$ and rows $\sum_{i=1}^{n} x_{ij}$.

$$\sum_{j=1}^{8} \sum_{i=1}^{12} x_{ij} = \sum_{i=1}^{12} \sum_{j=1}^{8} x_{ij} = 432. \tag{1}$$

Satisfaction of the condition (1) indicates the correct completion of the summary table of expert answers (Table 2) [10].

Table 2. Summary table of experts' answers

Experts' numbers	Defectiveness parameters of road pavements, most significantly affecting indicators of their reliability $x_1, x_2, \ldots, x_8$								$\sum_{j=1}^{8} x_{ij}$
	1	*2*	*3*	*4*	*5*	*6*	*7*	*8*	
1	3	4	2	5	1	6	7,5	7,5	36
2	3	5,5	2	4	5,5	7,5	1	7,5	36
3	2	5	1	6	3	4	7,5	7,5	36
4	3,5	5	1	3,5	2	6	7	8	36
5	1	2,5	4,5	4,5	7	2,5	8	6	36
6	3	4	1,5	5	1,5	6	7	8	36
7	7	5	4	1	2,5	2,5	6	8	36
8	5	6	2,5	1	4	2,5	7	8	36
9	3,5	2	1	7,5	5	3,5	6	7,5	36
10	5	8	3	2	4	7	1	6	36
11	2,5	5	1	8	4	2,5	6	7	36
12	2,5	3	1	8,5	2,5	7,5	5	6	36
$\sum_{i=1}^{12} x_{ij}$	41	55	24,5	56	42	57,5	69	87	432

On the values of the sums of the columns $\sum_{j=1}^{8} x_{ij}$ we make a conclusion about the degree of influence of the considered defining parameters, which are distributed in this way:

1. X_3 $\sum_i X_{i3} = 24{,}5$; 3. X_5 $\sum_i X_{i5} = 42$; 5. X_4 $\sum_i X_{i4} = 56$; 7. X_7 $\sum_i X_{i7} = 69$;

2. X_1 $\sum_i X_{i1} = 41$; 4. X_2 $\sum_i X_{i2} = 55$; 6. X_6 $\sum_i X_{i6} = 57{,}5$; 8. X_8 $\sum_i X_{i8} = 87$.

Thus, among the investigated defectiveness parameters of road pavements [11], the most essentially influencing indices of their reliability [12], the greatest influence is exerted by the determining parameter X_8, and the lest – X_3.

We will evaluate the degree of consistency of the interviewed experts using the concordance [13] coefficient according to formula (2), provided that the ranks in the matrix (see Table 2) have equal rank values for different parameters u the same specialist [14].

$$W = \frac{S}{\frac{1}{12} n^2 (m^3 - m) - n \sum_{i=1}^{n} T_i}, \quad (2)$$

$$\text{where } S = \sum_{j=1}^{m} \left[\sum_{i=1}^{n} X_{ij} - \frac{1}{m} \sum_{j=1}^{m} \sum_{i=1}^{n} X_{ij} \right]^2; \quad (3)$$

$$T_i = \frac{1}{12} \sum_{j=1}^{m} (t_j^3 - t_j); \quad (4)$$

t_j - the number of ranks j - th parameter in each row of the rank matrix;

m - number of experts;

n - number of ranks;

X_{ij} - evaluation of j - th rank of i - th expert.

We determine T_i (i =1, 2,…12) by the formula (4) and S by the formula (3). For the first line, rank 8 occurs twice. In this case

$T_1 = \frac{1}{12}(2^3 - 2) = 1/2$.

For the second row, ranks 5 and 6 occur twice. Thus we obtain

$T_2 = \frac{1}{12}(2^3 - 2) + (2^3 - 2) = 1$.

For the third row, rank 7 occurs twice. Then

$T_3 = \frac{1}{12}(2^3 - 2) = 1/2$.

For the rest of the lines, by analogy, we find

$T_4 = T_6 = T_7 = T_8 = T_9 = T_{11} = 0,5$; $T_5 = T_{12} = 1,0$; $T_{10} = 0$.

The value S is found from formula (3)

$$S = (41 - \frac{1}{8}432)^2 + (56 - \frac{1}{8}432)^2 + ... + (87 - \frac{1}{8}432)^2 = 2289,5$$

We determine the concordance coefficient according to formula (2)

$$W = \frac{(41 - \frac{1}{8}432)^2 + (56 - \frac{1}{8}432)^2 + ... + (87 - \frac{1}{8}432)^2}{\frac{1}{12}12^2(8^3 - 8) - 12(0,5 + 1,0 + 0,5 + ... + 1,0)} = 0,43$$

The significance of the difference in the concordance coefficient from zero we test with the help of the Fisher's z- criterion [15] by the formula:

$$Z = \frac{1}{2}\ln\frac{(n-1)W}{(1-W)}, \quad (5)$$

and the number of degrees of freedom γ_1 and γ_2 by formulas

$$\gamma_1 = (m-1) - \frac{2}{n}; \quad (6)$$

$$\gamma_2 = (n-1)\gamma_1. \quad (7)$$

To do this, we find by the formula (5) a numerical value Z :

$$Z = \frac{1}{2}\ln\frac{(12-1)0,43}{(1-0,43)} = 0,87$$

and degrees of freedom γ_1 and γ_2 according to formulas (6) and (7)

$\gamma_1 = (8-1) - \frac{2}{12} \approx 7$ и $\gamma_2 = (12-1)7 \approx 75$.

At the significance level $\alpha = 0,05$ for $\gamma_1 = 7$ and $\gamma_2 = 75$ from Table. VI of work [15] we find that $Z_\alpha = 0,35$.

Since the obtained Z=0,87 is greater than $Z_{\alpha} = 0,35$, then with probability $P \geq 0,95$ then with probability it can be asserted that there is a non-random agreement on the strength of the influence on the quality and indicators reliability covers of the of motor road [16], selected indicators [17] among experts.

An evaluation of the difference in the influence of the investigated indicators and the significance of the influence in the analysis of the investigated indicators is carried out by comparing the variances between the S_1^2, with the residual dispersion S_{res}^2, which are determined by formulas [8].

$$S_1^2 = \frac{1}{(m-1)} \sum_{i=1}^{n} \sum_{j-1}^{m} (\overline{X}_j - \overline{X}_i)^2 = \frac{1}{(m-1)} \left[n \sum_{j=1}^{m} (\overline{X}_j)^2 - mn(\overline{X}_{ij})^2 \right] \tag{8}$$

$$S_{res}^2 = \frac{1}{(m-1)(n-1)} \sum_{i=1}^{n} \sum_{j-1}^{m} (X_{ij} - \overline{X}_i - \overline{X}_j + \overline{X}_{ij})^2 = \frac{1}{(m-1)(n-1)} \left[\sum_{i=1}^{n} \sum_{j=1}^{m} (X_{ij})^2 - m \sum_{i=1}^{n} (\overline{X}_i)^2 - n \sum_{j=1}^{m} (\overline{X}_j)^2 + mn(\overline{X}_{ij})^2 \right]. \tag{9}$$

where X_{ij} - evaluation of j - th parameter i - th expert;

m - number of parameter;

n - number of experts.

Using formulas (8) and (9), we calculate $S_1^2 = 26,1$ and $S_{res}^2 = 3,1$. The average values of ranks according to the defining parameters are $\overline{X}_1$= 3,42, $\overline{X}_2$= 4,58, $\overline{X}_3$= 2,04, $\overline{X}_4$= 4,67, $\overline{X}_5$= 3,5, $\overline{X}_6$= 4,79, $\overline{X}_7$= 5,75, $\overline{X}_8$= 7,25.

The average values, by definition of the experts, are:

$\overline{X}_1$= 4,5, $\overline{X}_5$= 4,5, $\overline{X}_9$= 4,5
$\overline{X}_2$= 4,5, $\overline{X}_6$= 4,5, $\overline{X}_{10}$= 4,5
$\overline{X}_3$= 4,5, $\overline{X}_7$= 4,5, $\overline{X}_{11}$= 4,5
$\overline{X}_4$= 4,5, $\overline{X}_8$= 4,5, $\overline{X}_{12}$= 4,5.

The general average rank by definition of experts $\overline{X}_{ij}$ =4,5.

The significance of the difference between variances S_1^2 and S_{res}^2 is verified by the Fisher's z-criterion using formula (10).

$$Z = \frac{1}{2} \ln \frac{S_1^2}{S_{res}^2}. \tag{10}$$

As a result, we obtain

$$Z = \frac{1}{2} \ln \frac{26,1}{3,1} = 1,06.$$

According to the number of degrees of freedom (which were obtained earlier) $\gamma_1 = 7$ and $\gamma_2 = 75$ at $\alpha = 0,05$ according to Table 5, we find $Z_{\alpha} = 0,35$.

Since $Z > Z_{\alpha}$, then with a probability of $P \geq 0,95$ it can be argued that the difference in the influence of the parameters' under study is statistically significant [18], and their effect on the quality and indicators reliability of road covers of the of motor road is most significant [19].

Define the structure of the influence of defectiveness parameters of road pavements, which most significantly affect their reliability indicators. The groups of indicators can be found by evaluating the difference in the average values of the ranks between the parameters.

Let us consider an evaluation of a difference in average values of ranks between parameters [20]. The ranks for each of the parameters form random samples [21], each of which has its own average $\overline{X}_j$ and the mean square deviation S_j, defined by the formula

$$S_j = \sqrt{\frac{1}{n-1}\sum_{i=1}^{n}(X_{ij}-\overline{X}_j)^2}, (j=1, 2,..., m); (i=1, 2,..., n) \quad (11)$$

We determine the mean square deviations of the parameters bu formula (11):

S_1=1,59, S_3=1,23, S_5=1,76 S_7=2,37

S_2=1,64, S_4=2,55, S_6=2,10, , S_8=0,81.

The experts gave the highest average rating to parameter X_8 with $\overline{X}_8$=7,25 and S_{x_8} =0,81. Further, the value of X_7 with $\overline{X}_7$=5,75 and S_{x_7} =2,37 follows.

The significance of the difference in the mean values of the ranks between the parameters is checked by the Student's t-criterion [22] by a successive comparison of the average values of the ranks of the parameters:

$$t_{\overline{X}k-\overline{X}l} = (\overline{X}_r - \overline{X}_l)/S_{\overline{X}k-\overline{X}l} \quad (k, l=1, 2,...,m;\ k \neq l) \quad (12)$$

$$S_{\overline{X}k-\overline{X}l} = (S^2_{\overline{X}k} + S^2_{\overline{X}l})^{1/2} \quad (13)$$

$$S_{\overline{X}j} = \frac{1}{\sqrt{n}} S_j \quad (j=1, 2,..., m). \quad (14)$$

In this case, the number of degrees of freedom $\gamma = 2(n-1)$. (15)

If $t_{\overline{X}k-\overline{X}l} > t_\alpha$, taken from tables (see, for example, [22], for t_α =0,05, the t_α values are presented in Table 4.6b of [22], then with the probability $P \geq 1-\alpha$ it can be asserted that the difference in mean values is insignificant and these parameters about the average degree of influence are not different and they can be attributed to one group.

Using formulas (13) and (14), we find

$$S_{\overline{X}8-\overline{X}7} = \sqrt{\frac{0,81^2+2,37^2}{12}} = 0,72$$

$$S_{\overline{X}j} = \frac{1}{\sqrt{n}} S_j \quad (j=1, 2,..., m).$$

Using formulas (12) and (15), we respectively find the value of $t_{\overline{X}8-\overline{X}7}$ and the number of degrees of freedom γ:

$$t_{\overline{X}8-\overline{X}7} = \frac{7,25-5,75}{0,72} = 2,08;$$

$$\gamma = 2(12-1) = 22.$$

According to table 4.6b of work [22] for α = 0,05 and γ=22 we find t_α =1,72.

Since $t_{\overline{X8}-\overline{X7}}$ =2,08> t_{α} =1,72, then with probability $P \geq 0{,}95$ it can be argued that the difference in the means is statistically significant, therefore the parameters x_8 and x_7 can be attributed to different groups by the average degree of influence.

Similarly, we calculate the following values:

$$t_{\overline{X7}-\overline{X6}} = \frac{5{,}75-4{,}79}{0{,}91} = 1{,}05,\ t_{\overline{X6}-\overline{X5}} = \frac{4{,}79-3{,}5}{0{,}79} = 1{,}63,\ t_{\overline{X2}-\overline{X1}} = 1{,}76,$$

$$t_{\overline{X6}-\overline{X4}} = 0{,}13,\ t_{\overline{X5}-\overline{X1}} = 0{,}12,$$

$$t_{\overline{X6}-\overline{X2}} = 0{,}27,\ t_{\overline{X5}-\overline{X3}} = 2{,}35.$$

3. Outcomes

From the comparison of the calculated values of $t_{\overline{Xk}-\overline{Xl}}$ with $t_{\alpha=0{,}05}$ =1,72, we can distinguish the following groups: X_6, X_4, X_2, X_1; X_5, X_1; X_3.

As a result we have obtained that the parameter X_1 belongs to two groups.

However, comparing the $t_{\overline{X5}-\overline{X1}}$ and $t_{\overline{X2}-\overline{X1}}$ values, it is more preferable to relate a parameter X_1 to a parameter X_5.

The final structure of the influence of the defining parameter will be as follows: X_8 - the first group; X_7, X_6, X_4, X_2 - the second group; X_5, X_1 - the third group; X_3 - the fourth group.

According to the statistical analysis carried out on the degrees of influence, the determining parameter can be arranged as follows: X_3; X_5; X_1; X_2; X_4; X_6; X_7; X_8.

4. Conclusion

In general, the heuristic analysis made it possible to identify the most significant parameters X_4, X_6, X_7, X_8, affecting the quality of road pavements and their reliability indicators.

References

[1] Kurakina E V 2013 Expert characteristics of the road in road transport expertise *Modern problems of science and education* **5**.

[2] Hayati E, Majnounian B 2013 An expert-based approach to forest road network planning by combining Delphi and spatial multi-criteria evaluation *Environ Monit Assess* **185**(2)1767-76. doi: 10.1007/s10661-012-2666-1.

[3] Kokodeeva N E 2009 Evaluation of the quality of existing non-rigid-type road clothes taking into account the variation in humidity *Dorozhnaya Sfera* **19** 72-75.

[4] Stolyarov V V 1994 The design of highways taking into account the theory of risk (Saratov: Sarat. state. tech. University), part 2.

[5] Kıral I A, Kural Z, Çomu S 2014 Risk identification in construction projects: using the Delphi method. In *Proc. of 11th International Congress on Advances in Civil Engineering* (Istanbul Technical University).

[6] Outhred G P 1981 The Delphi method: a demonstration of its use for specific research types. In *Proc. of Construction & Building Research Conference* (Melbourne: RICS Foundation).

[7] Bazhanov A P, Saxonova E S 2018 The method of carrying out heuristic procedures in the selection and justification of the composition of the determining indicators of the quality of pavements that affect the reliability of roads. Preparation of a heuristic procedure and a survey of experts *Transport engineering* **5**(1) 18.

[8] Bazhanov A P, Saksonova E S, Kochetkov A V, Schegoleva N V 2018 The method of carrying out heuristic procedures in the selection and justification of the composition of the determining indicators of the quality of pavements that affect the reliability of roads. Statistical processing and formation of the results of an expert survey *Transport engineering* **5**(2) 14.

[9] *Methodological recommendations for the study of socio-psychological climate in teams* 2011 (Moscow: Center for emergency psychological assistance MZS of Russia).

[10] Yousuf M I 2007 Opinions through Delphi technique *Practical Assessment, Research & Evaluation University of Arid Agriculture* **12**(4).

[11] Kochetkov A V, Gladkov V Yu, Nemchinov D M 2013 Designing the structure of information support for the quality management system of road economy *Internet Journal of Science* **16**(3).

[12] Bliznichenko S S 1992 Simplified methodology for rapid assessment of the transport and operational state of the road network *Izvestiya Vuzov. Construction* **5-6** 126-130.

[13] Borodachev S M 2009 *Multivariate statistical episodes* (Ekaterinburg: USTU-UPI).

[14] Sidenko V M, Rokas S Y 1981 *Quality management in road construction* (Moscow: Transport).

[15] Fisher R A 1958 *Statistical methods for researchers* (Moscow: Gosstatizdat).

[16] *Online collection of technical regulations and policies* 2011 Use of the peer review method in the deployment of the quality function. Available at: https://studfiles.net/preview/1967369/page:5.

[17] *Online collection of technical regulations and policies* 2011 Industrial products and indicators of its quality. Available at: http://texnlit.ru/kachestvo/5.html/.

[18] Kobzar A I 2006 *Applied mathematical statistics* (Moscow: Fizmatlit).

[19] Jorge D, Ferreira A 2011 Road network pavement maintenance optimization using the HDM-4 pavement performance prediction models *International Journal of Pavement Engineering* **13**(1) 39-51.

[20] Emma K 2014 Matching GPS data to a road network In *Mathematical modeling and solution algorithms* (Gothenburg: Chalmers University of Technology).

[21] Arzhanukhina S P, Sukhov A A, Kochetkov A V, Yankovsky L V 2012 Organizational and economic mechanism of innovative activity of road economy *Innovative Regional Herald* **4** 40 - 45.

[22] Bolshev L N, Smirnov I V 1983 *Tables of mathematical statistics* (Moscow: Nauka).

IOP Conf. Series: Materials Science and Engineering **832** (2020) 012040 doi:10.1088/1757-899X/832/1/012040

Improvement of research methodology and reliability assessment of roads at design project and project documentation development stages

A P Bazhanov[1,2], E S Saksonova[2,3]

[1] Penza state university of architecture and construction engineering, 28 str. Titova, Penza, 440028, Russia
[2] E-mail: bajan_p@mail.ru
[3] E-mail: saksonovalena@yandex.ru

Abstract. The article touches upon the issues of the improvement of research methodology and reliability assessment of roads at engineering design and project documentation development stages. As a result of the study, the analytical expressions are obtained in order to improve the research methodology and assess the reliability of roads at the stages of the implementation of engineering project and the development of project documentation, using correcting coefficients, taking into account random errors in the numerical values of constituent quantities in the analytical expressions, which are functions of independent random arguments that determine the impact on these components of the magnitude of random weather and climate factors and conditions of vehicle movement during the operation of roads, as well as some corrections for non-linearity and the probabilistic dependence of these components in these analytical expressions. These analytical relations also take into account the obtained design estimates of the reliability of roads using statistical modeling procedures for calculating the reliability indicators of a specific category of road in a one-dimensional setting. In this work, the authors propose the use of a new probabilistic - statistical approach to the assessment of the reliability of roads, taking into account the risk of destruction of road structure during its usage. The results of the study allow concluding that the main provisions for the improvement of the research methodology and assessment of the reliability of roads at the stages of the implementation of their engineering design and the development of project documentation make a significant contribution. These provisions will contribute to a significant adjustment of design indicators of roads in the direction of their significant improvement.

1. Introduction

One of the popular solutions to the problem of research and reliability assessment of roads [1] at the stage of development of their pre-project documentation [2] was considered in work [3].

In the above mentioned work [3], the questions of substantiating a three-stage obtaining of generalized statistically uniform information on the same type of automobile roads were solved based on fairly rigorous methods [4] and models of the theory of pattern recognition [5], similarity and dimension theories [6], and a new predictive method of state analysis roads on the risk of their destruction and the actual service life to further using of the information received to study and assess the probability of failure of new road structures [7].

The following article is devoted to the improvement of the methodology for quantitative research and assessing the reliability of new modifications of roads at the stages of development of their engineering project [8] and working documentation [9].

2. Materials and Methods

As it was already mentioned, the methodology for studying and assessing the reliability of roads at the stage of development of pre-project design documentation [3] provides for a three-stage approach for obtaining a priori and forecast generalized statistically homogeneous information on analog roads and its subsequent use to calculate the reliability indicators of newly developed roads of the same type categories [3].

Obtained at the two initial stages of the methodological a priori generalized statistically homogeneous information on analog roads in the process of research and assessment of the reliability of roads, including a parametric description of their determining parameters $x_1, x_2, ..., x_N$ (for example, x_1 - elastic modulus, x_2 - strength, x_3 - flatness, x_4 - wear, etc.; N - number of determining parameters) the same types of analog roads, is the source for calculating the values included in the ratio and (or) the corresponding regression dependencies of the form [8]:

$$x_j = f\left(z_{1j}, z_{2j}, \ldots, z_{MN}\right) \tag{1}$$

where x_j is the values of specific determining parameters that affect the reliability indicators of roads;

z_{ij} - transport - operational characteristics [10] of structural elements (structural layers) of roads [11] ($i = \overline{1,M}$; M - number of transport - operational characteristics of road construction elements; $j = \overline{1,N}$; N - number of determining parameters).

Parameters z_{ij}, as a rule, are random variables subject to constant exposure to weather and climatic factors and traffic conditions of vehicles during the operation of roads (changing in pavement temperature, soil moisture of the road structure, characteristics of the heterogeneity of the structure of road materials, etc.).

Therefore, the defining parameters of the elements of the road structure x_j ($j = \overline{1,N}; N$ - the number of defining parameters) should also be considered as random variables that change randomly under the influence of random changes in the arguments $z_1, z_2, \ldots, z_N$, which are characterized by random errors. In the entire range of variation of random arguments included in relation (1), the determining parameters x_j in the general case change nonlinearly.

However, in a narrow range of random changes in the arguments z_{ij} ($i = \overline{1,M}; j = \overline{1,N}; M$ - the number of transport and operational characteristics of road construction elements that statistically significantly affect the j -th determining parameter of roads), the values of the determining parameters of roads x_j can be considered almost linear. Replacing the nonlinear function with a linear one, one can apply the apparatus of numerical characteristics to the last function, which was developed for linear numerical functions [12].

Knowing the numerical parameters z_{ij}, one can find the numerical indicators of the determining parameters x_j included in the relation of the form (1). In this case, an approximate solution to the final problem is obtained, which in most cases of practical design estimates of the reliability of roads is acceptable.

According to the above-mentioned estimates of mathematical expectations and variances of the determining parameters x_j (the index "j" is omitted), it can be calculated by the formulas [12]:

$$m_x = \varphi\left(z_1, z_2, \ldots, z_N\right) + \Delta_1 + \Delta_2 \tag{2}$$

$$\sigma_x^2 = \sum_{j=1}^{N} \left(\frac{\partial \varphi}{\partial z_j} \right)_{\bar{z}_j} \sigma_{z_j}^2 + \Delta_3 + \Delta_4 \tag{3}$$

In relations (2) and (3), the quantities $\Delta_1, ..., \Delta_4$ are corrections for the nonlinearity and the probabilistic dependence of the parameters $z_1, z_2, ..., z_N$ The values of these corrections can be calculated using the following formulas [12]:

$$\Delta_1 = \frac{1}{2} \sum_{j=1}^{N} \left(\frac{\partial^2 \varphi}{\partial z_j^2} \right)_{\bar{z}_j} \sigma_{z_j}^2 , \Delta_2 = \sum_{i<j} \left(\frac{\partial^2 z}{\partial z_i \partial z_j} \right)_{\bar{z}_i \bar{z}_j} \sigma_i \sigma_j \eta_{ij} \tag{4}$$

$$\Delta_3 = \frac{1}{2} \sum_{j=1}^{N} \left(\frac{\partial^2 \varphi}{\partial z_j^2} \right)_{\bar{z}_j} ; \sigma_{z_j}^4 + \sum_{i<j} \left(\frac{\partial^2 \varphi}{\partial z_i \partial z_j} \right)_{\bar{z}_i \bar{z}_j} \sigma_i^2 \sigma_j^2 \tag{5}$$

In formulas (4) - (6), instead of $\underset{i<j}{z}$, one can also use the notation $\sum_{i=1}^{N-1} \sum_{j=i+1}^{N}$, where index "$i < j$" means all possible pair combinations of parameters $z_1, z_2, ..., z_N$.

The determining parameter x_1 (for example, the measured modulus of elasticity of the road), which is a function of the constituent parameters $z_1, z_2, ..., z_N$, maybe in a probability dependence on another determining parameter x_2 (for example, the amount of wear of the road surface).

Let

$$x_1 = \varphi\left(z_1, z_2, \ldots z_{N_1}, \xi_1, \xi_2, \ldots, \xi_k \right) \tag{7}$$

$$x_2 = \varphi\left(z_1, z_2, \ldots, z_{N_2}, v_1, v_2, \ldots, v_l \right) \tag{8}$$

where $N_1 \le N_2$ (while N_1 arguments are common); k and l are the numbers of arguments of the form ξ_i and v_i, moreover, ξ_i and v_i are independent parameters. Then the pair correlation coefficient between the determining parameters of a particular type of road x_1 and x_2 is equal to

$$r_{x_1, x_2} = \frac{1}{\sigma_{x_1} \sigma_{x_2}} \left(a_{11} a_{21} \sigma_1^2 + a_{21} a_{22} \sigma_2^2 + \ldots + a_{1N} a_{2N} \sigma_N^2 \right) \tag{9}$$

The values $m_x, \sigma_x, r_{x_1, x_2}$ thus obtained are used at the stage of the engineering project in the calculated relationships (10) (11) for a point estimate of P probability P of the failure-free functioning of the road structure [13] and its lower confidence limit $\underline{P}_\gamma$ [14]:

$$\begin{aligned} P &= \prod_{j=1}^{N} Pj + (Pj_{\min} - \prod_{j=1}^{N} Pj) K_N = \\ &= \prod_{j=1}^{N} F(h_j) + \left[F(h_j)_{\min} - \prod_{j=1}^{N} F(h_j) \right] K_N, \end{aligned} \tag{10}$$

where $F\left(h_j\right)_{\min}$ is the minimum estimate $F\left(h_j\right)$;

$$K_N = \frac{2}{\pi C} \sum_{i \langle j} \arcsin r_{ij} ;$$

where r_{ij} is an estimate of the cross-correlation coefficient of the determining parameters $x_{ij}^{\lim}$ and x_{ij}^{n} (between the parameters x_j);

The index "$I<j$" under the sign of the sum means the number of all possible pair-wise combinations of quantities x_j with the number of such combinations $C = N(N-1)/2$ [14];

$$\underline{P}_{\gamma} = P\left[1-\sqrt{\left(1-\frac{\underline{P}_{j_{\min}}}{P_{j_{\min}}}\right)^2 + \left(1-K_N^2\right)\left(\prod_{j=1}^{N}\frac{P_j}{P}\right)^2 \sum_{j=2}^{N}\left(1-\frac{\underline{P}_j}{P_j}\right)^2}\right] \qquad (11)$$

or

$$\underline{P}_{\gamma} = P\left[1-\sqrt{\left(1-\frac{\underline{P}_{j_{\min}}}{P_{j_{\min}}}\right)^2 + \delta}\right], \qquad (12)$$

Where δ is the correction factor, which is defined as

$$\delta = \left(1-K_N^2\right)\left(\prod_{j=1}^{N}\frac{P_j}{P}\right)^2 \sum_{j=2}^{N}\left(1-\frac{\underline{P}_j}{P_j}\right)^2 .$$

The initially calculated relationships of the form (1), linking the value of the determining parameters x_j with the transport and operational characteristics z_{ij} in a number of cases has rather cumbersome expressions. This is especially true in cases where corrections of the form (4) - (6) are taken into account in the calculated ratios or the parameter values are determined from the corresponding systems of differential equations. This is mainly characteristic of the stage of development of design documentation of roads. At this stage, various verification and refinement calculations are usually carried out taking into account a priori data already available for this moment on new modifications of roads. In such cases, for updated design estimates of the reliability of the designed highway, it is advisable to apply statistical modeling methods. The essence of all methods of statistical modeling is reduced to multiple simulations of processes, as well as values of transport and operational characteristics z_{ij} and determining parameters x_j ($j=\overline{1,N}; N$ - the number of determining parameters for each specific element (structural layer) of the road structure) that underlie quantitative methods for assessing the reliability of roads.

The application of the statistical modeling method to the calculation of the reliability indicators of a particular type of road in a one-dimensional setting (i.e., when the failure model of the road structure as a whole or its individual critical elements is described by one determining parameter x, while one control point, one dangerous section is considered "limiting" element, one calculated case, etc.) is reduced to the multiple studies of the function of the health of the species.

$$\Delta x = x^{\lim}\left(\xi_1,\xi_2,\ldots,\xi_k\right) - x^{n}\left(v_1,v_2,\ldots,v_l\right) \geq 0 \qquad (13)$$

where $x^{\lim}\left(\xi_1,\xi_2,\ldots,\xi_k\right)$ - the limit value of the determining parameter x_j, which is a function of parameters $\xi_1,\xi_2,\ldots,\xi_k$ (k - the number of parameters that statistically significantly affect function $x^{\lim}$);

$x^{n}\left(v_1,v_2,\ldots,v_l\right)$ - normative (working) value of the determining parameter x_j, which is a function of parameters $v_1,v_2,\ldots,v_l$ (l - the number of parameters affecting function x^{n}).

The methodology for calculating the reliability indicators of roads can be based on two fundamentally different methodological approaches.

In the first approach, the values of function $\Delta x = x^{\lim} - x^{n}$ of random arguments $\xi_1, \xi_2, ..., \xi_k$, $v_1, v_2, ..., v_l$ are modeled, having the corresponding distribution laws (distribution density) $\varphi(\xi_1), \varphi(\xi_2), ..., \varphi(\xi_k), \varphi(v_1), \varphi(v_2), ..., \varphi(v_l)$.

Then the method provides for the calculation of the number m of values Δx in relation (13), which are less than or equal to zero, which corresponds to the failure model of the considered element of the road structure. Moreover, assessment $\hat{P}$ of the probability of failure of the road structure is found as

$$\hat{P} = 1 - m/n, \tag{14}$$

where m is the number of implementations of the value Δx with $\Delta x \leq 0$; n is the total number of implementations.

The lower confidence limit $\underline{P}_\gamma$ (often set as an indicator of reliability in the design documentation) of the probability of uptime according to the total amount of implementations n, the number of implementations m ($\Delta x \leq 0$) and confidence probability 1 [15] can be calculated using the well-known relation [16]:

$$\sum_{m=0}^{k} C_n^m \underline{P}_\gamma^m \left(1 - \underline{P}_\gamma\right)^{m-n} = 1 - \gamma, \quad m \leq k \leq n, \tag{15}$$

where C_n^m is the number of possible permutations of *m*, the values of the function Δ_x (with $\Delta_x \leq 0$) with the total number of implementations *n*; $\underline{P}_\gamma$ - lower confidence limit of rating $\hat{P}$ with confidence level γ.

Solving equation (15) for some fixed values of $\underline{P}_\gamma$, γ and *m*, we can obtain the volumes of the required number of implementations to confirm the values required for the design documentation of the probability of failure of the road structure. The calculation results are presented in tables 1 and 2.

The values of required implementations

$$n = n\left(\underline{P}_\gamma, \gamma, m\right).$$

Table 1. Probabilities of failures.

m	$\underline{P}_\gamma$	γ 0,90	0,95	0,99	m	$\underline{P}_\gamma$	γ 0,90	0,95	0,99
0	0,90	25	30	45	0	0,90	65	75	100
1	0,95	75	95	130	1	0,95	185	210	265
2	0,99	520	620	840	2	0,99	1540	1700	1700

* Note: in the table 1 numbers, rounded to the nearest digit "5"

Table 2. Values of required implementations.

$\underline{P}_\gamma$	$0{,}9_{(2)}$	$0{,}9_{(3)}$	$0{,}9_{(4)}$	$0{,}9_{(5)}$	$0{,}9_{(6)}$	$0{,}9_{(7)}$
n	$3{\cdot}10^2$	$3{\cdot}10^3$	$3{\cdot}10^4$	$3{\cdot}10^5$	$3{\cdot}10^6$	$3{\cdot}10^7$

* Note: Here $0{,}9_{(2)} = 0{,}99$; $0{,}9_{(3)} = 0{,}999$; $0{,}9_{(4)} = 0{,}9999$; $0{,}9_{(5)} = 0{,}99999$ etc.

From the analysis of tables 1 and 2, it follows that with an increase in the number m (with $\Delta x \leq 0$), the volumes of the required number of implementations n sharply increase. And to confirm sufficiently high values of probabilities P_γ, which is typical for the levels of reliability of the road ($P_\gamma \geq 0{,}9999$) even with $m = 0$, a very large number of implementations of n are necessary. It is known that with a large number of implementations (statistical modeling of the estimation values $\hat{P}$), systematic errors can accumulate in the formation of random variables. This circumstance leads to the need for duplicate (parallel) calculations in order to eliminate such errors. And with rather cumbersome initial calculated dependencies of the form (13), unreasonably large expenditures of machine time may be required.

In this regard, in many practical situations, a different formulation and solution of the problem associated with obtaining design estimates of the reliability indicators of the road structure by statistical modeling are more preferable. The methodology for solving this problem is as follows.

According to the initial analytical or regression dependencies of the species

$$x^{\lim} = x^{\lim}\left(\xi_1, \xi_2, \ldots, \xi_k\right) \tag{16}$$

$$x^{n} = x^{n}\left(v_1, v_2, \ldots, v_l\right) \tag{17}$$

as well as the well-known laws of the distribution of parameters $\xi_1, \xi_2, \ldots, \xi_k$, $v_1, v_2, \ldots, v_l$ a statistical sample of random variables $\eta_i = x_i^{\lim} / x_i^{n}$ is modeled ($i = \overline{1, n}; n$ is the number of realizations of the order of several hundred). Then, a suitable distribution density function $\varphi(\eta)$ is selected for the obtained statistical sample. Then the probability of failure-free operation of a particular element of the road structure or its important "limit" element can be calculated as [14]

$$P = (2\pi)^{-1/2} S_\eta^{-1} \int_1^\infty \exp\left(-\frac{(\eta_i - \bar{\eta})^2}{2S_\eta^2}\right) d\eta = F\left(\frac{(\bar{\eta}-1)}{S_\eta}\right) = F(\eta) \tag{18}$$

Under the normal distribution law of parameter η, an estimate $\hat{P}$ of probability P can be calculated as

$$\hat{P}(\eta > 1) = \int_1^\infty \varphi(\eta) d(\eta) \tag{19}$$

where $F\left(\hat{h}\right)$ is the standardized function of the normal distribution of parameter $\hat{h} = \frac{(\bar{\eta} - 1)}{S_\eta}$, tabulated in [15]. Moreover, in work [15] function $F\left(\hat{h}\right)$ is tabulated to $\hat{h} = 4{,}0$, and in work [17] – to $\hat{h} = 6{,}0$.

The lower confidence limit of the point estimate (19) can be calculated by the formula [18]

$$\underline{\hat{P}}_\gamma = F\left(\hat{h} - \frac{h_\gamma}{\sqrt{n}} \sqrt{1 + \hat{h}^2 / 2}\right) \tag{20}$$

Here h_γ is the quantile of the normal distribution, determined by the value of the confidence probability γ ($h_\gamma = 1{,}645$ for $P = 0.95$) using statistical tables [16].

In formulas (19), (20), estimates of the mean values and mean square deviations of parameter η are defined as

$\bar{\eta}=\frac{1}{n}\sum_{i=1}^{n}\eta_i, S_\eta=\left[\frac{1}{n-1}\sum_{i=1}^{n}\left(\eta_i-\bar{\eta}\right)^2\right]^{1/2}, \quad \eta_i = x_i^{\lim}/x_i^{n}$ where n is the number of realizations by which values $\bar{\eta}$ and S_η are calculated by the method of statistical modeling according to the initial relation (13).

In order to solve this problem in a multidimensional setting, i.e. for a multidimensional model of type

$$\left.\begin{array}{c} P\left(x_1^{\lim}-x_1^{n}=u_1\rangle 0\right) \\ P\left(x_2^{\lim}-x_2^{n}=u_2\rangle 0\right) \\ \dots\dots\dots\dots\dots\dots \\ P\left(x_N^{\lim}-x_N^{n}=u_N\rangle 0\right) \end{array}\right\}$$

the principle side of the question remains the same.

At the same time, M failure-free condition is simulated at the same time and the probability of failure-free operation of the road and their lower confidence limits are determined by the relations (10) - (12).

3. Results and Discussion

The above analytical expressions for improving the research methodology and assessing the reliability of roads at the stages of their engineering project and developing project documentation are based on two initial methodological steps for obtaining a priori generalized statistically homogeneous information on analog roads in the process of researching and evaluating their reliability [3].

These analytical expressions were obtained taking into account the introduction of correction factors Δ_i that take into account the presence in the analytical expressions of random errors of numerical values of component quantities z_{ij}, which are functions of independent random arguments ξ_i, v_i, which determine the influence of random weather and climate factors on these components (change in coating temperature , soil moisture of the road structure, characteristics of the heterogeneity of the structure of road materials, etc.) and the conditions of movement of vehicles during the operation of roads, as well as the presence in these analytical expressions of some corrections for non-linearity and the probabilistic dependence of these components.

In addition, these analytical relationships are associated with obtaining design estimates of the reliability of roads using statistical modeling procedures for calculating indicators of the reliability of a specific category of road in a one-dimensional formulation(i.e., when the failure model of the road structure as a whole or its individual critical elements is described by one determining parameter x, in this case one control point, one dangerous section of the "limit" element, one design case are considered, etc.).

Improving the methodology for studying and assessing the reliability of roads at the stages of completing their engineering project and developing project documentation at the third methodological stage of obtaining predictive information on analog roads in the process of research and evaluation of their reliability includes the use of risk theory formulas at this stage and applied in road construction statistical quality control [3].

Obtaining predictive information taking into account the influence of random arguments $\xi_1,\xi_2,\dots,\xi_k, v_1,v_2,\dots,v_l$ (fluctuations in the moisture content of the sub-grade, the quality of pavement and the temperature of the coating over time) on the strength characteristics of road pavement, which is further used to study and assess the likelihood of a failure of a given road structure, is carried out according to the results of a technical analysis of monitoring and predicting the rate (risk) of destruction of the road structure and its actual service life.

The analysis of monitoring and forecasting the rate (risk) of the destruction of the road structure and its actual life was carried out on the basis of obtaining predictive information on the study of the strength characteristics of pavement of capital and lightweight types with improved coating. It was also established:

- the effect of fluctuations in the moisture content of the soil of the sub-grade W_t on the parameters of the actual life of the road t_a and the risk of destruction of the road structure r (fig. 1);
- the impact of the quality of pavement on parameters t_a and r (fig.2);
- the effect of coating temperature over time (fig.3).

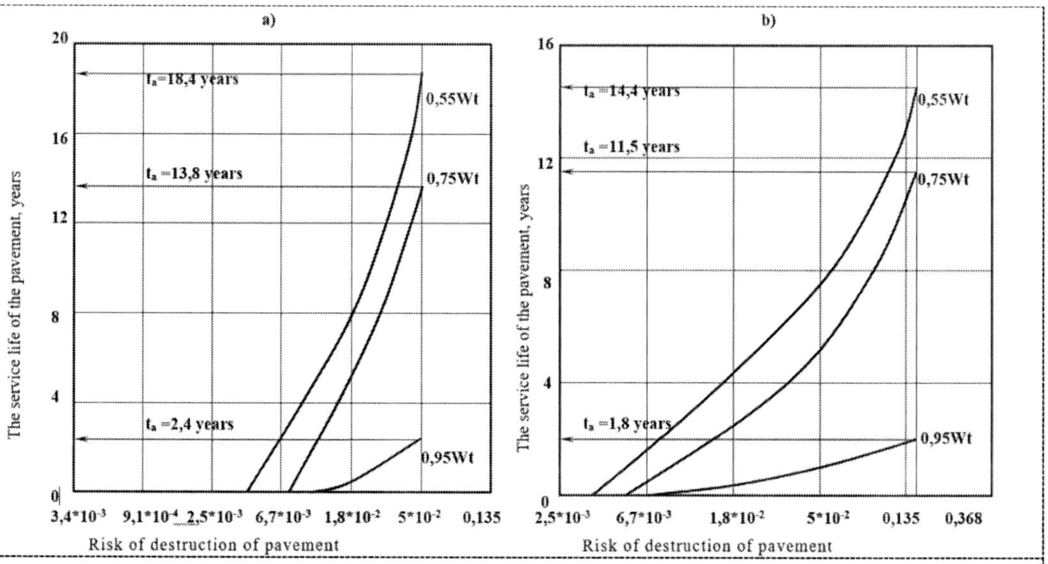

Figure 1. The effect of fluctuations in soil moisture in the sub-grade on the service life and the risk of destruction of pavement: a) capital type with improved coating; b) lightweight type with improved coating.

A comparative analysis showed that:

- with an increase in the soil moisture of the sub-grade (at the same value of the coefficient of variation, reduced to the calculated temperature of the equivalent modulus of elasticity of the pavement C_V^{eq}), the actual service life of the pavement is reduced, and the likelihood of their destruction increases (see fig. 1);
- with a decrease in the coefficient of variation of the equivalent modulus of elasticity (at the same soil moisture), the actual service life of the pavement increases, and the probability of their destruction, on the contrary, decreases (see fig. 2);
- the higher the capitality of the pavement, the longer their service life (see fig. 1 a, b) and the less intense their rate of destruction (see fig. 2 a, b).

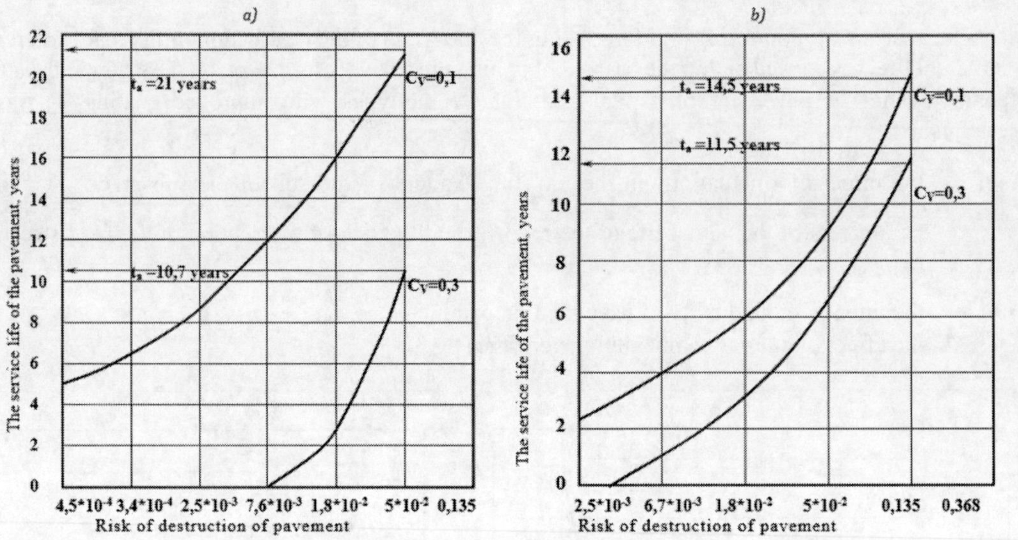

Figure 2. The impact of pavement quality on durability and risk of destruction: a) pavement of capital type with improved coating; b) lightweight pavement with improved coating.

4. Conclusion

As a result of the study, analytical expressions were obtained to improve the research methodology and assess the reliability of roads at the stages of the implementation of their engineering project and the development of project documentation, which are based on two the initial methodological stages of obtaining a priori generalized statistically uniform information on analog roads in the process of research and evaluation of their reliability at the stage of development of pre-project documentation. A key feature of the obtained analytical expressions is the presence of correction factors in them that take into account random errors in the numerical values of parameters z_{ij} related to the impact on these parameters of constant random weather and climate factors (change in coating temperature, soil moisture of the road structure, characteristics of the heterogeneity of the structure of road materials, etc.) and the conditions of movement of vehicles during the operation of roads, as well as the presence in these analytical expressions of some corrections for nonlinearity and the probability dependence of these parameters.

Improving the methodology for research and evaluating the reliability of roads at the stages of completing their engineering project and developing project documentation at the third methodological stage of obtaining predictive information on analog roads in the process of studying and evaluating their reliability consists in using risk theory formulas at these stages and the applied in road construction statistical quality control.

The graphical dependences obtained above of the influence of fluctuations in the moisture content of the soil of the subgrade on the service life and the risk of destruction of pavement, the influence of the quality of pavement on the service life and risk of destruction, as well as the temperature of the coating can be used as predictive calculation information in a quantitative study of the reliability indicators of new modifications of roads and their structural elements at the stages of designing roads, when a large number of different options for calculating structural solutions using computer-aided design (CAD) systems for automated assessment tasks.

The new approach presented by the author to improving the methodology for research and evaluating the reliability of roads at the stages of their engineering project and developing project documentation in the form of a sequential solution of a number of functional tasks is the basis for constructing human-machine interactive procedures for evaluating the reliability indicators of new modifications of roads. These procedures are based on the Monte Carlo approach to assessing the reliability indicators of roads at the stages of the implementation of their engineering project and the development of project documentation.

Using in the process of improving the methods of research and assessing the reliability of roads at the stages of the engineering project and developing project documentation, the probabilistic-statistical approach to assessing the reliability indicators of roads, as well as a new approach to analyzing the rate (risk) of their destruction of pavement in time during the operation of roads, which underlies the determination of their reliability, will contribute to a significant adjustment of design data performance towards their significant improvement.

References

[1] Stolyarov V V 1994 *Road design taking into account the theory of risk* (Saratov: Saratov state Technical University)

[2] Fedotov G A, Pospelov P I 2010 *Research and design of roads* (Moscow: Higher school) vol. 1.

[3] Bazhanov A P 2019 Development of a methodology for the study and assessment of the reliability of roads at the stages of their design *Transport facilities* **6**. Available at: https://t-s.today/PDF/19SATS118.pdf

[4] Iyudu K A 1962 *Optimization of automation devices by reliability criterion* (Moscow: Soviet radio).

[5] *Manual of Federal Highways Administration* 2012 Design consistency module engineers manual. interactive highway design model engineer's manual.

[6] Oguara T M 2010 A management model for road infrastructure maintenance *In Proc. of 19th Assembly of the Council for the Regulation of Engineering in Nigeria.*

[7] Stolyarov V V 2005 Risk theory formulas based on the normal distribution law - Problems of transport and transport construction: collection of scientific papers (Saratov: Saratov State Technical University) vol. 2, pp. 3 - 14.

[8] Bazhanov A P 1999 The methodology for assessing the mechanical component of the reliability of sensor equipment for rocket and space technology at the stage of development of schematic and technical projects. *Standards and Quality* **4** (in appendix) 38-45.

[9] Bazhanov A P 2019 General issues of assessing the reliability of roads *IOP Conference Series: Materials Science and Engineering* **632** 012012.

[10] Silyanov V V, Domke E R 2009 *Transport and operational qualities of roads and city streets* (Moscow: Publishing Center "Academy").

[11] *Guidelines for the design of non-rigid pavements under the impact of heavy traffic flow* 2015 Sectoral road methodological regulations 218.056-2015 (for highways I-II categories) (Moscow: LLC Dorinzhservis).

[12] Wentzel E S, Ovcharov L A 1988 *Probability Theory and Its Engineering Applications* (Moscow: Nauka).

[13] Sudakov R S, Chekanov A N 1972 To the question of calculating multidimensional normal integrals in reliability problems *Technical cybernetics* **I** 17-19.

[14] Volkov E B, Sudakov R S, Syricin T A 1989 To the question of calculating multidimensional normal integrals in reliability problems In *Fundamentals of the theory of engine reliability* (Moscow: Mechanical Engineering).

[15] Bolshev L N, Smirnov N V 1983 *Tables of mathematical statistics* (Moscow: Science).

[16] Smirnov N V, Dunin–Barkovsky I V 1965 *A course in probability theory and mathematical statistics for technical applications* (Moscow: Science).

[17] Ya B Shor, F I Kuzmin 1968 *Tables for analysis and reliability control* (Moscow: Soviet Radio).

[18] D Cowden 1961 *Statistical methods of quality control* (Moscow: Fizmatgiz).

IOP Conf. Series: Materials Science and Engineering **832** (2020) 012041 doi:10.1088/1757-899X/832/1/012041

Digital technologies and new mobility

N O Bludyan[1,2,3]

[1] Motor transportations, Moscow Automobile and Road Construction State Technical University (MADI), 64, Leningradsky prospect, Moscow, 125319, Russian Federation
[2] Administration, the Association for the advancement of the transport industry "Transport Association of the Moscow Agglomeration" (The Association "TAMA"), 2, Denisa Davydova, Moscow, 121170, Russian Federation
[3] Corresponding author email: np-tama@mail.ru

Abstract. New types of mobility for people started to appear with the development of digitalization in various fields of transport industry. The article discusses the development trends of digital technologies in transport sector. The current situation of the use of digital technologies in transport is observed. Forecasts of the development and introduction of new technologies in digitalization are given. The author presents the overview and forecasts of the need for regulation of transport, taking into account modern technologies as well as the overview of new types of mobility for remote cities where it is not possible to create scheduled routes. The analysis and review of such directions of transport as sharing are presented. The advantages and disadvantages brought by car sharing in cities are substantiated. The analysis of the policies of different cities in stimulating people to transition to general economy and sharing cars is presented. As a result the overview of the advantages and disadvantages of carpooling system is reflected. Necessary legal regulation of carpooling and control over this type of activity are proposed. The article touches upon the issue of the fight against mass motorization through the use of vehicle sharing.

1. On-Demand Mobility

Currently, a new method of organizing the system of urban passenger transport – on-demand mobility - is widely discussed in the professional environment. On-demand passenger transport offers a more flexible service to passenger traffic, as it is based on the number of incoming requests for transportation. This is a user-oriented type of public transport, which differs from the traditional route transport by flexible routes based on the presented demand [1].

An example of on-demand mobility is “Bwcabus” from Wales (UK). This company is focused on serving primarily the rural population and people living on the outskirts. Local bus routes are being reduced in order to save money, which causes isolation and the inability of people who do not have personal vehicles to carry out social and labor trips. This service allows people, after free registration, to order buses that provide them with communication with other main public transport and other cities, and access to basic services in the nearest cities at the right time [2].

On-demand transportation in urban traffic is relevant during non-peak periods, as well as in areas with low passenger traffic. Thanks to such a system, passengers will be able to request boarding and alighting at any non-prohibited place through a mobile application. Such a concept is planned on night bus routes instead of fixed-time traffic. Such a system will provide a better and more individualized bus service with the possibility of reducing the waiting time [1].

The studies by specialists from Sydney have shown the unprofitability of on-demand mobility. According to experts, this is due to the fact that small capacity buses are used and this forces to hire more drivers. They also noted the unwillingness of people to order transportation through the app in advance, as it causes discomfort, unlike a trip by private transport [3].

Successful projects and startups working on the principle of "on-demand" transportation can be highlighted. In Oxford, for example, passengers can call a PickMeUp minibus to any street in the eastern part of the city. Such a service provides travel cheaper than in a taxi, and no more expensive than transport following fixed routes. With the new technology, it is possible to determine where people are and provide them with transportation services, thereby creating new mobility and opportunities for people [4].

The cost of owning private vehicles, increasing the time of traffic congestion, restrictions on expanding road infrastructure, the need to save resources and reduce emissions - all this contributes to the development of on-demand mobility. It is predicted that in the near future a large proportion of the population in the city will come to usage public vehicles [5].

2. Car-sharing

Car-sharing is a group car use. This service is a form of short-term car rental. It is suitable for people who need a car for a short period of time. The peculiarity of the service is that drivers need to pay only for the time of actual use of the car, and the cost depends on the duration of use and class of the car.

Speaking about the value of car-sharing, it should be noted that this service can be used 24 hours a day. The development of specialized parking in different parts of the city makes it possible to take, leave such a car next to work or place of residence.

As of August 2017, Moscow car-sharing is the most dynamically developing free-float market in the world. The very first operator of car sharing in Moscow and Russia, the company “Anytime” appeared in 2012. By mid-November 2018, the car-sharing market in Moscow was over 13.5 thousand cars, and the total number of trips per day was 60 thousand. The largest operator of the car-sharing market in Moscow is "Yandex.Drive", the size of the rolling stock exceeds 6500 car-sharing cars. Belcacar and Delimobil are also among the largest domestic car-sharing companies. For comparison, according to analysts at ING Bank, as of October 2018, the total fleet of car sharing operators in 13 European countries combined amounted to only 15 thousand cars. According to the Moscow Department of Transport, 300 new cars are added to the fleet of car-sharing operators in Moscow every week. According to the data from the investment bank JPMorgan, the goal of the Moscow government is to reach the index of 25 thousand cars. In this case, one car-share will have 500 residents and according to this indicator, Moscow will be able to catch up with world leaders in this area: Toronto (one vehicle to 498 inhabitants), Madrid (500), Stuttgart (515) and New York (525). According to analysts of PwC consulting company, by 2025 the Moscow car-sharing fleet will reach 30 thousand cars [6].

Ongoing research and analysis of car-sharing activities in the urban environment shows that car-sharing has become part of an integrated public transport network. The interchangeability of car-sharing and conventional public transport in the event of a failure of one of them [7] is traced.

The German experience shows that car-sharing has become an important factor for sustainable urban mobility. To enhance the positive effect of car sharing, more users are required in such a system, then personal cars will be more actively displaced. This can be done with the involvement of additional users, such as women, the elderly, the population living on the periphery of cities [8].

Studies in Portugal have shown once again that the sharing of vehicles contributes to more efficient and sustainable mobility (with fewer vehicles per capita) and to the development of a sustainable transport system. In addition to the listed advantages it is possible to add such important fact as reduction of consumption of physical and economic resources at users of this system. The studies suggest a reduction in energy consumption and negative impact on the environment due to the refusal of people from personal transport in favor of car sharing. This is expected to result in a 35-47% reduction in energy consumption and a 35-65% reduction in $CO2$ emissions when hybrid vehicles are used in co-located environments. The number of owners of personal vehicles is expected to increase from 1 to 6 per car sharing vehicle [9].

According to experts, carsharing is one of the most promising services in the future. Using the systems of this service, it is possible to efficiently use unmanned vehicles, which, according to forecasts of Frost and Sullivan and RBK, will already occupy 40% of the global passenger car market by 2025 [10].

Surveys of car-sharing customers have shown that after joining the car-sharing scheme, they are much more likely to walk, cycle and use public transport than before. Muhammad's study in Switzerland showed that the use of mobile transport for car sharing users increased by 25% after they joined mobility, the Swiss provider of car-sharing [11].

A customer survey in Brussels showed that 60% of Cambio customers are holders of public transport subscriptions (STIB) and 22% have become new holders of subscriptions since they registered with Cambio [11].

In Quebec, thanks to the joint proposal "duo: car + bus" from RTC, the local public transport operator, and the car-sharing company Communauto, the number of monthly public transport tickets purchased by car-sharing clients increased by 45% after the launch of the proposal [11].

In addition, cooperation with carsharing provides a more dynamic image of public transport, and a more flexible and broad offer of mobility. These arguments clearly show that car sharing has become an important element of a sustainable transport strategy. In Zurich, Switzerland, 6% of residents are car sharing customers [11]. At the moment, this is the goal that large cities and agglomerations of the world should strive for.

3. Carpooling

Carpooling (ridesharing) represents joint use of vehicles for movement at different distances. It is assumed that the driver's trip was planned by himself and would have been carried out in any case, but he is looking for fellow travelers willing to share the costs. Examples of ridesharing are companies (services) BlaBlaCar, karzoo (France), mitfahrgelegenheit and mitfahrtzentrale (Germany), carpool (Belorussia), carpoolworld (America), etc. At present, we can observe a trend in the development of the paradigm of "sharing economy", with the transition from ownership of goods to joint use. Greater success among transport sharing is worth noting on Internet platforms that allow planning common trips, as well as finding travel companions for car owners and finding routes for passengers, such as Blablacar and Carpooling. Middle-aged people are more likely to use new digital resources, but older people unfortunately still use traditional transport systems [12]. In Europe, the concept of carpooling has been developed and promoted through internet and social media. The Internet has also helped spread the idea of real-time car sharing with a mobile device. However, there are still no clear algorithms for creating routes in applications that take into account all the interests of drivers and passengers [13].

In order to achieve compromise solutions that will ensure good compliance of the offer for both drivers and passengers of carpooling, constantly adjusting algorithms are necessary. Scientists from Poland have proposed algorithms that are applicable even in the BlaBlaCar application, presented the task of unification and the procedure of its solution. The task was reduced to the construction of an optimal route, taking into account the combination of "interests" for passengers and drivers [13].

Vehicle sharing is the foundation for a sustainable and safe mobility environment using fewer cars to transport more people. One of the main problems is drivers' confidence in passengers and vice versa. Greece has proposed a solution to this problem by using mechanisms to assess the reputation of users, which could rank users according to their behavior according to the assessments and feedback of other users. Special attention is paid to the stability of such a system from attackers and the preservation of personal data [14].

The draft federal law provides for the introduction of a new type of mobility in the Russian Federation - "joint trips" (carpooling, riddling) [15]. The main purpose of this type of mobility is to reduce the number of cars on the road network of large agglomerations and cities, reduce the intensity of traffic flows and the likelihood of traffic jams and congestion, reduce the amount of harmful emissions from vehicles into the atmosphere, etc. Taking into account this fact and the lack of practical experience in implementing such projects, as well as understanding the possibility of turning "carpooling" into an illegal sector of transportation in contrast to the existing system of transportation by passenger taxi, at the initial stage it is necessary to provide strict conditions and restrictions on market entry in this new transportation system.

1. Based on article 2 of the draft law, it is proposed to use M1 category vehicles for transportation not related to business activities. The M1 category includes vehicles with 8 seats in addition to the driver's seat. With a high degree of probability, it can be assumed that minibuses falling under the M1 category will become the prevailing vehicle for the provision of joint travel services. This circumstance, in turn, will lead to commercialization and entrepreneurship in this areaThis will especially have a negative impact at the initial stage of implementation of the system, when an effective system of control and supervision has not yet been established. In order to facilitate and prioritize the use of classic cars with no more than 4 seats in the trips under consideration, (in addition to the driver's seat), it is proposed to limit the number of passengers in one trip to three passengers. This will ensure the comfort of the trip of two passengers in the passenger car and one passenger next to the driver.
2. In article 2 of the draft law, it is proposed to clearly fix the name of aggregators of information on transportation that is not related to entrepreneurial activity as follows: "aggregator of information on services for non-commercial transportation".
3. We consider it necessary to legislatively prohibit the carriage of passengers under contracts of non-commercial chartering on interregional routes. Otherwise, we will create all the legal conditions for the prosperity of the illegal segment of commercial transportation in inter-regional communication by minibuses and sedans (for example, the Tula-Moscow route). Transportation on adjacent inter-regional routes will fully satisfy the needs of real carpooling.
4. It is proposed to identify only citizens of the Russian Federation with national category B driver's licenses as freighters that provide transportation services under non-commercial charter agreements. In this regard, delete from paragraph b) of article 7 the paragraph that begins with the words "or an international driver's license ..." (p. 11, 12 of the bill).
5. The conditions and requirements for the financial relationship between the owner of the aggregator and the charterer are not clear from the bill. On the basis of paragraphs g) and h) of article 5 the owner of the aggregator may impose on the shipowner and the charterer a commission is not acceptable to these participants. Clear legal regulation of the conditions of financial relations between the parties is required. The operational experience of the passenger taxi sector confirms the need for such regulation.

   5.1. Article 4 of the draft law proposes to prescribe a mechanism for generating the amount of compensation to the driver from the passenger for non-commercial transportation, taking into account the actual expenses of the driver, thereby:

   - Showing the passenger that the driver does not carry out business activities;
   - "Clear" the services for finding fellow travelers from illegal entrepreneurs;
   - Facilitating the control of supervisory authorities.

   5.2. In part 3 of article 4 of the draft law it is proposed to prescribe the procedure of control over the collection by the charterer of a fee for the use of a vehicle under a non-commercial chartering agreement, as well as the liability of the charterer for the collection of a fee unreasonably exceeding its maximum amount established by the Government of the Russian Federation.

   5.3. Part 9 of Article 4 of the draft law provides for a number of methods of payment for services under non-commercial charter contracts, and therefore this part of Article 4 is proposed to be revised as follows: "Payment for services under the non-commercial charter contract shall be made in the following order:"

6. We consider it inappropriate to allow the charterer carrying out the transportation of individuals by a vehicle which he has on loan. This right will lead to the widespread spread of uncontrolled

relations. It is proposed to edit paragraph 6 of article 2 in order to exclude the right to use rental vehicles.

6.1. Part 2 of article 9 of the bill makes changes to part 2 of article 8 of the Federal law "on protection of the rights of legal entities and individual entrepreneurs in the exercise of state control and municipal control", providing for the expansion of the list of works and services approved by the Government of the Russian Federation as part of activities in respect of which legal entities, individual entrepreneurs are notified of the commencement of certain types of entrepreneurial activity.

- In this regard, we consider it necessary to supplement Part 1 of Article 6 of the bill with the following words: “in the manner prescribed by the Federal Law “On the Protection of the Rights of Legal Entities and Individual Entrepreneurs under State Control (Supervision) and Municipal Control”.

7. In our opinion, the number of trips by one charterer during the day cannot exceed two only if provided that the second trip is on the same route, but only in the return message. In other cases, the number of trips should be limited to one trip. (Article 7, paragraph 2).
8. In paragraph c) of part 1 of article 7 of the draft law, it is proposed to determine the total driving experience of a driver with a size of at least 3 years (by analogy with a passenger taxi). Clarification of the concept of “seniority” in the driver of a personal vehicle is required, as well as a system for monitoring the length of the driving experience.

   8.1. In part 2 of article 7, it is proposed to establish a mechanism for controlling and blocking trips in excess of the proposed maximum. It is advisable to oblige the aggregators to automatically block the registration of driver trips in excess of the norm, and the norm should be clearly prescribed taking into account the long time intervals (number per month, year).

9. Article 3 (paragraph f)), as well as paragraph 3 of Article 4, introduces the concept of “maximum amount of a fee for using a vehicle”. From the considered edition of the bill its physical meaning is not clear. I consider it necessary to supplement article 2 with the concept of “maximum board size” and give a unit of measurement.
10. Paragraph 3 of Article 4 of the draft law defines two types of expenses. It is not clear why other costs are not taken into account, such as special liquids, tire wear and tear, depreciation of the vehicle, etc.

4. Conclusion

The mechanism of control and responsibility for a similar type of search for companions on the basis of any other online resources that are not essentially non-commercial transport aggregators is not clear. There are cases of mass use of social networks and messengers by "grey" carriers to search for passengers.

References

[1] DoorsAreClosing 2017 *LTA to trial on-demand bus services,* Blog on Land Transport Guru. Retrieved from: https://landtransportguru.net/lta-trials-on-demand-bus-services/

[2] European Network for Rural Development 2016 *An on-demand rural bus service in rural WalesAvailableat,* Project report. Retrieved from: https://enrd.ec.europa.eu/projects-practice/demand-rural-bus-service-rural-wales_en

[3] O’Sullivan M 2018 Sydney’s new on-demand buses test the public’s appetite *The Sunday Morning Herald* Retrieved from: https://www.smh.com.au/national/nsw/sydney-s-new-on-demand-buses-test-the-public-s-appetite-20180718-p4zs5d.html

[4] Topham G 2018 Oxford buses turn to Uber-style apps in on-demand experiment *The Guardian* Retrieved from: https://www.theguardian.com/business/2018/jun/29/oxford-buses-turn-to-uber-style-apps-in-on-demand-experiment

IOP Conf. Series: Materials Science and Engineering **832** (2020) 012041 doi:10.1088/1757-899X/832/1/012041

[5] Matus J, Heck S 2015 Understanding the future of mobility *Techcrunch* Retrieved from: https://techcrunch.com/2015/08/08/understanding-the-future-of-mobility/

[6] Posypkina A, Balenko E, Balashova A 2018 The Moscow city car-sharing market if the one of the fastest growing *RBC Media* Retrieved from: https://www.rbc.ru/technology_and_media/27/09 /2018/5bab94a69a79474169e307c1?from=newsfeed

[7] Tyndall J 2019 Free-floating carsharing and extemporaneous public transit substitution *Research in Transportation Economics* **74** 21-27.

[8] Giese F, Nobis C 2016 The impact of carsharing on car ownership in german cities *Transportation Research Procedia* **19** 215–224.

[9] Baptista P, Melo S, Rolim C 2014 Energy, environmental and mobility impacts of car-sharing systems. Empirical results from Lisbon, Portugal *Procedia of Social and Behavioral Sciences* **111** 28–37.

[10] Ol'hovikov D E 2018 Intelligent transport systems of Russia *Information Magazine* **7** 19–20.

[11] UITP 2011 *Becoming a real mobility provider Combined Mobility: public transport in synergy with other modes like car-sharing, taxi and cycling,* UITP Position paper. Retrieved from: https://www.uitp.org/sites/default/files/cck-focus-papers-files/FPComMob-en.pdf

[12] Guercio E, Borean C, Zenezini G, Gargiulo E, Giannantonio R 2015 Dynamic ride sharing service: are users ready to adopt it? *Procedia Manufacturing* **3** 777–784.

[13] Żak J, Maciej H, Grzegorz F 2019 Multiple criteria optimization of the carpooling problem *Transportation Research Procedia* **37** 139–146.

[13] Salamanis A, Kehagias D D, Tsoukalas D, Tzovaras D 2019 Reputation assessment mechanism for carpooling applications based on clustering user travel preferences *International Journal of Transportation Science and Technology* 8(1) 38–46.

[14] Draft Russian Federal Law 2019 *On regulation of relations arising from the conclusion with the use of information and telecommunication network "Internet" contracts of carriage of passengers by road on request for purposes not related to the implementation of business activities, and on amendments to certain legislative acts of the Russian Federation.*

IOP Conf. Series: Materials Science and Engineering **832** (2020) 012042 doi:10.1088/1757-899X/832/1/012042

Improvement of organization and functioning of taxi transportation through the introduction of digitalization

N O Bludyan[1,2,4], R S Ayriev[3], D G Moroz[1]

[1] Motor transportations, Moscow Automobile and Road Construction State Technical University (MADI), 64, Leningradsky prospect, Moscow, 125319, Russia
[2] Administration, Association for the advancement of the transport industry "Transport Association of the Moscow Agglomeration" (Association "TAMA"), 2, Denisa Davydova, Moscow, 121170, Russia
[3] Apparatus at the manual, GUP Mosgortrans, 22/21, Raushskaya naberezhnaya, Moscow, 115035, Russia
[4] Corresponding author: np-tama@mail.ru

Abstract. The article provides a brief retrospective of the development of taxi activity. The key characteristics of the changes in the quality of services provided, including the average time flow, average check, etc. The carried out researches and the analysis allowed determining relevance and expediency of usage of technologies for digitalization of the taxi market to solve "classical" problems of the system how to organize transportation by taxis. The possibility of using the Unified Regional Navigation and Information System of Moscow (ERNIS) was analyzed, as well as registers of issued permits on passenger and luggage transportation by taxis in Moscow and Moscow region to solve the problems of location of taxi stands and determine the optimal taxi car. The studies and analysis made it possible to determine the relevance and appropriateness of existing taxi cabs and the future ones, taking into account the requirements for rational location, in terms of ensuring a quick approach for passengers, as well as reducing the time for car delivery. The article concludes that it is necessary to revise the approach to solve "classical" problems how to improve the functioning of the taxi market.

1. Introduction

The first mention of a taxi in Moscow was in 1907. The formation of the Soviet taxi began in 1924, when the Moscow city Council decided to buy 200 taxi cars (minutes No. 35 of the meeting of the Presidium of the Moscow Council of RK and KD dated December 23, 1924) [1]. The first GAZ-A taxi cars produced in the USSR appeared in 1933.

The current stage of taxi development in the Russian Federation begins with the adoption on April 21, 2011 of Law No. 69-FL "On Amending Certain Legislative Acts of the Russian Federation". One of the most important achievements in passenger taxi transportation in recent years has been the reduction of the average delivery time to 4-6 minutes. This indicator is one of the most important in taxi transportation. Taxis in the capital have one of the leading positions in the world in terms of service level. The legalization and transformation of the taxi market has reduced travel prices and made taxis more attractive for Muscovites. These changes were made possible through the digitalization of the industry.

According to the data of the Department of transport and development of road infrastructure of Moscow, the average taxi delivery time decreased by 6 times compared to 2011 (figure 1). The cost of taxi services in the capital decreased by 2.4 % compared to 2017. For 2018, the average check for a taxi ride in Moscow amounted to 473 rubles. There has been a decrease in the average check for several years in a row due to an increase in the share of short trips. According to forecasts for 2019, the average check will be 495 rubles.

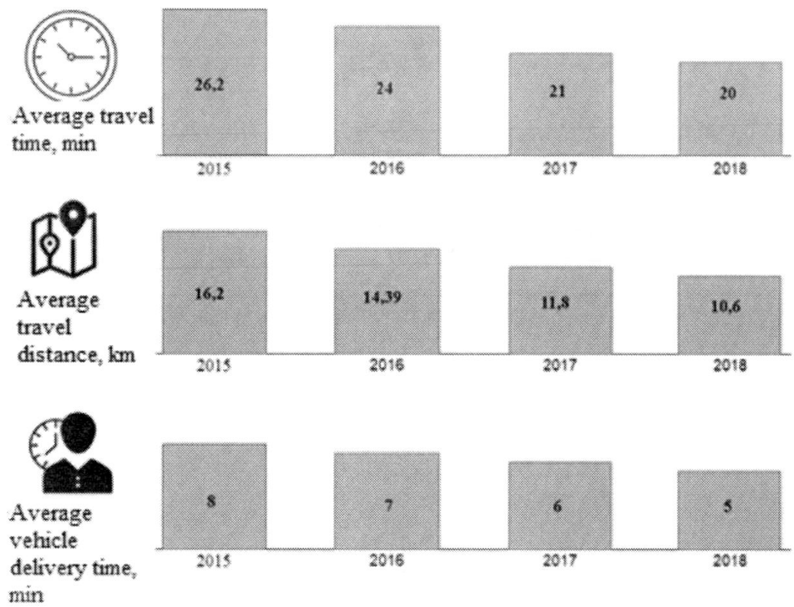

Figure 1. Dynamics of changes in car taxi rides

Taking into account the current trends in technology development, we can conclude that it is necessary to use the possibilities of digitalization of the industry to solve the classical problems of development and improvement of the system, organization and management of taxi transportation [2, 3, 4].

2. Planning for the organization of taxi stops based on geographic information

One of the "classic" tasks facing the system of organization of transportation by taxi cars is the issue of minimizing idle mileage, from the point of view of economic entities and minimizing the waiting time and emissions of exhaust gases into the environment. One of the approaches to solve this problem can be the optimization of the network of locations of taxi stands.

On the basis of the modern level of digital technologies, it is expedient to use large amounts of data obtained with the help of the GLONASS/GPS system when solving the problem of taxi parking arrangement. For the purposes of urban planning, it is possible to use the data transmitted in accordance with the Regulations on Information Interaction from the largest passenger taxi dispatcher orders to the Unified Regional Navigation and Information System of Moscow (ERNIS) [5,6].

The analysis of statistical data on the number of transferred statuses free/occupied allowed determining the most frequent points of start and end of taxi trips by districts on average per day (Table 1).

IOP Conf. Series: Materials Science and Engineering **832** (2020) 012042 doi:10.1088/1757-899X/832/1/012042

Table 1. Top 10 districts of Moscow to begin travel by taxi

1.

№	Area	Number of trips started, u.	% of the total number	Total %
1	Tverskoj	8009	5.85	36.52
2	Presnenskij	7249	5.3	
3	Taganskij	5477	4.01	
4	Basmannyj	5231	3.83	
5	Hamovniki	4410	3.22	
6	Krasnosel'skij	4373	3.19	
7	Ramenki	4146	3.03	
8	Dorogomilovo	3846	2.81	
9	Meshhanskij	3597	2.63	
10	Danilovskij	3576	2.61	
11-144	Other areas (average)	1484	1.86	63.48

2.

№	Area	Number of trips started, u.	% of the total number	Total %
1	Presnenskij	7166	5.39	35.35
2	Tverskoj	6768	5.09	
3	Basmannyj	5631	4.24	
4	Taganskij	4537	3.41	
5	Ramenki	4255	3.21	
6	Hamovniki	4152	3.13	
7	Krasnosel'skij	3820	2.88	
8	Danilovskij	3584	2.7	
9	Dorogomilovo	3531	2.66	
10	Ljublino	3479	2.62	
11-144	Other areas(average)	1478	1.11	64.65

In order to solve the problem of using geoinformation data, it is necessary to determine the following information structure according to the starting and ending points of the route of passenger taxis (Table 2).

Table 2. Information on the start and end points of travel by taxi

№	Name	Data type	Note
1	Date and time	Date	Date and time of data creation (to the nearest millisecond)
2	Latitude	Numerical	In the WGS84 projection, decimal degrees accurate to 3 decimal places (+ - 50 m)
3	Longitude	Numerical	In the WGS84 projection, decimal degrees accurate to 3 decimal places (+ −50 m)

At the first stage of planning it is necessary to divide the data set containing information on the starting and ending points of the passenger taxi route into blocks of hours with an interval of one hour in the section of the district. The second step is to construct ranked tables (Table 3) on the number of boarding/disembarkation points of passengers.

Table 3. Top 5 coordinates for the period from 00 to 1 a.m..

№	Coordinates	Number of orders per hour
1	55.758 37.612	27
2	55.761 37.620	23
3	55.756 37.614	22
4	55.759 37.611	18
5	55.757 37.613	17

The final step is an analysis of the potential and modelling of the specific location of the universal taxi stations, taking into account the changes in the concentration of taxi cars in the context of hours of day, days of the week and seasonality.

3. Determining optimal taxi car

Taxi companies regularly face the problem of choosing the type of vehicle when renewing their rolling stock. Often the choice is made through subjective assessment of the company's managers and the possibility of more profitable purchase of cars of a particular manufacturer.

The analysis of ERNIS data and the register of issued permits can help solve the problem of choosing the optimal type of vehicle, similarly to the problem of choosing locations for taxi cabs,

The analysis of 19621 vehicles for which "Permits" were obtained by the Department of Transport and Road Infrastructure Development of Moscow in 2017 shows that the vehicle brands were distributed as follows (Table 4, Figure 2):

Table 4. Distribution of permits by car brands

Car model	Hyundai Solaris	Ford Focus	Škoda Rapid	Volkswagen Polo	Škoda Octavia	Kia Rio	Ford Galaxy	Others
Quantity, units	4179	3041	2148	1863	1774	1539	1386	3691

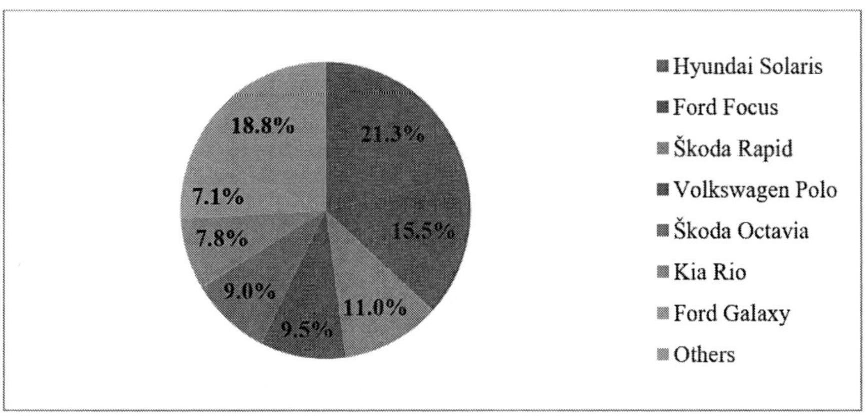

Figure 2. Distribution of passenger taxis by brand

IOP Conf. Series: Materials Science and Engineering **832** (2020) 012042 doi:10.1088/1757-899X/832/1/012042

The analysis of the fleet of passenger taxis used in the city of Moscow will allow identifying the 4 most popular brands / models, which account for about 60% of the entire sample (Figure 3).

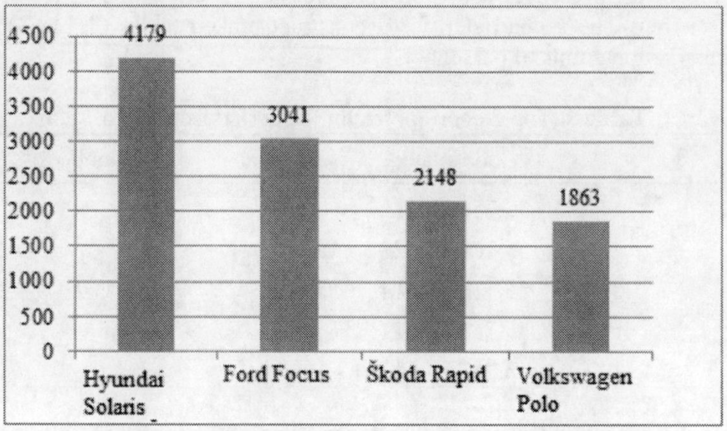

Figure 3. The most popular passenger taxi brands

Of the 19,621 vehicles for which permits were received in the Department of Transport of the city of Moscow in 2017, the most popular vehicle brand is Hyundai Solaris (4,179 vehicles), which is 21.2%. The second most popular vehicle brand is Ford Focus 15.5%, followed by Skoda Rapid (10.9%) and Volkswagen Polo (9.5%) [7].

On the other hand, for the selected car brands it is necessary to understand the interest (attractiveness) of passengers in each of them. Such information can be obtained by analyzing the data sets from the ERNIS system. The totality of information about the choice of taxi rolling stock allows determining the most accurate definition of the most popular taxi car for all market participants.

Table 5. Car parking areas for taxis

Offered parking areas for taxi cabs	Existing taxi parking areas
Mohovaja street, Manezh;	There's an active parking lot
The intersection of Tverskaya street with the Gazetniy and Nikitsky lanes;	There's an active parking lot in the immediate vicinity: Tverskaya Street, 3
Crossing of Bol'shaja Dmitrovka Street and Kamergersky Lane;	*No active parking, pedestrian zone*
Neglinnaja Street, TSUM department store;	There is an active parking lot
Slavjanskaja and Staraja Squares;	There is an active parking lot in the immediate vicinity: Lubyansky passage, 25s2
Tverskoj zastavy square, Belorussky railway station;	There's an active parking lot
The intersection of Teatralny passage and Nikolskaya street;	There's an active parking lot (the sign 3.27 with the inscription 8.4.14 "Except for taxis" is set)
Triumphal square, 2, Mayakovskaya metro station;	There's an active parking lot
The intersection of Novoslobodskaya street and Sushevsky Val.	*No active parking*

4. Conclusion

The use of ERNIS databases and information on issued permits makes it possible to apply fundamentally new methods and approaches to solving various problems in the field of passenger transportation by taxi cars, which were previously unavailable. In particular, for the Tverskoy district of Moscow, potential taxi parking areas have been identified based on and analysis of the data (Table 5).

Similarly, by means of the industry's digitalization, the tasks of choosing the optimal taxi car are solved, both for its owner and passenger.

The proposed approaches and methods for solving the "classical" problems of improving the functioning of the taxi market are impossible without a significant level of digitalization of the industry. It should be especially noted that the market of taxi services is currently one of the most advanced in terms of implementation of the concept of digital transformation of the transport industry.

Acknowledgments

The author expresses his deep gratitude to Professor Ostroukh A.V., an expert in the field of digital technology, for his great contribution to the creation of the electronic textbook on the discipline "Materials Science".

Reference

[1] Moskovskiy Transport LLC 2017 *Taxi transportation in Moscow*. Retrieved from: http://transport.mos.ru/common/upload/docs/1492676062_taxi_2017.pdf.
[2] Bludyan N O, Moroz D G, Hejfic P I 2015 Problems of state regulation and organization of taxi transportation in Moscow agglomeration *In the World of Scientific Discoveries* **6**(66) 243-251.
[3] Bludyan N O, Ayriev R S 2014 Agglomeration approaches to solving transport problems *Transport Enterprise* **8** 2-7.
[4] Bludyan N O, Moroz D G, Hejfic P I 2014 Territorial planning of taxi transportation in urban agglomeration *Automation and Control in Technical Systems* **8**(1-1) 62-73.
[5] Decree of the Government of Moscow 2017 *On measures to improve the organization of taxi services in the city of Moscow*, decree No. 9-PP from 24.01.2017. Retrieved from: https://www.mos.ru/authority/documents/doc/35928220/.
[6] Order of the Transport Department of Moscow 2017 *On approval of the Regulation on information interaction of dispatching services for passenger taxis in Moscow with the Unified Regional Navigation and Information System of Moscow*, order N 61-02-76/7 from 06.03.2017. Retrieved from: http://www.consultant.ru/cons/cgi/online.cgi?req=doc&base=MLAW&n=177335&dst=100001#06527366221921675.
[7] Bludjan N O, Moroz D G, Shestopalov N Ju 2018 Assessment and selection of vehicle type for taxi transport *MADI Newsletter* **3** 115–121.

IOP Conf. Series: Materials Science and Engineering **832** (2020) 012043 doi:10.1088/1757-899X/832/1/012043

Estimation of Passenger Load for Urban Passenger Transport on the Route

V N Bogumil and M J Duque

Moscow Automobile and Road State Technical University (MADI), 64, Leningradsky ave., Moscow, 125319, Russia

E-mail: v_bogumil@mail.ru, majoduque1990@gmail.com

Abstract. This article deals with the quality of passenger service of urban passenger transport of large cities and megalopolises. The quality of passenger service is estimated mainly by the passenger load, that is the number of passengers in a vehicle. The main content of the article is a description of the method of exponential forecasting used to forecast passenger load estimation on the public transport route with high passenger traffic and low predictability of passenger flow behavior. In order to assess the possibility of using analytical methods to predict the passenger load, it is suggested to forecast the load at a critical stop of the route.

1. Introduction

Public passenger transport plays a very important role for the population in Russia and in the world. Particular emphasis on the development of urban transport systems has a direct impact on the perception of system reliability, user comfort and traffic management [1, 2]. However, to control the perception of passengers and respond to territorial development, the efforts of state and administrative management and an effective political action are required to achieve the implementation of complex urban projects based on transport-oriented developments.

In this regard, the administration in large cities and megapolis pays much attention to urban passenger transport, improving transport services to the population. With the aim of improving the quality of public service, automated urban passenger transport control systems have been developed [3, 4], which manages and regulates passenger transport on urban and suburban routes.

The system is implemented on the basis of the use of computer equipment, communication between the organizational elements of the system, the transfer of information and control actions to perform the tasks of managing passenger traffic faced by the system. Controlled vehicles must be equipped with mobile communication and navigation units, the characteristics of which must be agreed with the administration of the cities [5].

To control the quality of the transport process, we suggested using an exponential forecasting method for passenger load estimation, which can be used by dispatch control system to compare the passenger load with the regulations that have been previously established and studied [6–12].

2. The use of the method of exponential forecasting to estimate passenger load on the public transport route with high passenger traffic and low predictability of passenger flow behaviour

Currently, the situation which is characterized by low predictability of passenger flow behavior and high traffic intensity of vehicles along the route [13] is typical of many main arterial routes of large cities and metropolis. In this case, it is necessary to use models with a high degree of adaptation to changing characteristics in the transport process, which are much higher than regression models. In this case, the source of information for the forecast should be only a group of the latest weighted data of the passenger load on the route.

3. Forecast of passenger load at a critical stop, close to the start of the route

We consider the situation when the vehicle of the previous trip passed a critical stop of the route before the next vehicle started the trip. This situation occurs, for example, on routes for which the critical stop is the starting point "A" of the trip as shown in Figure 1. In the city of Moscow, routes starting from the metro station are a typical example. The situation arises at "rush hour", a similar situation arises on routes where the critical stop is located so close to the starting point "A" that the vehicle performing the current trip has time to go through the critical stop before the next vehicle leaves for the trip. In this case, the most effective forecasting method is the method based on the use of data on the loading of the passenger at the critical stop of the current trip in the forecast for the next trip. On the basis of new data on the loading of the cabin, obtained at each next step, the model parameters are adjusted over time, that is, they are adapted to new, constantly changing conditions. Thus, the model constantly uses new information about the loading of the cabin and adapts to it. These methods can be used when predicting a small number of steps forward.

The process of adaptation to new conditions is as follows. Sets the initial state of the model. Initial data are the actual passenger load at a critical stop in the first two trips operated in the early morning when the loading of the cabin does not exceed the permissible value. Based on these data, a forecast is made for passenger load at a critical stop for the next (third) trip. After passing in the next (third) trip of a critical stop, actual data on the passenger load is obtained and it is determined how much the result obtained by the model differs from the actual value. The forecasting error is fed back to the model input through feedback and is used in accordance with its logic to calculate the forecast of passenger load for the next (fourth) trip.

Our analysis showed that in this case it is efficient to use the exponential smoothing model [14, 15]. The exponential smoothing of the series is carried out according to the recurrence formula:

$$S_t = \alpha x_t + \beta S_{t-1}, \tag{1}$$

where S_t – forecast of passenger load, expressed in the form of an exponential average at time t;

α, β – smoothing parameters; α=constant, β=constant; $0 < \alpha < 1$, $\beta = 1-\alpha$. These values are obtained from stochastic analysis of initial data.

A graphic illustration of such a case is shown in Figure 1.

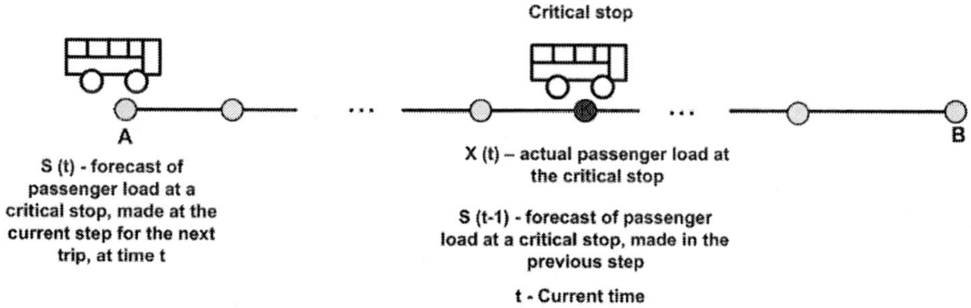

Figure 1. The scheme for generating the forecast for passenger load at the critical stop of the trip using actual data and the forecast passenger load

Expression (1) can be transformed as follows:

$$S_t = \alpha x_t + (1-\alpha)S_{t-1} = S_{t-1} + \alpha(x_t - S_{t-1}). \tag{2}$$

The model corresponds to "compensating" changes that are used for forecasts for the next moment

IOP Conf. Series: Materials Science and Engineering **832** (2020) 012043 doi:10.1088/1757-899X/832/1/012043

of time. The whole process is repeated at each next step.

As an example, the data obtained in the city of Moscow from Route 26 "Metro University-Metro Oktiabrskaya" (tramway) has been used to estimate the passenger load at morning peak, at the critical station "Lesteva Street" using formula (1). Figure 2 shows the actual passenger load and the estimation of passenger load.

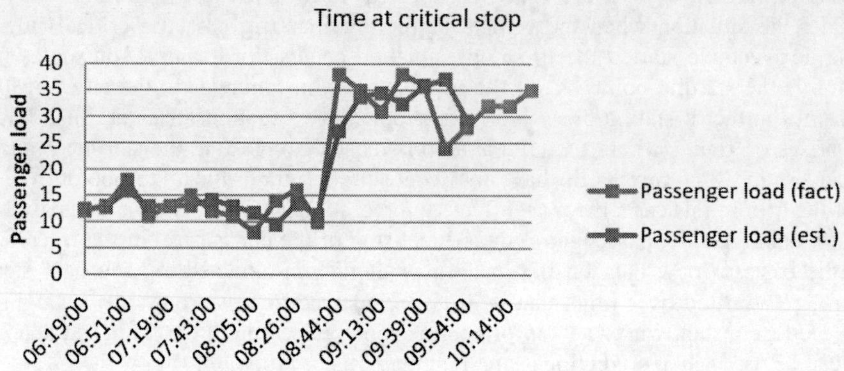

Figure 2. The actual passenger load and the estimation of passenger load at the critical station "Lesteva Street" using formula (1)

The analysis of the obtained results shows that in the case that the factual data are not found drastic changes, the estimate is accurate, which means that the estimate does not differ significantly from the factual data in real time. In the event of drastic changes in passenger flow, the estimate is less accurate, this is because there is a long interval of time between the time of departure of the tram for which the estimation will be made and the time at which factual data for estimation is taken at the critical station. To address this deficiency, a new solution is proposed, which is described below:

4. Formation of the forecast of passenger load at a critical stop according to the actual data of the vehicle that passed the critical stop and data on the loading of the passenger compartment of the next vehicle

Let´s consider the situation that occurs on routes with a high frequency of vehicles, and in which a critical stop is located relatively far from the starting point "A". In this case, more than one vehicle may be on the route section to the critical stop inclusively (Figure 2).

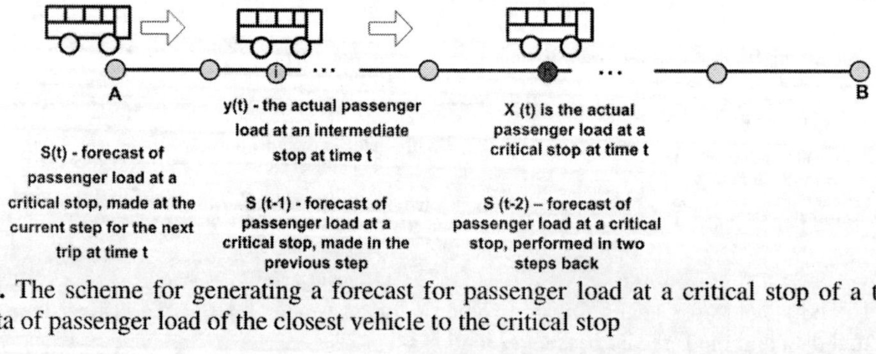

Figure 3. The scheme for generating a forecast for passenger load at a critical stop of a trip using actual data of passenger load of the closest vehicle to the critical stop

Figure 3 shows that the vehicle closest to the critical stop (k) has passed an intermediate stop (i). The route section to the k-th (critical) stop is divided by this vehicle into two sections:

1st section: from the first stop of the route to the i-th stop of the route inclusively;

2nd section: from the (i + 1) stop of the route to the k-th (critical) stop of the route inclusively.

At time t, when the next vehicle passes a critical stop with number k, data on the passengers' boarding / aligning on the section from the initial stop to stop with number i inclusive are received later than the data for the same section of the route received by the vehicle at the critical stop k at time t.

In accordance with the forecast formula (1), considered for the previous case, when there were no vehicles on the section between "A" point to the critical stop, the x(t) value characterized the actual passenger load that had just passed the critical stop. In [13], it was shown that the load of passengers in a vehicle that has departed from a i-th stop at time t ($c_i(t)$) is equal to the algebraic sum of the number of passengers boarding and alighting at all previous stops, including the last stop passed:

$$c_i(t) = \sum_{j=1}^{i} b_j(t_j) - \sum_{j=1}^{i} a_j(t_j), \tag{3}$$

where $b_j(t_j)$– number of passengers boarding at j-th stop at time t_j;

$a_j(t_j)$– number of passengers alighting at j-th stop at time t_j.

In accordance with the general approach to the forecasting of dynamically changing characteristics (in this case, such a characteristic is the load of the passenger compartment at a critical stop), the best solution is to use the latest information on the passenger boarding and alighting at the stops from vehicles currently on the route to predict the passenger load for the vehicle at point A, and have not yet set off on a trip (see Figure 3). We accept this statement as a working hypothesis. The hypothesis propose that instead of the actual amount of vehicle's passenger load, which has passed a critical stop in the forecast formula, as the x_t value, we will use the algebraic sum of the number of passengers boarding and alighting in the first section plus the algebraic sum of the number of passengers entering and leaving in the second section, which will be used in the forecast formula (3), the number of passengers in the vehicle at the time of passing a critical stop together.

Data in the first section should be obtained from the vehicle that passed the i-th stop at time t. Data in the second section should be obtained from the vehicle that has passed a critical stop k at time t. In this case, formula (3) after transformation, takes the following form:

$$S_t = \alpha\left[\sum_{p=1}^{i}\left(b_p^1 - a_p^1\right) + \sum_{p=i+1}^{k}\left(b_p^2 - a_p^2\right)\right] + \beta S_{t-1}, \tag{4}$$

where St – forecast of the passenger load at a critical stop k of the vehicle located at starting point A (first stop) at time t;

S_{t-1} – forecast of the passenger load at the critical stop k of the vehicle located at the starting point A, made in the previous step;

b_p^1, a_p^1 – the number of passengers boarding and alighting, respectively, at the p-th stop of the first section;

b_p^2, a_p^2 – the number of passengers boarding and alighting, respectively, at the p-th stop of the second section;

α, β – smoothing parameters; α=constant, β=constant; $0 < \alpha < 1$, $\beta = 1 - \alpha$.

The sequence of steps for generating a forecast at the current time t for the considered case is as follows.

1. Wait for information about the next vehicle passing the k-th (critical) stop of the route and passenger load
2. Get information on the number of passengers boarding / alighting on the first section of the route from the 1st stop to the i-th stop inclusively for the vehicle that has passed the i-th stop.
3. Get information on the number of passengers entering / leaving on the second leg of the route after the i-th stop until the k-th stop inclusively for the vehicle that has passed the k-th stop.
4. Perform, according to formula (4), the forecast for passenger load at the k-th (critical) stop of the route for a vehicle located at the initial stopping point "A".

5. Conclusions

The article reviews the possibility of evaluating the quality of passenger transport according to the passenger load in the vehicles of urban passenger transport. The exponential method has been used for estimating passenger load. The actual passenger flow data of the tramway working on Route 26 "Metro University-Metro Oktiabrskaya" have been used as the initial data to estimate passenger load. It has been shown that in the case when the factual data are not found drastic changes, the estimation is accurate, which means that the estimation does not differ significantly from the factual data in real time. In the event of drastic changes in passenger flow, the estimation is less accurate, this is because there is a long interval of time between the time of departure of the tram for which the estimation will be made and the time at which factual data for estimation is taken at the critical stop. To address this deficiency, a new solution was proposed that consists of using the latest information on the passenger boarding and alighting at the stops from vehicles currently on the route to predict the passenger load for the vehicle at point A, and have not yet set off on a trip.

References

[1] Epifanov V, Obshivalкin M and Lukonkina K 2018 Management of quality and security level of transportation in the system of regular passenger motor sport *Thirteenth Int. Conf. on Organizat. and Traffic Safety Managem. in Large Cities* ed. *SpbOTSIC* **36** 141–8 (Moscow: Transportation Research Procedia)

[2] Shang H Y et al 2019 *Bus timetabling considering passenger satisfaction: An empirical study in Beijing* (Beijing: Computers & Industrial Engineering)

[3] Efimenko D B et al 2018 Priority transfer management algorithm, based on interaction of the public transport dispatch systems information ad traffic lights control *Transp. and Telecommunic.* **19** 315–324 DOI: 10.2478/ttj-2018-0026

[4] Ryzkov A I and Zyuzin P B 2016 *Urban public transport development in Russia: Trends and reforms* (Moscow: National research university Higher School of Economics)

[5] Vlasov V M, Efimenko D B and Bogumil V N 2018 *Application of digital infrastructure and telematic systems in road transport* (Moscow: INFRA-M)

[6] Vlasov V M et al 2017 Development of Standards and their Harmonization with International Standards as a Necessary Condition of Normative and Technical Support in Construction and Development of Intelligent Transport Systems in Russia *Transport. Research Proc.* **20** 690–694

[7] *Transit Capacity and Quality of Service Manual* 3rd ed. 2013 (Washington: The national academies of Sciences, Engineering, and Medicine)

[8] *Highway capacity manual* 2010 (Washington: Transportation Research Board of the National Academy of Sciences)

[9] Russian Interstate Standard 27815-88 *Buses, General, Construction safety requirements*

[10] Russian Interstate Standard R-51004-96 *Transportation services. Passenger traffic. Nomenclature of quality indicators*

[11] Order of the Ministry of Transport of Russia 2017 No. NA-19-p *"On the approval of the social standard of transport services for the population during the transport of passengers and baggage by road and urban land electric transport"*

[12] Set of rules 42.13330.2011 *Urban planning. Planning and development of urban and rural settlements*

[13] Bogumil V N and Duque M J 2019 *Telematics in organizational and production structures* Methodical guidelines (Moscow: MADI)

[14] Prokofiev A B Shajmatov E V 2005 *Using regression analysis to study the management processes of complex systems (*Samara: SGAU)

[15] Lukashin Yu P 2003 *Adaptive methods of short-term forecasting of time series* a textbook for universities (Moscow: Financy i statistika)

IOP Conf. Series: Materials Science and Engineering **832** (2020) 012044 doi:10.1088/1757-899X/832/1/012044

Modeling of the management strategy of the research process on the basis of a digital model

V S Borovik[1] and A V Borovik[2]

[1]Volgograd Scientific and Technical Center, 6, St. Grushevskaya, Volgograd, 400074, Russia.
[2]Volgograd State Technical University, 1, St. Akademicheskaya, Volgograd, 400074, Russia.

E-mail: borovikv@mail.ru

Abstract. An important problem of efficiency growth of scientific and technological development by means of a centralized management of the research process is analyzed. The digital modeling of scientific problem solution was studied as an efficient tool of a precise, targeted management of the scientific search. It was shown that one of the priority goals of digitalization consists in a significantly exact and operational numerous calculation of forecasted results under the conditions of innovative development on the basis of use of factors which in reality reflect the conditions and means of system functioning. The theoretical justification of the strategy of a scientific problem modeling is provided. On the basis of the analysis of retrospective information of results of previous research with the help of a multi factor correlation and regressive analysis the digital modeling was carried out which allows creating the management strategy of a scientific problem solution. An example of optimization of the scientific search management strategy based on matrix transformations is given.

1. Introduction

Science is the most efficient sphere of capital investments. The world experience shows that the profit from capital investments into the science makes up 200 % [1]. Every year the expenditures on R&D in the world grow. For the last 50 years the volume of new knowledge grew about two or three times and at the same time the amount of the information reflecting the results of research grew eight or ten times and the volume of funds which are allocated for R&D increased more than 100 times. For example, in the USA the share of the expenditures in the Gross Domestic Product in 2018 made up 2.74 %, in 1950 this ratio was 1 %. China in 2018 spent 2.12 % of its GDP on science and in 1996 this ratio was 0.6 %. Russia has the 34th rank in the world with 1.1% of expenditures on R&D [2–5].

The research shows that neither the gross expenditures for R&D, nor the expenditures for R&D calculated per a researcher do not give any country the result which can provide the "leading positions in the world science". The volumes of the accumulated investments into R&D, development of scientific schools and R&D experience can be an indicator of these processes. These factors explain, for instance, the leading positions of scientists from Great Britain (70 Nobel Prize Winners) which spends money volumes on R&D comparable with Russia [5]. Science has become a specific human activity. The global organizations sneaked into the scientific institutions. A system of administration and management of science appeared [1, 6–9]. The problem consists in the incompliance of the existing management system of scientific research with the requirements of scientific and

IOP Conf. Series: Materials Science and Engineering **832** (2020) 012044 doi:10.1088/1757-899X/832/1/012044

technological development.

2. The analysis of conceptual foundations of strategy development of the research process

The opinion of specialists of the institute of management problems of the Russian Academy of Sciences is the following: "When we speak about the science management, we should understand that the R&D are the most important management process. Without a literate management of scientific research it is impossible to construct a new digital economy" [6].

The goals of efficiency growth of funds allocated for R&D are connected with the rationing of the work of scientific workers what allows justifying the planning of research and the expenditures on them. The work time standards are one of the most important stimuli and must be found in the foundation of the encouragement system of an efficient work of scientific workers. However the question of rationing of scientific and research work in the scientific community still remains open and disputable as it is possible to assess objectively the justification of the established ratios only taking into account its specific traits [10–12]. The presidium of the Russian Academy of Sciences plans the solution of problems connected with the development of suggestions for criteria development and growth of efficiency of the Russian science [13].

The solution of a set of goals which are connected with the efficiency growth of R&D determined the development of the corresponding standard. The Standard in the clause 4.4 provides "the choice of direction of research with the goal of determination of an optimal variant of research direction on the basis of condition analysis of the problems under analysis including the results of patent research and comparative analysis of variants of possible solutions taking into account the results of forecasted research which were carried out in analogous problems" [14]. Judging from the wording of the focal point of the All Union State Standard it is extremely difficult to determine with a sufficient exactness the optimal variant of research direction.

According to the opinion of the specialists, today an important problem is not only and not so much the use of a computer as a tool of research as the use of informational technologies for the efficiency growth of scientific activity management. New technologies can become a serious support in such areas as planning and the analysis of results of scientific activity [15].

Since 2008 in the Saint Petersburg State University, the informational and analytical system of the R&D maintenance has been working. The information about all current scientific projects which are being implemented by researchers and subdivision of the university is put into this database. The registration in this system is obligatory for all scientific projects. With the help of this system the following tasks are completed. They are: the planning of scientific and research work; record of results of scientific and research activity; analysis and forecast of the activity results; management of scientific research [16].

A possible variant of the creation of working strategies in the Russian Federation on the basis of forecasting are studied and the Presidium of the Russian Academy of Sciences points out that "In the world there exist about 20 models of the economic development but their results vary. It is important to come to a consensus and to adjust all the parameters which are used at the creation of these models. The methodic basis of mathematic modeling is not justified. That's why the mathematic models give various results. The problem consists in the absence of the tools for the decision making. 99 % of Russian strategies cannot be realized as they do not contain quantitative assessments which are presented in forecasts" [17].

To unambiguously determine the priority direction of the research process and its automation, digital models are needed that can reasonably justify the research strategy, taking into account not only quantitative, but also qualitative parameters.

The goal of research is the develop of the model of a unambiguous definition of the management strategy of the research process at the solution of a scientific objective under the conditions of the innovative development taking into account the real conditions on the basis of a digital model [18].

The objectives are:

1. To study the opportunity of quantitative assessment for the justification of the choice of the

strategy for a scientific task solution on the basis of a digital model.

2. To develop the strategy of a scientific task's completion on the basis of a digital model.

3. To solve the task of management strategy optimization of scientific search.

3. Example of the strategy determination of a scientific problem solution on the basis of a digital model

One of the first stages of the fulfillment of a research work is the analysis of the results of the research that has been accomplished earlier. In the result of such an analysis the factors x are identified which influence the target indicator Y. In a general case the challenge consists in the achievement of a desirable result of a target indicator by means of the influence on the factors to achieve. As a rule it is spoken about the search of an extreme point (maximum, minimum). For instance, it can be the increase of the carrying capacity of the road surface or, for instance, the reduction of the accident rate on highways [19]. The statistical information about the materials of research which reflect the values of factors and target indicators which were obtained by a number of researchers with the help of multi-factor correlation and regression analysis can be presented with the help of a function of type: $Y = C_0 \prod_{i=1}^{n} x_i^{\alpha_i}$. Here Y – target indicator; x_i, $i=\overline{1,n}$ factors influencing Y; α_i, $i=\overline{1,n}$ - index of power which characterizes the "input" of x_i into Y; C_0 – coefficient [20]. Let us analyze for instance the digital model:

$$Y=1.231\, x_1^{1.331}\, x_2^{1.502}\, x_3^{0.432}\, x_4^{0.231}\, x_5^{0.753}. \quad (1)$$

It is worth emphasizing that the obtained digital model, for instance (1), describes the hyper surface of a specifically studied process what gives the opportunity of realization on this surface of various operations which are connected with the optimization of the strategy of the solution of a specific scientific problem. In order to visualize the solution of the problem let us limit with a three dimensional model which includes the factors x_1 and x_2 as the factors having the biggest influence on Y. Let us use the simplified digital model of the research process $Y=1.231\, x_1^{1.331}\, x_2^{1.502}$ by means of attributing of various values for x_1 and x_2 and we will get the calculated values of Y. The obtained valued allow constructing a three dimensional graphical model, for instance, figure 1, illustrating the surface of the process under analysis with the help of which the search of management strategies can be carried out, for instance, of a maximum Y. The curves *1Y* and *2Y* are isoquant curves which connect the calculated points with the similar values of target indices.

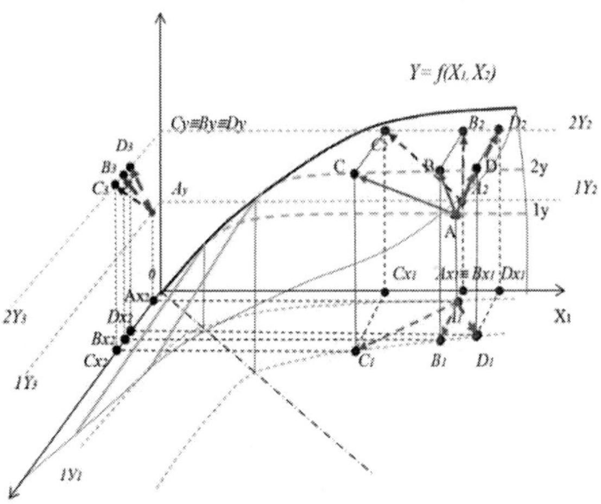

Figure 1. Visualization of management strategy

IOP Conf. Series: Materials Science and Engineering **832** (2020) 012044 doi:10.1088/1757-899X/832/1/012044

The curve $2Y$ connects the calculated points with the similar numerous values of a target indicator, the value of which it is necessary to achieve in the result of scientific research. As we can see from figure 1 one and the same value $2Y$ can be achieved by an eternal set of the combination of factors x_1 and x_2. For instance, the level of the target indicator $1Y$ obtained in the result of research is characterized by its projections A_1, A_2 and A_3. For the achievement of the target indicator $2Y$ from point A an eternal set of variants exists. Let us take three variants, for instance B, C and D. Every point is characterized by projections B_1, B_2, B_3, C_1, C_2, C_3, D_1, D_2, D_3 and Bx_1, Bx_2, Cx_1, Cx_2, Dx_1, Dx_2, which determine the parameters of the factors at the achievement of the target indicator $2Y$. As we can see in Figure 1, the achievement of the target indicator $2Y$ is accompanied by various combinations of factor values.

As we can see in Figure 1 the achievement of the target indicator $2Y$ is accompanied by various combinations of factors. For example, $OCx_1 < ODx_1$, but $OCx_2 > ODx_2$. Consequently, in the existing system under analysis the achievement of $2Y$ can be obtained by means of various values of x_1 and x_2. In dependence on real conditions (economic, technical, organizational and others) the corresponding values of x_1 and x_2 are selected. From the mathematical point of view the optimal variant is AB as AB is a perpendicular from the point A to overlying isoquant to a tangency in point B and is the shortest distance between the isoquants.

The exemplification of the development strategy of a scientific problem solution in progress can be seen well in Figure 2 in projections with numerical marks. The variants of management strategy can be presented by an infinite set.

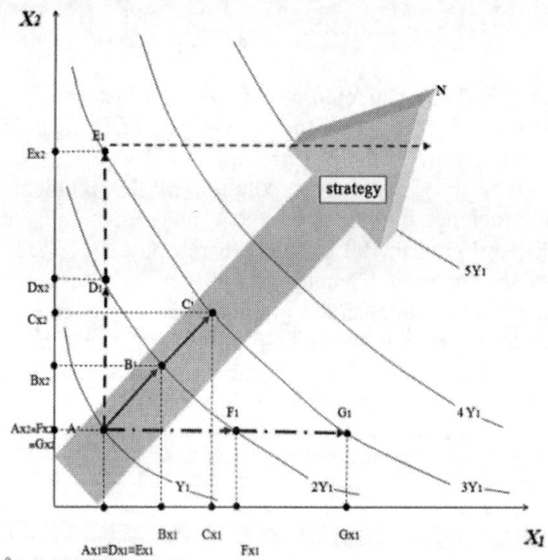

Figure 2. Visualization of strategies in projections with numerical marks

Let us analyze three of them. For example, the variant $A_1Д_1E_1$ at constant values of the factor x_1 and a constant growth of x_2 finally provides the achievement of a goal N.

Analogously the strategy can develop in directions $A_1F_1G_1$. However the direction in $A_1B_1C_1$ obviously and convincingly prove that an efficient strategy has another direction by means of the movement in the shortest distance between the isoquants. It is worth mentioning that a digital model which is being implemented in the projections with indices also reflects qualitative characteristics of a strategy. For example, on the basis of the distance between the isoquants in the same scale at two various strategies it is possible to calculate indirectly the time costs for their realization. The smaller the distance between them, the surface of the digital model has greater steepness and the more intense the influence of factors on the target indicator.

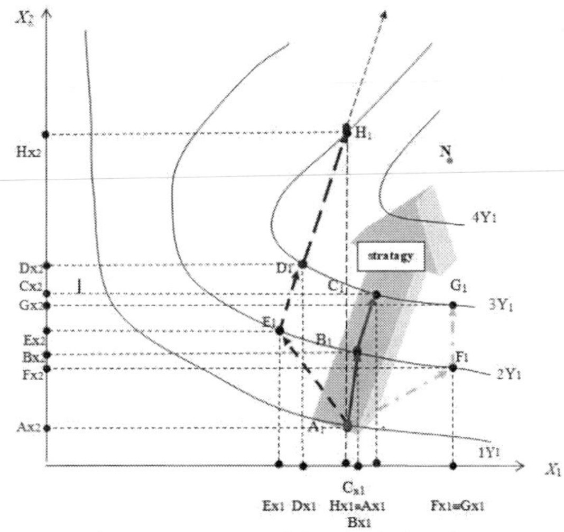

Figure 3. Visualization of consequences of the deviation from an optimal strategy

It is worth exemplifying the consequences from the deviation of the chosen management strategy. For example, in Figure 3 the operational management $A_1E_1D_1$ keeps the direction of the strategy but the dislocation takes place as a result of a significant reduction of the value of factor x_1 what we can see in the parameters Ex_1 and Ex_2. The implementation of the management strategy will not lead to the achievement of the target indicator which is shown in point N. After the achievement of the index $3Y_1$ despite a further growth of factors x_1 and x_2 (Hx_1 and Hx_2), the value of a target indicator will reduce what is seen in the position of point H_1.

The visualization of the strategy on the basis of a three dimensional graphical model can not be found a complete enough as here only two factors x_1 and x_2 of the function (1) out of five are used. For a more detailed determination of the strategy it is indispensible to analyze other combinations of factors.

It is worth mentioning that the direction of the strategy in $A_1B_1C_1$ (Fig. 2 and 3) is determined not only by the shortest distance between the isoquants but also by the real conditions of use of factors' combination which should be optimal.

4. Conclusions

It is proved that on the basis of the digital model it is possible to construct the trajectory of management strategy of a scientific search what will create the prerequisites for the automation of the research activity, for a significant efficiency growth of scientific and technological development. The digitalization is a tool in the management of the scientific search which supposes the use of all management functions: planning, organization, record and control. The digital model is formed on the basis of the set of data types and transformation rules which determine the set of rules of its construction and as a subtype of the general class of the transformation of basic models. The use of a digital model is showed which is formed on the basis of the set of statistical information which reflects various retrospective results of researches of the correlation of target indicators and factors. The visualization of a digital model allows exemplifying the construction of variants of management strategy of a problem's solution and determining an optimal variant as the shortest distance between the isoquants of the surface which is described by a digital model. The prerequisite of the model exactness is the information redundancy in respect to the model of a single object. It is proved that one of the first order digitalization goals consists in a considerably precise and prompt numerical calculation of the forecasted results of the scientific search under the conditions of the innovative development, on the basis of the use of the factors which reflect in reality the conditions and means of the scientific problem solution. On the basis of the retrospective information of the results of previous

IOP Conf. Series: Materials Science and Engineering **832** (2020) 012044 doi:10.1088/1757-899X/832/1/012044

research with the help of multi factor correlation and regression analysis the digital modeling was carried out which allows creating the management strategy of a scientific problem solution.

References

[1] Ryzhakova A V and Manakhov S V 2013 Assessment of the effectiveness of research work in a university *Bulletin of the Russian University of Economics G.V. Plekhanov* **12** 32
[2] Ranking of countries in terms of research and development costs 2019 *Institute for Statistical Studies and Economics of Knowledge, Higher School of Economics* Retrieved from: https://www.vestifinance.ru/articles/104411?page=10. Accessed 18.07.2019.
[3] *R&D Financing in Russia* 2019 Retrieved from: https://helpiks.org/3-21533.html. Accessed 22.07.2019
[4] *The national economy of the USSR in 1990* (Moscow), pp 9, 307
[5] Podosokorskiy N I 2019 *In terms of spending on science as a percentage of GDP, Russia ranked 35th in the world* Retrieved from: https://philologist.livejournal.com/9605867.html. Accessed 22.07.2019
[6] Scientific and Practical Conference 2017 *"Problems of R&D Management – 2017"* Institute of Management Problems named after V.A. Trapeznikov of the Russian Academy of Sciences Retrieved from: http://www.sib-science.info/ru/conferences/fano-gov-v-01112017. Accessed 18.07.2019.
[7] Shchedrovitskiy G P 2019 *Science Search Strategy. Forum Fund named after Shchedrovitsky G P* Retrieved from: entries|archive|friends|userinfo. Accessed 18.07.2019.
[8] Dzyun T 1983 *Theory of science* Translated from the Japanese (Moscow: Nauka Publ.) 192 p
[9] Riehl Alois Adolf 2011 *The theory of science and metaphysics from the point of view of philosophical criticism* (Moscow: "Librokom" Publ.) 448 p
[10] Feoktistova O A 2014 Rationing of research work: methodological approaches *Internet J. "Science of Science"* 5(24) Retrieved from: http://naukovedenie.ru.
[11] Kushnir A B 2010 Features of creative work in matters of rationing *Bull. of the Res. Instit. of Labor.* 23(34) 64–6
[12] Feoktistova O A 2014 Research Cost Planning: A Design Approach *Research Financial Institute Financial magazine* 1
[13] The meeting of the Presidium of the RAS and members of the Academy 25 June 2019 Retrieved from: https://scientificrussia.ru/articles/vak-ekspertiza-otsenka-deyatelnosti-nauchnye-sovety-i-drugie-temy?utm_source=pulse_mail _ru&utm_referrer=https%3A%2F%2Fpulse.mail.ru. Accessed 03.05.2019
[14] GOST 15.101-98 *Procedure for the implementation of scientific research*
[15] Emelyanov A I 2019 *10 theses on automation of science management* Retrieved from: https://habr.com/ru/post/123102/. Accessed 03.05.2019
[16] *Information and analytical system for supporting research* Retrieved from: http://philosophy.spbu.ru/3597. Accessed 03.06.2019.
[17] *Mathematical modeling of the Russian economy* Meeting of the Presidium of the Russian Academy of Sciences 11 June 2019 Retrieved from: https://scientificrussia.ru/articles/matematicheskoe-modelirovanie-ekonomiki-rossii?utm_source=pulse_mail_ru&utm_referrer=https%3A%2F%2Fpulse.mail.ru. Accessed 05.06.2019.
[18] Borovik V S, Borovik V V and Prokopenko Yu E 2013 Model for managing the implementation of the new technology based on the production function *Economic analysis: theory and practice* 42(345) 25–30
[19] Silyanov V V and Domke E R 2016 *Transport and operational qualities of roads and city streets* (Moscow: Academy), 352 p
[20] Borovik V S and Lukin V A 1995 Justification of measures to ensure traffic safety *Car roads* **3–4** 12–3

IOP Conf. Series: Materials Science and Engineering **832** (2020) 012045 doi:10.1088/1757-899X/832/1/012045

InfraBIM Open paradigm as the driver of informatization of the road sector in Russia

V N Boykov[1], A V Skvortsov[2] and V A Gurev[1]

[1]Moscow automobile and road construction state technical university (MADI), 64, Leningradsky ave., Moscow, 125319, Russia
[2]Tomsk State University, 36, Lenin str., Tomsk, 634050, Russia

E-mail: boykovvn@mail.ru

Abstract. The status of development and implementation of road infrastructure information modeling in Russia is reviewed. The InfraBIM Open paradigm is the basis for formation of information modeling processes practiced in Russia. The paradigm also solves the issue of software hybrid replacement allowing the use of both domestic and imported software packages. The current state of normative documentation regulating information modeling in Russia is analyzed. The evaluation of three BIM pilot projects is given, also the list of software packages and their modules having capabilities for implementation of road information modeling processes is provided.

1. Introduction

The OpenBIM paradigm is formulated and supported by the International consortium BuildingSmart. The paradigm is a universal approach to collaborative design, construction and operation of buildings and structures based on usage of open standards and working processes [1]. Specialists engaged in informational support of the life cycle of transport infrastructure note that there are significant differences in the formation of models for industrial and civil construction objects and those for transport infrastructure. In this regard while sharing the general principles of the OpenBIM paradigm transport engineers formulate another InfraBIM Open paradigm.

The principles and mathematical apparatus of creating information models have long been developed and tested, but the question of supporting the interoperability of these models at all stages of their life cycle remains open. One of the most promising approaches to ensuring the interoperability of models is the support of the open exchange format IFC developed by the BuildingSmart consortium and proposed for the entire professional world community [2].

Support for the IFC format does not solve the whole complex of problems for the implementation of BIM methodology, as it affects all aspects of engineering practice: organizational, technical, technological and to some extent even ideological.

In Russia, the issue of BIM implementation at the state level had been discussing and resolving since March 2014 when this topic was first discussed at a meeting of the Presidium of the presidential Council for modernization and innovative development of the economy [3]. The meeting was dedicated to construction, and BIM implementation topic was designated as one of the goals of innovative development of the construction industry at the state level. Over the past 5 years, we have come a long way from theoretical discussions on BIM to the creation of the first regulatory documents [4], the development of the first pilot projects [5] and the evolution of the domestic software [6].

With regard to transport infrastructure, both experts of railway industry [7] and specialists in motor roads made certain steps in the development and implementation of InfraBIM Open whose achievements we will discuss next.

2. Normative documentation

At the initial stage of BIM development in the Russian Federation, foreign experience was studied including normative documents in the field of information modeling. In 2016 the state company "Avtodor" developed the first regulatory industry document in the field of BIM: standard of organization STO AVTODOR 8.6-2016 "Organizational and technological maintenance of road information modeling on all stages of road lifecycle". This document made it possible to carry out the first pilot projects and to accumulate some experience for the further evolution of the road information modeling process.

In 2016-2017, on the instructions of the Federal road Agency (FDA) "Rosavtodor", a number of research works in the field of BIM were performed. As a result, we received road industry methodological document named ODM 218.3.105-2018. "Methodical recommendations on interaction of developers of design and working documentation on pilot projects for the construction, capital repair and reconstruction of roads with the application of BIM technologies", approved and put into effect by the decree of FDA "Rosavtodor" from 05.06.2018 No. 2084-r.

Development of Methodical recommendations pursued the following purposes:

- Improving the quality of project documentation and reducing the time of construction (reconstruction, repair) of roads through the implementation of the paradigm "Information modeling in the life cycle of roads".
- Improving the management system of the road network and improving the efficiency of capital investments at all stages of the life cycle of roads by introducing information modeling technology into engineering and management processes.
- Motivation of participants in the road construction process to the formation of the market of information modeling technologies and the creation of appropriate organizational, regulatory, technical and technological foundations.

The introduction of these Methodical recommendations will allow to conduct in 2019 a series of pilot projects at a new level of quality which, of course, should enrich the theory and practice of road information modeling at the stages of design and construction of roads.

Currently, also on the instructions of the FDA "Rosavtodor" there is developing two preliminary standards (PNST) to expand and detail the processes of roads information modeling. Preliminary standards should regulate the development of both the road information model and its individual components.

Development of industry-wide regulations and standards in the field of InfraBIM does not exclude the need to develop internal regulations (BIM-standards or STO), which should take into account the specifics of the work and human resources potential of a particular legal entity, as well as the application of international standards in the field of BIM [8–10].

3. Pilot projects

One of the first projects from the viewpoint of information modeling is the project "Feasibility study on the connecting road from M-4 "Don" to A-105, access to the airport "Domodedovo" (designer is "GorKapStroy" LLC). The purpose of this work is the formation of the road information model at the "Feasibility study" stage for the subsequent transfer of this model to the "Design" stage and, as a consequence, a significant time reduction for the development of project documentation.

For the first time, the mapping of geological data in the form of a 3D model was tested (Fig. 1) and cross-sections as part of the 3D-model of the road significantly increased the visibility of geological information in the road model.

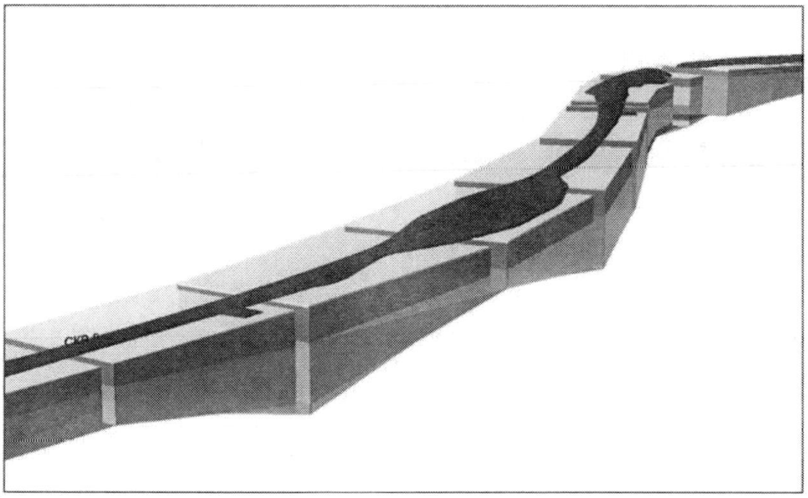

Figure 1. 3D model of the geological structure in the area of the right-of-way

In addition, the technology of detailed 3D study of overpasses, including spans, supports and foundations (pile grillage) was first tested at the feasibility study stage (Fig. 2).

In the development of the road information model the libraries of typical elements were used: reinforced concrete structures; metal structures; models of road signs (including pillars and foundations); fences; lighting, etc.

Figure 2. 3D-display of overpasses, including the underground part of the structures

The prepared documentation of the feasibility study consisting of drawings, explanatory note and the road information model was transferred to the Customer for its subsequent detailing at the "Design" stage.

The processes of road information modeling were implemented with most completeness and depth in the project "Far Western bypass of Krasnodar" (Fig. 3).

IOP Conf. Series: Materials Science and Engineering **832** (2020) 012045 doi:10.1088/1757-899X/832/1/012045

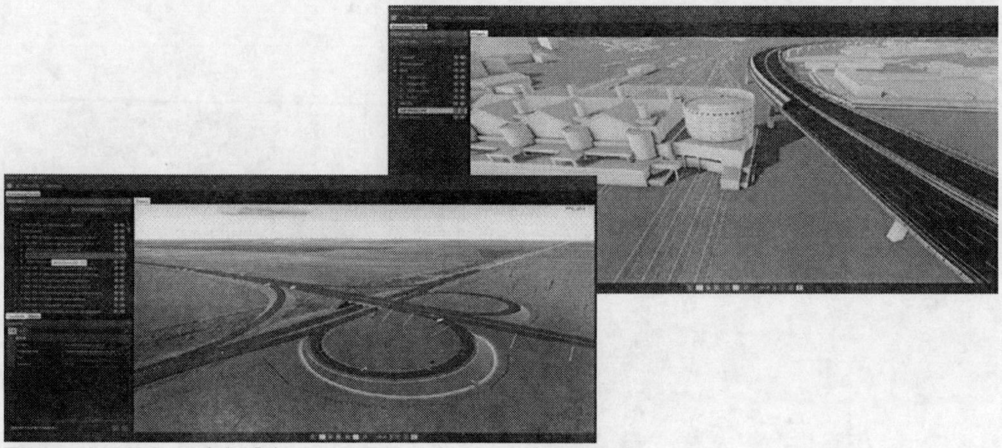

Figure 3. BIM-model of the Far Western bypass of Krasnodar

This project carried out by the engineering group "Stroyproekt" can be considered unique due to two circumstances. First, several information models were consistently developed: Territorial planning, Engineering surveys, pre-Design model and Design model. Secondly, the domestic software package S-Info for life cycle management of transport infrastructure facilities has been tested.

4. Software products

Both domestic and foreign software products are used in the practice of road designing in the Russian Federation.

Following the principles of OpenBIM concept and solving the issue of hybrid substitution software allowing the use of both domestic and imported packages, we propose a hypothetical range of software implementing InfraBIM at the design stage of highways.

As you can see from the list, these are all fairly well known software products. The software package S-Info ("S-Info" LLC, Saint Petersburg, Russia) is the only new in this line, but its appearance makes the line of Russian software products relatively self-sufficient.

S-Info is the first domestic software package that allows you to assemble a consolidated road information model from the elements developed in various software modules. S-Info can be used to generate queries, transmit, coordinate, synchronize, and view assembled model data, serving as a common data environment (CDE), and to address application engineering and management tasks in the design, construction, and operation of highways.

- **Engineering survey**: CREDO software package (Credo Dialogue LLC, Belarus), TopoCAD (Adtollo AB, Sweden)…
- **Road design**: IndorCAD/Road (IndorSoft, Russia), Robur-Dorogi (Topomatik, Russia), Credo-Dorogi (Credo Dialogue, Belarus), Autodesk Civil (Autodesk, USA), …
- **Design of culverts and bridges**: IndorCulvert & IndorBridge modules (IndorSoft, Russia), Renga (Renga Software, Russia)+Revit (Autodesk, USA), …
- **Design and calculation of pavements**: IndorPavement, (IndorSoft, Russia), Robur–Dorozhnaya Odezhda (Topomatik, Russia), Credo-RADON (Credo Dialogue, Belarus), …
- **Estimate calculations**: ABC-4 (NPP ABC-N, Russia»), 1C:Smeta (TsSP Erikos, Russia), 5D Smeta (NTC Gektor, Russia), …
- **BIM model assembly**: S-Info (S-Info, Russia), Autodesk Navisworks (Autodesk, USA), Tekla BIMsight (Tekla, Finland), …
- **Common Data Environment (CDE)**: S-Info (S-Info LLC, Russia), IndorBIM-Server (IndorSoft, Russia), ProjectWise (Bentley, USA), Pilot-ICE (Askon, Russia)…

- **Scheduling**: S-Info (S-Info, Russia), Synchro (Bentley, USA), Microsoft Project (Microsoft Corp., USA), …

As for the list of open exchange formats ensuring the interoperability of information models during their life cycle, here first it is necessary to talk about the support of the IFC format proposed by the buildingSMART consortium. However, the range of engineering and management tasks within InfraBIM is so wide and diverse that it requires support for other formats such as LandXML, CityGML and FBX.

5. Conclusions

The issue of road information modeling during their life cycle is solving at the practical level in Russia since 2016, including development of both regulatory documentation and domestic software products, implementation of pilot projects. Educating and professional development of engineers remain a weak point in this chain of information modeling implementation processes. The leading transport universities of the Russian Federation are engaged in solving this problem: Moscow automobile and road construction state technical university and the Russian university of transport.

6. References

[1] Boykov V N, Skvortsov A V and Sarychev D S 2018 Digital highway as industry segment of digital economy *Transp. Rossiysk. Fedetat.* **2** 96–100

[2] *Industry Foundation Classes Release 4* (IFC4 Official Release) Retrieved from: http://www.buildingsmart-tech.org/ifc/IFC4/final/html/ (accessing date: 10.01.2016)

[3] Korol M V 2014 BIM: information modelling — digital century of construction industry *Stroymetall* **39**, 26–30

[4] Sarychev D S and Skvortsov A V 2017 Drafts of Russian standards and regulations in the field of BIM for roads *SAPR I GIS avtomobilnykh dorog – CAD & GIS for roads* **1(8)** 44–8

[5] Boykov V N, Neretin A A and Skvortsov A V 2015 Probation of road information models at the stage of project execution *SAPR I GIS avtomobilnykh dorog – CAD & GIS for roads* **2(5)** 30–6

[6] Boykov V N 2014 Information technology for road lifecycle support *SAPR I GIS avtomobilnykh dorog – CAD & GIS for roads* **1(2)** 1, 6–7

[7] Kuprijanovskij V P, Sinjagov S A and Dobrynin A P 2016 BIM – Digital Economy. How to achieve the success? A practical approach to the theoretical concept. Part 1: Approaches and main advantages of BIM *Int. J. of Open Inform. Technol.* **4(3)** 1–8

[8] BS 1192-4:2014. Collaborative production of information. Part 4: *Fulfilling employers information exchange requirements using COBie – Code of practice* (Draft). 2014 45 p

[9] BS 8536-2:2016. Briefing for design and construction. Part 2: *Code of practice for asset management (Linear and geographical infrastructure)* 2015 98 p

[10] BIM e-Submission Template Guides. BCA *Building and Construction Authority* 2016

IOP Conf. Series: Materials Science and Engineering **832** (2020) 012046 doi:10.1088/1757-899X/832/1/012046

Evaluation of the impact of anti-alcohol policy in the regions of the Russian Federation on road traffic accidents

A B Chubukov[1], V T Kapitanov[1], V V Silyanov[1] and O Yu Monina[2]

[1]Moscow Automobile and Road State Technical University (MADI), 64, Leningradsky ave., Moscow, 125319, Russia
[2]FSIRDI for Road Safety of MIA of Russia, 17, Poklonnaia str., Moscow, 121170, Russia

E-mail: silyanov@bk.ru

Abstract Based on the earlier works performed by the authors, in which the role of socio-economic factors in road traffic accidents was noted, an analysis was carried out, models were developed. On their basis the analysis is developed and models are carried out. Regression equations allow highlighting the main factors affecting the number of deaths in accidents caused by drunk drivers in the regions of the Russian Federation.

1. Introduction

The order of the Government of the Russian Federation from 08.01.2018 № 1-p approved the Strategy of traffic safety in the Russian Federation for 2018–2024. This determined the main activities of various authorities till 2024 by achievement of reduction of social risks up to 4 deaths per 100 thousand of the population, and by 2030 – the zero death toll in the road accidents (RA).

The main tasks for the implementation of the Strategy are related to the improvement of the road safety management system, the change in the behaviour of its participants, the modernization of the street-road network, as well as the change of organizational and legal mechanisms for the admission of vehicles and their drivers to participate in road traffic.

The adoption of management decisions in the field of road safety makes it necessary to develop models reflecting the connection of accidents with the state of the DCRE complex ("driver-car-road-environment"). Statistical modelling techniques are used to evaluate this relationship [5, 7, 9, 10]. At the same time, regions of the Russian Federation have been selected as objects of research, which provides an opportunity to obtain a large volume of statistical information and significant conclusions.

2. Methods

In order to analyze, develop and justify a set of measures aimed at improving road safety, methods of correlation and regression analysis and worked-out approaches to the construction of a regional accident model are used [18].

In this article, the number of deaths in road accidents is taken as a criterion of the state of road traffic accidents.

Different options were considered to approximate the number of deaths in road accidents, but the best accuracy characteristics provide dependence [18]:

$$y = y_1 + \sum_{i=1}^{n} \overline{a_i} \cdot x_i + a_0,$$

where $a_0, \overline{a_i}$ $(i = 1,..., n)$ – parameters;

y – number of deaths;

n – number of parameters;

x_i – indicators of the process under consideration $(i = 1,..., n)$.

For the regions of Russia, the relationship between the number of deaths in road accidents y_1 with the population $\overline{x}_1$ and number of vehicles $\overline{x}_2$ for the 2015–2018 period is as follows:

$$y_1 = 0.427676 \cdot \overline{x}_1^{0.593329} \cdot \overline{x}_2^{0.296665},$$

for the period 2014-2017, this formula is almost the same

$$y_1 = 0.419425 \cdot \overline{x}_1^{0.60419} \overline{x}_2^{0.302095}.$$

There is a satisfactory match between the calculated and actual values of the considered accident indicator (rank correlation coefficient is 0.870906). However, additional indicators are used to improve the accuracy of calculations and to understand the nature of the process. For this purpose, an array of data was formed from 2015 to 2018, characterizing the socio-economic, natural and climatic state of the regions, as well as the measures taken to influence road users. 340 data sets relating to all subjects of the Russian Federation were considered. Later, the Republic of Crimea, the Chechen Republic and the city of Sevastopol were removed from consideration, as the statistics related to them introduce significant distortions in the results of the calculation, forming statistical emissions. Then 328 sets of indicator values (82 regions) were analyzed.

The following procedure for modelling road traffic accidents has been carried out.

A list of indicators is formed and a criterion (number of deaths in an accident) is determined. Correlation analysis is performed. For construction of regression models, indicators are selected, which correspond to the rank coefficient of Spirman correlation exceeding the threshold value at significance level 0.05 (for data array of dimension 328 the threshold value is 0.1095) [7].

Based on the results of the rank correlation analysis (rank correlation of Spirman coefficient values are given below in parentheses for data (2014–2017)/(2015–2018)), it is proposed to use the following indicators (factors) [1]:

- environment conditions, in points (0,440019/ 0,436966; x_9) [15];
- average annual temperature in the capitals of the constituent entities of the Russian Federation, degree Celsius (0,308329/0,311892; x_{10})[16];
- density of public roads of federal, regional or inter-municipal and local importance with hard surface for the constituent entities of the Russian Federation, km of roads per 1000 km2 of territory (0,406005/ 0,390925; x_{11}) [6];
- investment potential of Russian regions, share in all-Russian potential, % (0,790042/0,79297; x_1) [14];
- investment risk of Russian regions (–0,57448/–0,59530968; x_3) [14];
- level of socio-economic status (0,608389/0,630639; x_5) [2];
- final quality of life score, rating score (0,518697/0,51704; x_4) [4];
- number of registered administrative offences in the field of road safety, thousands (0,786226/ 0,806142; x_7) [3];
- number of registered administrative offences by means of automatic photo and video recording, thousands (0,671661/0,6754; x_8) [3];
- age of vehicles fleet (–0,35013/–0,34021; x_{13}) [12];
- annual sales of petrol in the regions (0,7936113; x_6). Data on the volume of gasoline sales in the regions correspond to 2015 [13].

The paper [19] also used indicators: level of economic development and sale of alcoholic beverages (beer). Due to the change in statistical accounts, these indicators are excluded from consideration.

In a number of works of the authors [1, 8, 11, 19] an attempt was made to assess the impact of the situation related to alcohol consumption on road traffic accidents. The main problem encountered is the choice of an adequate criterion for assessing the situation. In particular, the following were considered:

- the number of deaths due to alcohol;
- beer sales volume;
- sales of alcoholic beverages (vodka and other products) and beer;
- number of patients registered in medical and preventive institutions with the diagnosis of alcoholism and alcoholic psychosis.

Satisfactory correlation of the number of deaths in road accidents was noted only with the volume of beer sale. As for the sale of other alcoholic products, there seems to be a high share of counterfeiting in this part of the alcoholic market (according to experts, about 70 % [17]), as well as fake excise stamps [17] Therefore, there is a lack of correlation between official statistics and road traffic accidents.

In connection with the above, the result of the expert and analytical study "Rating of Sobriety of Regions-2018", carried out by the federal project "Sober Russia" together with "The Russian Television and Radio Broadcasting Company" (RTR), is of interest. The rating was the result of analysis, synthesis and systematization of regional and federal statistics. The rating used 6 criteria:

- A1: "Number of deaths by major class and individual causes of death (alcohol poisoning)";
- A2: "Number of patients registered in medical and preventive institutions diagnosed with alcoholism and alcoholic psychosis";
- A3: "Number of offences committed by intoxicated persons (unit)";
- A4: "Regional volume of all alcoholic products sold in litres of pure alcohol";
- A5: "Number of offences related to illegal production and trafficking of ethyl alcohol and alcoholic products;
- A6: "Strength of regional anti-alcohol legislation, – calculated on the basis of A6.1. sub-indicators "Number of hours of prohibition of sale of alcohol per day" and A6.2. "Number of days in the year when the sale of alcoholic products is completely prohibited".

In order to increase objectivity and representativeness, sales volumes of all types of alcoholic products for which official statistics were available were taken into account. At the same time, each of the types of alcoholic products was studied separately, and then was integrated into one criterion A4: "Sale of alcoholic products in litres of pure alcohol per capita". Sales volume for the following types of alcoholic products was taken into account:

- vodka;
- alcoholic beverage products with content of ethyl alcohol up to 25 %;
- alcoholic beverage products with ethyl alcohol content above 25 %, cognac and cognac beverages;
- cognac and cognac drinks;
- wine (natural);
- sparkling wines (champagne);
- a wine beverage produced without adding ethyl alcohol;
- a wine beverage produced with the addition of ethyl alcohol;
- liqueur wines, fruit and others;
- cider, poire, honey;
- beer;
- beer-based beverages.

The values of A1, A2, A3 and A5 were used as per 100,000 population for each subject of the Russian Federation. The value of indicator A4 was used as per capita of the subject of the Russian Federation, and the sub-indicators A6.1. and A6.2. were used in absolute form.

The coefficient of rank correlation of sobriety of regions rating x_2 and the number of deaths in road accidents is –0.32935, which is higher than the threshold value.

In this paper, a significant number of indicators are used for modelling, including interrelated ones, which makes it difficult to interpret the results. In many multivariate statistical processing tasks, the key figures that change most when an object changes are generally of interest. Therefore, it is a question of the possibility to reduce the dimension of the problem being solved. For this purpose, the method of main components [10] is applied, using which projections of observation points are built on directions of the largest data spread. Of the total number n, the principal components leave the most significant, that is, making the maximum contribution to the dispersion. The work [10] notes that the use of key components is most useful in situations where indicators share a common physical nature. On this basis, the indicators below are divided into three groups – socio-economic, administrative and other. At that algorithm of construction of main components is used, which is given in [10].

As a result, it is necessary to note the high values of the coefficient of rank correlation of gasoline annual sales in regions with a list of socio-economic indicators (investment risk of Russian regions is -0.64941283, final assessment of quality of life is 0.6139, investment potential of Russian regions is 0.829505, level of socio-economic situation is 0.715485), which is the basis for inclusion of this indicator in the mentioned list.

For the list of socio-economic indicators, the main components are:

$$y_1^1 = -0.442 \cdot x_3 + 0.501 \cdot x_4 + 0.403 \cdot x_1 + 0.504 \cdot x_5 + 0.371 \cdot x_6;$$

$$y_2^1 = 0.568 \cdot x_3 - 0.234 \cdot x_4 + 0.531 \cdot x_1 - 0.116 \cdot x_5 + 0.573 \cdot x_6;$$

$$y_3^1 = 0.228 \cdot x_3 + 0.118 \cdot x_4 + 0.622 \cdot x_1 + 0.121 \cdot x_5 - 0.729 \cdot x_6;$$

$$y_4^1 = 0.530 \cdot x_3 + 0.015 \cdot x_4 - 0.370 \cdot x_1 + 0.762 \cdot x_5 - 0.022 \cdot x_6;$$

$$y_5^1 = 0.387 \cdot x_3 + 0.825 \cdot x_4 - 0.177 \cdot x_1 - 0.370 \cdot x_5 + 0.042 \cdot x_6.$$

The relative fractions of the total dispersion due to one, ..., five main components are 0.677, 0.826, 0.935, 0.970, 1.0, respectively. Acceptable accuracy is achieved by the first three components used below.

For the considered indicators characterizing the law enforcement practice of the State Automobile Inspection, the main components are as follows:

$$y_1^2 = x_7 + 0 \cdot x_8;$$

$$y_2^2 = x_7 + 0.001 \cdot x_8.$$

The relative fractions of the total variance due to one, two principal components are 0.999, 1.0, respectively. The first component was used.

As a result, the regression equation (statistical model) for the accident death toll y in the region is:

$$y' = 1.3731368 \cdot y_1 - 7.9096146 \cdot x_9 + 4.54522087 \cdot x_{10} + 0.9232238 \cdot x_2 - 0.00472717 \cdot x_{11} + 0.548287544 \cdot x_{13} + 14.51177757 \cdot y_1^1 - 21.302621275 \cdot y_2^1 - 96.755847 \cdot y_3^1 - 0.0175580478\ 6 \cdot x_7 + 31.9557$$

Moving from the main components to the key figures, we get the regression equation in expanded form:

$$y' = 1.3731368 \cdot y_1 - 7.9096146 \cdot x_9 + 4.54522087 \cdot x_{10} + 0.9232238 \cdot x_2 - 0.00472717 \cdot x_{11} + 0.548287544 \cdot x_{13} - 40.5744 \cdot x_3 + 0.838 \cdot x_4 - 65.6456 \cdot x_1 - 1.9224 \cdot x_5 + 63.7125 \cdot x_6 - 0.0175580478\ 6 \cdot x_7 + 31.9557 \quad (1)$$

The correlation coefficient of actual values of the death toll in road accident in territorial subjects of the Russian Federation in 2018 and calculated values (using formula (1)) is equal 0.88881944, the Spirman coefficient of rank correlation is equal to 0.875255744.

Figure 1 graphically shows the estimated and actual accident deaths.

Weight in the % of composed values (components) in a formula received as the component relation to the death toll, at summation on regions (i.e. values for Russia): a component with population and number of vehicles is 136; socio-economic indicators (including gasoline) is 45.4 (investment risk of

Russian regions is 4.51; final assessment of life quality is 15.6; investment potential of Russian regions is 32.6; socio-economic status is 34.3; gasoline – 10.5); law enforcement practice of the State Automobile Inspection (considered indicators) is 8.71; environment conditions is 12.3; average annual air temperature is 5.75; sobriety rating is 13.7; road density is 5.1; the age of the vehicles fleet is 3.1. At the same time it must be kept in mind that depending on the sign of the component, the role of the factor in the accident (increase or decrease) differs in the formula.

Let us note that the weight contribution of undescribed (unknown) indicators included in the free term in the regression equation is within 13.3 %, which is quite good with such a large amount of data.

It should be mentioned that socio-economic indicators (investment potential, level of socio-economic situation of the region, annual sales of gasoline), sobriety rating, natural conditions and average annual air temperature, density of roads and others have the most significant impact on accidents within the framework of the macro approach. Since formula (1) includes the sales of petrol, transit flows and actual vehicle use are taken into account.

A similar approach was used to model the number of deaths caused by drivers driving vehicles with signs of intoxication y'' [3]. The significant variation of regional statistics in 2015–2018, as well as the relatively small value of this indicator, leads to the need to average its values by year when building the model. With this in mind, the model is as follows:

$$y'' = 0.289 \cdot x_1^{-0.50478} \cdot x_2^{-0.25239} + 4.30823 \cdot x_9 + 0.61 \cdot x_2 - 0.02243496 \cdot x_{11} + 0.356928 \cdot x_{13} + 1.491247 \cdot y_1^1 - 6.8342156 \cdot y_2^1 - 16.039655 \cdot y_3^1 - 0.000973949 \cdot x_7 - 39.85758 \quad (2)$$

The factors of rank correlation of the considered indicator and model parameters are equal: natural conditions (x_9) is 0.363736; sobriety rating (x_2) is -0.10896; road density (x_{11}) is 0.19348; age of vehicle fleet (x_{13}) is 0.30402; main components for the list of socio-economic indicators y_1^1, y_2^1, y_3^1 – are 0.578662, 0.50325, 0.56612, respectively; law enforcement practice of the State Automobile Inspection (x_7) is 0.681336; the number of road fatalities according to population and number of vehicles y_1 is 0,761247; average annual air temperature (x_{10}) is 0.016998.

The analysis showed that the population, the number of vehicles, the sobriety rating (48.52 %), natural conditions (36.07 %), the level of socio-economic situation of the region (–25 % – including gasoline), density of roads (–13 %) have a significant impact on the criterion under consideration. The impact of the remaining indicators is relatively small. In particular, the law enforcement activity of the State Automobile Inspection (the considered indicator) is –2.6 %.

Correlation coefficient of actual and calculated values is 0.802353, coefficient of rank correlation is 0.7867.

Figure 2 shows the estimated (formula (2)) and actual (4-year average) values of the fatalities number caused by drivers driving vehicles with signs of intoxication.

It should be noted that there is a close relationship between the total number of deaths in road accidents $\overline{x}$ and the number of deaths in road accidents caused by drivers in the state of intoxication and who refused to undergo a medical examination for the state of intoxication [3] $\overline{y}$ in the region (the correlation coefficient of actual and estimated values is 0.823949, the rank correlation coefficient is 0.793005). The corresponding relationship is as follows:

$$\overline{y} = 0.137345257 \cdot \overline{x} + 11.6808778 \quad (3)$$

and the results of the calculations are presented in figure 3, the estimated and actual (average for 4 years) values of the deaths number in road accidents due to the fault of drivers in the intoxication state, as well as those who refused the medical examination.

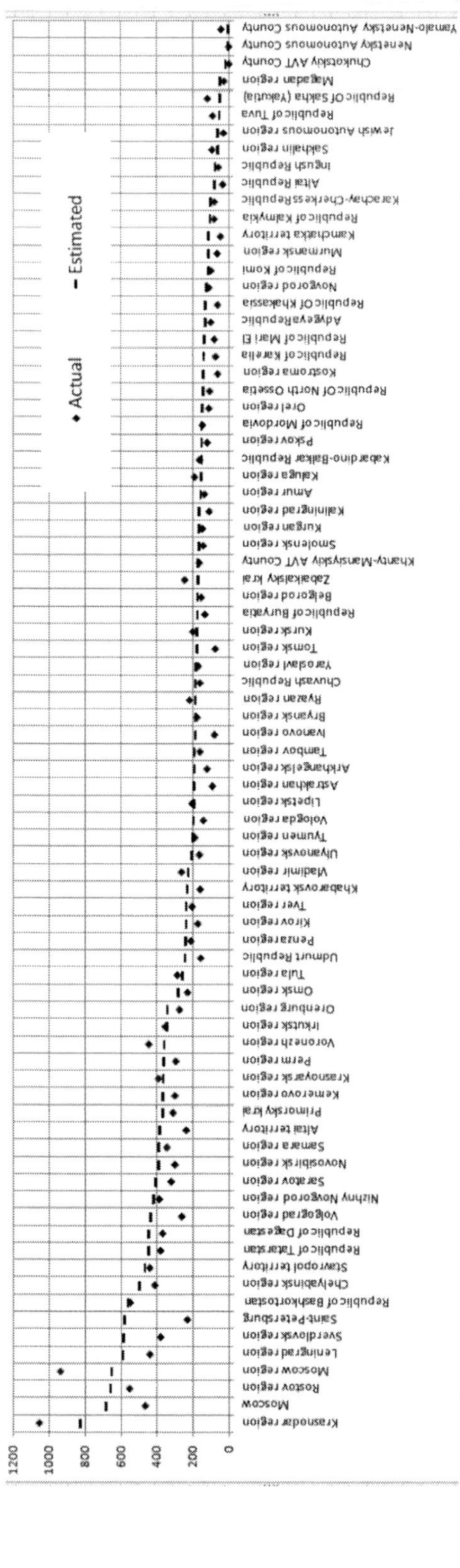

Figure 1. Estimated and actual values of the number of deaths in road accidents in the constituent entities of the Russian Federation in 2018

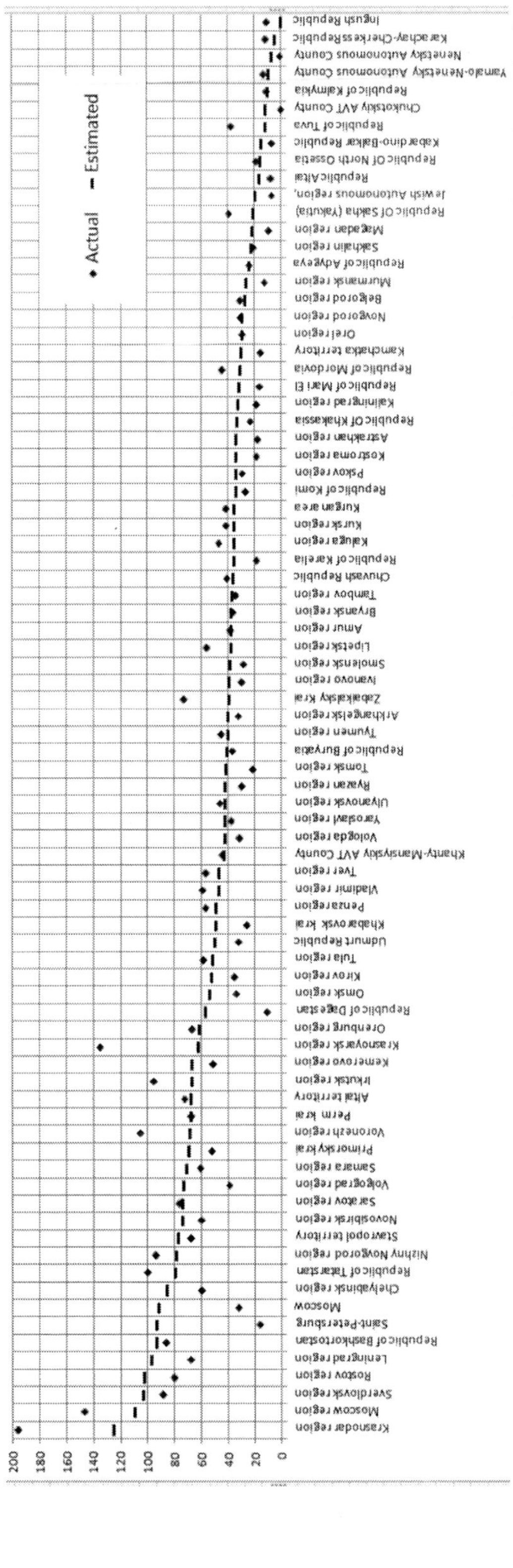

Figure 2. Estimated and actual values of the number of deaths in road accidents caused by drivers driving vehicles with signs of intoxication

IOP Conf. Series: Materials Science and Engineering **832** (2020) 012046 doi:10.1088/1757-899X/832/1/012046

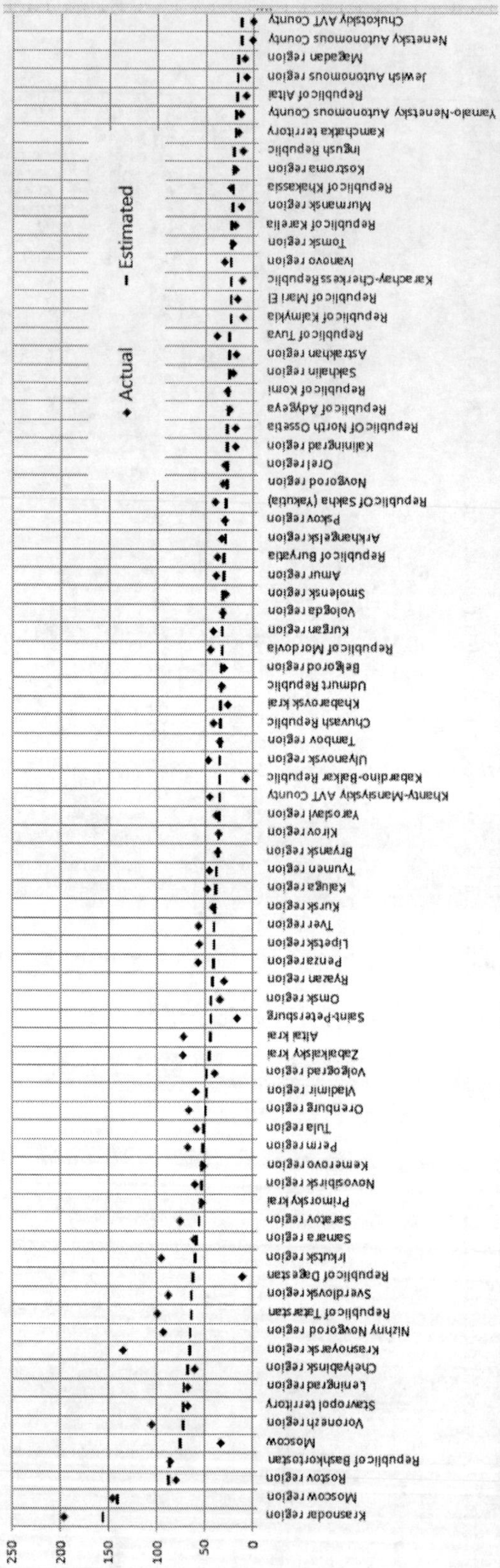

Figure 3. Estimated (formula (3)) and actual values of the deaths number in road accidents caused by drivers driving vehicles with signs of intoxication

IOP Conf. Series: Materials Science and Engineering **832** (2020) 012046 doi:10.1088/1757-899X/832/1/012046

3. Conclusion

In order to intensify anti-alcohol work, the federal project "Sober Russia" offered expert, legal, analytical and information support in the development of measures to protect the life and health of citizens within the framework of agreements with the regions of the Russian Federation. The agreements can contribute to the activation of new approaches aimed at protecting the life and health of Russian citizens, including the field of road safety.

The presented results of the study, according to the authors, can be used to assess the role of alcohol in road traffic accidents in relation to a specific region, as well as in Russia and to make appropriate decisions.

References

[1] Chubukov A B, Kapitanov V T, Monina O Yu and Silyanov V V (2014). Calculation of the Number of Fatalities in Road Accidents on the Basis of Socio-Economic Indicators *Sci. and Technol. in the Road Sector* **3**

[2] *Rating of investment attractiveness of Russian regions* Retrieved from: http://www.riarating.ru (viewed on 10.08.2019).

[3] *Official site of the State traffic inspectorate* Retrieved from: http://www.gibdd.ru.

[4] *Income level, final rating score of regions on quality of life, level of economic development, demographic situation* Retrieved from: 1prime.ru/ratings/20131217/773219432.html (viewed on 25.07.2019).

[5] Dubrov A M et al 2011 *Multidimensional statistical methods* (Moscow: Finance and Statistics) 352 p

[6] *Federal State Statistic Service* [Rosstat] 2018 Retrieved from: http://www.gks.ru (viewed on 10.08.2019)

[7] Kobzar A I 2006 *Applied Mathematical Statistics* (Moscow: Fizmatlit) 816 p

[8] Chubukov A B, Kapitanov V T, Monina O Y and Kosheleva P I 2012 Qualitative and quantitative analysis of accidents in the Russian Federation *Sci. and technol. in the road industry* **4**

[9] Ivchenko G I and Medvedev Yu I 1984 *Mathematical Statistics* (Moscow: Higher School) 248 p

[10] Ayvazyan S A 1989 *Applied Statistics.Classification and Reduction of Dimensionality* (Moscow: Finance and Statistics) 607 p

[11] Chubukov Al, Kapitanov V, Silyanov V and Monina O 2016 Simulation of region mortality rate in road accidents identifies the relationship of the vehicles which have arrived from other regions *Sci. J. of Transport.* Especial issue no. 07 International cooperation Journals MADI–SWJTU–UTC 129–136 (ISSN 2410-9088) (Moscow – Chengdu – Hanoi) Retrieved from: http://lib.madi.ru/sjt/index.html

[12] Retrieved from: https://www.autostat.ru (viewed on 27.07.2019)

[13] *Oil products and natural gas market in Russia* results 2015 Retrieved from: http://www.fuelbroker.ru/upload/iblock/ (viewed on 27.07.2019)

[14] *Rating agency "Ekspert"* Retrieved from: https://raexpert.ru/rankingtable/region_climat/2018/tab03 (viewed on 27.08.2019)

[15] *Classification of regions of the Russian Federation by natural conditions* Retrieved from: http://www.geoteka.ru/text.html?ident=usl (viewed on 20.08.2019).

[16] SNiP 23-01-99. *Construction climatology* 2000 (Moscow: Gosstroi of Russia)

[17] *National sobriety rating of subjects of the Russian Federation* 2018 Retrieved from: http://www.trezvros.ru/ (viewed on 20.08.2019)

[18] Chubukov A, Kapitanov V, Silyanov V and Monina O Approaches to assessing status and prediction of road traffic accidents rate *Transportation Research Procedia* (in print)

[19] Kaptains V T, Chubukov A B, Silyanov V V and Monina O Y 2017 Forecast of the number of deaths in road accidents based on socio-economic indicators *Sci. and technol. in the road industry* **4(82)** 4–6

IOP Conf. Series: Materials Science and Engineering **832** (2020) 012046 doi:10.1088/1757-899X/832/1/012046

[20] Chubukov A, Kapitanov V, Silyanov V and Brannolte U 2016 Simulation of Regional Mortality Rate in Road Accidents *12th International Conference "Organization and Traffic Safety Management in large cities"* SPbOTSIC-2016, 28–30 September (St. Petersburg) pp 112–4 Retrieved from: www.sciencedirect.com

The development of basic algorithms for processing navigation data in dispatching control systems of road transportation of goods and passengers

N A Filippova and V M Belyaev

Moscow Automobile and Road State Technical University (MADI), 64, Leningradsky ave., Moscow, 125319, Russia

E-mail: umen@bk.ru, belyaev-v@mail.ru

Abstract. The development of basic algorithms for processing navigational data in dispatching control systems of road transportation of goods and passengers on specified routes is discussed in this article. Algorithms for processing navigation data with the use of GIS-technologies are described. The features of the "linking" of navigation marks to the zones of control points represented in the spatial model of the route by polygons formed by means of specialized editors of geographic information systems are described. The algorithm of recognition of flights performed on a given route based on the analysis of the actual data on the passage of checkpoints of the route by the vehicle, as well as the algorithm of recognition of liberty mileage is considered in detail. The description of the algorithm for comparing the planned and actual indicators of transport work developed using the theory of sets is presented. A block diagram of an algorithm for a step-by-step evaluation of the results of performing transport work is presented on the basis of the application of the procedure of stepwise narrowing of the set of admissible alternatives.

1. Introduction

The main feature of the navigation control of vehicles performing transportation of goods or passengers on specified routes is the use of specialized spatial models of objects on the route, called virtual checkpoints.

If you define the real objects of the planned tasks in the form of an ordered sequence of virtual checkpoints (CKPT), located on the route, such checkpoints can be represented as a set of so-called geometric primitives, which are polygons. This approach to the formation of CKPT is to apply a polygon on an electronic geographical map of the area where the object of interest is located. At receipt in dispatching system of each navigation mark from the controlled vehicle the fact of its occurrence in each of the polygons representing checkpoints is checked. The process of determining the CKPT, which belongs to the navigation mark (NM), is "linking" NM to the CKPT. It is assumed that polygons cannot intersect, so any geographic point (x; y) in area G cannot belong to two or more polygons at the same time. In practice, this restriction is used to simplify subsequent information processing algorithms. If it is necessary to change the type of zones, this change is reduced to a dynamic indication of the type of checkpoint for a particular job [1, 2].

In the considered system, the described algorithm is based on the search for intersections of the coordinates of NM with the boundaries of CKPT. CKPT consist of primitives: triangle, rectangle or circle. If the CKPT contains more than one primitive, this CKPT is called a complex checkpoint (CCKPT), and if CKPT consists of one primitive, it is called a simple checkpoint (SCKPT). If it is not specified how many primitives a CKPT consists of, such a CKPT should be referred to a complex checkpoint [3].

The zone of the complex control point is the area of imposing at least one of the primitives of which the CCKPT consists. The result of "linking" the NM to a specific CKPT is the Boolean number.

Previously, it was noted that the checkpoint zones cannot overlap each other, so they cannot be "not linking" to any CKPT or "linking" to only one checkpoint from the list of registered in the system.

Set-off of checkpoint is called fixing a particular CKPT in the system. The main idea of this offset is the detection of "false" checkpoints encountered on the route [4].

The approach is based on determining the time of finding the vehicle in the zone of the checkpoint.

2. Description of the algorithm for singling out flights from the actual events of the vehicle

The process of obtaining a set of actual flights from the actual events of the vehicle is called the process of *singling out* the actual flights. A feature of singling out is the need to determine the first and last actual flights in the shift interval to form the correct values of production and empty runs. The event refers to the shift, where more than half of the event took place (in time). [5]

In this case, the empty run per shift is equal to:

$$L_x = l_{x1} + \sum_i l_{xi} + l_{x2} \,. \tag{1}$$

This can be expressed as follows:

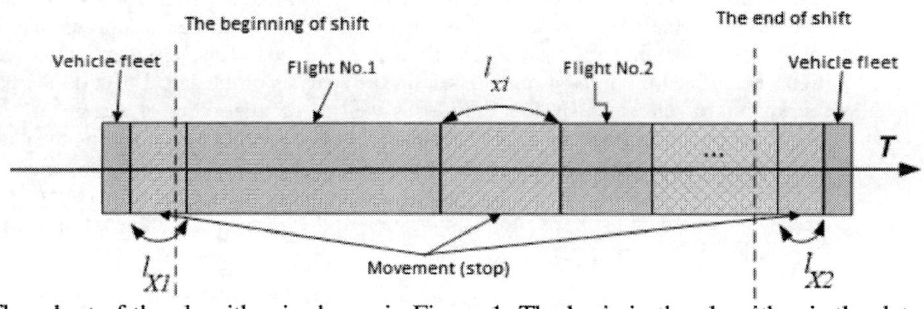

A flowchart of the algorithm is shown in Figure 1. The basis in the algorithm is the determination of the start and end time for the interval of calculated events. This is due to the possible coincidence of checkpoint in actual flights with planned flights, related to different shifts, and as a result, the wrong result of comparing the planned and actual indicators of flights is obtained [6].

3. The algorithm for comparing planned and actual transportation figures indicators

The algorithm is used to control the route movement of freight or passenger vehicles performing transport work, which plan is presented in the form of a set of flights on specified routes. The purpose of the comparison is to provide adequate, complete information to the dispatcher about the progress of the transport process in real time to solve the problems of dispatching control and regulation of the transportation process [7].

The algorithm is applicable for monitoring and accounting of transport work in the presence of information about the planned tasks (flights) in the system. The input data should contain: information about the planned flights, including the times of the start of flights, the order and the points of the route, as well as information about the actual flights, which is obtained in the system by the algorithm for determining the navigation marks obtained from the navigation blocks. This information includes the correct management of reference information on mobile units, drivers, control points of the system, linking radio stations to mobile units. The output data are the compared planned and actual flights [8].

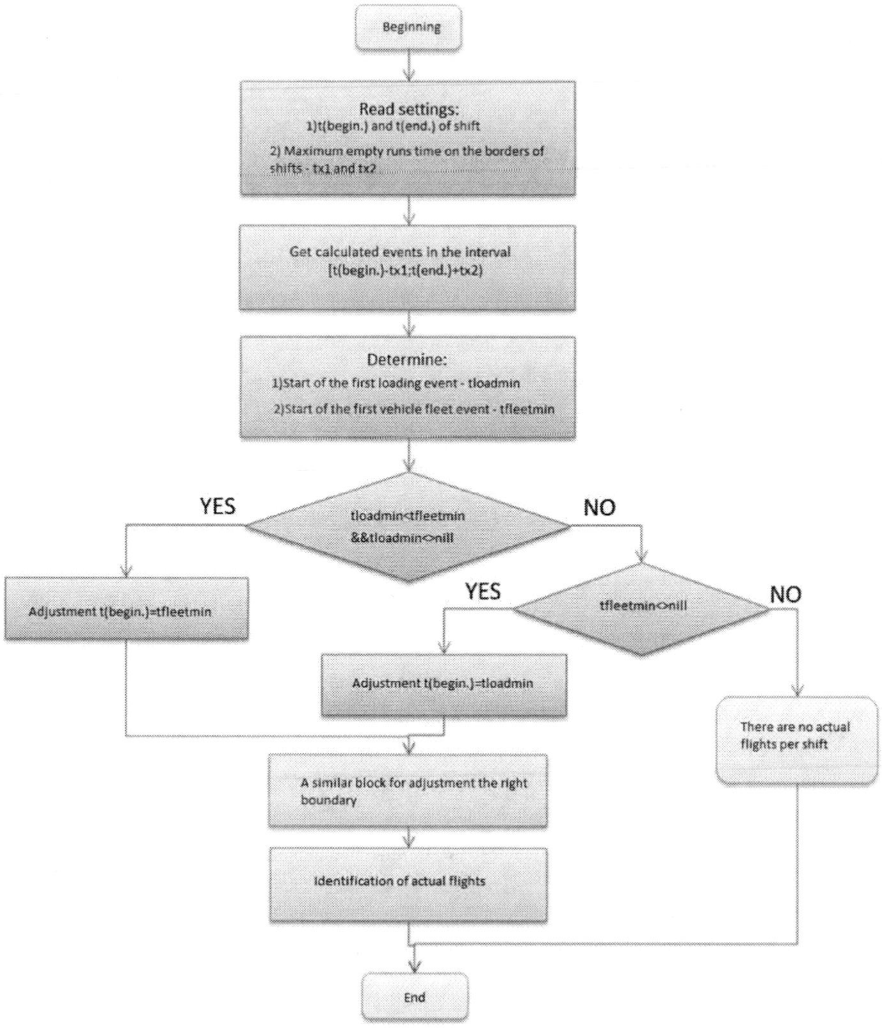

Figure 1. A flowchart of the algorithm for obtaining empty runs

Formalization of the process of movement of the vehicle in flight.

Initial data: information from the planning subsystem about the planned flights for each vehicle separately, ranked by the start time of each flight. Flights cannot overlap in time. We will designate a "scheduled flight" as a sequence of checkpoint, which the vehicle should visit in the process of transport work. In general, the term "flight" will be interpreted as a sequence of CKPT. There is at least one type of checkpoint in the flight of any type: type "fleet". Therefore, we take the designation of the CKPE type "fleet" as a $CKPT_F$ [9].

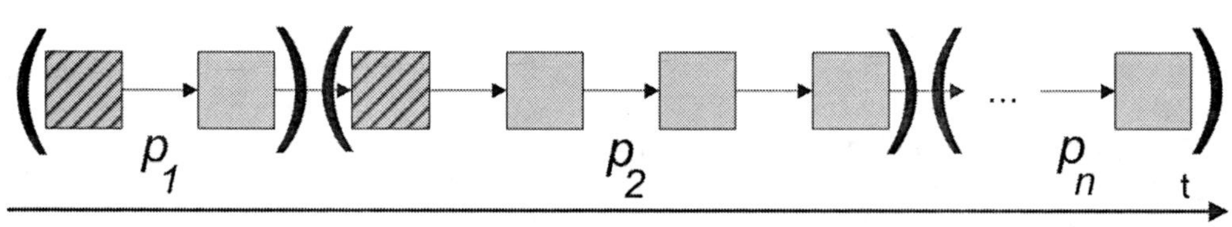

Figure 2. Sequence of scheduled flights

The figure shows the scheduled flights:

$$P = \{p_1, p_2,..., p_n), \text{ n-the number of scheduled flights} \quad (2)$$

Here and further, the shaded areas correspond to objects of the "fleet" type, not shaded - "special object" items.

Each $p_i \in P, i = \overline{1,n}$ is characterized by a set of attributes, the most significant of which are $t_{pibeginning}$ and t_{piend}. Each flight represents a sequence of checkpoints [10].

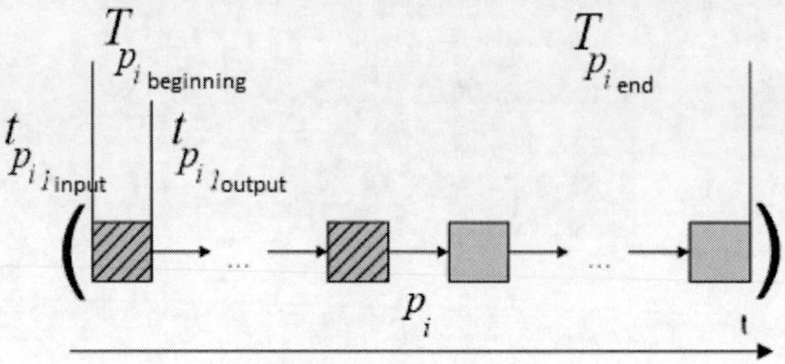

Figure 3. Time characteristics of a scheduled flight

The duration of the scheduled flight T_{pi} is:

$$T_{pi} = T_{piend} - T_{pibeginning} \quad (3)$$

where:

$$T_{piend} = t_{pioutput}, l - \text{the number of checkpoints in shift} \quad (4)$$

$$T_{pibeginning} = t_{piinput}, \quad (5)$$

There may be scheduled flights with the same CKPT, so in general we have:

$$p_i \cap p_j \neq \{\}, i, j = \overline{1,n} \quad (6)$$

Similarly, actual flights F are described.

With optimal planning of transport work in an ideal case, the vehicle should perform flights in the established order. Due to the fact that in the process of transport work the sequence and the set of checkpoints can change (operational planning on the one hand and disruptions in the process of transport work on the other), we define the following options (groups) of actual flights:

Table 1. Groups of flights

№ group	I group	II group	III group
Interrelation	$F\{f\} \cap P\{p\} \neq \{\}$ $F\{f\} \cap P\{p\} = \{F\{f\}, P\{p\}\}$	$F\{f\} \subseteq P\{p\} \neq \{\}$ $P\{p\} \subseteq F\{f\} \neq \{\}$ $F\{f\} \cap P\{p\} \neq \{\}$ $F\{f\} \cap P\{p\} \neq \{F\{f\}, P\{p\}\}$	$F\{f\} \cap P\{p\} = \{\}$

Method of comparing planned and actual flight information

To determine the degree of compliance of the work performed, it is possible to identify a number of criteria, which include: similarity criterion CKPT (k); the criterion of the minimum absolute deviation of the start time of each planned flight from each flight performed (t); the criterion of compliance with the sequence of passing checkpoints planned and completed flights (v). To solve the problem of determining the degree of correspondence between the actual and planned work, possible alternatives

$X = \{x_p\}(p = 1,2,...,N_x)$ representing objects whose quality and efficiency are estimated by the vector index should be determined:

$$w(x_p) = (w_1(x_p), w_2(x_p),..., w_n(x_p)), \tag{7}$$

where $w_1(x_p), w_2(x_p),..., w_n(x_p)$ – private similarity functions;

Expressions for particular functions for this case of similarity of flights:

$$K_\beta = w_1(x_p) = (\sum_\alpha k_\alpha)_\beta \rightarrow extr, \alpha = \overline{1,r} \tag{8}$$

$$T_\beta = w_2(x_p) = (\sum_\alpha t_\alpha)_\beta \rightarrow extr, \alpha = \overline{1,r} \tag{9}$$

$$V_\beta = w_3(x_p) = (\sum_\alpha v_\alpha)_\beta \rightarrow extr, \alpha = \overline{1,r} \tag{10}$$

Let object x_g be preferable (better) to the object x_h (denoted $x_g \succ x_h$) by $w_k, k = 1,2,3$ criterion if a strict $w_k(x_g) > w_k(x_h)$ inequality is satisfied for the direct measurement scale and $w_k(x_g) < w_k(x_h)$ for the inverse measurement scale. In this case, the problem of multi-criteria ranking (ordering) of objects evaluated by a set of indicators $w_k, k = 1,2,3$ can be posed. It is necessary to find such an ordering of objects $x_{i1} \succ x_{i2} \succ x_{i3}$ satisfying condition $w_k(x_{i1}) > w_k(x_{i2}) > w_k(x_{i3}), k = 1,2,3$.

The described method of systematic optimization is reduced to the search for an extremum of a particular objective function on the set of admissible solutions obtained in the previous step. Figure 4 shows a flowchart of the main stages of assessing the performance of transport work based on the use of a step-by-step procedure for a variety of acceptable alternatives [11].

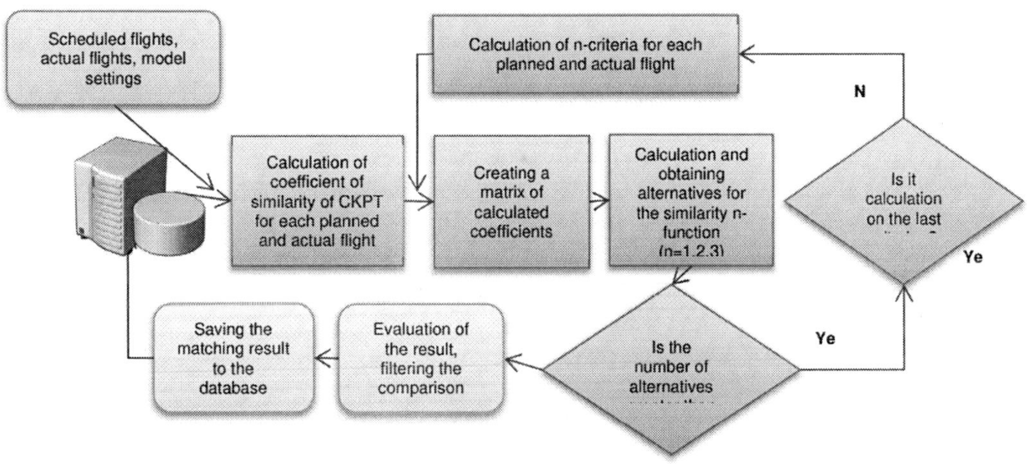

Figure 4. The procedure of step-by-step narrowing of the set of admissible alternatives

4. Conclusion

The article describes the basic algorithms for processing navigation data coming from vehicles performing transportation of goods or passengers on specified routes. The peculiarity of the considered algorithms is that they are equally applicable for the control of freight and passenger traffic, provided the description of the route in the form of an ordered sequence of checkpoints. Another feature of the

IOP Conf. Series: Materials Science and Engineering **832** (2020) 012047 doi:10.1088/1757-899X/832/1/012047

presented algorithms is the use of geoinformation technologies for the formal description of the boundaries of checkpoints.

The developed algorithms provide a formal basis for the development of specialized software for automated dispatching control systems for cargo and passenger transportation.

References

[1] Vlasov V M, Prikhodko V M, Zhankaziev S V and Ivanov A M 2011 *Intelligent transport systems in the highway complex* (Moscow: OOO "Meiler")

[2] Vlasov V M, Efimenko D B and Bogumil V N 2014 *Information technology in road transport: a textbook for students. Institutions of higher education. Education* (Moscow: Publishing Center "Academy")

[3] Efimenko D B, Terentyev A V and Karelina M Yu 2017 Methods of using areas as methods of optimization of motor transportation processes *J. of the "Bulletin of Civil Engineers"* **6(65)** 291–4

[4] Filippova N A 2015 *Methodology of organization and functioning of cargo delivery systems in the northern regions* monograph under. ed. Belyaeva V M (Moscow: Publishing House Tekhpoligratsentr)

[5] Filippova N A 2015. Information support of the transportation process of a multimodal system *Science of Krasnoyarsk* **1**

[6] GOST R 54027-2010 *Global navigation satellite system. Systems of dispatching control of cargo motor transport. Requirements to architecture, functions and applications of system of dispatching management by transport of construction materials on the hourly charts*

[7] Filippova N A, Belyaev V M, Shilimov M V, Koshkarev P P and Odinokova I V 2017 The Analytical Test of Methodological Approaches to the Increasing the Level of Automation of the Basic Functions of the Car Dispatching of the Cargo Delivery to Northern Regions of the Russian Federation *Int. J. of Applied Engineer. Res. (IJAER)* **12(21)**

[8] Bogumil V and Efimenko D 2013 Urban Transport Dispatch Control System Helps to Increase Intelligent Transport Systems Effectiveness *Proc. of the 11th European transport congr.* pp 20–5 (Prague 19–20 September)

[9] Bhouri N, Balbo F and Pinson S 2012 An agent-based computational approach for urban traffic regulation *Artifact Intelligence* 139–147 DOI 10.1007/s13748-012-0011-0

[10] Kammoun H M, Kallel I, Casillas J, Abraham A and Alimi M A 2014 Adapt-Traf: An adaptive multiagent road traffic management system based on hybrid ant-hierarchical fuzzy model, *Transport. Research Part C* **42** 147–67

[11] Qian Runhuaa, Cong Huaa, Zhao Ruiling and Li Yuanxing 2013 *Design Scheme of Public Transport Comprehensive Dispatching MIS based on MAS* 13th COTA Int. Conf. of Transportation Professionals (CICTP 2013)

[12] Tlig M and Bhouri N 2011 A Multi-Agent System for Urban Traffic and Buses Regularity Control *Procedia Social and Behavioral Sciences*, **20**, 896–905

IOP Conf. Series: Materials Science and Engineering **832** (2020) 012048 doi:10.1088/1757-899X/832/1/012048

BIM of transport infrastructure – practical aspects of data collection for DTM creation

V N Gulin, A A Neretin and S P Paudyal

Moscow Automobile and Road State Technical University (MADI), 64, Leningradsky ave., Moscow, 125319, Russia

E-mail: guvl@yandex.ru

Abstract. Road industry is moving towards the "Digital road" concept, and BIM of transport infrastructure is an integral part of it. Objects of transport infrastructure have linear structure required a special approach to their accurate coordinate description for proper digital terrain model creation. Approaches of railway and motor road experts to this issue is described. Implementation of mobile laser scanning and aerial photo survey using unmanned aerial vehicles for digital terrain model creation is discussed. Strengths and weaknesses of those two data collection methods are analyzed.

1. Introduction

Building Information Modelling (BIM) is a process involving various tools and technologies for generating and managing information on construction project across its lifecycle. Worldwide many countries are making efforts at the state level on developing and implementation of BIM standards. According to [1], the USA and the Great Britain are a bit ahead of other countries in this process [2–5]. It is important that main part of developed BIM standards cover mainly requirements for industrial and civil construction while BIM standards for infrastructure construction projects are under development. The article is focused on BIM in relation with transport infrastructure, i.e. railways and motor roads.

General trend in Russia today is moving towards so-called digital economy. There are publications describing what it means for railway [6–7] as well as motor road [8] industries. Although railways and motor roads differ by their organization, operation, methods of maintenance, etc., there is one common goal for both industries within "digital road" conception – implementation of BIM approach for all stages of the road life cycle.

Important part of BIM infrastructure project is Digital Terrain Model (DTM) of the territory where the road is located. Such DTM should have sufficient resolution to visualize all basic elements with required accuracy and detail. We would discuss practical aspects of DTM creation taking into account Russian specifics connected with using national coordinate systems.

Rail and motor roads have a narrow linear structure; their length may exceed hundred and even thousand kilometers. Nevertheless, BIM projects normally start at design stage and cover shorter lengths (few dozens of kilometers or less) provided those particular BIM projects are to be combined all together at later stage during road operation.

There was a number pilot BIM projects in Russia using different surveying methods of data collection for DTM creation. Among traditional surveys using GNSS and optical instruments modern technologies of mobile laser scanning and aerial photo survey by unmanned aerial vehicles (UAVs)

have been used for data collection what had significant impact on the result from point of accuracy and productivity.

Another important issue for assembling separate local BIM projects into bigger one is unified coordinate description of long linear object such as rail or motor road. For this purpose, there should be a number of known points along the road determined in some coordinate system(s). Selection of coordinate system also does matter. Existing national coordinate systems of Russian Federation such as SK-42, SK-95 or regional MSK (local coordinate systems) have some limitations to describe long linear object properly. Both railway and motor road industries undertaken some actions in this respect as described below.

2. Ensuring the unity of coordinate description

As already mentioned national coordinate systems of the Russian Federation such as SK-42, SK-95 or regional MSK do not allow one to describe long linear object properly, because those systems use mapping projection Gauss-Kruger having significant distortion in lengths at the borders of mapping zones. If the length of the project territory exceeds several dozens of kilometers, distortion may be at sub-meter level or even more. The distortion level also depends on object orientation. In particular, distortion in North-South direction is minimal while in West-East one it is maximal. In addition, the points representing SK-42, SK-95 and regional MSK coordinate systems at Earth's surface (reference GGS points) have not sufficient positioning accuracy. Moreover, the density of existing reference GGS points is not enough in some regions for solving surveying tasks at high level of accuracy. All that together does not allow getting precise DTM of a long linear object using the current reference GGS point's setup.

It is possible to overcome the described problem with SK-42, SK-95 and regional MSK by using new national global geodetic coordinate system GSK-2011, but density of reference points determined in GSK-2011 is not enough yet.

That situation is not acceptable for transportation industry, especially for design, construction and operation of high-speed railways. As a result, since 2008 Russian Railways (RZD) are developing and implementing their own corporate high accuracy coordinate system (VKS) [9]. In 2015, RZD developed and implemented Code of Practice [10] describing general requirements, rules of creation and use of the VKS. According to [10] VKS consists of the following components:

- Differential GNSS sub-system.
- Geodetic reference network (OGS).
- Communication segment.
- User segment.

Differential GNSS sub-system consists of:

- Permanent GNSS reference stations network.
- Network (computation, control) center.

The geodetic reference network consists of:

- Base stations – geodetic centers to install permanent or temporary GNSS reference stations; distance between base stations is within 50 km.
- Main control points – geodetic centers to install temporary GNSS reference stations; distance between main control points is within 4–5 km.
- Intermediate points – geodetic centers to install surveying equipment during surveys; distance between intermediate points is within 250–750 m.

Communication segment includes:

- Fixed communication channel – used to deliver satellite information from reference stations to network center.
- Mobile radio channel – used to deliver differential corrections from network center to mobile user equipment.

The user segment includes GNSS equipment of VKS users.

IOP Conf. Series: Materials Science and Engineering **832** (2020) 012048 doi:10.1088/1757-899X/832/1/012048

RMS error of the relative position of adjacent OGS points should not exceed 8 mm in plan and 5 mm in height (in the Baltic system of heights 1977) [10]. OGS points had been determined in different coordinate systems – ITRF2008, regional MSK, as well as special railway local coordinate systems.

It is important to note that accuracy requirements to operation of railways differ from those to motor roads. In particular, in motor roads segment there is no necessity to maintain road surface in design position over road lifecycle, also there is no need in monitoring high-accuracy position of moving vehicles in real time. At the same time, implementation of modern surveying technologies as well as development of preliminary BIM standards by the state company "Avtodor" resulted in understanding necessity of creation the corporate reference surveying network (named VOGS) having sufficient density of high-accuracy reference points along the roads comparatively to existing GGS setup [11]. Later general definitions and requirements to VOGS have been included into organization standard [12].

According to [11–12] VOGS consists of the following components:

- Network of frame stations.
- Network of benchmarks (or control points).

Frame stations ensure high-accuracy connection of VOGS to global coordinate systems ITRF2008 (international) and GSK-2011 (Russian). Centers of frame stations have to be permanently mounted at buildings with open sky view and availability of AC power connection to enable 5-days continuous GNSS observations; distance between frame stations is within 300 km.

Benchmarks have to be mounted by pairs into concrete foundations of road structures; the distance between points in pair is within 800 m while distance between pairs is within 30 km.

RMS error of the relative position of adjacent VOGS points should not exceed 20 mm in plan and 25 mm in height. VOGS points have to be determined in different coordinate systems – ITRF2008, GSK-2011 and regional MSK.

Currently VOGS segments had been created at several highways (M-1, M-4, M-11) operated by the state company "Avtodor" with total duration of approximately 2000 km.

Both VKS of RZD and VOGS of the state company "Avtodor" generate high-accuracy coordinate reference for effective implementation of advanced data collection methods for DTM creation.

3. Data collection methods

There is a number of methods enabling productive data collection. For big territories, technologies of airborne laser scanning and digital photography provide maximal productivity of the survey, but for quite small territories (few dozens of km) which we are talking about in the relation with BIM modelling at motor road design stage those methods are not reasonable from economical point of view. Alternatively, mobile laser scanning (MLS) method looks like most effective for such applications. MLS technology is developing rapidly resulting in permanent increasing of density, range and productivity of the survey. That is why MLS is widely accepted by the road industry specialists.

Another measurement technique to be discussed is air photo survey using UAVs. That method is also developing very fast demonstrating big potential for BIM data collection applications. It is obvious that every measurement technique has both strong and week points. Let us analyze MLS and UAV methods in more detail.

3.1. MLS survey

MLS technology has a number of advantages (strong points) comparatively to traditional surveying techniques [13]. Among those are:

- Rapid collection of enormous volumes of highly dense accurate geo-referenced data.
- Possibility of acquiring data with survey-grade accuracy at traffic speeds.
- Possibility of acquiring data in the form of 3D point clouds.
- Possibility of acquiring both laser-based and image-based measurement data.

In [14] the analysis of possible MLS applications for different stages of motor road lifecycle is provided. According to conclusions of this analysis, it is reasonable to use MLS for the following stages of road lifecycle:

- Design – engineering surveys.
- Construction – building control, executive survey.
- Operation – monitoring of road infrastructure, monitoring of road pavement.

In 2015, the company "Geoproektizyskaniya" LLC performed MLS of M-4 highway section to make DTM for subsequent repair of the pavement at this road section [15]. Part of DTM combined with the design alignment model is shown in Fig. 1.

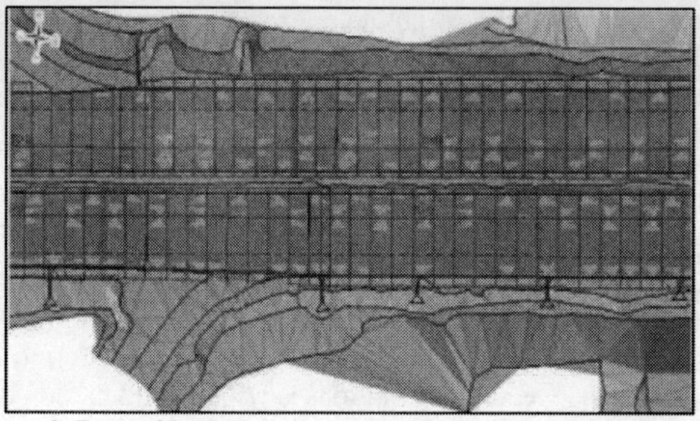

Figure 1. Part of DTM combined with design alignment model

According to [15], application of MLS technology in this project allowed to reduce milling volumes by 40.3 % resulting in savings of 4.35 million rubles.

Among weak points of MLS technology, we should mention dependence from weather conditions (rain, snow, water at road surface, etc.) and presence of obstacles not allowing to "seeing" surface behind them. Because the laser-scanning system itself is located not so high above the road surface, such obstacles as noise protection structures, other vehicles, etc. could reduce the range of MLS effective operation.

3.2. UAV survey

Aerial photo survey using UAVs is the data collection method developing rapidly from year to year. A wide range of different options for UAV's configuration make this method extremely flexible. Depending on the requirements, there may be selection from multiple options – airplane or helicopter type of vehicle, type of engine affecting the range of operation, resolution of photo camera, type of onboard GNSS equipment, etc.

Aerial photo survey using UAVs has following advantages:

- Rapid collection of big volumes of geo-referenced data.
- Flexibility to select proper vehicle configuration depending on range and accuracy requirements.
- Significant cost savings comparatively to operated aerial surveys.
- Easy operation, easy remote pilot education.
- Possibility to map territories from the top avoiding limitations for terrestrial surveys (such as obstacles for visibility, etc.).

In [16] there is description of the road construction project where the UAV's data collection method was tested for monitoring of construction process. In this test, cost-effective helicopter-type UAV DJI Phantom 4 Pro having reasonable camera resolution was used. The tested road segment is 600 meters in length; it was surveyed twice with 3-weeks interval (Fig. 2).

For DTM accuracy estimation a set of control points measured by GNSS and electronic total station has been used. Accuracy estimation demonstrates RMS of 2.5 cm in plan and 3.0 cm in height [16].

Figure 2. 3D-view of road job site

In general, among weak points of UAV's technology there are: dependence from weather conditions (rain, snow, strong wind, etc.), complicated procedure of getting flight permission, restrictions for flights in some areas (for instance, near airports), safety issues of flights over operated highways and living districts (in case of UAV's brake and drop), etc. Nevertheless, aerial photo survey using UAVs has very positive perspectives for data collection applications in relation with BIM of transport infrastructure, especially in combination with other data collection methods such as MLS.

4. Conclusions

Both railway and motor road industries are moving towards implementation of the "Digital road" concept. BIM of transport infrastructure is an integral part of this concept. It turn, detailed and accurate DTM is an integral part of the road information model. By this reason, particular efforts should be made to ensure the unity of linear objects coordinate description. Modern technologies of MLS and aerial photo surveys using UAVs can improve significantly productivity and accuracy of DTM creation process. Due to weaknesses of every single technology, it seems reasonable to combine MLS and UAVs data collection methods for DTM creation.

References

[1] Skvortsov A V 2016 The review of international regulation in BIM sphere *CAD & GIS for roads* **2(7)** 4–48
[2] *National BIM Standard – United States™ V3* 2015
[3] *PAS 1192-2:2013 Specification for information management for the capital/delivery phase of construction projects using building information, Modeling* 2013 68 p
[4] *PAS 1192-3:2014 Specification for information management for the operational phase of assets using building information modelling* 2014 44 p
[5] *PAS 1192-5:2015 Specification for security-minded building information modelling, digital built environments and smart asset management* 2015 38 p
[6] Rozenberg E 2017 Development of digital technologies for Russian Railways *Global Railway Review* **5** 71–5
[7] Rozenberg E and Ozerov A 2017 Russia's digital railway begins to take shape *Int. Railway J.* **57.4** 38–40
[8] Boykov V N, Skvortsov A V and Sarychev D S 2018 Digital highway as industry segment of digital economy *Transport Rossiyskoi Fedetatsii* **2** 96–100
[9] Tony O, Gelfgat A, Suvorov A and Samratov U 2009 About creation of high-precision coordinate system for design, construction and operation of the high-speed railway

Moscow – Saint-Petersburg based on global navigation satellite systems GLONASS/GPS/Galileo *Inzhenernye izyskaniya* Feb. 48–55

[10] SP 233.1326000.2015 *Railway transport infrastructure, High-accuracy coordinate system* Retrieved from: http://docs.cntd.ru/document/1200124325 (accessed 12 September 2019)

[11] Gulin V N, Mironov S A and Neretin A A 2015 The problem of providing unified coordinate description of the objects in road industry *CAD & GIS for roads* **1(4)** 75–82

[12] STO AVTODOR 8.6-2016 *Organizational and technological maintenance of road information modeling on all stages of road lifecycle* 2016 48 p

[13] Guan Haiyan, Li Jonathan, Cao Shuang and Yu Yongtao 2016 Use of mobile LiDAR in road information inventory: A review *Int. J. of Image and Data Fusion* **7(3)** 219–42

[14] Motuz V O and Sarychev D S 2014 Applications of lidar mapping and 3D models in road lifecycle *CAD & GIS for roads* **1(2)** 12–5

[15] Alekseenko N N 2016 Application of Laser Scanning Technology in Different Branches and on Different Stages of the Lifecycle of Objects *Proc. of Moscow State Univer. of Civil Engineer.* **2** 62–73

[16] Sukonnikov O G, Neretin A A and Guryev V A 2017 Analysis of UAV's implementation for geodetic control of constructing and exploiting highways *CAD & GIS for roads* **2(9)** 44–8

The procedure of statistical evaluation of accident rate equation for roundabouts

A N Kloyan, A M Byrgonytdinov, O A Fedoseeva, G B Lyalkina and L V Yashmanova

Perm National Research Polytechnic University, 29, Komsomolsky ave., Perm, 614990, Russia

E-mail: Anyutka.sh@mail.ru

Abstract. Road Design Manuals propose the procedure for determining the accident rate at modern roundabouts by the probable number of major accident of types on specific sites of a roundabout per year. An assumption is made that the regression equation may not take into account the design features of modern roundabouts, climatic conditions, as well as driving style in the Russian Federation. The paper presents a step-by-step method of the regression equation validation to assess the accident rate at the approaches to the roundabouts on the basis of other statistical methods. The validation involves the statistical evaluation of the sample correlation coefficients between the accident equation and the factors affecting it. Within the work the values included in the equation were non-dimensionalized, necessary assumptions were introduced, and the features of performing correlation analysis in the particular case were analyzed. The procedure proposed enables to determine the reliability of the equation for determining the accident rate at the approaches to a modern roundabout in the conditions of the Russian Federation, taking into account climatic conditions, design features, the state of the existing road transport network, as well as driving style.

1. Introduction

The problem of reducing road accident rates is urgent worldwide [1–3]. According to the World Health Organization (WHO) report the road traffic accidents are the major cause of deaths for children and young adults aged 15–25 years. Improving road safety, aimed at protecting people's lives and health, is the main priority of the Russian Federation transport policy [4, 5].

The main percentage of traffic accidents occurs at road intersections, due to the concentration of the largest number of conflict points. While studying the world practice on ways of improving road safety at road intersections, the modern roundabouts were found to be among the most popular and safest types of road intersection.

The Russian Federation is only beginning to increase the number of modern roundabouts therefore the design standards are being updated[1]. However, the adaptation of many aspects of design and reconstruction for the conditions of our country requires significant labor costs. For example, until now no one has studied in detail the methodology for calculating accidents at modern roundabouts.

To evaluate the road safety conditions[1] at modern roundabout, it is recommended to use the procedure designed by the Department of Main Roads of Australia in 2006. The procedure is given in detail in the road design manual and allows determining the accident rate for the major accident types on specific parts of a roundabout per year [6]. The regression equation proposed by the authors, however, may not take into account the design features of modern roundabouts, climatic conditions, as well as driving style in the Russian Federation. Therefore, in order to specify the adequacy of the proposed procedure, the regression equation requires additional checking for reliability.

2. Characteristics of regression equation

In the Design Standards the intersection accident rate is characterized by the traffic safety indicator K_a, as the number of accidents per 10 million cars that passed through one intersection. According to the procedure [6], it is recommended to assess the safety of a modern roundabout considering the accident rate indicators at the approaches including entry and exit curves to the roundabout by studying the number of road accidents per year.

The procedure [6] gives the empirical equation for an estimation of accident rate on approaches to a roundabout expressed by the formula:

$$A_1 = 1.81 * 10^{-18} * N_r^{1.39} * N_k^{0.65} * V_a^{4.77} * n^{2.31}, \quad (1)$$

where N_r – a number of approaching vehicles in one lane, (veh./day); N_k – the average annual daily traffic on the circulating carriageway adjacent to the approach (veh./d); V_a – 85th percentile speed on the entry curve (km/h); n is the number of lanes on the particular roundabout approach.

Equation (1) is a regression equation, where A_1 – the response function of the factors N_r, N_k, V_a and *n*.

The work, however, does not give clear evidence [6] at what values of the significance level and at what confidence intervals this equation can be used for exponents. Concerning this in order to validate the proposed equation (1), it is necessary to verify its validity on the basis of other statistical methods [7, 8]. First of all, it is necessary to perform a statistical evaluation of the sample correlation coefficients between the accident rate equation and the factors affecting it.

Previously, before using the empirical equation for assessing the accident rate on the approaches to the roundabouts, it is necessary to carry out the procedure of non-dimensionalizing the values included in it.

It allows moving from a physical problem to an abstract mathematical problem. This technique enables to apply formal mathematical methods for solving equations. In addition, non-dimensionalizing the quantities provides a way for moving from physical quantities, which can have very large or very small values in modulus, to their dimensionless analogues, which will take values of the order of units. This, in turn, ensures a much slower error accumulation.

In formula (1) the value which requires the procedure of non-dimensionalizing is V_a – the 85th percentile speed on the entry curve (km/h).

The most suitable value for its non-dimensionalizing is the traffic speed limits on the particular section $V_{\text{огр}}$. Dividing the value V_a by $V_{\text{огр}}$, we obtain the value *V*, which is dimensionless and optimally determines the accident rate under consideration. Then formula (1) will be as follows:

$$A_1' = 1.81 * 10^{-18} * N_r^{1.39} * N_k^{0.65} * V^{4.77} * n^{2.31}, \quad (2)$$

where $V = V_a (V_{\text{огр}})^{-1}$.

Take the logarithm of equation 2, to get its linear form. The expression will be as follows:

$$\log(A_1') = \log(1.18) - 18 + 1.39\log(N_r) + 0.65\log(N_k) + 4.77\log(V) + 2.31\log(n),$$ i.e. equation

$$\log(A_1') = -17.93 + 1.39\log(N_r) + 0.65\log(N_k) + 4.77\log(V) + 2.31\log(n), \quad (3)$$

It is necessary to note that response function $Y = \log(A_1')$ has a linear form relative to factors

$$X^{(1)} = \log(N_r),\ X^{(2)} = \log(N_k),\ X^{(3)} = \log(V),\ X^{(4)} = \log(n).$$

As a result, the equation linking the accident rate indicator *Y* at the roundabout with explanatory variables $X^{(i)} (i = 1,2,3,4)$ will take the following form:

$$Y = -17.93 + 1.39X^1 + 0.65X^2 + 4.77X^3 + 2.31X^4. \quad (4)$$

It is necessary to note that the analysis of multiple correlation relationships (statistical relationships between more than two variables) is associated with the need to measure the strength of relationship between the dependent variable *Y* and independent variables [8], as well as with emerging difficulties in interpreting paired correlation coefficients Y, $r_{(Y,X^i)}$ between Y and X^i, due to the possible indirect influence of others on this pairing (obviously not taken into account in the calculation of the independent variables). Therefore, in the general case, when selecting regression equations the number of independent variables may increase $X^{(i)}$ *(i = 1, 2, 3,..., k)*.

Multiple correlation analysis requires the introduction of such indicators of the statistical relationships that would give the correlation degree evaluation of each of the pairs (Y, X^i) for the variables we are interested in. These indicators must be "cleared" of indirect influence of other variables, that is, they would be calculated provided that the values of the other variables are fixed.

The search for a multiple regression equation begins with the evaluation of paired correlation coefficients for many factors $X^{(i)}$ *(i = 1, 2, 3,..., k)* and the resulting response function Y, $r_{(Y,X^i)}$.

3. Validation procedure of the reliability of the accident rate assessment

sAssuming that the multidimensional distribution law of the random variables X^1, X^2,..., X^k is normal, we denote by $r_{Y,X^1} = r_{Y,X^1(X^2,...,X^k)}$ a sample partial correlation coefficient between the response function Y and the variable X^1 for constant values of the variables X^1, X^2,..., X^k. Similarly (if $i > 1$) through $r_{(Y,X^i)}$ we will define the sample value of the partial correlation coefficient between Y and another factor X^i, at constant values of the remaining factors $X^{i-1}, X^{i+1},...,X^k$ when $i \geq 2$ [8].

While studying the properties of the selective partial correlation coefficient $r_{Y,X^i(X^1,X^{i-1},...,X^k)}$ *k*-th order we assume, as usual [8] that it is statistically distributed in exactly the same way as a normal (paired) sample correlation coefficient between the same variables Y and X^i. Naturally, it is necessary to introduce a correction for the number of degrees of freedom of the sampling, reducing its volume in *k* units, i.e. to take the appropriate number of degrees of freedom equal to $n_i - k$, where n_i – the size of each of the analyzed sample sets of values $X^{(i)}$ *(i = 1, 2,..., k)*.

To determine the accident rate, the peculiarities of its calculation on the approaches to the roundabout, as well as to validate the empirical relationships with necessary assumptions we propose the procedure enabling statistically reliable at a given α to identify the strength of relationship between the dependent variable Y and multiple independent variables $X^{(i)}$ *(i = 1, 2,..., k)*.

The main steps of calculation include:

- the calculation of the sample correlation coefficients $r_{Y,X^i(X^1,X^{i-1},X^{i+1},...,X^k)}$ for each of the *k* independent variables X^i;
- analysis of the statistical properties of sample correlation coefficients $r_{Y,X^i(X^1,X^{i-1},X^{i+1},...,X^k)}$;
- evaluation of multiple correlation coefficient $R_{Y,\overline{X}}$ between the dependent variable Y and the vector $\overline{X} = \left\{\overline{X}^1, \overline{X}^2, ..., \overline{X}^k\right\}$.

3.1 Step 1

Calculation of the sample (empirical) value of partial correlation coefficients is performed on the basis of the recurrent ratio [8]:

IOP Conf. Series: Materials Science and Engineering **832** (2020) 012049 doi:10.1088/1757-899X/832/1/012049

$$r_{YX^1} \equiv r_{YX^1(X^2,\dots,X^k)} = (r_{YX^1} - r_{YX^{1+1}} r_{XX^{i+1}})((1 - r^2_{X^1X^{i+1}})(1 - r^2_{YX^{i+1}})^{-1/2}),$$

$$r_{YX^i} \equiv r_{YX^i(X^{i+1},\dots,X^k)} = (r_{YX^i} - r_{YX^{i+1}} r_{X^iX^{i+1}})((1 - r^2_{X^iX^{i+1}})(1 - r^2_{YX^{i+1}}))^{-1/2}, (i = 1, 2, \dots, k). \quad (5)$$

In particular, equation 5 includes four dependent variables X^1, $X^2, \dots, X^4$ therefore requires calculation of the four sample correlation coefficients.

3.2 Step 2

When studying the statistical properties of the sample partial correlation coefficient of order k we assume that it is distributed in the same way as the usual (paired) sample correlation coefficient between the same variables. It is necessary to introduce an amendment, reducing the sample volume in k units, i.e. accept it to be equal to $n - k$. For small values of n and r, close to ±1, you should take into account that the value $\hat{r}$ is a biased evaluation of their theoretical r values.

Knowledge of the probability distribution law $\hat{r}$ enables to determine the adequate value of the sample correlation coefficient for a statistically valid conclusion about the presence of correlation between the studied variables.

For small values of $|r|$ there is a relatively good degree of approximation of the normal distribution, as a result a simple criterion for testing hypotheses about the lack of correlation between the studied variables is accepted:

$$H_0 : r = 0.$$

The approximate distribution of statistics according to the Student's probability distribution law with $n-2$ degrees of freedom is taken into account together.

It follows that the hypothesis H_0 is rejected, with probability of a mistake being equal to α following the inequality (6):

$$(|\hat{r}|(n-2-k)^{1/2})(1-\hat{r}^2)^{-1/2} > t_{\alpha(2)^{-1}}(n-2-k), \quad (6)$$

where $t_q(m) - 100q\%$ –point t distribution with freedom degrees m.

Using the Fisher transformation (7), confidence intervals are calculated for the true value of the correlation coefficient:

$$z_{1,2} = \operatorname{arcth}(\hat{r}) \pm (u_{\alpha/(2)^{-1}})(n-k-3)^{-1/2} - \hat{r}(2(n-k-1))^{-1}, \quad (7)$$

it follows that the true value of the correlation coefficient r with confidence probability $1-\alpha$ is concluded within:

$$\operatorname{th} z_1 < r < \operatorname{th} z_2$$

where $\operatorname{th} z$ – the hyperbolic tangent of the argument z, calculated by the formula:

$$\operatorname{th} z = (e^z - e^{-z}) / (e^z + e^{-z}). \quad (8)$$

Finding z is performed according to the table (Appendix 1 [8]), where the extreme columns show the values $|r|$, and the corresponding values $|z|$ are between them. It should be remembered that the signs of the argument and the function are the same.

It follows that the values of correlation coefficients r_{Y,X^i} with probability $p = 1 - \alpha$ will lie within the confidence interval:

$$(\operatorname{th}(z - m_z); \operatorname{th}(z + m_z)).$$

The value of the hyperbolic tangent from the argument is determined using statistical tables [8].

3.3 Step 3

The multiple correlation coefficient $R_{Y,\overline{X}}$ is used as a measure of the statistical relationship strength between the dependent variable Y by the set of independent variables $X^{(i)}$ $(i = 1, 2, \dots, k)$.

The calculation of the multiple correlation coefficient $R_{Y,\overline{X}}$ is performed using partial selective correlation coefficients $r_{Y,X^i(X^{i+1},...,X^k)}$ according to formula (9):

$$R^2_{Y,\overline{X}} = 1-(1-r^2_{YX^i})(1-r^2_{YX^{i+1}(X^i)})(1-r^2_{YX^{i+2}(X^i,X^{i+1})})...(1-r^2_{YX^k(X^i,X^{i+1},...,X^{k-1})}). \qquad (9)$$

To assess the statistical significance of the difference in the value of the multiple correlation coefficient $R_{Y,\overline{X}}$ from zero with a given significance level α we consider the following hypothesis H_0:

$$H_0 : R^2_{Y,\overline{X}} = 0.$$

Then the competing hypothesis H_1 has the form:

$$H_1 : R^2_{Y,\overline{X}} \neq 0.$$

In order to test the hypothesis H_0 about the insignificance of the multiple correlation coefficient $R_{Y,\overline{X}}$ we use the Fisher criterion. The experimental value F^{op} of the Fisher criterion is calculated by the formula:

$$F(\hat{R}) = \hat{R}^2_{Y,\overline{X}}(1-\hat{R}^2_{Y,\overline{X}})^{-1}(n-k-1)k^{-1} \qquad (10)$$

and at a given level of significance, it is compared with its critical value $F^{kr} = F(k, n-k-1)$ determined according to the statistical tables [7, 8].

4. Discussion of the results

Within the work, the analysis of the regression equation is performed, necessary assumptions are introduced and the features of the correlation analysis implementation in a particular case are analyzed.

On the basis of the conducted research using multiple and partial correlation coefficients the step-by-step procedure for the correlation analysis of the linear dependence of the accident rate at approaches to a modern roundabout (A_i) from the independent variables is developed and presented. The independent variables include a number of approaching vehicles in one lane (N_r); the average annual daily traffic on the circulating carriageway adjacent to the approach (N_k); 85th percentile speed on the entry curve (V_a); the number of lanes on the particular roundabout approach.

Correlation analysis of the equation for finding the accident rate at the approaches to modern roundabouts according to the proposed method enables to determine the reliability of this equation in the conditions of the Russian Federation, taking into account climatic conditions, design features, the state of the existing road transport network, as well as driving style.

References

[1] Gates T J and Maki R E 2001 *Converting old traffic circles to modern roundabouts* (Michigan: Michigan State University)
[2] Laureshyn A, Svensson A and Hyden C 2010 Evaluation of traffic safety, based on micro-level behavioral data *Accident Analysis and Prevention* **42** 1637–46
[3] *Roundabouts* 2010 An Information Guide 2rd ed (Washington: National Research Council)
[4] Burgonutdinov A M, Jushkov B S and Okuneva A G *Organization and safety of traffic on roads* (Perm: Perm National Research Polytechnic University)
[5] Lipnitsky A S 2009 *Modern roundabouts* (Irkutsk: Irkutsk national research technical university)
[6] *Roundabouts* 2006 Road planning and design manual ch 14 (Australia: Department of Main Roads) pp 67–87
[7] Lyalkina G B and Berdyshev O V 2014 Mathematical processing of the experimental results: training manual *Sovremennye probl. nauki i obraz.* **3** 180
[8] Aivazian S A and Mkhitarian V S 2001 *Probability theory and applied statistics* vol 2 (Moscow: Unity)

IOP Conf. Series: Materials Science and Engineering **832** (2020) 012050 doi:10.1088/1757-899X/832/1/012050

Methods of determining the correction coefficients of specific norms of spare parts consumption depending on fleet age

V A Maksimov[1], A A Zavgorodniy[2,3], N V Pozhivilov[1] and G A Krylov[1]

[1]Moscow Automobile and Road State Technical University (MADI), 64, Leningradsky ave., Moscow, 125319, Russia
[2]Research Centre, Mosgortrans, 22/21, b 1, Raushskaya naberezhnaya, Moscow, 115035, Russia

E-mail: zavgorodniy.aa@mail.ru

Abstract. The article considers changes in requirements for spare parts for maintenance and ongoing repair of urban buses taking into account their mileage from the beginning of operation. The research was carried out with the purpose of development of a technique that would help one to determine coefficients of correction made to specific norms of consumption of spare parts for various age groups of buses on the basis of changes in their failure flow. Correction factors were calculated for the four selected age groups of buses. In addition, the article considers the issue of reducing the annual mileage of the public transit bus while increasing its service lifetime. The methodology was tested on taking LiAZ-529222 and LiAZ-621322 urban buses as an example, operated in Moscow. As a result of the performed calculations, values of coefficients of correcting the consumption of spare parts have been obtained for the considered modifications of buses in each age group.

1. Introduction

An integral part of the technical operation of motor transport is the planning for the necessary spare parts. The number of spare parts required can be effectively assessed based on reliability. Not only do the reliability indicators depend on the operating time but also on factors such as the operating environment (town, metropolis, intercity, etc.), natural and climatic conditions (e.g. dust, humidity, temperature, etc.), which may worsen or increase reliability. These factors must be taken into account in order to determine the exact requirements for spare parts [1, 2].

Applying the norms of spare part consumption established by the manufacturer to the conditions of a particular transport company is carried out by their adjustments of their correction with coefficients K_1, K_2, K_3, according with the Regulation on maintenance service and repair of a fleet of motor transport. Correction factor K_1 takes into account the influence of operating conditions, coefficient K_2 takes into account the type and modification of the vehicle, coefficient K_3 shows the climatic conditions influence [3].

During fleet operation its reliability constantly decreases. The probability of failure-free operation of units and assemblies is reduced, operating time intervals are reduced and the failure flow is increased which leads to an increase in the required number of spare parts. This leads to an increase in time required for ongoing repair, increasing fleet downtime and the spare parts and repair costs.

To sum up, it is required to correct norms of the consumption of spare parts by factor of correction K_4 taking into account fleet mileage when calculating consumption of spare parts for the transit agency operating urban buses.

2. Organization of research

This method is developed to determine the coefficient of correcting the norms of spare parts consumption depending on the fleet age.

This technique was developed on the basis of generalization of theoretical and experimental researches carried out on the basis of the SUE "Mosgortrans" and implemented on the LiAZ-529222 and LiAZ-621322 urban bus modifications.

The purpose of the research was to calculate and determine the coefficient of correcting the norms of spare parts consumption depending on the fleet age by the example of LiAZ-529222 and LiAZ-621322 buses.

The subject of the research is operational failures of LiAZ-529222 and LiAZ-621322 buses.

Definition of the normative framework is based on the statistical information of failures and calculated indicators of fleet reliability under the actual conditions of the technical base and specific fleet modification.

Collection of failures data and actual mileage of LiAZ-529222 and LiAZ-621322 buses were carried out on the basis of the branch of SUE "Mosgortrans" in Moscow.

The implementation of the method is realized in several stages:

- collection, processing and analysis of statistical data of failures on the example of LiAZ-529222 and LiAZ-621322 bus modifications;
- calculation of reliability of LiAZ-529222 and LiAZ-621322 buses;
- determining the coefficient of correction for the norms of spare parts consumption for the ongoing repair cases depending on the operating time since the beginning of the bus operation.

Maintenance and current repair of the controlled buses for the entire period of research are carried out in accordance with the requirements of the "Guidelines for maintenance and repair of buses", "Typical technologies for maintenance and repair of buses", recommendations of the "Regulations on maintenance and repair of buses" approved by the SUE "Mosgortrans".

The system of keeping the technical condition of the controlled buses in working order for the whole period of research did not contradict the recommendations of the manufacturer.

The controlled buses were refueled with fuel, oils and other technical fluids in accordance with the operating manual. The controlled buses operated as usual in the course of the studies.

The current repairs were carried out as required after the inspection and diagnosis work and during maintenance work.

The following restrictions were imposed on the collection of data of bus failures:

- all available LiAZ-529222 and LiAZ-621322 buses at the time of the survey were included in the sample;
- the sample did not take into account failures and repair operations performed by the manufacturer during the warranty period on these modification of buses.

The frequency of observations of the monitored buses is carried out daily.

The sample size for obtaining reliable material was determined by standard methods. In the case of the nonparametric method (when the type of distribution law is unknown), the sample size for the estimation of the unknown reliability law was determined by the following formula [4]:

$$N = \frac{\ln(1-\beta)}{\ln P(x)} \qquad (1)$$

where β is the confidence level;

$P(x)$ the required probability of failure-free operation during some operating time x.

In case of failure and malfunction of the bus, the driver filled out an application for current repair and submitted the bus to the area of current repair for inspection (diagnostics). During the repair the

IOP Conf. Series: Materials Science and Engineering **832** (2020) 012050 doi:10.1088/1757-899X/832/1/012050

foreman recorded the actual time of the beginning and the end of the repair, the list of repair operations, causes of malfunction, etc. in the observation report.

Subsequently, the statistics collected were entered into the SAP software database and presented in MS Office Excel spreadsheets. The spreadsheets contained structured information on the following key criteria:

- modification of the bus;
- garage number of the bus;
- date of failure and malfunction;
- mileage at the time of failure or malfunction;
- actual repair time;
- name of repair operations;
- mechanic's conclusion, etc.

For the purpose of the further analysis and systematization of failures on certain intervals of operating time, the data were consolidated in the form of the table 1.

Table 1. Example of the table of the bus failure flow in the specified operating time intervals

Interval, thousand km	0–50	50–100	100–150	150–200	200–250	…	750–800
N_t							
$\sum r_i$							
$\omega(t)$							

where Δt is the interval of operating time (mileage), thousand km;

N_t – number of observed buses in the interval, units;

$\sum r_i$ – number of observed failures in the interval, units;

$\omega(t)$ – estimated value of the failure flow parameter.

To specify the observations of the monitored buses and a more detailed study of changes in the failure flow under the conditions of compressed timing of observations it is advisable to break the intervals of operating time on the mileage Δt = 50 thousand km. It was taken into account that the necessary number of intervals should be within the range from 8 to 12 [4].

Based on mileage data, the number and distribution of observed buses were calculated and carried out in the specified intervals (Δt = 50 thousand km). The different number of controlled buses in the observation intervals is caused by the uneven age fleet structure.

Similarly, the distribution of failures within the specified intervals for observed buses was performed.

The failure flow is the ratio of the number of failures of the object being restored for a sufficiently small amount of its operating time to the value of this operating time [1, 5].

$$\omega(t) = \frac{\{m(t+\Delta t)-m(t)\}}{\Delta t} \tag{2}$$

where Δt – operating time (mileage);

$m(t)$ – the number of failures that occurred from the initial point of time before reaching the operating time t;

the difference $m(t + \Delta t) - m(t)$ is the number of failures on the operating time (mileage) Δt.

Statistical evaluation of the failure flow parameter $\omega(t)$ in each of the operating time intervals is determined by the formula 3, the data are entered in Table 1 separately for each of the bus modifications.

$$\omega_t = \frac{\sum r_i}{N_t * \Delta t} \tag{3}$$

where $\sum r_i$ – the number of observed failures in the interval;

N_t – the number of buses observed in the interval;

Δt – operation time interval value.

In order to compare the operation of fleet in different conditions (different from the nominal ones), the standards of maintenance and repair of fleet are adjusted to the specific conditions of operation. The following types of correction are distinguished:

- resource (at the federal, sectoral and intrasectoral levels) – to create comparable working conditions for motor transport enterprises;
- operational (at the intra-industry and economic levels) – to ensure efficient use by the motor transport enterprise of its human and material resources.

Adjustments are made by changing:

- the quantitative value of the maintenance and current repair standards;
- list of maintenance operations;
- the ratio between maintenance and current repair work by including typical, frequently recurring current repair operations in maintenance.

At the same time, it is important to take into account the availability of reliable information of failures and costs of current repairs and preventive maintenance.

Determination of correction coefficients, in accordance with the recommendations of the "Regulations" [3], is carried out for the enlarged groups of fleet with a mileage interval of 0.25 L_R (0-200 thousand km) to the service life:

- Age group 1 – mileage from the beginning of operation from 0 to 200000 km;
- Age group 2 – mileage from the beginning of operation from 200001 to 400000 km;
- Age group 3 – mileage from the beginning of operation from 400001 to 600000 km;
- Age group 4 – mileage from the beginning of operation from 600001 to 800000 km.

L_R is a mileage of units before to resource depletion (Table 2), determined on the basis of experience in the operation of LiAZ buses in the Russian Federation [3, 5–8]. The resources of the bus units may be achieved submitted that the frequency of maintenance is observed, with the use of fuels and lubricants and operating materials recommended by the manufacturer.

Table 2. Unit mileage rates before resource depletion LiAZ-529222 and LiAZ-621322 buses, thousand km

Category of operating conditions	Engine, MAN D0836 LOH	Transmission, ZF Ecomat 4 6HP504C	Front axle RL-85A ZF	Rear axle AV-132/87, AV-132 II /87 ZF
III	800	800	800	800

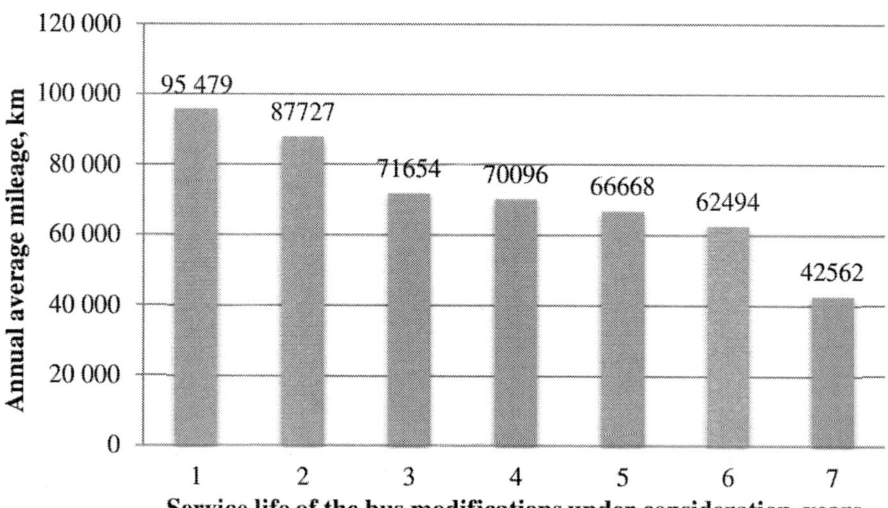

Figure 1. Histogram of the effect of the service life on the annual average mileage of LiAZ buses

The correction coefficient of specific norms of spare parts consumption defines increases in mathematical expectation of number of expected failures and corresponding quantity of spare parts in the considered interval of mileage (0.25 L_R) since the beginning of operation.

The mileage-correction factor is calculated in relation to the average value of the failure flow at the middle of the interval (0.25 L_R), relatively the first group. The value of the failure flow for the first age group is taken as 1.

Calculation of correction coefficients depending on the fleet age is based on the average annual mileage by year of operation.

Histogram (Figure 1) shows the effect of the service life (year of operation) on the annual average mileage of the LiAZ-529222 and LiAZ-621322 bus modifications.

3. Processing and analysis results

All LiAZ-529222 and LiAZ-621322 buses, which are on the balance of the company, were under observation during the period of statistical data collection from 01.08.2017 to 31.12.2018.

In the conducted studies, the sample volume for LiAZ-529222 buses was 107 units, and for LiAZ-621322 – 32 units which meets the specified conditions (see the formula (1)).

During the observation period, the statistics of failures of LiAZ-529222 and LiAZ-621322 buses were obtained in the amounts of 1148 and 319, respectively.

In accordance with Formula 1, a sample size of at least 28 shall be used to study the reliability characteristics of the safety-related components and assemblies (maximum reliability level).

The sample size for obtaining reliable material was determined by standard methods, according with the formula (1).

$$N = \frac{\ln(1 - 0.95)}{\ln 0.9} = 28$$

where 0.95 is the confidence level (in the field of road transport operation is in the range from 0.8 to 0.95);

0.9 – the required probability of failure-free operation during some operating time (in the field of road transport operation it is usually assumed to be 0.9).

Tables 2 and 3 buses included data on the number of observed buses in the operating time interval and the number of failures for the LiAZ-529222 and LiAZ-621322.

Further, the calculation of the failure flow for each of the intervals operation was performed according to the formula 3. For example, for the interval of 350-400 thousand km for the LiAZ-529222 bus modification:

$$\omega_t = \frac{\sum r_i}{N_t * \Delta t} = \frac{315}{73 * 50} = 0.086$$

where 315 – the number of recorded failures in the interval of 350–400 thousand km for the modification of the LiAZ-529222 bus;

73 – the number of observed LiAZ-529222 buses in the interval of 350–400 thousand km;

50 – operating time interval.

Similarly, calculations are performed for other intervals.

Estimated values of the failure flow of LiAZ-529222 and LiAZ-621322 buses for all intervals are presented in Table 3 and 4.

Table 3. The value of the LiAZ-529222 bus failure flow according to specified operating time intervals

Mileage interval, thousand km	150 200	200 250	250 300	300 350	350 400	400 450	450 500	500 550	550 600
N_t	2	2	12	44	73	82	51	20	3
$\sum r_i$	6	6	41	120	315	396	254	105	13
$\omega(t)$	0.060	0.060	0.068	0.054	0.086	0.096	0.099	0.105	0.086

Table 4. The value of the LiAZ-621322 bus failure flow according to specified operating time intervals

Mileage interval, thousand km	150 200	200 250	250 300	300 350	350 400	400 450	450 500	500 550
N_t	7	17	26	22	16	7	4	1
$\sum r_i$	19	46	64	76	59	32	20	3
$\omega(t)$	0.054	0.054	0.051	0.073	0.076	0.091	0.100	0.060

The correction coefficient of spare part consumption norms depending on age was calculated on average value of failure flow at the middle of an interval from the beginning of operation with a step of operating time of 0.25 L_R (0–200 thousand km). The data were entered in Table 5.

Based on the lack of statistical data on failures for the 4 age groups (mileage from the beginning of operation from 600 thousand km to 800 thousand km), the value of the coefficient was forecasted by the already known values of the coefficient for the three previous groups.

Calculation of correction coefficients depending on the age of the fleet is based on the average annual mileage by year of operation (Table 6, Figure 2).

Table 5. Values of the correction coefficient of specific norms of spare parts consumption (K_4) depending on the mileage from the beginning of bus operation

Bus modification	Mileage interval, thousand km			
	0-200	200-400	400-600	600-800
LiAZ-529222	1.00	1.12	1.62	2.05
LiAZ -621322	1.00	1.20	1.58	1.99

Table 6. Values of the correction coefficient of specific norms of spare part consumption (K_4) depending on the age of the buses

Bus modification	The age of buses, years						
	1	2	3	4	5	6	7
LiAZ-529222	0.85	0.94	1.02	1.11	1.21	1.30	1.37
LiAZ -621322	0.87	0.96	1.05	1.14	1.23	1.32	1.38

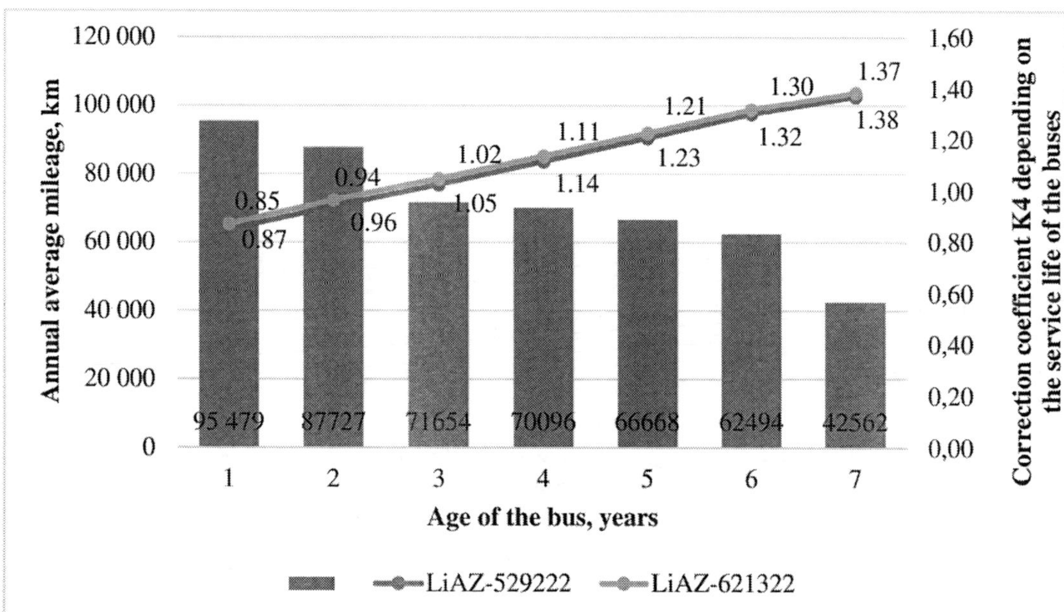

Figure 2. Histogram of correction coefficient values of specific norms of spare part consumption (K_4) depending on the age of buses

4. Conclusion

As a result of the research, the technique of experimental determination of correction coefficient of specific norms of spare part consumption (K_4) depending on mileage from the beginning of operation has been offered. Figure 3 shows a graph for comparing the values of the correction coefficient of specific norms of spare parts consumption according to various methods and regulatory approved documents [3, 8] for the class of LiAZ buses.

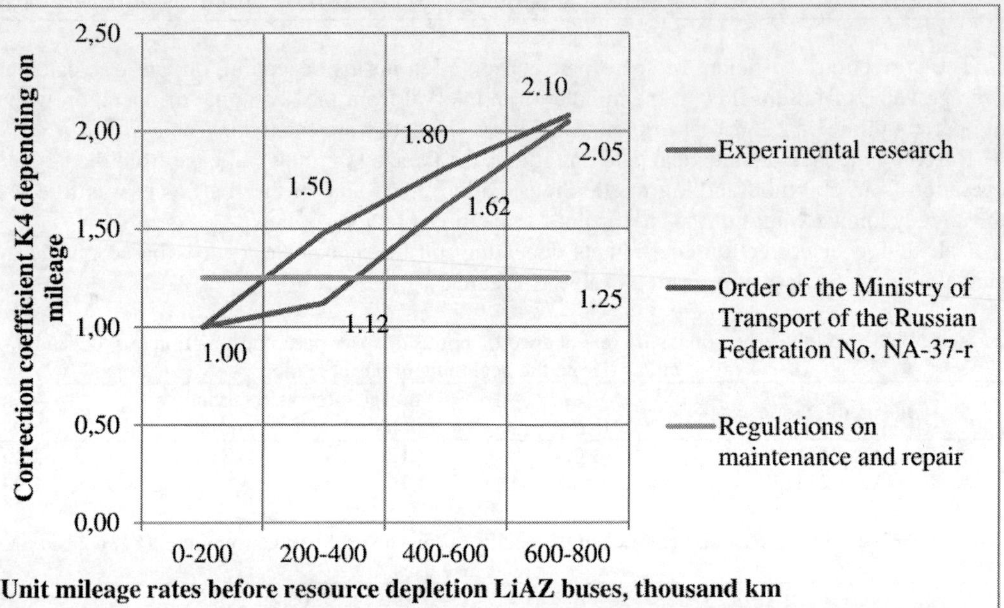

Figure 3. Values of the correction coefficient K_4 according to different methods

Technique testing is performed in real conditions of the operating branch of the SUE "Mosgortrans" by the example of buses LiAZ-529222 and LiAZ-621322, representing a significant share of the fleet of the company.

As a result of the performed calculations, the values of the correction coefficient of specific norms of spare parts consumption (K_4) depending on mileage and time from the beginning of operation have been received, taking into account the dynamics of change of average annual mileage of the considered modifications of buses at the enterprise of SUE "Mosgortrans".

It is necessary to adhere to a single rational system of management and planning of spare parts and materials consumption in order to increase the level of fleet reliability, improve the system of supply in the branches of SUE "Mosgortrans".

The obtained values of the correction factors are currently applied within the framework of the developed unified management and planning system of spare part consumption in the bus branches of SUE "Mosgortrans".

References

[1] Kuznetsov E S, Boldin A P and Vlasov V M 2001 *Technical operation of cars: textbook for universities 4th edited, revised and supplemented* (Moscow: Nauka) p 535

[2] Behzad Ghodrati 2003 *Product Support and Spare Parts Planning Considering System Reliability and Operating Environ*ment (Lulea University of Technology) p 96

[3] *Ministry of Motor Transport of the RSFSR* 1988 Regulations on Maintenance and Repair of fleet of Motor Transport (Moscow: Transport) p 78

[4] Boldin A P and Maksimov V A 2012 *Fundamentals of scientific researches: textbook for the higher education institutions* (Moscow: Publishing Center "Academy") p 336
[5] *Reliability in technique. Basic concepts. Terms and definitions* Resolution of the USSR Gosstandart Committee on Standards of 15.11.89 No. 3375 1989 GOST 27002-89 Interstate standard
[6] *LiAZ-529222 bus. Operation manual* 2012 Likino-Dulevo: LLC "Likinsky Bus Plant" p 407
[7] *LiAZ-621322 bus. Operation manual* 20131 Likino-Dulevo: LLC "Likinsky Bus Plant" p 373
[8] *On the introduction of the Methodological Recommendations on the calculation of economically justified cost of transportation of passengers and luggage in urban and suburban traffic by public motor and urban land-based electric transport* Order of the Ministry of Transport of the Russian Federation No. NA-37-r dated April 18, 2013

Innovative approach to integrative estimation of driver's aggressiveness and comfort

M P Malinovsky, E S Smolko and I R Nasibulov

Moscow Automobile and Road Construction State Technical University (MADI), 64, Leningradsky ave., Moscow, 125319, Russia

E-mail: ntbmadi@gmail.com

Abstract. Modern on-board intelligent systems of vehicles perform many tasks aimed at improving the safety and comfort of the driver and passengers, as well as the safety of the cargo carried. One such task is to prevent traffic accidents due to driver's inattention, drowsiness, or loss of consciousness. In order to fulfill this task various driver attention, monitoring systems are applied. Authors of the article note that the statistics does not count how many accidents and emergency situations occur due to aggressive or dangerous driving. The authors have invented a new on-board device called Comprehensive Integrator of Driver's Aggressiveness and Comfort (CIDAC), intended to train the driver to avoid the aggressive driving style by the method of scientific observation. CIDAC evaluates the driver's behavior based on indirect indicators, such as longitudinal, lateral, and vertical acceleration. Construction of the CIDAC device includes an Arduino controller and a three-axis accelerometer. CIDAC warns the driver in cases of aggressive driving by flashing the signal lamps when acceleration exceeds the threshold meaning. Thus, CIDAC increases the safety of cargo and passenger transportation by providing a training effect for the driver.

1. Introduction

The statistics claims that up to 25 % of the total amount of road accidents in the Russian Federation occur because of carelessness, drowsiness, or loss of consciousness of the driver due to fatigue, stress, neglect of his own health, as well as mistakes due to reassessment of his driving skills. In order to solve the aforementioned problem, various driver attention monitoring systems (DAMS) and similar devices are becoming more widespread.

We can distinguish the following types of on-board devices designed to monitor the condition of the driver:

1. Direct DAMS that directly records the physiological parameters of the driver's condition using biometric sensors (pulse, pupil condition, head position, sweating, skin conduction, respiratory rate).

2. Indirect DAMS that determines the condition of the driver by indirect signs (leaving the lane without turning on the turn signal, inadequate oscillating taxiing, prolonged lack of impact on the controls, ignoring traffic lights, analyzing the time of day and duration of the trip).

3. Vigilance control that periodically gives a light and sound signal. If the driver does not press the button, it takes measures to stop. It is widely used on railway locomotives.

4. Tachograph intended for registration of speed, fuel consumption and work schedule of the driver on a paper or electronic medium.

5. Remote DAMS that transmits data about the driver's condition to the dispatch center by signals from the above devices, as well as video cameras installed in the cabin.

The task of DAMS is to prevent and fix such driver states as inattention, drowsiness, stress, loss of consciousness. For each of those, different monitoring tools are effective. For example, indirect monitoring can be used to record an aggressive state, distract attention and drowsiness, while lateral acceleration of the vehicle can be used as a parameter for analyzing involuntary deviations from an ideal trajectory.

The simplest direct monitoring parameter that allows you to successfully identify all of the listed driver conditions is the pulse. For example, with a pulse of less than 70 beats per minute the driver is in a state of monotony, a range of 70 to 85 beats per minute corresponds to functional comfort; values above 85 beats per minute correspond to stress. Electroencephalogram and electrocardiogram are highly informative characteristics of the mental and physical activity of a person, but their use is limited by the complexity of mounting equipment on the driver. The skin-galvanic reaction allows identifying stressful situations and fatigue, however, this parameter is dependent on the personal characteristics of each driver, therefore it often filters out significant information, which leads to false alarms. Oculography quite effectively reveals drowsiness and distraction, however, its information content drops sharply at high levels of mental stress. In addition, there were cases when the driver fell asleep with his eyes open. It is also possible to use a head position sensor.

The first studies of mental tension using portable devices were conducted in the USSR in 1971 and were based on the registration the skin-galvanic reaction of the driver and oculography. The first indirect DAMS was patented by "Nissan" in 1977 [1]. However, in connection with the introduction of anti-lock braking system (ABS) in 1978 and other active safety systems, it did not receive due attention. The development of DAMS and active safety systems has long been parallel and independent of each other. Only 30 years later, Volvo introduced the first serial DAMS, which was called Driver Alert Control (DAC), and consisted of a video camera that monitored the trajectory of the vehicle and its position on the road, as well as a steering wheel speed sensor. In the second decade of the 21st century, it was understood that the synthesis of control systems gives a synergistic effect to improve road safety [2]. An integrated system, such as a preventive motion control system (PMCS), can prevent more accidents than each system individually. DAMS has become part of this complex.

2. Problems of modern DAMS

The main reason for the lack of effectiveness of active safety systems is a fragmented approach to the review of processes in the car. A modern DAMS must interact with other active safety systems (or other PMCS modules). If the driver's aggressive state is detected, it is advisable to limit the fuel supply by acting on the automated engine control system, and if the driver faints, to stop the vehicle within the lane with help of the lane-keeping assist and collision avoidance system [3–5].

Initially, DAMS included the following main functions:

1. Prevention of falling asleep with the help of sound, visual, and tactile signals.
2. Alarm signal when driver's distraction.
3. Warning by signals from temperature sensors inside the cabin when overheating above + 27 °C, which reduces the driver's ability to quickly and accurately assess the traffic situation, increases reaction time, leads to dull attention, and increases the number of errors.
4. Checking for alcohol intoxication with the possibility of displaying recommendations on the display and forced engine blocking.

However, no one even tries to count how many accidents and, moreover, emergency situations occur due to aggressive or dangerous driving [6–8]. Aggressive driving style is manifested as follows:

1. Sharp acceleration and braking, abrupt releasing the clutch.
2. The excess of the estimated speed in corners with a given curvature or when performing the lane-changing maneuver.
3. Speeding when moving over bumps.

IOP Conf. Series: Materials Science and Engineering **832** (2020) 012051 doi:10.1088/1757-899X/832/1/012051

Dangerous driving is characterized by the intentional creation of critical or emergency situations, including failure to provide priority.

A technical solution to reduce the number of the aforementioned cases is especially actual for the following reasons:

1. Aggressive driving by public transport drivers (buses, taxis) often leads to micro-injuries of the passengers' musculoskeletal system that no one records.
2. The safety of transported goods, especially fragile and liquid.
3. Comfort of passengers carried by private transport.
4. Increased durability of the vehicle chassis.

Therefore, modern DAMS should be supplemented with the following functions:

1. Assessment of driving style in terms of comfort and safety.
2. Checking the adequacy and mental state of the driver, assessing his danger.
3. Checking the driver's readiness to take control with the possibility of performing a minimum risk maneuver or automatic stop when driving on a vehicle with a high level of automation [9, 10].
4. Remote monitoring of the physiological and mental state of the driver through the intelligent transport system (ITS) [11].

3. CIDAC

One of the elements of modern DAMS should be a Comprehensive Integrator of Driver's Aggressiveness and Comfort (CIDAC), which forms the so-called "driver behavior profile" and implements the aforementioned functions using the following evaluation parameters:

1. Total acceleration and jerk (the first time derivative of acceleration) in three axes.
2. The number of cases of violation of the high-speed mode, driving to a prohibitory traffic signal, determined by either a system for recognizing traffic signs and traffic signals [12], or ITS.
3. The number of responses of various active safety systems.

CIDAC belongs to the group of indirect DAMS, that is, it evaluates the driver's behavior based on indirect indicators [13, 14].

The purpose of CIDAC is to increase the safety of freight and passenger transportation and the durability of the chassis of vehicles by reducing longitudinal, lateral and vertical overloads. In accordance with this, CIDAC solves three problems:

1. The limitation of longitudinal acceleration during acceleration and braking.
2. Improving the stability of curvilinear motion by limiting lateral acceleration, predicting the trailer jerk by angular speed of the steering wheel.
3. Prediction of the behavior of the trailer suspension along the first axle of the tractor.

CIDAC has a double character of action:

1. Informative and training.
2. Preventive due to interaction with other modules of PMCS.

For professional drivers of buses and commercial vehicles, it is possible to provide a penalization system for aggressive driving style on the part of the trucking company according to the integral indicators of acceleration and jerk.

4. Application of jerk for determination of aggressive driving style

As experimental studies have shown, the loss of passenger stability in the subway car is most likely at the time of acceleration change, that is, under the action of jerk [15]. Jerk is widely used in astronautics and in rail transport to assess human overloads. When designing highways, jerk is used to calculate transition curves [16]. Vertical jerk has long been used to assess the vehicle ride [17]. The use of longitudinal jerk to assess the smoothness of gear shifting is justified [18, 19]. Similarly, we can evaluate the impact of the lateral jerk during curvilinear movement. At the same time, it is advisable to use instantaneous jerk in the algorithms of active safety systems, and the accumulated (total) jerk can serve as an integral criterion for assessing the overloads acting on the driver, passengers and cargo. Studies show that the most difficult case for passengers is overload on all three axes at the same time.

In case of alternating overload when calculating jerk, it is necessary to add two acceleration amplitudes modulo (fig. 1), for example, 0.5g–(–0.5)g=1g.

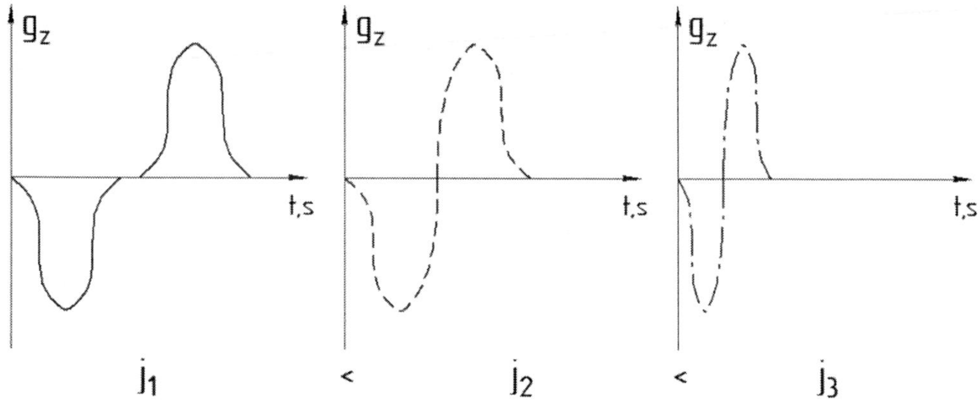

Figure 1. The dependence of jerk on the nature of acceleration

Two typical situations can be distinguished, covering most of the possible cases of curvilinear motion:

1) "U-turn" maneuver.

2) "Lane changing" maneuver.

The task is to tell the driver which trajectory is the best for passengers and cargo in terms of reducing lateral overloads. To do this, it is advisable to use the following motion parameters: the final speed of the maneuver, the total time of the maneuver, the integral acceleration along three axes, the average acceleration and the accumulated jerk.

There are three ways to perform the "U-turn" maneuver: with early, neutral and late apex, as well as two special cases – with shutter speed and with an additional turn of the steering wheel. The U-turn maneuver has two phases:

1) entrance to the turn (from the start of the steering wheel to the trajectory apex);

2) exit from the turn (from the trajectory apex to the return of the steering wheel to the neutral position).

At a constant speed, the most comfortable by jerk is neutral apex. Offset apex in any direction increases jerk. Acceleration does not depend on apex.

With equally slowing motion, the late apex is the most comfortable for acceleration, which allows the driver to find a balance between reducing lateral acceleration at the entrance due to the expansion of the trajectory, and at the exit due to the reduced speed. However, the smallest jerk is observed at neutral apex, since the acceleration between the phases does not change. The least comfortable, both by acceleration and jerk, is an early apex.

In the case of braking at the entrance and acceleration at the exit, neither acceleration nor jerk depend on the method of passing the turn, however, the final speed of the maneuver comes to the fore. The maneuver time does not depend on the apex, but depends only on the distance traveled and the average speed.

The lane changing maneuver has a different number of phases on the steering wheel and on the path. The former phases are:

1) entrance (steering wheel to the left);

2) counter-movement (steering wheel to the right);

3) exit (steering wheel left to the neutral position).

The latter phases are:

1) entrance (before the inflection);

2) exit (after bending).

When performing “lane changing”, the trajectory has an inflection, where the increment of the lateral coordinate changes sign, therefore, there are two apexes.

As experimental studies have shown, an early kink when performing the lane changing maneuver is preferable. The fact is that when the steering wheel is counter-running, a “suspension bounce” occurs, the jerk from which folds with the side jerk, increasing the likelihood of buckling – skidding or tipping over. To date, there is no mathematical model that adequately describes this phenomenon.

"Lane changing", as a rule, is performed either at a constant speed, or when engine braking, or coasting. When “lane changing”, the values of the radii affect the length of the maneuver.

Vertical jerk is very indicative of assessing the aggressiveness of the driver when moving over road irregularities, for example, artificial. However, with a decrease in vertical overloads, it is necessary not only to limit the speed, but also to take into account its value at which resonant vibrations of the sprung masses occur [20, 21]. Longitudinal jerk can help train the driver to gently engage the clutch.

On-board diagnostic is widespread nowadays [22–24].The selected parameters can also be used to diagnose shock absorbers and clutch in order to detect their wear or malfunction (fig. 2). Based on this data, the PMCS or the autonomous control system can make recommendations for replacement of those.

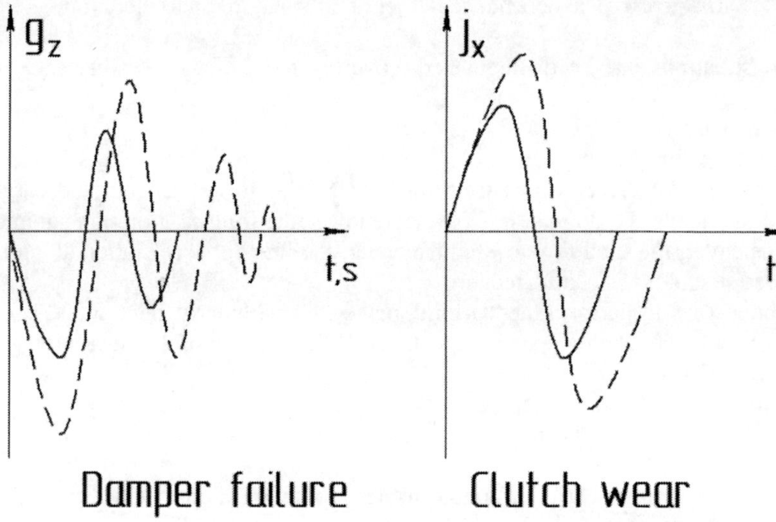

Figure 2. Application of jerk for testing dampers and clutch

5. Construction of CIDAC

In order to justify the link between the evaluation parameters and the driving style, the authors have invented a new on-board device. The device construction includes (fig. 3): housing 1; display 2; signal lamps 3, 4, 5; set-up buttons 6, 7, 8, 9; power lamp 10; port for MicroSD-card 11; power connector 12; toggle switch 13. Inside the housing 1 we placed the Arduino controller and a three-axis accelerometer (fig. 4).

The device is portable and supposed to be mounted on the dash-board inside the cabin in the driver’s field of view in accordance with the image of the axes on the housing 1. For the device to work correctly, the X axis must coincide with the longitudinal axis of the vehicle, the Y axis with the transverse axis of that, and the Z axis should be vertical.

The signal lamp 10 informs that the device is turned on. The device is connected to the power source through a cable, which is inserted into connector 12. After turning on the toggle switch 13, a self-test is performed, including the following items:

1. Checking the presence of a MicroSD-card in the slot 11.

2. The signal lamps 3, 4, 5 briefly (for 1 second) turn on to check their serviceability.

3. The horizon is determined to exclude errors in the installation of the device (manually with showing the current installation angles on the display or automatically, taking into account the deviation of the initial coordinates).

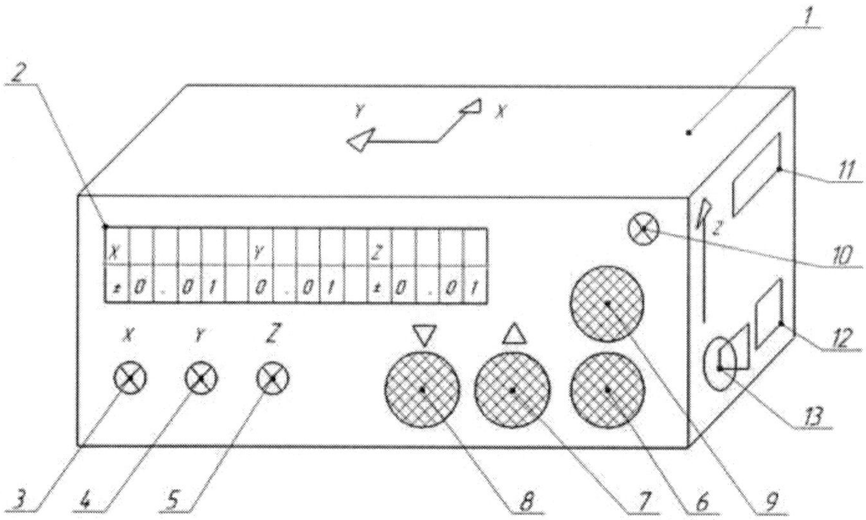

Figure 3. CIDAC, general external view: 1 – housing; 2 – display; 3, 4, 5 – signal lamps of an overload on axes X, Y, Z respectively; 6 – button SETUP; 7 – button UP; 8 – button DOWN; 9 – button MODE; 10 – power lamp; 11 – port for MicroSD-card; 12 – power connector; 13 – toggle switch

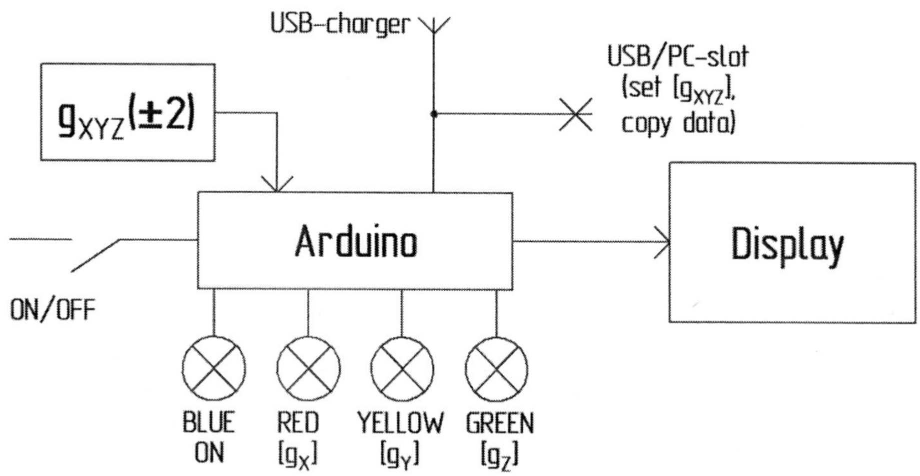

Figure 4. CIDAC, internal structural scheme

During the diagnosis, the power lamp 10 is in a blinking mode, then the device enters the recording mode, and the lamp 10 is lit continuously. During movement, the signal lamps 3, 4, 5 light up when the specified threshold acceleration values are exceeded along the X, Y, Z axes, respectively. Thus, the driver is taught to drive gently applying the method of scientific observation [25, 26].

The device is configured as follows. Button 9 is used to select the display mode between MAX and AVE. In the MAX mode the maximum acceleration values for the last 1...2 s are displayed, in the AVE mode a running average over the last 1...4 s are shown.

The transition to the setting of threshold values at which the signal lamps light up, and the selection of the adjustable axis is carried out by pressing the button 6. The display shows the current threshold values of accelerations on each axis, while the editable value flashes. The setting is made with buttons 7 and 8 in increments of 0.05 g. After completing the settings, the menu closes automatically when inactive for 5 s or by pressing the 9 button.

When the toggle switch 13 is turned off, the readings are reset to zero, and the data recorded during the last session is saved in a separate file on the MicroSD-card inserted in the slot 11.

6. Conclusion

The following problems need to be solved while working with CIDAC:

1. Initial determination of the horizon.
2. Removing sensor noise (normalization).
3. Selecting the optimal differentiation step using the moving average function (balance between accuracy and false response). The differentiation step significantly affects values of jerk.
4. The choice of threshold values and critical values. At threshold, the signal lamps should turn on; at critical, PMCS should be automatically triggered.

CIDAC setup is quite a subtle task, since the range of acceleration fluctuations along all axes during normal vehicle movement lies within 0.1...0.3 g, and during normal landing of a plane the vertical overloads do not exceed 0.5...0.6 g. Meanwhile, when jumping, a person develops acceleration in excess of 1 g.

The use of CIDAC will increase not only the safety of cargo and passenger transportation, but also passenger comfort and cargo integrity, while providing a training effect for the driver, increasing his skill level, and on-board diagnostics of suspension and transmission elements.

References

[1] Yanagishima T 1977 Patent US 4017843

[2] Malinovsky M, Vorobyev A and Zabudsky A 2018 Interaction of subjects having different control paradigms in a common transport space *Transportat. Res. Procedia* **36** 472–9

[3] Ivanov A M, Shadrin S S, Kristalniy S R and Popov N V 2019 Possible scenarios of autonomous vehicles' testing in Russia *IOP Conf. Ser. Mater. Sci. and Engineer.* **534(1)** 012001 DOI 10.1088/1757-899X/534/1/012001

[4] Ivanov A M, Kristalniy S R, Popov N V, Toporkov M A and Isakova M I 2018 New testing methods of automatic emergency braking systems and the experience of their application *IOP Conf. Ser. Mater. Sci. and Engineer.* **386(1)** 012019 DOI 10.1088/1757-899X/386/1/012019

[5] Dygalo V, Salykin E and Kotov V 2019 Methods of development and adaptation of engineering microprocessor engine control unit in real time to ensure the possibility of research work in various modes *Int. Conf. on Industr. Engineer., Applicat. and Manufact.* 8743053 DOI 10.1109/ICIEAM.2019.8743053

[6] Pegin P and Pegina O 2018 A method to assess accident psychological severity in drivers *Transportat. Res. Procedia* **36** 562–6

[7] Azemsha S, Kapski D and Pegin P 2018 A method for assessing the automobilization impact on population morbidity *Transportat. Res. Procedia* **36** 18–24

[8] Chubukov A, Kapitanov V, Monina O, Silyanov V and Brannolte U 2017 Simulation of regional mortality rate in road accidents *Transportat. Res. Procedia* **20** 112–24

[9] Malinovsky M, Vorobyev A and Pakhomov S 2018 Social aspect of anthropogenic adaptation to autonomous vehicles *Transportat. Res. Procedia* **36** 480–6

[10] Sidorov K M, Yutt V E, Grishchenko A G and Golubchik T V 2018 Practical implementation of

the concept of converted electric vehicle with advanced traction and dynamic performance and environmental safety indicators *IOP Conf. Ser. Mater. Sci. and Engineer.* **315(1)** 012026 DOI 10.1088/1757-899X/315/1/012026

[11] Zhankaziev S 2017 Current trends of road-traffic infrastructure development *Transportat. Res. Procedia* **20** 731–739 DOI 10.1016/j.trpro.2017.01.118

[12] Dygalo V, Lyashenko M and Potapov P 2019 Ways for improving efficiency of computer vision for autonomous vehicles and driver assistance systems Int. Conf. *Industrial Engineering, Applications and Manufacturing* 8743060 DOI 10.1109/ICIEAM.2019.8743060

[13] Mayboroda O V and Kalenov G K 2010 Increase of active safety of road traffic *Sci. J. of Transportat.* **2** 31–7

[14] Majboroda O V 2012 Patent RU 113218 U1

[15] Malinovsky M 2011 Application of jerk in active safety systems *Sci. J. of transportat.* **3** 62–6, 48–9

[16] Kostsov A 2018 Results of studies on traffic volume at left-turn exits of grade-separated intersections *Transportat. Res. Procedia* **36** 347–51

[17] Dubrovskiy A, Aliukov S, Keller A, Dubrovskiy S and Alyukov A 2016 Adaptive suspension of vehicles with wide range of control *SAE Techn. Papers* DOI 10.4271/2016-01-8032

[18] Aliukov S, Keller A and Alyukov A 2015 Dynamics of overrunning clutches of relay type *SAE Techn. Papers* DOI 10.4271/2015-01-1130

[19] Aliukov S, Keller A and Alyukov A 2017 Vibrations and properties of inertia continuously variable transmissions *Lecture Notes in Engineer. and Computer Sci.* **2** 702–6

[20] Keller A V, Gorelov V A and Anchukov V V 2015 Modeling truck driveline dynamic loads at differential locking unit engagement *Procedia Engineer.* **129** 280–7 DOI 10.1016/j.proeng.2015.12.063

[21] Keller A, Aliukov S and Anchukov V 2017 Mathematical model of the truck for investigation of stability and control of movement *Lecture Notes in Engineer. and Computer Sci.* **2** 711–6

[22] Revin A, Dygalo V, Boyko G, Lyaschenko M and Dygalo L 2018 Methods of monitoring the technical condition of the braking system of an autonomous vehicle during operation *IOP Conf. Ser. Mater. Sci. and Engineer.* **315(1)** 012020 DOI 10.1088/1757-899X/315/1/012020

[23] Dygalo V and Zhukov I 2018 The thermal loading estimation of the friction pairs of a vehicle automated brake system *IOP Conf. Ser. Mater. Sci. and Engineer.* **386(1)** 012012 DOI 10.1088/1757-899X/386/1/012012

[24] Solntsev A A, Feofanov S A, Guliy V V 2019 The research of the operation of the lithium-iron-phosphate battery of an electric vehicle in high and low temperatures *Systems of Signals Generating and Processing in the Field of on Board Communications* article № 8706728 DOI 10.1109/SOSG.2019.8706728

[25] Malinovsky M, Solntsev A, Juravleva A and Makovski S 2017 Estimation method of nonresident vehicle fleet inflow influencing road traffic safety in megalopolis *Transportat. Res. Procedia* **20** 751–5 DOI 10.1016/j.trpro.2017.01.121

[26] Kotov A and Pospelov P 2017 Engineering tools and methods of estimation of traffic capacity using mobile video monitoring *Transportat. Res. Procedia* **20** 347–54 DOI 10.1016/j.trpro.2017.01.044

Digital transformation infrastructure and transportation logistics systems

A G Nekrasov[1] and A S Sinitsyna[2]

[1]Moscow Automobile and Road Construction State Technical University (MADI), 64, Leningradsky ave., Moscow, 125319, Russia
[2]Russian University of Transport (MIIT), 9b9, Obrazcova str., Moscow, 127994, Russia

E-mail: tehnologistic@mail.ru

Abstract: The paper considers an integrated approach to the management of the life cycle of the transport and logistics infrastructure, basing on the concept of digital transformation and logistics engineering. The issues of digital transformation of the transport infrastructure are of particular importance in the context of the transition to the concept of Industry 4.0. This implies the existence of new infrastructure requirements, both in terms of costs and from a strategic approach to increasing business productivity, taking into account the sustainability criteria and response time to failure situations. As the basis of a highly efficient integrated TLS, there is a coordinated interaction of enterprises within the stages of the life cycle "design – operation – operational control – check and corrective actions – check of the control system". New approaches have been recommended that provide highly efficient support for global transport systems, not only through the use of digital technologies, but also logistic engineering methods.

Key words: Industry 4.0; complex organizational and technical objects; digital transformation; system life cycle; transport and logistics system; land-based transport and logistics infrastructure of the new generation; proactive management; cyber-physical systems; digital information and communication technologies; logistic engineering.

1. Introduction

Digital economy and logistics, digitalization of transport and logistics systems (TLS) are now part of the innovation of business processes and in practice prove greater efficiency. Prospects for the development of digital technologies are determined by the implementation of the concept of Industry 4.0 and the formation of new principles, models and architecture of engineering to ensure the digital transformation of the transport industry and related markets.

An analysis of the structural elements of the transport infrastructure in the Russian Federation showed that one of the key areas of development, along with digital technologies, is the creation and implementation of highly efficient TLS. This problem is of particular importance in the transition to the concept of Industry 4.0, when there are high demands on infrastructure both in terms of costs and on the part of a strategic approach taking into account the criteria of sustainability and response time to failure situations. Digitalization of the transport and logistics sector transforms this market based on innovative tools related to the management of life cycle processes and land transport infrastructure [1–4].

One of the key activities of the government of the Russian Federation in the medium term is the development of the digital economy. This is indicated at the disposal of the Government of the Russian Federation of July 28, 17 N 1632-p, in which, according to the instructions of the President of the Russian Federation V.V. Putin, approved the program «Digital Economy of the Russian Federation» [5]. One of the most important sectors of the economy where new digital services (services) will be introduced is transport. It is important to note that digital services in the new economy are the main elements of the digital transformation.

Digital transformation is the transformation of the relevant business processes in the new economy, or the combination of operational and information technologies. If earlier information technologies were used mainly for the exchange of information between individuals, now and in the future they will be used in technological processes as a basis for decision-making by the program, and not by man.

The model of innovative economy of the XXI century states are complex organisational-technical objects (COTOs), oriented towards combination of informational technologies and integrated models of life cycle of different systems of an enterprise.

Traditional (functional) logistics, dealing with placement of production, warehouses, customers and modeling supply chains with the aim of optimizing costs and order time, in modern conditions implies refocusing to adaptive-digital methods. A new approach is required, the one based on the system mechanism of integrated management not only in supply chains, but also that of processes in an enterprise basing on the interaction of intelligent and the Internet technologies (the Intenet-of-things type).

2. Formulation of the problem and methods

The systems, means and complexes used are usually characterized by multi-dimensionality, multi-structuralism, indefiniteness of functioning, hierarchy, excessive elements and connections, variety of implementing functions and processes and mobility of components Higher complexity of the existing COTOs requires more controlled parameters charactering functioning processes. Research and practical experience show that managerial delays and mistakes as well as incorrect process monitoring can lead to serious negative consequences. All of this largely depends on the level of sustainability of logistical processes, which provide for the efficiency of the main elements of the system (production) for the whole life-cycle of both the production and infrastructure [6–7].

In modern conditions the infrastructure of the life-cycle of transportation-logistics system can provide for conditions of innovative development, forming the demand for robot-oriented loading and transportation equipment and modern technological solutions, including logistic engineering and transport construction.

For example, it is impossible to imagine system approach implementation in road and rail transport without logistic engineering in the field of transportation infrastructure operation. Effective functioning of any system requires integration, which will define interaction of different companies, logistic operators in the common system services.

As a result, a sustainable transport system can be formed, which is an organizational-technical object, for the control of which adaptive cyber-physical systems (CPS) should be used. In addition, for various transformations, you can use various analytical capabilities to quickly identify needs, implement the creation, expansion and reconfiguration of resources to create new products and services. In this case, adaptive decision-making methods and an integrated model of digital enterprise transformation can be used (figure 1).

It is important to bear in mind that lack of resources (caused by a number of objective and subjective reasons) makes it impossible to keep up the required operational level of modern COTOs, which are to be usable in cases of faults, accidents and disasters and thus are to be sustainable. For this purpose it is possible to recommend new methods of proactive management, which are accompanied by intentional reconfiguration of COTOs and provide a complex system forecast and higher efficiency, basing on a large array of electronic data. Proactive management connects a physical object (infrastructure) with COTO's life-cycle processes by means of digital information-communication technologies (ICT).

IOP Conf. Series: Materials Science and Engineering **832** (2020) 012052 doi:10.1088/1757-899X/832/1/012052

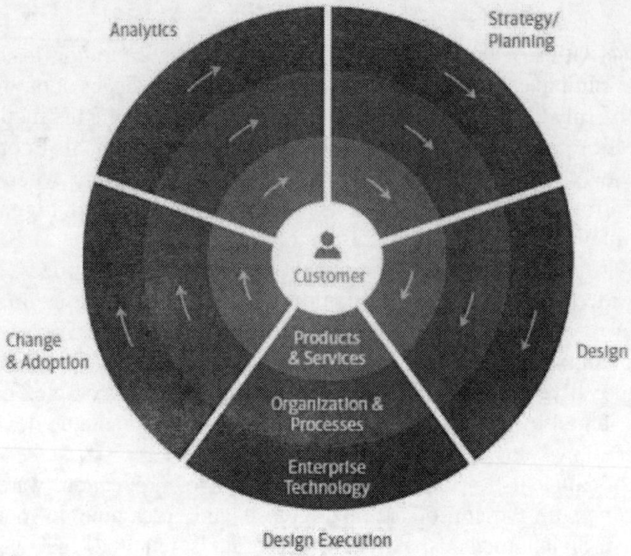

Figure 1. Complex model of digital enterprise transformation [8]

Proactive COTO infrastructure management unlike traditionally used reactive management (aimed at operative reaction and further avoidance of incidents) implies prevention of incidents basing on creating in a monitoring and managerial system principally new forecasting and preventing opportunities in formation and implementation of managerial activities basing on system (complex) modeling [9–11].

Now there are various types of organizing proactive monitoring and COTO management, including proactive technologies of managing structural dynamics of managing the objects concerned. These technologies include: changing ways and aims of COTO functioning, their content, operational sequence in various conditions, relocating elements and subsystems of COTOs, redistribution and decentralization of functions, tasks, management algorithms, information flows between COTO levels, using flexible (shortcut) technologies of COTO management, structural reconfiguration of COTOs in case of degradation.

3. Theoretical approach and implementation

Proactive (preventive) monitoring and management technologies in a COTO can be considered as prospective technologies in managing complexity of compound systems (transportation-logistic objects), thanks to which the objects remain sustainably operational at carrying out tasks in the forecast conditions to the required degree.

Modern development of international transportation and progressive transportation-logistics technologies implies using several transport modes with its interaction with different life-cycle stages that characterizes a transport system as capable of effective cooperation due to geographical and functional diversification and autonomy of separate supply chain stages. Introduction of modern technologies, including new through digital technologies in the near future will be aimed not only a cutting costs, unceasing reliability and safety of all stages, but also at mobile compatibility of transportation life-cycle processes.

High efficiency of transport is directly linked to both structure (processes) of enterprises, and the equipment and production assets used, which are employed at different life-cycle stages of systems. All this increases complexity of COTOs, which include transportation and transportation-logistics services [12–13].

Each transportation-logistics system is based not only o using new transportation technologies at a single enterprise of a network, dealing with receiving, handling and delivery to the final customer, but also on the creation of numerous models of sustainable interaction life-cycle. Integration of transportation-logistics processes and production assets included into infrastructure (terminals, warehouses, lifting and transportation equipment, rolling stock, etc.) provides for higher growth of the productivity of the whole system, and no single employees and workstations.

Thus methodology, technological solutions and tools of organizational interaction based on proactive management should be based on integrated new generation transportation-logistic systems (ITLS) combining subsystems of transportation-logistics services, supply stages and provide sustainability and client orientation at a more efficient level. That is one of the priorities of cyber-physical systems, provided for in the strategy of the Industry 4.0 [6, 14].

The formation and introduction of high-quality and complex system of freight handling based on the integration of transport network members is to be based on adaptivity, safety and sustainability. A highly effective ITLS is based on coordination and interaction of enterprises at the life-cycle stages "design – operation- operational control – check-up and correction – management system monitoring". This framework forms the life-cycle processes of proactive management. This direction is aimed at providing not only sustainability of transport, but also competitive infrastructure of Russian national economy. Possibly types of COTO stages based on structural dynamics methodology are presented in figure 2.

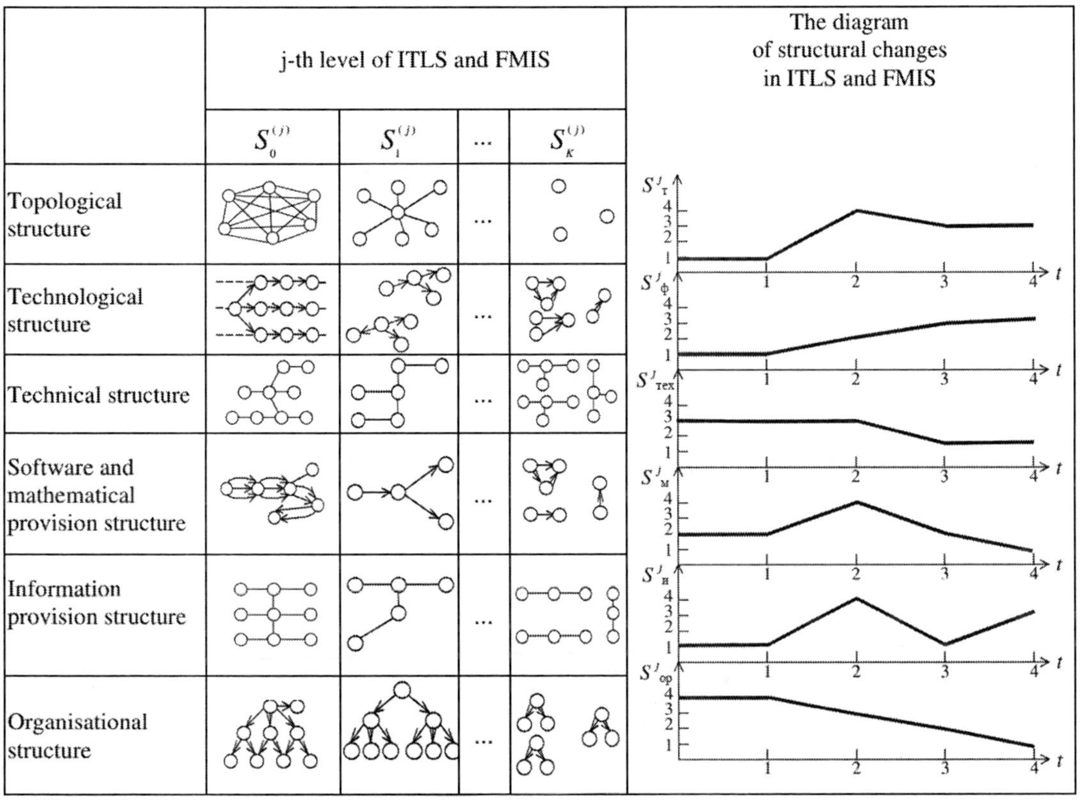

Figure 2. Structure types in structure dynamics

Thus, a task is set and solved not only to cut expenses at each stage of a supply chain, but also to carry out a complex assessment of transportation-logistics infrastructure. Due to this a balance should be found between stability (equilibrium) and reconfiguration of processes between different life-cycle models shown in picture 2 [1].

IOP Conf. Series: Materials Science and Engineering **832** (2020) 012052 doi:10.1088/1757-899X/832/1/012052

New technological framework known as Industry 4.0 provides for further integration of a network of machines that will be able to autonomously change production patterns, keeping up high efficiency. The increased integration of «intelligent» works into industrial and transport infrastructure will mean much lower energy consumption and higher labor productivity.

The creation of digital technologies M2M (Machine-to-Machine) and IoT (Internet of Things) will enable logistic companies to develop supply chain management systems and enterprise management systems via machine interaction technologies [15]. M2M-technology enables a remote device to transmit to the data processing center the data about a monitored object – COTO.

4. Results and discussion

However, the main challenge is not to increase information volume and automation via cyber-physical (robot-oriented) systems, but to form the management of self-governance, which is the most effective way to resist uncertainty (risks) of the environment), to support COTO's structures and functions in new models of life-cycle management. That is why the priority directions in digital transformation of the IT sphere are the strategy of adaptive and proactive systems, adaptive management and adaptive enterprise [16]. The priority of the range of tools in such systems and enterprises is improving the life-cycle management processes. Under these conditions the efficiency will first of all depend on its success in developing innovative technologies and adaptive enterprises (figure 3).

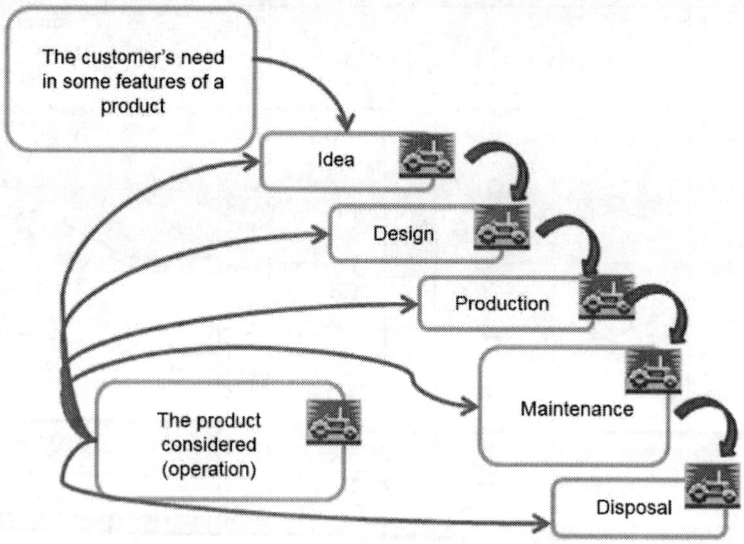

Figure 3. Infrastructure equipment life-cycle transformation

Thus, when considering the problem of transformation of integrated transportation-logistical systems in to the "digital" industry, it is necessary to create a new image and features of ITLS, which should comply with global changes taking place in the modern world, based on the Industry 4.0 strategy and logistic engineering. It will create the conditions for the adapting transportation-logistical infrastructure to the modern service-oriented transport sector in Russia.

5. Conclusion

The methods of digital transformation and logistics engineering, integrated with digital technologies in the 21st century, develop mainly as an interaction of the processes of CLS objects at various stages of the life cycle. This approach is global in nature and creates new opportunities for the digital transformation of the transport and logistics infrastructure (TLI) in the implementation of the concept of Industry 4.0.

The main driving force in global transport systems is the need for highly efficient supply and transportation processes based on the application of not only digital technologies, but also logistics engineering methodologies that ensure integrated management and growth of TLS productivity. This approach determines the current level of integration of ICT technologies, business within the framework of a holistic TLI.

Among the key factors of a balanced system of strategic business models and technologies, it is necessary to single out the management of the full life cycle of both the enterprise infrastructure itself and products, services and resources based on the enterprise architecture embedded in systems engineering and its applied discipline – logistics engineering.

The priority among business models of "knowledge economy" and logistical technologies should be given to the models of the management of the life cycle of the transport infrastructure, which have a long time-line.

One of the main ideas of life-cycle management systems in effective use of consistent system presentation and its environment in initially incompatible organizational-technical systems of «larger organization».

References

[1] Nekrasov A G and Sinitsyna A S 2017 Transformation of integrated transport and logistics systems in the digital industry *Logistics* **8** 36–41
[2] Kupriyanovsky V P, Namiot D E and Sinyagov S A 2016 Cyber-physical systems as the basis of the digital economy *Int. J. of Open Inform. Technol.* **4(2)** 19–25
[3] Nekrasov A G, Sokolov B V and Ataev K I 2017 *Management system for life cycle (transformation into digital infrastructure* (Moscow: Publishing house "Tehpoligraftsentr")
[4] Nekrasov A G and Sinitsyna A S 2017 Transport and logistics systems of the new generation in the digital economy *Power systems* **13** 11–23
[5] The program "*Digital Economy of the Russian Federation*". Approved by the order of the Government of the Russian Federation dated 28 July 2017 no. 1632-p Retrieved from: http://government.ru/
[6] Nekrasov A G and Sinitsyna A S 2016 Logistic Engineering as a Tool for the Integration of Logistic Systems *Logistics* **12** 40–5
[7] Oigman E G and Popov E V 1997 *Business reengineering. Reengineering organizations and information technology* (Moscow: Finance and Statistics) 336 p
[8] Kupriyanovskiy V P, Sinyagov S A, Dobrynin A P and Namiot D E 2017 Holistic Models of Transformation in the Digital Economy *Int. J. of Open Inform. Technol.* **5(1)** 26–32
[9] Kalinin V N and Sokolov B M 1995 A multi-model approach to the description of control processes of space vehicles *Theory and control systems* **1** 56–61
[10] Okhtilev M Yu, Sokolov B V and Yusupov R M 2015 Theoretical and technological foundations of the concept of proactive monitoring and management of complex objects *Sci.-Technical and Applied J. "Izvestiya SFU. Technical science" Section V Systems and control points* **1** 162–174 Retrieved from: http://izv-tn.tti.sfedu.ru/?page_id=1576
[11] Yusupov R M and Sokolov B V 2006 *Problems of the development of cybernetics and informatics at the present stage* Collection "Cybernetics and Informatics" (St. Petersburg: Publishing house SPbSPU) pp 6–21
[12] Mirotin L B, Omelchenko I N, Kolobov A A et al 2011 *Engineering logistics: logistic-oriented product lifecycle management* Textbook for high schools ed L B Mirotina and I N Omelchenko (Moscow: Hot line – Telecom) 644 p
[13] Kartashev A V, Nekrasov A G and Atayev K I 2016 *Life cycle management of complex high technology products in integrated supply chains* Monograph (Moscow : PrintUp) 324 p
[14] Meyer K and Davis S 2007 *Live organization* Trans. with English (Moscow: Publishing house "Good Book") 368 p
[15] Tsivlin A 2015 From M2M to IoT: evolution through synergy *Logistics* **12** 40–5
[16] Bir St 2005 *The brain of the firm* (Moscow: URSS) 315 p

IOP Conf. Series: Materials Science and Engineering **832** (2020) 012053 doi:10.1088/1757-899X/832/1/012053

Investigation of the overtaking and overtaken vehicles speeds on two-lane roads in the traffic conditions of the Russian Federation

P I Pospelov[1,3], A V Kostsov[1], D S Martyahin[1], D M Nemchinov[2]

[1] MADI, 64, Leningradsky avenue, Moscow, 125319, Russian Federation
[2] TransInzhProekt, 21, Avtozavodskaya street, Moscow, 115280, Russian Federation
[3] Email: pospelov@madi.ru

Abstract. The speeds investigation results of the overtaking, overtaken and moving on the opposite lane vehicles on two-lane roads in the conditions of the Russian Federation are presented in this paper. The investigations were carried out using unmanned aerial vehicles in compliance with generally accepted scientific methods. Based on the statistical processing of the measurement results, generalized vehicles speed data within the overtaking sections was obtained. It was noted that the presented investigations may be a rationale for improving the requirements for the design of roads on the territory of the Russian Federation. In addition, the obtained data may be used to develop measures to improve traffic safety on existing roads, which will have a generally beneficial effect on traffic safety.

1. Introduction

It has generally recognized that accidents in overtaking vehicles on two-lane roads cause some of the most severe consequences. The approaches to the designation of overtaking sight distances existing in the regulations of the Russian Federation are not adequately considering the modern traffic conditions on the country's road network. Therefore, the most important direction to improve the traffic safety on two-lane roads is investigating of modern traffic conditions in the overtaking areas in general and the traffic speed in such areas in particular.

The continuing vigorous motorization growth of the Russian Federation population sharply puts the task of ensuring road safety. Despite the overall decrease in the number of road accidents (Figure 1), the risk of accidents on the roads of the Russian Federation as of 2015 is 3-4 times higher than in countries with a high motorization level [10].

It has generally recognized that accidents in overtaking vehicles cause the most severe consequences - significant harm to the health and death of the victims. Studies conducted by German researcher fellows in the year 2016 have confirmed that accidents in overtaking lead to the most severe consequences on two-lane roads. At the same time, the main places of road accident concentration detected on-road sections with insufficient overtaking sight distance [5]. Thus, the implementation of measures to ensure overtaking in conditions of secured sight distance on two-lane roads is one of the primary tasks in preventing fatal crashes and severe injuries. The regulatory system of the Russian Federation establishes requirements for minimum values of sight distances sufficient for overtaking. Such values of sight distances are presented in Table 1.

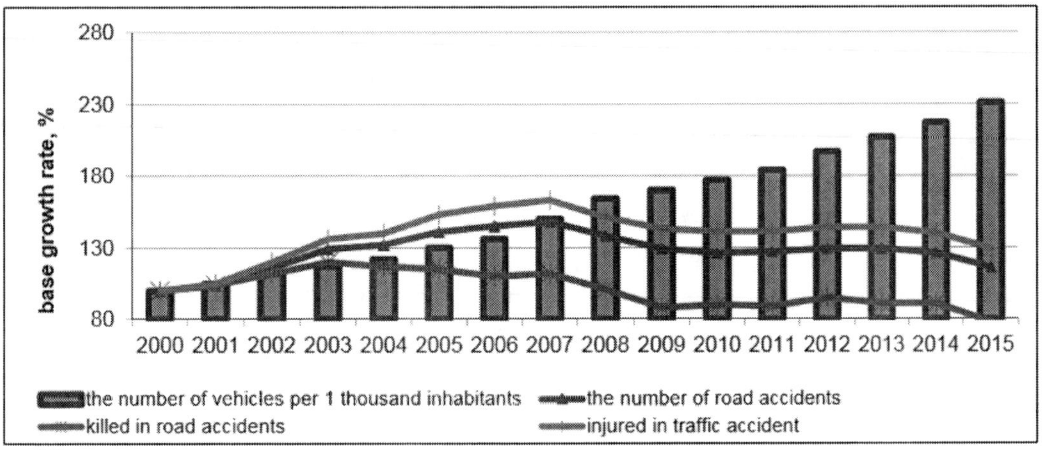

Figure 1. Trends in road accidents.

Table 1. Regulation of the sight distance in Russia.

Estimated speed, km/h	Sight distance at overtaking, m, in regulations of the Russian Federation	
	SNiP II-D.5-72 (1972) SNiP 2.05.02-85 (1985) SNiP 2.05.02-85* (2004)	CP 34.13330.2012 (2013), CP 34.13330.2012 rev.1 (2017)
120	-	800
100	-	700
80	-	600
60	-	500

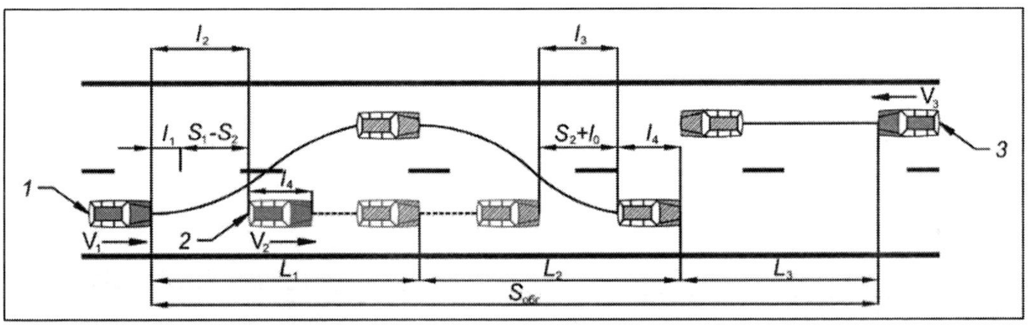

Figure 2. Sight distance determination scheme in overtaking

Considering the traffic conditions on the roads of the Russian Federation that have changed over the last decade, the authors performed research on traffic conditions on overtaking sections. The following overtaking model has adopted for the research work (Fig. 2) [11]:

a) The overtaking begins at the moment when the vehicle 1 approaches the overtaken vehicle 2 by a distance equal to the difference in their braking distances (S_1 -S_2) and the way the vehicle passes from the moment when the driver decides to overtake (L_1). Therefore, considering that the vehicle 1 is at a distance (L_2) from the vehicle 2 and knowing the speeds difference of the vehicles (V_1 -V_2), determine the path that the vehicle 1 will pass until it equals the vehicle 2 (L_1) on the horizontal road segment (i = 0):

$$L_1 = \frac{(l_1 + l_4) \cdot V_1}{V_1 - V_2} = \frac{V_1^2}{V_1 - V_2} + \frac{k_э \cdot V_1 \cdot (V_1 + V_2)}{2 \cdot g \cdot \varphi_1}, \tag{1}$$

where V_1 andV_2 - is the speed of the rear and front vehicle, respectively, m/s; φ_1 - is the longitudinal friction (adhesion) coefficient; $k_э$ - is the operational brakes state coefficient, assumed equal for both vehicles; l_4 - is the vehicle length, m.

b) The vehicle 1 returns to its lane in front of the vehicle 2 at a safe distance (l_3) equal to its braking distance (S_2), increased by the safety gap (l_0) and the length of the vehicle (l_4). Then the path passed by the vehicle 2 before returning to its lane (L_2) will be equal to:

$$L_2 = \frac{(l_3 + l_4) \cdot V_1}{V_1 - V_2} = \left[\frac{k_э \cdot V_2^2}{2 \cdot g \cdot \varphi_1} + l_0 + l_4\right] \cdot \frac{V_1}{V_1 - V_2}. \quad (2)$$

c) Overtaking is possible when the vehicle 1 returns to its lane at the time of meeting the oncoming vehicle 3 moving at a speed V_3. During the overtaking period, this vehicle passes the path (L_3) equal to:

$$L_3 = \frac{L_1 + L_2}{V_1} \cdot V_3, \quad (3)$$

Then the sight distance from the overtaking condition will be

$$S_0 = L_1 + L_2 + L_3 = (l_0 + l_1 + 2 \cdot l_4 + \frac{k_э \cdot V_1^2}{2 \cdot g \cdot \varphi_1}) \cdot \frac{V_1 + V_3}{V_1 - V_2}. \quad (4)$$

From the equation (4), within the adopted model, the sight distance of the oncoming vehicle depended mostly on speeds of the overtaking (V_1) and overtaken (V_2) vehicles, as well as the vehicle moving in the opposite traffic direction (V_3).

2. Investigation of overtaken and overtaking vehicles speeds on two-lane roads

In order to actualize the requirements for sight distances accepted by the regulatory system of the Russian Federation (Table 1), the authors have conducted studies aimed at determining the speeds of vehicles in such areas. Investigations have carried out on a network of two-lane roads in the Moscow region.

The investigation of traffic speeds has carried out using the aerial video method recording using unmanned aerial vehicles (UAV) (Figure 3) in compliance with specially developed [9] and generally accepted scientific methods [8], and subsequent processing of the record received using the software package Movavi Screen Capture Studio.

Figure 3. Operating moment during the measurement using UAV.

The obtained data of the overtaking (V1) and overtaken (V2) vehicle values have presented as the theoretical distribution curves according to Poisson's law corresponding to those obtained during the measurements. The smoothing of the experimental data was carried out by the methods described in [6]. Based on the statistical processing results of the obtained data array, cumulative curves have constructed containing data on the speeds of the overtaking (V_1) and overtaken (V_2) vehicles, and the vehicle moving in the opposite traffic direction (V_3), presented in Figure 4.

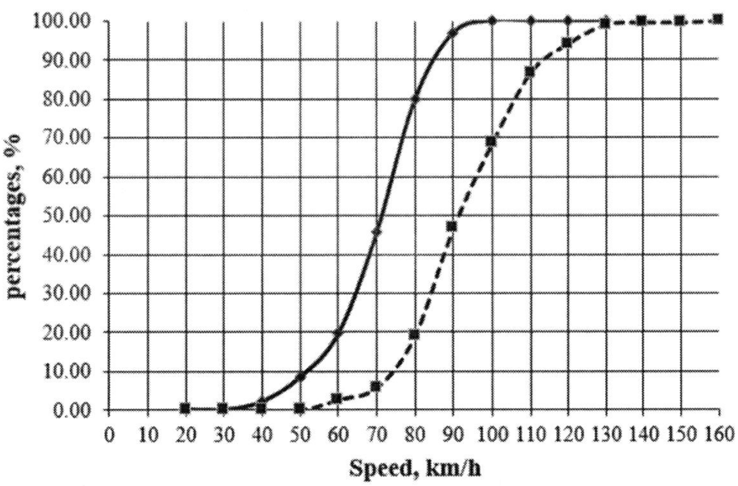

Figure 4. Frequency of accidents vs vehicles speed

3. Conclusion

The investigations conducted allowed establishing the values and ratios of the overtaking (V1) and overtaken (V_2) vehicle speed, as well as a car, moving in the opposite traffic direction (V_3) on the overtaking sections of two-lane roads in the conditions of the Russian Federation. The investigations presented in the paper may be used as the substantiation for the improvement the requirements to design roads in the Russian Federation territory. Moreover, the data obtained may be implemented for the development of measures to improve traffic safety on existing roads, which will have a generally beneficial effect on traffic safety.

References

[1] CP 34.13330.2012 2013 Automobile roads. Updated version of SNiP 2.05.02-85 * (Moscow) (in Russian).
[2] Revision No. 1 to the CP 34.13330.2012 2017 Automobile roads. Updated version of SNiP 2.05.02-85* (Moscow) URL: http://docs.cntd.ru/document/1200095524 (in Russian)
[3] SNiP II-D.5-72 1973 *Automobile roads. Design code* (Moscow: Stroyizdat) (in Russian)
[4] SNiP 2.05.02-85 1985 *Automobile roads* (Moscow: Gosstroy) (in Russian)
[5] Richter T, Ruhl S, Ortlepp J, Bakaba E 2016 Prevention of overtaking accidents on two-lane rural roads *Transportation Research Procedia* **14** 4140-4149.
[6] Zaks L 1976 *Statistical estimation* (Moscow, Statistics) p. 598 (in Russian)
[7] STSI 2018 Statistics URL: http://gibdd.ru/stat/.
[8] Silyanov V V, Domke E R 2008 *Road service quality of roads and city streets* (Moscow: Academy Publishing Center) p. 352
[9] Pospelov P I, Kostsov A V, Martyahin D S, Martyahin K Yu 2018 Application of unmanned aerial vehicles to the investigation of vehicle traffic modes *Vestnik MADI* **2**.

IOP Conf. Series: Materials Science and Engineering **832** (2020) 012053 doi:10.1088/1757-899X/832/1/012053

[10] Pugachev I et al 2016 Factor analysis of the organization and road safety system In *Proceedings of the Twelfth international scientific and practical conference Organization and safety of traffic in large cities*, St. Petersburg.
[11] Babkov V F 1979 *Design of roads* (Moscow: Transport) vol. 1, p. 367.

IOP Conf. Series: Materials Science and Engineering **832** (2020) 012054 doi:10.1088/1757-899X/832/1/012054

The formation of the digital transport space of Russia through the Platon's system

V I Prusova and M A Zhidkova

Moscow Automobile and Road State Technical University (MADI), 64, Leningradsky ave., Moscow, 125319, Russia

E-mail: archive@madi.ru

Abstract. The article is devoted to the issues of digitalization of the motor transport complex through the Platon's system. This system is designed to charge owners of vehicles with a permitted mass of more than 12 tons for travelling on federal highways. The article focuses on domestic experience in organizing the collection of tolls, calculating the tariff, the positive and negative sides of the implementation of this system. Based on the analysis, the authors draw conclusions about the prospects of this system in Russia and abroad.

1. Introduction

Road transport is a type of transport that carries goods and passengers on trackless tracks. The main areas of the increasingly expanding expedient use of automobile transport are the transportation and transportation of goods to main modes of transport, the transportation of industrial and agricultural goods over short distances, domestic transportation, transportation of goods for trade and construction, and the possibility of delivering goods "door to door". Over long distances, road transport transports perishable, especially valuable, requiring fast delivery, inconvenient for overloading by other means of transport cargo. At present, without a motor transport, the activity of any branch of the economy is impossible.

Therefore, it is quite natural that road transport needs the most modern technologies in order to be competitive not only in the domestic market for road transport, but also with a swipe to the world market; meet the need for undefined increasing volumes of cargo transportation, as well as ensuring the availability and quality of services provided [5]. One of the main points in the transition to the "figure" for freight is digital control. Paper invoices, declarations should be replaced by electronic payments, paperwork should be electronic. This will allow you to choose the most optimal delivery methods, depending on traffic congestion, to track the location and condition of the cargo and optimize staff work.

2. Freight Market Digitalization

2.1. Domestic experience

The digitalization of the freight transportation market, according to the authors, is primarily the safety of freight transportation, their convenience, accessibility for people and businesses. This also includes measures to reduce costs, expand transit traffic and export opportunities [1]. More recently, in Russia, there was a system for collecting fees from freight vehicles for using the Platon motorway. In our country, the Plato system has recently appeared. It can be deciphered as "pay per ton". It is intended to charge owners of vehicles with a permitted mass of more than 12 tons for travelling on federal roads in order to comply with Article 31 of Federal Law No. 257-FZ "On Roads and Road Activities in the Russian Federation". RT-Invest Transport Systems LLC is the sole executor of the project.

This system was launched on November 15, 2015. The toll was undefined 1.53 roubles for every 1 km of the way.

The system operator maintains a register, which is a structured database of users of the charging system, to organize the procedure for making a payment for damages. Registration of the vehicle and its owner in the registry of the charging system is performed by the operator free of charge.

2.2. Technical base

Payment for travel is carried out in two ways – using a route map or an on-board device. The calculation is made depending on the actual mileage of the vehicle.

Route map is a document formed on the basis of data on the planned route provided by the owner of the vehicle, and which is the basis for calculating the fee for damages in the absence of the on-board device [4].

The route map contains the following data:

- Route card number;
- state vehicle registration plate;
- the beginning of the validity period of the Route map with the date and exact time;
- date and time of issue of the Route map;
- validity period;
- the total length of sections of public roads of federal importance included in the route;
- description of the route in relation to sections of federal roads for the passage of which this document was issued;
- amount of payment.

Payment methods for route cards:

- from settlement record undefined Owner of a vehicle open in the Charging System;
- by bank / fuel card in the Personal Account and the Platon Mobile Application, in the Platon Self-Service Terminals, in the Information Support Centers of the Charging System users, through Qiwi e-wallet, as well as through the Internet bank and Sberbank Online mobile application;
- in cash at partner agent terminals: Qiwi, Sberbank, Moscow Credit Bank, Eleksnet, in the network of Euroset stores;
- using a mobile phone, using the payment service MOBI.Money [2, 3].

Ways to replenish an account:

- non-cash bank transfer according to the details of the Operator of the Charging System;
- by credit / fuel card in the Personal Account and Platon Mobile Application, in Platon Self-Service Terminals, in the Information Support Centers of the Charging System users, through Qiwi e-wallet, as well as through the Internet bank and Sberbank Online mobile application;
- cash in agent undefined partner terminals; Qiwi, Sberbank, Moscow Credit Bank, Eleksnet, in the network of salons Euroset;
- using a mobile phone, using the payment service MOBI.Money.

On-board device - an object of the Charging System, which is a technical device that allows using the GLONASS or GLONASS / GPS satellite navigation technologies to determine the route of the vehicle along federal highways [6].

The use of the device is convenient because in the presence of funds on the Owner's account of the vehicle, the calculation and debiting of funds from the account in the payment account is carried out automatically based on data received from the on-board device.

2.3. Rates

The initial tariff for road damage was set by the Government of the Russian Federation by resolution No. 504 of June 14, 2013 at the level of 3.73 roubles per km. On November 3, 2015, Decree No. 1191 of the Government of the Russian Federation "On Certain Issues of Collection of Payments for Compensation of Damage Caused by Public Highways of Federal Importance by Vehicles with a Permissible Maximum Mass of Over 12 tons", reducing coefficients were introduced, at that time the charge was 1.53 roubles per km of track.

From April 15, 2017, the tariff began to be instead of 1.53 roubles per 1 km, currently 2.04 roubles per km. Initially, they wanted to set the tariff at 3.06 roubles. The decision on a smoother increase was made on March 23 at a meeting of Russian Prime Minister Dmitry Medvedev with representatives of small and medium enterprises in the field of road transport.

2.4. Advantages and disadvantages

Despite all the advantages of the new system for collecting tolls of heavy trucks on federal roads, carriers call it unfair and note that it is not a payment or a fee fixed in the Tax Code. In fact, this is a fee for driving on federal roads, which in theory should be restored from the means of a paid transport tax, logistics say.

"With the introduction of a duty for driving on highways, the rate and tax base for the transport tax remained unchanged. Until the end of last year, a privilege was applied to logistics companies, but since January 2019 it has been cancelled, and, in fact, freight transport owners pay for road repairs twice; in the form of payments to "Plato" and in the form of a transport tax".

Then, changes in the system of weight control became a great stress for carriers – a network of automatic posts (automatic weight control system) (ASVGK) was introduced on federal and regional routes undefined in the country since October 1, 2016. "The vehicle drove through this frame, the sensors recorded even a slight excess – get a fine". Specialists in this field and the authors of the article join believe that Plato only managed to remove carriers from the gray zone. But a qualitative change in the technical condition of the roads did not occur. Until now, the Plato system is controversial; it is still believed that the Plato is a double-edged sword. On the one hand, it is heavy trucks that cause damage to roads – there are about 50 million cars in the country, and their owners pay transport taxes and fuel excise taxes, replenishing road funds at different levels. Of these vehicles, more than 1 million vehicles are heavy vehicles, which are payers in the Platon system. According to various estimates, the damage caused by heavy trucks to roads compared to cars differs thousands of times. "Platon" allows you to compensate for the loss of roads from carriers [5].

"On the other hand, not defining the very principle of the Plato system raises a lot of questions. It is not very clear why, for this, it was necessary to attract some investment capital (on a concession basis, RT-invest transport systems – Regional Economy). In my opinion, it was enough to make Plato a state structure that could do everything at its own expense and receive money directly. And not like it is now – of the money that comes to compensate for the damage to roads, only a part is left to compensate for the damage, and the other part goes to pay off some investments of a private organization." – Alexey Nosov says.

According to Finam JSC, in 2015, 1.397 billion roubles were received to the Russian budget through the fees of Plato. When comparing the data of the Ministry of Transport of the Russian Federation with the financial statements of the operator of the system available at Spark-Interfax, the result was approximately the same: the revenue of RT-invest transport systems LLC, that is, most likely, the year the revenue system began operating undefined were divided equally [7, 8]. In 2016 and 2017, the operator's revenue was about 10 billion roubles per year, which looks like a fixed fee. Budget revenues in 2016 amounted to 16.434 billion roubles and in 2017 – 19.836 billion roubles. In 2018, the budget already received 23.182 billion roubles from Platon, and from the beginning of 2019, another 15.829 billion roubles. The reporting of RT-Invest Transport Systems LLC for 2018 is not yet available, but it can be assumed that the revenue should have been the same 10 billion roubles.

IOP Conf. Series: Materials Science and Engineering **832** (2020) 012054 doi:10.1088/1757-899X/832/1/012054

2.5. Summary

The results of the implementation of the Platon's system. In the state system "Platon" in July 2019 more than half a million users were registered (501913). The number of vehicles is 1.198 million. There are 50.774 km of federal roads in the system. In the road fund of Russia collected 77 billion roubles. The funds received from Platon are spent on improving road transport infrastructure, financing repair and road construction works, and maintaining the state of highways of federal importance. Road tolls are not levied on passenger vehicles, military equipment and trucks used for its transportation, as well as on vehicles operated by emergency services: ambulance, fire department, police, Ministry of Emergencies, etc. almost 80 billion roubles were collected for road repair and bridge construction for the entire period since 2015, according to Platon's operator, RTITS (RT-invest transport systems). On these funds undefined over 2 thousand km of roads were repaired. More than 30 bridges were restored and built in 40 regions of the country, according to the Ministry of Transport. So, 500 million roubles were allocated for the construction of a bridge across the Volga River in the city of Dubna in the Moscow Region. Another 930 million roubles – for the construction of the Frunze's bridge in Samara, 625 million roubles – for the construction of a bridge over the river Katun (Gorny Altai). 380 million roubles were allocated for the construction of the bridge across the Cheptsa River in Udmurtia, and more than 330 million roubles were allocated for the reconstruction of the bridge across the Terek River on the Ischerskaya-Grozny highway. It is also planned to build and reconstruct bridges across the Cheptsa River (two overpasses in Udmurtia in the Balezinsky and Kezsky districts), Sosna (the city of Livny, Oryol Oblast), Zusha (Mtsensk), Oka (Oryol Oblast), and the Samara River (Samara Oblast). Also in 2018, funds from the Platon system were used to co-finance the repair of the federal highway R-243 Kostroma – Sharya – Kirov – Perm. The Ministry of Transport of Russia and Rosavtodor made such a decision undefined on the basis of the choice of the carriers themselves in the framework of the project "Map of the killed roads". More than 2.5 billion roubles were transferred to the Siberian Federal District for the repair of federal roads and bridges over four years. In particular, 750 million roubles were allocated to the Omsk Region, including 625 million roubles – to repair the bridge across the river Katun, 250 million roubles – for road repairs in the Altai Republic, 135 million roubles – for the reconstruction of the bridge over the river. In the Krasnoyarsk Territory, 82 million roubles are to restore the bridge across the river Alley in the Altai Territory and another 63 million roubles are to repair the bridge in Buryatia (in November 2018, the Republic of Buryatia became part of the Far Eastern Federal District). Also, the Siberian region this year received 670 million roubles as compensation for the shortfall in income from the deduction for the transport tax (according to the Tax Code of the Russian Federation, owners of 12-ton heavy trucks registered in the Platon's system were actually exempted from the transport tax if the amount of the fee for using federal roads exceeded or was equal to it. The norm undefined acted until the end of 2018). These funds were used to build roads, bridges and overpasses. In particular, the Novosibirsk region of this amount offset 108 million roubles. As representatives of the state system noted, freight forwarders themselves determine which roads should be repaired at the expense of the funds transferred by them to the Russian road fund. This happens during an open vote in the framework of the project of the All-Russian Popular Front "Map of killed roads". For example, 700 km of federal roads of Russia were repaired, including more than 80 km of the R-258 Chita-Ulan-Ude highway repaired in Siberia.

3. Conclusion

Starting July 1, 2019, the Platon's state system fare for heavy trucks on federal highways increased by 14 kopecks – up to 2.04 roubles per km. From February 1, 2020 it will amount to 2.2 roubles, and in 2021 2.35 roubles. In turn, the Ministry of Transport of the Russian Federation presents this not as an increase, but as a "tariff reform". "The government met the carriers and last year refused to increase the tariff, and in 2019 decided to make it a phased one – for three years. The applied approach with stretching the increase in the tariff for accumulated inflation will increase the number of roads and bridges being repaired due to additional funds from the Platon's system, which will positively affect traffic safety. If we talk about commercial structures, then the Platon's system will ultimately prove to be more profitable than its absence due to more expeditious cargo delivery and savings on depreciation of vehicles". In 2019, according to the original plan, the fare was supposed to undefined make up 3.73 roubles. Experts call the spending of funds collected from carriers not transparent enough, and the increase in tariff unreasonable. According to freight forwarders, it is necessary to improve Plato. In particular, legal companies point to unscrupulous competitors who shy away from using the system and receive an influx of additional customers due to low prices and higher profits than those from bona fide entrepreneurs. "Before raising the system's tariff, it is necessary to ensure high fundraising – at least 95 %, so that legal freight carriers do not suffer from the actions of gray competitors." According to the data of the Platon's operator, RTITS, at the moment about 90 % of heavy trucks in Russia are registered in the system. This means that approximately one in ten heavy trucks in the country is still not equipped with an on-board Plato device. Currently, there are proposals to introduce "Plato" in the Tax Code. However, this measure with undefined the prospect of criminal penalties and high fines may not have a positive effect on everyone. "The first category is those who simply have nothing to pay. The second category is those who simply do not pay. And the third option: a person calculates what is more expensive – fare or the amount of fines? Only if the amount of fines exceeds this value, then people will go to pay using the Platon's system. Digitalization is not a separate move, operation, it is a whole technology, and therefore it is quite obvious why there are so many problems in the development and implementation of the Plato system. But they are overcome. And the authors hope that in the near future the Platon's system will cover most federal highways and the Platon's experience will be used abroad. As shown by the successful practice of using it in India.

References

[1] Zhidkova M A and Shpilkina T A 2018 Digital decisions in the transport system of the country *Financial management in the conditions of new industrial revolution: trends and prospects* 96–8

[2] Landsman A Ya 2016 Problems of modernization of a mobile terminal system of financial service on transport *Car. Road. Infrastructure.* **3(9)** 8

[3] Landsman A Ya 2017 Stability of a mobile service system of financial service on transport// *Car. Road. Infrastructure* **1(11)** 15

[4] Politkovskaya I V and Hvichiya D T 2019 Digital technologies and financial performance of the enterprises of the transport industry *Bull. of the Moscow automob. and road state techn.l univer. (MADI)* **1(56)** 126–33

[5] Prozumentova V V, Beznovskaya V V and Prusova V I 2016 Risks of an entry of transport companies into the foreign markets *Motor transportation enterprise* **12** 49–52

[6] Prusova V I, Beznovskaya V V and Anosova A V 2019 Digital transformation of the Russian economy: problems and prospects *Economy and business: theory and practice* **4-3** 124–9

[7] Shpilkina T A, Zhidkova M A and Rybyakova O I 2018 Rol of digital technologies in the strategy of development for the country 2024 *Modern information technologies in education, science and the industry the X International conference, the VIII Int. competiti. of scie. and sci. and method. works, the IV Competit. "Sci. School Community"* Coll. of works Ed T V Piryazeva and V V Serov 116–8

[8] Shpilkina T A, Zhidkova M A, Politkovskaya I V, Kazitskaya N V and Rybyakova O I 2018 Digital economy: instruments of development and their influence on a financial system of the country *Modern information technologies in education, science and the industry the XI Int. conf., the IX Int. competit. of sci. and sci. and method. works* Ed T V Piryazeva and V V Serov 188–91

IOP Conf. Series: Materials Science and Engineering **832** (2020) 012055 doi:10.1088/1757-899X/832/1/012055

Software modeling application for verification of transportation planning engineering hypotheses

N Shchegolova[1], **V Talalai** [2], **C Gorshenina**[1], **D Smirnova**[3]

[1] Saratov State Technical University Gagarin U.A., 77, Polytechnic, Saratov, 410054, Russian Federation
[2] Moscow Automobile and Road Construction State Technical University, 64, Leningradsky prospekt, Moscow, 125319, Russian Federation
[3] Astrakhan State University, 20A, Tatishcheva, Astrakhan, Russian Federation

Emails: shegoleva123@mail.ru, talalay@bk.ru, ekaterina_84_07@mail.ru, dsmirnovad@mail.ru

Abstract. The organization of traffic is a complex of engineering and organizational arrangements aimed at maximizing the use of traffic flows opportunities represented by the road geometric parameters and condition. The article deals with an example of using software modeling for creating a digital twin of a complex intersection in Saratov in order to verify the engineering hypotheses concerning the effective transportation planning. The current traffic situation in the PTV VISSIM program is simulated. The transportation planning engineering hypothesis proposed and simulated the solution options of changing the working traffic light duration at the intersection for verifying and widening the carriageway by adding a lane. The analysis of two models options allowed determining the most effective transportation planning measure in a given transport hub of the city with a high level of traffic delays, which is actually the duration change of the working traffic light at the intersection. The software modeling application is confirmed for creating a digital twin complex intersection in Saratov.

1. Introduction

Road transport put down deep roots in modern life. It provides a large volume of traffic in all spheres of human activity. The global car fleet is steadily increasing. Saratov took the 17th place among the major cities of Russia according to the number of officially registered cars. There are 253,000 registered cars in Saratov. According to Russian and international practice the growth of number of cars provide both the positive impact on the economy and social city development and the negative one associated with a large number of motor vehicle collision (MVC), fatalities, injuries , huge material damage. These problems are of great interest to the transportation planning experts whose task is to ensure traffic safety [1]. For solving these problems, many methods are used, including software modeling. Their use helps to significantly relieve the teacher and students, to free up additional time for analysis of complex or specific moments, allow making a creative element in the study of the subject.

2. Methods of creating a multimedia textbook

We consider the application of digital twin of a complex road intersection in Saratov, performed in the program PTV VISSIM. The software package allows the transportation planning engineering hypotheses verifying [2].

We use the following hypothesis: the identification of an acceptable transportation planning solution in a given transport hub of the city with high level of traffic delays [3].On the basis of such multimedia technologies, it is possible to create electronic educational resources (EER) for educational materials containing images, texts accompanied by sound, video, animation and other visual effects.

3. Materials and methods

The acceptable solution for transportation planning is considered: change the duration of a traffic light at an intersection or widen the carriageway by adding a lane. To verify the hypothesis, we selected a complex intersection of «Sokolova» and «Murmansk travel» streets in Saratov (Fig. 1), as it is a high traffic and busy transport way of the city. The traffic is provided in two directions, four lanes on the «Sokolova» street in one direction and two lanes on the «Murmansk travel» street.

Figure 1. Intersection of «Sokolova» street and «Murmansk travel» street in Saratov: (a) - the scheme of the intersection "Google Maps" (Internet resource); (b) - the general view of the intersection.

We make a diagram of the intersection and determine the traffic intensity at the intersection.

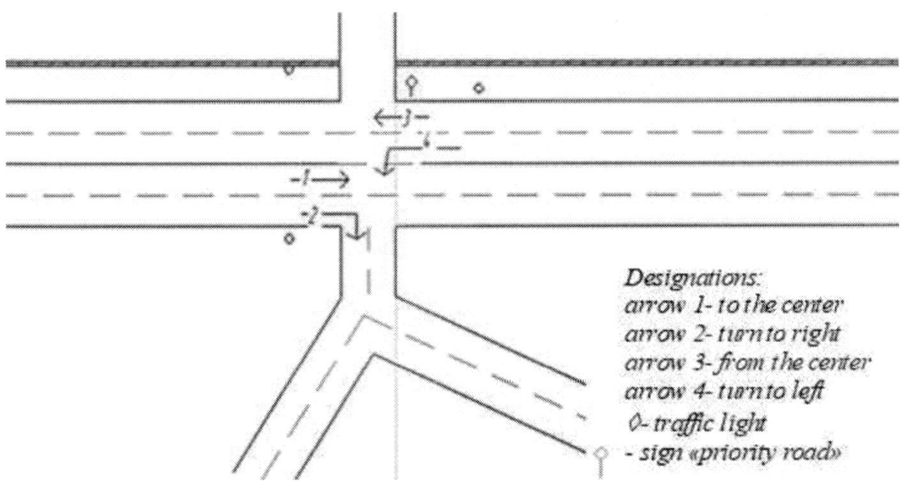

Figure 2. Scheme of the intersection of «Sokolova» and «Murmansk travel» streets in Saratov

IOP Conf. Series: Materials Science and Engineering **832** (2020) 012055 doi:10.1088/1757-899X/832/1/012055

The inhabitants of the city are used to the road network constant congestion on the considered segment.

Consider the current single-phase of stage-by-stage movement of cars, which allows providing the separation of conflicting flows in time [4-6]. The «Sokolova» street traffic is organized in 2 phases, it is necessary to let the traffic flows pass in straight and turning directions 3 and 4 in first phase. The direct and turning directions 1 and 2 are just as the parallel flow 3 in the phase 2 in accordance with the scheme of Figure 2. The pedestrian traffic through the roadway is organized by pedestrian traffic signals with a button for pressing which has the duration of working traffic signals positive go signal for 30 seconds. The traffic signals cycle is 116 seconds.

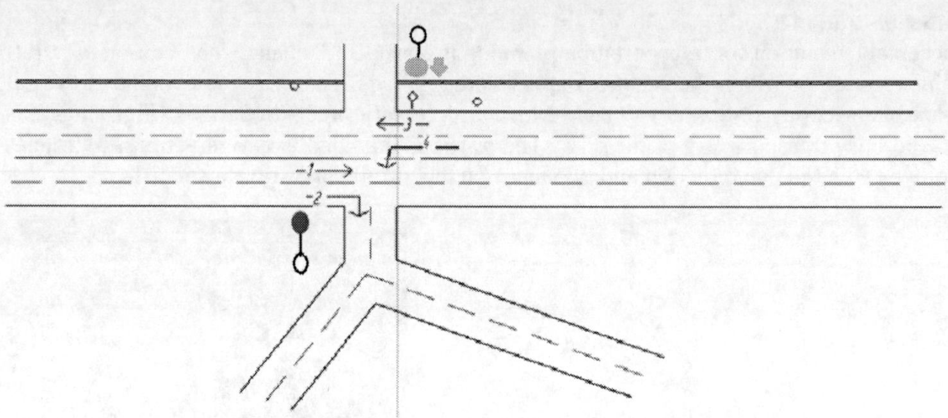

(a)

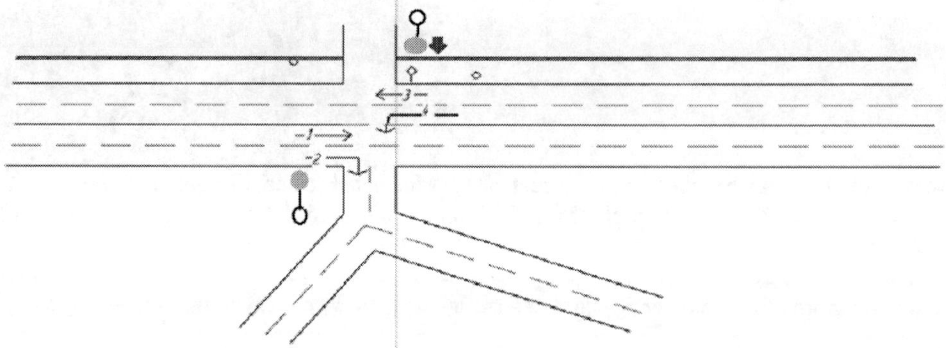

(b)

Figure 3. Single-phase of stage-by-stage movement of cars (a) - the first phase of regulation; (b) the second phase of regulation.

We present the Intensity plot of the traffic flow by the hour from 6.00 am to 12.00, determined by the monitoring (Fig. 4).

It is necessary to determine the traffic flow composition for creating the digital twin and the engineering hypothesis verification (Fig. 5).

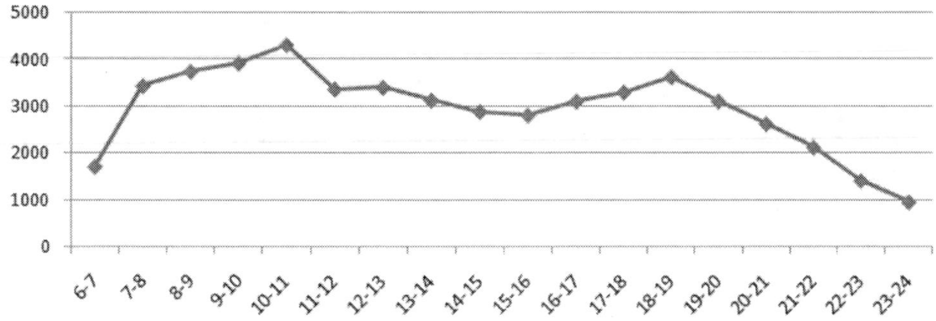

Figure 4. Intensity plot of traffic flow according to the hour of monitoring

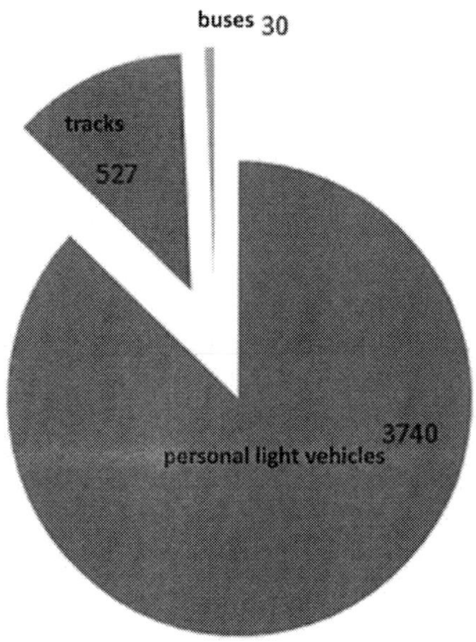

Figure 5. Composition chart of the traffic flow during rush hour (10.00 am to 11.00 am).

We simulate the current traffic situation in the PTV VISSIM program, the result is presented in Figure 6.

We formed traffic congestion on the «Sokolova» street seen in Figure 6; therefore, we can speak about the low capacity at the intersection under review.

It is necessary to simulate the proposed solutions for verifying the transportation planning engineering hypothesis.

4. Discussion

PTV Vissim is a software package that allows simulating the traffic flow movement, as well as pedestrian traffic [8-11].

In general, the PTV Vissim provides the ability to simulate traffic, test the transportation planning engineering hypotheses, etc.

Figure 6. Current situation model

We formed traffic congestion on the «Sokolova» street seen in Figure 6; therefore, we can speak about the low capacity at the intersection under review.

It is necessary to simulate the proposed solutions for the verification of transportation planning engineering hypothesis.

5. Conclusion

1. The software package PTV VISSIM is acceptable for the creation of a digital twin complex intersection in Saratov, as the model reiterated the actual traffic situation.
2. The models analysis allows confirming the hypothesis about the possibility of solving traffic congestion by the transportation planning in a given transport hub of the city with a high level of traffic delays.
3. The analysis of two models options allowed determining the most effective transportation planning measure in a given transport hub of the city with a high level of traffic delays, which is actually the duration change of the working traffic signals at the intersection. The most effective traffic signals cycle will be equal to 90 seconds. At the same time, the optimal working positive go signal duration of the foot traffic signals can be considered 10 seconds, because traffic of pedestrian is not intensive.

Reference

[1] Shchegoleva N V 2010 The road traffic safety problems In *Proc. of international scientific-practical symposium "The socio-economic problems of residential construction and the solutions during the way out of crisis period»* pp. 308-311 (Saratov: Saratov State Technical University)
[2] Gorshenina C Yu 2017 In *Proceedings of the II International scientific-practical conference Science publication* edited by Zhankazieva S V, pp. 334-341 (St. Petersburg: Publishing of Saint-Petersburg State University of Architecture and Civil Engineering)
[3] Shchegoleva N V, Gusev V A, Vorozhejkin M A 2016 The congestion formation in traffic flow *Technical regulation in transport construction* **5**(19) 3 Retrieved from: http://trts.esrae.ru/38-204
[4] Konoplyanko V I 2007 *Organization and road safety* (Moscow: High school) p. 383
[5] Pugachev I N, Gorev A E, Oleshchenko E M 2009 *Organization and road safety: studies and mauals* (Moscow: IC "Academy") p. 272.

[6] Rabchinsky A I, Gudkov V A, Kravchenko E A 2013 Organization of transportation services and the safety of the transport process (Moscow: IC "Academy").

[7] Kochetkov A V, Andronov S Yu, Shchegoleva N V, Valiev Sh N, Talalay V V 2018 Industry risk management system in the technical regulation of transport construction *Building materials* **5** 61-67.

[8] Zhankaziev S V, Vorobyev A I, Morozov D Y, Novikov A N, Kulev A V 2017 Efficiency of operation and functioning of the system of an indirect transport flow regulation and control *International Journal of Applied Engineering Research* **12** 3645-3652.

[9] Zhankaziev S V, Vorobyev A I, Morozov D Y, Novikov A N, Kulev A V 2017 Definition of accuracy of qualitative correspondence matrixes for indirect traffic flow control and regulation *International Journal of Applied Engineering Research* **12** 3653-3658.

[10] Ostroukh A, Nuruev Y, Ephimenko D, Zhankaziev S, Moroz D 2016 Automated dispatching control system of the mobile concrete batching plants *ARPN Journal of Engineering and Applied Sciences* **11** 6733-6736.

[11] Maksimychev O I, Ostroukh A V, Pastukhov D A, Nuruev Y E-O, Karelina M Y, Zhankaziev S V *2016* Automated control system of road construction works *International Journal of Applied Engineering Research* **11** 6441-6446.

IOP Conf. Series: Materials Science and Engineering **832** (2020) 012056 doi:10.1088/1757-899X/832/1/012056

An overview road data collection, visualization, and analysis from the perspective of developing countries

V V Silyanov[1], J I Sodikov[2], R Kiran[3] and A I Sadikov[2]

[1]Moscow Automobile and Road State Technical University (MADI), 64, Leningradsky ave., Moscow, 125319, Russia
[2]Tashkent Highway Engineering Institute, 20 Amir Timur street, Tashkent, Uzbekistan
[3]Lonrix Limited, Unit 1.48 Tawn Place, Pukete, Hamilton, New Zealand

E-mail: osmijam@gmail.com

Abstract. Road data collection went through dramatic changes over the course of technology enhancement. It started with simple paper sketches of pavement distresses, then manual and automatic road data collection, current trends are mobile phone-based road data collection and in near future connected/autonomous vehicles based road data collection will dominate. The evolution of road data collection methods shows how technology is shaping the way how data is collected, stored and analyzed. Even though the technology is rapidly progressing, most of the developing countries are still practicing manual (visual) road data collection practices. The paper overviews the advantages and disadvantages each approach describes data quantity requirements proposes an approach for road asset management systems in developing countries. It also highlights how to visualize road data by utilization open-source GIS software and briefly overviews turnkey solutions for road asset management.

1. Introduction

It's a well-known fact that roads are blood vessels of an economy which carry loads and passengers across short or long distances safe and quick. To keep roads safe and in good condition, highway engineers have to know what is current road condition which requires the collection of road data. Various road data types can be collected such as road inventory data, road condition data, cost data, climate data, and historical data. It depends on the road agency's policy on how much data would be efficient to make a proper decision. Some developed countries collect over 100 items related to road data. But in developing countries, there are challenges like lack of sophisticated computerized road survey vehicles, human capital to utilize hi-tech equipment and advanced pavement performance evaluation models, poor telecommunication systems, paper-based document exchange, and others. In the beginning, highway engineers collected road data by visually inspecting roads, some developing and third world countries still practicing paper-based road visual inspection. These include making sketches, taking notes and photography. Initial attempts to visually map pavement distress and its unification carried out by [1–2]. As road networks increased over time there was a need to collected road data by utilizing vehicles. Besides equipment based road data collection reduces rater's error and increases safety. During this period when vehicle-based road data collection systems were started widely utilize, data collection system vendors offered automatic and manual data recording systems [3–4]. There were several papers related to the automated collection of various pavement distress. For instance, automatic rut depth data collection [5], automatic skid resistance data collection [6], and one of the key pavement distresses such as pavement roughness collected manually and automatically [7–9]. Around the 90s and beginning 2000s road data collection system vendors started offering video recording data [10]. An increase in computing power led to incorporation simultaneous data collection of pavement distresses. Vendors like ARAN, DYNATEST, SIRANO, SCRIM, ROMDAS, and others started providing vehicles equipped with sensors, video logging, and GPS systems. ROMDAS utilized relatively inexpensive responsive type roughness measurement tool (bump integrator) with GPS and video logging specifically for developing countries [11]. Availability low-cost smartphones led to investigation of road data collection using smartphone acceleration sensor which behaves like bump integrator used for measuring road roughness [12–14].

2. Road data collection

It is a key element for pavement performance evaluation based on the quantity and quality of the data highway engineers who decide to minimize total cost of maintenance activities and works. Road data collection passes through several development stages such as in the beginning highway engineers used manual paper-based approach which is labor-intensive, low accuracy, requires skilled rater and limited data can be collected. The second stage of the development started in 1980s when vehicle-based road data collection systems gradually used by road agencies. Vehicle-based road data collection systems developed according to computer computing power and nowadays they have number of sensors for collecting roughness, rutting, skid resistance, odometer, GPS, video logging for road inventory and pavement crack detection. The cost of modern vehicle-based road data collection system depending on functionality may go up to 1 million US dollars. It requires highly skilled operator, data has high accuracy and detailed data. Smartphone road data collection initiated by using a built-in accelerometer to measure vehicle vibration which in turn can be converted into roughness index. Advantages this method of data collection is low cost of smartphone, no specific requirements for operator, acceptable accuracy and disadvantage is limited data can be collected. The future of road data collection is connected/autonomous vehicles equipped with high-resolution LIDAR cameras, GPS and numerous sensors. The advantages are zero cost of equipment, detailed data, high accuracy, and no operator are required. By utilizing deep learning algorithms data collected by connected vehicles can be automatically categorized and visualized on map which can be accessed by any highway engineer. Besides data continuously updated in real-time (Fig. 1). The evolution of road data collection methods shows how technology is shaping the way how data collected, stored and analyzed. Even though technology is rapidly progressing, most developing countries are still practicing manual (visual) road data collection practices. The reasons are lack of highly skilled operators, over-complicated road data collection systems, tons of pavement condition data, sophisticated pavement performance models. Therefore, there is a need for developing a simplified approach road asset management in developing countries.

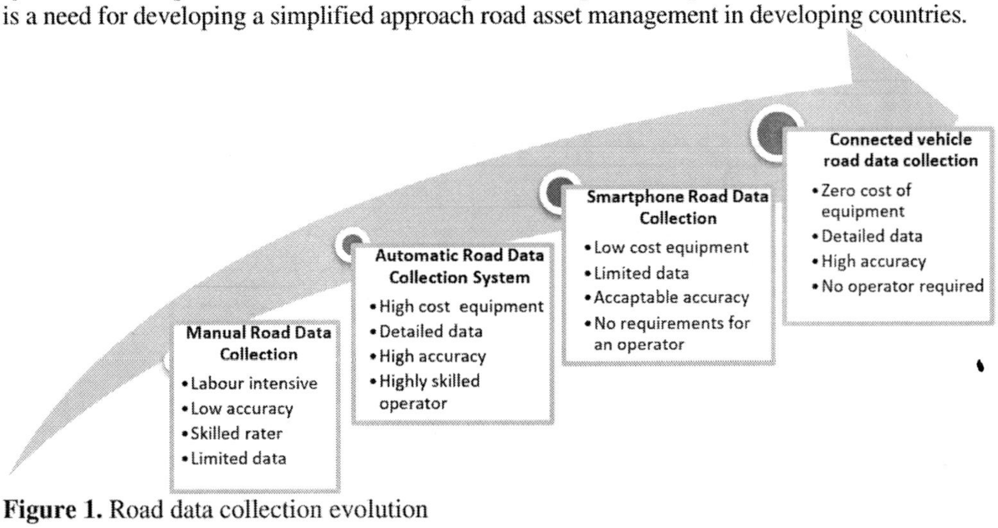

Figure 1. Road data collection evolution

3. Road data visualization and analysis

In Uzbekistan, there is more than 43 000 km of public roads, in total, there is more than 183 000 km. A small fraction of network around 10 percent of the road network has pavement condition data which is collected by automated road data collection systems. The rest of the network partially surveyed by visual inspection in the spring or autumn seasons. These visual inspections are paper-based and used for quick fixes or follow the strategy "patch the pothole first". To collect data across the whole network, a low-cost smartphone data collection has to be implemented. Availability acceleration sensor, GPS, photo and video recording in a smartphone make it a low cost and handy tool for measuring vehicle vibration in 3 axes (x,y,z) which in turn can be converted into roughness indices, GPS assists visualize various road attributes such as roughness index, photos and other road characteristics on map.

Two levels of analysis are proposed to optimally manage the road network in Uzbekistan. The first level is the network level where road agency manages the whole budget and entire road network. When financial resources are limited it's impossible acquire all pavement condition data across the road network. At network level road agency needs to prioritize road works by regions. This can be achieved by weighting each region by suitable factors such as traffic volume, road condition, population or concentration industrial companies/agricultural centers. The other approach is to employ highway functional classification. Highway functional classification serves not only for identifying what function is playing certain road sections but also importance of that section. As one of the performance indicators AADT (Annual Average Daily Traffic) or VKMT (Vehicle Kilometers Traveled) plays significant role in budget spendings. Importance of each highway functional class: arterial, collector and local was defined by factors such as traffic safety, speed, riding quality and environment using analytical hierarchy process [15]. After prioritization regions by taking into account highway functional classification, the next step is to prioritization by road section in each region. Figure 2 shows pavement management system framework which sets relationship between data and budget at network and project-level analysis. At network level of analysis there is relatively limited data available because it's practically impossible to collect data up-to-date all conditions and inventory especially in developing countries. Therefore, road agency has to develop a simplified tool to prioritize road works by regions. The analytical hierarchy process (AHP) can be used to prioritize road works at network level and project level. There is quite a large number of research papers dedicated on application of AHP in pavement management and optimization [16–25]. After prioritization by region, next step is to find road sections in each region that need to be repaired first. There are several approaches to optimize road repair works but one of the simple approaches is optimization using dynamic programming for detailed explanation reference [26].

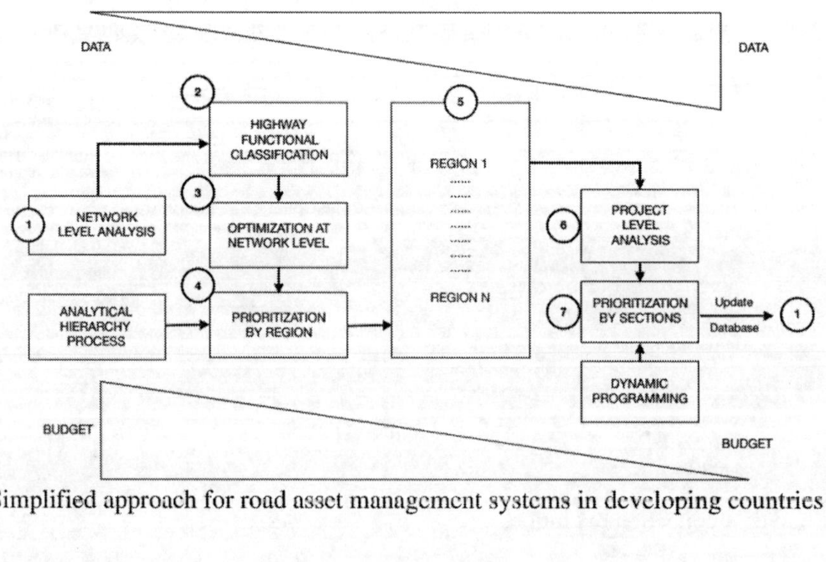

Figure 2. Simplified approach for road asset management systems in developing countries [27]

Developing countries are facing challenges not only in data collection but also in data storing and visualization. Moving from paper-based to digital technology requires extra technical skills and experience. Besides, there are two options open source and proprietary software. From developing countries' standpoint, open-source is most suitable because it's free to use, distribute and customizable. But again it demands extra knowledge to customize road organization needs. Nevertheless, open-source GIS software QGIS provides quite a variety tools that can be used out of box to visualize road attributes (Fig. 3a and 3b).

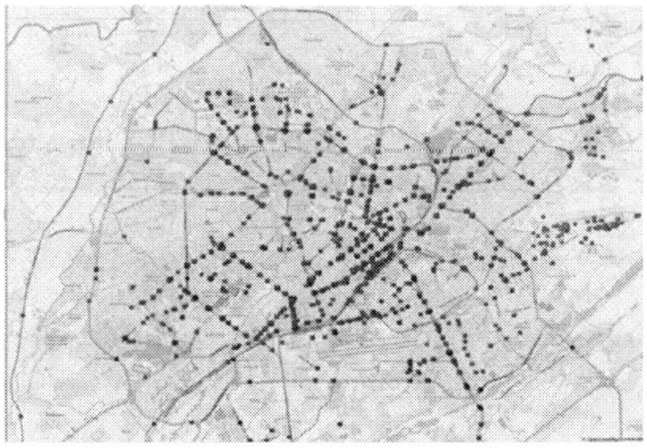

Figure 3a. Traffic lights visualization in Tashkent city

Figure 3b. IRI roughness index visualization in Tashkent city

There are turn-key solutions for road asset management. No asset management system can function in isolation. Successful implementation relies on three basic components – processes, people, and technology. Implementation requires that the agency must have a management mindset. Institutional policies and organization protocols must support a business process that can nurture the system. The technology adopted has to be effective and efficient. It must house the asset data and allow the required interrogation, analysis, and reporting. A suitable support and maintenance function is vital for sustainability. It is also important to recognize that no system is static. There will always be opportunities for refinements and improvements based on both the agency practice and technological progress. The framework acknowledges the importance of all three resource pillars noted above. As such, it provides a platform for the agency to implement management and auditing processes. At the same time – the flexibility of the framework, coupled with our highly energetic development team, is optimized for organizations with energetic asset management champions who have ideas of their own. Customized modules can easily be added at a fraction of the cost involved in configuring a more abstract commercial-off-the-shelf (COTS) system. JunoViewer is a framework for building modern, web-based asset management systems [28]. The JunoViewer platform is ideal for situations where clients have very specific, localized needs for reporting, workflow, and field inspection tools. It contains the following main modules and feature sets:

- Browser-based central web application and web-enabled data architecture.
- Store network asset inventory and condition data.
- Reporting and data analysis features.
- Centralized processing, storage, and streaming of digital content such as photos and videos.
- Develop and optimize budgets and Forward Work Plans (FWPs) for individual local
- governments.
- Analyze FWPs and budget requirements at the central and local government levels.
- Facilitate central management and auditing of FWPs.
- Perform graphical analysis and reporting of network condition and condition trends.
- Model and analyze future network conditions under different budget optimization scenarios.
- Facilitate GPS enabled offline field inspections with seamless sync-to-cloud server technology.

- Facilitate offline smartphone apps for inventory and inspections of "off-pavement" assets.
- A 100 % customizable framework to meet all asset management requirements.

JunoViewer is built using modern web-based protocols supported by Microsoft technologies. The application is 100 % browser-based and truly zero footprints. It requires no installation of software on the user's computer or plugins in the browser – operations that would be challenging in a distributed government system. JunoViewer shines when it comes to making data trends visible so that it can drive informed decisions. Optimized for fast analysis of time series data, the framework can provide a wide array of graphs – either for individual segments, groups of segments, or for entire networks. Deterioration modeling is a core feature of the framework and facilitates budget and treatment optimization using established approaches such as the HDM models or later, more innovative approaches such as those based on Heuristics and Fuzzy Logic analysis techniques. The modeling architecture exists at two tiers. At the lower level is a suite of looping algorithms to optimize the allocation of funds based on optimal treatments. At the higher level is a modeling setup that the user can control. At this level, the user has full control over the model and can specify deterioration parameters and equations as well as the treatment selection algorithm. The day-to-day management of forwarding Works Plans (FWPs) is an integral part of the framework. An innovative Forecast View allows network managers and central planners to quickly see where and what work is planned. The FWP Summary view can be used to quickly see a summary of future costs and forecasted network conditions. JunoViewer makes it easy for you to create and store inventories and condition data for any asset type. Once an asset class and its attributes have been defined, the table structures are automatically created. The video analysis views enable the user to view their network video, paired with a number of other features, such as a map and strip map. Google StreetView is enabled on the map so the user can interrogate a specific location further in detail. In addition, scatter plot and trend graphs are available (fig. 4).

In conclusion, road agencies depending on long-term strategies, available financial resources, and human capital establish what type of data collection system is suitable for them, what data analysis and visualization tools fit company needs, what optimization method is the best option to succeed in minimization total cost. All these questions need to be answered individually but the key is to utilize available low-cost technologies such as the smartphone data collection method which indeed provides adequate accuracy at the minimal data collection cost. Smartphone acceleration sensors, GPS, photo, and video capturing capabilities make them attractive low-cost alternatives in comparison high cost and sophisticated automatic vehicle-based road data collection systems.

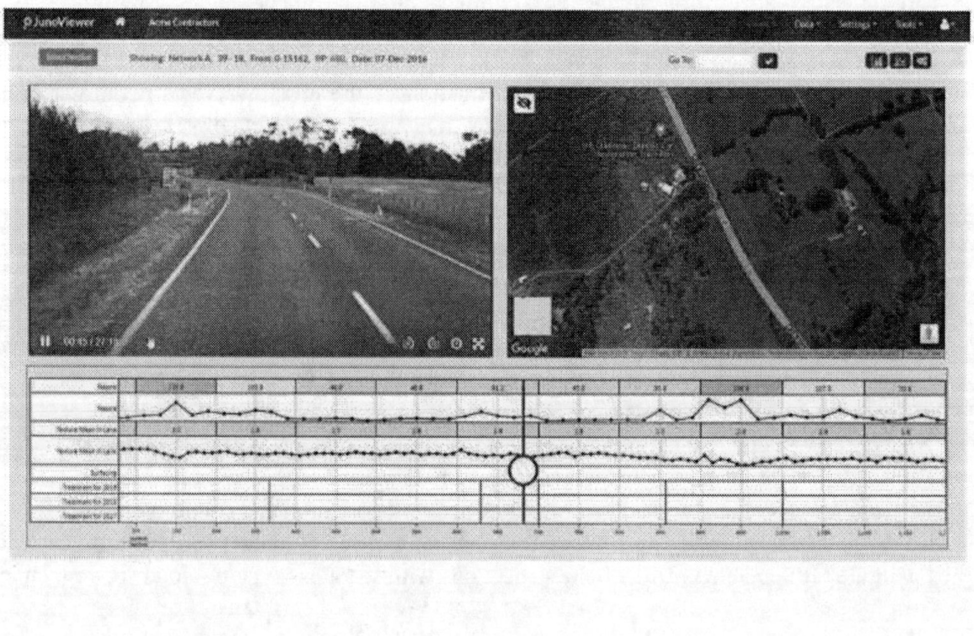

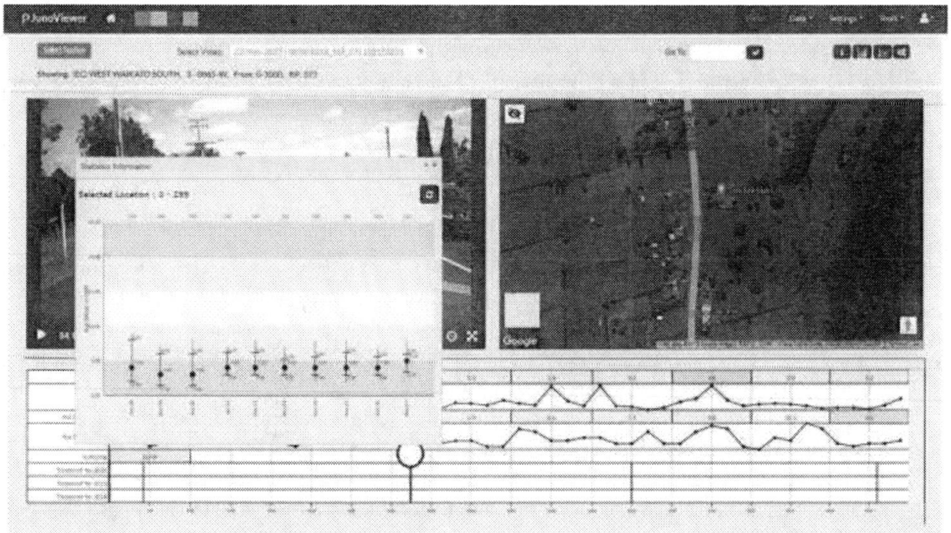

Figure 4. JunoViewer video analysis screenshot

References

[1] *Special Report 61E: The AASHO Road Test Report 5 Pavement Res.* 1962 (Washington, D.C.: HSB, National Research Council)

[2] Hveem F N 1958 Types and Causes of Failure in Highway Pavements *Bull. Highway Res. Board* **187** pp 1–51

[3] Birulja A K 1966 *Operation of highways* (Moscow: Transport) 324 p

[4] Benson K R, Elkins G E, Uddin W and Hudson W R 1988 Comparison of Methods and Equipment to Conduct Pavement Distress Surveys *In Transportation Res. Record* **1196** (Washington, D.C.: TRB, National Research Council) pp 40–50

[5] Moore R K, Clarks G N and Gisi A J 1993 Comparison of Pavement Surface Distress Measurement Systems *In Transportation Res. Record* **1410** (Washington, D.C.: TRB, National Research Council) pp 11–18

[6] Wadel G, Georges S, John E H and Kazuhiko M 1993 Obtaining Rut Depth Information for Strategic Highway Research Program Long. Term Pavement Performance Sites *In Transportation Res. Record* **1410** (Washington, D.C.: TRB, National Research Council) pp 19–24

[7] Tracy C C, James C Mc and Prianka S 1993 Maintenance Skid Correction Program in Utah *In Transportation Res. Record* **1410** (Washington, D.C.: TRB, National Research Council) pp 32–40

[8] Sanjay A, Khaled K and Al-Suleiman T I 1993 Consistency of Roughness and Rut Depth Measurement Collected with 11 South Dakota Road Profilers *In Transportation Res. Record* **1410** (Washington, D.C.: TRB, National Research Council) pp 41–51

[9] Zhu J J and Rajeev N 1993 APPARE: Personal Computer Software for Automated Pavement Profile Analysis and Roughness Evaluation *In Transportation Research Record* **1410** (Washington, D.C.: TRB, National Research Council) pp 51–58

[10] Bertrand C B 1993 Automated Versus Manual Profilograph Correlation *In Transportation Res. Record* **1410** (Washington, D.C.: TRB, National Research Council) pp 67–79

[11] Maccarley C A, Need, D and Nieman R L 1993 Video Cameras for Roadway Surveillance: Technology Review, Test Methods, and Results *In Transportation Res. Record* **1410** (Washington, D.C.: TRB, National Research Council) pp. 80–88.

[12] Sodikov J, Tsunokawa K and Ul-Islam R 2005 Road Survey with ROMDAS System: A Study in Akita Prefecture *Saitama University Departmental Bull. Paper* 149–51

[13] Jiménez L A and Matout N 2014 A Low Cost Solution to Assess Road's Roughness Surface Condition for Pavement Management *Presented at 93rd Annual Meeting of the Transportation Research Board* (Washington, D.C.)

[14] Allaire F and Hanson T 2017 Potential of road roughness data from smartphones as an input to spring weight restriction decision-making *Paper prepared for presentation at the Innovations in Pavement Management, Engineering and Technologies Session of the Conf. of the Transportation Association of Canada* (St. John's, NL)

[15] Viengnam D and Hiroyuki O 2013 Estimation of Road Roughness Condition from Smartphones under Realistic Settings *13th Int. Conf. on ITS Telecommunications*

[16] Silyanov V V and Sodikov J I 2017 Highway Functional Classification in CIS countries *Transport Infrastructure and Systems. Proc. of the AIIT Int. Congress on Transport Infrastructure and Systems*

[17] Azis I J 1990 Analytic Hierarchy Process in benefit-cost framework: A post-evaluation of the Trans-Sumatra highway project *Eur. J. of Operational Res.*

[18] Hagquist R F 1994 High-precision Prioritization Using Analytic Hierarchy Process Determining State HPMS Component Weighting Factors *Transportation Research Record* **1429** 7–14

[19] Masami S 1995 Analytic Hierarchy Process (AHP)-Based Multi-Attribute Benefit Structure Analysis of Road Network Systems in Mountainous Rural Areas of Japan *Int. J. of Forest Engineer.* **7(1)**

[20] Ramanathan R 2001 A note on the use of the analytic hierarchy process for environmental impact assessment *J. of Environmental Management* **63** 27–35

[21] Kim K and Bernardin V 2002 Application of an Analytical Hierarchy Process at the Indiana Department of Transportation for Prioritizing Major Highway Capital Investments *J. of the Transportation Res. Board* **1816** 266–78

[22] Kengpol A 2002 The Design of a Support System (DSS) to Evaluation the Investment in New Distribution Centre Using The Analytic Hierarchy Process (AHP) *Capital Investment Model and Transportation Model*

[23] Taslicali A and Ercan S 2006 The analytic hierarchy & the analytic network processes in multicriteria decision making: a comparative study *J. of aeronautics and space technol.* **2(4)**

[24] Kang S and Lee S M 2006 AHP-based decesion-making process for construction of public transportation city model: case study of Jeju, Korea *Joint Int. Conf. on Computing and Decision Making in Civil and Building Engineer.*

[25] Sarfaraz A, Vedagiri P and Krishna Rao K V 2017 Prioritization of pavement maintenance sections using objective based Analytic Hierarchy Process *International Journal of Pavement Research and Technology* **10** 158–70

[26] Sodikov J I and Silyanov V V 2015 Road asset management systems in developing countries: Case study in Uzbekistan *Sci. J. of Transportation* **6** (Moscow–China–Vietnam) Retrieved from: https://www.researchgate.net/publication/272415247_road_asset_management_systems_in_developing_countries_case_study_uzbekistan

[27] Sodikov J I, Silyanov V V and Shahidov A F 2018 Optimization of pavement maintenance activities considering vehicle operating cost *Int. Cooperation Issue of Transportation* **9** Retrieved from: https://www.researchgate.net/publication/333449423_optimization_of_pavement_maintenance_activities_considering_vehicle_operating_cost

[28] Sodikov J I 2019 *Forecasting and management of the road assets system* Monograph. (Tashkent) p. 205

[29] *JunoViewer framework outline* Retrieved from: http://www.junoviewerweb.com/junoviewer-features.html

Research on the integration of intelligent transport systems

B S Subbotin and E U Ahmetzhanova

Moscow Automobile and Road Construction State Technical University (MADI), 64, Leningradski ave., Moscow 124319, Russia

E-mail: onti@nich.madi.ru

Abstract. In article top trends and the directions of application of the intellectual transport systems (ITS) are considered and analyzed. Now, proceeding from the analysis of world practice, it is possible to draw a conclusion that application of the intellectual transport systems as systems which provide adaptive and expected management of safety and the traffic organization in the transport and road system of regions will lead to the growth of transport services quality level. Development of the intellectual transport systems methodologically is based on system approach as the intellectual transport systems are formed as systems, but not separate services. Methods of creation of the intellectual transport systems are based on the principles of operating transport systems modernization and reengineering.

1. Introduction

Modern intelligent transport systems combine communication, telematics and information management technologies. Their goal is the search and implementation of the most effective traffic management solutions to improve traffic safety and reduce stress levels.

The multilevel system of influences on the traffic flow forms intelligent transport systems.

The first level is controlling the movement of an individual vehicle. The second level is to ensure coordination of the movement of several vehicles moving together in passing and oncoming directions. The third level consists of a traffic management network (city, highway, etc.).

At each of the indicated levels, two main functions are introduced: monitoring and application of control action by intelligent transport systems. In addition, it is necessary to implement the coordination function between all three levels of intelligent transport systems.

2. The relevance of introducing intelligent transport systems in a megalopolis

The trend of modern cities is to increase the level of motorization, as a result, the presence of parking spaces on the roadway, which inevitably leads to a decrease in the level of road safety, as well as the throughput of the road network. An integral part of the transport system of large cities are intersections. Monitoring traffic and pedestrians refers to the management of transport systems [1].

The construction and use of automatic traffic control systems on the street-road network of large cities allow to improve indicators [2, 3]. The task of traffic regulation is not only to increase safety, but also to simultaneously increase the operational effect of a particular production process [4]. Understanding that the digital transformation of roads as a basis for life safety, social and economic development, becomes relevant and timely.

The trend of modern cities is to increase the level of motorization, as a result, the presence of parking spaces on the roadway, which inevitably leads to a decrease in the level of road safety, as well

IOP Conf. Series: Materials Science and Engineering **832** (2020) 012057 doi:10.1088/1757-899X/832/1/012057

as the throughput of the road network. An integral part of the transport system of large cities is intersections. Monitoring traffic and pedestrians refers to the management of transport systems [1].

The construction and use of automatic traffic control systems on the street-road network of large cities allow improving indicators [2, 3]. The task of traffic regulation is not only to increase safety but also to simultaneously increase the operational effect of a particular production process [4]. Understanding that the digital transformation of roads as a basis for life safety, social and economic development, becomes relevant and timely.

After analyzing the experience of using indirect traffic management systems in other countries, it was revealed that in the USA much attention is paid to control technology, namely to such components: message structure, frequency of information and the impact of this structure on decisions made directly by the driver, within the framework of such basic components as:

- detection;
- perception;
- readiness for action [5].

The main goals of intelligent transport systems: improving the quality of life of the population, improving road safety, developing intermodal transport, international transport corridors, reducing the negative impact of transport on the environment, reducing transportation costs in the cost of services, choosing the best logistics support for the provision of transport services, improving quality and competitiveness creating favorable conditions for private investment, supporting the adoption of operational decisions on rule, the implementation of the transit potential, improving the efficiency of the transport system [6]. The most important of them are presented in Figure 1.

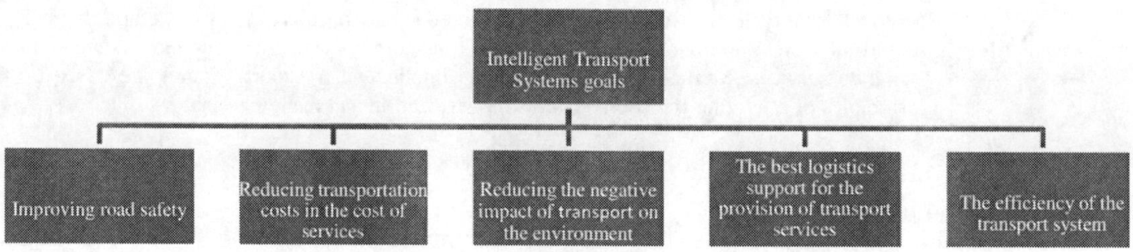

Figure 1. The most important of intelligent transport systems goals.

Since the integration of the subsystems of indirect control of traffic flows affects the efficiency of the road network of large cities, there is a need to develop and improve this area.

Intelligent transport systems are most actively developing in the central part of Russia; in Moscow, such systems help reduce traffic congestion, and as a result, improve the environment due to reduced transport downtime, and can also optimize individual public passenger transport routes. The subsystem of informing about situations that present difficulties in traffic, slowing down the flow of traffic is an effective way to warn drivers of a traffic jam [5]. Intelligent transport systems also received their distribution in the constituent entities of the Russian Federation: Ryazan, Kaluga, Leningrad and Ulyanovsk regions [7]. Thus, if drivers and passengers are informed about the traffic situation in a timely manner, indirect losses from downtime can be avoided, which leads to such an adverse effect as a decrease in the profit of individual enterprises associated with transport.

There are many aspects in the direction of movement towards intelligent transport systems; they can be classified according to the following criteria: comfort and energy efficiency.

Given the demand by the international community for the dynamic development of transport infrastructure, the tasks of increasing energy efficiency and safety of both individual vehicles and the transport system as a whole are relevant and promising [8]. Currently, the development of smart energy-efficient systems for road transport has a positive trend.

During operation, for every 1,000 kilometers, the car requires an energy consumption of approximately 240…350 k /h. Such energy costs are associated with the consumption of operational materials, spare parts, and technological processes during maintenance and repair. When carrying out major repairs, costs range from 96 to 131.2 kW/h per engine.

Analytical estimates show that, depending on the type, one passenger car requires 100.000...150.000 k /h of energy consumption for the period of its life cycle (from manufacture to disposal); truck – 150.000...500.000 kW/h; the bus – 300.000...600.000 kW/h. An average passenger car consumes approximately 9.3 tons of fuel during its operation, and its production costs another 2 tons of fuel [8].

As one of the promising methods for generating energy, generators operating on the piezoelectric effect are considered. For example, using the kinetic energy of automobile transport, it is proposed to use overhead strips of electric-generating road pavement, which allow the piezoelectric transducers of the wheels of the car (inside the pavement) to generate a sufficient amount of electric energy to provide autonomous lighting of the route [9]. As a result, one kilometer of the roadway can produce up to 500-kilowatt hours [10, 11].

To optimize the processes, "New Energy Technologies" has developed a device for collecting the kinetic energy of transport and converting kinetic energy into electricity. The piezoelectric generator is designed as a speed bump.

3. Management of the movement of an individual vehicle

Internal (autonomous) systems are based on information received from a directly intelligent vehicle. They are widely used in cars and trucks to increase active safety. These include various systems increasing stability, examples in Figure 2.

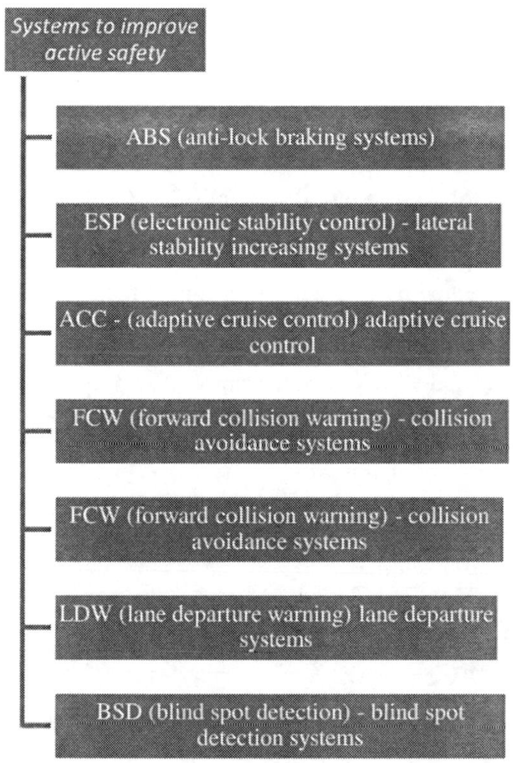

Figure 2. Systems to improve active safety

One of the effective options for the implementation and development of telematics systems in road transport is the introduction of satellite navigation technologies to solve the main technical problems of transport management [12]. The stimulation of work in the field of navigation support in Russia was promoted by the Decree of the Government of the Russian Federation of March 7, 1995 No. 237 "On the implementation of work on the use of the global navigation satellite system GLONASS in the interests of civilian consumers." In 2010, the Order of the Ministry of Transport of the Russian Federation (dated March 9, 2010 No. 55) was issued "On Approving the List of Types of Automobile Vehicles Used for the Carriage of Passengers and Dangerous Goods to Be Equipped with GLONASS or GLONASS / GPS Satellite Navigation Equipment". This order noted that vehicles are equipped with GLONASS, GLONASS / GPS equipment in order to increase the safety and efficiency of road transport of passengers and dangerous goods.

Currently, there is a change in the status of a transport unit from an independent subject of traffic, towards a predictable subject of transport and information space, which is a distinctive feature. Therefore, the development of a telematic complex of road infrastructure is a priority in this area. Leading car manufacturers are exploring new use cases and telematics business models, such as user-based insurance and personal car sharing.

Over the past decade, scientists and engineers in the automotive industry have achieved good results in the development of devices and systems that ensure the safety of pedestrians and other unprotected road users. Modern cars are able to detect not only pedestrians but also vehicles while activating the brake system with the highest possible braking performance.

For example, Google developed an unmanned vehicle. This system uses information collected by the Google Street View service, video cameras, a LIDAR sensor that is installed on the roof, radars in the front of the car, as well as a sensor connected to one of the rear wheels, which helps in determining the position of the car on the map. [13]. In the process of testing cars with this system in 2010, it was found that the car drove about 1,600 km completely autonomously, 225.308 km with partial human participation [13]. This direction has been developed in subsequent years. In 2016, an agreement was signed between Google and Fiat Chrysler on cooperation, the purpose of which was the release of one hundred self-governing Chrysler Pacifica [13].

The desire to minimize accident rate is also characteristic of manufacturers of components and systems of commercial vehicles.

4. Ensuring coordination of the movement of several vehicles

All road users, as well as infrastructure, need to exchange a large amount of data in real-time. Based on them, an alert algorithm is generated. Despite the complexity of the task, such a communication system – ITS Connect was invented by Toyota. The Japanese manufacturer has prepared a system of interaction with other road users, and moreover, with objects of transport infrastructure. The automaker introduced the innovation system in October 2015. At the heart of the system is a cloud network that interacts with cars and allows them to transfer data. For example, ITS Connect adjusts the work of cruise control, warns drivers about approaching ambulances and police, helps to maneuver at intersections equipped with communications.

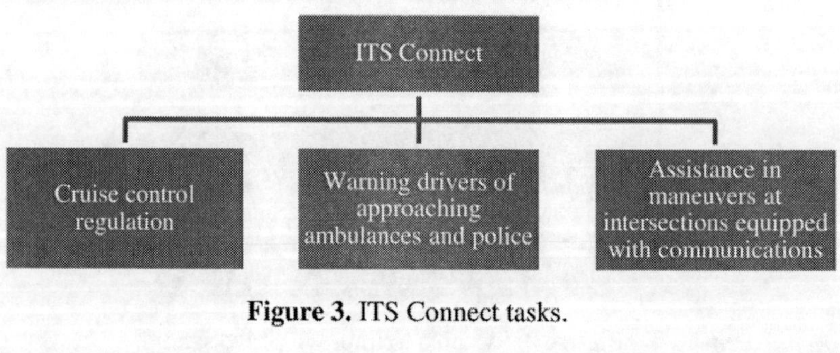

Figure 3. ITS Connect tasks.

Together with the Swedish Ministry of Transport, Volvo has created a project to combat accidents on icy roads. The essence of the project lies in the fact that a car falling on how many sections, through internal mobile communications Volvo Cars transmits information about it to the data center, where then this information is transmitted to nearby cars. In addition, information about the icy section of the road is sent to the road service. The data obtained allow road services and contractors to better plan and maintain roads in the winter, as well as respond quickly if necessary.

5. Management of network traffic flows

The introduction of new technologies in the traffic management system will create centralized monitoring of the road situation. Currently, saturation with the intellectual technology of traffic control systems is necessary to ensure the interconnected functioning of vehicles in the stream. Multi-program and adaptive systems of connected traffic control in megacities and adjacent highways that can receive, summarize, analyze, manage, communicate information to all road users about the state of the system, for example, these include road signs with automatically changing information, special dynamic information boards and etc. The analysis of large volumes of data allows for preventative maintenance of equipment and improved territory planning. Intelligent infrastructure requires the integration of several technologies, such as IoT, Big Data, etc. The collected data will allow you to adjust the development of the territory [14]. IoT devices are a key element of the smart city concept. Built-in sensors collect useful data for analysis and obtaining the necessary information. Free exchange of information between complex urban systems will be carried out in real-time; The integration of data analysis will minimize accidents and unintended consequences.

The work of the intelligent traffic management system is aimed at maximum unloading of intersections and improving the safety of urban traffic. The intelligent traffic control system collects information about the movement of vehicles through the city's video fixation points, then analyzes the data and takes measures to optimize traffic automatically or semi-automatically [6]. Changes in control actions at one particular intersection inevitably necessitate changes in the characteristics of traffic flows at neighboring locations. Therefore, systemic flow control is required.

Despite the complete automation of traffic control, the analysis of information and the adjustment of the operating modes of objects are sometimes carried out manually. In the future, it is necessary to reduce the impact on the operation of these systems.

According to the forecasts of the Massachusetts Institute of Technology, such a sign of a smart city as joint mobility should appear. Based on an empirical study, it is established that in conditions of heavy traffic, two phases of the traffic flow can be distinguished: "synchronized flow" and "wide moving traffic jam". A three-phase model was also proposed [15]. An autonomous car, according to the present in the navigation system, from a computer, tablet or smartphone will take citizens whose routes cross, to the office, to business meetings, to the hospital, to study and return for passengers with a similar route in the evening. The only thing that will unite passengers is the route. An experiment was conducted by Senseable City Lab. During the year, they tracked 150 million taxi rides in New York. Movement models have been identified and developed efficient car-sharing systems. Experts concluded that the development of transport sharing technologies could reduce the number of taxi rides by 40 %. As a result, the length of traffic jams and the number of harmful emissions will decrease. According to experts, the adoption of the concept of general mobility by cities, four out of five private vehicles would be superfluous, and minimizing the number of cars would lead to the release of parking spaces [14].

The introduction of intelligent transport systems throughout the country is local and uneven, since the systems are not interconnected, fragmented, without proper scientific and conceptual elaboration. There is no systematic approach. A comprehensive application of technologies is needed, which will improve the level of road operation [16].

6. Conclusion

Transport problems are inseparable from problems such as the planning of city infrastructure. [17]

At the present stage of development of road transport complexes, intelligent systems are actively being introduced into traffic flow control systems, vehicle designs, and interactive systems for informing road users and payments for services provided by the transport system to its participants.

The volume and pace of implementation, the effective deployment of communication and the competent stimulation of consumers to new technologies play a significant role for the effective interaction of intelligent transport systems with other projects, for example, related to road construction.

The integrated application of technologies can increase the level of road operation and optimize the costs of servicing individual vehicles, minimize the costs associated with the capacity of the road network.

References

[1] Popkov V K 2011 On modeling city traffic systems with hypernetworks, *Autom. Remote Control*, **72(6)** 1309–18
[2] Mitrofanova M A 2017 A new system for efficient traffic management, *Molodoy uchenyy*, **4(138)**, 669–71
[3] Vorob'yev A I and Subbotin B S 2012 Analysis of the methodological basis for the construction of subsystems of indirect and directive management of traffic flows, *Vestnik MADI* **4(31)** 49–53
[4] Golberg Ya M 1936 Automation of the production process in road transport, *Automation and Remote Control*, **5**, 91–8
[5] Vorobyev A I, Subbotin B S and Vlasenko G V 2013 Analysis of technologies for indirect control of traffic flows by the example of foreign experience, *Vestnik MADI* **4 (35)** 83–9
[6] Basyrov R R and, Valiev T R 2016 The main capabilities of managing intelligent systems of traffic flow, *Aktual'nyye napravleniya nauchnykh issledovaniy XXI veka: teoriya i praktika* **5-3(25-3)**, 193–8
[7] The III International forum and exhibition "Intelligent Transport Systems of Russia". Retrieved from: http://itsrussiaforum.ru/2018/ru/results.php (date of the application: 06.03.2019)
[8] Shatrov M G, Ivanov I E, Prishvin S A, Matyukhin L M, Dunin A Yu and Ereshenko V E 2013 *Heat engineering: a textbook for students. institutions of higher prof. education* (Moscow: Academia)
[9] Gerasimov E M and Tretjak L N 2014 New approaches to designing of the highways illumination systems *The fundamental researches* **3-1** 22–7
[10] Roshani H, Dessouky S, Montoya A and Papagiannakis A T 2016 Energy harvesting from asphalt pavement roadways vehicle-induced stresses: A feasibility study, *Applied Energy* **182(15)** 210–8
[11] Wu M S, Huang H H, Huang B J, Tang C W and Cheng C W 2009 Economic feasibility of solar-powered led roadway lighting *Renewable Energy* **34(8)** 1934–8
[12] Vlasov V M, Efimenko D B, Bogumil V N and Konin I V 2016 *Industry requirements for the design and implementation of telematics systems in road transport: a training manual* (Moscow: MADI)
[13] Fomenko A D and Nikolayenko D V 2017 Overview of Existing Intelligent Transport Systems Technology Solutions *Sovremennyye tendentsii razvitiya i perspektivy vnedreniya innovatsionnykh tekhnologiy v mashinostroyenii, obrazovanii i ekonomike* vol 3 **1(2)** 281–4
[14] City of the future: what concepts are turned into reality. Retrieved from: https://iot.ru/gorodskaya-sreda/gorod-budushchego-kakie-kontseptsii-voploshcheny-v-realnost (date of the application: 06.03.2019)
[15] Kerner B S, Klenov S E and Wolf D E 2002 Cellular automata approach to three-phase traffic theory, *J. Physics A: Math. General* **35(47)** 9971–10013
[16] Dmitriyev I I and Kirillov A M 2017 Smart roads and Intellectual transport system *Construction of Unique Buildings and Structures* **2 (53)** 7–28

[17] Gribova V V, Shamray N B and Fedorishchev L A 2017 Traffic modeling flows in a developing urban infrastructure with a software suite for creating interactive virtual environments *Autom. Remote Control* **78(2)** 235–46

Digital object-oriented control models in automobile-road complex systems

A V Terentyev[1], M Yu Karelina[1], T Yu Cherepnina[1], D A Linnik[3] and V A Demin[2]

[1]Department of Machinery Parts and Theory of Mechanisms, Moscow Automobile and Road State Technical University (MADI), 64, Leningradsky Ave., Moscow, 125319, Russia
[2]NOZ-TL, Moscow Automobile and Road State Technical University (MADI), 64, Leningradsky Ave., Moscow, 125319, Russia
[3]Department of Engineering and Technical Operation of Cars, Yanka Kupala State University of Grodno, 22, Ozheshko str, Grodno, 230023, Belarus

E-mail: aleksej.terentev.67@bk.ru

Abstract. The extensive development of the structures of the automobile-road complex requires optimization of the processes of their integration into the regional and global transport infrastructure. In order to effectively manage unstable traffic flows, automated control systems are developed that use complex algorithms based on the principles of artificial intelligence. The basis of this approach is the concept of transferring the concept of a control object from a generalized or aggregate traffic stream to a separate control element: a separate vehicle, a separate passenger, a separate consignment. The completed management process in the system of the road complex, which meets the current level of development of scientific and technological progress and productive forces, requires the use of digital object-oriented management models based on the creation of a unified communication and computer network for managing information flows as well as the development and application of control models physical objects of movement in the transport network, determining the optimal trajectory of movement of the latter. To form a control system based on object-oriented models, it is necessary to develop specialized analytical methods for multi-criteria evaluation of possible options. It is of fundamental importance that the developed methods can investigate the field of probabilities of changes in informational situations of the environment and are applicable to factors of any degree of uncertainty.

1. Introduction

The development of the structures of the road complex and their integration into the regional and global transport infrastructure cannot be based solely on the mechanisms of regulation of the "free" market. In this case, each interested explant of the transport system, and these are cargo carriers, owners of transport and storage complexes, municipal authorities etc. optimize "their own transport logistics and economy", which leads to such a transport situation when the transport infrastructure does not satisfy everyone. An alternative to this approach is the use of mechanisms for synchronous or integrated optimization of the parameters of traffic flows and elements of the transport infrastructure: parking lots, interchange hubs, cargo terminals etc., which will reduce the total mileage of the rolling stock, more efficiently use the power of both the road network and infrastructure of the transport and logistics system as a whole as well as to differentiate in time and space and optimally redistribute traffic flows of goods and passengers. Currently, the operation of control systems in the road complex is based on outdated ideas about the control object - the traffic flow. In this case, as a control object, a traffic flow has the following properties:

- non-stationary state (daily, weekly and seasonal unevenness),
- stochasticity (differences in the goals of the participants in the movement and their behavior),
- sustainability (periodic nature of travel),
- inertia,
- interconnectedness (presence of internal dependencies between flow parameters),
- insufficient manageability.

In order to effectively manage such an unstable object as a traffic stream, transport science today is developing more and more new approaches. Control systems have gone the way of development from local hard one-program regulation on separate sections of the transport network to automated control systems (ACS) using complex algorithms and principles of artificial intelligence. The next stage in the development of ACS are those operating on the principle of feedback with the flow or adaptive systems using various self-learning algorithms, systems that simulate the behavior of living organisms [1] and systems based on neural network technologies and artificial intelligence [2, 3].

When designing these systems, there is a tendency towards a transition to the so-called "Active Traffic Management" [4, 5]. This approach assumes a more intense impact on the behavior of traffic flows based on modeling and predicting changes in its parameters, which requires full integration of all local control systems. Coordination of control actions is considered an attribute of a complete management process. This level occurs when individual elements of the system begin to develop control actions on the basis of incoming information in the interests of the entire system as a whole. The stage of coordinating control actions requires maximum efforts to overcome organizational and technical barriers, since, as a rule, the creation of a unified communication computer network is required. At this stage, each participant must correlate their decisions with the overall management concept. That is, the concept of a control object is transferred from a generalized or cumulative traffic stream to a separate control element: a separate vehicle, a separate passenger, a separate consignment of cargo. Today, the integration of transport network components is carried out by transmitting information about the interaction between individual vehicles and transport network elements. In one study, more than 30 connections between the components of the transport network are highlighted [6]. Therefore, the completed process in any control system of the road complex that meets the current level of development of scientific and technological progress and productive forces requires the use of digital object-oriented control models:

- creation of a unified communication and computer network for managing information flows;
- development and application of object-oriented models for managing physical objects of movement in the transport network, forming optimal trajectories of movement of the latter.

2. Problem Statement

When considering issues of interaction of objects related to each other by a number of diverse relationships, it is necessary to develop algorithms that are able to maintain integrity relations between objects [7]. Modern high requirements for the reliability and security of information systems lead to the fact that all the main decisions in the field of organizing information exchange processes have a client-server architecture. This approach allows us to restrict access to confidential information by concentrating all the functions of direct data management in a single center. As a rule, various database management systems (DBMS) are used as server software. A significant number of DBMSs are based on the relational principle [8], when tables or stored procedures are the means of interaction. The main disadvantage of the relational interaction model is that the designed system as a set of tables is very often difficult for analyzing and understanding of processes. In turn, with the increasing complexity of the system, this leads to the inability to fully monitor all the necessary places and procedures for implementing changes, and the process of developing the system itself becomes poorly managed or uncontrollable. An alternative to this approach (relational data model) is an object-oriented approach. When using an object-oriented approach, a program should not only describe objects and their properties in the form of criteria and the relationships between them in the form of goal setting, but also describe their interaction (methods) in the form of operations on objects.

3. Materials and Methods

A clear advantage of the object-oriented approach is the conceptual proximity in any subject area to an arbitrary structure and purpose of the system. At the same time, attribute conversion mechanisms or methods should allow building derivative objects and structures on the basis of basic ones, thereby creating a model in a more complex subject area with the necessary properties, providing the possibility of constantly analyzing and making changes if necessary. In this case, objects and methods are polymorphic, which makes the software more versatile and flexible [9]. Despite the clear advantages of the object-oriented approach, object-oriented control systems are still not widely used [10]. At the same time, hybrid object-relational DBMSs are widely used, which partially use object-oriented principles of working with data, but at the same time, data storage is performed during the implementation of the relational model (Table 1).

Table 1. Architectural solutions for information systems

Solution option	Solution description
Direct work with the control system	As an automated workstation (AWS) a specialized application is used that directly accesses the control system (CS). This scheme is used for small application systems; the implementation of this approach for complex databases makes it difficult to develop the system
Using an intermediate layer	As an AWS specialized browsers are used that do not access the CS directly, but some kind of intermediate layer, which then accesses the CS directly. The use of the intermediate layer as a link between the AWS and the CS enables the AWS not to depend explicitly on the CS and increases the possibility of developing systems
Using a server application	The AWS directly refers to a specialized application (server), which then accesses the CS directly or using an intermediate layer. This 3-level architecture makes it possible to implement logic on the server, thus increasing the security of working with data, increasing scalability

The given review of options for possible architectural solutions in the development of an information system that provides simultaneous access to information of many users allows you to create a scheme for using technologies with an intermediate layer on the server. When applying the object-oriented approach to implement the interaction between objects (classes), an event model is implemented. The scheme of this model is shown in Figure 1.

The main result in the interaction using the event model is defined in the following. The event handler object (B), the data and state of which depend on the state of the object (A), implements a subscription to the event of a state change or change of the object (event source). When an event occurs (when the state or data changes), object (A) generates an event that causes the execution of the corresponding handler object (B). The use of the event model for the implementation of object relationships in order to maintain the integrity of relationships in the control system allows us to perform the relationship between implicit objects, thereby simplifying the procedures for adding new and changing existing relationships. This mechanism is good enough if the objects are fixed once and fixed all the time the application exists or until they are in demand. But when forming control systems intended for simultaneous work with a sufficiently large number of interconnected objects, when objects are stored in the CS and loaded for work only when performing direct actions with them, it is unacceptable. In this case, additional tools are needed in the form of interaction control subsystems that implement event-based models of interaction of objects in the intermediate layer.

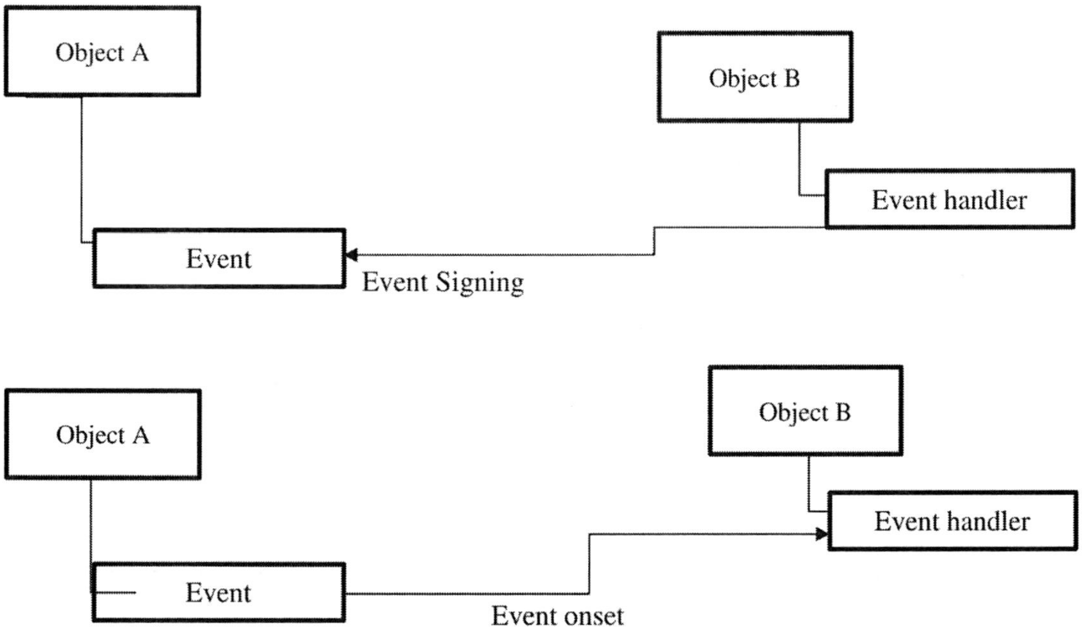

Figure 1. Event interaction scheme

Naturally, when developing a control system that implements an event-based model of interaction between objects within an intermediate layer on a server, support for events of two types is required:

- events that occur when performing actions with any object of the system – object events.
- events that appear when any particular point in time occurs – temporary events.

Event handlers also need to be divided into two types:

- synchronous handlers, when the processing of an event occurs immediately at the moment of its occurrence, that is, the same transaction in which the event occurred (happened);
- asynchronous handlers, when the event handler is queued and implemented during system downtime, for example, in a separate transaction.

The use of asynchronous handlers will reduce the time it takes to complete a transaction and, accordingly, will reduce the response time of the CS to user requests. It is also possible to increase the overall performance of the CS by implementing additional processes, including on individual computers, when invoking asynchronous processors.

The main elements of the object interaction control system under consideration are the queue of asynchronous event handlers for events and the storage of event subscriptions. When implementing a new subscription to an event, information about it is added to the subscription repository. Each record may contain certain information about the source object:

- event of the source object;
- type of subscription – synchronous or asynchronous;
- handler object;
- method of the handler object that must be called for to process the event.

When any event occurs, all subscriptions to the source object event are searched for in the repository, then, depending on the type of subscription, either a direct access to the handler object is made, or an entry is added to the queue of asynchronous handlers. If the asynchronous handler is accessed during system downtime, this increases its performance. It is important to note two main circumstances necessary when implementing the subsystem for managing interactions between objects:

1. Technological. Any object that is an event handler or source must have a unique identifier, using which one can load the desired object from the database into the application memory. The identifier should allow storing links to objects in the subscription description, which are event handlers and sources.
2. Scientific and methodological. The method of the processor object to be implemented must be formed, formalized and programmed as an analytical decision-making model based on information about the state or current state of the source object, for example, in the form of tables of values of the efficiency of actions in a multi-criteria goal setting structure.

4. Discussion of Results

Any level of tasks of information systems, the solution of which is aimed at increasing the efficiency of their functioning by increasing the completeness and efficiency of information about the control objects and the systems themselves, and also at improving the quality of generated control actions, requires optimization of control decisions for participants in the movement based on object-oriented modeling (OOM). The object-oriented approach involves the formation of an abstract representation of the structure of any system in the form of a hierarchy, which is necessary to study the functional interactions of its objects and their effects on the entire system as a whole [11]. The stages of the process of building a system hierarchy are as follows:

- The boundaries of the system are determined based on the initial information state (degree of certainty) for the given environmental conditions of the system functioning in the form of a formalized model of the source database based on possible solutions depending on the criteria being designed.
- The goals of the system are fixed as an assessment of the highest level of the hierarchy based on the significance of the interaction of various levels, and not directly dependent on the objects of the system at these levels.
- Criteria are identified as measuring instruments of the control process that affect the solution of system tasks. Obviously, each criterion should reflect the main function at its level in decision-making in the system, and not a secondary one.
- A hierarchy of criteria is defined. The simplest hierarchy is linear, ascending from one level of elements to a neighboring level. In a nonlinear hierarchy, the upper level can be both in a dominant position with respect to the lower level, and in a dominated one (for example, in the case of information flow). In the mathematical theory of hierarchies, a method is used to assess the impact of a level on a neighboring upper level by composing the corresponding contribution (priorities) of the lower level elements with respect to the upper level element. This composition can spread up the hierarchy.
- A set of technical indicators or formalized values is developed for each criterion, ensuring the requirements for the functioning of the system.
- A method for solving the problem is described, which consists in determining the permissible values of the parameters that define not only the performance of the system and its objects, but also the effectiveness of functioning in general, taking into account the goal-setting.

In the developed hierarchical structure of the system, the question of the optimal number of objects that are subordinate to the senior object and (or) located at the same level is always relevant. The number of such determines the degree of controllability in the system: the more of them exist, the less is the controllability in the system. On the other hand, a large number of seniority levels in subordination are justifiably undesirable, as it leads to an increase in the timeframe for promoting information in the system. In this case, the solution to the controllability problem in the system is determined by the effectiveness of the mathematical apparatus aimed at finding the optimal solution. [12]. If we consider the management process as a series of specific organizational and administrative measures and technical measures for the development of existing systems of the automobile-road complex, then in order to ensure their joint work, at least three main groups of criteria must be taken into account:

- completeness and effectiveness of implementation of the main functional tasks of a control system;
- volumes and complexity of work;
- possibility of development in the process of further functioning.

The initial decomposition of the hierarchy of the CS structure based on OOM for the development of solutions is shown in Figure 2.

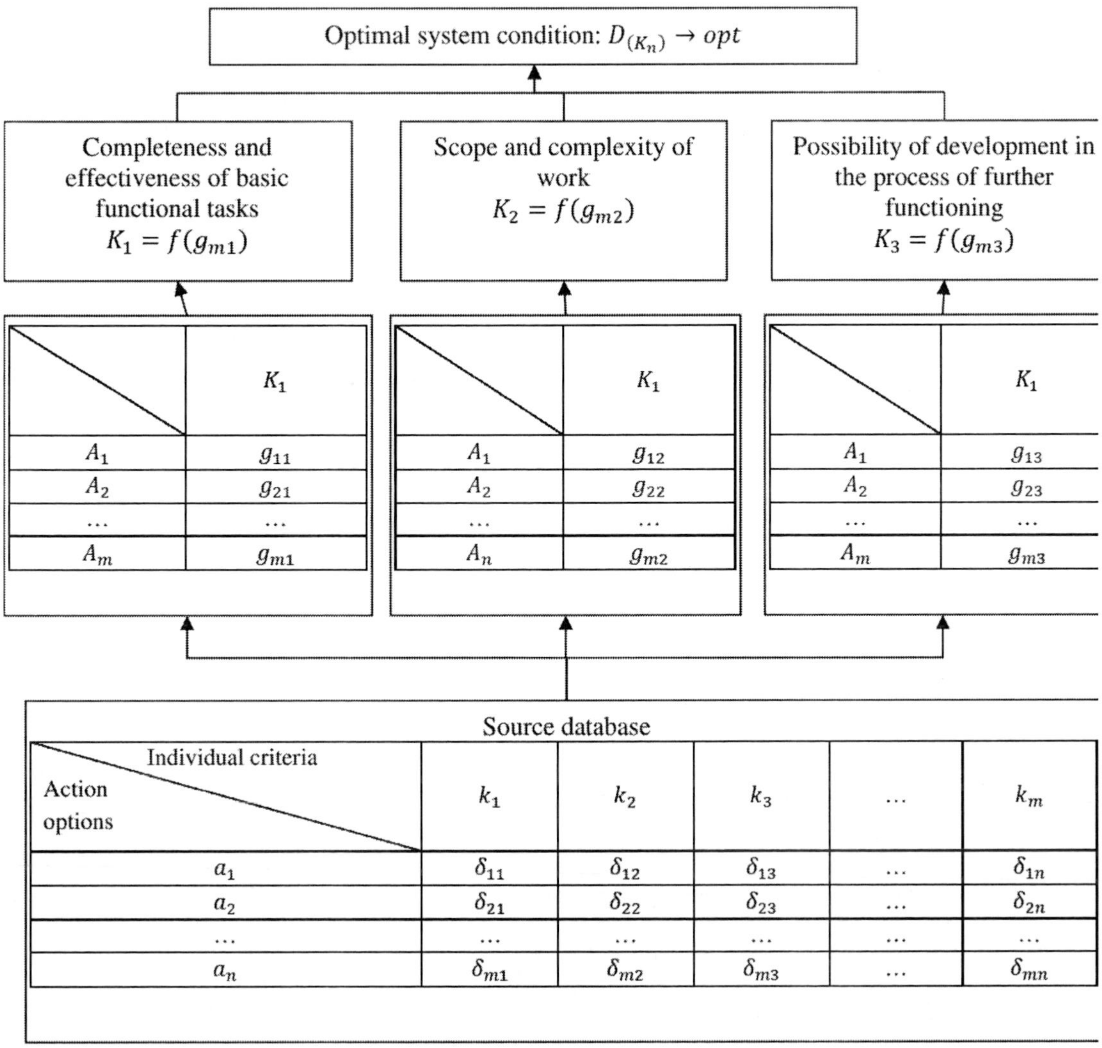

Figure 2. Decomposition of structure hierarchy of control system based on OOM

5. Conclusion

In the structure of the OOM-based control system presented in Figure 3, it is necessary to apply analytical methods for multi-criteria evaluation of possible options for operations, which allow operating with initial data and obtaining the optimal solution analytically. In real conditions, the effectiveness of an ACS functioning can be determined by a large number of criteria, in addition to those considered. Therefore, it is important that the developed methods investigate the field of probabilities of changes in informational situations of the environment and are applicable to factors of any degree of uncertainty [13], that is, rely on the Bayesian rule. In other words, if previously known information about a process exists, then it is possible to determine the probability of a new event in a given process more reliably.

$$P(A|B) = \frac{P(B|A)P(A)}{P(B)} \quad (1)$$

where $P(A)$ is the probability (a priori) of an event A; $P(A|B)$ is the probability of event A in case of event B or a posterior probability; $P(B|A)$ is the probability of event B in case of reality (occurrence) of event A; $P(B)$ – is the probability (full) of event B.

Application of the Bayesian rule will make it possible to remove objective difficulties in choosing a system for calculated cases that arise in the process of solving the multicriteria problem of evaluating actions in ACS under the condition of uncertainty [14]. Namely, a number of formal-logical methods is based on the Bayesian rule that allow one to analytically determine the probabilities of possible events when there is information about previous events or the state of the system as a whole [15].

References

[1] US Public Law 102-240 *Intermodal Surface Transportation Efficiency* Act of 1991
[2] US Public Law 105-178 *Transportation Equity Act for the 21st Century (TEA-21)* 1998
[3] *History of Intelligent Transportation Systems* U.S. department of transportation, report FHWA-JPO-16-329, 2016
[4] *Synthesis of Active Traffic Management Experiences in Europe and the United States* 2010 FHWA
[5] Active Traffic Management for Arterials 2013 *National Cooperative Highway Research Program*
[6] Building the ITI: *Putting the National Architecture into Action*, Mitretek Systems, FHWA, April 1996
[7] Romanov B L and Slobodetsky D Ya 2009 Interaction of objects in an object-oriented environment *Proc. of the Instit. for Syst. Anal. of the Russ. Acad. of Sci.* **45** 59–67
[8] Zaliznyak E 2005 *"Three Fat Men" DBMS occupied the market for life CNews* 15th August Retrieved from: http://www.cnews.ru/reviews/index.shtml?2005/08/15/184770
[9] Мейер Б 2005 *Object-oriented design of software systems* (Moscow: Russ. edition)
[10] GOST R ISO 14813-1-2011 *Intelligent transportation systems. The scheme of building the architecture of intelligent transport systems* Part 1 *Service Domains for Intelligent Transport Systems, Service Groups and Services* (Moscow: Standartinform)
[11] Kuzin M V 2009 Simulation model of coordinated traffic flows in the city road network *Inform. Technol. of model. and control* **4(56)** 502–8
[12] Mesarevich M 1973 *Theory of hierarchical multi-level systems* (Moscow: Mir)
[13] Kornikov V V, Seregin I A and Khovanov N V *Bayesian model for processing non-numeric, inaccurate and incomplete information on weight coefficients* Retrieved from: http://inftech.webservis.ru/it/conference/scm/2000/session3/kornikov.htm/
[14] Pegat A 2009 *Fuzzy Modeling and Control* (Moscow: BINOM, Lab. of knowledge)
[15] Karelina M Yu, Arifullin I V and Terentyev A V 2018 Analytical determination of weight coefficients in a multi-criteria evaluation of effectiveness of vehicles *Bull. of the Moscow Automobile and Road State Techn. Univer. (MADI)* **1(52)** 3–9

IOP Conf. Series: Materials Science and Engineering **832** (2020) 012059 doi:10.1088/1757-899X/832/1/012059

Development and testing of a conveyor for detecting various types of vehicles when transporting agricultural products from the field

I A Uspenskiy, G K Rembalovich, I A Yukhin, D S Ryabchikov and A S Stepashkina

Ryazan State Agrotechnical University named after P.A. Kostychev, 1, Kostycheva Str., Ryazan, 390044, Russia

E-mail: ivan.uspensckij@yandex.ru rgk.rgatu@yandex.ru rds_62@mail.ru

Abstract. Research purpose: development of a technology that will detect various types of vehicles in the image. To achieve this purpose it is necessary to solve the following objectives: Highlighting requirements for the technology being developed; Development of the technology for finding an object in an image; The choice of methods and algorithms for the allocation of objects for the development of the technology; The choice of methods and algorithms that allow detecting objects of a certain class; Implementation of the developed technology in the software system. The scientific novelty of the work lies in the application of previously known methods, which have been shown to be effective in other studies, to a new object of study, namely, to detect in images vehicles of various classes used for the transportation of agricultural goods.The practical value of the expected results lies in creating the software based on the developed technology, which makes it possible to detect vehicles of a certain class with a high degree of probability in a static image. The development of such software will automate part of the business processes and reduce labor costs. The following methods are used as the main ones in the developed technology: the HOG method (histograms of oriented gradients) and the support vector method (SVM).

1. Introduction

In modern conditions of technology development, people are increasingly striving to automate processes. Automation is understood as a process where previously performed functions are transferred by a person to some equipment and automatic devices. Automation surrounds us everywhere: from domestic to world-class issues.

Undoubtedly, high-tech devices have already been developed and are widely used, which make it possible to collect the necessary data on the parameters of the movement of vehicles from harvesting sites to storage units. For the operational management of freight transport during loading, it is necessary to know its carrying capacity in advance. Determining the class of the vehicle and its load capacity in an automated way will reduce human labor costs. However, all existing special instruments for measuring the intensity of the traffic flow have a high price. Therefore, in practice, the manual method of determining the class of vehicles is often used.

To reduce capital and labor costs, a decision was made to develop a technology for detecting various types of vehicles in the image [1].

The task of finding an object in an image is complicated by a number of factors: distortion of an object in an image, overlapping of objects with other objects, stray signals in the image, a variety of objects in the shape and color of objects of the same class, etc.

When creating automated systems several widespread approaches to search for objects in the image are used.

Depending on the method of obtaining the source image for solving the problem of finding objects, the following approaches are used.

Modeling the background. This approach is applied when a stable and slightly changing background, i.e. an image comes from a fixed camera.

Modeling an object. The approach is applied in cases where the background regularly and substantially changes. To solve some problems, it is appropriate to apply both approaches together, this can significantly improve the results.

2. The choice of the sequence of the main stages of the developed technology

The work of Andrey Vedaldi and Andrew Zisserman was chosen as a model for the technology to be developed [2].

In their work, the researchers set a goal: the discovery of objects of a certain category. Pedestrians, cars, traffic signs on the city streets were identified as search objects.

The source data for the program are images that depict the target search objects. At the output of the program the target objects are marked with a bounding box.

The problem of finding an object in an image is the search for the object by the detector, regardless of the location of the object, its scale in the image, as well as lighting, color, etc.

The search for objects in the image in the technology developed by the researchers is to search for objects in a training sample. That is, before detecting an object in the image, a statistical analysis of the image is performed. And then, to detect the object in the image, the obtained model of the object is applied with the help of a sliding window to all parts of the image at all possible scales [3].

Now let's pass to a more detailed study of the developed technology.

3. Basics of detection

The target objects of the search are vehicles of two classes: cars and trucks.

Two classes of cars were chosen as the initial data for training, and street pictures with cars were selected as test images.

To search for objects in the test image, the sliding window method will be used, based on the construction of the HOG feature extraction method (histograms of directed gradients).

In MATLAB, a training sample for practical use is loaded with the m-file loadData.m. The loadData (targetClass) function takes a targetClass argument that points to an object of the interest class.

The LOADDATA function loads data for training. LOADDATA (TARGETCLASS) loads data, configuring it for training. TARGETCLASS - the target class - is the vector of one or more labels. If several labels are specified, then several classes are combined into one. LOADDATA (TARGETCLASS, NUMPOSIMAGES, NUMNEGIMAGES) indicates positive and negative examples.

Figure 1 shows the elements of the training sample related to two classes of objects. On the left side of each figure there are examples of objects of one class (type) of cars and on the right side of each figure there is an average image of cars of one class.

Consider the first part of the program.

When working out the example1.m script, variables are loaded into the working space:

- trainImages – a list of training sample image names;
- trainBoxes – 4 × N array variable of bounding rectangles having coordinates $[x_{min}, y_{min}, x_{max}, y_{max}]$.
- trainBoxImages – the name of the object enclosed in the bounding box,

- trainBoxLabels – the label of the object for each bounding box (one of the indices in target class targetClass);
- trainBoxPatches – an array variable of image sections, one for each training object (in RGB format).

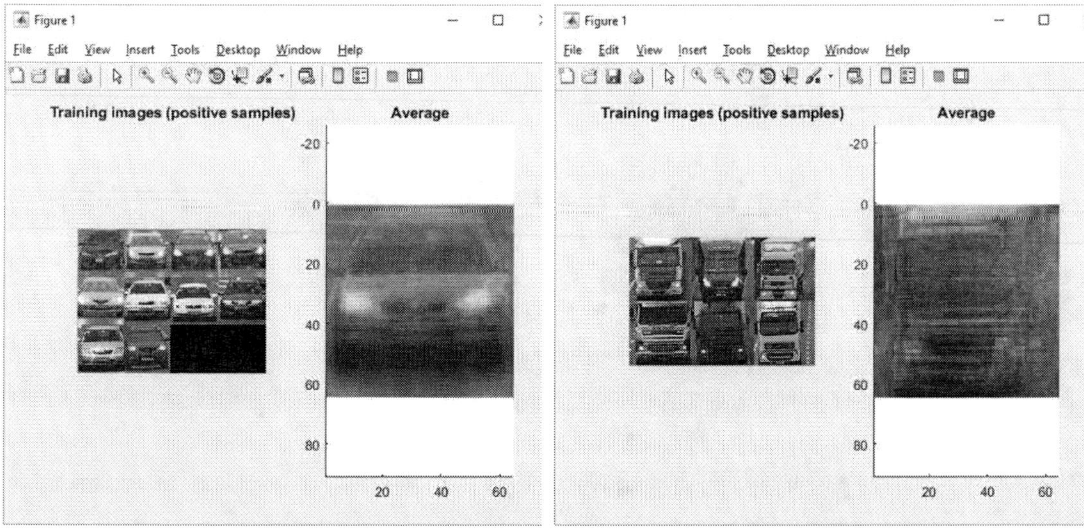

Figure 1. Elements of the training set of cars of two classes

A similar set of testImages, testBoxes variables was created for checking test images.

In step 1.1, an image is created that visualizes the complete training list of objects of one class and their average value.

Figure 2 shows the image obtained in step 1.1. The figure shows cars. All elements belong to the same class, on the right side of the image there is an average image obtained from the sample elements of this class.

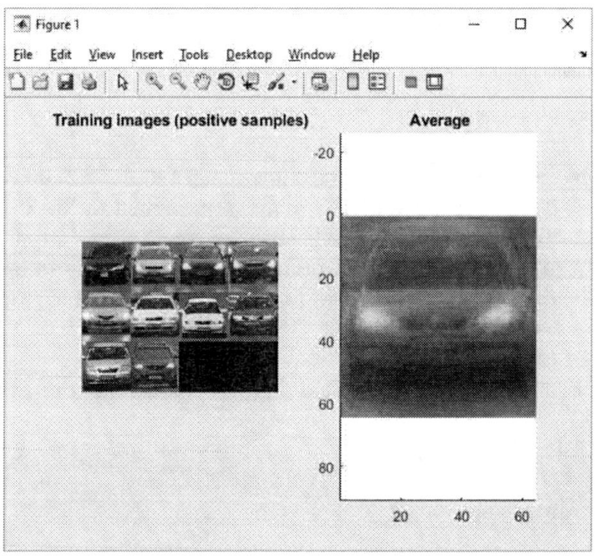

Figure 2. The image visualizing the complete list of training for objects of one class (cars) and their average value

The same actions are performed to obtain an averaged value for a training sample consisting of trucks (Figure 3).

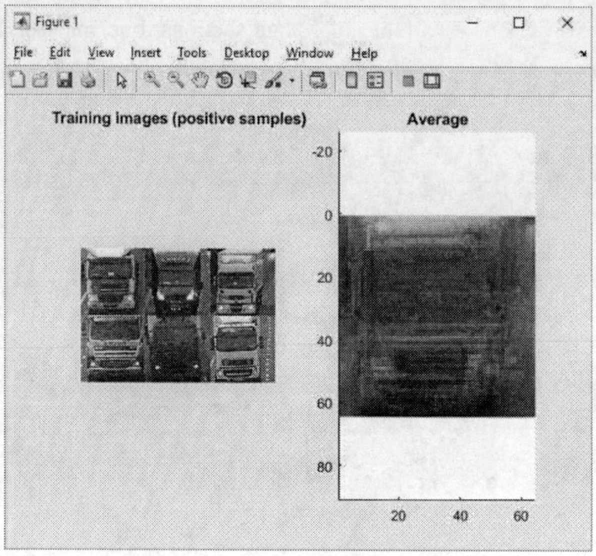

Figure 3. The image visualizing the complete training list of objects of one class (trucks) and their average value

4. Extracting HOG features from training images

In most cases, detectors work on top of a layer of low-level objects. In this case, a HOG detector (histograms of oriented gradients) is used. To obtain a HOG model of an object, objects are extracted from parts of the images corresponding to the existing training examples. This action is performed in the next cycle:

```
hogCellSize = 8 ;
trainHog = {} ;
fori = 1:size(trainBoxPatches,4)
trainHog{i} = vl_hog(trainBoxPatches(:,:,:,i), hogCellSize) ;
end
trainHog = cat(4, trainHog{:}) ;
```

HOG is computed by the VLFeatvl_hog (doc) function. This function takes as a parameter the pixel size for each HOG cell. It also accepts an RGB image represented in MATLAB as a w × h × 3 array variable (extracted as a fragment of trainBoxPatches patches). The output is the size of the array variable: w / hogCellSize × h / hogCellSize × 31. One array variable is extracted for each training sample, then the array variables are combined to form a 4D array variable (fourth dimension).

5. Building a model of the HOG pattern

A simple object model can be obtained by averaging the characteristics of the objects in the training sample.

This is done in the following way:

```
w = mean(trainHog, 4) ,
```

where *w* – the average value of HOG.

The model can be visualized by rendering w, like an array variable of HOG objects. Rendering (Engl. Rendering - "visualization") is a term used in computer graphics, denoting the process of obtaining images from a model using a computer program.

In this case, the operation is performed using render option vl_hog:

```
figure(2) ; clf ;
imagesc(vl_hog('render', w)) ;
```

The result of the program during this step is shown in Figures 4 (a car) and 5 (a truck).

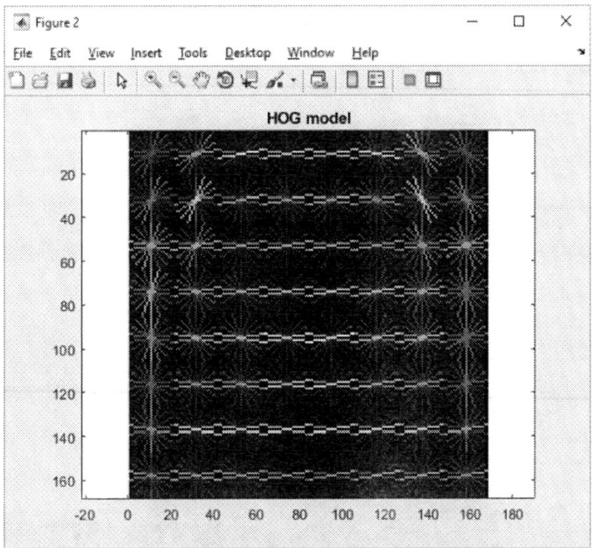

Figure 4. The result of building a model of the HOG pattern of a car

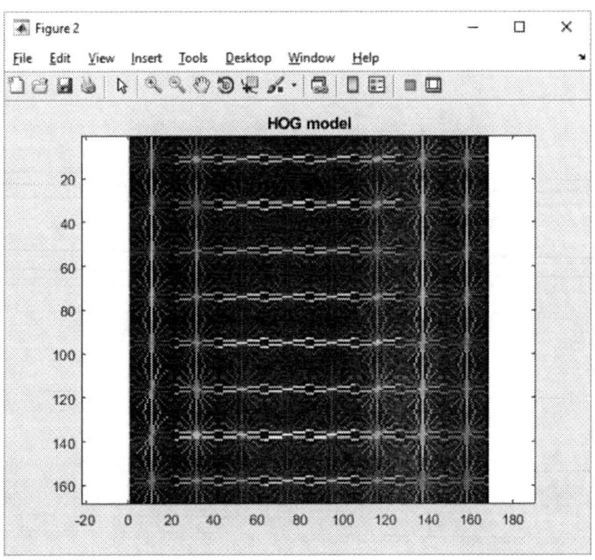

Figure 5. The result of building a model of the HOG pattern of a truck

IOP Conf. Series: Materials Science and Engineering **832** (2020) 012059 doi:10.1088/1757-899X/832/1/012059

6. Applying the HOG model to a test image

The model is compared with the test image by: extracting the HOG model of the image and convolution of the model according to the resulting map of objects:

```
im = imread('data/test.jpg');
im = im2single(im) ;
hog = vl_hog(im, hogCellSize);
scores = vl_nnconv(hog, w, []);
```

The first two lines read an example image and convert it to one format. The third line computes the HOG model using vl_hog described above. In the fourth line, the HOG convolution from pattern w is compared with the HOG model of the test image, returning a response map.

The result of the program at this step is shown in Figures 6 and 7.

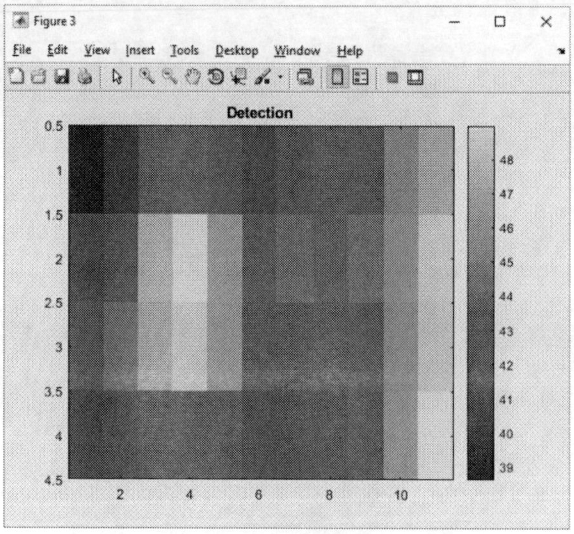

Figure 6. Car image response map

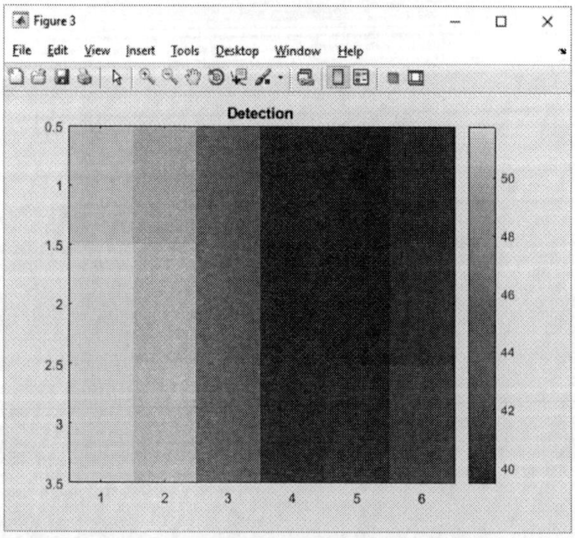

Figure 7. Truck image response map

On this map, the image area is highlighted in yellow, where the HOG model of the training sample examples is most identical to the HOG model of the test image area, the opposite value is shown in blue.

7. Selecting an object in the image

After receiving the system response map, the image area is found where the maximum response is observed and the bounding box of the image section containing the corresponding HOG model is calculated.

The maximum response is:

```
[best, bestIndex] = max(scores(:);
```

It is worth noting that bestIndex is a linear index in the range [1, M], where M is the number of possible filter locations.

The lower left coordinate of the bounding box (h_x,h_y) is calculated in MATLAB using function ind2sub:

```
[hy, hx] = ind2sub(size(scores), bestIndex).
```

Coordinates (h_x,h_y) are calculated in units of HOG cell values. Converting values to pixel coordinates is as follows:

```
x = (hx - 1) * hogCellSize + 1;
y = (hy - 1) * hogCellSize + 1.
```

The size of the model pattern in the coordinates of the HOG cells can be calculated in several ways and one of them is as it follows:

```
modelWidth = size(trainHog, 2) ;
modelHeight = size(trainHog, 1).
```

Thus, one gets enough information to calculate the bounding box:

```
detection = [ x - 0.5 ;
  y - 0.5 ;
  x + hogCellSize * modelWidth - 0.5 ;
  y + hogCellSize * modelHeight - 0.5 ;].
```

A bounding box spans a portion of the image including the exact size of the HOG model. In MATLAB, pixel centers have integer coordinates, and the pixel borders are ± 0.5.

According to the results of step 1.5, the area of the vehicle of one of the classes is selected on the test image as a bounding box of rectangles (Figures 8 and 9).

The result of the program is an object enclosed in a bounding box on the test image. However, the search for the object is carried out correctly, provided that the size of the object in the image matches the size of the HOG model, that is, it does not scale.

8. Image scalability and application of SVM reference vector methods

The implementation result in the first part of the program gave a positive result, however, the algorithm described above does not work well when changing the scale of an object in the image.

In the second part, we expand the detector for searching for objects in cases of zooming and improve the algorithm for searching for an object in the image by using the support vector method. To get started, data for cars wil be uploaded:

N = 10; % Number of images

```
trainBoxPatches = single(zeros(64,64,3,N));
fori = 1:N
trainBoxPatches(:,:,:,i) = ...
imread(['data/LittleCars/Car00' int2str(i - 1) '.jpg']);
end
```

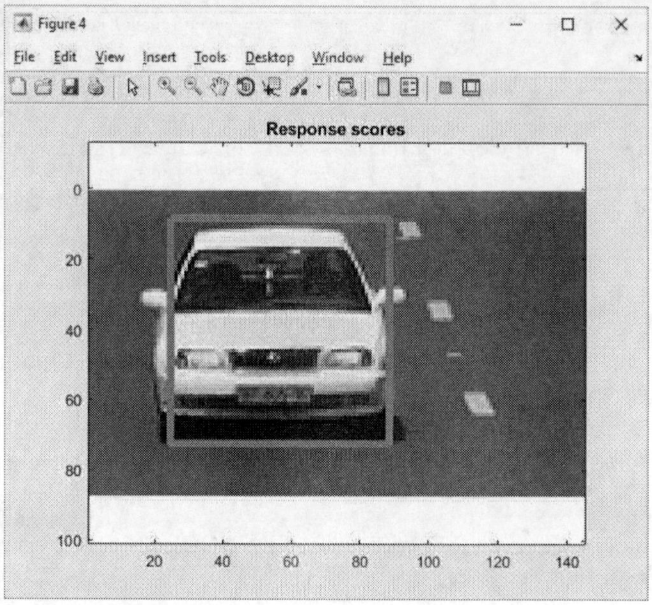

Figure 8. The result of detecting a car in the image using the construction of the HOG model

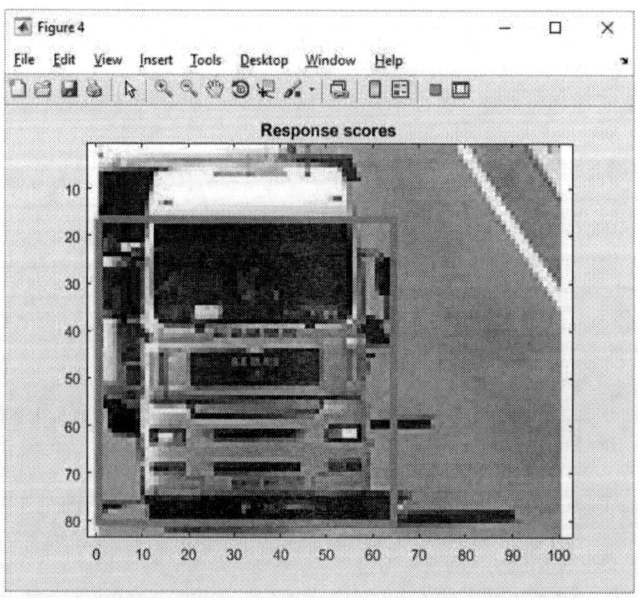

Figure 9. The result of detecting a truck in the image using the construction of the HOG model

For trucks:
N = 6; % Number of images

```
trainBoxPatches = single(zeros(64,64,3,N));
fori = 1:N
trainBoxPatches(:,:,:,i) = ...
imread(['data/Truck/truck00' int2str(i - 1) '.jpg']);
end
```

Just as in the example described above, a model of the HOG pattern (simple model) is built. Since the models will be identical to the models built in the example above (Figures 12 and 13), they are not given.

9. Multiscale detection

In most cases, the size of the objects in the image differs from the size of the desired pattern. To solve this problem, the image is scaled in all directions and objects are searched continuously.
Набор искомых масштабов определяется следующим образом:

```
% Scale space configuraiton
minScale = -1 ;
maxScale = 3 ;
numOctaveSubdivisions = 3 ;
scales = 2.^linspace(...
minScale,...
maxScale,...
numOctaveSubdivisions*(maxScale-minScale+1)).
```

Given *w* model defined in part 1 and using detectAtMultipleScales function, objects are searched on several scales:

```
detection = detectAtMultipleScales(im, w, hogCellSize, scales).
```

As in Part 1, based on a comparison of the HOG convolution from *w* pattern and the HOG model of the test image, a response map is constructed at different scales (Figure 10).

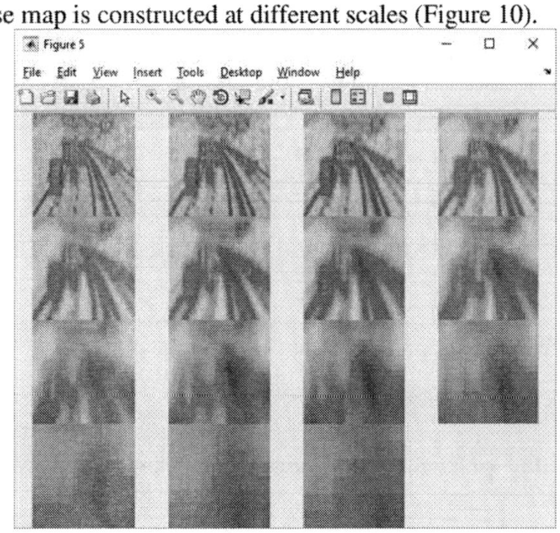

Figure 10. Making a response map

IOP Conf. Series: Materials Science and Engineering **832** (2020) 012059 doi:10.1088/1757-899X/832/1/012059

Based on the comparison of HOG models, objects are searched on the test image. The result of the program at this stage is shown in Figure 11.

Figure 11. Simple detector output

As a result of the operation of a simple detector, the image area is highlighted on the test image, where in fact the desired object is located.

10. Acquisition of positive and negative training results

Now that there is positive and negative data, you can apply the SVM model for the developed algorithm. Function vl_svmtrain will be used for this.

The function requires that the data be combined into a matrix of dimension D × N, where D is the size of the object, and N is the number of training points, i.e.:

```
% Pack the data into a matrix with one datum per column
x = cat(4, pos, neg) ;
x = reshape(x, [], numPos + numNeg).
```

A vector of binary labels is constructed, where value +1 is assigned for positive points and –1 for negative points:

```
% Create a vector of binary labels
y = [ones(1, size(pos,4)) -ones(1, size(neg,4))].
```

Set the value of parameter λ for the SVM algorithm, and use it instead of equivalent parameter C:

```
numPos = size(pos,4) ;
numNeg = size(neg,4) ;
C = 10 ;
lambda = 1 / (C * (numPos + numNeg)).
```

The construction of the HOG model is carried out using the SVM reference vector method (Fig. 12).

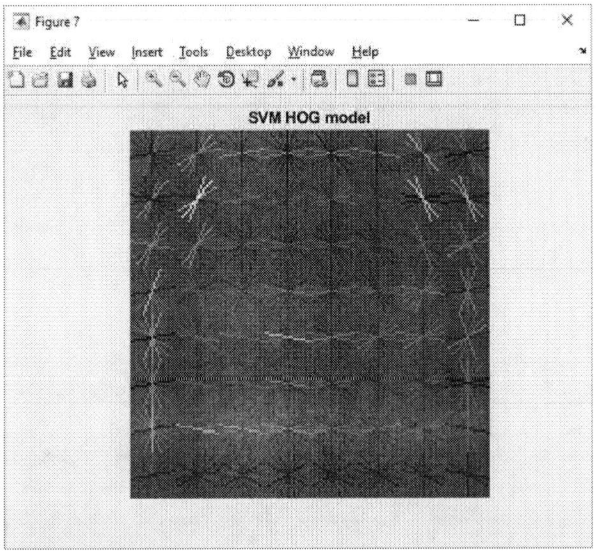

Figure 12. Building a HOG model using the SVG support vector method

Again, based on a comparison of the HOG convolution from *w* pattern and the HOG model of the test image, a response map is constructed at different scales (Figure 13). And the search for objects in the test image is performed (Figure 14 for the search for a car and Figure 15 for the search for a truck).

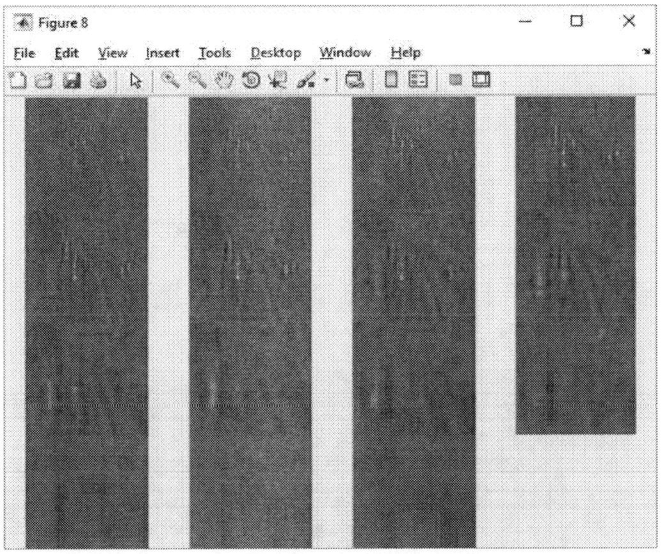

Figure 13. Making a response map at different scales

It can be seen from the results of the algorithm that the joint use of two methods (the oriented gradient method and the support vector method) makes possible to obtain more accurate results of determining the desired objects in the image at different scale values.

From the above example, one can see that the detector recognizes objects in those parts of the image where in fact they do not exist. Therefore, it is necessary to assess the adequacy of the detector. For this, the Pascal criterion VOC with the calculation of the average accuracy value (AP) is used.

Multiple testing of the developed technology made it possible to establish that even with a small training sample, the accuracy of detection of objects of a certain class is quite accurate.

In most cases, objects located in the foreground, i.e. had a greater system response, were detected. The use of only HOG method does not allow to detect objects in the image accurately, as with the simultaneous use of two methods (SVM + HOG). The combined use of two methods shows high efficiency, since it describes the parameters of objects more accurately.

Figure 14. The result of the object search using the oriented gradient method (HOG) and the support vector method (SVM) for the search for cars

Figure 15. The result of the object search using the oriented gradient method (HOG) and the support vector method (SVM) for the search for trucks

IOP Conf. Series: Materials Science and Engineering **832** (2020) 012059 doi:10.1088/1757-899X/832/1/012059

11. Conclusion

Based on the results of this work, the task of developing a technology for detecting various types of vehicles used for the transportation of agricultural goods in a transport stream using images was solved. The practical implementation of the technology made it possible to draw conclusions about the effectiveness of the developed technology, which allows it to work correctly even with a small training sample. The task set has been completely solved: according to the results of the work, a software product was implemented and its effectiveness was confirmed by tests.

Based on the results of the work, the following conclusions are made:

1. The technology for detecting various types of vehicles (trucks and cars) in a static image has been developed.

2. The developed technology for searching for objects in an image using the calculation of the HOG descriptor has shown its effectiveness only in images where the overall dimensions (in pixels) of the desired object coincide with the dimensions of the constructed HOG model based on the elements of the training sample.

3. The developed technology for searching for objects in the image using two methods (oriented gradients and reference vectors) has shown great efficiency in detecting the desired objects. This result was achieved due to the multiple scaling of the image and the search for the object due to the sliding window.

4. The developed technology makes possible to satisfy the requirements for the detection system of an object in the image.

References

[1] Viola P 2001 Rapid Object Detection using a Boosted Cascade of Simple Features *Accepted Conf. on Computer Vision and Pattern Recognit. (CVPR 2001)* pp 511–8

[2] *Object category detection practical* Retrieved from: http://www.robots.ox.ac.uk/~vgg/practicals/category-detection/index.html#part2.

[3] Hartley Richard and Zisserman Andrew 2004 *Multiple View Geometry in Computer* (Vision Cambridge University Press)

IOP Conf. Series: Materials Science and Engineering **832** (2020) 012060 doi:10.1088/1757-899X/832/1/012060

Method for determining integral damage from road accident

L V Zhivoglyadova[1], V G Zhivoglyadov[2] and V V Silyanov[3]

[1]KVVU, 4, Krasina str., Krasnodar, 350063, Russia
[2]Academy of Science and Technology and social progress, 4, Kalinina str., Krasnodar, Russia
[3]Moscow Automobile and Road State Technical University (MADI), 64, Leningradsky ave., Moscow, 125319, Russia

E-mail: Zhivogladova @mail.ru, jivogljadov_vg@mail.ru

Abstract. In practice, the inspection of the scene of an accident, the conclusions, the collection of materials and execution are carried out, as a rule, by measuring electronic roulette, filming and visual methods without instrumentation, which does not allow to obtain the true (direct and indirect) causes that caused the accident, dimensions failures in the organization and management of traffic, both on the streets of its occurrence, and on the streets parallel to it, and the consequences of an accident, both on the transport involved in the incident and on other vehicles transport vehicles and their passengers affected by this incident. This article presents a method for determining the integral damage from a road accident taking into account the participating vehicles, drivers and passengers, as well as thousands of traffic participants (vehicles and passengers) who suffered damage due to a failure in the organization and management of traffic (factors, mileage), as well as road conditions that do not meet the requirements of traffic safety.

Each traffic accident (traffic accident) is associated with the restriction of traffic flows not only in the traffic lanes of the carriageway at the place of its occurrence, but also with an increase in the density of traffic flows, reducing their speed, and correspondingly increasing the travel time associated with speed and mileage, as well as higher cost of the final transport product [1, 2, 5]. On the street where the accident occurred, as a rule, a traffic jam (oncoming traffic) condition occurs. The oncoming streams of this street, leaving (right and left) through intersected (transverse) streets to parallel streets, where there is an opportunity for their passage. With the distance from the street of the accident, the flux density – resorption decreases, reaches up to 2–3 km in the transverse and longitudinal directions (figure 1).

The causal relationship of an accident, as a rule, is based on material with reliability. The unsatisfactory technical condition of the transport (according to the traffic police of the Ministry of Internal Affairs of the Russian Federation) [4] accounts for 0.7...1.5 % of all accidents committed in the Russian Federation. For this reason, 17...27 % – in Germany, 13...21 % – France, 22 % of all accidents committed in these states were committed in the USA. The reason is lack of technical competence or lack thereof.

Road conditions, professionalism and law-abiding drivers of vehicles, the human factor (the driver's ethical condition, attentiveness in the process of driving) are the basic guarantors of ensuring traffic safety.

IOP Conf. Series: Materials Science and Engineering **832** (2020) 012060 doi:10.1088/1757-899X/832/1/012060

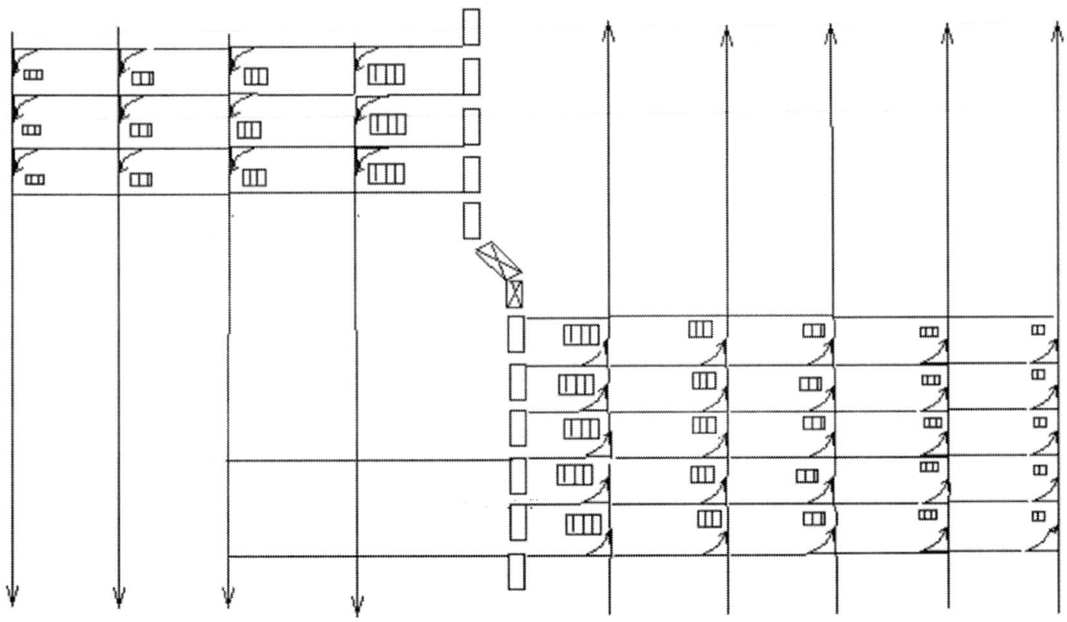

Figure 1. The scheme of redistribution of traffic flows from the street of the accident to the transverse and parallel streets (in the plan)

These parameters should be set during the inspection of the scene of the accident and the preparation of a conclusion on the true causes. The duty units of the traffic police of the territorial bodies are not always equipped with traffic control engineers, automotive engineers, and therefore these parameters are not available in the materials of road accidents.

The services of duty units of the traffic police traffic police inspection of places of road traffic accidents (DTP) and their execution is carried out, as a rule, by visual [3] method, video shooting, without proper instrumentation, which does not allow to collect and present reliable materials, characterizing their material involvement, preliminary and emergency situations of accident processes not only in the contours of the places of their implementation, but beyond them, as a failure of the traffic management system, the redistribution of traffic flows, which resulted in reprogram, congestion, and sometimes traffic accidents associated with the loss of time of passengers in vehicles due to traffic accidents and malfunctions. At the same time, often the establishment of the true cause of the accident, as well as related factors and the extent of the consequences, is not ensured. As a rule, a charge is brought against a driver or a pedestrian, while neither one nor the other is responsible for the organization of traffic, for the creation of road conditions that meet the requirements of road safety (road lighting, a visibility triangle at intersections, overloading the information field on lanes of carriageways, the degree of serviceability of the surface layer of the carriageway, coefficient of adhesion, provision with bends on curved roads, respectively, their radii, etc.). The consequences of an accident are usually limited to the loss of working capacity of the passengers participating in it and damage to vehicles. All this negatively affects the conduct of preventive measures and, accordingly, the reduction of casualties on the roads of cities and towns and beyond. Each accident causes damage to tens of thousands of vehicles due to failures in the organization and management of traffic and passengers of these funds, which is the focus of this scientific study.

Despite the significance and mass character of the size and consequences of road traffic accidents, damage to road traffic from a significant part of the Russian population is not established and is not reflected in the statistics. On the inspection schedule, the list of materials and instrumentation that ensure their reliability, registration of incidents on the basis of software and methodological principles aimed at preventing emergencies in the aspect of emergency safety, allowing to establish direct and indirect causes of the accident, as well as the amount of integral damage, allowing at the same time to determine and develop targeted (priority and promising) – long-term comprehensive measures, standards are not developed us. There is no standard for the competence (qualification) of an executor who has the right to draw up accidents, produce opinions, and conclusions.

Since traffic is a multifunctional, multi-factorial, multi-parameter and multi-level process that inspects the traffic police, the device must be equipped with competent engineers who know the theory of movement of transport and pedestrian flows and the functioning of the road transport complex [5].

The formula for determining the damage (integral) caused by a traffic accident to a society, a city (district), road facilities, traffic, ecology can be presented as follows:

- for one i-th accident

$$Y = \sum_{1}^{365}\sum_{i=1}^{I}\sum_{k=1}^{K}\sum_{1,2,3}\sum_{z=1}^{Z}\sum_{l=1}^{L}\sum_{m=1}^{M}\sum_{t=1}^{T}\sum_{j=1}^{J}\sum_{h=1}^{H} [((\lambda_1[(q^{\lambda}{}_1\cdot\delta^{\lambda}{}_1) + [(t^{\lambda}{}_1\cdot\delta^{\lambda t}{}_1)]+[(\mu^{\lambda}{}_1\cdot(b^{\lambda}{}_1\cdot\delta^{\lambda}{}_1)) +$$
$$+ (\mu^{\lambda}{}_2\cdot(b^{\lambda}{}_2\cdot\delta^{\lambda}{}_2))+\ldots+(\mu^{\lambda}{}_n\cdot(b^{\lambda}{}_n\cdot\delta^{\lambda}{}_n))]+ [(\mu^{p}{}_1\cdot(b^{p}{}_1\cdot\delta^{p}{}_1)) + (\mu^{p}{}_2\cdot(b^{p}{}_2\cdot\delta^{p}{}_2)) +\ldots+ (\mu^{p}{}_n\cdot(b^{p}{}_n\cdot\delta^{p}{}_n))] +$$
$$+ [(\mu^{p}{}_1\cdot(t^{p}{}_1\cdot\delta^{p}{}_1)) + (\mu^{p}{}_2\cdot(t^{p}{}_2\cdot\delta^{p}{}_2)) + \ldots+ (\mu^{p}{}_n\cdot(t^{p}{}_n\cdot\delta^{p}{}_n))] + [(y_k{}^{1,2,3} + y^{\lambda}{}_z + y^{\lambda}{}_l + y^{\mu}{}_z + y^{l}{}_m + y_j + y_h + y_c)],$$

where 1, 2, 3 – number of the knee (gender, genealogy);

$y_k{}^{1,2,3}$ – damage caused to the k-th relative 1, 2 and 3 the knee (clan), thousand rubles;

$y^{\lambda}{}_z$ – damage caused to the λ-th vehicle associated with damage to its technical condition and delays Z due to the i-th accident, as well as congestion situations, thousand rubles;

$z = 1, 2, \ldots, Z$ – the duration of the delay, h;

$\lambda = 0.1; 0.2;\ldots; 0.7$ (λ_{max}) = 360, 720,…2520 (λ_{max}) – traffic flow, TE/s, TE/h;

$k = 1, 2,\ldots, K$ – the number of the relative, of the 1, 2, 3-th knee;

$q^{\lambda}{}_1$ – a list of malfunctions (consequences) in the first vehicle that arose as a result of the i-th accident;

$\delta^{\lambda}{}_1$ – the cost of the cost of eliminating technical malfunctions (consequences) that arose in the first vehicle related to the i-th accident, thousand rubles;

$i = 1, 2, \ldots, I$ – accident number;

$t^{\lambda}{}_1$ – the duration of the preservation of the place of the first accident, h;

$\delta^{\lambda t}{}_1$ – the cost of one hour to save the accident scene, thousand rubles;

$\mu^{\lambda}{}_1$ – the number of drivers in the first vehicle participating in the accident;

$b^{\lambda}{}_1$ – a list of injuries, damage caused by the drivers of the first vehicle of the i-th accident;

$\delta^{\lambda}{}_1$ – the cost of loss of health (injuries) and damage, thousand rubles;

$\mu^{\lambda}{}_2$ – the number of drivers in the second vehicle participating in the i-th accident;

$b^{\lambda}{}_2$ – a list of injuries and damage caused by the driver of the second vehicle in the i-th accident;

$\mu^{\lambda}{}_n$ – the number of drivers in the n-th vehicle participating in the i-th accident;

$b^{\lambda}{}_n$ – a list of injuries and damage caused to drivers of the n-th vehicle in the i-th accident;

$\delta^{\lambda}{}_2$, $\delta^{\lambda}{}_n$ – the cost of treating injuries and damage caused by drivers of the second and n-th vehicle in the i-th accident, thousand rubles;

$\mu^{p}{}_1$, $\mu^{p}{}_2$, $\mu^{p}{}_n$ – the number of passengers in the first, second and n-th vehicle participating in the i-th accident, people;

$b^{p}{}_1$, $b^{p}{}_2$, $b^{p}{}_n$ – a list of injuries and damage caused to passengers of the first, second and n-th vehicle involved in the i-th accident;

$\delta^{p}{}_1$, $\delta^{p}{}_2$, $\delta^{p}{}_n$ – the cost of treating injuries and damage caused to passengers of the first, second and n-th vehicle participating in the i-th accident, thousand rubles;

$t^{p}{}_1$, $t^{p}{}_2$, $t^{p}{}_n$ – the duration of the delays of passengers in the first, second and n-th vehicle involved in the i-th accident, h;

$\delta^{p}{}_1$, $\delta^{p}{}_2$, $\delta^{p}{}_n$ – the cost of an hour of passenger delays in the first, second and n-th vehicles participating in the i-th accident, thousand rubles;

$y^{\lambda}{}_l$ – damage to the n-th vehicles related to their mileage within the l-extent, thousand rubles;

l=1, 2, …, L – the length of the mileage of the n-th vehicles associated with the i-th accident, km/TE;

λ =1, 2,…, λ_{max} – the intensity of the movement of vehicles on each lane of the carriageway in the j-th direction associated with rerun at the i-th accident, TE/h, TE/s;

y^{μ}_{z} – the delay value of the μ-th passenger in vehicles participating in the i-th accident, as well as those related to the rerun of other vehicles in this accident, where μ means the delay value of the μ-th, i.e. passenger number, hour/pass., (z =1, 2, …, Z);

y^{t}_{m} – damage caused to the t-legal and m-th individuals (owners of vehicles), thousand rubles;

y_j – damage caused by the failure of traffic organization on the lanes of the carriageway of the j-th direction to the city (district) road traffic, utilities, including the carriageway of the roads and the environment, thousand rubles;

j =1, 2, 3 – number of the object that suffered damage, respectively, traffic, urban (district) economy, ecology;

y_h – damage caused to productivity (of the final transport product) and associated with the loss of working time of h-persons involved in road accidents, registration, visiting emergency rooms, hospitals, etc. – disability, thousand rubles;

h= 1, 2, …, H – number of the person involved in the commission, execution of an accident that has lost its working capacity, and sometimes death;

y_c – costs of investigative, investigative and judicial proceedings (actions) on the i-th accident, thousand rubles;

c =1, 2, 3 – number of the event, respectively: inquest, investigative and judicial.

On the roads of municipalities of cities and rural settlements and between them in the Russian Federation, about (on average) 200 thousand accidents occur each year with serious consequences, about 30 thousand with a fatal outcome and more than 200 thousand people received various degree of injury. At the same time, unjustified material and moral damage is caused to road users, road and public utilities, the environment and society as a whole (table 1).

Not all employees of the duty department of the traffic police traffic police have comprehensive knowledge in organizing and managing the processes of traffic and pedestrian flows, ensuring road conditions within the requirements of regulatory installations, the functioning of aggregates and systems, trouble-free driving, and the assessment of movement as vehicles (any models), pedestrian flows, and the safe movement of passengers on all types of urban passenger transport.

Table 1. The number and consequences of road accidents in traffic in the Russian Federation [4]

Years	Number of accidents	Consequences of accidents	
		Dead people	Wounded people
2008	210000 218322 [3]	29936	270883
2009	210500 203603 [3]	26084	257062
2010	211000 199431 [3]	26567	250635
2011	199868 224035 [3]	27957	251848
Change in 2011 to 2010, %	5,3	5,2	0,5

It is necessary to acquire knowledge not only of the content of standards and normative settings in the field of traffic safety, but also to solve the problems of their implementation in production in vertical and horizontal planes.

The lack of in-depth knowledge among traffic police officers does not allow, not only to evaluate, but also to ensure the collection of reliable information at all stages of the accident. In parallel, other employees study and confirm the facts of the consequences of an accident (the extent of damage to the vehicles of their passengers and pedestrians and other vehicles of their passengers who lost time in traffic jams, reruns) due to the redistribution of flows due to the accident.

It is advisable to develop a program-methodical complex of regulated solutions to problems associated with establishing the true causes of the accident and the size of the injuries to road users, and material damage to road traffic and utilities.

In the last century, there was a standard for determining braking distance:

$$L = K{\cdot}V^2/(254{\cdot}(\varphi \pm i)),$$

where $K = 1{\div}2{,}5$ – the coefficient of wear of the brake system;

V^2 – square of the speed of movement, km²/h²;

$254 = 2{\cdot} 3.6^2 \cdot 9.81$ (3.6^2 – km /h to m/s conversion, 9.81 m/s² – body gravity acceleration);

$\varphi = 0.1 \ldots 0.3$ – coefficient of adhesion of the roadway;

i – descent or ascent.

At the end of the last century, the standard was adjusted:

$$L = V^2/(26{\cdot}j_{ust}),$$

where V – the speed of movement, km/h,

V^2 – square of the speed of movement, km²/h²; $26 = 2 \cdot 3.6^2$;

j_{ust} – steady-state acceleration 5 or 5.8 m/s².

These formulas are empirical. The kinematic energy formula is used:

$$E= \frac{1 \cdot m \cdot V^2}{2},$$

where m – the mass;

V^2 – square of the speed of movement, km²/h².

According to two standards, these formulas did not provide the required reliability, one of the key parameters characterizing the serviceability of the vehicle's brake system under loading conditions and without it, the state of tires and the value of adhesion coefficients, which directly affect the length of the braking distance. Despite the importance of this system, there are no monitoring systems for estimating the length of the braking, from both the state of the braking system, the degree of its loading, the speed of movement, from the nose of tires and the coefficient of adhesion.

Conclusion

In our opinion, it would be justified by the Ministry of Transport of the Russian Federation to find the opportunity to allocate allocations in the order of fulfilling the priority tasks for the development of:

- systems for monitoring the state of the brake system, such as the braking length, depending on the degree of loading of vehicles (cargo, passenger, passenger), tire tread wear, speed, both on horizontal sections of roads and on slopes;
- a software and methodological complex that optimizes the solution of tasks related to the design of accidents (establishing a pre-emergency, emergency and post-emergency situation), determining the severity of the injured and the extent of damage both to the participants in the accident and to other traffic flows and their passengers due to mileage and delays in congestion associated with this accident;
- registration of an accident, inspection of vehicles, establishment of direct and indirect causes of the accident should be carried out by specialists (engineers) in road transport or traffic management.

It would be right for the Ministry of Transport of the Russian Federation to take on the mission to streamline the training of transport drivers at the level of standards, in particular: to take into account the drivers of buses, trams, trolleybuses and minibuses, to transport a large (18–200 people) number of passengers. This work is carried out on intense routes with high traffic intensity, training should be carried out according to specially developed programs for at least 1.5 years, which, during the movement, will exclude not only conflict situations, but also prevent traffic accidents.

IOP Conf. Series: Materials Science and Engineering **832** (2020) 012060 doi:10.1088/1757-899X/832/1/012060

Given there is mass profession (the driver of the subway, tram, trolley-bus, driver-master of the bus, minibus taxi and taxi driver), (driver-master of the truck (carrying capacity is 5–12 tons), the driver of a car), we develop and publish the state standard for these professions and their training in schools of passenger transport on the basis of universities, institutes, in schools of freight and passenger transport on the basis of universities, institutes and technical schools (colleges).

References

[1] Feofilova A A 2012 *On models for the dynamic transfer of traffic flows to free sections of the UDS in the event of congestion situations Organization and safety of traffic in large cities* (St. Petersburg: Publishing house of SPb state architect-building-it Univ.) pp 211–7

[2] Zhivoglyadov V G, Zhivoglyadova L V and Vladimirov V F 2013 The analytical base for calculating and evaluating the final transport product VINITI RAN *Transport: sci., technol. Managem. Sci. Inform. Coll.* **5** 47–80

[3] Kravchenko E A, Kravchenko A E, Barseghyan A R, Evich O M 2013 Analysis of the production of buses and factors affecting the accident rate on the road network in Sochi VINITI RAN *Transport: sci., technol., managem. Sci. Inform. Collection* **3** pp 56–61

[4] Oleshenko E M and Kravchenko P A 2012 Innovative elements in regional programs for road safety *Mater. X Int. sci.-pract. conf. "Organization and road safety in large cities:* resources and capabilities" (St. Petersburg) pp 37–48

[5] Zhivoglyadov V G 2005 *The theory of traffic and pedestrian flows* (Rostov-Don: Universities North-Caucasus region Engineering) 1082 p

IOP Conf. Series: Materials Science and Engineering **832** (2020) 012061 doi:10.1088/1757-899X/832/1/012061

Analysis of reliability and sustainability of organizational and technical systems of railway transportation process

V A Olentsevich, Yu I Belogolov, N N Grigoryeva

Irkutsk State Transport University, 15 Chernyshevsky Street, Irkutsk, 664074, Russia

E-mail: olencevich_va@irgups.ru

Abstract. The increase in the throughput and processing capacity of railway stations in many cases is solved through organizational and technical measures, which makes it possible to solve this problem only partially, while limiting the possibility of increasing the volume of work in the future. Private funds should become investors in the creation of specialized points for bulk staging and preparation of station cars. Car owners and manufacturers, leasing companies, and rolling stock operators are interested in creating this type of stations. The layout of the stations specialized for the car staging should ensure the implementation of a complex of processing operations: reception, sorting, dispatch, technical inspection, all types of maintenance and repair. The article presents the dynamics of the throughput capacity of the East Siberian Railway. It analyzes the indicator un-der consideration with parallel and non-parallel schedules, which determined the main factors affecting the throughput capacity of railway stations and sections of the East-Siberian Railway and the degree of their significance. To determine the main ac-teal reasons for reducing the throughput capacity of railway stations and improving the quality of use of the indicator in question, the Ishikawa diagram was constructed. It provides a systematic approach to determining the root causes of the problem under consideration based on data on the throughput capacity of railway stations in Russia in 2013-2015. The analysis of the Ishikawa diagram identified the main factors: mismatch in the level of development of the federal railway network; errors in the management of car fleets; excess of car fleets; non-uniform arrival of car traffic volumes; decrease in the quality of car fleet management; private car fleet growth.

1. Introduction

The possibility of the increase in the throughput and operational capacity indicators of railway stations and hubs is currently solved in most cases, primarily through organizational and technical measures that do not require significant financial investments. Reconstructive measures related to capital investments in the construction of new fixed assets, to the construction and technical equipment of the arrangement of structural units of the industry are generally not to be considered. This technical and economic approach solves the problem only for today, while limiting the possibility of increasing the volume of work and activities of most railway stations in the future.

In the face of strong and open opposition to the accelerated development of Russia by the West and economic stagnation, the creation of a modern transport infrastructure can give an additional impetus to the development of the country. This area will most likely stimulate the development of the real sector of the economy, small and medium-sized businesses and result in the development of new areas of residence. For stable and rapid development of the country, investments in the railway transport system should be made at a faster pace.

According to statistics, the current situation is such that at least 80% of railway stations in the trunk lines do not have the sufficient track arrangement [1-4]. The experience of other countries shows that it is necessary to strengthen the technical equipment of transport structural units during a decrease in freight traffic or until the time when the load on the equipment of the stations approaches a critical level [5-7].

The relevance of the problem under study consists in the excess of the rolling stock, which leads to an increase in the load on the railway transport system and to a decrease in the throughput capacity of railway stations and hubs. The problems of insufficient capacity of the public and private tracks in loading and unloading areas become exposed.

2. Throughput capacity and its research

The dynamics of the throughput capacity of the East Siberian Railway (hereinafter referred to as the ESR) by sections: Zaudinskiy – Ulan-Ude, Ulan-Ude – Mysovaya, Tulun – Nizhneudinsk, Tayshet – Yurty is represented in Figs. 1 and 2.

The dynamics of the throughput capacity of railway hauls with the parallel and non-parallel schedule is represented in Tables 1 and 2, and also in Figures 1, 2, 3 and 4.

The analysis of Table 1 and Figure 1 allows concluding that the indicator in question with the parallel schedule in the out direction has an unstable dynamics in the studied areas. In 2014, no changes occurred in relation to the base at the sections. In 2015, compared with the previous one, the throughput capacity of the railway hauls increased only in two of four sections: in the Zaudinsky - Ulan-Ude section by 12.409% (or 17 trains) and in the Tulun – Nizhneudinsk section by 11.382% (or 14 trains), while in the other two sections the indicator decreased by 20.13% (or by 31 train).

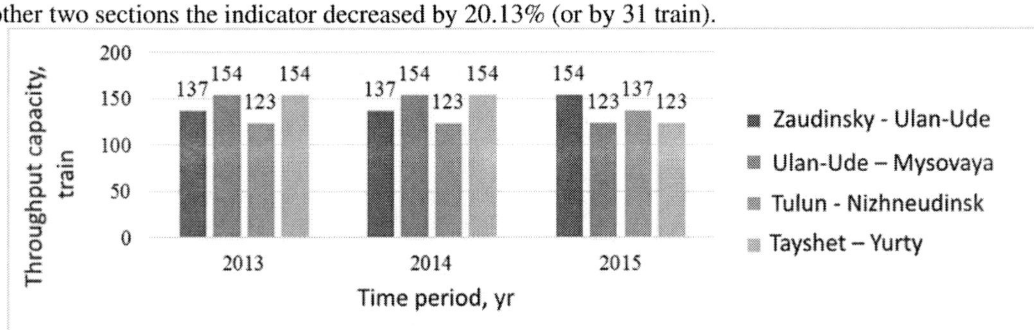

Figure 1. Throughput capacity of railway hauls with a parallel traffic schedule in the out direction.

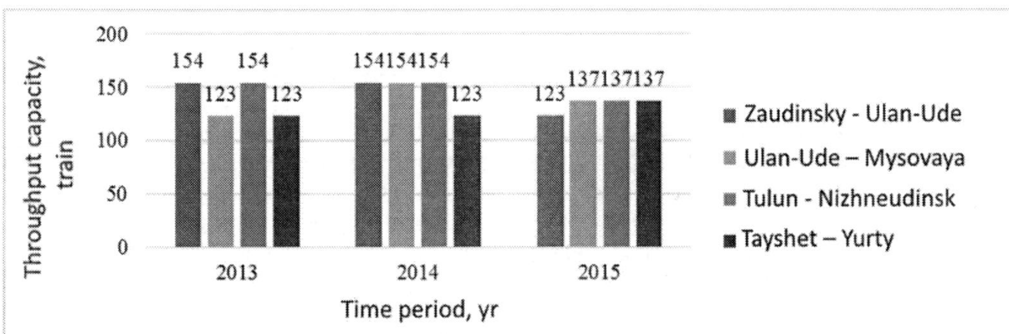

Figure 2. Throughput capacity of railway hauls with a parallel traffic schedule in the return direction.

Using the data listed in Table 1 and Fig. 2, we can conclude that in 2014 relative to 2013, the throughput capacity with a parallel traffic schedule in the "return" direction underwent a change only in the Ulan-Ude - Mysovaya section, namely, the indicator grew by 25.203% (or by 31 trains). In 2015, compared with 2014, the throughput capacity at the three sections under consideration decreased: by 20.13% (or 31 trains) at Zaudinsky - Ulan-Ude, by 11.039% (or 17 trains) at Ulan-Ude - Mysovaya and Tulun - Nizhneudinsk. Despite this, the indicator grew by 11.382% (or 14 trains) in the Taishet - Yurty section.

Table 1. Throughput capacity of railway hauls with a parallel traffic schedule in the out direction.

Section name	Year 2013		Year 2014				Year 2015					
	actual		actual		deviation by 2013,%		actual		deviation by 2013,%		deviation by 2014,%	
	out	return	out	return	out	return	out	return	out	return	out	return
Zaudinsky - Ulan-Ude	137	154	137	154	100	100	154	123	112.409	79.870	112.409	79.870
Ulan-Ude - Mysovaya	154	123	154	154	100	125.203	123	137	79.870	111.382	79.870	88.961
Tulun - Nizhneudinsk	123	154	123	154	100	100	137	137	111.382	88.961	111.382	88.961
Tayshet - Yurty	154	123	154	123	100	100	123	137	79.870	111.382	79.870	111.382

Table 2. Throughput capacity of railway hauls with a non-parallel traffic schedule.

Section name	Year 2013		Year 2014				Year 2015					
	actual		actual		deviation by 2013,%		actual		deviation by 2013,%		deviation by 2014,%	
	out	return	out	return	out	return	out	return	out	return	out	return
Zaudinsky - Ulan-Ude	89	92	96	118	107.865	128.261	117	85	131.461	92.391	121.875	72.034
Ulan-Ude - Mysovaya	104	86	100	106	96.154	123.256	93	105	89.423	122.093	93.000	99.057
Tulun - Nizhneudinsk	58	77	60	96	103.448	124.675	99	104	170.690	135.065	165.000	108.333
Tayshet - Yurty	116	80	123	90	106.034	112.500	89	101	76.724	126.250	72.358	112.222

The analysis of Figure 3 shows that the throughput capacity of railway hauls with the non-parallel schedule in the out direction has been changing non-uniformly for three years. So in 2014, in relation to the base in the Zaudinsky - Ulan-Ude section, the indicator grew by 7.865% (or 7 trains), as well as in the Tulun - Nizhneudinsk sections by 3.448% (or 2 trains) and Tayshet - Yurty by 6.034% (or 7 trains). In 2015, compared with the previous one, the throughput capacity of the railway hauls increased only in two of four sections: in the Zaudinsky - Ulan-Ude section by 21.875% (or 21 trains) and in the Tulun - Nizhneudinsk section by 65% (or 39 trains). In sections such as Ulan-Ude – Mysovaya and Tayshet – Yurty, the index decreased by 7% (or 7 trains) and by 27,642% (or 34 trains) respectively.

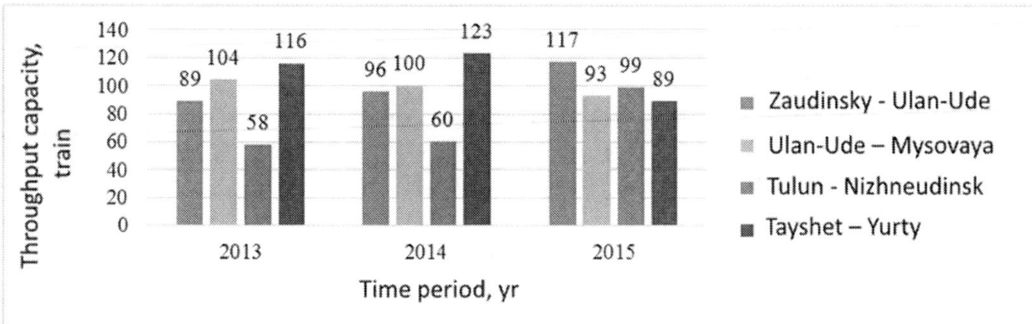

Figure 3. Throughput capacity of railway hauls with a non-parallel traffic schedule in the out direction.

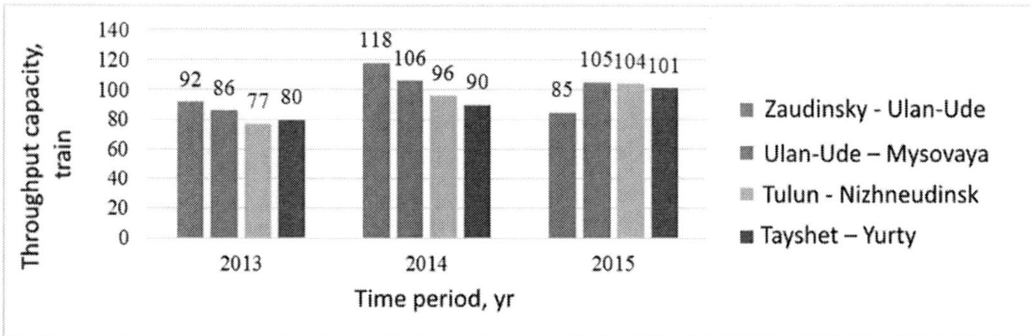

Figure 4. Throughput capacity of railway hauls with a non-parallel traffic schedule in the return direction.

Having analyzed Table 2 and Figure 4, we can conclude that in 2014, in relation to the base, the throughput capacity of the railway hauls with the non-parallel schedule in the "return" direction increased in all sections: in the Zaudinsky – Ulan-Ude section by 28.261% (or 26 trains), Ulan-Ude - Mysovaya - 23.256% (or 20 trains), Tulun - Nizhneudinsk - 24.675% (or 19 trains), Taishet - Yurty - 6.034% (or 10 trains). In 2015, compared to 2014, in the Zaudinsky - Ulan-Ude and Ulan-Ude - Mysovaya sections, the indicator under consideration decreased by 27.966% (or 33 trains) and 0.943% (or 1 train), respectively. As for the rest of the sections, vice versa, the throughput capacity increased, namely, in the Tulun - Nizhneudinsk section by 8.333% (or 8 trains) and in the Taishet - Yurty section by 12.222% (or 11 trains).

Table 3 and Figures 5, 6 and 7 show the dynamics of the throughput capacity of the stations of the considered sections.

The analysis of Table 3 and Figure 6 showed that the throughput capacity of stations in 2014 compared to 2013 increased in two sections: Zaudinsky - Ulan-Ude - 4.673% (or by 5 trains), Tayshet - Yurty - 46.269% (or by 31 trains), and in the Zaudinsky - Ulan-Ude and Ulan-Ude - Tulun sections, the indicator in question decreased by 28.205% (or 33 trains) and by 20% (or 21 trains) respectively. In 2015, compared to the previous one, in the Zaudinsky - Ulan-Ude section, the throughput capacity increased by 2.381% (or 2 trains), while in the remaining sections the indicator decreased: Ulan-Ude - Mysovaya – by 1.19% (or 1 train), Tulun –Nizhneudinsk – by 11.607% (or 13 trains), Taishet – Yurt – – by 30.612% (or 30 trains).

IOP Conf. Series: Materials Science and Engineering **832** (2020) 012061 doi:10.1088/1757-899X/832/1/012061

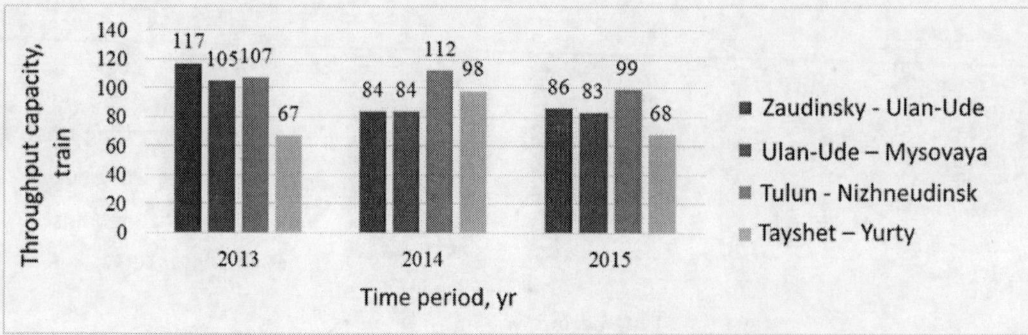

Figure 5. Throughput capacity of the stations in the out direction.

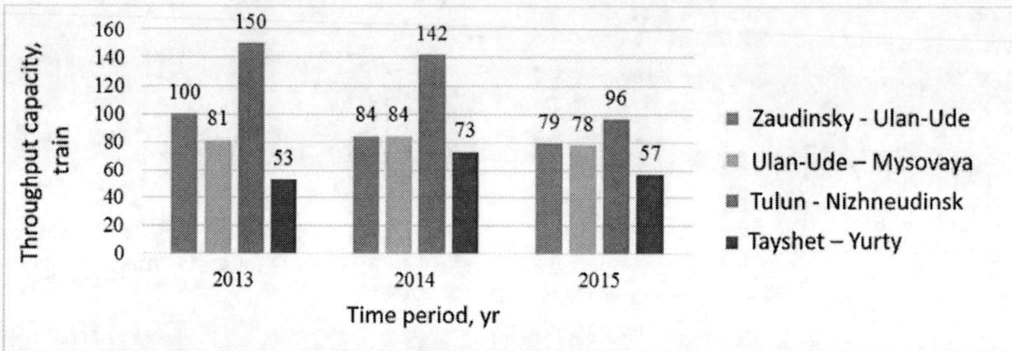

Figure 6. Throughput capacity of the stations in the return direction.

The study of Table 3 and Figure 7 allowed concluding that in 2014, with respect to the base, the throughput capacity of stations in the Zaudinsky – Ulan-Ude and Tulun – Nizhneudinsk sections decreased by 16% (or 16 trains) and 5.333% (or 5 trains), respectively. Along with this, the indicator increased in two other sections: Ulan-Ude - Mysovaya - 3.70% (or by 3 trains), Taishet - Yurty - 37.736% (or by 20 trains). In 2015, in relation to 2014, the throughput capacity rate decreased in all sections: Zaudinsky - Ulan-Ude - 5.952% (or by 5 trains), Ulan-Ude - Mysovaya - 7.143% (or by 6 trains), Tulun - Nizhneudinsk - 32.394% (or by 46 trains), Taishet - Yurty - 21.92 % (or by 19 trains).

Table 3. Throughput capacity of railway hauls with a non-parallel traffic schedule.

Section name	Year 2013		Year 2014				Year 2015					
	actual		actual		deviation by 2013,%		actual		deviation by 2013,%		deviation by 2014,%	
	out	return	out	return	out	return	out	return	out	return	out	return
Zaudinsky - Ulan-Ude	117	100	84	84	71.795	84.000	86	79	73.504	79.000	102.381	94.048
Ulan-Ude - Mysovaya	105	81	84	84	80.000	103.704	83	78	79.048	96.296	98.810	92.857
Tulun - Nizhneudinsk	107	150	112	142	104.673	94.667	99	96	92.523	64.000	88.393	67.606
Tayshet - Yurty	67	53	98	73	146.269	137.736	68	57	101.493	107.547	69.388	78.082

Traction power supply is the supply of electric energy to the electric rolling stock. To implement traction power supply in electric railways, electrical installations and devices are used, which include traction substations and traction networks belonging to either the railway or other transport or industrial enterprises. Table 4 and Figs. 8 and 9 show the dynamics of the throughput capacity of traction power supply devices.

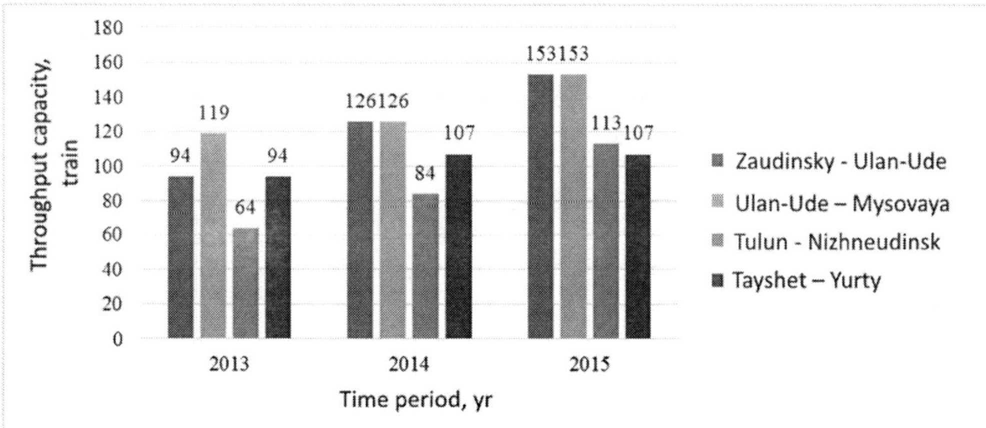

Figure 7. Throughput capacity of traction power supply devices in the out direction.

The analysis of Table 4 and Figure 8 is indicative of the positive dynamics of the throughput capacity of traction power supply devices in the "out" direction for the duration of the entire period under consideration. So in 2014 compared to 2013, the indicator grew in the Zaudinsky - Ulan-Ude section by 34.043% (or 32 trains), Ulan-Ude - Mysovaya - 5.882% (or 7 trains), Tulun - Nizhneudinsk - 31.25 % (or 20 trains), Taishet - Yurty - 13.83% (or 13 trains). In 2015, compared to the previous year, in the Zaudinsky - Ulan-Ude and Ulan-Ude - Mysovaya sections by 21.429% (or 27 trains), Tulun - Nizhneudinsk - by 34.524% (or 29 trains) and there were no changes in the Taishet - Yurty section.

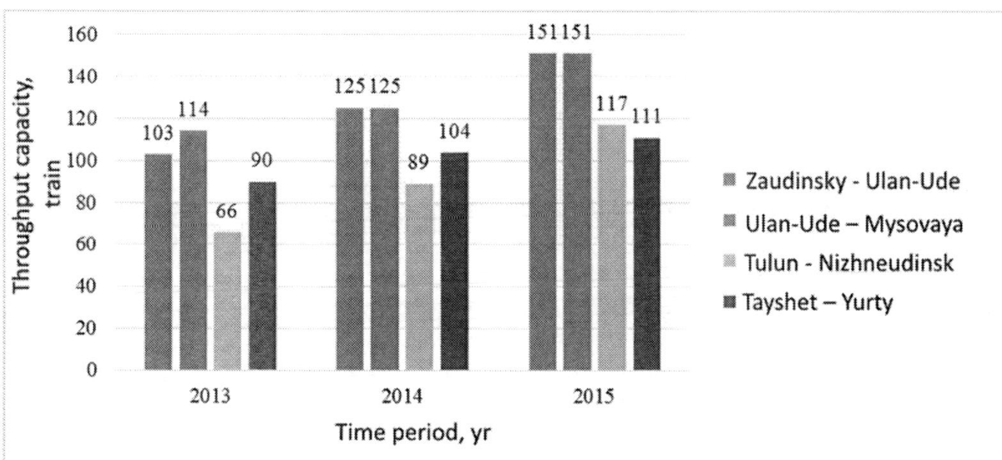

Figure 8. Throughput capacity of traction power supply devices in the return direction.

Using the data in Table 4 and Figure 9, we can conclude that the throughput capacity of the traction power supply devices in the return direction tends to increase. For three years, the indicator in question increased in the Zaudinsky - Ulan-Ude section by 46.602% (or 148 trains), Ulan-Ude - Mysovaya - by 28.571% (or 37 trains), Tulun - Nizhneudinsk - by 77.273% (or 51 trains), Tayshet - Yurty - by 23.333% (or 21 trains).

Based on the obtained data on the throughput capacity of the locomotive depot and equipment devices, we can conclude that during the period under review there were no changes in the indicator values.

The analysis of the dynamics of changes in the indicator under study identified the main factors, affecting the throughput capacity of railway stations and sections of the ESR and the degree of their significance, a more detailed description of which is given below.

3. Main reasons for the throughput capacity reduction

When describing railway systems, it seems appropriate to use combinatorial structures and their determinant and stochastic estimates [8-12]. We will use a different approach [13-18].

To determine the main actual reasons for reducing the capacity of railway stations [19-24] and improving the quality of use of the indicator in question, the Ishikawa diagram was constructed (Fig. 9). It provides a systematic approach to determining the main causes of the problem under consideration based on data on the capacity of railway stations in Russia in 2013-2015.

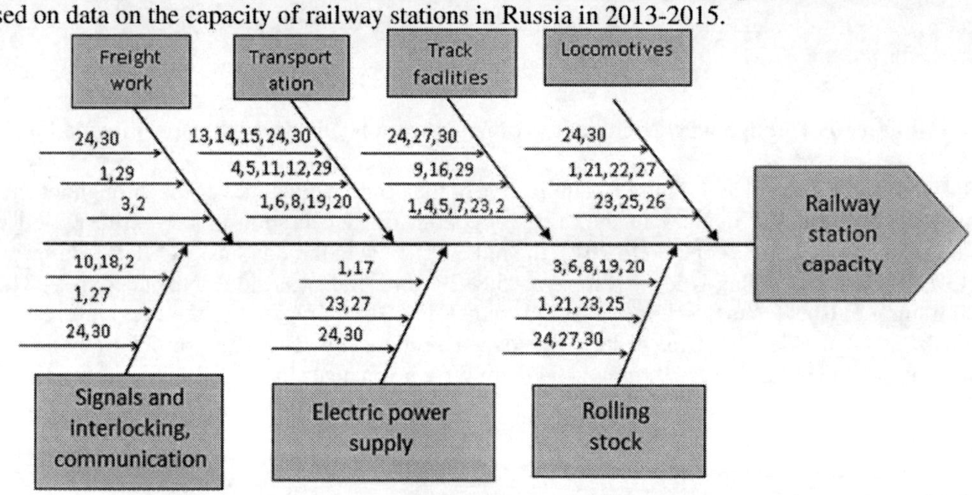

Figure 9. Factors affecting the throughput capacity of railway stations.

The following aspects are identified as the main reasons influencing the capacity of the ESR railway stations: mismatch in the level of development of the railway network (1); lack of promising centralized transportation planning (2); errors in the management of car fleets (3); insufficient available throughput capacity of the polygon network (4); insufficient processing ability of holding sidings (5); excess of car fleets (6); track possessions provided to perform the works on the track overhaul (7); non-uniform arrival of car traffic volumes (8); track profile (9); means for the movement of trains (10); norms of weight and length of trains (11); unpaired movement of trains (12); ratio of passenger and freight train speeds (13); the ratio of running, civil and section speeds (14); maximum permissible speed and its limitations (15); arrangement of intermediate color light signals (16); electrification of railways (17); signaling and communication (18); reduced quality of car fleet operation management (19); private car fleet growth (20); low quality and violation of technology for depot repair and overhaul of rolling stock (21); reduced quality of train inspection (22); high physical depreciation and aging of fixed assets (23); reduction in the number of employees (24); increase in warranty mileage of loaded and empty cars (25); train delay of more than 1 hour due to malfunction of the locomotive (26); failures in the operation of technical means and equipment (27); poor track maintenance (28); inconsistency of actions of third parties and transport companies (29); low labor and technological discipline, insufficient professional level of personnel, poor knowledge of and non-compliance with regulatory documentation, omissions in the organization of preventive work (30).

The insufficient available throughput capacity of the main transportation directions and the low operating capacity of holding sidings, as well as the excess of car fleets are the consequences of a not very effective strategy for the development of railway transport after the collapse of the USSR. If at the time of the work of the State Planning Commission, rail transport developed in close coordination with the needs of the country's economy, then with the renouncing of centralized planning, especially during the period of a sharp decrease in freight traffic in 1991-1998, there was no strategic planning for the development of the transport network, especially railways. This led to a mismatch between the needs of the economy and the real situation.

Traditionally, the main way to increase the track-carrying capacities of polygons and directions is to increase the mass of the train. This approach has proven effective. For many years it has been widely used in practice. Such an operational indicator as locomotive performance is traditionally one of the main indicators of the operational work quality. Under the conditions of the general locomotive fleet, its significance is indisputable. However, in modern practice, interruptions in the regularity of operation pace of key stations often lead to stops in train traffic volumes [25–28], which leads to a decrease in local speed and locomotive productivity, while the fleet of locomotives sharply increases and, coincidently, there is a shortage of them at the locomotive shift points. The transition to the operation of the locomotive fleet in long circulation journeys seriously increases the price of errors in adjusting the locomotive fleet. The use of technology for handling empty trains with a length of 100 conditional wagons leads to the emergence of oddness. As a result, on the Trans-Siberian Railway there is traditionally an excess of circulation on the eastern borders and, vice versa, a lack of circulation on the western ones.

As the analysis of the Ishikawa diagram shows, from the perspective of the throughput capacity of railway stations, one of the main most significant factors can be distinguished: an excess of car fleets; errors in the management of car fleets; low level of development of the federal railway network; non-uniform arrival of car traffic volume; decrease in the quality of car fleet management; private car fleet growth.

Even a small excess of the rolling stock on private tracks, certain train oversupply in sections leads to disruption of interaction in operational work. Factors associated with congestion in traffic flows and certain oversupply of car fleets in sections and private railways, with the impossibility of sending trains from stations and (or) supplying arriving wagons to stations, as a rule, become the primary cause of the car oversupply at station track capacities.

Since 2009, there has been an increase in the share of the car fleet located on the railways of the Eastern polygon, which is associated with a redistribution of freight traffic. With the growth of the operating fleet located in the common infrastructure, the following main quality indicators deteriorate in the network railways: car turnover; downtime of transit (with and without the yard operation) and local cars; local speed; average daily mileage of locomotives; the number of "abandoned" trains and the train-hours of their delays [1, 29-31]. When establishing operator companies, the registration of the rolling stock was often carried out nominally, i.e. without checking the capacity of the home station layout.

According to the procedure used in railway transport until 1993, any enterprise that wanted to buy freight cars was under an obligation to build additional access roads, both at the loading station and at the unloading station. Now the total fleet is almost 1.1 million cars, but there are no staging tracks for them. It is the rejection of the previous rules that largely explains the current difficulties with the location of rolling stock on the tracks of the railway network stations. The excessive car fleet occupies thousands of kilometres of station tracks. According to various estimates, it is up to 4-5 thousand kilometres. Railway track congested by empty cars of the private fleet of most of the East-Siberian Railway stations interferes with operational work. Private tracks adjacent to the stations are not able to accommodate all empty cars that have been freed from cargo or awaiting loading, so the bulk of unloaded cars are formed at loading and unloading stations [31-32].

4. Conclusion

In our opinion, to create specialized points for bulk staging and preparation of station cars is the main way to improve the quality of the polygon operational work in the conditions of excess fleet circulation and limitations on the railway network capacity. It is private funds that should invest in the creation of such points. Car owners and manufacturers, leasing companies, and rolling stock operators are interested in creating this type of stations. The layout of the stations specialized for the car staging should ensure the implementation of a complex of processing operations: reception, sorting, dispatch, technical inspection, all types of maintenance and repair.

References

[1] Zheleznov D V 2012 Establishment of stations specialized for bulk staging and preparation of cars as the primary way to improve the quality of polygon operational work under conditions of excess fleet circulation and capacity limitations *The Bulletin of the Russian State University of Railway Transport* **2** pp 142-150

[2] Vdovin K N, Feoktistov N A, Sinitskii E V, Gorlenko D A, Durov N A 2015 Production of high-manganese steel in arc furnaces. Part 1 *Steel in Translation* Vol. 45 **10** pp 729-732

[3] Vdovin K N, Gorlenko D A, Feoktistov N A 2016 Characteristics of excess phase in cast high-manganese steel *Steel in Translation* Vol. 46 **7** pp 484-488

[4] Vdovin K N, Feoktistov N A, Gorlenko D A 2016 The effect of the cast high-manganese steel primary structure on its properties *Materials Science Forum* Vol. 870 pp 339-344

[5] Spiess H, & Florian M 1989 *Optimal strategies. A new assignment model for transit networks. Transportation Research Part B: Methodological, 23 B (2)* 83-102

[6] Yolkin K S, Yolkin D K, Kolosov A D, Ivanov N A, Shtayger M G 2018 Technologies, which allow to reduce an impact of metal silicon production on the environment IOP Conference Series: Materials Science and Engineering 411 012028

[7] Kuzmin O V 2006 *Combinatorial methods of modeling discrete distributions* (Irkutsk: Irkut. un-ty Publ.) p 138

[8] Kuzmin O V, Orkina K P 2006 Building error-correcting codes using a Pascal-type triangle *The Bulletin of the Buryat State University* **13** pp 32-39

[9] Bulatov Y N, Kryukov A V, Suslov K V 2017 Multi-agent technologies for control of distributed generation plants in the isolated power systems *Far East Journal of Electronics and Communications* Vol. 17 **5** pp 1197-1210

[10] Solodusha S, Suslov K, Gerasimov D 2016 Applicability of Volterra integral polynomials in the control systems of electric power facilities *International Conference Stability and Oscillations of Nonlinear Control Systems (Pyatnitskiy's Conference)* 7541227

[11] Suslov K V, Solonina N N, Smirnov A S 2011 Smart Grid: A new way of receiving primary information on electric power system state *IEEE PES Innovative Smart Grid Technologies Conference Europe* 6162654

[12] Kolosov A D, Gozbenko V E, Shtayger M G, Kargapoltsev S K, Balanovskiy A E, Karlina A I, Sivtsov A V, Nebogin S A 2019 Comparative evaluation of austenite grain in high-strength rail steel during welding, thermal processing and plasma surface hardening *IOP Conference Series: Materials Science and Engineering* **560** 012185

[13] Kargapol'tsev S K, Nachigin V A, Frolov V F 2013 Algorithmic support for the assessment of transportation process targets *Modern technologies. System Analysis. Modeling* **4 (40)** pp 152-156

[14] Sivtsov A V, Sheshukov O Y, Tsymbalist M M, Nekrasov I V, Makhnutin A V, Egiazar'yan D K, Orlov P P 2019 Steel Semiproduct Melting Intensification in Electric Arc Furnaces Using Coordinated Control of Electric and Gas Conditions: II. On-Line Control of the State of the

Charge and Melt Zones in Electric Arc Furnaces *Russian Metallurgy (Metally)* **6** pp 565-569
[15] Sivtsov A V, Sheshukov O Y, Tsymbalist M M, Nekrasov I V, Makhnutin A V, Egiazar'yan D K, Orlov P P 2018 Steel Semiproduct Melting Intensification in Electric Arc Furnaces Using Coordinated Control of Electric and Gas Conditions: I. Heat Exchange and Structure of the Electric Arc Furnace Laboratory *Russian Metallurgy (Metally)* **12** pp 1108-1113
[16] Nekrasov I V, Sheshukov O Y, Metelkin A A, Sivtsov A V, Tsymbalist M M 2016 Slag conditions in electrosmelting: A review *Steel in Translation* Vol. 46 **6** pp 435-442
[17] Sivtsov A V, Sheshukov O Y, Tsymbalist M M, Nekrasov I V, Egiazar'yan D K 2015 The Valve Effect of an Electric ARC and Problems in Controlling Electric-ARC Furnaces *Metallurgist* Vol. 59 **5-6** pp 380-385
[18] Nekrasov I V, Sheshukov O Y, Metelkin A A, Sivtsov A V, Tsymbalist M M, Egiazar'yan D K 2015 Ensuring Consistent Foamability in Electric-Furnace Slags *Metallurgist* Vol. 59 **3-4** pp 300-304
[19] Sheshukov O Y, Nekrasov I V, Sivtsov A V, Tsymbalist M M, Egiazar'yan D K, Metelkin A A 2014 Dynamic monitoring of slag oxidation and thickness in the ladle-furnace unit *Steel in Translation* Vol. 44 **1** pp 43-46
[20] Abrosimov N V, Ageev A I, Adushkin V V, Akimov V A, et al. 2015 *The security of Russia. Legal, socio-economic and scientific-technical aspects. Scientific foundations of technological safety* (Moscow)
[21] Konstantinova M V, Balanovskiy A E, Gozbenko V E, Kargapoltsev S K, Karlina A I, Shtayger M G, Guseva E A, Kuznetsov B O 2019 Application of plasma surface quenching to reduce rail side wear *IOP Conference Series: Materials Science and Engineering* **560** 012146
[22] Guseva E A, Kargapoltsev S K, Balanovskiy A E, Karlina A I, Shtayger M G, Gozbenko V E, Konstantinova M V, Sivtsov A V 2019 Comparative evaluation of corrosion resistance of wheel and rail steels in various media *IOP Conference Series: Materials Science and Engineering* **560** 012181
[23] Grechneva M V, Balanovskiy A E, Gozbenko V E, Kargapoltsev S K, Karlina A I, Shtayger M G, Karlina Yu I, Govorkov A S 2019 *IOP Conference Series: Materials Science and Engineering* **560** 012142
[24] Shtayger M G, Balanovskiy A E, Kargapoltsev S K, Gozbenko V E, Karlina A I, Karlina Yu I, Govorkov A S, Kuznetsov B O 2019 *IOP Conference Series: Materials Science and Engineering* **560** 012190
[25] Kopylova T A, Mikhailov A Yu 2016 Evaluation of the functioning of intermodal nodes of urban passenger transport *In the collection: Transport systems of Siberia. The development of the transport system as a catalyst for the growth of the state economy. The International scientific and practical conference* pp 528-532
[26] Balanovskiy A E, Shtayger M G, Karlina A I, Kargapoltsev S K, Gozbenko V E, Karlina Yu I, Govorkov A S, Kuznetsov B O 2019 Surface hardening of structural steel by cathode spot of welding arc *IOP Conference Series: Materials Science and Engineering* **560** 012138
[27] Gozbenko V E, Khomenko A P, Kargapoltsev S K, Minaev N V, Karlina A I 2017 Creating of the alternative lubricants and practice of their use *International Journal of Applied Engineering Research* Vol. 12 **22** pp 12369-12372
[28] Chriqui C, Robilland P 1975 Common bus lines *Transportation Science* **9** pp 115-121
[29] Khomenko A P, Gozbenko V E, Kargapoltsev S K, Minaev N V, Karlina A I 2017 Comparative analysis of simulation results and test of the dynamics of the wheelset *International Journal of Applied Engineering Research* Vol. 12 **23** pp 13773-13778
[30] Kargapoltsev S K, Khomenko A P, Gozbenko V E, Minaev N V, Karlina A I 2017 Development of new lubricants for reducing the wear of the elements of the path and running parts of rolling stock *International Journal of Applied Engineering Research* Vol. 12 **22** pp 12362-12368

[31] Lebedeva O A, Kripak M N 2016 Modeling freight transportation in the transport network *The*

Bulletin of Angarsk State Technical University **10** pp 182-184
[32] Sazonov V N, Zagitov E D 2008 Russian Railways need modern effective solutions, not lightweight theories *World Railways* **4**

IOP Conf. Series: Materials Science and Engineering **832** (2020) 012062 doi:10.1088/1757-899X/832/1/012062

Improving the educational process to address issues of digital technologies development in the road industry

G S Merentsova

Altai State Technical University after I.I. Polzunov, 46, Lenin Ave., Barnaul, 656038, Russia

E-mail: adio-06@mail.ru

Abstract. The features of the educational process in training personnel for the road industry at Altai State Technical University are considered, taking into account the development of the digital economy. Much attention has been paid to key technologies that provide digitalization for the road industry. The set of disciplines necessary for obtaining knowledge in digital technologies for the further activities of university graduates is analyzed. When studying bachelor's and master's programmes, future specialists acquire knowledge of modern methods in computer-aided design of highways, geographic information systems in construction, information modeling and intelligent transportation systems. During the educational process, an understanding of the importance of implementing an automated control of equipment in road construction works using onboard information management system is also formed. The directions of scientific research are described in which students studying Master's Programme in Highway Design Construction and Management work. These studies make extensive use of digital technology. The analysis of road traffic accidents for identifying hazardous sections is given, as well as the analysis of the current road state with the development of measures for improving the transport and operational road condition. Current research areas using digital technologies are in demand in modern conditions.

1. Introduction and Problem Statement

Specialist training is of great importance in the strategy of digital economy development. Currently, in training specialists for the road industry, much attention should be paid to the use of digital technologies in road construction. In this case, it is necessary to provide training in modern methods of road design and in the use of digital technologies for road safety and road operation.

Obtaining knowledge in developing digital terrain models and digital models of situations that are used in the road design is of great importance..

The three-dimensional road design is also relevant for providing visual smoothness and clarity of the road. When designing a road in three-dimensional space (3D design), a virtual (digital) road model is created, on the basis of which you can check the geometric consistency of the model.

Future specialists in the road sector must have an understanding of GIS technology, which is an essential element of the modern economy. Regarding this, the necessary technologies and software tools have been developed at present. The road industry also has an understanding of information technology importance.

In the field of higher road education, more attention must be paid to computer-aided design and geographic information systems in the educational process. These technologies have to become an integral tool of road engineer activities for university graduates.

Information modeling and intelligent transportation systems (ITS) are the key technologies for the road industry digitalization. Intelligent control systems must be actively implemented on the roads.

Students need to obtain an understanding of the feasibility of installing an automated management system (AMS) for the appropriate equipment. They also must acquire knowledge about the advantage of automated management systems based on 3D technology that will improve the quality of performed work and increase the efficiency of using equipment with savings in material and labor costs, providing a quick return on investment.

2. Features of the educational process, using modern methods of digital design and effective executing road construction works.

The high school must take into account the needs of the modern economy. Currently at the Altai State Technical University after I.I. Polzunov in the curriculum of the Department of Road and Airfield Construction, there are disciplines directly related to informatization of the road sector, where elements of the digital economy are used including the advantage of its application in the road industry.

Thus, the bachelor's programme includes the following disciplines: Automated Road Design, Geographic Information Systems in Road Construction, and the Master's programme in Highway Design, Construction and Management includes the disciplines:

- Theoretical Aspects of the Geographic Information Systems Development;
- Theory of Modeling Flow Movement;
- Methodological Aspects of Intelligent Transportation Systems Development;
- Special Issues of Road Design;
- Modern Methods of Road Design.

When teaching these disciplines, much attention is paid to modern methods of highway design. Students have an understanding that the design of such complex technical structures as highway requires a comprehensive analysis of a variety of factors to ensure the required transport and operational characteristics of the road and traffic safety. Therefore, the choice of optimal technologies and tools used for design plays a decisive role.

For the development of the digital economy in the road industry, it is very important to obtain relevant knowledge in this direction during training.

When studying the discipline Modern Methods of Road Design, students get acquainted with digital terrain modeling that is used to construct earth surface cuts in a given direction, to find coordinates and elevations of points for creating plans, longitudinal and transverse profiles, etc. A digital terrain model (DTM) consists of:

- digital elevation model;
- digital situation model;
- digital model of the geological and hydrogeological structure of the area;
- digital model of environmental parameter distribution, etc.

Working with CREDO software package students get acquainted with digital elevation model (DEM), digital model of the situation (DMS), digital geological model (DGM).

The digital elevation model is formed on the basis of points, structural lines, and elevation contours. In this case, a triangular network is created, with the help of which the heights of points with known coordinates can be calculated, sections in a given direction are constructed, the relief is displayed using elevation isolines, legend symbols of slopes and cliffs.

Creating a digital terrain model, students get acquainted with the recommended sequence of work. Horizontals in the form of broken lines are used to visualize the road carriageway for various layout types. There are two methods to analyze the relief: constructing a cut along any broken line drawn along the formed surface, and the second method implies viewing the relief in the three-dimensional perspective.

Students study the recommended work order when creating a digital model of the situation. The digital model of the terrain situation includes areal objects (settlements, platforms, individual buildings), linear objects (roads, watercourses, power lines) and point objects. The information on the terrain prepared in the form of a digital model should also include land-use boundaries, status and characteristics of existing infrastructure objects, such as roads and railways, utilities, buildings and structures, as well as data on natural objects – such as wetlands, rivers and other water bodies, forest massifs and relief.

Students get acquainted with AutoCAD Civil 3D. This system is an innovative solution. Its obvious advantage is a dynamic design environment, built on the basis of the well-known and widely used AutoCAD platform. Using an automated design system AutoCAD Civil 3D allows you to perform any stage of design including construction, reconstruction and repair of roads [1].

Structurally, the design process in AutoCAD Civil's 3D should include the following steps:

- preparing a digital terrain model;
- defining the road route in the plan and profile;
- 3D modeling of the road;
- calculating the volume of the work and creating the output documentation.

Students begin to understand that the digital elevation model (DEM) is the basis on which the entire dynamic model of the road project is built. The digital elevation model is used to create longitudinal profiles of linear structures, and is the target object for determining project slopes and profiling.

At the design stage, several options are being worked out for passing the road route with an assessment of technical and economic indicators and taking into account the features of the terrain and other obstacles. Three-dimensional modeling of a road begins with the definition of typical transverse profiles (structures).

Calculating the earthwork volume is carried out along the axes of the cross-sections and surfaces of black marks (earth) and surfaces of black marks (earth).

In the process of training, attention is paid to improving geographic information technologies in the road sector using digital aerial photography of roads and mobile laser scanning. For planned high-altitude justification of GIS works, GLONASS-based satellite measurement technologies are being implemented.

GIS remains an integral part of the overall process, playing a dominant role in the operational phase of roads. The process of operating roads in the Russian Federation is based on a diagnostic procedure. When diagnosing roads, the main attention is paid to variable parameters (evenness, adhesion, strength, defects of the carriageway and to other structural elements of the road). Based on the diagnostic results, road sections that do not meet regulatory indicators are determined. A set of measures to improve transport and operational conditions is being appointed at these sections. Much attention is paid to studying capabilities of modern GIS using the ArcGIS software package as an example.

Students acquire skills in solving practical problems using geoprocessing operations in GIS, as well as creating and maintaining spatial databases in modern GIS. In practical classes, they solve the task of investigating the territory in the GIS for selecting places for road design and operational measures. The training process uses the Arcview3.2 / ArcGIS technical documentation package.

The mathematical basis of the map is studied, taking into account the basic methods of mathematical cartography implemented in modern GIS packages.

Thus, geographic information systems are a tool for engineering and road management and play a significant role in the implementation of information modeling.

In the master's programme, considerable attention is paid to studying methodological aspects of the developing intelligent transportation systems (ITS), which allow automated search and adoption of the most effective ways to control transport and road complex of a region, specific vehicle or group of vehicles.

The key technologies for digitalization of the road industry are information modeling and intelligent transportation systems (ITS).

When studying the discipline Modern Technologies for the Mechanization of the Road Construction and Operation, students get acquainted with modern road equipment. The task of the road equipment management systems is the formation of the surfaces of each road layer (subgrade and pavement). For operation of such systems, it is necessary to use 3D digital models representing the surfaces of the structural layers of the subgrade and pavement, each of which requires a certain type of road equipment at each stage of the work [2].

Currently, modern road equipment is being introduced at domestic engineering enterprises. Thus, some machines are equipped with an intelligent leveling system that allows accurate positioning of the excavator bucket or the dozer blade. It is planned to create a new bulldozer with a full fledged onboard information and control system, which will provide control over the functioning of all units, will fully automate the control of the machine and its working equipment, including remote control. This will increase the level of productivity, and the operator of the equipment will be able to control several machines remotely at the same time.

3. Research areas for using digital technologies in road construction, road safety and road operation

The research work in using geographic information technologies was carried out in the following areas:

- analyzing traffic safety and accident rate on highways;
- analyzing operational conditions of roads and developing measures to improve transport and operational conditions of roads;
- analyzing road lighting conditions and developing measures of their improvement;
- analyzing current road traffic conditions and their improvement for road users with disabilities.

In carrying out these research works, digital terrain models and digital situation models were used, as well as the necessary software systems.

One of the areas of scientific research carried out by master's students in the course of work on their theses is the use of geoinformation technologies to identify foci of traffic accidents concentration in the city. To solve this problem, in the GIS an information model was developed which allow creating interactive accident maps. Using this map, an analysis of high road accident concentration spots was carried out to develop recommendations indicating measures reducing the number of accidents. Such studies give an approach to solving the problem of ensuring road safety and the effective organization of the road network in settlements.

To accomplish these tasks, the ArcGIS software package produced by ESRI was used for it has the most complete range of functions for creating and analyzing spatially distributed information. Using GIS the most hazardous, in the sense accident rate, road sections were identified

The interactive accident map constructed by means of ArcGIS represents a visual model of accident foci in the urban area.

The created geodatabase of traffic accidents has its own independent value and assumes annual updating data for the actualization of the interactive accident map [3].

Research in this direction allows solving one of the most important tasks of road and transportation management.

For analyzing accident data on the regional and federal roads, the geographic information system of the web version of ArcGIS, ArcGIS Online (ESRI), was used. ESRI software can be applied for a wide range of tasks related to spatially distributed objects and analysis of spatially distributed data. It is successfully used in the areas of cadastre, land management, utility management, surveying and others. The choice of the ArcGIS Online system is due to the fact that it is, first of all, an instrumental GIS that works with many formats of geographical data, to which attribute (descriptive) data can be easily connected.

When conducting studies on traffic safety, a digital topographic base was used, corresponding to the locality, consisting of spatial data sources in shapefile format and including the necessary geographic layers.

In ARcGIS Online, an interactive thematic map was created that allows to visualize data on traffic accidents specified in geographic coordinates, view information about them and search for accidents by some of their features or by their spatial position or combining spatial and logical conditions..

In addition, the GIS allows to perform visual and quantitative analysis of the road features, that is, with the help of GIS we obtain complex information about the object, which allows to make more informed decisions on the road maintenance and management.

Using information on the number of accidents, it is possible to classify sections of roads according to the degree of danger and show this on the map with the help of symbols (the symbol class "Gradient coloring" was used).

Usining the created interactive accident map, you can view information on any accident and on various road sections. In addition, as in any GIS, it is possible to search for accidents or road sections using spatial-logical queries. You can also add various Internet resources to the created map.

This map can be updated by adding new roads, new data on accidents. The map is a road accident analysis tool. Using this map in GIS, you can monitor the situation on the roads, and also use it as the basis for making management decisions to reduce road accidents.

The resulting GIS accident map provides comprehensive information about road and traffic accidents, and gives a possibility to search for objects of interest, all that makes the content of the accident analysis more complete and functional.

One of significant importance is a research work carried out by master's students by means of GIS technology which allows analyzing road accidents to identify accidentally hazardous road sections and develop recommendations for reducing accident rate. The expediency of using GIS technologies in analyzing the current state of roads and their operational conditions for the following development of measures improving their transport and operational conditions is established. When conducting these studies, digital technologies were used, which allow performing spatial analysis of various road features and situations. Working on of digital solutions for people with disabilities on the roads is also noteworthy. For them, digital applications and modern means of communication are extremely useful.

The main objectives of these studies are the implementation of a comprehensive analysis of the causes and conditions causing road traffic accidents and the subsequent assessment of the effectiveness of decisions made, including the use of mathematical modeling methods. On this basis, proposals were prepared on priority areas in the field of road safety.

A number of studies were devoted to the problem of analyzing the state of the road surface in urban conditions, as well as in the districts of the Altai Krai. These studies were performed in the ArcGIS program.

Based on the results of the visual analysis, the collected data was reflected in the updated maps of a particular area in a visual format. On these GIS maps, roads with minor and significant defects were highlighted with corresponding symbols. As a result of the studies, measures were proposed to eliminate defects on specific roads, which are currently taken into account when repairing problem sections of roads, both in urban conditions and in several regions of the Altai Krai.

Studies were also conducted to analyze the lightning of streets and roads in the city of Barnaul. The use of the ArcGIS program allowed us to present the results of our studies in a clear, systematic form. Based on the data GIS-maps were created. At the same time, sections of roads and streets with poor lighting were revealed. Recommendations have been developed on optimizing the lighting system of road sections, as well as on the effective use of LED lighting lamps. GIS-maps were obtained showing the state of illumination in the studied settlement.

The master's theses also discuss digital solutions for people with disabilities (with hearing and vision problems). For them, digital applications, modern means of communication are extremely useful. When performing such studies, a state analysis was carried out to improve the traffic conditions of road users. In the course of these studies, existing means are analyzed that ensure the safety and comfort of movement for people with limited mobility in the urban environment. At the same time, an experimental analysis of the state of existing conditions for ensuring the safety and comfort of movement of these population groups is performed by means of GIS.

IOP Conf. Series: Materials Science and Engineering **832** (2020) 012062 doi:10.1088/1757-899X/832/1/012062

As a result of these studies, an interactive GIS map is created, which can be further developed when new objects appear for road users with disabilities. Based on the studies, recommendations have been developed on creating new facilities that increase the safety and comfort of people with disabilities in a specific urban environment.

Scientific research carried out using digital technologies in the above areas is relevant in modern conditions. Its implementation is necessary for the road industry.

4. Conclusion

To train qualified road industry specialists possessing knowledge of digital technology the following problems need to be addressed:

- acquiring knowledge of modern methods in the road design;
- studying the importance of digital models in the issues connected with road building equipment management with a computer-aided 3D design system.

Of the significant importance is a research work carried out by master's students by means of GIS technology which allows analyzing road accidents to identify the most hazardous road sections and develop recommendations for reducing their accident rate. The expediency of using GIS technologies in analyzing the current state of roads and their operational conditions for the following development of measures improving their transport and operational conditions is established. When conducting these studies, digital technologies were used, which allow performing spatial analysis of various road features and situations. Working on of digital solutions for people with disabilities on the roads is also noteworthy. For them, digital applications and modern means of communication are extremely useful.

References

[1] Kruglov S 2011 Highway Design in CAD AutoCAD Civil 3D *CAD and Graphics* **8** 1–3

[2] Gulin V N 2015 Digital Models for Control Systems of Road-Building Machines *CAD and GIS for Roads* **1(4)** 56–9

[3] Mikhailidi I M and Kishko D I 2019 GIS-Based Approach for Detecting Sites of High Accident Rate in the Industrialny District of the City of Barnaul *Polzunov Almanakh* **2** 1–3

Contemporary paradigm of cargo motor transport vehicles service

S Ozornin

Transbaikal state university, 30, Aleksandro-Zavodskaya Str., Chita, 672039, Zabaykalsky kray, Russia

E-mail: s.ozornin2013.s@ya.ru

Abstract. The results of the study of the organization of cargo vehicles operation by the transport processes mechanization in the environment, complicated by the climatic and industrial factors, are presented. The integrated technological process of using freight vehicles as a new modern paradigm of their operation during the life cycle is described.

1. Introduction

The problem of *cargo motor transport vehicles* (hereinafter – CMTV) maintenance should be considered as an important component of the strategy of creating a competitive national market and improving the marketability of the Russian automotive industry on the international market. In developed countries, the CMTV service has completely become an independent sphere of the economy.

Modern marketing and dealer services of automakers have formed their service concept, which provides for the services rendering in the maintenance of the CMTV for the entire duration of operation. This concept is supported by the orientation of large enterprises to increase the service sector, which stimulates the formation of their own internal grouped service points (for example, information technology structures that serve certain units), as well as the creation of subsidiaries or the purchase of the enterprises already providing the CMTV services. It does not change the basic marketing purposes, but reprioritizes general marketing decisions in the direction of the service, offered to the CMTV owners, or building an effective system of the CMTV service. This allows attracting more and more new customers and expanding the system of regular customers. The total set of services, provided during the CMTV operation, becomes a criterion for securing an order in those industries where competition in price, quality and reliability is no longer possible, where all enterprises use high-tech production methods. In general, the agreement on the service provision is now one of the most important conditions for the CMTV purchase and operation.

The need for the development and implementation of more economical modern methods and methods of CMTV operation, allowing obtaining high economic results of the operational enterprises, is becoming increasingly important. Undoubtedly, reducing the costs associated with the CMTV use contributes to the efficiency enhancement of any production, as well as the competitiveness of transport operations.

At the origins of the science of the machine service, there have been some prominent Soviet scientists: academician B.S. Svirschevsky, a honored worker of science and technology of the RSFSR, doctor of Technical Sciences, Professor G.V. Vedenyapin, Professor N.Ya. Govorushchenko, N.S. Zhdanovsky, G.V. Kramarenko, P.T. Frolov, A.M. Scheinin. A significant contribution to the development of science was made by Professor V.I. Balovnev, V.A. Zorin, A.P. Kryvshin, E.M. Kudryavtsev, A.N. Maksimenko, A.K. Morozov, B.I. Filippov and others.

Through the joint efforts of these scientists and production specialists, the science of machine operation was formed, which became a powerful factor in ensuring the high-performance use of equipment for various purposes in the national economy.

Rational machines operation, allowing increasing productivity and efficiency, has become a system of machine service, implying a certain scheme of implementation of their life cycle (Figure 1). This approach can be considered the first paradigm of machine operation, which was formed in the 30s – 50s of the twentieth century.

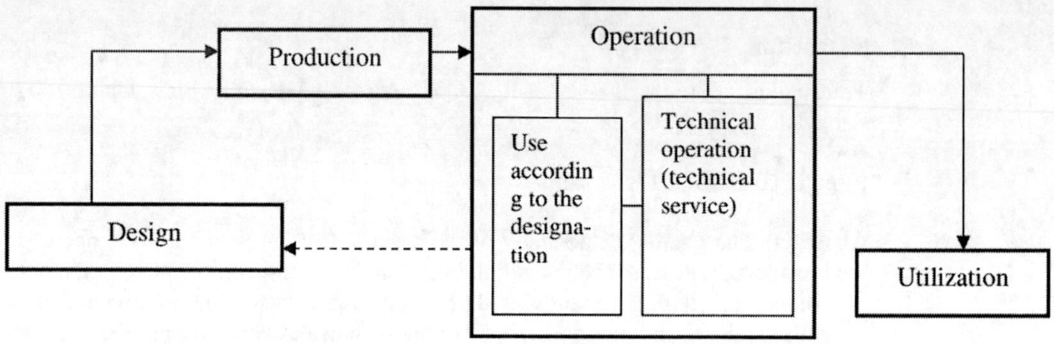

Figure. 1. Scheme of the machine life cycle realization.

The paradigm (from the Greek *paradeigma* – example) is a conceptual model of problem formulation and solution, predominating in the academic circles during a certain historical period, some clearly defined concept, style of thinking and methodology. It should be recognized that this is a special way of organizing scientific knowledge, which sets a particular vision of the world and, accordingly, examples and models of setting and solving research problems [1]. The paradigm shift is generally seen as a scientific revolution, which is now clearly visible in various fields of science, production and education. The science of machinery operating occupies a worthy place specified by its history, adequately responds to the many demands of production and human community, having powerful potential. The characteristic feature of the present time is globalization. It is the process of integration of mankind into a complex interconnected single planetary system, a qualitatively new level of human activity, based on the latest achievements of science and technology, the introduction and development of market economy.

In fact, currently, the CMTV operation is an integrated technological process (ITP CMTV), which includes technological processes of the CMTV use for their intended purpose, technological processes of CMTV technical operation (maintenance and repair processes, i.e. technical service), as well as technological management processes [4].

This approach makes it possible to form a system of freight vehicle operation for ITP CMTV. It should be noted that this system should include a CMTV producer, which plays the role of a CMTV supplier in the area of operation. At the same time, they need to participate in technological processes of maintenance service, carrying out so-called original service maintenance (OSMCMTV). The interactions between the CMTV manufacturer (PCMTV), operator (OCMTV), the operation control structure (SCOCMTV) in the sphere of the CMTV operation (SOCMTV) are shown in Figure 2.

Thus, the scope of the CMTV operation is considered as a single flexible high-performance complex, ensuring the implementation of transport works and the provision of industrial infrastructure services to other participants of the integrated technological process of the CMTV. In modern conditions, the field of the CMTV operation should assume additional tasks, which, in comparison with the traditional operation process, dictate new requirements for the development of the entire infrastructure, technical equipment and technology of all CMTV units and services [4].

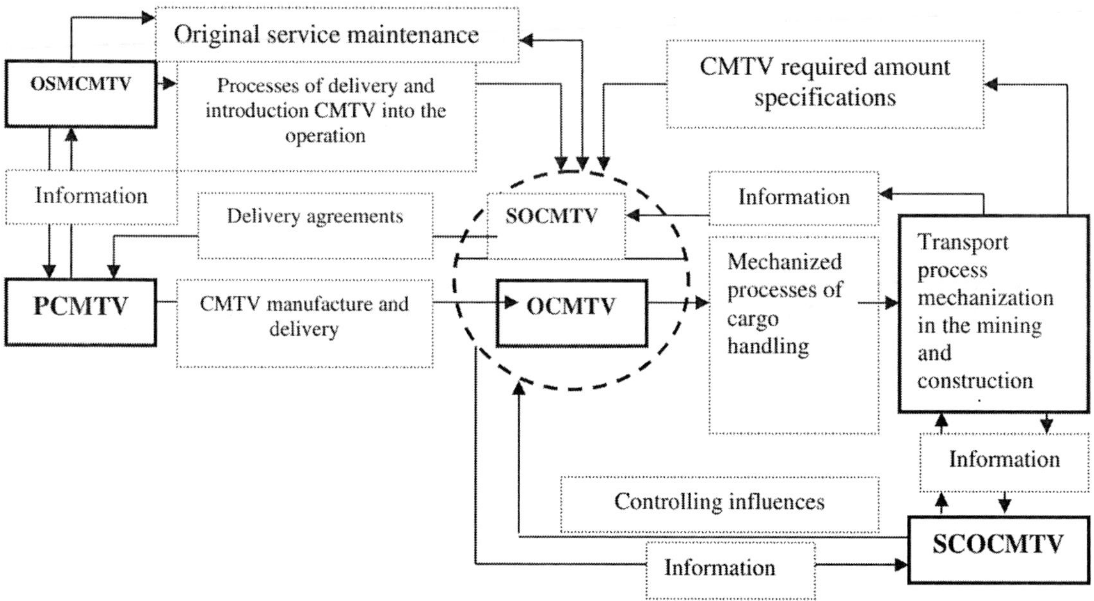

Figure. 2. Diagram of the cooperation of the cargo motor transport vehicles operator (OCMTV), the cargo motor transport vehicles producer (PCMTV), the original service maintenance system of the cargo motor transport vehicles (OSMCMTV) and the operation control structure of the cargo motor transport vehicles (SCOCMTV) in the operation sphere of the cargo motor transport vehicles (SOCMTV).

Any enterprise of the automotive industry, operating within the framework of the service concept, should use a systematic approach, first of all, to solve the problems of market competition. This involves the study of needs and demand; the analysis of the internal and external environment of the enterprise, the development of goods and programs of their production; pricing and marketing policy evaluating; rendering various kinds of services related to sales and consumption, including the organization of after-sales services and out-of-use goods utilization.

The observed steady trend of increasing demand for technical services is supported by the automotive industry enterprises, which have the value of products added by services. At the same time, the CMTV owners have the advantage of differentiated benefits owing to additional services. The activities primarily associated with science-intensive services take on greater importance. They include comprehensive services to the CMTV owners, consulting services, engineering and system integration services, project management organisation, special software of the CMTV owners, training and retraining of personnel, maintenance of installations, financial decisions and so on. The organization and expansion of knowledge-intensive service business is central to the strategic plans of many companies in the automotive industry [6]. The increasing complexity of the CMTV designs leads to a growth in the need for comprehensive solutions to technical problems, for example, in the supply of "turnkey", assistance in the management of systems and etc.

The purpose of this study is to improve the CMTV exploitation efficiency and safety and the competitiveness of the transport operations by optimizing the processes and costs associated with the use of modern CMTV for their intended purpose and technical service.

The scope of the CMTV operation cannot be considered in isolation from the environment in which they operate. Since the resource provision of the CMTV operation field is carried out from the external environment, its functioning should be connected with it systematically. At the same time, the direction of the CMTV operation is responsible for the external environment, performing transport work for the customers – consumers of services of operating enterprises. The field of operation also works for itself, performing the CMTV maintenance and repair.

For the analysis of the CMTV operation system effectiveness a process approach, in the terms of which a complex organizational and technological system as a set of processes, structurally reflecting the CMTV life cycle, can be seen as follows. "CMTV Design – CMTV Production– CMTV Operation – CMTV Utilization" ("Des. CMTV – Prod. CMTV – OCMTV – UCMTV"). In this case, the concept of «operation of CMTV» includes the processes of their intended use and technical service processes (including OSMCMTV) [5, 7, 8].

To describe the structure of the complex organizational and technological operation system of the CMTV, cooperation and interaction of its constituent subsystems, a synergetic approach is applied. In terms of the approach, the considered system is not local and between its elements there is an interaction organized, which is perceived as the information transmission with an infinite velocity (first of all, the quantum mechanical inequality by Jhon Bell reminds about this). The set of the processes, accompanying its functioning, structurally reflects the CMTV life cycle [5, 8–11, 13–16].

With due regard to the process approach, the CMTV life cycle is divided into three levels:

1 – the level of the CMTV design and production (LDPCMTV).

2 – the level of the CMTV operation (LOCMTV).

3 – the level of the CMTV utilization (LUCMTV).

The CMTV operation technological processes are represented by the sequence: "Breakingin – Intended Use – Original Service – Technical Service" *("BrInCMTV – IUCMTV – OSCMTV – TS CMTV")*.

The scope of the CMTV operation, considered as a single flexible high-performance complex, should ensure the implementation of transport operations on a commercial basis, as well as the provision of commercial services of the technical service production infrastructure to other participants in the production process of the CMTV operation.

The effectiveness of the CMTV use is a set of two concepts: the CMTV performance and the cost of their purchase and operation.

The parameters ensuring maximum performance of the CMTV are called rational, and the parameters that provide maximum efficiency are optimal.

Finding the best solution, that achieves the maximum productive capacity (maximum profit) or a minimum cost of funds to guarantee the operation and operational reliability maintenance, is called optimization of the CMTV use.

In practice, it is impossible to strive for maximum results and minimum costs at the same time. It is necessary to get the most results, provided that the costs do not exceed certain values, or get the minimum cost, provided that the results are not less than the specified ones.

The task of the CMTV use optimizing is extreme (finding the points of the greatest and least values). Notably, it has the only most effective solution in this confluence of facts and circumstances. As the most complete criterion of optimality (efficiency) of the CMTV use, it is accepted to apply the criterion of specific reduced costs [5, 6].

The information support of the technological processes of the CMTV technical service is the most important element of the unified operation technological process. The sources of information are the results of statistical processing of data on the CMTV individual elements reliability and their operational reliability in general. In addition, the information about the CMTV technical condition should appear in the result of the data analysis obtained from the CMTV diagnosing on the service companies fixed stations. The speed operation modes and levels of load impact on the CMTV should be monitored by the diagnosis and registration equipment built into the CMTV construction design [5, 6].

Formed as a result of the analysis of the production and technological system functioning (in a set of processes), structurally reflecting the CMTV life cycle [5, 7], the system is transformed as follows:

"Performance Specification Development – CMTV Design – CMTV Production – CMTV Operation – CMTV Utilization" ("PSD – DCMTV – PCMTV – UCMTV "). In this case, the block of "Performance Specification Development" is introduced into the system, taking account of all requirements for the design and operation of the new CMTV. With the consideration of the activities of the structures, designing and producing machines and ensuring their implementation and delivery to the location of the user (operator), it is necessary to consider the above-mentioned production processes as *the production-sales-maintenance-service system* (PSMSS): "CMTV Manufacturer – CMTV Sales Dealer – CMTV Owner – CMTV Operator – CMTV Service Representative" ("MCMTV – SDCMTV – OwnCMTV – OCMTV – SRCMTV") [5, 7].

The CMTV life cycle levels should also be defined, introducing their notation conventions: *LPSDCMTV* – the level of performance specification development of the CMTV design; *LDM1CMTV* – the design level of the CMTV; *LDM2CMTV* – the production level of the CMTV; *LOCMTV* – the operation level of the CMTV; *LUCMTV* – the utilization level of the CMTV.

The technological processes of the CMTV operation are as follows: *BrInCMTV* – the CMTV breakingin; *InUCMTV* – the intended use of the CMTV; *OSMCMTV* – the original service maintenance of the CMTV; *TSCMTV* – the CMTV technical service.

Natural-climatic and other factors characterizing the surrounding (external) environment have a significant impact on the CMTV technological processes. The level and characteristics of the operating conditions impact on the CMTV technical condition, causing its changes, are taken into account by the indicator "the cumulative value of the load impact on each CMTV", in which, according to the principle of the "blackbox", both the nature and the magnitudes of the specified cumulative value of the load impact are hidden. Modern built-in means of control of operating conditions and changes in the CMTV technical condition allow the monitoring mode to reveal the uncertainty of load effects and the CMTV technical state changes, which are concealed by the "black box". The theoretical justification and the examples of the practical use of these means are given in the dissertation of I.E. Berdnikov, written under the supervision of the author.

The theory of self-organization in complex, open, nonequilibrium and nonlinear systems of any nature is considered to be synergetic [8, 12]. Since the considered system of the CMTV operation consists of a large number of elements, it is complex and self-organizing. Besides, it is open, nonequilibrium and nonlinear. With an increase in non-equilibrium above a certain limit, it turns into an unstable state, in which there is a consistent behavior of the system elements and the transition of the system to a qualitatively new state with an ordered structure. Therefore, it can be attributed to the category of *self-organizing synergetic systems*, the way out of the instability of which is a jump due to the rapid restructuring of the system elements. However, the choice of one of the possible states of such a system is always random [5, 8–11].

In the process of the CMTV operation, the most common problem situations are associated with *the development of sound management decisions* (DSMD). This applies both to the problems of using the CMTV for their intended purposes (commercial operation), and the problems connected with the control and assessment of the CMTV technical state, constantly changing in the course of their operation due to stress effects. The changes in the technical condition should be compensated by preventive maintenance measures (PMM), and in cases of failures – by repairing (R).

To assess the effectiveness of the DSMD, the model, called "white box", is acceptable, in which the intangible component of the system – information is handled. In the process of the DSMD, the entities carry out operations only with information. To solve the problems, not only the resources of the company operating or servicing the CMTV can be used, but also the favorable conditions of the external environment as a specific resource for the development of the operational system.

The confluence of resources and problems allows for a variety of resource use cases to be considered for each problem. For example, a certain resource may be required to solve some problems, but predominantly the required synergetic effect is provided by a combination of resources. The gaps (contradictions) between the goals of the CMTV operation and their current technical condition allow to identify the problems, and the goals, in turn, are the quantitative indicators, which can be used to evaluate the developed alternatives and make a grounded decision. The prerequisites for the new approaches search in the CMTV operation organization are created by the theory of technocratic discourse and awareness of the scale and importance of technical reality, determining literally all aspects of human life in the modern world.

It should be borne in mind, however, that the CMTV operation is a multi-faceted area of expertise, services and organizations, consisting in the task-specific and full employment of the CMTV potential, especially when it comes to commercializing these processes.

Since the scope of the CMTV operation should be considered as a single flexible high-performance complex, ensuring the implementation of transport works on a commercial basis and the provision of services of industrial infrastructure of technical service to other participants in the production process of the CMTV operation also on a commercial basis. The goals included in this activity and the methods of achieving them change depending on the level at which the work is carried out. However, at all levels, the purpose of the CMTV operation is:

1) to improve the service for the CMTV use and their parks for transport operations, taking into account their specific features in order to obtain an acceptable level of income;

2) to tighten control over the CMTV operation modes in the course of their functioning in order to obtain an acceptable level of income;

3) to technologically maintain and restore the CMTV performance;

4) to provide CMTV with the high quality maintenance materials;

5) to organize routes of cargoes handling;

6) to save, store the CMTV during the periods when they are not used for their intended purpose.

These areas of expertise and services simultaneously point to the causal factors that determine the CMTV operation effectiveness and are controllable.

In this case, it is necessary to remember the subjective factors affecting the CMTV use effectiveness. This may include the driver, managing the CMTV work, as an element of the system "Man – CMTV – Environment", the effectiveness of the implementation of the above tasks largely depends on his skills and experience.

Currently, in the period of the new scientific areas development, such as the control theory, reliability theory, technical diagnostics, probability theory and its section – Queuing theory and etc. – theoretical foundations of the CMTV operation are significantly enriched.

Thus, the CMTV operation should be an integrated technological process (ITP CMTV), which includes the technological processes of using the CMTV for their intended purpose, the management technological processes, as well as the technological processes of technical operation (maintenance and repairing) of the CMTV. All these processes in the market management mechanism, for the most part, are entrepreneurial, so the goal of most of them is to make a profit. In this regard, it should be recognized that the third synergetic paradigm of cargo vehicles operation has already been established.

References

[1] Vedenyapin G V 1973 *General method of experimental research and experimental data processing* (Moscow: Kolos) 199 p

[2] Govorushchenko N Ya 2011 *Systems engineering of motor transport (calculation research methods)* (Kharkiv: KhNADU Publishing house) 292 p

[3] Zhdanovsky N S, Alliluev V A and Mikhlin V M 1973 *Diagnostics of tractor engines using electronic devices* (Leningrad: LSKhI Publishing house) 127 p

[4] Ozornin S P 2017 *Modern paradigm of transport and technological machines operation: monograph* (Chita: Transbaikal SU Publishing House) 195 p

[5] Volkov V P, Matejcik V P, Nikonov O I, Komov P B, Gritsuk I V, Volkov Yu V and Komov E A 2013 *Integration of cars technical operation in the structures and processes of intelligent transport systems: monograph* (Donetsk: Knowledge) 400 p

[6] Komov A B, Komov P B and Gritsuk I V 2010 *Improving reliability of the vehicle transmission. Monograph* (Donetsk: DonIZhT Publishing House) 144 p

[7] Brekhovskikh S M, Prasolov A P and Solinov V F 1995 *Functional computer systematics* (Moscow: Mechanical Engineering) 551 p
[8] Ozornin S P and Berdnikov I E 2017 Modeling the processes of the technical service system of transport and technological machines to improve its functioning quality. *Systems. Methods. Technologies*. **2(34)** 13–20 https://doi.org/10.18324/2077-5415-2017-2-13-20
[9] Haken G 2003 *The mysteries of Nature. Synergetics: the doctrine of the interaction* (Moscow, Izhevsk: Institute of Computer Researches) 320 p
[10] Kapitsa S P, Kurdyumov S P and Malinetsky G G 2001 *Synergetics and forecasts of the future* (Moscow: Editorial URSS) 288 p
[11] Prigozhin I R and Kondepudi D 2002 *Modern thermodynamics. From heat engines to dissipative structures* (Moscow: Mir) 464 p
[12] Prigozhin I R and Stengers I 2003 *Time, chaos, quantum. To a solution of the time paradox* (Moscow: Editorial URSS) 240 p
[13] Borodkin L I 2005 *History and synergetics. Research methodology* (Moscow: KomKniga) 184 p
[14] Friedel R 2007 *A Culture of Improvement* (London: England) 600 p
[15] Pribram K H 2002 *The Corsini Encyclopedia of Psychology and Behavioral Science* vol 3, W E Craighead, Ch B Nemeroff eds (Wiley) 1952 p
[16] *Convergence of Knowledge, Technology and Society: Beyond Convergence of NanoBio-Info-Cognitive Technologies*, M C Roco, W S Bainbridge, B Tonn, G Whitesides (eds). (New York, London: Springer) 2013
[17] Jean-Paul R, Claude C and Brian S 2017 *The Geography of Transport Systems* (New York: Routledge) 440 p

IOP Conf. Series: Materials Science and Engineering **832** (2020) 012064 doi:10.1088/1757-899X/832/1/012064

Digitalization of the English language correspondence course for engineering students

T Y Polyakova and O V Gorbacheva

Moscow Automobile and Road Construction University (MADI), 64, Leningradsky Pave., Moscow, 125319, Russian Federation

E-mail: 89169451288@yandex.ru

Abstract. The paper is devoted to the problem of digitalization of a foreign language course for engineering students. Realizing the advantages of the educational process digitalization in engineering universities, MADI Foreign Languages Department launched the project aimed at the development of digital materials for teaching English. The course is addressed to the students of the University correspondence department at the level of baccalaureate. The digital English materials can be used as an integral part of the traditional course providing flexibility of self- study in any place with an Internet connection, self-paced learning and immediate assessment and feedback. The digital materials are intended to help correspondence students to overcome the main difficulties they have while studying English and to improve their learning outcomes. The paper gives the analysis of the definitions of the terms "digitization", "digitalization", and "digital transformation", shows the difference between them, describes the aims and content of the digital materials, their structure, and the main language skills they develop.

Keywords: digitalization, English correspondence course, engineering students.

1. Introduction

The modern world is changing fast due to the development of new techniques and technologies. Digitalization is one of the most relevant innovations of our life. A lot of the countries pay special attention to this process. In 2018, Russia launched the national project "Digital Economy" and its implementation will take five years from 2019 to 2024. The key objectives of the national project are to increase funding for the development of digital economy, to create a stable and secure information and telecommunication infrastructure for high-speed transmission, processing, and storage of large amounts of data available to all organizations and households, to use mainly domestic software by state bodies, local governments and organizations [1]. But even now in our professional activities and everyday life we see more and more examples of digitalization.

Discussing digital solutions, as a rule, we use the terms "digitization", "digitalization", and "digital transformation". The analysis of these terms definitions show the difference between them. According to Gartner's IT Glossary, "Digitization is the process of changing from analog to digital form, also known as digital enablement. In another way, digitization takes an analog process and changes it to a digital form without any different-in-kind changes to the process itself." [2]. In other words, it is a technical process, the results of which facilitate our life. Thanks to the fact that a great number of documents, files, photos are in a digital form now, it is not necessary to spend hours looking through a huge manual for actual information in order to repair a device or a car, we do not need thick photo albums to store them in the bookshelves or wardrobes. Many of us cannot imagine their lives without smartphones that help us organize events, remember important dates, business meetings or current affairs.

If by "digitization" we mean the process of digital enablement, the term "digitalization" denotes "the use of digital technologies to change a business model and provide new revenue and value-producing opportunities; it is the process of moving to a digital business." [3]. Thus, through digitalization we are finding the ways to apply digital technologies most effectively. You can use them in various fields such as medical care, for instance, when nurses use portable vein finders for blood tests, or when auto mechanics use a laptop to troubleshoot a car, or when teaching students with methods of distant education, etc. Digitalization penetrates our personal life: we chat instead of making telephone calls. The number of smartphones in use increased from 23,000 in 1980 to 7,2 billion in 2015 which makes approximately one smartphone per person on Earth [4]. Digitalization increases the speed of people's work and makes it more efficient.

Finally, the third term in question is "digital transformation" that "refers to anything from IT modernization (for example, cloud computing), to digital optimization, to the invention of new digital business models. The term is widely used in public-sector organizations to refer to modest initiatives such as putting services on-line or legacy modernization." [5]. Thus, according to the author the meaning of the term "digital transformation" is closer to the meaning of the term "digitization" than to "digital business transformation." In reality, the process of digital transformation requires specific organization of activities and consequently specific competences from professionals.

Speaking about digital solutions, we should bear in mind that all the three processes are closely connected and related. First, we digitize information (digitization), second, we digitalize processes and roles that make up the operations of a business or any other activity (digitalization) and, finally, we digitally transform the activity and its strategy (digital transformation). Each stage is quite necessary but taken separately it is not sufficient. The nature of digitization and digitalization processes is technological, and the nature of digital transformation is not. The latter concerns the customer whoever he or she is: a patient, a passenger, or a student. In the paper, we use the term "digitalization" in the narrow meaning as discussed above and traditionally in its wide meaning covering all the three processes [6].

Digitalization has become an integral part of all spheres of modern society, including the system of higher education. In Russian engineering universities, digital solutions are being introduced at various levels: in administration, educational environment, research and educational process.

Digital transformation in the field of administration results in the growth of efficiency in the interaction between all the departments of the university, the decrease of time losses in the process of decision-making.

Digitalization also contributes to the creation of innovative educational environment. The universities modernize the equipment of the lecture- and classrooms, do a lot to develop digital libraries and digital campuses, renovate IT resources, work out training simulators, virtual laboratories, use more and more e-textbooks, multimedia teaching materials including videos, cartoons, or/and online presentations [7].

In the sphere of research up-to-date software, hardware and equipment are being used. Digital technologies provide unprecedented opportunities for scientific communication and international cooperation.

As far as educational process is concerned, first of all universities are aimed at training specialists ready to implement digital solutions in various fields of professional activity. So educators are busy with determination of the core competences necessary for digitization, digitalization, and digital transformation in different fields of economy, design of basic models of digital competences development and methods of their implementation, providing efficient personal trajectory of students' achievements. For this purpose, universities offer courses of training and retraining specialists in the field of digitalization [8].

The establishment of university expert digitalization centers stimulate further digital transformation of the educational process that offers a lot of advantages to students within formal, informal, open, additional, life-long education. Among them, it is necessary to mention the following:

- accessibility for the students who live far away from the university (possibly in another town) without waste of time for transportation;
- flexibility of self- study in any place on condition of an Internet connection;
- self-paced learning having the opportunity to skim the material you know and to focus on the material requiring more time and effort to grasp;
- the opportunity to have immediate assessment and feedback;
- accessibility for those with restricted mobility or busy work schedule.

At the same time, specialists in the field of higher education underline that a modern university should have a combination of both modern information technologies and real direct communication between students and teachers, experts or scientists. One of the ideas suggested is to replace the standard set of theoretical lectures with on-line products, increasing the number of contact hours, aimed at securing the material and improving practical and project activities of learners.

Understanding the unique possibilities of digital technologies, the educators are beginning to use them wider and wider in one course and at the same time for a broader scope of disciplines [9]. The discipline "Foreign language" is among them.

2. Problem

Teaching a foreign language in engineering universities is aimed at the development of foreign language communicative competence necessary to use a foreign language in the professional activity. The discipline has a rich background connected with wide implementation of various technical teaching aids. Long ago, in the middle of the 20th century the specific peculiarities of the subject made it necessary to use audial, visual and audiovisual aids. Besides using the traditional chalkboards, charts, models, slides, demonstrations, flannel graphs, flash cards, pictures, photographs, foreign language teachers were perhaps the first to bring the radio, a record player and a tape recorder in the classroom. The most popular modern teaching aids in use at present are English language labs, computers, overhead projectors, power point presentations, Internet, on-line dictionaries, on-line encyclopedias, VCDs, DVDs, DVD players.

Some time ago, from 10 to 15 years, a great variety of on-line English schools and courses were developed and offered to consumers. Some private schools are quite successful and popular among learners now. The courses use different online-platforms and try to attract as much students as possible. For example, the school "Englex" first started in 2011, and now it has over 14000 graduates [10]. Tutors use "Skype" as the way of communication with learners. "Lingualeo" provides teaching on the interactive interface basis [11]. As the school provides most of the teaching materials free of charge, they have in total about 18 billion visitors and learners. But some schools are not so popular for some reasons and quite often they have to stop their activity. However, it is evident that for many reasons digitalization of the foreign language courses makes them very attractive for learners.

The analysis of the theory and practices of foreign language teaching shows that introducing digital innovations in this process it is necessary to preserve an optimal combination of digital teaching materials and real face-to-face communication between teachers and students. Therefore, MADI Department of Foreign Languages had to solve a number of problems for digitalization of the course. It was necessary to answer the following questions:

- Who are the target audience for digital teaching materials?
- What are the main difficulties of learners that the digital materials will help to overcome?
- What part of the course is most appropriate for digitalization?
- What is the form and the structure of digital materials?
- Do foreign language teachers need any specific skills for working out digital teaching materials?

The Federal State Educational Standards (FSES) for Higher Education determine precisely the learning outcomes for graduates of engineering baccalaureate [12]. The discipline "Foreign Language" is a compulsory subject that is included in the basic part of the "Humanitarian, social and economic cycle." At the end of the course students should acquire general cultural competence in the sphere of a foreign language (either English, or German, or French), that is formulated in many FSES as follows: "the ability to communicate in oral and written forms in the Russian and foreign language for solving tasks of interpersonal and intercultural interaction" [13]. In order to acquire this competence a student should have sufficient foreign language knowledge, skills and abilities. Among them, the graduate should know:

- the basic vocabulary necessary for oral and written interpersonal and intercultural communication;
- grammar typical for oral and written intercultural communication.

The student should have the following skills:

- reading skills necessary to understand polytechnic, intercultural and professionally oriented texts with the help of a dictionary;
- speaking skills necessary for producing a logically structured monologue within the topics studied;
- listening skills necessary to understand a monologue and a dialogue within the topics studied.

The student should have the following abilities:

- the ability to use four types of reading (critical, skimming, scanning and survey types of reading);
- to produce a monologue and a dialogue within the topics studied.

The requirements mentioned above do not cover all the skills and abilities necessary but even their enumeration shows that achieving such teaching objectives takes a lot of time and effort from the students. According to the Curriculum, engineering students of a daytime department have from 200 to 350 contact hours of "Foreign Language" and at the same time the opportunity to participate in some extra-curricular work as well. Therefore, the majority of students of the daytime department demonstrate rather good results of learning at the final exam. Students of the evening department also have enough number of contact hours and their results are quite satisfactory. On the contrary, the engineering students of the correspondence department study the discipline also for two years and have the same workload, but according to schedule, they have only 16 contact hours per a term. In spite of a great difference in schedule and training time, according to the educational program all the requirements are the same for students of daytime, evening and correspondence departments. We must admit that students achieve the required level of communicative competence having to do a lot of self-study work and nevertheless some of them fail in the exam.

For that reason, the Department of Foreign Languages took the decision that the target audience for the digital materials is engineering students of MADI correspondence department. As most correspondence programs are those of baccalaureate and about 90% of students of this department study English we decided to start with digitalization of teaching materials for learning English in correspondence baccalaureate.

The results of the final English examination show that learners have difficulties with proper pronunciation of the vocabulary, the meaning of the words and their usage in oral and written speech. The students demonstrate insufficient level of grammar knowledge, interpretation of grammar structures in oral and written speech. Reading skills are also to be improved in order to provide better understanding of written texts. While listening students show rather low level of comprehension. Besides, the correspondence students have a lot of problems of non-linguistic character. They are young people who work long hours and sometimes their work prevents them from attending classes. Some of them live far from the University.

Digital English learning materials are expected to be effective in overcoming the difficulties mentioned above and some others for two reasons. Firstly, the correspondence students are representatives of Generation Z who were born at the beginning of the 21th century, and in contrast to the previous generation, they never knew the world without access to the Internet, social networks, computer games or virtual communication. They are not "Digital Immigrants" but they are "Digital Natives" and digital technologies are an integral part of their life. They use them most effectively [14]. Secondly, digital materials organize and control learners' self-study work. Students can carry out tasks any time they want, stop at any moment if necessary and continue studying when they again have time for that. It gives freedom of choice and planning their lives. Once the task is ready, they can send it to their tutor for checking via e-mail.

The implementation of the project has three stages: digitization of teaching materials, their digitalization and digital transformation of the educational process. The work at these three stages requires implementation of modern digital technologies from a teacher of a foreign language. This situation is a new challenge that sets the task of continuous professional development of the teacher. Before designing the digital materials it was necessary for the English teacher to attend the course "Information systems, electronic educational environment, the fundamentals of on-line course design" provided by MADI Institute of Professional Development. The course gave the opportunity to improve digital technology skills of teachers and apply them for teaching materials design. Besides, at each stage of the project, it was possible to consult IT specialists to solve the problems.

3. Materials and methods

The purpose of the English course for correspondence students is to prepare them for using a foreign language in their future professional activities, both in written and oral communication. In addition, the acquired knowledge can serve as a basis for further self-education. The selection of the content of training is based on the survey professional foreign language needs of engineers [15].

The teaching content and methods are presented in the textbook "English for Engineers" [16] and the digital materials were worked out to supplement it. The textbook is the main part of the educational set of materials. Its goal is to provide a basic course of English for engineering students of the first and second years of study. The audio supplement (CD) is used in parallel with the textbook, and the manual includes additional texts with tasks for developing various types of reading at the final stage of training. The textbook is aimed at the development of learners' reading, speaking, listening and writing skills.

As one of the main objectives of the English course in non-linguistic universities remains teaching reading. The textbook develops four types of reading – critical, skimming, scanning and survey types of reading, the choice of which is determined by the task set when working with the original literature: authentic general scientific articles, monographs, regional and cross-cultural literature, technical documentation, materials of polytechnic and popular science, information that determines our behavior in everyday life.

As far as the development of oral speech skills is concerned, the textbook prepares students to implement the prepared monologue in the form of a presentation or report, which, as recent studies show, is the most typical for the sphere of professional communication. An attempt is made to prepare students to participate in elementary dialogue (acquaintance, presentation, expression of request, etc.) and polylogue (conversation or discussion), which implies the necessity of developing the ability of expressing their own opinion, approval or disapproval of other people's statements, requesting information, etc. The textbook provides the ability to perceive oral texts of a polytechnic nature, presented at an average pace.

In addition, the textbook prepares students for written communication in English. The plan provides the formation of skills necessary to fill out various questionnaires and forms, conducting informal correspondence and writing certain types of business letters. Along with this material, the skills of fixing information extracted from the written text (in the form of annotations and abstracts) and translation of texts in the specialty from English into Russian are traditionally developed.

The textbook has two parts. The first part "English in Practice" contains 13 Units. They have identical structure and consist of the following sections:

- Language Material

- Listening
- Speaking
- Reading
- Writing.

Though the sections are differentiated according to the main type of speech activity in focus, the tasks of each section involve the interrelated development of reading, listening, speaking and writing skills, which is natural for verbal communication in real life.

According to the University foreign language program, correspondence students study three Units for the first term. That means that supplementary digital materials should correspond the teaching content of the three Units. Before preparing the digital materials, the Foreign Language Department undertook a pilot survey and interviewed quite a number of correspondence students in order to find out their preferences and wishes concerning digital teaching materials. There were two possible approaches: to give a detailed grammar description in a written form or to explain vocabulary, grammar rules imitating the presence of the tutor. The vast majority of the respondents said that they preferred to see a tutor explaining grammar rules on the screen of their computers instead of reading explanations of language material themselves. Taking this information into account and the recommendations received form IT specialists, we began the process of digitization on the platform of iSpring Suit which is an addition to Microsoft PowerPoint. It is a rather new, but very successful program which serves for creation of various courses. With the help of iSpring Suit, it is easy to construct your own course of studying by adding interactive interface, tests, presentations and different educational simulators.

4. The Discussion of the Results

Following the Unit structure, it was necessary to begin with the first section "Language Material" that contains subsections “Vocabulary” and “Grammar”. They involve working on language material, which is a kind of means of implementing speech communication.

Working on language material begins with the introduction and consolidation of “Vocabulary”. The textbook activates about 1300 words and phrases that belong to the most frequent general scientific vocabulary. Each Unit has approximately 45 words. These words are digitized and recorded in the digital course separately, one by one. The student is able to listen to any word by clicking on a chosen vocabulary item. The word can be played at a normal pace or in slow motion.

New words are also introduced in sentences. Various exercises are offered to activate new words. Two-purpose exercises are of particular note: subtraction a word from a row, training skills of rapid reading on the one hand and increasing the repeatability of lexical material on the other hand.

Particular attention is paid to the expansion of the potential vocabulary of students. In this regard, in each section there are exercises for recognition and correct interpretation of international words. There is a series of exercises aimed at mastering the word-building elements. The 19 elements included in the textbook are the most typical of the sphere of professional communication of future engineers. This series of exercises ends with a small text containing previously studied derivatives of words built on worked word-formation models, meaning of which each student must deduce on the basis of the root and affix known to him.

There are also exercises to develop language guess, exercises on the formation of skills to use the dictionary, crosswords etc. Repeated use of the inserted and fixed words in other sections of the Units provides the frequent occurrence of vocabulary necessary for mastering active and passive vocabulary.

While creating exercises it was necessary to take into consideration the peculiarities of Generation Z. It is well-known that it is very difficult for representatives of Generation Z to hold the attention for a long time. According to the latest research [17], it became clear that the continuation of students' attention can be focused on one task for about eight seconds and after that they have to switch to another activity. When designing digital materials, it was necessary to introduce new educational material in small portions by dividing it into blocks using visual images in the form of tables, graphs and drawings. At the same time, each portion of the new material should be followed by its primary fixation. Exercises should be not too big and different in the form of execution. It is advisable to ensure the use of new material in all types of speech activities: listening, speaking, reading and writing, providing multiple repetition of the studied material in new situations. Some thematic videos can be created for the introduction of new educational material. So digitizing exercises for Vocabulary, we divided one rule into several portions. It is also very important to give exercises for drilling grammar rules after each part.

Designing exercises, it is necessary to take into account that a student should hear the pronunciation of all the words and phrases within the program as well as the texts. It is compulsory for further reproduction of sounds without any errors. Besides, it is possible improve traditional education by transferring it into a digital lecture with the teacher illustrating all the material in a digital format. Watching such lectures students will have the opportunity to stop the video at any time they want, rewind it, play it again as many times as they need, and thus repeat grammar rules and structures of each of the Units and learn the rules properly. Acting in this way, students will have enough time to absorb new information, fulfill all tasks and exercises and learn the words by heart without any haste. With the digital course, we can compensate the lack of academic hours at correspondence department without losing any detail of the educational program.

The first section of the Unit also provides work on "Grammar". The basis for the selection of grammatical material is the data of linguistic and statistical studies, revealing the most typical for the scientific and technical style of the structure. Students have the opportunity to get acquainted with the meaning of selected grammatical phenomena (illustrated with the help of a small written text), with their form (tables) and consolidate the skills formed in a series of communicative exercises. It should be noted that in the section "Students' Material" one can find grammatical comments, allowing a student, if necessary, to study or repeat a particular grammatical phenomenon.

Instructions for the exercises in this section are presented in Russian, so that students' attention could be focused on the phenomenon being studied. In the digital course, all tasks to the exercises are also given in Russian language. We can take as an example one of the first grammar rule, put it into a table, divide it into three different parts and give exercises for drilling after each part (table 1–3).

Table 1. Affirmative sentences with the verb “to be” in the Present Indefinite Tense /The first part/.

person/number	Sing.	Plur.
1 p	I am a student	We are students
2 p	You are a student	You are students
	He is a student	
3 p	She is a student	They are students
	It is a dog	

Table 2. Affirmative sentences with the verb “to be” in the Past Indefinite Tense /The second part/.

person/number	Sing.	Plur.
1 p	I was a student	We were students
2 p	You were a student	You were students
	He was a student	
3 p	She was a student	They were students
	It was a dog	

Table 3. Affirmative sentences with the verb “to be” in the Future Indefinite Tense /The third part/.

Person/number	Sing.	Plur.
1 p	I will be a student	We will be students
2 p	You will be a student	You will be students
	He will be a student	
3 p	She will be a student	They will be students
	It will be cold	

Exercises that follow each of the three parts should be short and different in the form of execution. It is obvious that drills for each part should include all types of activity such as listening, speaking, reading and writing.

Section "Listening" presents assignments to two texts so that students have the opportunity to listen to. Listening to the texts is preceded by brief introductory information and an instruction. Introductory information, on the one hand, introduces the student to the situation, creates conditions for the emergence of students listening motivation, on the other hand, it facilitates oral comprehension of the text, as it orients them to a certain content of the text and provides the necessary language support.

Executions of the tasks for the first text requires, as a rule, its general understanding. The control of understanding is achieved by performing question-and-answer exercises, with students being pre-familiarized with the questions. The information extracted when listening to the text can be used by students in subsequent assignments.

At the final stage of working on the text and for correct understanding, as well as to identify obscure fragments and eliminate the causes of this misunderstanding, students should write out all the unknown words.

The textbook and the digital course provide the gradual complication of the language material of the texts for listening. It is also recommended to gradually increase the rate of presentation of texts.

In accordance with modern requirements, the section "Speaking" provides tasks for the development of monological speech skills on the given topic. The list of words and phrases provides the necessary language support. Students can use them as samples to build sentences. Such a list of questions provides a monological statement consisting of 6–7 or more sentences. Particular attention should be paid to the students' expressing their own opinions. Correspondence students can send digitally typed audio of their speech as it was mentioned above.

There is an extra opportunity the digital course offers correspondence students: they can record their monologue and send it to the teacher. In this way, their speaking skills on a particular topic or theme can be checked and controlled by the teacher. There are six oral tasks in this section and thus six different topics should be recorded by the correspondence students. They are the following:

- The school I went to.
- My Institute.
- City traffic of future.
- The scientist I'd like a new street to be named after.
- One of the greatest inventions of the mankind.
- Choosing a tour route.

The student's monologue should take at least two minutes. He or she type their speech in the special program of the digital course and afterwards press the button and send it to the teacher.

In the section "Reading" there are four texts: A, B, C, D, each of them has a specific methodological task. In the selection of text material the main criterion was the informative value of texts and their compliance with the interests of the correspondence students of technical universities. Texts content data of particular interest to students were obtained with the help of questionnaires.

Most of the texts are samples of the original English and American literature. In some cases, they have been adapted for better understanding and reduced in order to facilitate the process of learning for the part-time students.

Text A is considered to be the main text for studying, and therefore it has the most developed system of tasks. First of all, there is a brief introduction to the situation that precedes the execution of the tasks. Before reading the text, students are offered the tasks aimed at developing probabilistic forecasting, which means predicting the whole content or its elements on the basis of what has already been perceived. This foresight is expressed in the textbook in the form of hypotheses about the most likely content of the text. Multiple choice exercises are often used for this purpose.

In addition, these tasks create a natural motive for the first reference to the text in order to test learners' own hypotheses, which provides practice in fast types of reading, usually skimming, which takes from two to five minutes. Then students are given the task to find fragments of the text of a certain content. As a rule, these passages are the most saturated in terms of activated language material: vocabulary and grammar. For this purpose, it is supposed to use search reading.

The discovered fragments are intended for practice in the learning form of reading, which involves a complete and accurate understanding of the text. The means of verification is written translation typed by correspondence students and send for the checking to the teacher.

The main requirement for the translation is its adequacy to the original. Particular attention should be paid to the language means, formalizing semantic links.

Based on the importance of reading aloud for the development of speaking skills (pronunciation training, the formation of automatic pronunciation, etc.), special attention is given to the development of reading mechanisms in the textbook. For each text A, a student can find a task to read the fragment aloud. Since reading aloud is practiced on passages that must previously be read silently and understood, for this purpose one of the fragments selected earlier for practicing the student's reading is chosen. Students are given the opportunity to get acquainted with a sample of the sound of this fragment. It is presented in a recording of a digital course. Then the student independently solves two problems: at the first stage – to achieve the correct reading of the text, at the second – to increase the speed of its pronunciation.

Text B is designed to develop the skill of "fast reading" and contains no more than 3–4 % of new words, the translation of which is given in brackets in the text. This makes it possible not to interrupt the reading process and not to waste time searching for the meaning of a word in the dictionary. It is planned to increase the rate of reading, ranging from 70 words per minute in the first lessons to 100 words per minute in the last.

The main task of working on text B is to achieve the understanding of the basic facts contained in the text, which is checked by means of a multiple choice test and other types of tasks.

The main task of working on text C is to develop the ability to subtract certain information from the text. To implement the task, the text is divided into fragments and provided with pre-text questions.

Within the digital course for correspondence students we make some changes for easier transformation of it into a digital format. The texts were recorded and transferred to a digital format so that the students could clarify all difficulties in pronunciation and have the chance to playback texts as many times as they need. Thus, they can read and listen to the text at the same time, and pronounce all the sounds, words and phrases correctly, without any mistakes. Besides, they can make a self-checking, for example, read a sentence, then play it and find out if his or her pronunciation corresponds to the original.

For the correspondence students there is also a section in the digital course with the tasks of writing business letters, postcards, filling different information and registration forms, landing cards and customs declarations. Below there is an example of the form for filling in.

Table 4. Registration form

Registration Form
Full name
Age
Nationality
Occupation
Home address
Length of course
Number of hours per day
Course starting date
Accommodation required or not required
Signature

The process of digitalization results in the digital English training materials. Digital transformation requires recommendations for their implementation in the educational process. The Foreign Languages

Department organizes a workshop for all the teachers on the methods of using digital materials. Teachers have to acquire new skills in a digital sphere. Later on digital English and other foreign languages teaching materials will be worked out for the students of different levels of high engineering education of daytime and evening departments.

5. Conclusions

Digitalization has become an integral part of all spheres of modern society, including the system of higher education. Digital solutions introduced in the educational process offer a lot of advantages. In Russian engineering universities, educators introduce them for a broad scope of disciplines. The discipline "Foreign language" is among them.

MADI Foreign Languages Department worked out digital English teaching materials for engineering students of the University correspondence department. This digital English course is intended for the development of foreign language communicative competence including reading, writing, listening, and speaking skills. The course is aimed at overcoming learners' difficulties and organizes their self-study work. The digital English course give correspondence students the opportunity of flexibility of self- study in any place with an Internet connection, self-paced learning and immediate assessment and feedback. The course provides students' involving in the learning process. They can listen to the texts or words, read or watch grammar material explained by the teacher, type the results of their homework and send them to the teacher.

The processes of digitization, digitalization, and digital transformation used for the development of the English correspondence course can be applied for the creation of similar courses for the students of other levels of higher education including daytime and evening departments.

6. Acknowledgement

We would like to thank the head of a digital course development at MADI, Professor and Doctor of Engineering S. V. Aleksakhin who leads the course of training specialists in the sphere of digitalization which is called "Information systems, electronic educational environment, the fundamentals of on-line course design." And also we appreciate his team for the productive collaboration. They provided us with all the necessary information within the digital course, supported and assisted us at any time we needed help.

References

[1] Passport of the national program "*Digital economy of the Russian Federation*" 2018. Retrieved from: http://government.ru/info/35568/

[2] *Gartner's IT Glossary, word digitization.* Retrieved from: https://www.gartner.com/it-glossary/digitization/

[3] *Gartner's IT Glossary, word digitalization.* Retrieved from: https://www.gartner.com/it-glossary/?s=digitalization

[4] Muro M, Liu S, Whiton J and Kulkarni S 2017 *Digitalization and the American Workforce* (Washington: Metropolitan Policy Program at Brooklings). Retrieved from: https://www.brookings.edu/wp-content/uploads/2017/11/mpp_2017nov15_digitalization_full_report.pdf

[5] *Gartner's IT Glossary, word digital transformation.* Retrieved from: https://www.gartner.com/it-glossary/digital-transformation/

[6] Bloomberg J 2019 *Digitization, digitalization, and digital transformation: confuse them at your own peril (Forbes Media LLC).* Retrieved from: https://www.forbes.com/sites/jasonbloomberg/2018/04/29/digitization-digitalization-and-digital-transformation-confuse-them-at-your-peril/#739d22c72f2c

[7] Ainslee J 2018 *Digitization of Education in the 21st century.* Retrieved from: https://elearningindustry.com/digitization-of-education-21st-century

[8] Irniger A 2017 *Difference between Digitization, Digitalization and Digital Transformation,*

IOP Conf. Series: Materials Science and Engineering **832** (2020) 012064 doi:10.1088/1757-899X/832/1/012064

Systems, Applications and Products. Retrieved from: https://www.coresystems.net/blog/difference-between-digitization-digitalization-and-digital-transformation

[9] Mavlyotova G 2018 Digitalization of modern institutions of higher education *Economical security and quality* **3(32)** 5–7. Retrieved from: https://cyberleninka.ru/article/v/tsifrovizatsiya-v-sovremennom-vysshem-uchebnom-zavedenii

[10] *Online-school "Ingles"*. Retrieved from: https://englex.ru/

[11] *English language learning Service.* Retrieved from: https://lingualeo.com/ru

[12] Order of the Ministry of education of Russia from 12.03.2015 No. 201 "*On approval of Federal state educational standard of higher education in the direction of training 08.03.01 Construction (undergraduate level)*" (Registered in Ministry of justice of Russia 07.04.2015 N 36767). Retrieved from: http://fgosvo.ru/080301

[13] Ministry of education of Russia from 14.12.2015 N 1470 "*About the statement of Federal state educational standard of higher education in the direction of training 23.03.03 Operation of transport and technological machines and systems (undergraduate level)*" (Registered in Ministry of justice of Russia 18.01.2016 N 40622). Retrieved from: http://fgosvo.ru/230303

[14] Polyakova T Yu 2018 Peculiarities of the Modern Generation of Students and their Influence on the University Foreign Language Training Development *Vestnik of Moscow State Linguistic University* **2** 43–54. Retrieved from: http://libranet.linguanet.ru/prk/Vest/2_796.pdf

[15] Polyakova T Yu and Karelova D G 2018 Engineering Students' Needs in Foreign Language Studying in Russia, Teaching and Learning in a Digital World *Proceedings of the 20th International Conference on Interactive Collaboration Learning*, vol 2, New York: Springer, pp 481–90. Retrieved from: https://link.springer.com/chapter/10.1007%2F978-3-319-73204-6_53

[16] Polyakova T Yu and Sinyavskaya E V 2016 *English for Engineers* (Moscow: publishing center "Academy»)

[17] Kirilova E 2015 *Generation Z and its place in history. Theory of generations. Generations X, Y and Z.* Retrieved from: https://fb.ru/article/210221/pokolenie-z-i-ego-mesto-v-istorii-teoriya-pokoleniy-pokoleniya-x-y-i-z

IOP Conf. Series: Materials Science and Engineering **832** (2020) 012065 doi:10.1088/1757-899X/832/1/012065

Maintenance and repair management taking into account specialization, cooperation and unification within a united bus company consisting of several branches

N V Pozhivilov[1], V A Maksimov[1], G A Krylov[1] and A A Zavgorodniy[2]

[1]Moscow Automobile and Road State Technical University (MADI), 64, Leningradsky ave., Moscow, 125319, Russia
[2]Research Centre, Mosgortrans, 22/21, b. 1, Raushskaya naberezhnaya, Moscow, 115035, Russia

E-mail: poj-nikita@mail.ru

Abstract. The issue of maintenance and repair management of fleet in a bus transit agency consisting of several branches is presented in the article. Technical base development takes place due to increasing of effectiveness as a result of specialization and cooperation of maintenance and repair, as well as using maximum potential, reducing and optimal costs allocation. The minimum total reduced cost was taken as a criterion for the efficiency of production placement, focusing on the costs associated with the choice of specialization and subsequent cooperation of maintenance (mainly the costs of transportation of units and vehicles between the branches). The management approach of cooperating in the issue of maintenance and repair of buses is aimed at the distribution of a particular type of work between the technical bases of different branches of the united transit agency. The applied method is based on the common approach to the solution of the "transport problem" and adapted specifically for the united transport company involving two and more separate branches. The method was tested on the united agency operating urban buses in Moscow, aiming to allocate fleet maintenance with regard to its needs and resources.

1. Introduction

The main goal of the development and improvement of the technical base of a transit agency is to increase the efficiency of the technical base based on specialization and cooperation in maintenance and repair of fleet due to the maximum use of the existing technical potential, reduction and rational use of capital investments.

The directions that determine the form of organization of the technical base include [1–6]:

- concentration of the technical base;
- specialization in the types of maintenance and repair;
- cooperation of different technical departments of transit agency;
- creation of a centralized service for fleet maintenance and repair.

Concentration of a technical base shall lead to a concentration of the production funds necessary for ensuring fleet technical readiness and creation of conditions of high-performance human labor within the company [3].

IOP Conf. Series: Materials Science and Engineering **832** (2020) 012065 doi:10.1088/1757-899X/832/1/012065

Distribution of the scope of work on maintenance and vehicle repairs leads to the creation of new facilities at a higher organizational, technical and technological level, while the number of departments, shops and sections in the overall structure of the united enterprises is reduced, but their grade, staffing and technical equipment are significantly increasing.

A transit agency with incomplete maintenance or repair works is suitable only when the organization of some technical activity is economically unreasonable, i.e. when the establishment and support of the full program of technical base requires significant capital investments, the use of expensive equipment, large production areas.

Distribution as a condition for improving the technology and organization of production would not be acceptable with any consolidation company, but one that is accompanied by a decrease in the diversity of its products or service. This can be achieved by concentrating technologically uniform production or service.

The main goal of the specialization is to create such a concentration of uniform service at the company that allows for the most efficient use of high-performance equipment and technology. Specialization is characterized by the similarity of the technological process, the equipment, the qualification and occupational structure of the staff.

Two or more agencies can work in cooperation providing each other with mutual services for the maintenance and repair performance. Cooperation in such cases will make it possible to increase the degree of utilization of fixed assets and reduce the total costs of the technical requirements.

There is internal and external cooperation [1]. Internal cooperation shall take into account the cooperation of the most labor-intensive technological processes within the framework of individual transit agencies. External cooperation involves the vehicles repair and its units at other specialized shops and technical bases.

The article discusses the advisability of creating a concentration of specialized service for fleet maintenance or repair due to the internal cooperation of several branches (various territorial agencies) of a united transit agency. Territorial sites are subsidiaries which operate the fleet of city buses that provide passenger transportation on a fixed network of regular routes, and also have its own technical base, whose potential is sufficient to complete the main scope of the maintenance and repair of buses.

2. Different approaches to creating the conditions for specialization, distribution and cooperation of fleet maintenance and repair at transit agencies

Labor productivity is increased and costs are reduced as a result of concentration and specialization.

According to the expert assessment, the following preferable sequence of concentration of types of repair and preventive activities was identified [7]:

- Maintenance – the most preferable for concentration.
- Engine repair.
- Repair of fuel equipment; aggregate repair and tire repair.
- Repair of electrical equipment and gearbox repair
- Ongoing repair and daily maintenance – less preferable for concentration.

Types of repairs are specified from highest priorities to the least preferable for the types of work concentration and centralization.

The total effect of all organizational and technical measures caused by the specialization should be taken into account when evaluating the effectiveness of specialization of works, sites and zones separately as well as in general.

Specialized posts have the highest level of mechanization of work and the level of throughput, but they can carry out technological operations of a limited range. Therefore, specialized posts are organized at transit agencies with a large amount of fleet [6].

There are 4 types of specialization in automobile transport agencies [2]:

- Specialization by the type of the vehicle provides for maintenance and repair of a transport of a specific type of land transport, for example, buses, trolley-buses and trams.
- Specialization by brands when vehicles of the same brand are being repaired.

- Specialization by structural elements when one enterprise concentrates maintenance of similar parts regardless of the type and brand of vehicles. For example, repair of diesel fuel equipment, hydraulic or automobile electrical equipment, crankshafts, gearboxes, etc.
- Specialization by maintenance processes for the implementation of a specific type of work: general mechanical works, diagnostic, tire, etc.

Specialization and concentration are limited since as the concentration increases the cost of part transportation to the maintenance place increases significantly. In addition, the organization and production technologies have a significant impact on this process [8]. In addition, the level of specialization and concentration affect each other being in a correlation and influence transportation costs [9].

It is recommended to take a minimum total reduced costs as a criterion for the efficiency of service management. Meanwhile, it is needed to take into account operating costs, the necessary capital investments connected with the required changes as well as the costs connected with the choice of specialization option and subsequent cooperation (mainly, transportation costs of units and vehicles).

The rational allocation of different shops and labor force in branches of the united transit agency consists of optimal allocation of specific types of maintenance and repair works, ensuring a minimum of the total reduced costs.

The distance of an economically viable transfer of a bus from the i-th place (for example, company location) to centralized maintenance can be calculated as follows [5]:

$$l_{ij} \leq \frac{C_{kai} - C_{kaC}}{2S^a(1 - \beta_{ij}^a)}$$

where C_{kai} – specific reduced costs for the k-th type of bus maintenance at the j-th placement point (site);

C_{kaC} – the same in a centralized maintenance;

S^a – prime cost of 1 km of the bus run;

β_{ij}^a – the capacity utilization of the run when driving a bus from the i-th to the j-th point and back.

3. Methodology for the rational distribution of the scope of bus maintenance by branches at the united bus company

Given the limited capital investments as well as the fact that reconstruction and technical re-equipment are carried out under the conditions of existing transit agency, it is advisable to carry out specialization stage-by-stage with the priority development of those organizational and technological forms that provide the maximum efficiency increase. At the same time, the stages of specialization of technical base should be coordinated in such way that the previous stage creates technological and resource prerequisites for the effective development of subsequent stages. The overall effect of specialization should be maximum.

The analysis of actual and normative fleet downtime allows the company to identify the specific processes of maintenance that cause excessive downtime, establish the causes and outline a way to reduce downtime and justify the specialization of the various territorial technical bases (located in different branches) of the united transit agency during the development.

In general, an integrated approach for creation of the technical base system in a united transit agency based on specializations and cooperation in bus maintenance provides the following main steps [5]:

The choice of organizational and technological forms of a rational system for centralized fleet maintenance.

Determination of the main types of specialization for each element of the technical base.

Establishment of a rational structure and optimal capacity of technical base of united transit agency.

Determination of the scope of work performed by different branches within the cooperation.

Development of an approach to create a centralized system for the supply, maintenance and repair of technological and engineering equipment.

The achievement of this goal shall meet the needs for the complete and necessary maintenance and repair work of the fleet in order to achieve the planned value of the coefficient of technical readiness with rational costs for the operation of technical base branches.

This task of optimization of the fleet maintenance system can be solved by following the five steps:

I step. Initial data formation

The following initial data is necessary for optimal concentration and cooperation:

- Data for calculating the required labor and production capacity needed to provide the fleet maintenance in a united transit agency.
- Specific costs for the maintenance of technical equipment for each base of the united transit agency.
- Specific costs for bus and its units transportation between different technical bases of the united transit agency.

The initial data can be implemented as shown in table 1 (in the case of the united transit agency consisting of the three branches with their own technical bases as an example).

Table 1. Initial data for the task of finding an optimal allocation of the value of maintenance work

Branches of united transit agency	Specific costs for bus transportation, rub.			Specific costs for one maintenance, rub.	The maximum capacity of annual amount of maintenance, qt.	Planned annual amount of maintenance by specific branch, iqt.
	Between 1 and 2	Between 1 and 3	Between 2 and 3			
№1	C_{12}	C_{13}	-	C_1^M	T_1^H	T_1
№2	C_{12}	-	C_{23}	C_2^M	T_2^H	T_2
№3	-	C_{13}	C_{23}	C_3^M	T_3^H	T_3
In total					$\sum T_3^H$	$\sum T_3$

where C_{12} – specific costs for bus transportation between branch no. 1 and branch no. 2.
C_{13} – specific costs for bus transportation between branch no. 1 and branch no. 3.
C_{23} – specific costs for bus transportation between branch no. 2 and branch no. 3.
C_1^M, C_2^M and C_3^M – specific costs for one maintenance on branch no. 1, branch no. 2 and branch no. 3 respectively.
T_1^H, T_2^H and T_3^H – the maximum possible annual amount of maintenance potentially possible on branch №1, branch no. 2 and branch no. 3 respectively.
T_1, T_2 and T_3 – planned annual amount of maintenance by specific branch for buses operated on branch №1, branch no. 2 and branch no. 3 respectively.

II step. Identifying the possibility of optimization

The minimum prerequisite reflecting the opportunity to provide maintenance of the united transit agency:

$$\sum T_i \leq \sum T_i^H. \tag{1}$$

Only if (1) is satisfied, it is advisable to proceed to the third step.

III step. Formation of restrictions allowing for the performance of all the maintenance works on a full scale

The total amount of maintenance of buses operated at branches no. 1, no. 2 and branch no. 3:

$$\begin{cases} x_1 + x_2 + x_3 = T_1 \\ y_1 + y_2 + y_3 = T_2 \\ z_1 + z_2 + z_3 = T_3 \end{cases} \tag{2}$$

where x_1 , x_2 and x_3 are the volumes of maintenance for the fleet operated on branch no. 1 and are made on branch no. 1, branch no. 2 and branch no. 3, respectively.

y, y_2 and y_3 are the volumes of maintenance for the fleet operated on branch no. 2 and are made on branch no. 1, branch no. 2 and branch no. 3, respectively.

z_1 , z_2 and z_3 are the volumes of maintenance for the fleet operated on branch no. 3 and are made on branch no. 1, branch no. 2 and branch no. 3, respectively.

Then the maximum amount of maintenance of buses performed at the technical base on branch no. 1, branch no. 2 and branch no. 3:

$$\begin{cases} x_1 + y_1 + z_1 \le T_1^H \\ x_2 + y_2 + z_2 \le T_2^H \\ x_3 + y_3 + z_3 \le T_3^H \end{cases} \quad (3)$$

The cost of performing the maintenance at the technical base on branch no. 1:

$$\sum C_1^M = C_1^M \cdot x_1 + (C_1^M + C_{12}) \cdot y_1 + (C_1^M + C_{13}) \cdot z_1 .$$

The cost of performing the maintenance at the technical base on branch no. 2:

$$\sum C_2^M = (C_2^M + C_{12}) \cdot x_2 + C_2^M \cdot y_2 + (C_2^M + C_{23}) \cdot z_2 .$$

The cost of performing the maintenance at the technical base on branch no. 3:

$$\sum C_3^M = (C_3^M + C_{13}) \cdot x_3 + (C_3^M + C_{23}) \cdot y_3 +\cdot C_3^M \cdot z_3 .$$

IV step. Target function generation (minimum of total costs).
Target function has to be represented as follows:

$$\sum C_1^M + \sum C_2^M + \sum C_3^M \to min. \quad (4)$$

It is necessary to find x_i , y_i , z_i that satisfies the target function (4) and condition restrictions (2) and (3).

V step. Finding an optimal solution using the transport problem approach.
The optimization problem was solved by constructing a support plan and its improvement by the method of potentials. Eventually, such x_i , y_i , z_i were found when constrains (2), (3) are implemented as well as target function (4) is achieved.

Analytical manual solution of a transport problem of a large scale (by the number of "consumer-suppliers") can be difficult due to the large number of actions necessary to obtain an optimal plan. It is advisable to use special software or conduct calculations online on the Internet using one of the many web sites to automate the calculation [9].

VI step. Analysis of eventual results
The analysis of the obtained results has to be carried out and the decision of the optimal allocation of work between different technical bases of transit agency has to be made. When deciding in favor of the optimization there should be the model of organizational measures aimed at systematic and rational embodiment of the resulting model optimization.

4. Collecting initial data for the calculation of the rational allocation of fleet maintenance by the example of the maintenance zone of the united transit agency in Moscow

Approbation of the methodology was carried out at the united transit agency including three branches with their own technical bases. The transit agency operates urban bases on regular routes in Moscow. The goal was to obtain the system of rational allocation of maintenance by shares on three technical bases. The special point of the task was that such type of work had been stopped on the 1st technical base and the calculation according to the methodology should give the results meeting such constrains and propose the solution when the eventual allocation of shares of the fleet maintenance between two other technical base are optimal. The economic effect is not compulsory, but preferable in this specific task.

The initial data was collected by the following stages.

On the 1st stage the costs on bus transportation between technical bases were calculated (table 2).

Table 2. Costs on bus transportation between technical bases (branches) of united transit agency in Moscow

Bus transportation	The length of the transportation route, km	Fuel consumption, lit.	Fuel consumption costs, rub.	Full operational costs of the transportation, rub.	
				One way	Return
between **no.** 1 and **no.** 2	6.5	2.1	82.7	330.8	661.6
between **no.** 1 and **no.** 3	10	3.24	127.7	510.8	1021.6
between **no.** 2 and **no.** 3	11	3.6	141.8	567.2	1134.4

Further consumptions are made:

- Fuel consumption of bus LiAZ-529222 – 32,4 lit/100 km.
- Average cost for 1 lit of fuel – 39,4 rub.
- Fuel consumption is 25 % of full operational costs.

On the 2nd stage the planned annual amount of maintenance of buses operated on three branches has to be calculated. The data is recommended to be presented on a year according to annual maintenance program and fleet write-off and supply (table 3).

On the 3rd stage the maximum capacity of maintenance by technical bases of all branches (2nd and 3rd branches in our case) was calculated. The data was provided by the transit agency based on the previous period of work.

Obtained data (*I step. Initial data formation*) allow us to carry out a calculation according to the proposed methodology and allocate the full amount of maintenance by shares on different technical bases.

Table 3. Initial data for optimal allocation of maintenance by technical bases of three branches of a united transit agency in Moscow

Branches of united transit agency in Moscow	Specific costs for bus transportation, rub.			Maintenance		
	between no. 1 and no. 2	between no. 1 and no. 3	between no. 2 and no. 3	Specific costs of one maintenance of bus, rub. C_i^M	The maximum capacity of annual amount of maintenance, qt. T_i^H	Planned annual amount of maintenance by specific branch, iqt. T_i
№1	662	1 022	–	53 053	0	1 187
№2	662	–	1 134	54 114	1 602	898
№3	–	1 022	1 134	51 992	1 958	1 230
		In total			3 560	3 315

5. Rational distribution of amount of bus maintenance by branches at the united transit agency in Moscow

II step. Identifying the possibility of optimization

The minimum prerequisite to reflect the opportunity to provide maintenance of the united transit agency was examined to be as follows:

$$3\,315 \leq 3\,560 \tag{5}$$

(5) is satisfied, it means that the goal is achievable and there are no obstacles to go further to the III step.

III step. Restrictions allowing performing maintenance in full scale.

The total amount of maintenance of buses operated at branches no. 1, no. 2 and branch no. 3:

$$\begin{cases} x_1 + x_2 + x_3 = 1\,187 \\ y_1 + y_2 + y_3 = 898 \\ z_1 + z_2 + z_3 = 1\,230 \end{cases}. \quad (6)$$

Then the maximum amount of maintenance of buses performed at technical bases on branches no. 1, no. 2 and branch no. 3:

$$\begin{cases} x_1 + y_1 + z_1 \leq 0 \\ x_2 + y_2 + z_2 \leq 1\,602 \\ x_3 + y_3 + z_3 \leq 1\,958 \end{cases}. \quad (7)$$

The cost of performing the maintenance at the technical base on branch no. 1:

$$\sum C_1^M = 53053 \cdot x_1 + (53053 + 662) \cdot y_1 + (53053 + 1022) \cdot z_1 .$$

The cost of performing the maintenance at the technical base on branch no. 2:

$$\sum C_2^M = (54114 + 662) \cdot x_2 + 54114 \cdot y_2 + (54114 + 1134) \cdot z_2 .$$

The cost of performing the maintenance at the technical base on branch no. 3:

$$\sum C_3^M = (51992 + 1022) \cdot x_3 + (51992 + 1134) \cdot y_3 +\cdot 51992 \cdot z_3 .$$

IV step. Target function (minimum of total costs)

Target function is presented as:

$$\sum C_1^M + \sum C_2^M + \sum C_3^M \to min. \quad (8)$$

It is necessary to find x_i , y_i , z_i that satisfies target function (8) and condition restrictions (6) and (7).

V step. Finding an optimal solution using the transport problem approach.

The solution of the transport problem is reduced to the online data input [9], analysis and solution of the problem.

Optimal plan is performed as follows:

$$X = \begin{bmatrix} 0 & 459 & 728 \\ 0 & 898 & 0 \\ 0 & 0 & 1230 \end{bmatrix}.$$

VI step. Analysis of eventual results.

The eventual result obtained according to the initial data and the proposed methodology is the allocation of shares of the fleet maintenance at the technical base of every branch of the united transit agency:

1. Planned amount of maintenance of buses operated at branch no. 1 is 1 187 of whose amount shall be:
   - 459 (around 39 %) amount of maintenance is optimally conducted at technical base of branch no. 2;
   - 728 (around 61 %) amount of maintenance is optimally conducted at technical base of branch no. 3.
2. Planned amount of maintenance of buses operated at branch no. 2 is 898 and the entire scope of work (100 %) is optimally conducted at the in-house technical base of branch.
3. Planned amount of maintenance of buses operated at branch no. 3 is 1 230 and the entire scope of work (100 %) is optimally conducted at the in-house technical base of branch.

The goal has been achieved because the whole scope of the fleet maintenance can be fully completed using the proposed allocation by just two technical bases at branch no. 2 and no. 3. Moreover, operational costs after the optimization do not exceed the operational cost before it. However, economic effect is not mandatory in the specific observed case; a small saving can be achieved (175.23 against 175.51 million rubles) that made the result profitable.

6. Conclusion and further recommendation

The article presents the topical issue of an optimal allocation of the fleet maintenance and repair works at a united transit agency operated urban buses through concentration, specialization and cooperation. Literature research shows that such issue is relevant as more and more transportation companies aim to reduce their operational costs and use rational approach of maintaining the fleet at an appropriate level of reliability.

Minimum of the total operational cost of a united transit agency has been taken as a criterion of optimality taking into consideration the full scope of the fleet maintenance.

The method of optimal allocation of maintenance at a united transit agency has been developed and implemented on transportation company operated buses in Moscow. The summary or the proposed methodology is as follows:

- initial data collection that include transportation costs between branches, specific costs of a single maintenance procedure for bus, annual amount of maintenance, planned annual amount of maintenance by specific branch;
- identifying the possibility of optimization;
- obtaining an optimal allocation through calculating the minimum specific cost and compliance with restrictions that guarantee a fully performed maintenance at the entire fleet of a united transit agency.

The proposed method has been implemented at a united transit agency operating urban buses in Moscow and consists of three branches with their own technical bases. After applying the method the optimal allocation of the shares of maintenance between branches has been found: relocation of the full scope of fleet maintenance from branch no. 1 to branch no. 2 (39 %) and to branch no. 3 (61 %), whereas the whole scope of the fleet maintenance from branches no. 2 and no. 3 is recommended to perform on their in-house technical bases.

References

[1] Won Il Lee 2018 The effect of external technology cooperation and internal relation on innovative behavior in technology intensive organizations *International journal of engineering & technology* **7(4.4)** 26–9

[2] Kravchenko I N 2005 *Fundamentals of designing operational enterprises* (Moscow)

[3] Kournikov I P, Kuznetsov E S 1989 *Development of the technical base of road transport* (Moscow)

[4] Haedo C and Mouchart M 2017 A stochastic independence approach for measuring regional specialization and concentration *Regional Science* **97(4)** 1151–68

[5] Napolsky G M 1993 *Technological design of motor transport enterprises and service stations* (Moscow: Transport)

[6] Kuznetsov E S, Voronov V P and Boldin A P 1991 *Maintenance of automobiles* (Moscow: Transport).

[7] Bergman A K 1975 *Efficiency and methods of formation of organizational and production structures of associations in urban passenger transport* (Moscow: Dissertation)

[8] Pozhivilov N V, Maximov V A, Krylov G A and Zavgorodniy A A 2019 Methodological approach to the rational distribution of work on maintenance and repair of fleet of the united motor transport enterprise *Problems of technical operation and car service of rolling stock of automobile transport* ed L L Zimanov (Moscow: MADI) pp 23–6

[9] Aiginger K, Rossi-Hansberg E 2006 Specialization and concentration: a note on theory and evidence *Empirica* **33** 255–66

[10] *Minimum Transportation Cost Calculator.* Retrieved from: https://www.easycalculation.com/ operations-research/minimum-transportation-least-cost-method.php.

Innovative development of the automobile and road complex

I N Pugachev, Yu I Kulikov and S V Telnova

Pacific National University, 136, Tikhookeanskaya str., Khabarovsk, 680035, Russia

E-mail: ipugachev64@mail.ru

Abstract. The issues of the system approach to the automobile and road complex strategic development in accordance with the departmental project of the Ministry of Transport of the Russian Federation "The Digital Transport and Logistics" have been considered. They are a part of the national program "The Digital Economy of the Russian Federation" in connection with the Strategy of the Spatial Development of Russia for the period up to 2024. The national program is based on the transport modes infrastructure ensuring the areas interconnection and requiring the gradual modernization, digitalization and integration into one digital services ecosystem both for the transport individual modes and in the aggregate. The subjects of the study have become the automobile and road infrastructure objects that are the material and technical resources of transport services for the cargo, passengers and luggage transportation. However, the highway transportation in Russia holds the leading position in the cargo and passengers transportation total volume, where a significant proportion corresponds to the cities and agglomerations transport system with the deployment of cargo and passenger terminals, fuel stations, carriers and consumers of transportation services. The relevant national projects Passports being the background of the research and approved by the Presidium of the Presidential Council for the strategic development and the national projects contain the achievement of goals, targets and tasks in the format of instructions from the President of the Russian Federation to the Government. The technique of every national project (program) implementation in the relevant field at the first stage requires the conceptual strategic development, the digital technologies and the platform solutions reflected in the work.

1. Introduction

The keynote in the context of the national transport policy is the Decree of the President of the Russian Federation of 07.05.2018, No. 204 "On the national goals and strategic objectives of the national development up to 2024". They include the digital economy, safe and high-quality roads, science and education, which together determine the scientific, educational, technical and technological breakthrough as well as the future development of the transport sector. This ensures the change in the motor vehicles production structure in all the segments in accordance with the "Strategy of the development of the automotive industry of the Russian Federation for the period up to 2025", approved by the Decree of the Russian Federation Government of 28.04.2018, No. 831-r. The unmanned motor vehicles manufacture of passenger cars, light commercial and goods-carrying vehicles, buses, and advanced types of innovative transport on the territory of the Russian Federation include electric vehicles and gas-powered vehicles running on compressed natural gas.

The manufacture of unmanned motor vehicles is the highest achievement of the automotive industry in the implementation of the intelligent transport systems. The world practice of the unmanned motor vehicles production and traffic testing has identified five levels of vehicle traffic automatic performance (autonomy), the requirements for the innovative transport infrastructure modernization, ensuring the passage of the unmanned motor vehicles, and the need to develop a concept of road safety with the unmanned vehicles on public roads. At the same time, the digital transformation of the road infrastructure ensures the implementation of the departmental project of the Ministry of Transport of the Russian Federation "The Digital Transport and Logistics" as a part of the national program "The Digital Economy of the Russian Federation".

2. Features of the automobile and road complex digitalization

The national spatial development strategy proposed by the President of the Russian Federation in the Address-2018, the strict implementation of the national projects require the gradual modernization, digitalization and integration into one digital services ecosystem both for the individual transport modes and for all the participants in the transport process in the aggregate.

The development and implementation of the digital transport economy must comply with the national project "The Digital Economy of the Russian Federation" approved by the Decree of the Russian Federation Government of 28.07.2017, No. 1632-r. The program document is based on the idea that the digital economy is an economic activity where the digital data become the key production factor allowing to form the information space with the transport infrastructure characteristics, the transport services consumers demand and the carriers offers. The target result of the digital transport economy should be a balance between the supply and demand for the transport services, the goods logistics and the passengers delivery based on the space monitoring of transport with the services provision to transport consumers.

The current practice of the national transport complex operating shows that the automobile transport is the most popular and susceptible to the digital economy in commercial terms. Especially it concerns public transport, which mainly forms the convenience and accessibility of the urban environment, acting as a digitalization indicator of a harmonious spatial development, connecting cities and agglomerations in various types of the transport communication depending on the route schedules. At the same time, the automobile and road complex digitalization has its own distinctive features in the automotive and road sector.

The motor vehicle fleet digitalization provides for the availability of the motor vehicles electronic passports circulated since July 1, 2018, the insurance policies of motor vehicles owners in a digital form, the installation of the emergency response indicators for the accidents "ERA-GLONASS".

It should be noted that the space group as a part of JSC "GLONASS" provides monitoring of the motor vehicles registered by all the indicators, the vehicles contour digitization and the effectiveness of the interaction between the participants in the transportation process.

The public road transport digitalization is determined by the transportation specifics. The general-purpose transportation is carried out by public buses in various types of communication depending on the route schedules. The schedule is made for each stopping point of the regular transport route provided for a mandatory bus stop. At the same time, buses are equipped with the satellite navigation monitoring equipment that provides the transportation dispatching and regularity, as well as the interactivity of bus traffic on routes for mobile digital services (smartphones and tablets) of the passengers, which allows the passengers to plan the trips. Moreover, it is proposed to introduce the transport cards and their validation everywhere (the ability to use a single card during transfers) [1].

An alternative to public transport in the digital circuit is the individual order systems in the mobile digital applications "Yandex. Taxi" and car sharing as subcategories of public transport. Any type of car sharing in the strategic development is an alternative to the global motorization, the benefits of which are outweighed by the growing burden of responsibility and ownership costs, which foreshadows the stage of declining the global motorization and opening the unmanned transport era with a widespread introduction of the intelligent transport systems [2].

The digitalization of the urban public transport is of particular importance in the strategic development. It involves the use of the unmanned vehicles and the image recognition technologies (identification) of the passengers getting on public transport with an automatic fare payment from electronic wallets.

The cargo transportation digitalization is based on the cargo registration in a digital circuit congestion and the road traffic of freight rolling stock carrying heavy, oversized and dangerous cargoes. It is also based on the traffic interactivity of the motor vehicles that carry out the cargo terminal transportation in the main domestic and international routes along the certified transport corridors. The small consignments logistics delivery by light commercial vehicles in cities and agglomerations is determined by the digital interactions between the consignors and consignees. In this case, the electronic document management is combined in a single cargo digital circuit.

In accordance with the national project for the safe and high-quality roads, it is planned to introduce new technical requirements and standards of the roads arrangement based on the digital technologies and aimed at eliminating the accidents concentration. The introduction of the automated and robotic technologies of the traffic management and monitoring the compliance of the traffic rules are envisaged as well as the use of the new mechanisms for developing and operating the road network, including the infrastructure mortgages use, the life cycle contracts, the best technologies and materials. It is proposed to equip the "smart system" of safe and high-quality roads with indicators built into the road infrastructure for the automatic registration of congestion and the road traffic of freight rolling stock carrying heavy cargo in the payment system "Plato", illegal carriers, illegal parking and other violations.

An example of the innovative technologies is the construction of the Crimea main road - the Federal highway "Tavrida" with 3D-model technology. It provides the unmanned automatic mode operation of graders using a special equipment with an electronic map of the road. The electronic sensors work in the satellite monitoring system of the graders operation program and on-board computers that ensures the ideal profiling of the roadway by leveled indicators (marks) of evenness, slopes, transverse and longitudinal slopes of the road surface, taking into account the terrain. The asphalt laying is carried out by loaders ensuring homogeneity of asphalt mixture. Many sensors are built into the asphalt rollers that monitor the temperature of the asphalt mix and its compaction value. The georadar and laser inspections are carried out during the road operation to assess the condition and thickness of the pavement layers. In case of the violations, the scan results are formatted as digital models with the possibility of integration into the environment of design, reconstruction or repair of road sections [3].

During the construction of the road "Bypass of Khabarovsk 13–42 km" the expert supervision is done by unmanned drones. The drones transmit the panoramic images of the construction to computer monitors that make it possible to identify the technological flaws in the blading operation and compaction of the roadbed.

The creation of a digital circuit on roads based on the use of the electronic platform of the Etalon system enables to provide the digital interaction of the Federal Road Agency Rosavtodor with the subordinate organizations and applicants in the format of digital workflow and the presentation of technical requirements and conditions for placing advertising structures in the roadside area and joining road service facilities.

The road transport is not only an independent type of the transport system of the Russian Federation, but also an integral part of the urban, industrial and interacting transport modes in the multimodal cargo and passengers' transportation. Therefore, the conceptual approaches to the strategic development of the transport sector, formed as a result of the discussion at the state, government and departmental levels, have been determined of some complex directions:

1) the digital optimization of the multimodal direct and mixed (combined) cargo transportation with several transport modes through the transport and logistics hubs and in cross-border traffic requires the unification and integration of a digital telemetry platform using on-board computers for compatibility, trust, interaction and coordination of transport modes in logistics business processes that will streamline the work and responsibility of the operators;

2) the creation of a single digital integrated platform for the multimodal passenger transportation will allow the passengers to choose the best route for the transportation by any means of transport and in any combination, to purchase a single ticket with a guaranteed level of comfort and safety, the passengers loyalty considered;

3) the Association of the Digital Transport and Logistics with the Federal Centre of Competence should become an integrating and coordinating center of the transport complex digital transformation for all the industry digitalization participants under the Ministry of Transport of the Russian Federation; the Ministry has announced a competition to develop the concept of creating a digital platform of a transport complex of the Russian Federation, the project, as we think, being the foundation in the field of artificial intelligence and big data for the implementation of digital logistics.

The implementation of the digital transport economy is possible with the consolidation of science, education, business and government, which requires the creative development of the existing pass-through digital platforms and technologies and the creation of new ones. The environment necessary for the platforms and technologies developing, the effective interaction of the road transport market subjects includes the regulatory framework, the information structure, personnel and education, the information security and interoperability, in accordance with the state policy in the field of various transport modes [4–15].

3. HR Training

1. The key to the implementation of the departmental project "The Digital Transport and Logistics" is training the relevant industry personnel. The commonality of the transport problems and the unity of the digital performance indicators make it possible to form the digital circuit and technologies for the cargo and passenger transportation of the individual transport modes followed by the integration of the digital services into a single digital platform both for the individual transport modes and in the aggregate. This circumstance makes it possible to train and retrain the personnel in the educational program specialization of "The Digital Transport and Logistics" with an industry component (specializing in transport modes and its infrastructure facilities, training directions and educational programs being considered, a list of professions, positions and qualification requirements for them).

2. The basis for the digital transport staffing is the national project "Education". The introduction of the pass-through training, starting from the early vocational guidance of the secondary school students, implemented in the open online lessons "Proektoriya" and to the subsequent training of the skilled workers, middle and senior managers in accordance with the selected professional competences implemented by the project "Ticket to the Future" will allow to prepare a highly-qualified industry personnel provided that the platform for "The Modern Digital Educational Environment" would be created. This platform will ensure the high quality and availability of education of all types and levels. At the same time, the educational programs for the transport infrastructure digital transformation should be coordinated with the Federal Centre of Competence under the Ministry of Transport of the Russian Federation and meet the requirements of the national projects [16].

4. Conclusion

The paper has investigated the features and capabilities of the strategic implementation for the digitalization of the automobile and road complex in accordance with the requirements of the national projects.

5. References

[1] Pugachev I N, Kulikov Yu I, Markelov G Ya, Lipsitz L M and Boreyko A E 2018 Features of the Digital Economy in the Transport Sector *Transport of the Russian Federation* **3(76)** 34–6

[2] Pugachev I N, Markelov Yu Ya and Burkov S M 2013 *Organization of Information and Technological Support. Methodology of Infrastructure Research on the Example of the city of Khabarovsk* (Khabarovsk: Pacific National University Publishing House)

[3] Eremin R A 2017 Innovative Technologies in Road Facilities and Urban Economy *Roadway*

Network **6** pp 104–108

[4] Pugachev I N, Kulikov Yu I, Markelov G Ya and Kondratenko T E 2017 *Development Strategy of the Transport Systems of Russian Cities* (Khabarovsk: Pacific National University Publishing House)

[5] Pugachev I, Kulikov Y, Markelov G and Sheshera N 2017 Factor Analysis of Traffic Organization and Safety Systems. *Transportation Research Procedia, 12th International Conference "Organization and Traffic Safety Management in Large Cities"*, (St. Petersburg, Russia) vol 20 pp 529–535

[6] Kulikov Yu I, Pugachev I N and Markelov Yu Ya 2013 *Innovative Doctrine of Road Transport Development* ed Kulikov Yu I (Khabarovsk: Publishing house of the Pacific National University)

[7] Pugachev I N, Kulikov Yu I and Yarmolinsky A I 2018 Innovative Technologies for Digital Transformation of Automobile and Road Complex in Strategic Development *Far East. Roads and traffic safety*: *Int Collection of Research Papers* ed Yarmolinsky I vol **18** (Khabarovsk: Pacific National University Publishing House) pp 36–9

[8] Pugachev I N and Kulikov Yu I 2018 Strategy for the innovative development of road transport Road Transport of the Far East-2018 *Proc of the IX Int Scient Conf* (Khabarovsk-Vladivostok, ed P P Volodkin (Khabarovsk: Pacific National University Publishing House) pp 238–41

[9] Pugachev I N, Kulikov Yu I, Markelov G Ya, Kondratenko T E and Shpakov V N 2016 *Methodology for the development and implementation of the concept of road safety and the program of activity in the territory of the subject (on the example of Khabarovsk Territory) (Monograph)* (Khabarovsk: Pacific National University Publishing House)

[10] Pugachev I N, Kulikov Yu I and Markelov G Ya 2018 Khabarovsk: Features of the Regional Organization of Road Safety *Transport Security and Technologies* **2(53)** pp 204–9

[11] Pugachev I N, Kulikov Yu I and Markelov G Ya 2018 Strategic development of the automobile and road complex in the digital environment *Scientific readings in the memory of Professor M. P. Danilovsky: Proc of the Eighteenth National Scientific and Practical Conf*: in 2 vol (Khabarovsk: Pacific National University Publishing House) vol 1 pp 336–8

[12] Pugachev I N, Kulikov Yu I and Sedyukevich V N 2015 Indicators of qualitative functioning of the transport complex of the Russian Federation *Science & Technique* **3** 51–60

[13] Pugachev I, Yarmolinsky A and Kulikov Yu 2018 Current trends in development of the transport infrastructure of large cities of the Far East, Russia *Transportation Research Procedia 36, Thirteenth Int Conf on Organization and Traffic Safety Management in Large Cities* (SPbOTSIC) pp 622–6

[14] Pugachev I, Kamenchukov A and Yarmolinsky V 2018 Evaluation of road repair efficiency in terms of ensuring the traffic quality and safety *Transportation Research Procedia 36, Thirteenth Int Conf on Organization and Traffic Safety Management in Large Cities* (SPbOTSIC) pp 627–33

[15] Kamenchukov A, Pugachev I, Yarmolinsky A, Vasilyev A and Kulikov Yu 2019 Improving the Criteria for Determining the Dates of Repair of Highways *J. of Applied Engineer. Sci.* **17(1)** 81–6

[16] Pugachev I N and Kulikov Yu I 2019 Training – the key to the implementation of the departmental project "Digital Transport and Logistics" *Proc of the IV International Scientific-Practical Conference Transport Planning and Modeling* (SPbGASU) pp 67–70

Formation of fuel consumption norms for cargo transportation in Vietnam

A N Rementsov, V A Egorov, Fan An Vu and D V Antonov

Moscow Automobile and Road State Technical University, 5-47, Ak. Ilushina str., Moscow, 125319, Russia

E-mail: rementsov@yandex.ru, yegorov@tea.madi.ru, anhvumta@gmail.com

Abstract. The unified transport system of Vietnam includes road; rail; river; sea and air transport. Road freight transport has a particularly important role and provides for the delivery of goods, especially in large cities. Transport problems and the problem of improving the quality of freight transport in Viet Nam are classified as difficult and still not resolved. The paper analyzes the role of road freight transport in the economy. Here we propose a refined approach to determining the rationing of fuel consumption of trucks and its features in Vietnam conditions. By assessing the influence of various factors on fuel consumption, it is possible to increase the accuracy of standardization and reduce fuel consumption.

Keywords: transport, road freight transportation, operation of trucks, rationing of fuel consumption, fuel consumption of vehicles.

1. The role of road transport for the development of the economy of the Socialist Republic of Vietnam (SRV)

During the years of existence of the SRV, a unified transport system (UTC) has been created in the country, which includes all types of transport: automobile; railway; river; sea; air. The main indicators of the efficiency of the transport system are the volume of cargo transportation, cargo turnover.

Visual representation of the using different types of transport in the transportation of goods for the period 2010–2017 is shown in Fig. 1.

The advantages of automobile transport are the possibility of direct delivery of goods from the warehouse of the sender to the warehouse of the recipient without transshipment operations at short distances, the possibility of centralized delivery and export of goods from stations, sea and river ports. When transporting large cargoes over long distances, preference is given to rail and sea transport.

Statistics show that in freight turnover, the first place is taken by sea transport, and the second place is occupied by road transport; in 2017, maritime transport provided a cargo turnover of 140,442.9 million tkm, while road transport – 63,803.5 million tkm [10].

According to fig. 1 it may be noted that in Vietnam there is an increasing trend in the volume of traffic by all modes of transport. Road freight transport plays a special role in meeting the demand for transport services. Carrying out a significant volume of freight transportation, road transport should essentially become the main carrier, which is facilitated by its objective market advantages.

Currently, SRV road freight transport plays an important role in the country's economy, and has a leading role in the transportation of goods in the country. Every year road transport carries about 75 % of all cargo in the country.

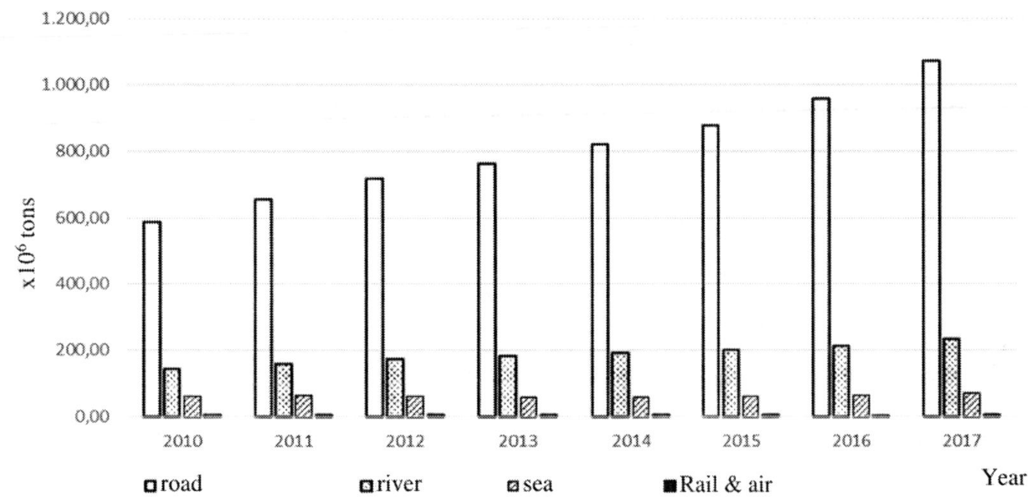

Figure 1. The share of types of transport in the transportation of goods in Vietnam

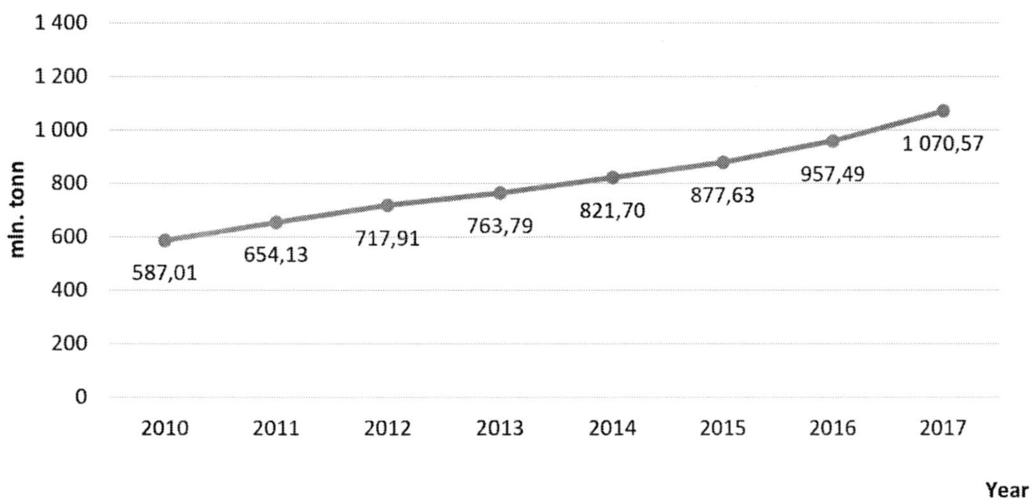

Figure 2. Dynamics of the volume of cargo transportation by road

The scope of application of road transport in Vietnam is very wide. It performs most of the short intra-district transportation, delivers goods to railway stations, seaports and river piers and delivers them to consumers. In the Northern regions and the Central plateau, where there are almost no other types of land transport, it carries out long-distance inter-district transport. Road transport has a flexible tariff policy, as there are many small motor transport enterprises in the country with different types of ownership.

In addition, it is known that one of the main tasks of road transport is the timely, high-quality and complete satisfaction of the needs of the national economy and the population in transportation, increasing the economic efficiency of their work. With regard to road transport, this can be achieved by improving the efficiency of the use of rolling stock, and energy (fuel).

It should be noted that this task can be achieved not only through the purchase of new vehicles, but also by improving the methods of maintenance and repair, as well as the development of methods for rationing the fuel consumption of vehicles.

Thus, in our opinion, the automobile transport of Vietnam is the most flexible and mass mode of transport. Mass motorization has a significant impact on the development of territories and settlements, on the processes of trade and consumption, on the formation of entrepreneurship. This accelerated development of road transport is due to the following main factors:

- About 80 % of production, transport and distribution structures, as well as the majority of settlements do not have access roads other than automobile, which in these conditions determines the uncontested use of automobile transport.
- Automobile transport has the following advantages: high speed of delivery and the possibility of door-to-door and just-in-time delivery, high cargo safety, mobility, maneuverability, accessibility, the possibility of small-lot and container transportation, the possibility of transporting different types at the same time and in small batches.

These features allow us to consider it as the most market-oriented mode of transport. So, all these and other functions put the development of road transport in the category of the most important state tasks, make it an integral part of the economy of Vietnam as a whole.

2. Methodological approach to determining the fuel consumption rates of trucks in Vietnam

The number of vehicles used in Vietnam in 2017 amounted to 2.411.292 pieces. In particular, the number of trucks was 1.004.499 pieces. The annual growth rate of motor vehicles was 11.17 % for the period from 2005 to 2017, and the growth of trucks was 12.35 % (Fig. 3) [3, 4].

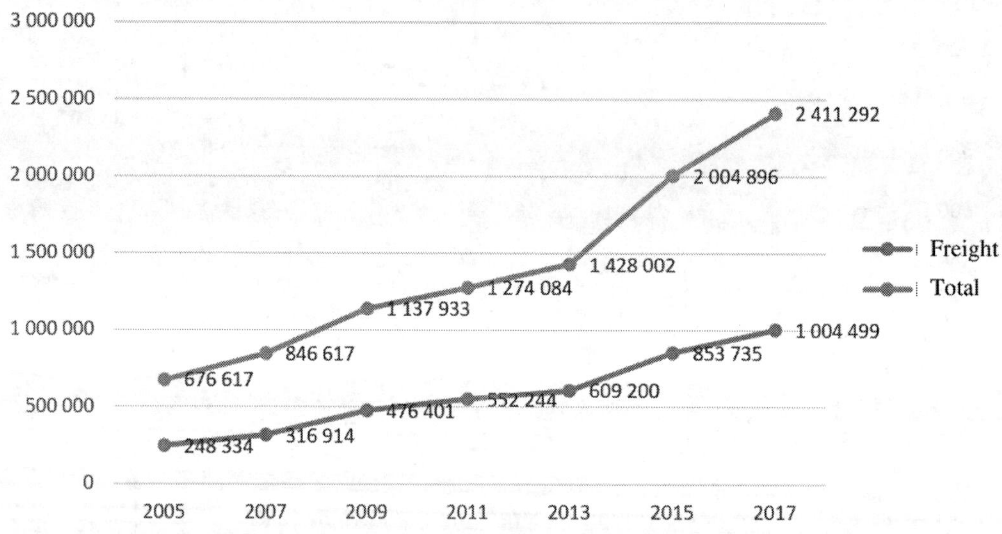

Figure 3. Dynamics of growth in the number of freight vehicles in Vietnam

The demand for transportation is constantly growing and, accordingly, the number of vehicles increases, which leads to an increase in fuel consumption. By the end of 2017, fuel consumption in road transport amounted to about 16 million tons, according to expert estimates, fuel consumption in 2020 in Vietnam will be 22.4 million tons, and in 2025 – 29.9 million tons. However, at present, domestic oil production meets approximately 40 % of the fuel demand and 60% must be imported.

The above-mentioned facts point to the need for resource saving in the field of fuel consumption. Thus, in addition to decisions to modernize and improve the quality of the vehicle fleet by applying new technological advances and improving the quality of transport infrastructure, it is necessary to develop and implement a number of regulatory documents governing operating costs. To solve the ambitious task posed for trucks, it is necessary to develop a system of statistical monitoring in automobile transport and improve the methods of standardization and estimation of operational fuel consumption.

According to the results of an expert survey conducted by World Bank, in Vietnam fuel costs account for about 65 % of the total cost of transportation [8]. Due to the lack of state regulation in the field of rationing of fuel consumption of trucks, enterprises independently set fuel consumption rates for their trucks. The fuel rationing applied to all categories of motor vehicles is widely used by transport companies as follows.

The standard value of fuel consumption is calculated by the formula:

$$M_c = K_1 \cdot \frac{L}{100} + K_2 \cdot \frac{P}{100} + n \cdot K_3,$$

where Mc is the standard fuel consumption, l; L – vehicle mileage, km; K_1 – fuel consumption rate for running a vehicle in running condition without cargo (basic norm), l/100 km; P – the volume of transport work, t·km or passenger·km; K_2 – fuel consumption rate for transport work, l/100 t·km or l/1000 passenger·km; n is the number of stops, units; K_3 – fuel consumption rate for each stop, l.

The actual fuel consumption differs from the values obtained by this method (and most often do not correspond to reality). The calculations show that practically errors can be approximately 20–25 %. This shows that the application of this method is not fully suitable for rationing freight traffic in the real conditions of Vietnam at the present time [7].

In Vietnam, most transport enterprises are small and medium-sized private trucking companies that do not have the financial capabilities and human resources to conduct studies on rationing fuel consumption that require large volumes of measurements with a large scale for a long time. The lack of state regulation in the field of rationing the fuel consumption of trucks leads to a lack of coordination in the field of rationing fuel consumption; some problems arise with the management and planning of transport. In addition, the quality of transport infrastructure is low: small width of roads; there is no lane reserved for each type of vehicle, which leads to difficult traffic conditions, etc. Therefore, it is necessary to develop a methodology for standardizing fuel consumption that is close to the actual operating conditions, the advantages of which are unnecessary for long and expensive experiments, a minimum of time spent to obtain a result, etc., which is convenient for practical use by motor transport enterprises workers.

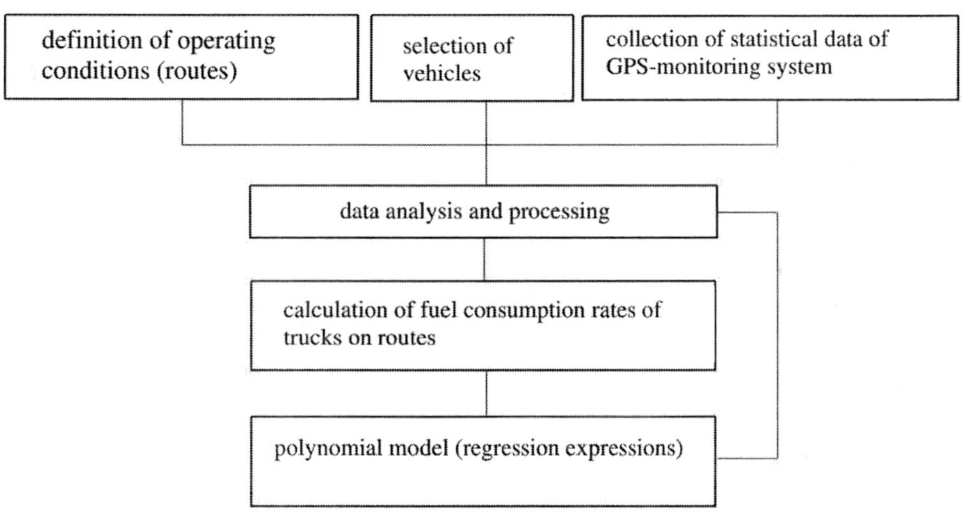

Figure 4. Block diagram of the methodological approach to the definition of rationing fuel consumption for trucks

IOP Conf. Series: Materials Science and Engineering **832** (2020) 012067 doi:10.1088/1757-899X/832/1/012067

According to the rules of the Ministry of Transport of Vietnam "on the installation of equipment for monitoring trips on vehicles carrying goods (passengers)" of 01.01.2017, it is necessary to install equipment for monitoring trips on trucks. Based on this, a refined approach to standardizing fuel consumption for trucks is proposed, which consists in calculating an individual fuel consumption rate calculated according to the theory of a car based on polynomial dependencies using GPS technologies (Fig. 4) [1–3].

The proposed approach allows us to take into account the following factors that significantly affect fuel consumption:

- road conditions;
- transport conditions;
- natural and climatic conditions for the operation of vehicles.

Thus, it is possible to increase the accuracy of standardization and reduce fuel consumption.

3. The main factors affecting the fuel consumption of trucks in Vietnam

When developing a more advanced fuel management system, it is very important to correctly take into consideration all the variety vehicle operating conditions. Therefore, the features of rationing the fuel consumption of trucks in Vietnam take into account the following main factors.

3.1. Air temperature

Vietnam is completely located in the tropics, and the country's climate is characterized by high rainfall, high humidity, and many sunny days. The average annual air temperature varies from 22 to 25 degrees Celsius. Annual temperature fluctuations in the North range from 15–20° in winter to 22.5–27.5° in summer; in the South-from 26–27° in winter to 28–29° in summer. Almost the entire territory of Vietnam receives more than 1000 mm of rainfall per year, on the hills-from 2000 to 2500 mm. In addition, Vietnam is one of the countries most affected by natural disasters. According to the UN, Vietnam is among the five countries that will be affected most by climate change. Because of its climate, the country is at risk of flooding, drought, salt water intrusion, landslides and flooding due to rising sea levels. These circumstances have seriously affected the transport infrastructure, as well as transport activity in general [8, 10].

3.2. Road conditions

The SRV territory extends in the meridional direction (the distance between the extreme northern and southern points is about 1650 km), and in the latitudinal direction its length is from 616 km in the north (from Mongkai to the Vietnam-Laos border) to 46.5 km in the central part (in Quangbin Province). More than three quarters of the territory is occupied by mountains and plateaus, the country has two large alluvial-delta plains – the Hongha River Delta in the north and the Mekong Delta in the south. These characteristics do determine the spatial distribution of inhabitants and the corresponding transport infrastructures.

Vietnam's road network is divided into six categories: national highways, regional roads, district roads, urban roads, rural roads, specialized roads. Currently, the total length of Vietnam's roads includes more than 200,000 km of highways with the main road Langshon-Hanoi-Ho Chi Minh city passing along the railway, 5 major highways and 104 national highways. As of the road surface, in Vietnam there are 113,989 km of roads with asphalt-concrete, cement – concrete and asphalt coating, which is 52 % of the entire road network (table. 3.1). This means that the condition of the road network does not meet the needs for transportation. Most of the district, rural roads have gravel and soil coating. Approximately 60 % of these roads have less than two lanes. In addition, with the rapid growth in the number of vehicles, traffic jams are increasing along the main transport corridors, especially in the vicinity of large cities such as Hanoi, Ho Chi Minh city and others [6, 9, 10].

Table 1. Characteristics of the Vietnam Road Network

Road category	Length of road, km	Length of roads by type of road surface, km				
		Asphalt-concrete	Cement-concrete	Asphalt	Grave	Dirt road
National	19.545	9.464	344	9.301	400	36
Regional	25.335	5.474	2701	14.488	1.926	746
District	64.254	2.762	12.581	11.092	19.957	17.862
Rural	92.452	4.616	18.442	10.226	24.907	34.261
Urban	10.535	2.465	2.776	3.750	976	568
Specialized	6.100	2.800	160	547	2.593	0
Total	218.221	27.581	37.004	49.404	50.759	53.473

3.3. Quality of operation of Motor Vehicles

Currently, there are about 1050 automobile transport companies in the transport performing freight transport in the country. Most of them are small and medium – sized enterprises. The privatization process has had a significant impact on road transport and as a result, the role of state-owned enterprises in providing transportation has decreased significantly. This has led to the formation of a market for transport services with a high level of competition. On the one hand, competition is a key element in increasing the ability to meet market requirements and reduce transportation prices on the other hand it can cause negative problems. So to reduce initial costs and operating costs, enterprises do not pay attention to the need to modernize the fleet of trucks, reduce the requirements for maintenance and repairing of rolling stock, overload vehicles, etc. This leads to increased fuel consumption, environmental pollution, an increase in the number of road traffic accidents [5].

3.4. Traffic conditions

Each type of transport has its own purpose in the national transport system. However, road transport due to its mobility and the ability to deliver goods "door to door" plays a dominant role in the implementation of freight transport in Vietnam. Road transport is able to meet the needs of the market for the transportation of various types of goods. The main ones are food and industrial products, construction materials.

According to statistics, up to 87 % of the cargo volume in Vietnam is transported within 200 km, which is mainly due to domestic trade concentrated around Hanoi and Ho Chi Minh city. In the internal transport model of Vietnam, road transport occupies a large part with short distance freight transportation, but this share decreases sharply with increasing transportation distance (Fig. 5) [10].

The characteristics of road freight transport by city, region and route are presented in tables 2–4 [10].

The analysis of road freight transportation performed by road transport enterprises to various destinations using vehicles of different capacity allows assessing the efficiency of the transport infrastructure as a whole.

Heavy trucks are often used for transportation in the industrial cities of Can Tho, Quignon, Danang, Hue, Haiphong, the central and southern parts of the country and at large distances from the cities of millionaires Hanoi, Ho Chi Minh.

The practice of freight transportation in Vietnam has its own characteristics:

- the average operating speed is 30–35 km/h and has some variation depending on the type of rolling stock;
- the utilization rate of the carrying capacity is of the greatest importance for vehicles performing transportation in the central region and on routes to cities with million-plus population. The percentage of rides of trucks returning without cargo is about 4 %; compared with the national average of 28 %;
- the average annual mileage also differs depending on the routes. The average annual mileage is of the greatest importance (63,000 km) for vehicles performing transportation on routes to cities with

million-plus population

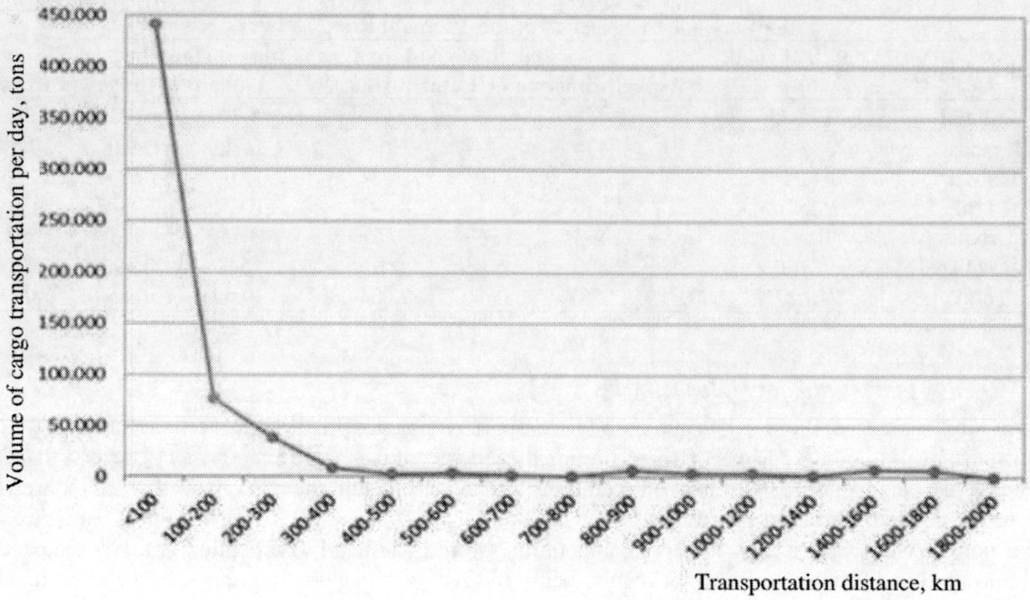

Figure 5. Trucking volumes by distance

Table 2. Characteristics of road freight transport in Vietnam by cities

	Cities	Cities of over one million	Industrial cities	Other cities	Countrywide
Average annual value	Age of trucks, year	16,5	17,1	17,3	16,6
	Truck weight, t (a)	7,6	12,9	8,7	8,0
	Cargo weight, t (b)	14,0	21,1	13,3	14,6
	Overloading, t (c)	8,3	8,5	5,8	8,3
	Cargo weight / Truck weight. (b/a)	1,8	1,6	1,5	1,8
	Overloading / Truck weight. (c/a)	1,1	0,7	0,7	1,0
	Number of times loading-unloading the truck	45,0	79,8	65,1	47,6
	Percentage of rides without cargo, %	27,3	40,2	28,5	28,2
	Transportation distance, km	499,4	453,2	392,5	493,7
	The duration of the trip, h	14,1	12,0	9,3	13,9
	Operational speed, km / h	33,7	35,9	32,1	33,8
	Total mileage, km	26750	31338	24582	27019

Table 3. Characteristics of road freight transport in Vietnam by region

	By region	Northern	Central	South	Countrywide
Average annual value	Age of trucks, year	17,1	17,1	15,9	16,6
	Truck weight, t (a)	5,8	12,0	10,4	8,0
	Cargo weight, t (b)	10,8	18,0	19,7	14,6
	Overloading, t (c)	6,5	6,1	9,9	8,3
	Cargo weight / Truck weight. (b/a)	1,9	1,5	1,9	1,8
	Overloading / Truck weight. (c/a)	1,1	0,5	1,0	1,0
	Number of times loading-unloading the truck	40,6	54,7	55,5	47,6
	Percentage of rides without cargo,%	27,1	3,5	31,0	28,2
	Transportation distance, km	393,8	780,9	643,5	493,7
	The duration of the trip, h	11,0	20,9	17,9	13,9
	Operational speed, km / h	35,8	39,8	30,8	33,8
	Total mileage, km	22858	34424	32216	27019

Table 4. Characteristics of road freight transport in Vietnam by route

	By route	To cities of over one million	road to international ports / border crossings	road to other cities / nearby rural areas	Countrywide
Average annual value	Age of trucks, year	17,4	16,9	16,0	16,6
	Truck weight, t (a)	10,0	9,0	6,6	8,0
	Cargo weight, t (b)	19,6	16,8	11,0	14,6
	Overloading, t (c)	11,1	8,7	5,9	8,3
	Cargo weight / Truck weight. (b/a)	2,0	1,9	1,7	1,8
	Overloading / Truck weight. (c/a)	1,1	1,0	0,9	1,0
	Number of times loading-unloading the truck	31,2	48,0	53,6	47,6
	Percentage of rides without cargo,%	3,9	30,9	34,7	28,2
	Transportation distance, km	1.516,2	343,4	226,0	493,7
	The duration of the trip, h	42,8	13,2	6,3	13,9
	Operational speed, km / h	31,7	34,0	34,6	33,8
	Total mileage, km	63253	22368	16426	27019

4. Conclusion

The analysis of transport work performed by different modes of transport shows that the leading operator of freight transport in Vietnam remains road transport. To improve the efficiency of freight transport by road, it is necessary to determine the rational structure of the fleet, improve the production technical base of transport enterprises, develop and implement a number of regulatory documents governing operating costs – tire costs, fuel costs, maintenance costs, etc.

Based on the analysis, the following conclusions can be drawn:

1. Due to geographical, social and economic conditions, road transport plays a leading role in the transportation of goods in the country. Goods transported by trucks are mainly food, construction and industrial goods.

2. Transportation by trucks is mainly carried out over short distances (less than 400 km) around two major economic centers: Hanoi in the North and Ho Chi Minh city in the South.

3. Small and medium private motor transport enterprises do not have the financial capabilities to modernize the truck fleet. In addition, experiencing serious competition, enterprises do not pay due attention to the maintenance and repairing works of rolling stock, and trucks, often carry out transportation with significantly exceeded cargo masses from the permitted one, have a low operating speed. This leads to rapid physical aging, deterioration of the technical condition of vehicles, increased fuel consumption, environmental pollution, an increase in the number of road accidents.

4. The main drawback of the current methodology of rationing fuel consumption in Vietnam is the lack of an objective assessment of the operating conditions of vehicles, which makes it impossible to assign exact allowances to the linear norm. The current methodology does not explicitly take into account a number of basic parameters of the transport process.

The above allows us to conclude that the improvement of the current system of rationing fuel consumption is an urgent task that allows more objectively taking into account the operating conditions.

References

[1] Borisov G V, Lelyavski K Y and Pachurin G V 2015 To the question about regulation of the consumption of liquid fuels in road transport *Fundamental res.* **3** 28–35

[2] Kuzmin N A 2010 The problem of rationing the costs of automotive fuels and lubricants in the Russian Federation *Motor transport enterprise* **8** 20–2

[3] Nueng, H M, Rementsov A N and Nguyen M T 2014 Organization of passenger traffic in Vietnam *Truck* **4** 20–2

[4] Fan A V and Rementsov A N 2018 Features of the operation of trucks in the conditions of Vetnam Problems of technical operation and car service of rolling stock of automobile transport *Sat. sci. articles* 193–200

[5] Fan A V and Rementsov A N 2018 Features of freight transport by road in Vietnam *Bull. of MADI* **4(55)** 20–3

[6] Nguyen Thi Thu Ha 2017 Phat trien ha tang giao thong nham nang cao hieu qua logistics tai Viet Nam *Giao Thong Van Tai* **5** 87–9

[7] Nguyen Manh Hung and Dinh Quang Toan 2016 Thuc trang loai hinh van tai da phuong thuc trong nganh Dich vu logistics tai Viet Nam *Giao Thong Van Tai* **10** 34–6

[8] Le Ba Thao 2018 *Viet Nam – Lanh tho va cac vung dia ly* (Ha Noi: NXB The Gioi) 609 p

[9] *Tong cuc thong ke, Nien giam thong ke* 2017 (Ha Noi: NXB Thong ke) pp 639–71

[10] *Vietnam Urbanization Review* 2011 Technical Assistance Report (Ha Noi: World Bank) pp 69–111

New digital technologies of training in the transport education

A F Smyk, T M Tkacheva and Yu A Portnov

Moscow Automobile and Road State Technical University (MADI), 64, Leningradsky ave., Moscow, 125319, Russia

E-mail: afsmyk@mail.ru

Abstract. The article analyzes the use of new digital technologies in transport education considering the Moscow Automobile and Road Construction State Technical University (MADI) as an example. The results Massive Open Online Courses (MOOC) and other electronic systems in the training of engineers using on specialty «Land transport and Technological Vehicles» are given. Positive opportunities of online courses using arise in the organization of student extracurricular activities and the implementation of a differentiated approach in training. Different content of online courses in the areas of training makes it difficult to use MOOC in the practice of general engineering education. The authors suggest using the Modular Object-Oriented Dynamic Learning Environment (Moodle) which is a free web application that provides to create websites for online learning. The experience of using the online resource «Information system MADI» («IS-MADI: Physics») to study Physics, created by MADI's teachers, is also considered. This experience makes to conclude that using of the Internet resource «IS-MADI: Physics» awakes interest of students to learn Physics. In a whole students begin to understand that knowledge of Natural Science disciplines gives them powerful tool to create and to exploit modern cars and roads, they begin understand importance of Physics.

1. Introduction

During various historical stages the economic development of Russia was inextricably linked with the transport system formation. In turn this system stimulated the development of primary vocational and higher engineering education. The current stage of innovative economy development and technological modernization is associated with digitalization, which means the widespread introduction of digital technology in all areas of activity

In transport education in the training of motor engineers in the direction of "Land Transport and Technological Vehicles", three forms of training are implemented: full-time, part-time education and distance learning. For all of these educational forms there are different models of the educational process organization as well as of training materials presenting. The experience of the Moscow Automobile and Road Construction State Technical University (MADI) in the use of new digital technologies for training of engineers motorists during the study of natural sciences in junior courses is considered as an example.

2. Problem statement

2.1. Models of electronic educational resources

Comparison of curricula of mechanical engineers training in the Moscow Automobile and Road Construction Institute (1933) and specialists on the specialty of "Land Transport and Technological

Vehicles" in the Moscow Automobile and Road Construction State Technical University (2013) showed that these curricula keep from the XIX century the tradition of Russian engineering education to give students fundamental physical and mathematical training [1]. A large number of hours are still allotted to the study of physics and mathematics. These hours are divided between class work and the student's self-work. During intramural educational process student's self-work takes up 50 % of all time, during part-time and extramural educational process it takes up to 80 % of all study time. Despite the advantages of the independent nature of educational activities, there is a problem of organizing and controlling this activity, which can be solved with the help of an information-educational environment.

The creation of an educational environment emphasizes the importance of learning, ensures the assimilation of comprehensive knowledge, skills, formation and development of competencies. The educational environment helps to implement all the functions of training, assimilation of all elements of the content of education. An important subsystem of the developing educational environment is the educational and technical environment of professional and personal self-development of students [2].

Modern young people are often referred to as "digital aborigines," and it is not difficult for them to obtain any information from the Internet. But information itself is not knowledge. It is the teacher's task to turn this information into knowledge. There is no unified approach to this path, every time the teacher is faced with the task of choosing different models of electronic educational resources, taking into account the tasks of training specialists of a certain specialty.

Today the educational space is filled with various open educational resources (online courses), there is a format of MOOC (Massive Open Online Courses), which allows to carry out network interaction of participants of the educational process. In our country there is a steady growing interest in MOOC, in the implementation of students "training in this format. The best universities of the Russian Federation offer on the platform "Open Education" their online courses of basic disciplines studied in Russian Universities [3]. Any student user can take an online course of interest from the leading Universities of the Russian Federation completely free of charge, as well as in the case of successful completion of training he can receive set-off by using these results in his university within the framework of his own educational program.

For extramural studies, MOOC seems to be the most effective. The experience of using MOOCs in training specialists on the specialty "Land Transport and Technological Vehicles" at the Siberian Federal University has shown that there are inconsistencies in the curricula of different Universities, and there is a difference in the volume of study of disciplines, which makes it difficult to integrate MOOCs with traditional forms of education [4]. Therefore, it is possible to use the results of MOOCs to set off individual modules of the academic discipline.

3. Materials and methods

3.1. IS-MADI: Physics

Another way to train physics was chosen at the Department of Physics MADI: it includes using well-known digital educational technologies and creating one's own web resource. At the MADI Department of Physics, the transition to digital forms of education is carried out using electronic presentations, including videos and materials from the Internet, electronic textbooks and teaching materials, the certified information system "IS-MADI: Physics" [5], and computer testing – input, current, and final. Students use electronic materials to carry out mini-projects, as well as to prepare reports for student conferences. Internet resource "IS-MADI: Physics" was created and developed by teachers of this Department to solve the problem of training automotive engineers in Physics and other disciplines of Natural Science.

An important part of this resource is the section "Library", which is available for students throughout the work with the information system. This section consists of textbooks, texts and presentations of lectures, environment descriptions of laboratory works, and guidelines for solving problems. A survey was conducted among the 2nd year students of the Faculty of Road Transport (specialty "Automotive engineering in transport technologies"). They systematically use "IS-MADI:

Physics" during two semesters. The aim of this survey was to determine the effectiveness of the Library section using by students. It turned out that 64 % of respondents constantly use lecture texts, 42 % use methodological recommendations for solving problems, 48 % use lecture presentations.

In the learning module there is a "Bank of tasks", which contains about 2500 questions and tasks to test the knowledge gained in the study of each topic, structured by levels of difficulty that students pass sequentially. "IS-MADI: Physics" was created as a game with several levels, each of which can be reached only after passing through the previous stage. At the first level assimilation by students of the basic concepts and definitions of the studied section of Physics training course is checked. At the second level it is required to solve a quantity of simple tasks with the choice of the answer. In case of successful execution of a task the system transfers the student to the following level, but if the number of the wrong answers or a limit of time is exceeded, the system returns this student on the first level. At the third level the student is offered to solve a number of problems of basic level of complexity.

The answer must be entered from the keyboard, taking into account the specified accuracy. If the number of incorrect answers or a limit of time is exceeded the system returns this student to the first step again. The quantity of questions, the number of the offered possible answers, time limits, quantity of admissible errors at each stage is set by the teacher based on the curriculum and the number of training hours. The important advantage of an information system is that it helps junior students to organize their self-work work in an optimal way.

Practice of pedagogical work shows that first-year students satisfactorily perform their self-work when it is specific and tied to a certain point in time. The use of "IS MADI: Physics" makes it possible to clearly define the necessary minimum sections for development and the time frame for their implementation. At the same time students are given the opportunity to pass sections of the information system any number of times at a convenient time for them and in any place where there is access to the Internet, using a personal computer or smartphone.

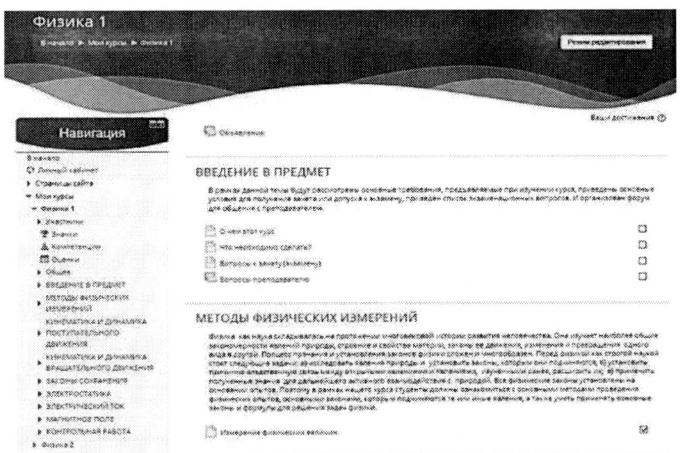

Figure 1. Moodle training page

3.2. Modular Object-Oriented Dynamic Learning Environment

A wide range of opportunities to ensure and control the self-work of students is provided by the electronic system Moodle (Modular Object-Oriented Dynamic Learning Environment). This system is a free web application that allows a teacher to create sites for online learning. By using the Moodle system, it is possible to create web pages where teachers can post texts, pictures, video and audio information. Also, Internet tools allow a teacher to place flash animation on his pages where he can create interactive laboratory work when describing certain physical phenomena. In this case, the teacher can use a linear learning mode that consists of a series of training pages (Fig. 1) [6], as well as create more complex modes that contains various ways or options for the student. Depending on the answer chosen by the student during testing or quiz, students can go to another page, return to the previous page or be redirected in a completely different way. Another important tool of the Moodle system is the ability to host all kinds of electronic content: hyperlinks to electronic resources, e-books, video and audio files.

To carry out an active interaction with students in the Moodle system, the "Video Conference", "Forum" and "Chat" modules are provided. The "Video Conference" module allows student to create links to web conferences. Using this module, student can specify the name and description of the event in the calendar, as well as groups and recording parameters of the online session. The Forum module allows participants to communicate in asynchronous mode. There are several types of forums, such as a standard forum where everyone can start a new discussion at any time and a "Question and Answer" forum where students must first reply to a message before they can see the answers of other students. The Chat module allows participants to have the opportunity of synchronous written communication in real time. Chat can be either a one-time event or repeated at the same time every day or every week.

4. Results and Discussion

The results of the «IS-MADI: Physics» used by first-year students (set 2018/2019 academic year) in the discipline "Ground vehicles," in total 110 people, are given (Fig. 2). First-year students solved problems in eight sections of Physics: Kinematics, Dynamics, and Conservation Laws in Mechanics, Electrostatics, DC Laws, Magnetism, Molecular Kinetic Theory and Thermodynamics. 90–92 % of the first year students received set-off in physics, as the result of work in the semester.

Approximately the same feature is observed in Fig. 3. The results of the use of the Internet resource «IS-MADI: Physics» by second-year students (set 2017/2018 academic year) in the same discipline "Ground vehicles," in total 98 people, are given (Fig. 3). They solved problems in six sections of Physics: Oscillation, Waves, Wave Optics, Quantum Optics, Thermal Phenomena, Atomic and Nuclear Physics/ It is shown? That the most difficult sections of Physics are Quantum Optics, Atomic, and Nuclear Physics. But in spite of difficulties 88–90 % of students pass exam successfully.

The use of the Internet resource "IS-MADI: Physics" makes it possible to draw a conclusion not only about the abilities and knowledge of students of the first and second academic year of the technical university, but also gives a cut of school training in Physics in the whole country, as MADI is attended by young men and girls from different cities of Russia.

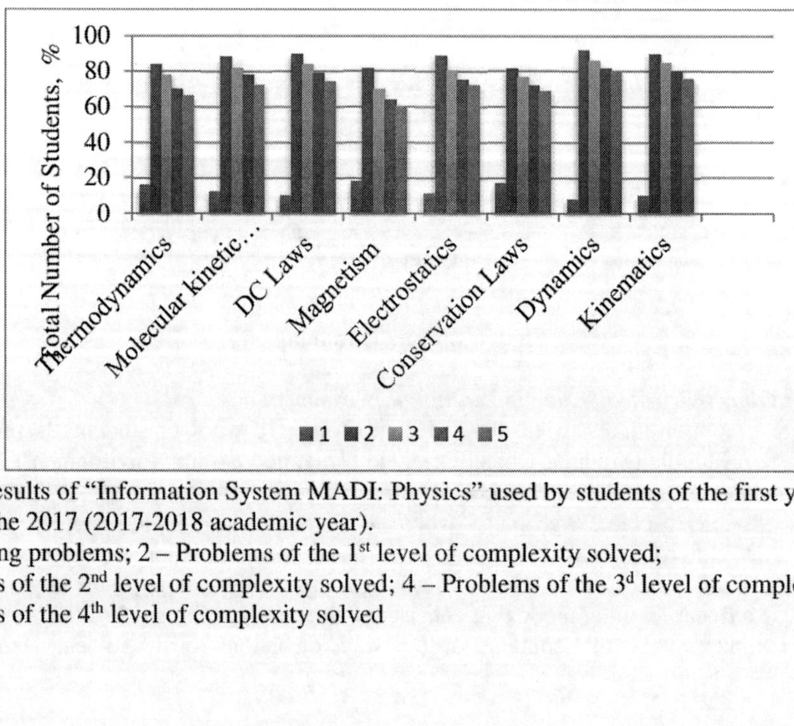

Figure 2. Results of "Information System MADI: Physics" used by students of the first year in the Autumn of the 2017 (2017-2018 academic year).
1 – No solving problems; 2 – Problems of the 1st level of complexity solved;
3 – Problems of the 2nd level of complexity solved; 4 – Problems of the 3[d] level of complexity solved
5 – Problems of the 4th level of complexity solved

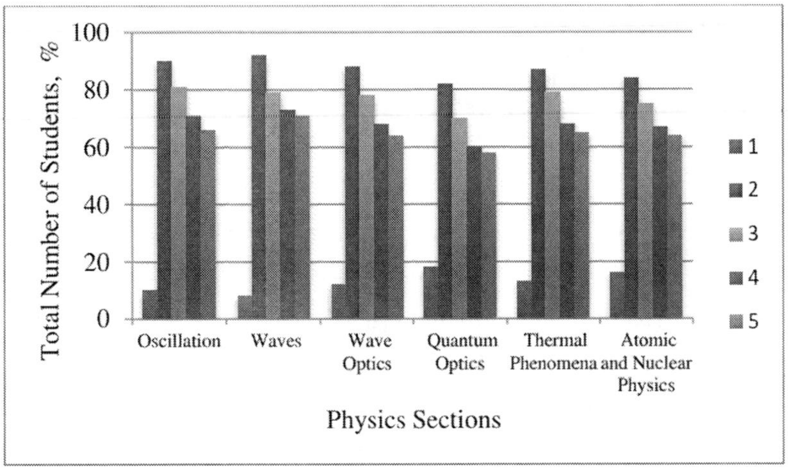

Figure 3. Results of "Information System MADI: Physics" used by students of the second year in the Spring of the 2018 (2017–2018 academic year).
1 – No solving problems; 2 – Problems of the 1st level of complexity solved;
3 – Problems of the 2nd level of complexity solved; 4 – Problems of the 3d level of complexity solved
5 – Problems of the 4th level of complexity solved

To increase the effectiveness of electronic forms of training, an interactive information educational environment is required. This thesis implies the use of training and monitoring programs. The compilation of such programs will ensure their successful application if these electronic forms take into account the individual abilities and capabilities of students, the teacher's ability to arrange training tasks according to degree of difficulty, absence of difficult tasks with large time costs, allowance to have a few percent of possible errors, and an opportunity to find answers to assignments and detailed explanations. All of these factors allow us to hope that students will be able to work independently. If at the same time there is a two-way communication with the teacher (it can also be electronic: e-mail, Skype or other instant messengers), then the probability of passing the exam or other control task will be close to 100 %.

It is possible to compare the merits of MOOCs and the information system "IS-MADI: Physics". Both tools are proposed to note the possibility of using interactive tasks with automated assessment of the result. In the information system "IS-MADI: Physics" mentioned above, such tasks and their assessment are included in the program. The course of the assignment is recorded by the program, and the teacher can track how many attempts to solve the task the student completed, how many of them were successful. At the end of the task, the program is evaluated as a percentage.

Examination with proctoring (as MOOC uses), that is exactly what we created in the information system "IS-MADI: physics": the task has limited the time of its execution to the specified time interval, the number of possible errors at each level of task execution and instant electronic assessment. Of course, during the task, the student is offered a countdown. However, in real time, students in this case are not monitored by the teacher, and verification of the identity of the person who sits at the computer and performs the task is also not provided. However, this information system "IS-MADI: Physics" is convenient for both students and teachers: an unlimited number of students can perform the same task at the same time. Despite the simultaneous use of this system, due to a large database of tasks (more than 2500) each student will have its own set of tasks.

5. Conclusions

In conclusion, we can draw the following statements:

1. Digital forms of training (on-line training) increase the effectiveness of training due to the speed, individualization of the educational process and the objectivity of student assessments.

2. The transition to digital forms of education is inevitable in connection with the widespread, more accessible and accelerating spread of electronic forms of interaction.

3. To increase the effectiveness of electronic forms of training, it is necessary to create an interactive information educational environment.

4. Accessibility, relative ease of use, very fast control and the ability to immediately receive encouragement for a well-completed assignment contributes to the growth of motivation to learn.

References

[1] Smyk A and Zimanov L 2017 Training of automotive engineers in the mirror of education reforms in the Russian Federation *Bull. of transp.* **3** 12

[2] Tkacheva T 2013 Educational environment as a factor of improving the quality of education: MADI The Tidings of the Baltic State Fishing Fleet Academy: Psychological and pedagogical sciences *Theory and methods of professional education* **23** 86

[3] Babaeva M and Smyk A 2018 Distance education: a historical path to MOOK *Higher Education in Russia* **4** 156

[4] Moskalev A, Serukova I and Dolgopolova M 2019 The use of mass open online courses in teaching physics to bachelors of general engineering areas of training *Bull. of Krasnoyarsk state pedagog. Univer.* **1** 26

[5] Smyk A, Spiridonov A, Bachtina E, Belkova Y and Spiridonova L 2016 Information system for effective study of physics *Phys. Educat. in univer.* **22** 97

[6] *MADI PKI Distance Learning System* Retrieved from: https:// cdo.madi.ru

IOP Conf. Series: Materials Science and Engineering **832** (2020) 012069 doi:10.1088/1757-899X/832/1/012069

Methodological approach to digitalization of management processes in automobile and road complex

A V Terentyev[1], M Yu Karelina[1], A A Pavlovskaya[2], I V Arifullin[3] and E A Karelina[4]

[1]Department of Machinery Parts and Theory of Mechanisms, Moscow Automobile and Road State Technical University (MADI), 64, Leningradsky Ave., Moscow, 125319, Russia
[2]Department of Logistics and Management Methods Yanka Kupala State University of Grodno, 22, Ozheshko str, Grodno, 230023, Belarus
[3]Department of Management, Moscow Automobile and Road State Technical University (MADI), 64, Leningradsky Ave., Moscow, 125319, Russia
[4]Department of Foreign Languages, Moscow State University of Technology «STANKIN", 1, Vadkovsky lane, Moscow, 127055, Russia

E-mail: aleksej.terentev.67@bk.ru

Abstract. In order to successfully accomplish the tasks of digitalizing control processes in the road complex integrated into a focused process, new generations of control systems — intelligent control systems – are being actively developed today. Intelligent control systems should be able to independently develop a goal, implement the decision-making process for an action, provide the necessary set of functions to achieve a goal, predict the values of system indicators as a result of actions and compare them with real data, providing feedback in the system and adjusting the control procedure. Currently, the methodological basis for making management decisions in control systems is the concept of scenarios based on the theory of expert assessments. The main disadvantage of this approach is the inability of the system to gradually adapt to changes in input information without completely reworking the structure of the initial requirements. For effective implementation of this task, it is necessary to develop a decision-making methodology in control systems based on formal logical methods of decision theory, namely, analytical methods for obtaining many effective plans, which should allow us to build algorithms that provide optimal solutions when implementing applied management tasks while having minimal information about the conditions of the research object.

1. Introduction

To successfully accomplish the tasks of digitalizing control processes in the road complex integrated into a focused process, new generations of control systems are actively developing today – intelligent control systems (ICS) [1]. In the future, ICS should be able to independently develop a goal, implement the decision-making process for an action, provide the necessary set of functions to achieve the goal, predict the values of system indicators as a result of actions and compare them with real data, providing feedback in the system and adjusting the control procedure [2]. However, ICS that implement the abovementioned functions are complex software systems designed for automatic decision-making in the operational management of complex systems and processes. It should be noted

IOP Conf. Series: Materials Science and Engineering **832** (2020) 012069 doi:10.1088/1757-899X/832/1/012069

that full-fledged ICS for automatic control are still under development. However, the implementation of individual functions of the ICS in the form of systems that allow us to evaluate and implement actions in changing environmental conditions is possible in the form of dynamic control systems (DCS). The DCS is a tool that allows for given initial data on the state of the system and changes in informational environmental conditions that are random in nature, to compare the parameters of the possible and real results of the action, to develop optimal solutions that help achieve the goal of the control system. To do this, the necessary database and the mathematical apparatus for solving the tasks should be formed in the DCS. The functions of the DCS are also the assessment of the results of solving the problem, the formation of the parameters of the possible result of the system, the adoption of the control decision, process modeling to evaluate the solution of the problem [3].

2. Problem Statement

Depending on the goal that the DCS faces, the knowledge base, algorithms for solving a problem, making a decision, developing a control may have a different idea, which, in turn, depends on the nature of solving problems. Accordingly, it is possible to form various types of DCS, depending on the approximation of their functionality to the ICS. The general structure of the DCS with the use of simulation methods for obtaining solutions (DCS of the first type) is shown in Figure 1. The model of this structure assumes that the actual data accurately reflect the described processes, only in this case a reliable solution to a problem can be obtained in accordance with the statement and specification.

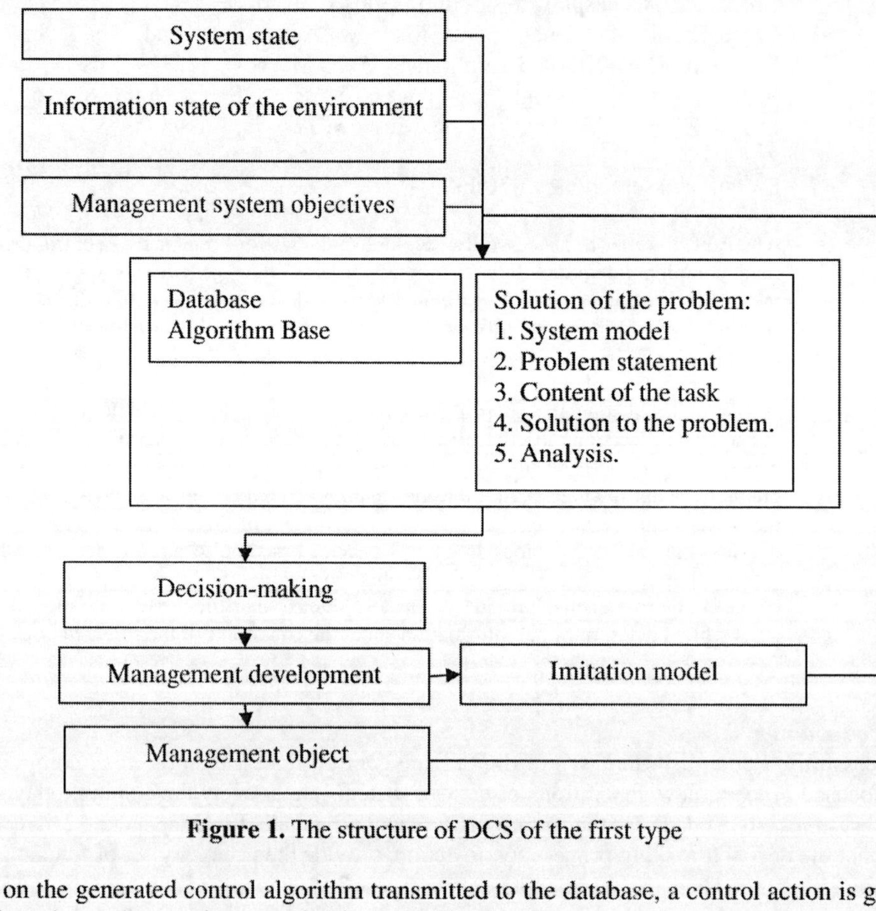

Figure 1. The structure of DCS of the first type

Based on the generated control algorithm transmitted to the database, a control action is generated. Before this action arrives at the control object, its effectiveness and consistency can be assessed using various methods used depending on the degree of uncertainty of information situations in the environment being studied. For ICS tasks, this is the most difficult case – the uncertain state of the research environment [4, 5].

As a rule, in this case it is possible to resort to the use of expert models based on the knowledge of experts – specialists in this problem area, or methods of decision-making theory. What is one of the main problems in the design of DCS of a higher level of approximation to the ICS (second type) is that the task of choosing the necessary apparatus for describing decision-making processes and building on its basis a decision-making model that is adequate to the state of uncertainty in the environment becomes important. It is possible to use the combined DCS of the first and second type, where the knowledge base combines the description in the form of rigorous mathematical formulas with expert information, as well as the mathematical methods of finding a solution with non-strict heuristic methods [6]. But when developing a DCS of this type, the following problems may arise:

- The determination of the knowledge base structure and the principles of its formation.
- The need to develop new and adequate use of well-known theories and methods to describe processes in the system.
- The need to develop ways to formalize and organize the process of implementing decisions.
- The need to develop algorithms and software.

It is important to note that a DCS must have the property of adapting to dynamic changes in the environment of a process under study, the ability to introduce new elements and relationships into the description of situations, change the rules and strategies for the operation of objects in the decision-making process and develop control, work with incomplete, fuzzy and contradictory information etc. [7].

3. Materials and Methods

All known decision-making methods imply an iterative application of synthesis, analysis and estimation procedures [8]:

- Synthesis includes the formation of a certain set of requirements described in the language of functioning. The synthesis is most effective while using both ascending and descending actions, and the results of applied research and the possibility of using known technologies are taken into account.
- Analysis of system solutions includes calculating and predicting the value of parameters depending on technical or informational characteristics. In all cases, information on physical processes and phenomena, technical information as well as economic information stored in databases is used. System analysis and operations research are necessary steps towards evaluating system design options, but the required adaptation of appropriate models and methods to the characteristics of the subject area is required. In general, the use of analysis is a necessary but not a sufficient component of the decision-making procedure on the choice of a project version of a system.
- Evaluation implies that each solution option (or alternative) is evaluated in comparison with other options, and also checked for compliance with the requirements of interested parties. Evaluation of each of the options is carried out after receiving information about its characteristics. Characteristic data not related to design decisions are external factors that are used in relation to all options to be evaluated. Each option is subjected to a final assessment with the selected optimal indicators, after which a final decision is made.

The iterative use of the synthesis – analysis – evaluation triad is a fundamentally important feature of decision-making methods. The application of one method or another begins with the recognition of the needs of interested parties and the determination of their requirements, which are then transformed according to certain rules to obtain an initial description of system solutions. In the future, the description of the system is refined and detailed; moreover, at lower levels of the system hierarchy, the system engineering process is already used recursively, which allows achieving a high level of concretization in the description of the system. Using the method "synthesis – analysis – evaluation" allows us to describe and build a system, providing a gradual reverse transition from the level of detailed description of the components to the general structure of the system under study.

IOP Conf. Series: Materials Science and Engineering **832** (2020) 012069 doi:10.1088/1757-899X/832/1/012069

The decision-making process consists of a number of stages (steps). At the initial stage, it is necessary to identify a problem, a problem situation or determine the subject area in which decision-making is required. The term "decision" has many semantic meanings. These may be actions to solve problems, leading to some kind of result, the consequences of which must be assessed. Under a solution we can also understand some objects or systems requiring evaluation. A solution is also different options, alternatives, opportunities regarding actions and objects. The solution is also called the process of finding the best option in a sense, and this option itself.

At the next stage, it is necessary to substantively describe the subject area, to identify the goals of decision-making and also the limitations. Next, we need to collect and analyze detailed information about the subject area. In some cases, it is necessary to build a model, determine the range of alternatives (options) of decisions, perform decision generation and formulate a decision-making problem.

At the next stage, decisions are made based on the selected (developed) methods. Evaluation of solution options, their comparison, classification, ranking and selection of the most preferred one are carried out. And finally, the final phase is related to the implementation of the decision and the assessment of its consequences.

In decision theory, subjects are identified that affect the different stages of decision making. First of all, this is the decision maker (DM), that person or group of people who makes the final choice. The owner of the problem, an expert as well as a decision consultant can be highlighted.

The properties of the subject area, the conditions for the functioning of decision support systems can be different. Depending on the completeness and initial data, decision-making problems can be well-structured, weakly structured, and unstructured. For well-structured tasks, data are formulated in quantitative form, for weakly structured tasks, they contain quantitative descriptions with the dominance of qualitative and uncertain factors, for unstructured tasks, data are presented in the form of a qualitative description of the initial factors and the interdependencies between them. The choice of decisions can be made in conditions of certainty, risk or uncertainty.

Under certainty, the initial data and consequences of each alternative solution are known.

Risk conditions occur if the probabilities of alternatives to solutions are known or the law of the distribution of their probabilities is known. Then the task is reduced to the choice of statistical solutions. In turn, risk conditions can be defined as stochastic uncertainty conditions generated by random factors that, when they appear massively, have the property of statistical stability and are described by some kind of probability distribution law [9].

Uncertainty conditions under which the probability distribution law for uncertain factors is unknown are defined as conditions of statistical uncertainty. In turn, conditions of statistical uncertainty are divided into two types: first, with known distribution parameters (mathematical expectation, variance, and other characteristics of a random variable); secondly, with unknown distribution parameters.

If uncertain factors are nonrandom, do not have statistical stability, and are not described by any law of probability distribution, then it is impossible to obtain reliable information about these uncertain factors. And the probability of decisions related to the impact of these factors cannot be determined. In this case, we speak of conditions of non-stochastic uncertainty.

The following reasons for the occurrence of non-stochastic uncertainty in the choice of solutions should be highlighted:

- incompleteness and insufficiency of information about all factors of the problem of decision-making;
- fuzziness, ambiguity or inconsistency of allocation and description of the factors of decision-making problems.

In addition, the number of factors, variables, signs, indicators and criteria that determine the subject area can vary significantly. Информация об этих факторах может быть полной или частичной. Information on these factors may be complete or partial. Factors can be independent of each other or dependent. The dependence may be consistent or inconsistent. Indicators can be quantitative or qualitative, evaluated on different scales.

Decisive for the choice of solutions are models for evaluating indicators and choosing alternatives for these solutions. The structures of the models for evaluating and choosing decisions can be different, they might differ in "connectedness" between factors, and hierarchical assessment. The structure of a model is largely determined by the goals of choice and the criterion of the efficiency of decisions. There are various languages for describing selection problems in the form of: quality criteria, selection functions, binary relations, axiom systems [10].

It is advisable to use the language of binary relations or systems of axioms if the choice of solutions cannot be made unambiguously and is represented by vectors of criteria and restrictions, depends on the system of preferences of the decision maker, as well as on the conditions and relevant information that he possesses [11].

The most widely used languages for describing selection problems are the languages of quality criteria (efficiency) and selection functions. These description languages are focused on the fact that the initial set of decision alternatives is uniquely defined. And for making decisions, criteria for the effectiveness of decisions or a set of rules are used.

Under the conditions of one or a small number of criteria, complete certainty of the subject area, the methods of decision theory are well developed.

However, in conditions of a large number of incompletely defined, both qualitative and quantitative, interacting indicators that determine the subject area and also the complexity of objects that require an assessment to make decisions, classical methods and models of decision theory are often not applicable.

In such conditions, the construction of a domain model and the development of methods for evaluating complex multi-criteria objects, which can greatly facilitate the subsequent choice of an option, are of great importance.

4. Discussion of Results

The main tasks of evaluating and choosing alternatives to solutions include the following:

- evaluation of alternative solutions;
- comparative assessment and ranking of many decision alternatives;
- definition of strategies for evaluating decision alternatives;
- determination of the values of particular indicators of decisions by given values of a generalized indicator;
- search for the values of particular indicators that provide the required values of the generalized indicator;
- search for the best solutions (values of particular indicators) providing the required values of a generalized indicator.

Single- and multi-criteria methods for evaluating and choosing alternatives to solutions can be distinguished [12, 13]. In the single-criterion case, the following are considered known:

- many alternatives to solutions (objects, options) – $a_j, j = 1 \ldots m$;
- assessment of alternatives according to the selected criterion (indicator) – $p(a_j)$;
- rule for choosing the best option:

$$\max_j(p(a_j)) \text{ or } \min_j p((a_j)).$$

In the process of solving the problem of choosing the best option, an alternative a^* is determined, for which:

$$p(a^*) = \max_j(p(a_j)) \text{ or } p(a^*) = \min_j p((a_j)).$$

Another case of the selection problem is the ranking of solution options in accordance with a variety of alternative estimates. More significant for practical use is the case of evaluating and choosing alternatives according to many criteria. In the multi-criteria case, each variant a_j can be associated with a vector of the form:

$$\{p_1(a_j), p_2(a_j), \dots, p_n(a_j)\},$$

representing option estimates by criteria (indicators): $p_1, p_2, \dots, p_n$.

Further, approaches to comparing and choosing solutions can be divided into two large groups: reduction and not reduction of many criteria to one [14].

Methods based on non-reduction of multicriteria assessment problems and choice of solution options to one criterion, set as their task comparison of options based on assessment vectors for all criteria. Among the important factors taken into account when solving this problem, one can indicate the weight (importance) of the criteria, as well as information on the preferences of the decision-maker. Often these methods are used in another setting: the search for optimal solutions on a set of parameters that determine the decision-making conditions [15].

Among the methods of this group the following can be identified:

- method of dominance;
- method based on global criteria;
- lexicographical ordering;
- methods of mathematical programming;
- threshold methods;
- methods of the theory of multisets etc.

As a rule, the application of these methods is due to the independence of the criteria and their full certainty.

In the case of using methods based on the reduction of multi-criteria assessment problems and the choice of solutions to one criterion, the choice problem is solved on the basis of constructing an integral (generalized) criterion. For this, various methods of aggregation, "convolution" of indicators are used, i.e. the construction of various general indicators, primarily additive and multiplicative.

An additive generalizing criterion (indicator) is obtained as a weighted sum of estimates by particular criteria (indicators):

$$p_{об}(a_j) = \sum_{i=1}^{n} w_i * p_i(a_j),\ j = 1, \dots, m,$$

where $p_i(a_j)$ is the assessment of the alternative a_j by the criterion p_i, w_i is the weight (importance) of the criterion p_i .

The multiplicative criterion is defined as:

$$p\,(a_j) = \prod_{i=1}^{n} p_i(a_j)^{w_i},\ j = 1, \dots, m,.$$

However, the above expressions do not always adequately reflect the characteristics of the tasks of assessment and choice, in particular, the interdependence, inconsistency of individual indicators.

The approach based on the convolution of many criteria into one is used in many methods of decision theory, among which the following can be distinguished:

- methods based on the theory of value, utility;
- methods of analysis of hierarchies;
- nonlinear convolution methods;
- methods of the theory of fuzzy sets [15].

The approach to multi-criteria assessment of complex objects, based on the reduction of many criteria to one, is attractive in that it reduces the complexity of such decision theory procedures as comparison, ranking, classification and choice of options (alternatives). At the same time, the demand for the search for effective assessment methods that reflect the characteristics of the subject area and the preferences of a decision maker or the knowledge of an expert increases significantly.

An analysis of existing methods allows us to conclude that they do not fully reflect the features of the subject area, which are the uncertainty as well as the strongly interacting different-quality indicators and also the complexity of the objects being evaluated.

5. Conclusion

In contrast to the first type of DCS designed to find the optimal solution and based on rigorous mathematical methods and optimization models, the second type DCS is mainly focused on solving difficult formalized problems in the absence of complete and reliable information (Figure 2).

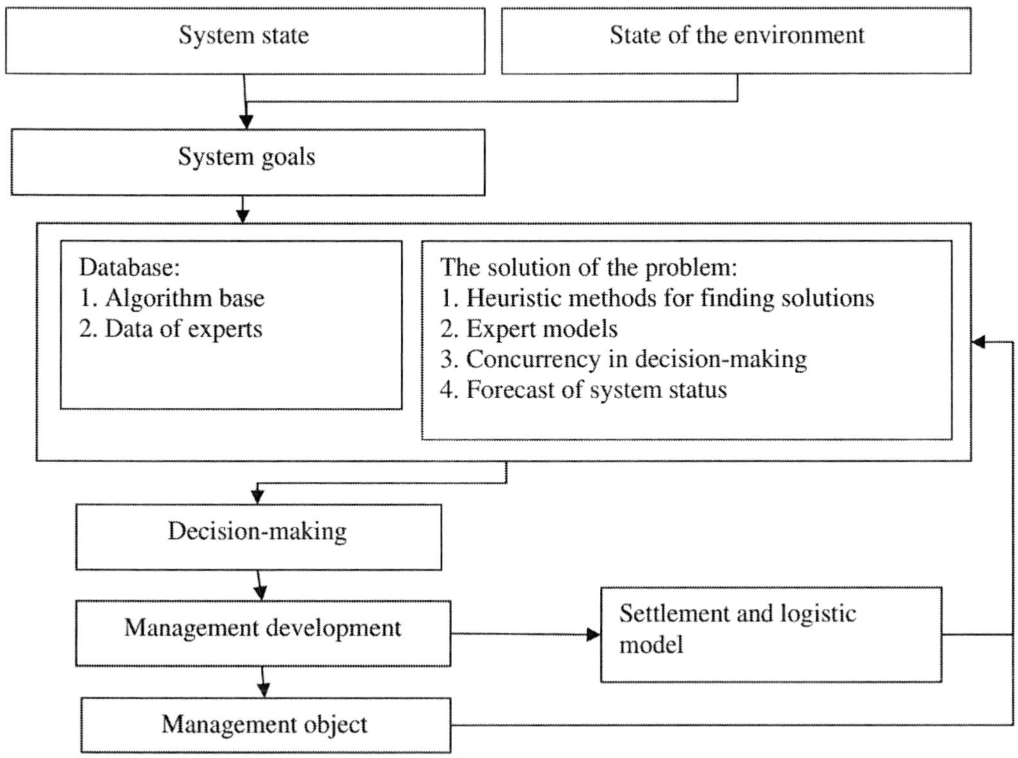

Figure 2. The structure of DCS of the second type

A DCS of the second type is designed to search for the optimal solution based on rigorous mathematical methods and optimization models (lack of reliable information). Naturally, one of the main problems in the development of a second-type DCS is the choice of an analytical apparatus for the algorithmization of decision-making processes and the construction on its basis of a mathematical model that is adequate to the studied area (semantically correct). However, as a rule, the main developments are carried out in the context of a deterministic interpretation of a single-criterion model with its inherent consistent decision-making scheme. As a result, the obtained models often have a significant share of subjectivity in decision-making and are inadequate to real practical processes characterized by non-determinism in the development of the situation in the research environment.

When determining the second type of DCS, it is necessary to focus on the computational-logical model for obtaining solutions. In this case, the information base should combine the description of processes in the form of a strictly mathematical apparatus with the information of experts, that is, form parallel branches of the search for a solution with their further synthesis. Therefore, mathematical methods for finding a solution are combined with non-strict heuristic methods, and the weight (significance) of one or another system action is determined taking into account further testing of the model in practice. When developing a second-type DCS, it is necessary to overcome a number of difficulties associated with the need to solve the following problems:

1) formation and updating of the database;

2) development of new and adaptation of known theories and methods for describing information processes in DCS;

3) development of formal ways of organizing the application of the developed DCS (methodological support for practical implementation);

4) development of algorithms and software with data processing and obtaining results in parallel processes for developing optimal solutions.

It can be stated that the second type of DCS cannot be fully attributed to the ICS, since it has all the disadvantages of hybrid systems. The above difficulties can be removed by integrating mathematical modeling methods into a set of effective plans into the DCS, which should allow to build algorithms that provide optimal solutions for the implementation of applied control problems having minimal information about the state of the research object.

References

[1] Solodkiy A I and Yenokaev V K 2015 Formation of rational options for the integration platform of local automated traffic control systems *Bull. of Civil Engineers* **6(53)** 190–4

[2] Solodkiy A I and Yenokaev V K 2016 Methods of choosing a rational option for integrating local automated traffic control systems *Innovations in transport and engineering* Proc. of the IV Int. sci. and pract. Conf. ed V.V. Maksarov (St. Petersburg: St. Petersburg Mining University) vol I pp 57–60

[3] Khovanov N V and Fedotov Yu V 2006 Models of accounting for uncertainty in the construction of composite performance indicators of complex production systems *Sci. reports* **28** 37

[4] Terentyev A V and Prudovsky B D 2016 Decision-making methods in the conditions of the uncertain state of the "external environment" *Transport Planning and Modeling* Proc. of the Int. Sci. and Pract. Conf. (Saint-Petersurg: SPSUACE) pp 145–9

[5] Kornikov V V, Seregin I A and Khovanov N V *Bayesian model for processing non-numeric, inaccurate and incomplete information on weight coefficients* Retrieved from: http://inftech.webservis.ru/it/conference/scm/2000/session3/kornikov.htm

[6] Karelina M Yu, Arifullin I V and Terentyev A V 2018 Analytical determination of weight coefficients in a multi-criteria evaluation of effectiveness of vehicles *Bull. of the Moscow Automobile and Road State Techn. Univer. (MADI)* **1(52)** 3–9

[7] Chong Y, Quek C and Loh P 2009 A novel neuro-cognitive approach to modeling traffic control and flow based on fuzzy neural techniques *Expert Syst. with Applicat.* **36** 4788–803

[8] Buslaev A and Novikov A 2003 *Probabilistic and simulation approaches to optimizing road traffic* (Moscow: Publ. House "Mir")

[9] Antonova A S and Aksenov K A 2012 Multi-criteria decision making under risk based on the integration of multi-agent, simulation, evolutionary modeling and numerical methods *Engineer. J. of Don* **4(2)** Retrieved from: http://www.ivdon.ru/magazine/archive/n4p2y2012/1466

[10] Saaty Thomas L 2008 *Decision Making with Dependencies and Feedbacks: Analytical Networks* (Moscow: Publ. house "LKI")

[11] Taha Hemdi A 2005 *Introduction to Operations Research* (Moscow: Publ. house "Williams")

[12] Chernorutzkiy I G 2005 *Methods of decision making* (St. Petersburg: BHV-Peterburg)

[13] Steuer R 1982 *Multi-criteria optimization: theory, calculations, applications* (Moscow: Science)

[14] Prudovsky B D and Terentyev A V Methods for determining the Pareto set in some linear programming problems *J. of Mining Instit.* **211** 89–90

[15] Pegat A 2009 *Fuzzy Modeling and Control* (Moscow: BINOM, Lab. of knowledge)

Taxi transport characteristics in Vietnam

Huy Truong Thai[1], Andrey Rementsov[2], Khac Minh Nguyen[3] and AnhTuan Le[4]

[1]Academy of air defense-air Force, town Son tay., Ha Noi, 154480, Vietnam
[2]Moscow Automobile and Road State Technical University (MADI), 64, Leningradsky ave., Moscow, 125319, Russia
[3]Thai Nguyen University of Technology, 666, Tich Luong Strs., Thai Nguyen, 24131, Vietnam
[4]Institute of Military Mechanical Engineering, 42, Dong Quan Strs., Ha Noi, 122000, Vietnam

Email: truongk25mta@gmail.com

Abstract: In recent years, economic growth and urbanization have been rapid in Vietnam, which has led to an increase in mobility in urban areas, especially in large cities. The government of Vietnam and the city authorities are trying hard to find the best solution, in which developing public transport is considered the most optimal solution to traffic urban problems. Hanoi and Ho Chi Minh City are the two special cities in Vietnam; they have applied the best policies for developing public transport and actually made a great leap forward in economy and tourism. However, the reality shows that the level of meeting the transport needs of urban people is still very low (about 14 % in Hanoi and 8 % in Ho Chi Minh). Obviously, there are objective and subjective reasons for this problem, such as: low quality of urban transport infrastructure, asynchronous development of urban transport network and the shortage of taxis and system bus. In particular, the management and monitoring of the operation of the means of public transport are very weak leading to poor quality transport services, reducing the attractiveness of the customer for the service of public transport. In this situation, the goal of increasing the efficiency of public transport to satisfy the needs of people in the large urban centers becomes a major and urgent problem. The article analyzed and evaluated the role of passenger transport in Vietnam's economy. Important factors that will be needed to improve the quality of passenger service of Hanoi's public transport system have been identified. The characteristics of managing and operating taxi cars of major passenger carriers are also analyzed and evaluated in this article.

Keyword: transports, passenger transports, the operation of passenger cars, transport infrastructures, taxi.

1. Introduction

Road transport plays an important role in the transport complex of Vietnam, constantly providing the activities of enterprises and organizations of all forms of ownership. The features and advantages of road transport are primarily determined by its high mobility and maneuverability, which allows you to quickly concentrate vehicles in the required quantity in the right place, and when the need passes, quickly transfer them to another place. In addition, road transport can receive goods and passengers directly at the place of their formation without intermediate loading and unloading operations and transfer passengers and deliver them to their destination "door-to-door", and therefore, without additional costs for these operations. This type of transport ensures the regularity and sequence of delivery of goods. However, the cost of transportation by road is very high and on average exceeds the similar indicators of river and rail transport. The high cost price is determined by the low carrying capacity and, consequently, the low productivity of the rolling stock and, in this regard, a significant share of wages in the total amount of operating costs. Reserves of cost reduction are mainly intensive factors – an increase in utilization rates of vehicle mileage, carrying capacity and average speed.

2. Main part

Road transport, as one of the types of land transport in Vietnam, has received the greatest distribution and occupies a leading position in the transportation of passengers. In Vietnam, road transport accounts for more than 77.4 % of the volume of transported goods and 94 % of the volume of passengers. Thus, it is the main carrier for the main sectors of the economy in table 1 [1].

Table 1. Statistics of volume goods transported by types of the transport in Vietnam in the period of 2007–2017

Year	Quantity (thousand tons)	Types of the transport				
		Roadway	Railroad	River way	Seaway	Air way
2007	596800.9	403361.8	9050.0	135282.8	48976.7	129.6
2008	653235.3	455898.4	8481.1	133027.9	55696.5	131.4
2009	715522.4	513629.9	8247.5	137714.5	55790.9	139.6
2010	800886	587014.2	7861.5	144227.0	61593.2	190.1
2011	885681.5	654127.1	7285.1	160164.5	63904.5	200.3
2012	961128.4	717905.7	6952.1	174385.4	61694.2	191.0
2013	1010413.9	763790.0	6525.9	181212.7	58701.6	183.7
2014	1078580.9	821700.0	7178.9	190600.0	58900.0	202.0
2015	1146895.7	877628.4	6707.0	201530.7	60800.0	229.6
2016	1255458.2	969721.0	5209.0	215768.2	64474.4	285.6
2017	1378999.8	1070572.0	5636.1	231950.8	70535.0	305.9
2018	1500918.0	1161632.3	6253.6	256361.3	76325.6	345.2

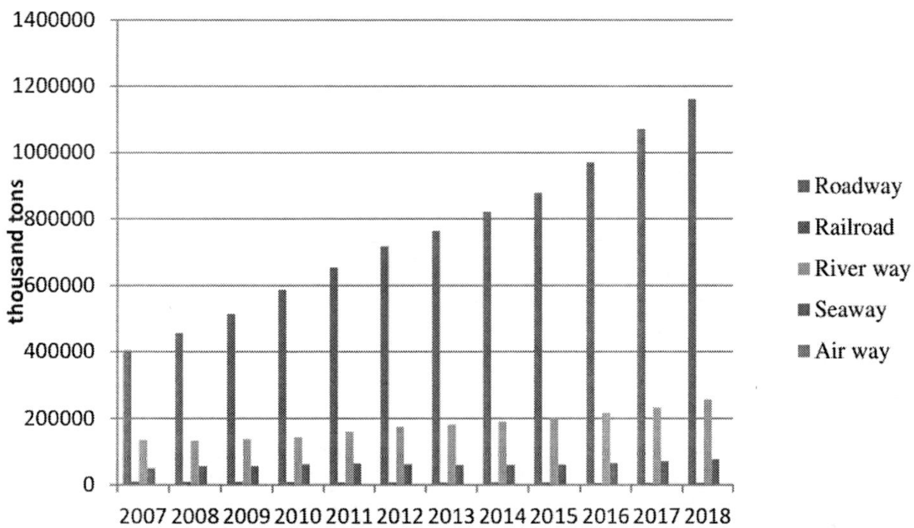

Figure 1. The growth of volume goods transported by types of the transport in the period of 2007–2017

Population growth and urbanization speed are very fast in Vietnam (96 million people in 2017) has led to demand for passenger transport in the major cities such as Hanoi, Da Nang, Hai Phong, Ho Chi Minh rising rapidly [2]. In recent years, each year more than 3401.9 million passengers are transported by the roadway (Table 2). In addition, this increase in transport volumes is supported by the quality of the roadway infrastructure and convenient road systems [1].

Table 2. Passenger volume statistics are transported by types of the transport in Vietnam in the period of 2007–2018

Year	Quantity (million people)	Types of the transport			
		Roadway	Railroad	River way	Air way
2007	1638.0	1473.0	11.6	144.5	8.9
2008	1793.5	1629.0	11.3	143.0	10.2
2009	2016.9	1843.6	11.1	151.3	10.9
2010	2315.2	2132.3	11.2	157.5	14.2
2011	2476.1	2306.7	11.9	142.4	15.1
2012	2676.5	2504.3	12.2	145.0	15.0
2013	2839.9	2660.5	12.1	150.4	16.9
2014	3056.8	2863.5	12.0	156.9	24.4
2015	3310.5	3104.7	11.2	163.5	31.1
2016	3623.2	3401.9	9.8	172.9	38.6
2017	3994.1	3760.0	9.6	180.0	44.5
2018	4347.5	4101.3	10.3	186.3	49.6

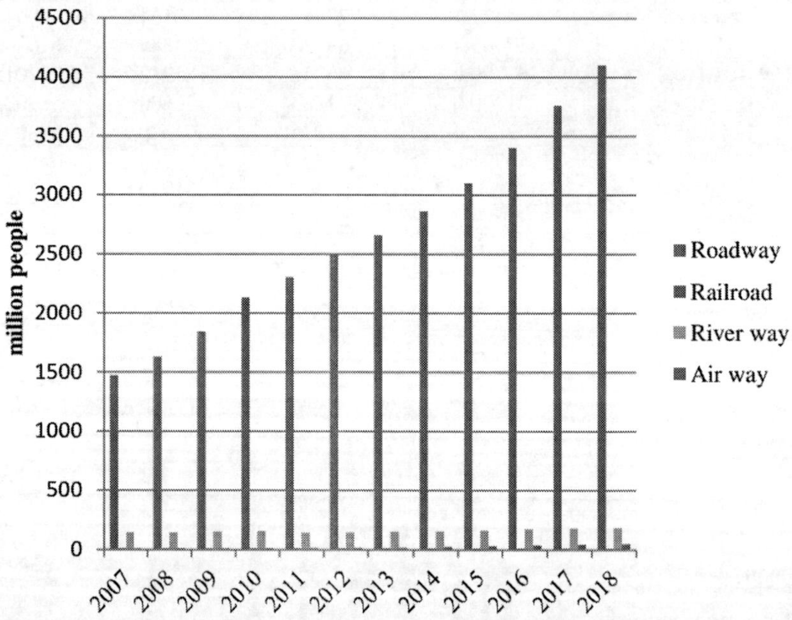

Figure 2. The growth in the number of passengers transported by types of the transport in the period of 2007–2018

Passenger transport in Vietnam is plentiful, including public transport, buses, taxis and personal vehicles. They are exploited in the city and suburbs. In the period from 2011 to 2020, according to the development plans of big cities such as Hanoi, Da Nang, Hai Phong and Ho Chi Minh, the main form of passenger transport is Public services [2]. By 2020, taxi transport should meet the needs of people with a growth rate of 15 and 20 % growth in 2025 in the table 3 [1, 2].

Table 3. Quantity characteristics of the passenger transport Hanoi in the period of 2011–2018

Year / Type of cars	2011	2012	2013	2014	2015	2016	2017	2018
Personal cars	116695	119832	123354	127536	132936	140672	149548	159123
Taxi	14632	16220	17880	19635	21563	23126	24665	26350
Buses	983	1016	1138	1269	1462	1589	1583	1936
Total	132310	137068	142372	148440	155961	165387	175796	187409
Rate of increase	3.3 %	3.6 %	3.7 %	4.0 %	4.8 %	5.7 %	5.9 %	6.2 %

The following factors show an increasingly important role of taxi transport in passenger transportation in Hanoi:

- In the future, the demand for passenger transportation will increase dramatically because most people use roadway transports as a daily means of transportation. This is evident in large cities and densely populated areas.

The analysis shows that there is a strong increase in the number of taxis in the roadway traffic systems in Hanoi

Currently, in the capital Hanoi Taxi 12 joint stock companies, 6 personal companies and the number of taxi vehicles contract is 26.350 units. The most popular taxis are branded Toyota Vios, Gran i10, Kia Morning. In 2018, the number of passengers using the taxi service was about 180.5 million. The average travel distance for a taxi is from 23000 to 25000 km. Taxi transport plays a special role in roadway traffics in Hanoi. Passengers in Vietnam use many different types of the transportation, bus service accounts for 8 %; taxi service accounts for 14 % and personal vehicles accounts for 8.5 % [3]. In 2018, passenger taxi transport increased by 12.1 % compared to 2017 and the number of taxi cars reached 26.350 units in figure 3.

So, currently in the capital of Hanoi there are 12 joint-stock companies, 6 personal taxi companies and contract cars in the amount of 26350 units. As a taxi, such car brands as Toyota Vios, Grandi 10, Kia Morning are often used. In 2018, passenger transport by taxi amounted to about 180.5 million passengers. Each year, the average mileage of one car is about 23000–25000 km. The general structure of passenger transportation in Hanoi is impossible to imagine without a taxi. Therefore, taxi transportation occupy a special place in the system of urban passenger transport. Currently, passengers in Vietnamese cities use various modes of transport. In general, bus services account for 8 % of passenger traffic, taxi – 14 %, private cars – 8.5 % [3]. In 2018, passenger taxi transport increased by 12.1 % compared to 2017, and the number of taxi cars amounted to 26.350 units (figure 3). Table 4 presents the main companies operating taxi in Hanoi [2].

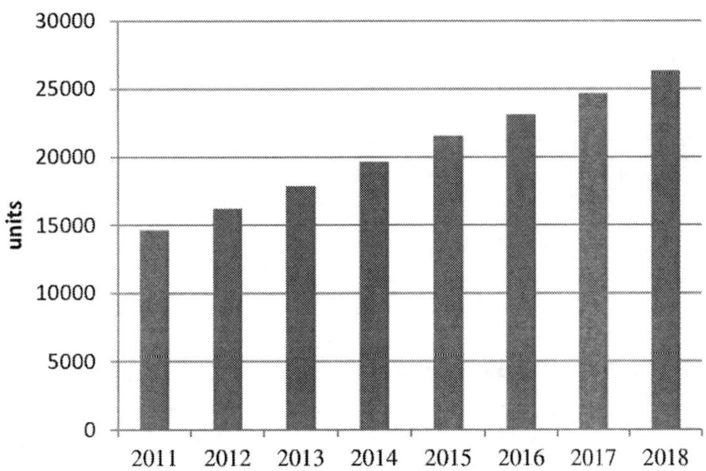

Figure 3. The number of taxis in the roadway traffic system in Hanoi

Table 4. Number of taxi cars of joint-stock and personal companies

Id	taxi company	Quantity (units)
1	Taxi Mailinh	4565
2	Taxi Group	2122
3	Taxi Noibai	1532
4	Taxi thudo	1230
5	Taxi Hanoi Tourist	1135
6	Taxi Airport	608
7	Taxi Thanglong	698
8	Taxi ABC	642
9	Taxi Vanxuan	684
10	Taxi Thekymoi	542
11	Taxi Grab	2250
12	Taxi Uber	2260
1	Taxi 54	322
2	Taxi 37	431
3	Taxi 12	453
4	Taxi 83	416
5	Taxi V20	363
6	Taxi CP	460
7	Contract cars in the suburbs	5637
Total		26350

According to the plan to develop passenger transport in Hanoi capital from 2020 to 2025, the rate of using personal vehicles such as motorbikes and personal cars will sharply decrease and the demand for public transport such as buses and taxis will increase rapidly [2]. In Hanoi, buses are a popular means of public transport, but the level and quality of exploitation is still very limited: some routes have not yet operated buses, the distance between the stations is quite far (more than 2 km), bus management process is quite outdated, the bus operating time is usually only up to 19 hours and a few activities up to 21 hours. Obviously, the bus does not meet the needs of people and the type of passenger transport by taxi is a very promising method to solve this problem. Advantages of this type of taxi transport are the maximum number of passengers is up to 9 people, the integration with software such as Grab, Uber helped taxi transportation become very comfortable. In the future need to develop transport by taxi towards the modernization of management, improve service quality and meet the needs of the people. By the above analysis, the demand for passenger and cargo transportation in major cities in Vietnam has achieved significant growth and promises to continue to increase sharply in the future. The question is how to manage and organize the operation of the taxi system to meet the needs of the people and suit the tropical climate conditions and underdeveloped infrastructure.

3. Conclusions

Analysis of taxi transport status in Hanoi shows that, in the context of increasing demand for passenger transport, and the goal of improving efficiency, further development of passenger transport is one of the major issues of economic and social Vietnam in the current period [6]. To solve this problem it is necessary to:

- reduce personal vehicle use and increase taxi and bus usage. Also, taxi services to ensure the reasonable cost of services and the segment so that people can use it;
- improve the quality of road infrastructure; constructing reasonable and convenient bus stops for operation;
- improve the quality of service delivery, enhance the comfort and safety for our customers;

- establish policies needed to encourage citizens to develop business in the field of passenger transport, more established taxi companies.

Obviously, in Hanoi city the level of public transport is very high and the demand for service quality continues to increase. To solve this problem, it is necessary to improve the organization of urban passenger transportation and taxi maintenance.

References

[1] *Management of the general statistics* Retrieved from: https://gso.gov.vn/default.aspx?tabid=721
[2] *Logistics* Retrieved from: https://logistics4vn.com/thuc-trang-loai-hinh-van-tai-da-phuong-thuc-trong-nganh-dich-vu-logistics-tai-viet-nam.
[3] Tran H T 2011 *Public Transport Issues in Hanoi* **4** 87–9
[4] Rementsov A N and Fan A B 2018 Features of cargo transportation by road transport in vietnam *Vestnik MADI* **55(4)** 96–102
[5] Nguyen H M and Rementsov A N. 2013 Organization of passenger traffic in Vietnam *Vestnik MADI* **35(4)** 8–13

Improving the performance of technical services personnel of car enterprises in digital ecosystems

V A Trubitsyn

Moscow Automobile and Road State Technical University (MADI), 64, Leningradsky ave., Moscow, 125319, Russia

E-mail: tva_va@mail.ru

Abstract. Based on the research, the article discusses the trends in the use of digital technologies and the creation of digital ecosystems in the road transport industry. Transport informatization has created the basis for a transition to a new level of functioning in the context of digital production ecosystems, which focuses personnel's attention on increasing labor productivity, accelerating interactions with consumers of transport and car maintenance units. However, the digital environment requires automotive personnel to improve existing and create new competencies related to digital literacy, information exchange, and increased communication skills. A digital ecosystem of the engineering and technical service of a car enterprise, a model of personnel competencies (by the example of a controller of the technical condition of vehicles) are proposed.

Key words: digitalization, digital ecosystem, transportation services, vehicle service, personnel, competency model, reliability of personnel.

1. Introduction

The use of digital technologies in transport is a complex system task that significantly affects the development of the transport industry. Considering that the advanced economies, including Russia, are gradually being drawn into the so-called fourth industrial revolution (Industry 4.0), the leadership of many road transport enterprises (RTE) and vehicle service stations (CSS) needs to review development strategies. 4.0" [1] provides for the massive introduction of digital elements and systems, the complete digitization of space, subjects and processes, elaborated digital transformation of all material resources and their integration into digital ecosystems to ensure the production and service level which satisfies consumers. Taking these trends in our country into account on the initiative of the Ministry of Transport, a digital platform of the transport complex is being formed, which will combine services and data arrays and become an ecosystem for all participants in the transport process. The digital platform will establish common standards, rules and regulations for information exchange on vehicles, transport infrastructure, etc. Since the main participants at the primary level in the new forming digital system are transport companies and service stations, they must substantially transform their activities in the light of new requirements caused by digitalization processes in the economy.

The creation of digital ecosystems in transport and service companies is possible only if there are well-established processes in place for obtaining and analyzing information on the achieved performance indicators based on modern services application and arrangement of effective exchange for the development of management decisions.

To date, a unified understanding of a number of terms of the "digital" economy has not been formed yet, nevertheless, some of them are used by scientists and specialists to explain the digitalization phenomenon. Accordingly, there is a need to cite the most frequently used terms as revised by the author [2]. "A digital (electronic) economy is an economy which characteristic feature is the maximum satisfaction of the needs of all stakeholders through the use of information, including personal one. This is possible through the development of information and communication and financial technologies, as well as the availability of infrastructure, which together provide the opportunity for full interaction in the hybrid world of all economic activity participants: subjects and objects of the process of creation, distribution, exchange and consumption of goods and services".

"The platform is a digital environment (hardware and software complex) with a set of functions and services that meets the consumer and manufacturer needs, as well as realizes the possibility of direct interaction between them".

"Digital ecosystem [3] is a representation of the social and technical system in the form of a combination of computer programs with distributed interaction and mutual use of agents for the exchange of knowledge in the evolutionary development".

2. Problem statement

At present, transport and service enterprises use automated information systems with software tools that allow for personnel, financial and warehouse accounting, processing of transport work documents, accounting of service works (maintenance) and repair (R) works performed, etc. However, the analysis of effectiveness of the information systems use in transport companies has shown that they are outdated, do not take into account and do not process a number of important data, especially indicators of the vehicle technical condition, personnel performance indicators in the maintenance and repair system of vehicles. So, the information is not collected on the reliability of the activities of auto mechanics, controllers of the vehicle technical condition, the principle of continuous monitoring of vehicle technical condition data, especially those that affect road safety are not respected, insufficient attention is paid to accounting of the reliability indicators of equipment. This altogether leads to the fact that the level of road traffic accidents (RTA) remains high due to vehicle malfunctions (according to statistics, up to 4 % of the total number of accidents, according to research estimates – up to 18 %), the costs in the engineering and technical service of a transport enterprise is about 40 % of the cost of transportation. These circumstances pose a problem, primarily in terms of maintaining the rolling stock in working condition, reducing road transport incidents due to technical malfunctions, reaching of optimal costs for transportation and vehicle services. Thus, the creation of a digital ecosystem based on modern digital services for transport enterprises remains an important task which implementation will significantly improve the management of processes to maintain the required level of vehicle performance, provide productive work of personnel, timely accounting and analysis of information on the parameters of activities of the road transport enterprises units.

3. Materials and methods

The range of human activities is continuously changing. New technologies appear, routine work is transferred to technical devices in increasing volume, and personnel is required to be creative, to find new solutions, master new types of activities, willingness to work in a team, etc. All this happens due to the changes in various sectors of the economy of our country, which, together with other advanced economies, is moving to the next level of development of new technologies based on digitalization, artificial intelligence, and cognitive technologies.

The author [2] notes "that with the advent of the Internet, a virtual world is being formed, full of new opportunities. The real and virtual worlds are interconnected and their merger forms a "hybrid" world, characterized by the ability to perform all vital actions in the real world through the virtual one" and that "digitalization is an objective, inevitable process and can not be stopped."

IOP Conf. Series: Materials Science and Engineering **832** (2020) 012071 doi:10.1088/1757-899X/832/1/012071

Based on these concepts, it should be expected that in the near future significant changes will occur in various sectors of the economy, new content will be carried out in management, science, security, and various sectors of the economy.

In this contest, digitalization in the transport sector is gaining momentum: ecosystems are being formed in the market for the transport of goods and passengers, in the organization of road safety, and logistics. Transport experts note that further development of the transport industry is impossible without introduction of digital technologies, self-driving vehicles, artificial intelligence. Digital Transport and Logistics Project which will reflect the best domestic achievements in creating a domestic digital platform and the deployment of information flows on its basis in the transport complex of the country is being developed [4].

The introduction of innovations in transport will require new personnel competencies, especially in terms of the development of digital culture, technologies as part of artificial intelligence, algorithms for interacting with partners and consumers in digital production ecosystems.

Building an ecosystem at the level of a motor transport or vehicle service enterprise and, particularly, a technical engineering service (TES), which maintains the performance of vehicles, involves the inclusion of all structural units and management personnel in this process. Such an ecosystem can be formed as the "TES Digital Ecosystem and Reliability". However, it should be borne in mind that the ecosystem will consist of two parts: internal and external. Building an internal ecosystem, it is necessary to ensure harmonious interaction with the external ecosystem through the appropriate choice of platform and services.

Fig. 1 shows a diagram of a digital production ecosystem of a vehicle enterprise (CE) with a developed technical engineering service (TES). General change management and the formation of the digital production ecosystem of ATP are carried out by the Director, and at the TES level, by the Chief Engineer (Technical Director). Operational service (OS), human resources department (HR), planning department (PD) and information technology (IT) department provide information to enterprise's management and communication with external environment. In order to make management and technological processes at the TES level more effective, it is necessary to develop digital tools in the form of programs and databases on a single platform, equip structural units with appropriate services that can function both in the internal and external production ecosystems.

The external digital ecosystem consists of many stakeholders, the main of which are: consumers, suppliers of spare parts and materials, labour market, tax authorities, banks, various forums (conferences, consumer sites, etc.).

The organization of work of a car enterprise TES digital transformation should begin with the creation of a transformation office. Production and technical department (PTD) is quite suitable for this role. At the same time, some employees of production control department (PCD), mechanical department (MD), material and technical supply (MTS) system, maintenance (M) and repair (R) system, and information technology (IT) department (temporarily) on a temporary basis are required.

Transformation office is created to develop a project for the digital transformation of the automotive engineering and engineering service, as a result of which organizational and information technology proposals for business processes based on the capabilities of new digital technologies should be prepared and implemented.

The development of a project, first of all, begins with an analysis of domestic and foreign experience in using digital technologies and an examination of solutions that are acceptable for improving business processes in a car enterprise TES. In this case, of course, changes will be required in operational service (OS) and planning department (PD) of the CE.

On the next phase, it is necessary to make a choice of digital technology that best meets the specifics of business processes in the TES on the basis of expert decisions and is compatible with the services of structural units of a car enterprise.

The next stage should be devoted to the analysis of the current situation in the TES of a car enterprise in such areas as management, organization of technological processes of vehicle maintenance and repair, reflection of processes in the information system, development of personnel competencies.

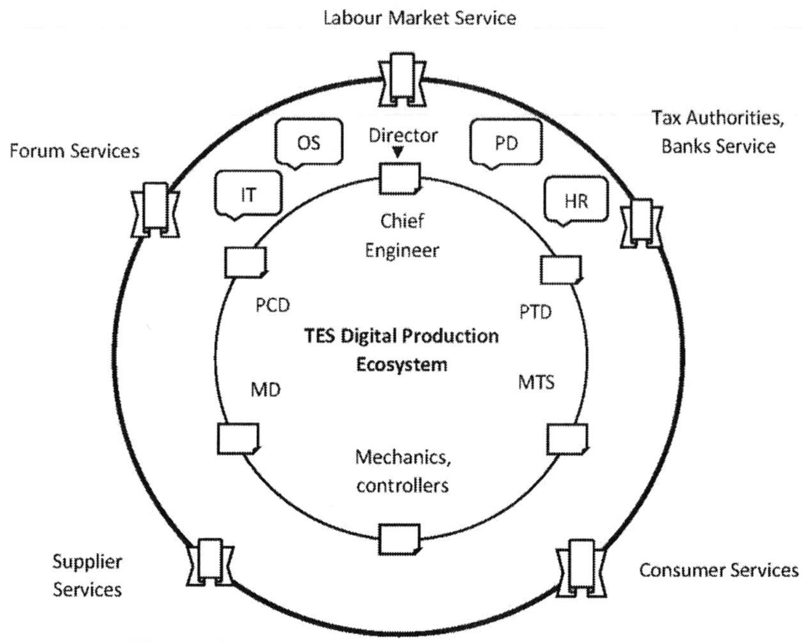

Figure. 1. TES Digital Ecosystem and Reliability

Such an approach will make it possible to form a roadmap for the implementation of the project, including digitalization of processes to maintain the required operability of vehicles, and improving the competencies of management and production personnel.

Digital transformation of the internal production ecosystem will require digitization of the storage resources of spare parts and materials, updating the fleet of diagnostic and technological equipment and installing chips on them, creating digital profiles of management and production personnel, and developing new competencies. In such a system, reliability indicators are determined for automobiles and for equipment and, which is very important, for personnel.

Digitalization of interactions of the production ecosystem with the external environment will allow the car enterprise to communicate with consumers, suppliers, the labour market and government bodies at a new level.

The implementation of the roadmap will require digitization of information on the performance indicators of all structural units of a car enterprise TES. The most complex are the performance indicators of Maintenance and Repair systems, in which auto mechanics (repair workers) work, using diagnostic and technological equipment to carry out the processes of determining and restoring technical condition of vehicles.

Fig. 2 shows the "TES Digital Ecosystem and Reliability" scheme, which allows to improve the interaction of Maintenance and Repair systems with the units of a car enterprise.

In the proposed scheme, the role of production and technical department (PTD) is increasing, which is additionally entrusted with the functions of coordinating the work of the TES structural divisions, receiving from them relevant digital information, processing it and presenting it to the chief engineer in a structured form to make management decisions. The controllers of vehicles technical condition transmit information through their mobile services to the PCD, which is processed and transmitted digitally to the PTD. After performing technological operations, the auto mechanics of the Maintenance and Repair systems also transmit information about the work for each vehicle that has undergone maintenance or repair through their mobile services. Data from the equipment chips comes to PTD automatically and is processed in real time. All departments (MD, MTS, PCD) that carry out their activities as part of the TES transmit certain information to the PTD. Thus, valid data is daily received by the TES management and is checked and communicated digitally to the enterprise's

General Manager referring major issues. If the achieved results are within the planned values, then operational regulation of production processes is not performed, otherwise the place of failure in the enterprise's system is determined and current management decisions are made to correct the situation. Particular attention is required to apply the competency model of management and operational personnel to work in digital ecosystems.

The competency model of a specialist is a combination of key competencies with specific results of their application, which allows them to successfully perform labor functions. Key competencies represent the specialists' competencies which are the most important for the enterprise at a certain stage of its development.

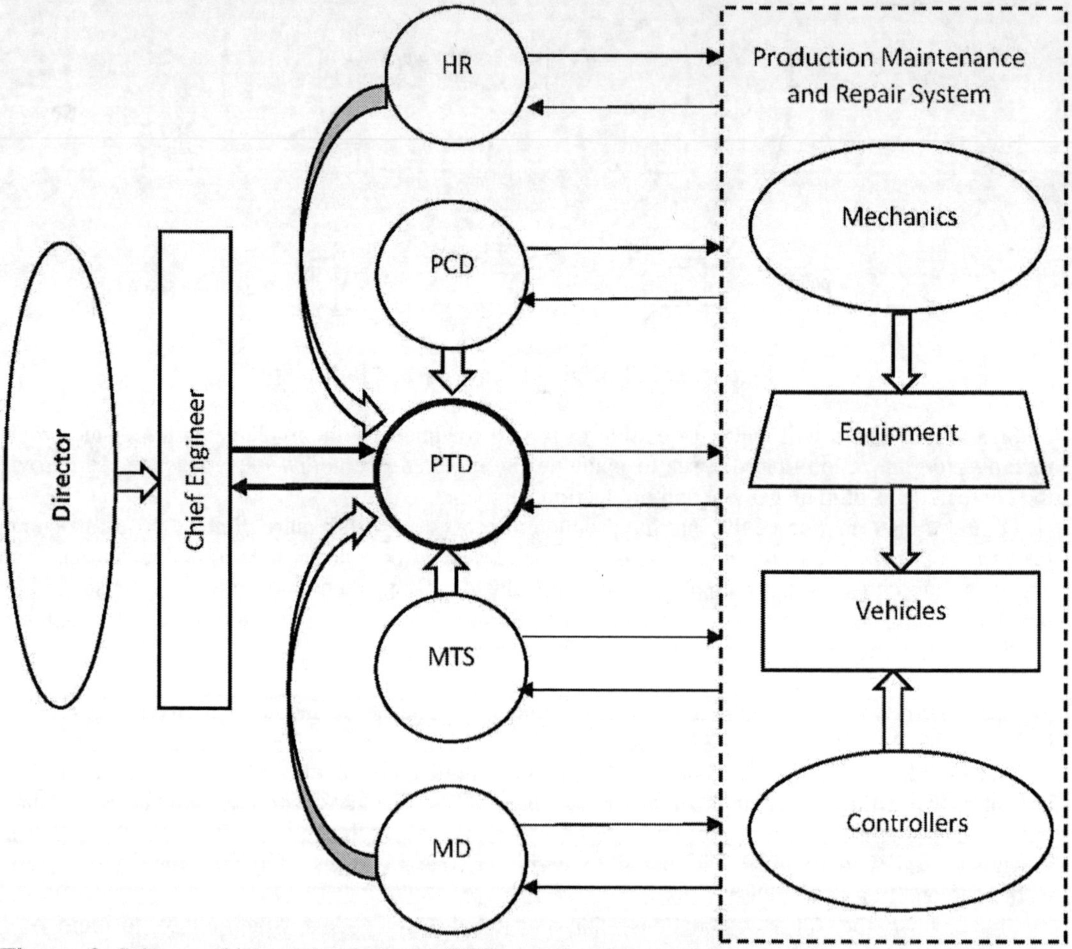

Figure. 2. Scheme of interaction between PTD and the maintenance and repair system with the units of a car enterprise as part of the "TES Digital Ecosystem and Reliability" model

The competency model (on the example of a controller of the technical condition of vehicles) can be represented by the following set of key competencies:

- ability to use digital technologies for work organization on preparation, calibrating, monitoring of readiness and use of technical diagnostics, measuring instruments, additional technological equipment;
- ability to perform maintenance of diagnostic devices, measuring instruments, technological equipment;

- ability to perform work on measuring, checking and analyzing the parameters of the technical condition of vehicles;
- ability to make decisions on the conformity of the technical condition of vehicles with requirements of traffic safety;
- ability to develop and digitize organizational and technological documentation for monitoring the technical condition of vehicles;
- ability to control and analyze the process of monitoring the technical condition of vehicles, the work of personnel to perform labor functions;
- ability to work in the digital ecosystem for recording, storing and updating data on vehicle monitoring and the operability of technical diagnostics, measuring instruments, and additional technological equipment.

4. Discussion of the results

The creation of digital production ecosystems in transport and vehicle services enterprises is possible using existing and developed digital platforms and services for transport, investments in the development of new competencies of management and operational personnel, new forms of information flow management in the Maintenance and Repair system and structural units of car enterprises. Previously, the authors [5] considered the possibilities of applying advanced information technologies in transport systems and justified not only the need for digitalization of management and technological processes, but also proposed models and algorithms that would increase the efficiency of personnel, reduce material and labor costs for vehicle operation. At the same time, the basic principles necessary for setting information systems for CE are formulated: "document revision, usage of common database by all units, automation of engineering decisions based on expert systems, implementation of information exchange between CE units through a computer network, real time information system operation, use of rolling stock control systems". These principles are in good agreement with the modern digital economy. However, they affect only the inner part of the CE ecosystem, and insufficient attention has been paid to the important part of the digital interaction with the external environment. The author [6] reasonably points out the technical capabilities available for car enterprises to electronically identify objects, transmit information based on automation-equipped working places (AWP), implement information technologies when organizing a supply chain.

The studies highlighted in [5, 6] confirm the fact that increasing the efficiency of car enterprises and their units can be achieved by expanding their functions in the provision and use of services in digital format, electronic processing of information on business process results, and analytics based on the use of artificial intelligence, connecting new consumers involved in the digital space.

The practical implementation of the elements of the digital production ecosystem of car enterprises has shown that the time for interaction with consumers on organizing the transportation of goods and passengers, servicing and repairing vehicles at service stations is significantly reduced. The amount of information processed automatically increases 5–6 times, and the time to receive analytical data and make management decisions decreases 3–4 times.

The direction of further research is seen in improving mathematical models and algorithmization of reliability value of management and production personnel [7] and units of the car enterprise, its technical engineering service in order to bring values into a format convenient for digitalization, and to increase the share of optimization tasks in the digital production ecosystem.

5. Conclusion

Thus, one of the main factors that can improve the efficiency of transport and vehicle service enterprises and their technical services is the creation of digital production ecosystems at enterprises. This should be promoted by the developed and proposed TES Digital Ecosystem and Reliability digitalization concept, updated personnel competency models, which are preferable for working in the new digital environment.

IOP Conf. Series: Materials Science and Engineering **832** (2020) 012071 doi:10.1088/1757-899X/832/1/012071

References

[1] *"Industria 4.0": sozdanie tsifrovogo predpriatiia* Vsemirnyi obzor realizatsii "Industria 4.0 za 2016 god" Retrieved from: http // www.pwc.ru/ru/technology/assets/global industri-2016 rus.pdf

[2] Keshelava A V, Budanov V G and Rumyantsev V Yu *Vvedenie v "Tsifrovuyu" ekonomiku 2017* pod obshch. ed by A V Kechelava (Moscow: VNII Geosistem)

[3] Avdeenko T V and Aletdinova A A 2017 *Digitalization of economy, based on improving expert knowledge management systems* (St. Peterburg: Spbstu.ru) pp 7–18

[4] *Transcript of the Report of the Minister of transport Yevgeny Dietrich at the final enlarged Meeting of the Board of the Ministry of transport of the Russian Federation on the results of the Ministry of transport of the Russian Federation for 2018, goals and objectives for 2019 and the planning period until 2021* 2019 p 18 Retrieved from: www.mintran.ru

[5] Postolit A V, Vlasov V M and Efimenko D B 2004 *Information support of motor transport systems* Textbook (Moscow: MADI) 242 p

[6] Oshchepkova E A 2012 *Information technologies in road transport* Textbook (Kemerovo: KuzGTU) 144 p

[7] Duchkov BA, Korolev AV, Smirnov BA 2002 *Fundamentals of engineering psychology* Textbook (Moscow: Academic project; Ekaterinburg: Business book) 576 p

IOP Conf. Series: Materials Science and Engineering **832** (2020) 012072 doi:10.1088/1757-899X/832/1/012072

ICT-method for control of virtual rank-trajectory for researcher on automobile and road construction industry

M.V. Yashina and M.E. Karaeva

Moscow Automobile and Road State Technical University (MADI), 64, Leningradsky Ave., Moscow, 125319, Russia

E-mail: karaeva96@gmail.com

Abstract. This article raises questions about the processes of bibliometric ICT-tools used by the scientific community in the automobile and road industry. This refers to scientific developments in the field of digitalization, in the operation of motor vehicles services, maintenance and design of highways, safety and traffic management. Modern ICT-methods can be implemented in the form of work with abstract and bibliographic databases, such as Web of Science, Scopus, Russian Science Citation Index, which show scientometric indicators of the publication activity of scientists, individual groups of researchers, organizations and journals. The following areas have been studied in the areas of scientometric databases: automotive industry, road complex and traffic flow management. Our study is devoted to the search for the above areas in the bibliography and the development of methods for providing a separate study to build its scientific activities and publication activity on an international scale. The study revealed that the above areas also include many other areas that, one way or another, affect the interests of other scientific fields. Due to the fact that the scope of scientometric databases is different, in order to study the scientific activities of scientists or journals, it is necessary to consider all indicators comprehensively in all databases. The rating trajectory of Doctor of Engineering Sciences, Professor M. Yashina, was also constructed, according to Web of Science, Scopus and Russian Science Citation Index.

1. Introduction

Nowadays there are various rating scientometric tools such as Web of Science, Scopus, Russian Science Citation Index (RSCI) and so on. They are available anywhere in the world and automatically calculate the rankings of individual researchers, scientific publications and universities. Unlimited use of Web of Science and Scopus by researchers requires a subscription from the organization in which the researchers operate, while RSCI is a public platform.

However, there are difficulties for different branches of knowledge. If for classical Sciences, such as mathematics, physics or chemistry, these tools have been developed for a long time and have a representative base, then for interdisciplinary newly formed topics this is not the case. This article discusses 27 areas that relate to the digitalization of development in the operation of road transport and service, maintenance and design of highways, safety and traffic management. These areas, one way or another, affect the interests of other scientific fields. Scopus service covers and indexes 7820 scientific journals, books, conference proceedings and industry publications.

IOP Conf. Series: Materials Science and Engineering **832** (2020) 012072 doi:10.1088/1757-899X/832/1/012072

Important and considerable merit of Thomson Reuters is building a search platform (or database) Web of Science (hereafter WoS), which now belongs to Clarivate Analytics. This platform is known and recognized all over the world. In 2014, the 50th anniversary of the search platform was celebrated. During this time, it has become a means for research activities and a means of information exchange in different scientific fields. The anniversary, which was celebrated by the company, in truth, is not even the anniversary of WoS, and Science Citation Index (SCI), which was created and used before WoS.

Today, the WoS database is a huge network with access to information about many things: people, countries, documents, subjects, organizations, etc. It is the most authentic and reliable source for combining information from scientific sources. Web of Science now has an updated legible user interface, increased content and web access for its subscribers. WoS service is widely used by owners of publications, readers and experts of scientific literature. Finding the right scientific literature before the SCI citation index was developed was very difficult. In the 50-IES of XX art. doctor of philosophy Eugene Garfield proposed the idea of creating a registration of cited sources in published materials of scientists. This innovation made it possible to register and track references to citation sources, establish links with other scientists/institutions and find works by other authors on similar fields [1].

The first issue of SCI was printed, then put on magnetic tape and on CDS. And after the social Science Citation Index (social Sciences) and Arts and Humanities Citation Index (Humanities and arts) appeared, they entered the SCIE database, which represents an extended version of SCI. And then there was WoS.

A similar situation is observed with the Scopus database. Scopus is a bibliographic and abstract database and tool for tracking the citation of articles published in scientific journals. The database indexes scientific journals, conference proceedings and serial books. The developer and owner of Scopus is Elsevier publishing Corporation. The database is available on a subscription basis through a web interface. The search device is integrated with the Scirus search system for web pages and the patent database.

As in the case of WoS, Scopus consists mainly of journals published abroad, the largest share of which are publications in the field of natural and exact Sciences. Perhaps the only Russian journal in which the results of scientific research in the field of pedagogy can be published is the journal of the Russian Academy of Sciences “Social Sciences”, published in the USA, in Minnesota.

Russian science citation index (RSCI) is a bibliographic database of scientific publications of Russian scientists. RSCI project has been developed since 2005 by "Scientific electronic library". With the help of the RSCI database, a scientist can find out the index of his citation. This index will be lower than the true one due to the "limited list and retrospective of processed journals, dissertations and dissertation abstracts". The manual also describes the procedure for determining the citation index for a scientist.

WoS was founded in 1963, while Scopus and RSCI in 1996 and 2005 respectively, resulting in different coverage of journals and publications.

2. Statement of problem

Various national projects have been launched in the Russian Federation. Of great importance is the National project – Safe and quality roads [2].

The National project "Safe and quality roads" sets goals:

1. The increase in the share of regional roads that meet regulatory requirements in their total length to at least 50 percent (relative to their length as of December 31, 2017), as well as the approval by the state authorities of the subjects of the Russian Federation of such standards on the basis of established at the Federal level requirements for road safety.

2. Reduction of the share of Federal and regional roads operating in overload mode in their total length by 10 percent compared to 2017.

3. Reducing the number of places of concentration of road accidents (accident-hazardous areas) on the road network twice compared to 2017.

4. Reduction of mortality as a result of road accidents by 3.5 times compared to 2017 – to a level not exceeding four people per 100 thousand population (by 2030 – the pursuit of zero mortality).

To achieve these goals, the following tasks are to be solved:

1. Bringing the proportion of regulatory-compliant roads in the largest urban agglomerations to 85 per cent of their total length.
2. Application of new mechanisms for the development and operation of the road network, including the use of infrastructure mortgages, life cycle contracts, the best technologies and materials.
3. Bringing the standard of transfer of tax revenues of budgets of subjects of the Russian Federation from excises on fuels and lubricants to 100 percent.
4. Implementation of a public information system for monitoring the formation and use of road funds at all levels (in 2019).
5. Creation of mechanisms of economic stimulation of safety of highways of regional and local importance.
6. Introduction of new technical requirements and standards of arrangement of highways, including on the basis of digital technologies, aimed at elimination of places of concentration of road accidents.
7. Implementation of automated and robotic technologies of traffic management and monitoring of compliance with traffic rules.
8. Strengthening the responsibility of drivers for violation of traffic rules, as well as increasing the requirements for the level of their professional training.

The national project fixed, in particular, the need to bring to the normative state of at least half of the regional roads. Since large amounts of funding are allocated for this purpose, scientific justification is needed to ensure the effectiveness of the investment.

As of the end of March 2019, last year the number of publications by Russian scientists in journals indexed by Scopus and Web of Science Core Collection (WOS CC) exceeded 98.000 and 75.800, respectively. Recall that measures to increase publication activity are taken, in particular, within the framework of the national project "Science". The national project formulated key indicators, among them-the place of the Russian Federation in the share in the total number of articles in the areas identified by the priorities of scientific and technological development [3].

It is required to develop indexable tools to determine the individual strategy to improve the scientific rating, taking into account changes in web-resources.

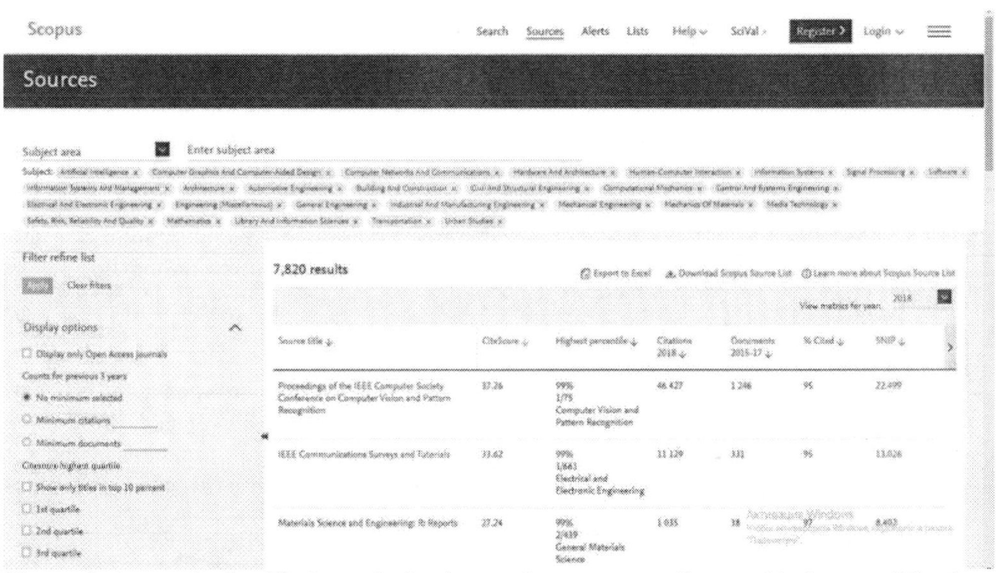

Figure 1. Areas of scientific journals, books, conference proceedings and industry publications indexed in Scopus

A total of 27 areas relate to the digitalization of development in road transport operation and service, maintenance and design of highways, safety and traffic management (Fig. 1). These areas, one way or another, affect the interests of other scientific fields. Scopus covers and indexes 7820 scientific journals, books, conference proceedings and industry publications.

Our research is devoted to the search for the above areas in the bibliography to the development of methods to ensure individual research to build its scientific activity and publication activity on an international scale

3. The algorithm of formation of the rank-trajectory

The main scientometric indicator, from which the rating of the scientist is formed, is the citation. In this case, to build the rank-trajectory of doctor of technical Sciences, Professor Yashina M.V. for 2014–2018, we used information from WoS, Scopus and RSCI broken down by year.

Previously, to study the publication activity of the author in WoS, it was possible to use the Researcher Id platform. However, in 2019 there was an integration of author profiles from Researcher Id to the new platform Publons, which allows to record the work of a scientist as a reviewer, expert and editor, to keep statistics, to offer their services in this field. It is worth noting that researchers form their own profile. Posts to the author's profile are not automatically backed up. Publication activity is represented by a histogram of publications and a citation graph (Fig. 2). According to the metric, the citation graph begins in 1968, although the histogram shows that the first publication was made in 1986, which may not be possible, because publications could not be cited before they were published. This is a good example of what information should be tracked and how it is displayed in the profiles of researchers.

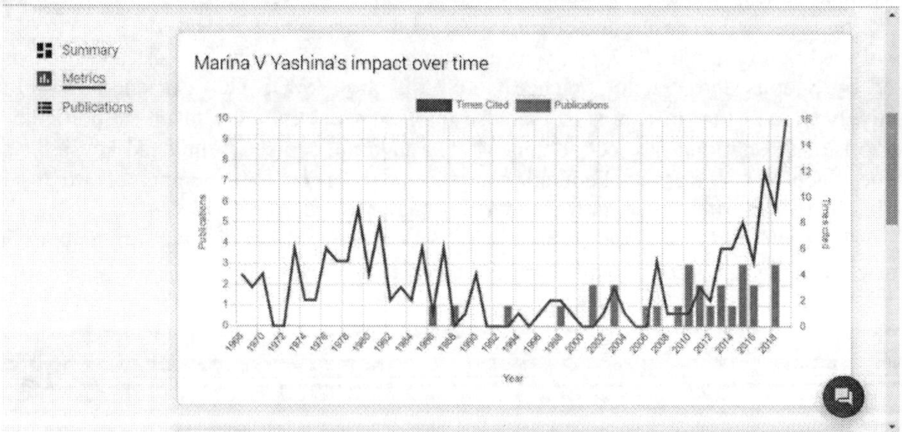

Figure 2. Statistics of publications and citations by year of Professor M. Yashina in Publons

Scopus automatically generates author profiles. However, if you find inconsistencies, you can send a request to the moderators of the site and they will correct all errors. In Scopus, publication activity is also represented by a histogram of publications and a graph of citations of the author (Fig. 3).

In RSCI it is possible to view various statistical reports on the activities of the researcher, including the report "distribution of citations by years of cited publications", which is represented by a diagram (Fig. 4).

On the basis of the available information, it is possible to make a joint schedule of how Professor M. Yashina's works for 2014–2018 were cited (Fig. 5). The graph fully illustrates how information in different bibliographic and abstract databases can differ. Based on the information obtained on the example of indicators of citation of works, it can be concluded that in order to give a real assessment of the scientific activity of not only the author, but also scientific publications and universities, it is necessary to consider the indicators in different databases comprehensively.

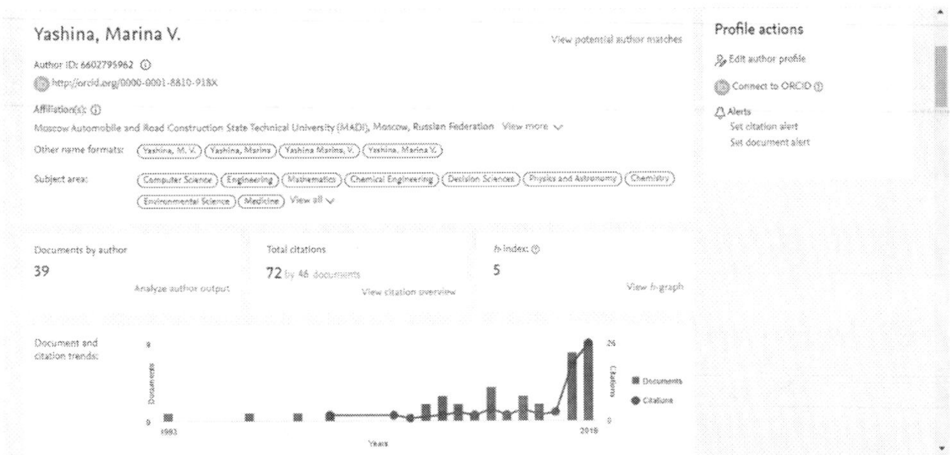

Figure 3. Statistics of the number of publications and citations by year Professor M. Yashina in Scopus.

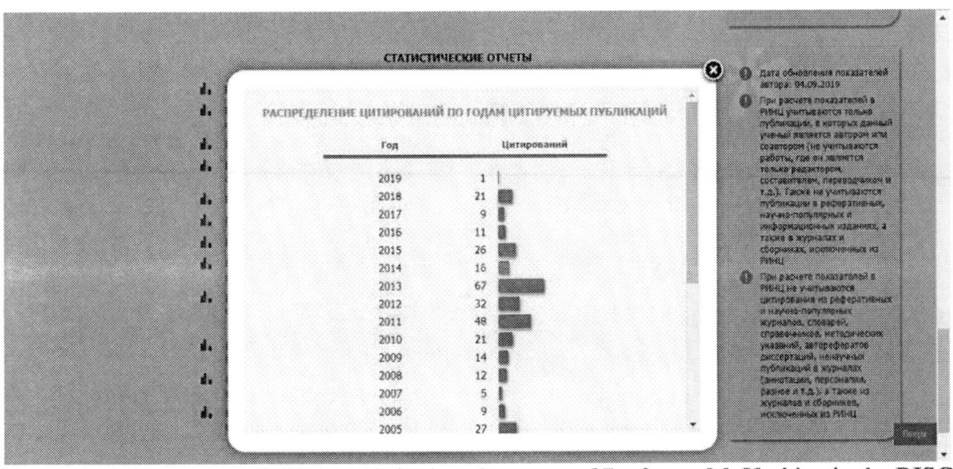

Figure 4. Statistics of the number of citations by years of Professor M. Yashina in the RISC.

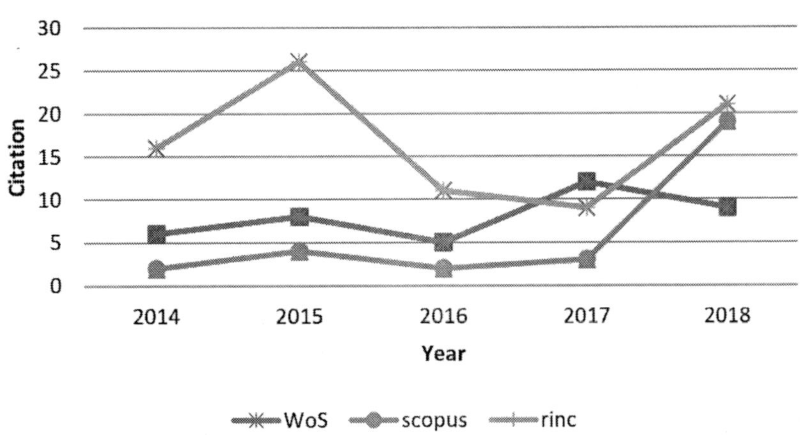

Figure 5. Statistics of the number of citations by year of Professor M. Yashina in Publons.

4. Conclusion

The article presents an algorithm for the formation of the rank-trajectory. It should be noted that for researchers to improve their rating and the rating of the organization in which the scientist works, it is necessary to publish their scientific works in those publications that are indexed by abstract and bibliographic databases.

References

[1] Retrieved from: http://123mi.ru/articles/Istoriya-sozdaniya-i-razvitiya-Thomson-Reuters

[2] Loktyushina E A 2014 Substantive and procedural aspects of the articles preparation for the database "Scopus" *Concept Special issue* **21** ART 14761 – 0,4 p l. Retrieved from: http://e-koncept.ru/2014/14761.htm. State reg. El no FS 77-49965 ISSN 2304-120X.

[3] Retrieved from: https://www.5top100.ru/news/100908/

[4] Retrieved from: https://cyberleninka.ru/article/n/soderzhatelnye-i-protsessualnye-aspekty-podgotovki-statey-dlya-bazy-dannyh-scopus.pdf

[5] *Natsionalnyy proyekt – Bezopasnye I kachestvennye avtomobilnye dorogi.* Retrieved from: https://ru.m.wikipedia.org/wiki/

[6] Dotkulova A S, Moseva M S and Yashina M V 2017 Features of information support of educational process in technical universities *T-Comm-telecommunicat. and transp.* **11(9)** 58–64

[7] Pechnikov A A and Chuiko Yu V 2007 Mathematical models of coordinated behavior of small Internet communities *Internet-mathematics: collection of works of participants of the contest of scientific projects on information search* ed P I Braslavsky (Yekaterinburg: Ural publ. House, UN-TA) pp 164-70

[8] Yashina M V and Nakonechnyy I I 2014 WEB Service for Data Extraction from Semi-structured Data Sources Proc. of the IX Int. Conf. on Dependability and Complex Systems DepCoS-RELCOMEX *Advances in Intelligent Systems and Computing* (Brunow, Poland, 30 June – 4 July 2014) vol 286 pp 499–510 Retrieved from: DOI 10.1007/978-3-319-07013-1 48

[9] Witkowski A Yu and Yashina M V 2015 Distributed server system of information collecting from moving objects *Telecommunicat. and inform. Technol.* **2** 8–10 Retrieved from: https://www.5top100.ru/news/100908/

[10] Yashina M V and Dotkulova A S 2017 Infocommunication methods of pedagogical measurements for assessing the rating of Junior students Proc. of XI Int. industry sci. and techn. Conf. *Information Society Technologies* pp 445–6

[11] Yashina M V and Moseva M S 2017 Methods of automation of methodological support in the University within the competence-oriented approach Proc. of XI Int. industry sci. and techn. Conf. *Information Society Technologies* pp 458–9

[12] Yashina M V and Prudnikova Yu V 2017 Analysis of information national scientometric system "ISTINA" Proc. of the Int. sci. and techn. Conf. *Telecommunication and computer systems-2017* pp 239–41

IOP Conf. Series: Materials Science and Engineering **832** (2020) 012073 doi:10.1088/1757-899X/832/1/012073

Methodology for making management decisions in digital economy systems

D V Zaitsev[1], S S Evtyukov[2], E A Karelina[3], V V Voropaev[4] and A O Kholodova[5]

[1]Department of Economics and Road Management, Moscow Automobile and Road State Technical University (MADI), 64, Leningradsky Ave., Moscow, 125319, Russia
[2]Department of Ground Transport and Technological Machines, Saint-Petersburg State University of Architecture and Civil Engineering, 4, 2nd Krasnoarmeyskaya St., St. Petersburg, 190005, Russia
[3]Department of Foreign Languages, Moscow State University of Technology «STANKIN", 1, Vadkovsky lane, Moscow, 127055, Russia
[4]Department of Logistics and Management Methods, Yanka Kupala State University of Grodno, 22, Ozheshko str, Grodno, 230023, Belarus
[5]LLC «LUKOIL-ENERGOSETI», 3-1, Pokrovsky boul., Moscow, 109028, Russia

E-mail: opferpriesterin@mail.ru

Abstract. For effectiveness assess of management decisions in modern economic systems, as a rule, single-criterion mathematical models are used, which have several disadvantages. The use of qualitative single economic efficiency criterion for complex production processes comparative evaluation in digital economy leads to a high proportion of subjectivity. Modern complex production, based on intelligent communication technologies, operates in specific information situations context that require the use of adequate mathematical tools for making effective management decisions. In such tasks, effective solution choice depends on external and internal environment state of the system being studied and implies the use a number of methods for solving multi-objective tasks. The main methods used in uncertainty situation cases due to information situation complexity are prior ranking or expert evaluation methods. Naturally, methods based on expert evaluation have significant degree of subjectivity. This disadvantage can be avoided if you solve the problem using scientifically based effectiveness evaluating methods of production processes not by one but by several relevant performance indicators.

1. Introduction

As a whole economy digitalization should be viewed as technological evolution natural product, that is, an evolutionary, not a revolutionary process. Next period purpose of digital economy development is to formulate the basic principles for using scientific results in intellectual environment objects, to develop scenarios for transition of individual industries to digital intelligent technologies, to create an infrastructure for transition period support, including scientific support for making new decisions. Achieving this goal implies a change in organizational and technological structure in the basic sectors of national economy under the influence of management systems unification based on digital information technologies, as an inevitable consequence of digitalization. This will lead to creation of a new composition of industries and to fundamental change in the information interaction principles

IOP Conf. Series: Materials Science and Engineering **832** (2020) 012073 doi:10.1088/1757-899X/832/1/012073

between social and technological systems that enter it. Uniting for all problem-oriented systems is infocommunication technology, whose role in the digital era is significantly increasing.

Purposeful transition processes regulation to digital technology is one of the most important management tasks in Russian economy. As you know, management is a function of the system, focused on preserving its basic quality and improving efficiency in the face of changes in environment state. In other words, the goal of control systems in the digital economy is to ensure transforming information process into targeted actions that take a managed object from its original state to a given or optimal state corresponding to scientific and technological process development level.

2. Purpose of the study

Management processes in modern complex information digital systems (IDS) are united by information conditions fundamental commonality which determined by:

- changing environmental conditions for the system under study;
- dynamic development of changing parameters process in system itself;
- lack of sufficient degree of certainty for the necessary information state when solving problems aimed at system efficiency improving;
- presence of larger number of goal setting criteria and a multitude possible solutions.

In general consideration of objects interaction conditions in IDS, related to each other by a number of diverse relationships (analytical links), it is necessary to develop algorithms that are able to maintain integrity relations between objects. Modern approaches to creation management information systems (IS) are based on relational principle implementing the subject-oriented approach (SOA) to managing procedures in the system, when the means of interaction are tables or stored procedures. In this case, the control is carried out by a certain set of objects through the determination of regularities, as a rule, of stochastic character from the accumulated statistical data. The main disadvantages of this relational interaction model in complex systems based on a subject-oriented approach are:

1. Projected system, as a tables set is very often difficult to analyze and understand processes.
2. Increasing system complexity, leads to inability fully or reliably track all the necessary places and procedures for changes implementation, and system development process is transformed into poorly managed or unmanaged.

An alternative to this approach is an object-oriented approach (OOP). In the object-oriented approach, the program should represent not only objects description and their properties in criteria form, and relations between them in the form of goal setting, but also the methods of their interaction (methods) in the form of operations on objects. A clear advantage for OOP is conceptual proximity in any subject area to an arbitrary structure and purpose of the system. At the same time, attribute conversion mechanisms or methods should allow building derived objects and structures on the basis of basic ones, thereby creating a model in a more complex subject area with the necessary properties, ensuring continuous analysis possibility and making changes if necessary. In this case, the objects and methods are polymorphic, which makes the software more versatile and flexible.

Despite the clear advantages of OOP, object-oriented control systems have not yet received widespread distribution. At the same time, hybrid object-relational management systems are widely used, which partly use object-oriented working principles with data, but at the same time the data storage representation is performed when implementing the relational model [1–3].

3. Research questions

Therefore, as management quality improvement in IDS, transition process from traditional management methods in economic systems to object-oriented analytical management models is natural, allowing management processes automation using modern digital technologies. Analytical solution of this problem is possible on dynamic multicriteria programming methods basis, which allow simultaneous accounting the significant number of indicators values in the solved extremal problem according to a number of criteria depending on IDS environment state conditions at a given time. Quality change process in IDS form a number of additional tasks that are mandatory for implementation. New tasks formation in any management system inevitably entails changes in its structure. The general theory of systems establishes that complex systems management structure, such as IDS, is hierarchical (multi-level) in nature. In hierarchical structure, the question of optimal elements number which subordinate to the senior element and on the same level is extremely important. The more such elements, the less manageable the system is, but on the other hand, the creation a larger number of hierarchy precedence levels is also undesirable because it will lead to a long process for passing information [4–6].

In the first case, in order to preserve system controllability, it is necessary to develop a mathematical apparatus aimed at finding an effective solution and allowing generation of algorithms for optimizing processes in IDS taking into account several criteria.

In the second case, the transition from level to level, as a rule, reduces the number of tasks to be solved, but their importance and complexity significantly increases. The criteria for junior tasks should be consistent with senior management level interests. Practical implementation of this method is often associated with considerable difficulties, because it is not always, having risen to a higher level, it is possible to formulate and solve a corresponding single-criterion optimization problem.

Based on the fact that the structure of vehicle's IDS should be built with minimum possible number of stages, and system's controllability should be provided with an adequate mathematical apparatus, it is advisable to determine system hierarchy and limits of the study. Hierarchy is an abstract representation of any system structure, necessary for studying functional interactions of its elements and their effects on entire system. The hierarchy can take various forms, each of which, as a rule, descends from a common goal down to sub-goals, further to impacts that affect these sub-goals, further away to strategies, and finally to outcomes that are results of these strategies.

4. Research methods

The stages of IDS hierarchy building process are [7]:

1. System objectives are recorded as a higher level assessment based on interaction of various hierarchy levels, and not directly dependent on the elements at these levels.
2. Formed management tasks defined by system environment.
3. Identify criteria that affect the task or tasks of management system. Each individual criterion should reflect the main, rather than a secondary function when governing decision making in system.
4. Criteria hierarchy is determined. The simplest hierarchy is linear, rising from one elements level to a neighboring level. In a non-linear hierarchy, the upper level can be both in a dominant position relative to the lower level, and in a dominated one, for example, in the case of information flow. In the mathematical theory of hierarchies, a method is used to assess level impact on a neighboring upper level by composing a corresponding contribution (priority) of lower-level elements with respect to an upper-level element. This composition may extend up the hierarchy.
5. Technical indicators set is being developed that ensure requirements for system functioning. Technical solution of the problem consists in determining the permissible system parameters values, which determine not only the efficiency of its elements, but also the efficiency of functioning in general, taking into account certain goals.

IDS hierarchy can be considered in various terms:

1. Interactions of its parts.
2. Structures in accordance with physical classification of its parts.
3. Functions in accordance with what the system elements functions are, what goals they should fulfill, parts of which goals of a higher order are these goals (leading to system overall goal).
4. The goals laid down in its structure from the point of view of external environment (the larger surrounding system), for which it is a subsystem [8,9].

5. Research results

IDS hierarchy formation should occur on basic principles of program-target approach [10]:

1) Hierarchy principal difference in IDS functions is several performance criteria imposition on one level. At the same time, the number of hierarchy levels remains unchanged (system flexibility

remains unchanged), but in order to preserve initial system control parameters, a multi-criteria task is being formed that requires an analytical solution.

2) IDS has several levels of different significance goals, therefore it is necessary to streamline them by building goals hierarchy.

3) systematization and ordering of identified ways to achieve IDS goals is carried out by building subsystems hierarchy.

4) identification and determination of IDS interaction order, taking into account information situation degree of uncertainty.

The presence in research a number of situations that have a certain degree of uncertainty requires, for its description, involvement certain mathematical apparatus, which would a priori include uncertainty occurrence probability. Nowadays, such an apparatus can be: probability theory; game theory (used to uncertainty describe generated by conflict and "players" antagonistic interests; statistical solutions theory (describes games with a passive environment or "nature", whose behavior is characterized by random variables with known or unknown distribution laws); theory of vague (blurred) sets.

Figure 1 shows methods classification for making management decisions.

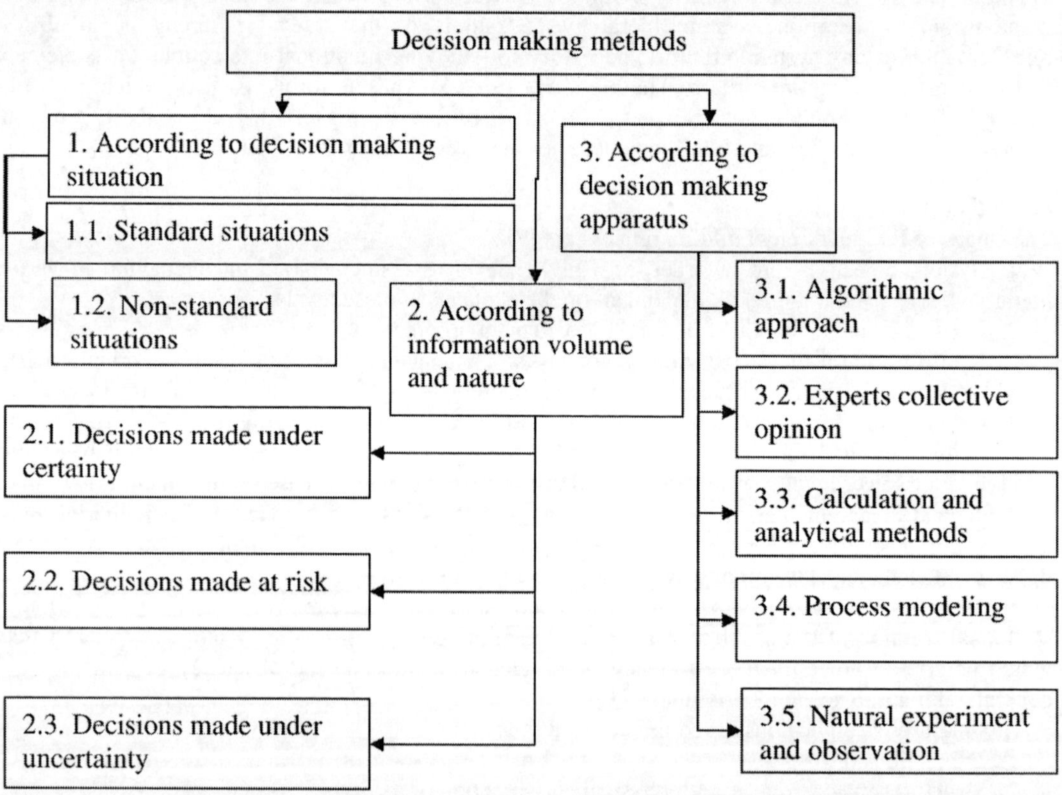

Figure 1. Decision making methods classification

In accordance with the above scheme, decision-making methods in IDS occur:

1) on decision-making situations - standard;
2) by the nature of information - taken in uncertainty conditions;
3) on decision-making apparatus - settlement and analytical.

To apply a particular mathematical apparatus, it is necessary to classify possible uncertainty gradation degrees or classify information situations I_i. All initial information (quantitative data, etc.) can be represented as a set $Q = \{q_1, q_2, ..., q_n\}$ of mutually exclusive SVS - C of their states Q. The information situation is defined as certain of uncertainty degradation degree by environment C of their states [11]. Following information situations can be singled out as the main ones:

I_1 – the first informational situation is characterized by a priori probabilities given distribution P_j on elements q_j of the set Q (in discrete form):

$$P_j = P\{Q = q_j\},\ \sum_{j=1}^{n} P_j = 1 \tag{1}$$

I_2 – the second information situation is characterized by a priori probabilities given distribution P_j with unknown parameters on the elements q_j of the set Q:

$$P(w) = \{P_1(w), P_2(w) ..., P_n(w)\},\ 0 \leq P_j(w) \leq 1, \tag{2}$$

where w – an undefined parameter from parametric set Q.

I_3 – the third informational situation is characterized by a given system of preferences on the prior probability P_j distribution of the set Q. The ordering of environment states $q_j \in Q$ implies the introduction of order relations for the components $P = \{P_1, P_2 ..., P_n\}$. For example, a simple order relation $q_1 > q_2 > \cdots > q_n$ is determined by setting the inequality:

$$P_1 \geq P_2 \geq \cdots \geq P_n \geq 0. \tag{3}$$

I_4 – the fourth informational situation is characterized by unknown probability distribution P_j on q_j elements from the set Q, on the one hand, and absence of active environment counteraction on the other. Such environment behavior is equivalent to "passive environment" behavior, studied in statistical solutions of "games with nature" theory. It should be noted that in this situation, distribution law ignorance does not exclude possibility of taking into account any information about elements of environment state (various kinds of restrictions, average and variance estimates, etc.)

I_5 – the fifth information situation is characterized by antagonistic environment interests.

I_6 – the sixth information situation is mixed and is determined by information elements presence that characterize "intermediate" environment behavior.

It is important that information situations I_2, I_3, can be reduced to the first I_1. In the case of information situation I_4, if "nature" states probability of the third factors group can be determined, the decision is made under risk. Further, the task of minimizing it should be carried out, that is, information situation also reduces to I_1. For I_4, when information lack situation or its inadequacy about environment can be considered as under conditions not opposed by "nature", but its behavior is unknown, the problem is solved under uncertainty conditions.

6. Conclusion

Systemic approach methodology implies the study IDS interaction with external environment, the change of which affects system parameters. System effectiveness as a whole is enhanced by successive series of operations performed, substantiated with help of a number of mathematical and economic methods. In each case, the methods development to ensure overall or local system effectiveness or its elements, we have to work with three factors groups:

1) group of defined factors;
2) group of controlled factors;
3) group of factors classified as undefined (unknown) conditions.

The most complex and close to reality is the third case. The nature of uncertainty for these variables (factors) may be different. Non-stochastic nature indefinite factors can be divided into two subgroups:

1) with known membership functions (the membership function defines a subset of the total allowable factor change range, reflecting the uncertainty degree);
2) with unknown membership functions.

The latter cases have the greatest degree of uncertainty. Usually an expert evaluation procedure is applied to them for ranges of changes in their values. Uncertainty of this type is called "nature" [12]. The presence of this type of uncertainty in IDS is explained by the following provisions:

1. Unclearly expressed end goal of the research in quantitative characteristics.
2. Insufficient knowledge at the stage of phenomena development which accompanying the system functioning process.
3. Targeted opposition presence from external environment.

In such tasks, the choice of solution depends on the state of "nature", and mathematical models are called "games with nature". In these cases, the methods of desired solutions obtained should be based on the rules of games and statistical decisions or can be obtained using vector optimization methods. Information situation I_5 requires involvement of game theory. Information situation I_6 requires involvement of decision theory and game theory.

References

[1] Tarasov R V, Makarova L V and Akzhigitova O F 2014 On the issue of applying expert methods in forecasting processes, assessing the quality level and making management decisions *Modern sci. res. and innovat.* **4** Retrieved from: http://web.snauka.ru/issues/2014/04/33142
[2] Jarratano D and Riley G 2007 *Expert systems: development principles and programming* (Moscow: OOO "I.D. Williams")
[3] Jackson P 2001 *Introduction to expert systems* (Moscow: "Williams" Publ. House)
[4] Saaty Thomas L 2008 *Decision Making with dependencies and feedbacks: Analytical Networks* (Moscow: LCI Publ. House)
[5] Pegat A 2009 *Fuzzy modeling and control* (Moscow: BINOM, Lab. of knowledge)
[6] Antonova A S and Aksenov K A 2012 Multi-criteria decision making under risk based on integration of multi-agent, simulation, evolutionary modeling and numerical methods *Engineer. Herald of the Don* **4(2)** Retrieved from: http://www.ivdon.ru/magazine/archive/n4p2y2012/1466
[7] Nogin V D, Protodyakonov I O and Evlampiev I I 1986 *Fundamentals of optimization theory* (Moscow: Higher School)
[8] Terentyev A V, Efimenko D B and Karelina M Yu 2017 Zoning methods as methods for optimizing road transport processes *Herald of civil engineers* **6(65)** 291–4
[9] Terentyev A V and Prudovsky B D 2015 Methods for determining Pareto set in some linear programming problems *Proc. of Mining Instit.* **211** 89–90
[10] Chernorutsky I G 2005 *Decision making methods* (St. Petersburg: BHV-Petersburg)
[11] Taha Hemdi A 2005 *Introduction to operations research* (Moscow: Publ. House "Williams")
[12] Wentzel E S 2001 *Operations research. Tasks, principles, methodology* (Moscow: Higher School)

IOP Conf. Series: Materials Science and Engineering **832** (2020) 012074 doi:10.1088/1757-899X/832/1/012074

Analysis of failures of urban buses in the conditions of the transit agency "Mosgortrans"

A A Zavgorodniy[2], V A Maksimov[1], N V Pozhivilov[1] and G A Krylov[1]

[1]Moscow Automobile and Road State Technical University (MADI), 64, Leningradsky ave., Moscow, 125319, Russia
[2]Research Centre, Mosgortrans, 22/21, b 1, Raushskaya naberezhnaya, Moscow, 115035, Russia

E-mail: zavgorodniy.aa@mail.ru

Abstract. The article presents the results of analysis of failures of low-floor urban buses LiAZ-529221, LiAZ-529222, LiAZ-621321 and LiAZ-621322 in the conditions of operation in Moscow. The research was based on the statistical data collected on failures of the bus modifications under consideration for the period from 01.08.2017 to 31.12.2018. As a result, the causes of failures of systems, assemblies and units, their seasonality, development of recommendations to improve the maintenance and repair system of LiAZ buses were identified. The research was carried out on the base of the branch of SUE "Mosgortrans".

1. Introduction

One of the most important issues with vehicle operation is to ensure the necessary level of reliability of the operated fleet [1]. Definition of reliability indicators and their further management at the expense of specific measures (organization of maintenance and repair, advanced training of personnel, modernization of production and technical base, etc.) has an impact on the durability and reliability of buses and their units [2].

In the process of fleet operation there is a decrease in its reliability, reduction in the probability of reliable operation of components and units and the operating time between failures. The aforementioned leads to an increase in the labor intensity of current repairs, downtime in repair and the cost of spare parts and maintenance.

The objectives of the research were to identify the most frequent causes of system failures, the seasonality of their occurrence in order to further develop recommendations on improvement of the system of maintenance and repair of the considered modifications of LiAZ buses. The solution of such tasks is primarily based on reliable and systematic information about failures, their operating time, spare parts costs and etc. [3].

2. Organization of research

The research was conducted at the testing site on the branch of SUE "Mosgortrans" in Moscow.

The following restrictions were imposed on the collection of data on bus failures:

- all LiAZ-529221, LiAZ-529222, LiAZ-621321 and LiAZ-621322 buses operated in the transit agency for the period of surveys from 01.08.2017 to 31.12.2018 were included in the quantity of 95, 107, 55 and 31 units, respectively;

- the research did not take into account failures and repair operations performed by the manufacturer during the warranty period on these buses.

The frequency of observations of the monitored buses was carried out daily.

In the case of failure of the bus the driver filled out an application for current repair and provided the bus to the area of current repair for inspection (diagnostics). During the repair the foreman recorded the actual time of the beginning and the end of the repair, the list of repair operations, causes of malfunction, etc. in the observation report.

Subsequently, the statistics collected were entered into the SAP software database and presented in MS Office Excel spreadsheets. The spreadsheets contained structured data on the following key criteria:

- modification of the bus;
- garage number of the bus;
- date of failure and malfunction;
- mileage at the time of failure or malfunction;
- actual repair time;
- name of repair operations;
- mechanic's conclusion, etc.

The analysis of failures for the whole period of observations was carried out for the following factors:

- the amount of failures of the controlled modifications of buses;
- the amount of recorded system failures and their share of the total amount of failures;
- distribution of failures by months and seasonality of their occurrence;
- total time of downtime of the fleet in case of system failures;
- average time of repair of failure on the system;
- the proportion of fleet downtime in repairs.

The special statistical sample limited by the calendar year for the period from 01.01.2018 to 31.12.2018 was formed in order to monitor correctly the seasonality of failures. It has allowed to complete an estimation of influence of a natural-climatic conditions on reliability of systems of the fleet in use.

Due to similar technical characteristics and construction features [4–7], in the initial processing the statistical data on buses LiAZ-529221, LiAZ-529222 and LiAZ-621321, LiAZ-621322 were combined according to their type of vehicle – high capacity (HC) and especially high capacity (HCV).

3. Processing and analysis results

The obtained statistical data on the distribution of failures on the systems of buses LiAZ-529221, LiAZ-529222 is presented in Table 1 and for LiAZ-621321, LiAZ-621322 in Table 2.

Thus, in the period from 01.08.2017 to 31.12.2018, the current repair complex registered 3135 and 974 failures of LiAZ-529221, LiAZ-529222 and LiAZ-621321 and LiAZ-621322 buses, respectively.

Later on, the causes of failures for individual systems and units were analyzed.

According to the obtained statistics, the largest number of failures on the bus LiAZ-529221, LiAZ-529222 and LiAZ-621321, LiAZ-621322, was found in the following systems (given is the average failure rate for all bus modifications):

The engine and his systems – 28.1 and 32.4 %:

- leak tightness of cooling system circuits (51 %) – replacement of thermostat tubes (11.6 %), expansion tanks (7.9 %), radiator tubes (6.2 %), etc. (25.3 %);
- coolant pump failure (19.1 %), thermostat failure (4.8 %), cracks in expansion tanks (5.5 %);
- burn-through of the exhaust silencer (62.5 %) and violation of the silencer attachment (37.5 %).

Brake system – 22.6 and 20.9 %:

- leakage of power accumulators (49.9 %) and brake chambers (4.6 %);
- jamming and excessive wear of brake mechanisms – supports (44,5 %), disks and pads (18,2 %);
- loss of tightness of brake system circuits – replacement of brake hoses (9.0 %) and brake pedals (5.3 %).

Table 1. Distribution of failures by systems and units of buses LiAZ-529221 and LiAZ-529222

Bus systems	Amount of failures, pcs.	Amount of failures, %	Average time of failure repair, m/h	Downtime in repair, %
Engine, including:	880	28.1	2.20	29.4
– Power system	55	1.8	2.35	2.2
– Exhaust system	81	2.6	2.34	3.2
– Cooling system	744	23.7	1.91	24.0
Transmission, including:	183	5.9	1.68	4.7
Suspension	136	4.3	1.28	2.9
Steering system	284	9.1	1.72	8.2
Brake system	707	22.6	2.54	30.4
Electrical equipment and devices	231	7.4	1.24	4.8
Cabin, bodywork and its equipment	58	1.9	1.36	1.3
Air conditioning and heating	265	8.5	1.34	6.0
Pneumatic system	324	10.3	1.58	8.6
Front axle	50	1.6	3.38	2.9
Rear axle	6	0.2	3.64	0.4
Wheels and tyres	11	0.4	1.70	0,3
In total:	3135	100.0	1.98	100,0

Table 2. Distribution of failures by systems and units of buses LiAZ-621321 and LiAZ-621322

Bus systems	Amount of failures, pcs.	Amount of failures, %	Average time of failure repair, m/h	Downtime in repair, %
Engine, including:	316	32.4	1.94	33.4
– Power system	19	2.0	1.66	1.8
– Exhaust system	25	2.6	2.29	3.2
– Cooling system	272	27.9	1.86	28.4
Transmission	39	4.0	1.51	2.9
Suspension	48	4.9	1.06	2.9
Steering system	71	7.3	1.77	7.1
Brake system	204	20.9	2.37	27.1
Electrical equipment and devices	86	8.8	1.12	5.4
Cabin, bodywork and his equipment	12	1.2	1.13	0.8
Air conditioning and heating	42	4.3	1.47	3.5
Pneumatic system	135	13.9	1.78	13.5
Front axle	13	1.3	3.33	2.4
Middle Axis	0	0.0	0.0	0.0
Rear axle	3	0.3	5.11	0.9
Wheels and tyres	5	0.5	1.00	0.3
In total:	974	100.0	1.93	100.0

Pneumatic system – 10,3 and 13,9 %:

- air tightness disorders of pneumatic system circuits (48.4 %) – replacement of main tubes and hoses of pneumatic system of air dryers (25.1 %) and compressor (23.1 %);
- failure of vacuum dehumidifiers (34.2 %);
- failure of floor level cranes (12.4 %).

The steering system is 9.1 and 7.3 %:

- leakage of circuits – replacement of power steering tubes (70.4 %);
- tight steering – power steering pump failure (12.7 %);
- lack of transverse (5.9) and longitudinal steering rods (6.8).

Electrical equipment and devices – 7.4 and 8.8 %:

- no battery charging – replacement of generators (22.4 %);
- blowing motor failure (20.2 %) and windscreen wiper drive motor failure (5.7 %), starter replacement (13.9 %);

IOP Conf. Series: Materials Science and Engineering **832** (2020) 012074 doi:10.1088/1757-899X/832/1/012074

- failure of controllers of turns (10.4 %).

Air conditioning and heating system – 8.5 % and 4.3 %:

- failure of cabin heaters (60.9 %);
- air-tightness breach of the passenger compartment heating system circuit (30.0 %);
- leakage of windscreen blower radiators (8.8 %).

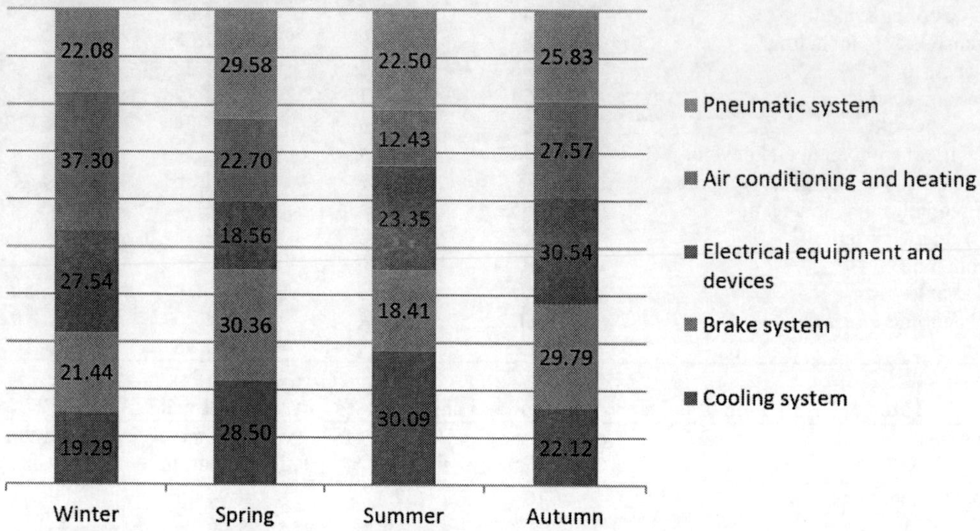

Figure. 1. Histogram of seasonality of LiAZ-529221 and LiAZ-529222 buses failure

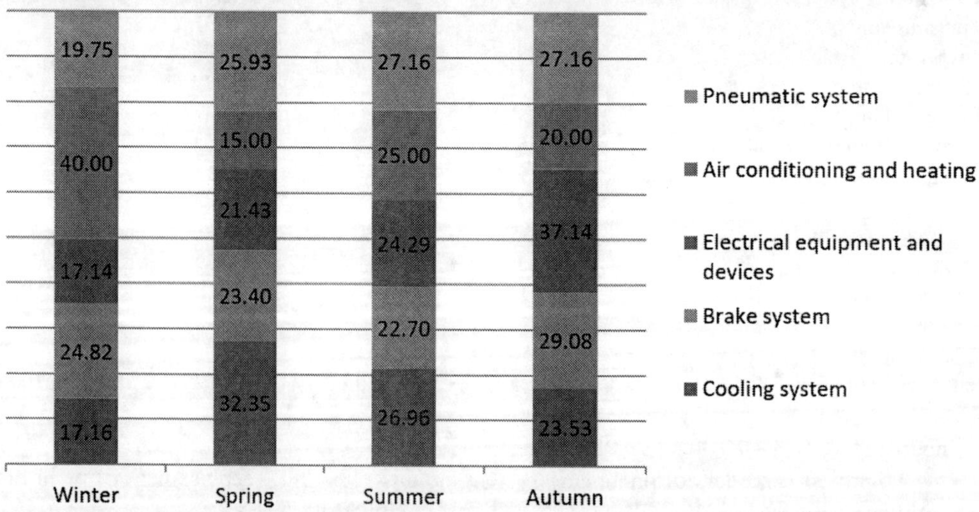

Figure. 2. Histogram of seasonality of LiAZ-621321 and LiAZ-621322 buses failure

4. Conclusion

On the basis of the conducted researches and the collected statistical data it is possible to formulate recommendations on development of fleet maintenance service system within the "Mosgortrans". In particular, to apply objective corrections to the existing system of maintenance service, and form a number of additional, preventive operations of maintenance service for individual systems and units.

IOP Conf. Series: Materials Science and Engineering **832** (2020) 012074 doi:10.1088/1757-899X/832/1/012074

The analysis of statistical data gives a general view of the quantitative indicators of reliability and statistics of failure distribution of different systems and units of buses LiAZ-529221, LiAZ-529222 and LiAZ-621321, LiAZ-621322.

The collected statistical data make it possible to initiate the base for developing integrated reliability strategy of the considered modifications of buses by systems and units for different age groups of buses. In addition, the collected statistical data are applicable in foundation of the nomenclature and planning of norms for spare part consumption, distribution of the scope of repair, calculation of the required scope of equipment, labor etc.

Reference

[1] Kuznetsov E S, Boldin A P and Vlasov V M 2001 *Technical operation of cars: textbook for universities. 4th edited, revised and supplemented* (Moscow: Nauka) p 535
[2] Bazhenov Y V 2006. *Fundamentals of the theory of reliability of machines*: manual (Vladimir: Vladimir State Unitary Enterprise) p 160
[3] Boldin A P and Maksimov V A 2012 *Fundamentals of scientific researches: textbook for the higher education institutions* (Moscow: Publishing Center "Academy") p 336
[4] *LiAZ-529221 bus. Operation manual* 2011 Likino-Dulevo: LLC "Likinsky Bus Plant" p 360
[5] *LiAZ-529222 bus. Operation manual* 2012 Likino-Dulevo: LLC "Likinsky Bus Plant" p 407
[6] *LiAZ-621321 bus. Operation manual* 2011 Likino-Dulevo: LLC "Likinsky Bus Plant" p 397
[7] *LiAZ-621322 bus. Operation manual* 2013 Likino-Dulevo: LLC "Likinsky Bus Plant" p 373

IOP Conf. Series: Materials Science and Engineering **832** (2020) 012075 doi:10.1088/1757-899X/832/1/012075

Application of digital technologies in higher education in "Materials Science" for transport complex

O V Chudina

Department of Construction Materials MADI, 64, Leningradsky Avenue, Moscow, 125319, Russian Federation

Email: Chudina_madi@mail.ru

Abstract. The article is devoted to the problem of digitalization of higher technical education in transport engineering and is aimed to improve the efficiency of independent work of students and improving the classroom work of the teacher with students through the visualization of lecture materials. The article presents a new electronic educational resource (EER) on the subject "Material Science", which is a multimedia textbook on the chapter "Theory of heat treatment" and "Practice of heat treatment" for visualization of complex hidden processes occurring in metals during heat treatment. Specific examples show how the usage of multimedia teaching methods in the educational process contributes to the visualization of invisible diffusion processes, which are accompanied by the formation of micro-and nano-sized structural components. The technology of open modular systems (OMS), used to create a software environment, its advantages and wider interactive possibilities in comparison with traditional software tools are described.

1. Introduction

Nowadays digital technologies are used in teaching various disciplines in higher and vocational education. The analysis of foreign and domestic experience of digitalization of education shows that the existing developments are mainly focused on the modernization of education and science through the creation of network services operating through the Internet. The application of interactive network resources is aimed on the solution of innovative scientific and technological problems, training of students and retraining of specialists through their interaction with remote software, multimedia and educational and scientific complex. In real time through the Internet technology, Workplace as a Service (WaaS), etc. are created conditions for remote access to the modern high-tech equipment, implementation of researches and experimental tasks, involvement of the student in the interested independent learning process. During the distance learning of students in natural Sciences, mainly the problem of remote access during laboratory and practical work on special equipment is solved [1,2].

There is no information about the application of digital technologies to describe the structure of materials and processes occurring in them. This is probably due to the fact that specialists in digital technologies do not have knowledges of material scientists, and, on the contrary, material scientists do not know how to use multimedia technologies to describe the hidden processes occurring in materials under the influence of various factors. Therefore, the creation of electronic educational resources (EER) in materials science is possible only with the direct interaction of programmers with material scientists.

The development of EER in materials Science is an urgent task of the modern educational process, as their application will significantly relieve the teacher and students, free up additional time for analysis of complex or specific moments, allows making a creative element in the process of studying the subject.

On the one hand, the development of an electronic educational resource (EER) is caused by the need to visualize the processes occurring in the metal during heating and cooling. On the other hand, the need to use multimedia materials is dictated by a number of factors that have been observed in the educational process in recent years. First, most students are very advanced users of computer technology and are willing to accept materials presented in electronic form. Secondly, students found, from their point of view, the easiest way to "study" the discipline – search for materials on the Internet; often they are not only poor quality, but also illiterate. The main objective of the proposed development is to increase the efficiency of independent work of students and improve the classroom work of the teacher with students through the visualization of lecture materials.

In higher technical schools imperative for engineering disciplines is "Material Science", which contains chapters on the science of metals and heat treatment of metals, which include basic concepts about the structure and properties of metals, methods of processing. Complex diffusion processes occurring in metals during heat treatment, plastic deformation, recrystallization, aging and other methods of exposure are accompanied by the formation of micro-and nano-sized structural components and are not visible to the unarmed eye. The description of these processes in textbooks or directly at the lecture does not solve the problem of assimilation of educational material by students, which creates an opinion about the complexity of the discipline.

In this regard, an urgent problem is the creation of a multimedia interactive electronic educational resource (EER) on the subject of "Material Science", corresponding to the educational standards of the Russian Federation for higher technical educational institutions.

The aim of this work is to develop an educational multimedia textbook on the chapters "Theory of heat treatment" and "Practice of heat treatment" for the visualization of invisible processes occurring in metals under thermal influence.

The work was carried out jointly by the departments of Technology of structural materials and Automated control systems MADI.

2. Methods of creating a multimedia textbook

In the development of the modules "Theory of heat treatment" and "Practice of heat treatment", the technologies of open modular systems (OMS) were used, which have wider interactive possibilities in comparison with traditional software tools, such as Adode Flash [3, 4]. With its help, all sections and subsections of the topic are placed in one window, which greatly facilitates the search for any plot or fragment of the lecture.

On the basis of such multimedia technologies, it is possible to create electronic educational resources (EER) for educational materials containing images, texts accompanied by sound, video, animation and other visual effects.

The basis of the software environment is a modular architecture, in which each module is an independent content and functionally complete educational resource designed to solve a specific educational problem. In accordance with the principle of separation of programs and data software environment is separated from the content (content) modules [5, 6].

The main innovative features of the developed EER are [7]:

- provision of all components of the research process: obtaining information, practical training, modeling. For comparison, the book only transmits information;
- implementation of active forms of interaction with the content of EER, due to its high interactivity and multimedia.
- a significant increase in the capacity to improve self-study. The developed EER allows "at home", i.e. outside the classroom, to implement such activities that were previously possible only at the University (laboratory experiment, current knowledge control, and much more up to the collective educational work of remote users).

IOP Conf. Series: Materials Science and Engineering **832** (2020) 012075 doi:10.1088/1757-899X/832/1/012075

The main advantages of open modular systems (OMS) in comparison with traditional software tools are [7]:

- no technical limitations, it is possible to use and distribute interactive, multimedia in global computer networks;
- unlimited system life cycle. CHI is a dynamically expanding resource that does not require significant improvement when changing the content or technical external conditions.

In addition to the positive qualities of OMS can be attributed:

- ability to distribute on local media. Selected e-learning materials from the OMS together with the software can be easily transferred to a CD, flash drive, etc.

3. Results and Discussion

The chapter "Theory of heat treatment" discusses the processes, that describe the phase transformations in steels during heat treatment and most requires the development of video or animation visualization techniques, as the processes occur in motion in nano-sized level. By moving the cursor on the thermometer, independently varying the temperature, the student can observe structural changes in the steels that occur during heating or cooling.

The module "Theory of heat treatment" on the topic "Diagram of the isothermal decay of austenite" provides a virtual laboratory work in which the student, using text prompts, independently heats the samples, cools them in isothermal furnaces and in water, then measures the hardness, draws graphs, determines the time of the beginning and end of the decay of austenite and then builds a diagram of the isothermal decay of austenite.

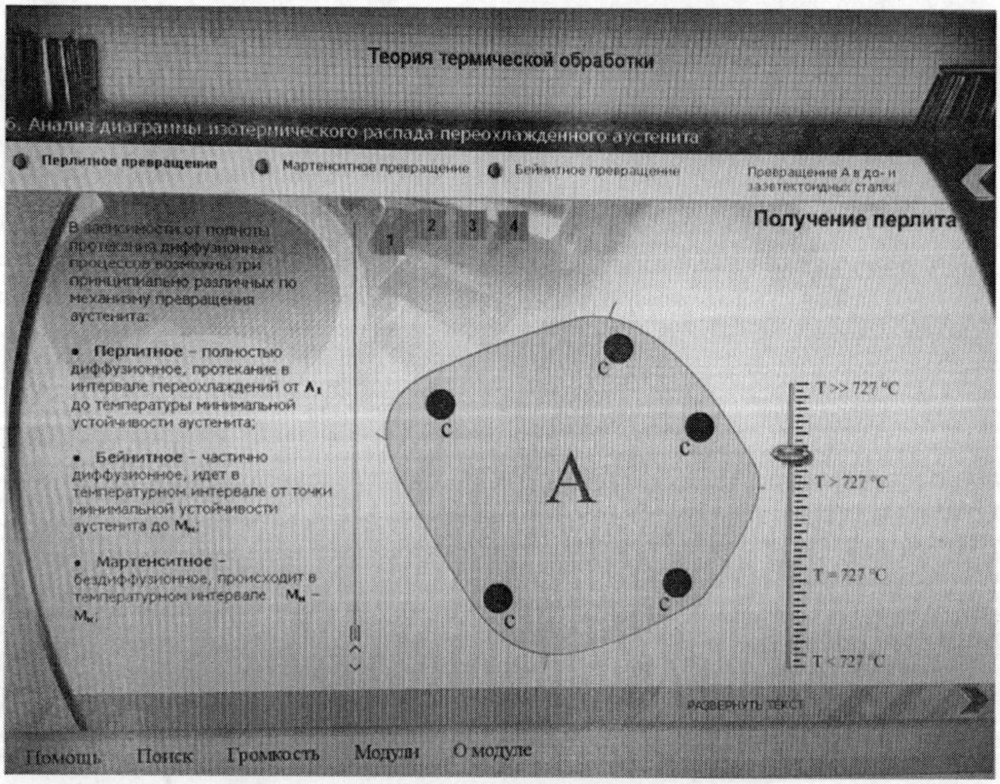

Figure 1. Visualization of the pearlite transformation.

In the subsection describing the transformation of austenite under slow cooling, the process of austenite decomposition into a ferrite-cementite mixture of plate structure is visualized. The mechanism of perlite transformation, consisting in the diffusion of carbon atoms to the grain boundaries of austenite, the formation of cementite particles, their growth deep into the grain and polymorphic transformation of austenite into ferrite is shown in dynamics (Figure 1).

In the subsection describing the martensitic transformation, it is shown how, with rapid cooling, austenite instantly acquires a needle structure. Structural changes occurring during vacation are also shown. At the same time on the left in the window you can read the text of the lecture material with a description of what is happening on the screen.

4. Conclusion

Traditionally the laboratory furnaces and a hardness tester are used in educational process in laboratory work. Technically, laboratory work is difficult, and in a real educational process for its full implementation is usually not enough time. In addition, in classical textbooks on materials science its description is absent. Therefore, the virtual laboratory work is the only method to acquaint students with the method of constructing a diagram of the isothermal decay of austenite. In the process of laboratory work the student chooses an isothermal furnace, conducts heating, cooling, cleaning of samples, measures their hardness, draws graphs and concludes on the structure and properties of steel (Figure 2).

In laboratory work on the topic "Practice of heat treatment" research method is used, in which a student independently conducts technological operations of heating and cooling of samples, measures their hardness after heat treatment, draws graphs and based on the measurements makes conclusions about the effect of a type of heat treatment on the structure and properties of metals and alloys.

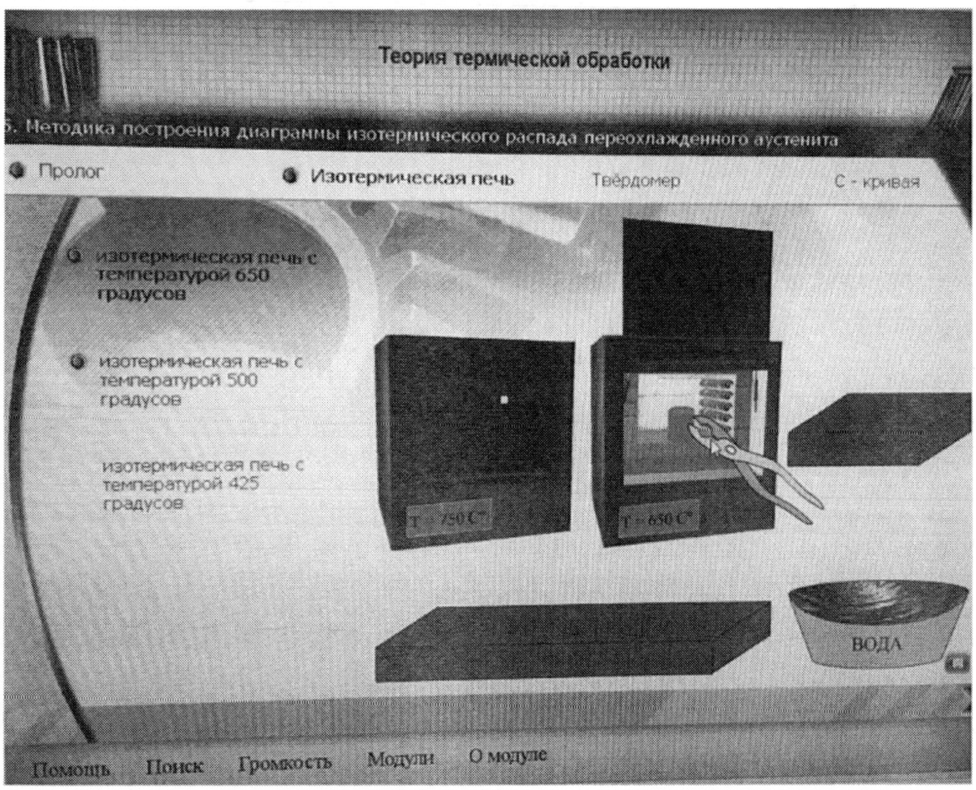

Figure 2. Virtual laboratory work "Diagram of isothermal decay of austenite".

Thus, the developed electronic educational resource that allows visualizing the complex hidden processes occurring in metals during heat treatment, increases the efficiency of a teacher during lectures, laboratory work, and facilitates the assimilation of the material by students during independent work on the discipline "Materials Science".

Acknowledgments
The author expresses his deep gratitude to Professor Ostroukh A.V., an expert in the field of digital technology, for his great contribution to the creation of the electronic textbook on the discipline "Materials Science".

Reference

[1] Mazuritsky M I, Soldatov A V 2014 Interactive network scientific and educational resources for natural science education *Higher education in Russia* **1** 80 -87.
[2] Safontsev S A, Mazuritsky M I, Konoplev B G, Boldyreva A M 2013 Features of development of electronic educational and methodical complex of natural science and technical disciplines *Informatization of education and science* **2** 111-120.
[3] Chudina O V, Ostroukh A V 2014 Development of electronic educational resource in materials science *Higher education in Russia* **1** 108-111.
[4] Chudina O V, Ostroukh A V, Snegirev I I, Malamut S A, Klimov P S 2011 Creating an electronic textbook for materials science *Young Scientist* **4** (27) 73-78.
[5] Ostroukh A V, Surkova N E 2011 *Electronic educational resources in professional education.* (Saarbrucken: LAP LAMBERT Academic Publishing) p. 103.
[6] Ismailov M I, Nikolaev A B, Ostroukh A V 2013 *Training and retraining of personnel of enterprises of industrial and transport complexes with the use of mobile technologies* (Saint Louis, MO: Publishing House Science and Innovation Center) p. 166.
[7] Ostroukh A V 2011 Experience in the development of new generation electronic educational resources for distance learning *In the World of Scientific Discoveries* **9**(21) 149-158.

IOP Conf. Series: Materials Science and Engineering **832** (2020) 012076 doi:10.1088/1757-899X/832/1/012076

Actual proposals for improving the transport and road network of the Kyrgyz Republic

D Glazunov

Kyrgyz-Russian Slavic University. 44, Kyev str., Bishkek, Russia

E-mail: glazunowdv67@mail.ru

Abstract. The main problems of improving the operation of road transport in the Kyrgyz Republic are considered, the problems of the road network affecting road safety. The intensity of traffic and the environmental safety of the cities of our country are noted not only as a result of the impact of compliance with measures to ensure road safety, but also reconstruction of highways. Measures have been developed to solve these problems.

Keywords: public transport, road safety, environmental safety, road transport infrastructure

1. Introduction

For several years, Kyrgyzstan has been trying hard to solve the acute problem of improving the operation of road transport. Considering that the entire road network was designed and built back in the seventies and eighties of the last century, it practically did not change and was not unified. For the Soviet republic, the street capacity developed at that time was sufficient in the field. There is small motorization, the road network of the city of Frunze, and now Bishkek, fully ensured normal operation not only during the day, but also during peak hours in the evening and in the morning.

Despite the fact that these years have been actively discussing these problems at all levels, not only in the media, but also in all possible information resources, the solution to this problem is progressing very hard. All road, design and other services involved in ensuring safety and improving traffic capacity try to find a way out of this situation, but for some reasons the solution to the problem does not move off the ground.

2. Statement of the research problem

In many cities and countries this problem is perfectly solved by public transport. In our city, all passenger traffic is carried out by 95 % using fixed-route taxis. A small number of regular buses do not actually affect street unloading. So, at present, Bishkek's public transport fleet has about 900 buses and trolleybuses. But if you take into account real data on the number of rolling stock, then we can draw the following conclusion. Bishkek City Municipal Passenger Motor Transport Company has more than 700 buses. But of these, about 130 buses are in operation, of which 50 % are new buses, and 50 % are old buses that have been repaired but are in a decadent state. More than 700 units are in a destroyed and malfunctioning state, although they are on the balance sheet of a car enterprise. The Bishkek trolleybus department sends more than 160 trolleybuses daily to the city streets. In addition, there are two bus routes in the city, which are provided by buses by a Chinese company. So, on bus route No. 35, passengers are transported by 18 buses, and on route No. 5 there are 12 buses.

IOP Conf. Series: Materials Science and Engineering **832** (2020) 012076 doi:10.1088/1757-899X/832/1/012076

The main transportation of passengers is carried out by private minibuses and private transportation companies. If several years ago there were more than 3,000 minibuses, now, for certain reasons, daily from 1,600 to 1,800 minibuses take to the streets of the city via 120 approved routes. The lack of buses in the morning and evening hours is now palpable.

Based on information from the State Registration Service under the Government of the Kyrgyz Republic, in 2014 about 1,050,000 units of motor vehicles were registered in our republic. World Bank (WB) experts involved in and developing the draft National Road Safety Strategy of the Kyrgyz Republic "Road Safety" indicate that "with statistics of 2190 deaths per million cars, the problem with road safety in Kyrgyzstan is one of the worst in the world" [1]. According to various estimates, the number of victims and disabled after an accident is unacceptably high. So, in 2017, 907 people died in road traffic accidents in the Kyrgyz Republic, 104 of them were children, and 938 people died in 2016. 1.622 children were injured in accidents. In 2015, there were 1.432 accidents [2].

According to the World Health Organization (WHO), at present, Kyrgyzstan is in second place among the countries of the WHO European Region in mortality per 100,000 population – 19.3 people. (2012), 19.2 hours (2010)" [1]. Injury is one of the causes of death in the age group of 5 to 29 years, and this indicates that we are losing the most efficient and promising population. Every year, mortality among children is 3–4.5 thousand children up to 14 years old – 30 %. More than 1,200 people die in traffic accidents each year, and up to 10,000 people are injured. Each year, these indicators reflect a certain increase in casualties, which has a very negative impact on the state of security not only in Bishkek, but throughout the republic. Despite this, Kyrgyzstan does not actually use the scientific data of the results of the analysis of road accidents, which are not only studied by scientists of the Kyrgyz-Russian Slavic University and KSTU named after I. Razzakova, but they also carry out monitoring, count and develop a certain measure leading to a decrease in road accidents. The Unified database on offenses is still underdeveloped, the installation and operation of technical means of monitoring offenses and traffic regulation have just begun.

According to the National Institute for Strategic Studies of the Kyrgyz Republic: "... to date, technical regulations for the construction, reconstruction and repair of roads have not been developed. Until now, regulation has been carried out on the basis of post-Soviet Construction Norms and Regulations and GOST, which, according to the conclusion of the Ministry of Justice of the Kyrgyz Republic, have only a recommendatory character" [1]. The studies carried out claim that there are no safety standards on roads at the legislative level, there are no methods for assessing the risks of road accidents, no regulations have been established for monitoring road safety, and this leads to a low level of safety. In addition, it is necessary to take into account the secondary costs of reconstruction and repair of roads, corruption in this area has not yet been eliminated. All the work on the creation of parking spaces, road fences, the installation of billboards is performed in the absence of technical regulations, not observing the land use requirements of the Kyrgyz Republic.

As a result of the calculations of the National Institute for Strategic Studies, the following conclusion was made. It is necessary to develop standards that meet modern requirements, taking into account the development of the leading countries of the world, so that in all structures associated with this problem, transport, infrastructure, law enforcement, technical, educational institutions need to have and implement targeted road safety programs.

The materials of these studies, regulatory and technical and legal acts, road transport, engineering infrastructure, culture on the roads, compliance with traffic rules remain at the level of unsolved problems. Our country needs urgent solutions to this problem and the adoption of priority measures by government bodies. Lack of solutions to road safety problems leads not only to economic damage but also to social disaster.

3. Proposed recommendations and urgent measures

Based on our analysis, it is necessary not only to recommend but also to seek the adoption of the following urgent measures:

IOP Conf. Series: Materials Science and Engineering **832** (2020) 012076 doi:10.1088/1757-899X/832/1/012076

1. At the legislative level, assess the situation, develop and adopt safety standards, develop modern risk assessment methods, and technical regulations for monitoring road safety. Since SNiP was approved in Soviet times, and it was difficult to reaffirm it, they made it a recommendation. But now I have to get a mandatory legal status. In addition, in the "Code of Administrative Responsibility of the Kyrgyz Republic" dated July 15, 2016 No. 117, there are a sufficient number of articles that specify the requirements for the operation of vehicles and emissions, for example, Article 173. Operation of vehicles and other mobile vehicles in excess of content standards pollutants in emissions [3].

2. Continue to introduce and install on the roads technical means of traffic regulation, video and photo-fixing equipment to all relevant organizations, including large Business facilities located along the roads.

3. To monitor traffic accidents not only in the city of Bishkek, but also throughout the Kyrgyz Republic. Introduce training programs and subjects in educational institutions.

In addition, it is necessary to pay attention that the inefficient development of urban transport leads not only to a decrease in the capacity of the road network, but also to major environmental problems in the capital, constantly poisoning the townspeople. For the purposes of Sustainable Development, ensuring mobility of the population, safe movement and a favorable, safe environmental environment in cities, it is additionally necessary to hold public hearings and take the following measures that must be taken into account, including in the spatial and planning regulation and in the general plan of the city:

1. To improve road safety in Bishkek and other large cities, increase the number of one-way streets that reduce traffic accidents by up to 50% in accordance with SniP.

2. To increase the number of buses and trolleybuses of urban public transport, reducing the number of minibuses accordingly.

3. During the construction of large supermarkets, shopping centers and other facilities where there is a large concentration of people in the estimated radius from each other, to take into account and enter, underground parking only where the road network allows pedestrian streets.

4. To reduce the number of high-rise buildings with public and commercial functions on the ground floors in the city center.

5. To increase the construction of preschool and school facilities evenly distributing the quality of education throughout the city and with equal access in order to reduce the forced movement of citizens in the city in search of the best schools and kindergartens.

6. To ensure continuous monitoring of the level of air pollution on all major highways of the city.

7. To carry out a strict policy to reduce the MPC of atmospheric air in Bishkek, where more than 82 % of emissions come from transport. Take drastic measures and pursue a tough policy that ensures an annual reduction of 15 %, which is quite realistic.

At present, atmospheric air quality, measured in various parts of Bishkek city, shows threatening indicators of maximum permissible concentration (MPC) of air, which are above regulatory requirements from 10 to 100 or more times in different areas of the capital. Especially difficult is aeration from the city center, where the MPC is always 8 times higher than that of other parts of the city, since the city is located in a foundation pit, in a bowl-shaped relief. In connection with the growth of the population of the capital from 500,000 to 970,000 people, without taking into account the unregistered population, the territory of the city itself increased by only 20–25 %, the standard level of landscaping fell sharply.

The city of Bishkek needs to be provided with a sufficient quantity of comfortable and high-quality public transport, into which citizens will transfer. In the city center, it is advisable to run electric or trolleybus lines, that is, green transport [4]. Such a project has been developed and is available to the authors of the project. Restore pedestrian walkways on all city streets and return them within the boundaries of the red lines, which are municipal lands.

In the city of Bishkek, urgently and tightly regulate the emission standards for cars on the roads and introduce large fines into the plans before banning the import of vehicles, since they are the main sources of air pollution, and when the climate changes that have already occurred create a level of environmental disaster for the health of citizens. Among the diseases in the city in the first place there is bronchopulmonary, which overtook cardiovascular highly dependent on the climate, natural and landscape conditions of the country.

In the cities of the Kyrgyz Republic, it is advisable to propagate the concept of "pedestrian city" accepted around the world. Pedestrian cities appeared in many European countries when the population deliberately abandons cars, except for special vehicles. This is a huge cost savings and the organization of eco-cities in the mountainous environment. Currently, in small and medium-sized cities of the republic there is no intra-urban transport, which needs to be supported. Alternative transport can be offered, from live transport to bicycles. It is necessary to use the experience of Tajikistan, where in villages police officers, district doctors, medical sisters give out bicycles as a transport.

Table 1. An approximate plan and assessment of reduction of harmful emissions with the following measures for the city of Bishkek

No.	Activity	Consumption Control	Accounting	Quality	Reduction of harmful emissions by years 2019	2022	2022–2025
Task 1: Combustive-lubricating materials. Reduction of harmful emissions							
1	Raise the duty on fuel and lubricants consumption by cars	Purchased and imported	+	+	15 %	30 %	45 %
2	Gasoline quality, alternative fuel sources	Purchased and imported	+	+	15 %	30 %	45 %
3	Determination of emission standards. Technical regulation, SNiP, regulations, etc.	Emission standards	+	+	Rate entry	Regulatory requirements	Regulatory requirements
4	Penalties	Visual fixation,	+	+	Budget replenishment	Budget replenishment	Budget replenishment
		Video Photo Capture	+	+	Budget replenishment	Budget replenishment	Budget replenishment
5	Incentive measures	Price tax support	Price tax support	Price tax support	Price tax support	Price tax support	Price tax support

Table 2. Transport in the cities of the Kyrgyz Republic

No.	Activity	Control	Accounting	Quality	Taxes	2018–2022	2022–2025
Task 2: Transport							
1	Private transport	+	+	+	Introduce a check-in tax to the city center	Paid parking	Reduced parking
2	Any public transportation	+	+	+	Tax free	Free parking	
3	The buses	+	+	+	Tax free	Free parking	
4	Trolleybuses According to the project proposal in the center of Bishkek	+	+	+	Tax free	Free parking	
5	Alternative transport	+	+	+	Tax free	Free parking	
6	Green transport	+	+	+	Tax free	Free parking	
7	Incentive measures	Price tax support	Price tax support	Price tax support	Price tax support	Price tax support	Price tax support

It is necessary to analyze and evaluate several parameters of the road transport infrastructure at the Institute of Road Transport or to invite experts at city halls. In this article, we offer a specific solution to the problems and an approximate plan of activities, which will depend on full-scale research by specialists. Monitoring, accounting, quality of consumption are the main for any tasks, the implementation of which will give tangible results in the coming years.

Table 3. Roads in the cities of the Kyrgyz Republic

No.	Activity	Monitoring			Necessary measures	2018–2022	2022-2025
		Control	Accounting	Quality			
Task 3: Roads							
1	Main streets. The highest category	+	+	+	Toll roads		
2	For the city of Bishkek, increase the number of one-way traffic on the streets, which reduces accidents by 50%.	+	+	+	-		
3	Restore pedestrian sidewalks on the streets and return municipal lands within the boundaries of red lines.	+	+	+	Penalties		
4	Bike paths	+	+	+	-		
5	Pedestrian streets	+	+	+	-		
6	parking	+	+	+	Environmental Duty		
7	Regulatory indicators	+	+	+	tax		
8	Schedule. Route	+	+	+	Calculation of passenger traffic		
9	Incentive measures for the installation of video-photo fixing equipment	+	+	+	Price tax support	Price tax support	Price tax support
10	Penalties	+	+	+			

Table 4. Road Safety in the Kyrgyz Republic

No.	Activity	Control	Accounting	Quality	Taxes	2018–2022	2022–2025
Task 4: Security. Reducing the number of accidents							
2	One way streets	+	+	+	-	50 % reduction in accidents	50 % reduction in accidents
3	Traffic lights	+	+	+	-	increased security	increased security
4	Traffic violation	+	+	+	-		
5	Underpasses	+	+	+	-		
	Incentive measures					Price tax support	Price tax support
	Penalties	+	+	+	+	+	+

Task 5: Mobility.
a. Ticketing.
b. Uniform Route Distribution.
c. Incentive Measures and Penalties.

Task 6: PWD.
a. In all objects of new construction and the organization of improvement, introduce the mandatory requirements noted in the acts of acceptance of work taking into account PWD.
b. Organization of special vehicles.

IOP Conf. Series: Materials Science and Engineering **832** (2020) 012076 doi:10.1088/1757-899X/832/1/012076

c. Public transport equipment for people with disabilities.
d. Incentive Measures and Penalties

Task 7: Create a project on the Concept for the Development of Public Spaces for Road Transport Infrastructure, which should take into account all the above tasks and the green frame.
a. Public spaces.
b. The types and protective measures of public spaces along the roads, I agree to SNiP, GOST, those regulations and the design of the urban environment.
c. Capitalization of facilities for self-sufficiency. Capitalization of facilities for self-sufficiency.
d. Incentive Measures and Penalties.

Hence, the following recommendations can be made:
1. The Ministry of Transport and Roads of the Kyrgyz Republic needs to develop a technical regulation that sets the basic development strategy. To decide who should research and monitor on an ongoing basis about urban transport for cities.
2. In this paper, the proposed activities, which include the basic requirements and measures to improve the transport and road facilities with tasks and an approximate plan of activities, will depend on full-scale research by specialists.

References

[1] National Institute for Strategic Studies of the Kyrgyz Republic *Scope, consequences and preventive measures of road traffic accidents in the Kyrgyz Republic* (Date of treatment 08.15.2018)
[2] Sputnik news agency. Retrieved from: https://ru.sputnik.kg/incidents/20180119/1037326719/skolko-chelovek-pogibli-v-dtp-v-2017-godu.html/ (Date of treatment 08.15.18.)
[3] *The Code of Administrative Responsibility of the Kyrgyz Republic* dated July 15, 2016 No. 117. Retrieved from: http://www.ecology.gov.kg/page/view/id/112 (Date of treatment 1.08.18)
[4] From the archive of the company "Mark Audit" 2015 *The project was developed by the Czech company for green transport for the city of Bishkek.* Unpublished material

Determination of Power and Kinematic Wheel Parameters when Rolling against a Drum

T A Balabina[1], D S Simonov[1], V R Rogov[1], A N Mamaev[2], A S Vorontsov[3]

[1]Department of Machinery Parts and Theory of Mechanisms, Moscow Automobile and Road State Technical University (MADI), 64, Leningradsky Ave., Moscow, 125319, Russia
[2]Moscow Polytechnic University, 38, B. Semyonovskaya Street, Moscow, 107023, Russia
[3]Department of Materials Science and Resource-Saving Technologies, Yanka Kupala State University of Grodno, 22, Ozheshko str, Grodno, 230023, Belarus
E-mail: dimsimonov94@mail.ru

Abstract. Car tire tests, including rolling resistance tests, are often performed on drum stands. Car tests on drum stands of various designs are also becoming more widespread: with a different number of drums, their locations and types of drives. However, in the literature there are no works describing the process of interaction of the wheel with the drum. In this regard, there is a need to consider the mechanics of the interaction of an elastic wheel with a drum and determine its kinematic and power characteristics, which is the task of this study. The mechanics of rolling the wheel against the drum is considered, taking into account the frictional interaction of the pair "elastic wheel – rigid drum" using the theory of preliminary displacement of the wheel elements in the zone of contact with the drum. This made it possible, inter alia, to obtain relationships for determining the rolling resistance caused by hysteresis losses in the wheel material, the tangential force acting in the contact, the moment on the wheel, the power of friction losses in the contact, the coordinates of the boundary of the adhesion and sliding sections in the contact of the wheel with the drum, relative loss of speed (slippage) of the wheel. The interaction of the wheel with two drums is also considered.

1. Introduction

Car tire tests, including rolling resistance tests, are often performed on drum stands. Car tests on drum stands of various designs are also becoming more widespread: with a different number of drums, their location and type of drive.

However, the process of rolling a wheel against a drum remains poorly understood – in foreign and Russian literature, there is no analysis of the mechanics of rolling a wheel against a drum, although this requires, with regard to the above, close study including the determination of both power and kinematic parameters of the wheel.

2. Main Part

Currently, issues related to the mechanics of rolling a wheel on a flat surface are considered quite widely [1–24].

The rolling mechanics of an elastic wheel against a drum are the same as when the wheel is rolling on a flat rigid supporting surface [1–5].

IOP Conf. Series: Materials Science and Engineering **832** (2020) 012077 doi:10.1088/1757-899X/832/1/012077

When a driven wheel is loaded, only with a normal load, due to imperfect elasticity of the material, there is a loss of internal friction in the wheel material (hysteresis), which causes the appearance of a moment of resistance M_f and the appearance of a rolling resistance force F_f – the longitudinal tangential force acting in contact of the wheel with the base in the opposite direction to the wheel rotation. A similar rolling resistance force also arises in a brake wheel, loaded in comparison with a driven wheel, with an additional braking torque M_t. The presence of this force leads to slipping of the elements of its treadmill relative to the base in the contact zone and to the loss of the angular velocity of the wheel.

When the drive wheel is rolling, the movement of which occurs under the influence of the torque M_k, a leading (traction) force arises in the contact, directed along the wheel rotation. As in the previous case, this force causes slippage of the treadmill elements of the wheel in the zone of contact with the base and to the loss of the linear velocity of the wheel axis.

The mechanism of occurrence of sliding of the surface elements of the wheel relative to the base is considered in detail in [1–5]. Using the scheme of the reversed mechanism "elastic wheel – rigid base", on the basis of the theory of preliminary displacement, it is shown that with steady rolling, the wheel surface elements entering the contact zone are not yet "prepared" for perceiving tangential forces and at the same time pressed to the base normal force, they begin to move without sliding, while receiving tangential displacements (directed opposite to rolling for the brake and driven wheels, and in the direction of rolling – for the driving wheel). As the interlocked elements of the wheel and the base move in the inverted mechanism in the contact zone, their tangential displacements increase, therefore, the tangential friction force acting between the interlocked elements also increases. In the place of contact, where the increased friction force reaches its ultimate in adhesion, a breakdown occurs and on the entire part of the contact located beyond the breakdown point, regardless of whether it is in the zone of decreasing or increasing normal pressures, sliding occurs.

With an increase in the loss of wheel speed and a corresponding increase in the tangential force acting in the contact, the slip zone increases, as well as the power of friction losses in the contact, which characterizes the wear rate of the treadmill and partly the wheel rolling resistance.

The tangential displacements of the points of the treadmill in the contact zone, due to the implementation of the tangential force in the contact, are determined [1–5] by the relationship:

$$U = (a - x)\left(\frac{\omega_{wh} r_{wh}^{f}}{V} - 1\right) \tag{1}$$

where a is the half-length of the wheel area with a rigid supporting surface, x is the distance from the beginning of the contact area to the point of the wheel in the contact zone, r_{wh}^{f} is the radius of free rolling, $V = \omega_{wh} r_{wh}$ is the speed of the wheel axis in the inverted mechanism, r_{wh} is the rolling radius of the wheel.

As applied to wheel rolling against a rigid drum, $V = V_d = \omega_d r_d$, where ω_d and r_d are the angular velocity and radius of the drum. As a result, the tangential displacements of the surface points of the elastic wheel, due to the implementation of tangential force in contact with the drum, in the adhesion section can be represented by the expression:

$$U = (a - x)\left(\frac{\omega_{wh} r_{wh}^{f}}{\omega_d r_d} - 1\right) = \xi(a - x) \tag{2}$$

where

$$\xi = \frac{\omega_{wh} r_{wh}^{f}}{\omega_d r_d} - 1 \tag{3}$$

ξ – relative speed difference.

With a known value of ξ, the ratio of the angular velocities of the wheel and drum will be equal to:

$$\frac{\omega_{wh}}{\omega_d} = (1 + \xi)\frac{r_d}{r_{wh}^{f}}. \tag{4}$$

Based on the proportionality of the tangential stresses (specific tangential forces) to tangential displacements, we can write, that the tangential stresses due to the implementation of the tangential force in the contact are:

$$q_t = \lambda U = \lambda \xi (a - x) \tag{5}$$

where λ is the tangential stiffness coefficient of the wheel, defined [1] as:

$$\lambda = \lambda_{wh} \frac{r_d}{r_d + r} = \frac{1.5qr}{a^3} \frac{1}{1 + r/r_d}. \tag{6}$$

Under a parabolic law, the distribution of normal pressures along the length of the contact area, the coordinate of the boundary of the adhesion and slip sections, determined from the equality $q_t = \mu q_n$, can be represented by the relationship [1–3]:

$$x_B = -a \pm \frac{\lambda \xi}{\mu q_{n_0}}. \tag{7}$$

The moment on the drum, due to the action of the tangential force, is equal to:

$$M_t = 2b \int_{-a}^{x_B} r_d q_t^{sl} dx + 2b \int_{x_B}^{+a} r_d q_t dx = \left[2b \int_{-a}^{x_B} q_t^{sl} dx + 2b \int_{x_B}^{+a} q_t dx\right] r_d = F_t r_d. \tag{8}$$

In the last formula, the value in square brackets equal to the algebraic sum of all the specific tangential forces in the contact is called the circumferential traction force (Figure 1):

$$F_t = 2b \left[\frac{\lambda \xi}{2} (a - x_B)^2 \pm \frac{1}{3} \mu q_{n_0} (2a^3 + 3a^2 x_B - x_B^3)\right] \tag{9}$$

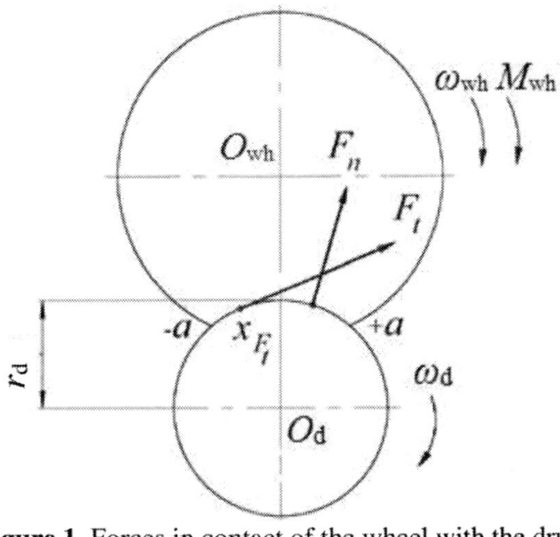

Figure 1. Forces in contact of the wheel with the drum

Substituting in (9) the expression $\xi = \pm \mu q_{n_0} (a + x_B)$, obtained from (7), after transformations, we obtain an equation whose solution gives a relationship for finding the coordinate of the boundary of the adhesion and slip sections:

$$x_B = a \left(1 - 2 \sqrt[3]{1 - \frac{F_t}{\mu F_z}}\right). \tag{10}$$

As a result:

$$\xi = \frac{\pm 1}{\lambda} 2 \mu q_{n_0} \left(1 - \sqrt[3]{1 - \frac{F_t}{\mu F_z}}\right) \tag{11}$$

or, taking into account the expressions for q_{n_0} [1 ... 5], (2) and (6):

$$\xi = \pm \frac{\mu a}{s} \left(\frac{1}{r} - \frac{1}{r_d}\right) \left(1 - \sqrt[3]{1 - \frac{F_t}{\mu F_z}}\right). \tag{12}$$

The last expression, if we do not take into account the saturation coefficient of the treadmill pattern s (for a wheel without a treadmill pattern s = 1), coincides with a similar formula obtained by G. Fromm [6, 7] (the difference lies only in the degree of radical: G. Fromm – square root), and then R.V. Virabov [1] for the friction gear, consisting of two cylinders.

With a known relationship between ξ, the ratio of the angular velocities of the elastic wheel and the rigid drum as a function of the traction force F_t and normal load realized in the contact in accordance with formulas (3), (11) and (12) can be represented as:

$$\frac{\omega_{wh}}{\omega_d} = \frac{r_d}{r_{wh}^f}\left[1 \pm \frac{2\mu q_{n_0}}{\lambda}\left(1 - \sqrt[3]{1 - \frac{F_t}{\mu F_z}}\right)\right] = \frac{r_d}{r}\left[1 \pm \frac{\mu a}{s}\left(\frac{1}{r} - \frac{1}{r_d}\right)\left(1 - \sqrt[3]{1 - \frac{F_t}{\mu F_z}}\right)\right]. \tag{13}$$

For small (compared with the adhesion limit $F_t^{max} = \mu F_z$) tangential forces, the last formulas can be simplified if the expression $\sqrt[3]{1 - F_t / \mu F_z}$ is expanded in a power series, then discarding second-order negligible values:

$$x_B = a\left(-1 + \frac{2}{3}\frac{F_t}{\mu F_z}\right) \tag{14}$$

$$\xi = \frac{a}{3s}\left(\frac{1}{r} + \frac{1}{r_d}\right)\frac{F_t}{F_z} \tag{15}$$

$$\frac{\omega_{wh}}{\omega_d} = \frac{r_d}{r}\left[1 - \frac{a}{3s}\left(\frac{1}{r} + \frac{1}{r_d}\right)\frac{F_t}{F_z}\right]. \tag{16}$$

In the latter expressions, the force F_t is positive for the drive wheel and negative for the driven and braking.

The friction power losses in the contact of the elastic wheel with the rigid drum, due to the implementation of the traction force in the contact, are determined by the same relationship as for the case of rolling the wheel on a flat supporting surface:

$$P_f = F_t \xi V = F_t \xi \omega_d r_d. \tag{17}$$

The point of application of the resulting tangential force in the contact (Fig. 3) is found from the relationship [8]:

$$x_{F_t} = a\left(1 - \frac{\mu F_z}{F_t}\right)\left(1 - \sqrt[3]{1 - \frac{F_t}{\mu F_z}}\right) \tag{18}$$

which for small tangential forces compared to the ultimate in adhesion can be represented in a simplified form, if we expand $\sqrt[3]{1 - F_t / \mu F_z}$ in a power series:

$$x_{F_t} = a\left(1 - \frac{\mu F_z}{F_t}\right)\frac{F_t}{3\mu F_z} = \frac{a}{3}\left(\frac{F_t}{\mu F_z} - 1\right). \tag{19}$$

The arm of the tangential force relative to the center of the wheel is found from geometric considerations (Figure 2):

$$h_\tau = O_d O_{wh} \cos\alpha_\tau - r_d = (r_a + r_d)\cos\alpha_\tau - r_d \tag{20}$$

where the angle

$$\alpha_\tau = x_{F_t}/r_d. \tag{21}$$

After the transformations we get:

$$h_\tau = r_a\left(1 - x_{F_t}^2/2r_a r_d\right). \tag{22}$$

As with the rolling of an elastic wheel along a flat supporting surface, in the case of the wheel interacting with a rigid drum, there is a redistribution of normal pressures along the length of the contact area and, as a result, a shift in the point of application of the normal reaction of the drum relative to the center of the wheel (Fig. 2).

The arm of the normal reaction of the drum relative to the center of the wheel is:

$$h_n = (r - h_\tau) F_t / F_n. \tag{23}$$

We consider the case [9], when an elastic wheel rolls against two supporting drums (Fig. 3). The wheels of a car, in particular, are installed in this way during testing on drum stands.

e' and e'' denote the distance between the center of the wheel and the centers of the left and right drums, respectively:

$$e'=r'_d+r'_a \qquad e''=r''_d+r''_a,$$

where r'_a and r''_a – the shortest distance from the axis of the wheel to the left and right drums.

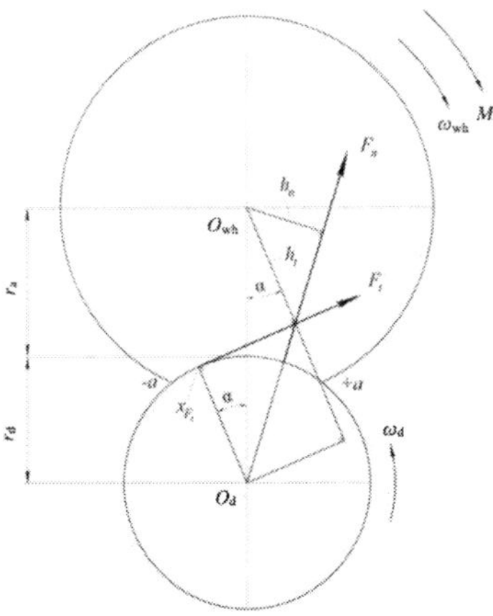

Figure 2. The arms of the forces acting on the wheel

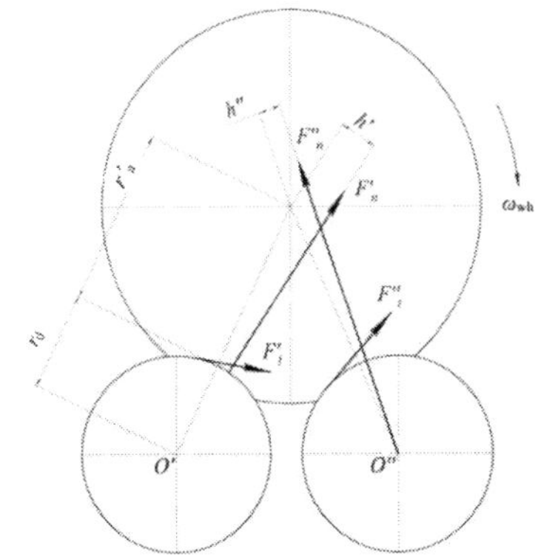

Figure 3. Forces acting on the wheel when rolling against two drums.

Under the action of the moment M_{wh} applied to the wheel, the normal reactions of the drums are shifted by the values of h' and h'':

$$h'=h'_n+h'_0 \qquad h''=h''_n+h''_0 \tag{24}$$

where h'_n and h''_n are the displacements of normal reactions due to the realization of tangential forces F'_{τ_0} and F''_{τ_0} in the contact; h'_0 and h''_0 are the displacements of normal reactions due to hysteresis in the tire material, which can be calculated as:

$$h_0^{\prime(\prime\prime)} = f_0 r\left(1+\frac{r}{r_d}\right)\frac{a^{\prime(\prime\prime)}}{a^{area\prime(\prime\prime)}} \tag{25}$$

where a' and a'' are the half-lengths of the contact area of the wheel with the drums under the action of normal forces F_n' and F_n''; $a^{area\prime(\prime\prime)}$ – the half-length of the contact area on the plane under the action of the forces $F'_z = F_n'$ and $F''_z = F_n''$.

The points of application of tangential forces are determined from formulas (18) (for small tangential forces, from (19)).

The angles γ_i, which determine the displacement of the point of application of the normal reactions of the drums, can be found from geometric considerations:

$$\gamma' = \arcsin\frac{h'_n + h'_0}{e'} \qquad \gamma'' = \arcsin\frac{h'_n + h''_0}{e''}. \tag{26}$$

It should be noted that with a rigid kinematic connection between two drums, the distances between the wheel axis and the axes of the drums, as well as normal reactions in the contacts, would be different. The arms of the circumferential and normal forces, hysteresis losses, power of friction losses in the contact, sizes of the contact areas will also be different.

When the wheel is mounted on two freely rotating unconnected drums, the normal reactions from the side of both drums and the dimensions of the contact pads are the same.

Under the action of the moment M_{wh}, the center of the wheel is located asymmetrically relative to the axes of the drums.

The rolling resistance of the driven wheel, as already noted, is determined by the moment of rolling resistance due to hysteresis losses in the wheel material. The value of this moment is more convenient to determine [10] through the hysteresis loss power $M_H=P_H/\omega_H$.

To determine the power of hysteresis losses in the material of an elastic wheel rolling along a rigid drum, we will take into account only the normal deformation of the wheel, which can be represented as the sum of terms (Figure 4): $W=W'+W''$.

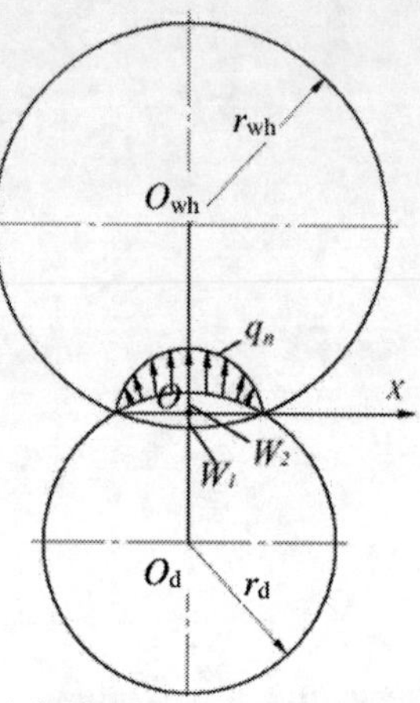

Figure 4. Normal stresses q_n and deflection w when the wheel is pressed against the drum.

Since:

$$W' = \frac{a^2-x^2}{2r} \; and \; W'' = \frac{a^2-x^2}{2r_d}$$

then:

$$W = \frac{a^2-x^2}{2}\left(\frac{1}{r}+\frac{1}{r_d}\right). \tag{27}$$

The hysteresis power loss can be found using the following relationship:

$$P_H = \beta_H \int_0^a q_n \left|\frac{dW}{dt}\right| dx2b \tag{28}$$

where β_H is the hysteresis loss coefficient; dW/dt – wheel deformation rate:

$$\frac{dW}{dt} = \frac{dW}{dx}\frac{dx}{dt} = \frac{dW}{dx}V_{ang.} = \frac{dW}{dx}\omega_{wh}r \tag{29}$$

Here, the x coordinate lies on the OX axis (Fig. 4); $\frac{dx}{dt} = V_{ang.} = \omega_{wh}r$ – since the change in the normal deformation dW/dt as the treadmill element moves deeper into the contact occurs at a speed equal to the peripheral speed of the wheel.

In view of (27):

$$\frac{dW}{dt} = -x\left(\frac{1}{r}+\frac{1}{r_d}\right)\omega_{wh}r. \tag{30}$$

As a result:

$$P_H = \frac{3}{16}\beta_H F_n a\omega_{wh}r\left(\frac{1}{r}+\frac{1}{r_d}\right). \tag{31}$$

With the relationship found for P_H, the moment of hysteresis in the tire material can be represented as:

$$M_H = \frac{P_H}{\omega_{wh}} = \frac{3}{16}\beta_H a F_n\left(1+\frac{r}{r_d}\right). \tag{32}$$

Then the arm shift of the normal reaction of the drum will be equal to:

$$h_0 = \frac{M_H}{F_n} = \frac{3}{16}\beta_H a\left(1 + \frac{r}{r_d}\right). \tag{33}$$

According to [4, 5]:

$$3\beta_H a^{area}/16 = f_0 r_{wh}^{f} \cong f_0 r$$

then

$$h_0 = f_0 r\left(1 + \frac{r}{r_d}\right)\frac{a}{a^{area}}. \tag{34}$$

Here, a^{area} is the half-length of the contact of the wheel with a flat rigid supporting surface at the same load F_n.

Knowing the arm h_0, we can find the relationship for the tangential force (it is the rolling resistance of the driven wheel), due to hysteresis:

$$F_{\tau_0} = F_n f_0\left(1 + \frac{r}{r_d}\right)\frac{a}{a^{area}}. \tag{35}$$

Since the ratio of rolling resistance to normal force $F_{\tau_0}/F_n = f_0^d$ is the coefficient of rolling resistance of the wheel, for the case of rolling of an elastic wheel against a rigid drum in this case:

$$f_0^d = f_0\left(\frac{r}{r_d} + 1\right)\frac{a}{a^{area}}. \tag{36}$$

For $r_d \to \infty$, relationships (32), (35), (36) lead to the expressions derived for the case of rolling of an elastic wheel along a flat rigid supporting surface. Comparison of these expressions with the above relationships leads to the conclusion that both the moment of hysteresis and the strength and coefficient of rolling resistance of the driven elastic wheel against the rigid drum, caused by hysteresis, increase in $a(1+r/r_d)/a^{area}$ times compared to rolling the same wheel on a flat rigid surface.

An increase in rolling resistance against the drum leads to a difference in the drag coefficients determined on the drum and when the wheel moves along a flat supporting surface.

When using the coefficients of resistance to lateral withdrawal and rolling resistance, obtained experimentally on a drum stand, for the case of movement of the wheel on a flat supporting surface, appropriate correction factors should be introduced.

The interaction of an elastic wheel with a rigid supporting surface is described in more detail in the references cited at the end of this article [1–24].

3. Conclusion

1. Relationships that make it possible to calculate the power and kinematic parameters of a wheel while rolling against one or two rigid drums are obtained.
2. The coefficient of rolling resistance of the driven wheel against a rigid drum increases in $a(1+r/r_d)/a^{area}$ times compared to rolling on a flat rigid surface.
3. An increase in rolling resistance against a drum leads to a difference in the drag coefficients determined on the drum and when the wheel moves along a flat supporting surface.
4. When using the coefficients of resistance to lateral withdrawal and rolling resistance, obtained experimentally on a drum stand, for the case of movement of the wheel on a flat supporting surface, appropriate correction factors should be introduced.

References

[1] Virabov R V 1982 *Traction Properties of Friction Gears* (Moscow: Engineering)
[2] Virabov R V and Mamaev A N 1980 Analysis of Kinematic and Power Relationships When Rolling an Elastic Wheel on a Rigid Base *Mechanics of machines* **57** 101–6

[3] Virabov R V and Mamaev A N 1978 Determination of Power Losses for Friction in the Contact of a Friction Pair – A Wheel with a Pneumatic Tire-Rigid Base *Infinitely variable transmission: interuniversity* **3** 61–7
[4] Virabov R V, Mamaev A N and Balabina T A 2010 General Issues of the Interaction of an Elastic Wheel with a Rigid Supporting Surface *AAI "Automobile and Tractor Engineering": Abstracts of the International Scientific and Practical Conference* (Moscow) pp 73–80
[5] Balabina T A and Mamaev A N 2014 The Rolling Mechanics of an Elastic Wheel on a Rigid Supporting Surface *Compilation: Engineering: trends, prospects and development technologies* pp 20–25
[6] Fromm H 1927 Berechnung des Schlupfes bein Rollendeformierbaren Scheiben *Math. Und mech. Bd.* **7**
[7] Fromm N 1928 Arbeitsverlust, Formanderungen und Schlupf bein Rollen von treibenden und gebremsten Radern oder Scheiben. Beitrag zur Analyse der Reibungsgesetze *Technische Physik* **9**
[8] Mamaev A N, Virabov R V, Portuguese V M and Chepurnoy S I 2010 Determination of Power and Kinematic Characteristics of an Elastic Wheel when Rolling against a Rigid *Drum Sat Proceedings of the international scientific conference dedicated to the 145th anniversary of MSTU MAMI* (Moscow: MSTU MAMI)
[9] Abuzov V I and Mamaev A N 2012 Rolling of an Elastic Wheel on Two Rigid Drums *Auto prom.* **10** p 19
[10] Mamaev A N 2010 Resistance to Rolling of a Driven Wheel on a Rigid Drum *Sat Proceedings of the international scientific conference dedicated to the 145th anniversary of MSTU MAMI* (Moscow: MSTU MAMI)
[11] Virabov R V and Mamaev A N 1983 The Effect of Toroidal Elastic Wheels on Uneven Wear across the Width of the Treadmill *Proceedings of universities. Engineering* **9** 94–7
[12] Virabov R V and Mamaev A N 1987 To the Question of the Smallest Value of the Coefficient of Rolling Resistance of an Elastic Wheel on a Rigid Horizontal Surface *Proceedings of the universities. Engineering* **10** 85–8
[13] Mamaev A N 1982 Determination of the Tangential Elasticity Coefficient of a Wheel with a Toroidal Treadmill Shape *Izvestiya Vuzov. Engineering* **10** 80–6
[14] Mamaev A N, Vukolova G S and Dmitrieva L N 1999 The Influence of the Form of the Adopted Law of the Distribution of Normal Pressures in Contact of the Wheel with a Rigid Base on the Calculated *Collection of scientific papers dedicated to the 60th anniversary of the reconstruction of MAMI* (Moscow: Publishing. MSTU MAMI) pp 112–8
[15] Virabov R V, Mamaev A N, Marinkin A P and Yuryev Yu M 1986 The Influence of the Rolling Mode of the Elastic Wheel on the Value of Lateral Force during Lateral Withdrawal *Bulletin of mechanical engineering* **1** 33–5
[16] Virabov R V and Mamaev A N 1980 Analysis of Power Relations when Rolling a Driven Elastic Cylindrical Wheel along a Curved *Path Mechanics of machines* **57** 105–12
[17] Virabov R V and Mamaev A N 1980 Investigation of Contact Phenomena during Curvilinear Rolling of a Toroidal Wheel *Proceedings of the universities. Engineering* **2** 33–8
[18] Virabov R V and Mamaev A N 1980 Determination of Forces and Moments Acting on a Toroidal Wheel during Curvilinear Rolling *Proceedings of the universities. Engineering* 3 30–4
[19] Mamaev A N, Sazanov I V and Nazarov Yu P 1990 Determination of the Power Characteristics of an Elastic Wheel during Rolling by Withdrawal along a Curved Path *Problems of tires and rubber-cord materials. Strength and Durability: Materials of the II All-Union Symposium* (Moscow: NIISHP) 50–6
[20] Mamaev A N 1999 The Influence of Tire Operating Conditions on the Wear of Their Treadmill *Truck* **9** 12–4
[21] Mamaev A N 1982 The Influence of the Design Parameters of Elastic Wheels on the Magnitude of their Deflection and the Dimensions of the Contact Area with a Rigid Base *Safety and*

reliability of the car: Interuniversity. Sat scientific works 203–11
[22] Mamaev A N and Alepin E A 1980 Determination of the Dimensions of the Contact Area and Deflection of the Wheel with a Rubber Tire with Static Pressing of the Wheel to a Rigid Base *Machine Science: Sat scientific works* **251** 82–5
[23] Mamaev A N 1982 On the Determination of the Hysteresis Loss Coefficient of Highly Elastic Rolling Elements *NIINavtoprom* **779**
[24] Balabina T A, Mamaev A N and Chepurnoy S I 2013 Determination of the Ratio of Camber and Toe Angles of Elastic Wheels, Providing the Least Rolling Resistance *News of MSTU "MAMI"* **1(15)**

IOP Conf. Series: Materials Science and Engineering **832** (2020) 012078 doi:10.1088/1757-899X/832/1/012078

Influence of Tire Tread Convexity on its Rolling with a Slip

M Yu Karelina[1], T A Balabina[1], T Yu Cherepnina[1], A N Mamaev[2] and A S Vorontsov[3]

[1]Moscow Automobile and Road State Technical University (MADI), 64, Leningradsky Ave., Moscow, 125319, Russia
[2]Moscow Polytechnic University, 38, B. Semyonovskaya Street, Moscow, 107023, Russia
[3]Yanka Kupala State University of Grodno, 22, Ozheshko str, Grodno, 230023, Belarus

E-mail: 2783974@gmail.com

Abstract. The tire tread (inflated) may have a convex (toroidal) shape. At present, there is practically no data on the influence of this structural parameter of the tire on its power and kinematic characteristics when rolling with a slip. The solution to this problem, given in this paper, is based on the consideration of contact phenomena in the friction pair "elastic wheel – rigid support surface". As a result, relationships that make it possible to determine the lateral force, the turning moment acting in the contact area, the friction loss in the contact area for wheels with different toroidalities as a function of the slip angle and the longitudinal tangential force realized in the contact area are obtained; conclusions about the influence of the convexity on the lateral force, the stabilizing moment and the power of friction losses in the contact area are drawn.

1. Introduction

Most modern wheeled-vehicle tires have a treadmill (tread) that is nearly cylindrical in shape. The contact patch of such tires with a solid supporting surface is close to rectangular. Along with these, there are tires in which (in the inflated state) the tread has a convex shape (toroidal).

As shown in [1], the toroidality (convexity) of the treadmill of the wheel affects the process of its linear rolling only with small longitudinal tangential forces acting in the contact area.

At the same time, the question of the influence of the toroidalness of the treadmill of an elastic wheel on its rolling with a slip and along a curved path remains unclear.

2. Problem Formulation

In this paper, we solve the problem of determining the power and kinematic characteristics of elastic wheels (which include tires) with a convex treadmill while rolling with a slip. That is, in the case when the velocity vector of the wheel axis makes an angle δ, called the slip angle, with the plane of rotation of the wheel, namely: relationships for finding the magnitude of the lateral force, the stabilizing moment and the friction-loss power in the contact area are derived.

3. Solution

To solve this problem, we will use the methodology and equations obtained earlier [1–7] for a wheel with a cylindrical treadmill (the shape of the contact patch is close to rectangular).

In order to simplify, we assume that the shape of the contact patch is close to elliptical (Figure 1), i.e., the coordinates of the contour line of the contact patch are related by:

$$\frac{a^2}{A^2}+\frac{y^2}{b^2}=1, \tag{1}$$

where A and b are the semi-axes of the ellipse, and the normal pressures in both the longitudinal and transverse sections of the contact patch are distributed according to a parabolic law [8]:

$$q_n = q_{n0}\left(1-\frac{x^2}{A^2}-\frac{y^2}{b^2}\right), \quad q_{n0}=\frac{2F_n}{\pi Ab}. \tag{2}$$

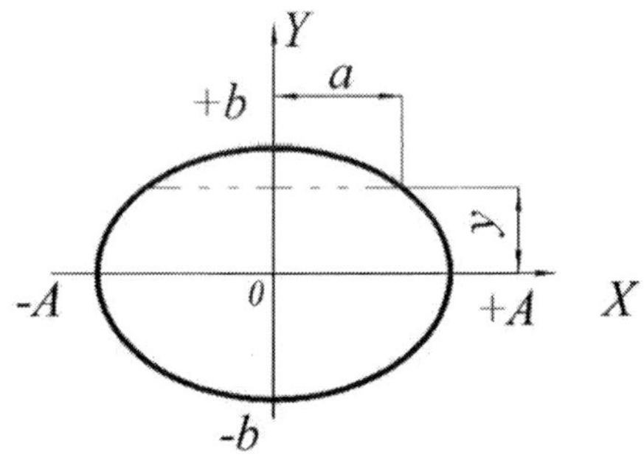

Figure 1. The shape of the contact patch of an elastic toroidal tire with a rigid supporting surface

As in the above studies, we believe that there are two zones in the contact patch – an adhesion section at the contact entry and a sliding section at the contact exit.

In the adhesion section, the longitudinal and transverse tangential displacements of the point of the wheel treadmill, due to the action of the longitudinal and transverse tangential forces, respectively, are defined as [5]:

$$u_x = \Delta V_x t = \xi_x(a-x), \tag{3}$$

$$u_y = \Delta V_y t = V_y t = (a-x)tg\delta \approx \delta(a-x). \tag{4}$$

The corresponding specific tangential forces will be equal to, respectively:

$$q_{tx} = \lambda_x u_x = \lambda_x \xi_x (a-x) \tag{5}$$

$$q_{ty} = \lambda_y u_y = \lambda_y \delta (a-x). \tag{6}$$

In these expressions, λ_x and λ_y are the tangential stiffness coefficients of the wheel in the longitudinal and transverse directions:

$$\xi_x = \frac{\Delta V_x}{V_x} = 1-\frac{r_{wh}^f}{r_{wh}} \tag{7}$$

relative loss of wheel speed:

$$r_{wh} = \frac{V_x}{\omega} \tag{8}$$

rolling radius:

$$r_{wh}^f = r_0 + \frac{a^2}{3r_0^2} \tag{9}$$

free rolling radius (when the longitudinal force in the contact area is zero).

In formula (5), the value of the relative loss of wheel speed is different in different longitudinal sections in the contact area, since in accordance with (7), the relative loss of speed depends on the radius r_{wh}^f, which is different for different longitudinal sections [1]. The tangential stiffness coefficient λ_x is the same for all longitudinal sections.

At some point of contact of the longitudinal section upon reaching the total tangential stress:

$$q_t = \sqrt{q_{tx}^2 + q_{ty}^2} = (a-x)\sqrt{\lambda_x^2\xi_x^2 + \lambda_y^2\delta^2} \tag{10}$$

grip limit value i.e. $|q_t| = \mu q_n$, a breakdown occurs and sliding begins. The coordinate of the boundary areas of adhesion and sliding in each of the longitudinal sections, taking into account expressions (2), can be found as:

$$x_0 = -a \pm \frac{A^2}{\mu q_{n0}} \sqrt{\lambda_x^2 \xi_x^2 + \lambda_y^2 \delta^2}. \tag{11}$$

The lateral force arising in the contact area can be calculated as:

$$F_y = \int_{-b}^{+b} t_y \, dy, \tag{12}$$

where

$$t_y = \int_{-a}^{x_0} q_{ty}^{sl} dx + \int_{x_0}^{+a} q_{ty} dx \tag{13}$$

linear lateral force; q_{ty}^{sl} is the transverse component of the tangential stress in the sliding section. To determine it, it should be taken into account that in the sliding section, the tangential stresses $q_{ty}^{sl} = \mu q_n$ coincide in the direction with the sliding velocity, and in this case:

$$q_{ty}^{sl} = \mu q_n \frac{V_y^{sl}}{V^{sl}}, \quad q_{tx}^{sl} = \mu q_n \frac{V_x^{sl}}{V^{sl}}. \tag{14}$$

To determine $V^{sl}, V_x^{sl}, V_y^{sl}$, we use the expression obtained in [3], which allows us to determine the average value of the longitudinal sliding velocity:

$$V_x^{sl} = \xi_x V_x, \quad V_y^{sl} = \delta V_x. \tag{15}$$

As a result:

$$q_{ty}^{sl} = \mu q_n \frac{\delta}{\sqrt{\xi_x^2 + \delta^2}}, \; q_{tx}^{sl} = \mu q_n \frac{\xi_x}{\sqrt{\xi_x^2 + \delta^2}}. \tag{16}$$

Taking into account the relationship obtained for q_{ty}^{sl}, and also taking into account the expression for q_{ty}, q_n and x_0 after some transformations, we get the following expression for linear lateral force:

$$t_y = \delta \left[\frac{\mu q_{n0}}{\sqrt{\xi_x^2 + \delta^2}} \cdot \frac{2a^3 + 3a^2 x_0 - x_0^3}{3A^2} + \frac{\lambda_y}{2} (a - x_0)^2 \right]. \tag{17}$$

Similarly, we find the longitudinal tangential force:

$$F_x = \int_{-b}^{+b} t_x dy, \tag{18}$$

where linear longitudinal shear force is expressed by the relationship:

$$t_x = \int_{-a}^{x_0} q_{tx}^{sl} dx + \int_{x_0}^{+a} q_{tx} dx = \xi_x \left[\frac{\mu q_{n0}}{\sqrt{\xi_x^2 + \delta^2}} \cdot\cdot \frac{2a^3 + 3a^2 x_0 - x_0^3}{3a^2} + \frac{\lambda_x}{2} (a - x_0)^2 \right]. \tag{19}$$

Due to the asymmetry of the distribution of longitudinal and transverse tangential stresses on the contact area, moments arise in the contact patch:

$$M_{tx} = \int_{-b}^{+b} t_x y dy,$$
$$M_{ty} = \int_{-b}^{+b} m_y dy, \tag{20}$$
$$M_z = M_{tx} + M_{ty},$$

where m_y is the elementary moment arising in the longitudinal section due to the asymmetry of the distribution of transverse tangential stresses:

$$m_y = \int_{-a}^{x_0} q_{ty}^{sl} x dx + \int_{x_0}^{+a} q_{ty}^{sl} x dx. \tag{21}$$

Applying the above formulas, for a wheel with a massive rubber tire (rubber layer 36 mm thick, diameter in the equatorial section $D = 2r = 260$ mm, wheel width $2B = 106$ mm, normal load $F_n = 3500$ N), figures 2–3 show the graphical relationships between F_y and M_z; illustrating the effects of wheel toroidality, slip angle and longitudinal traction force (traction coefficient $\psi_x = \frac{F_x}{\mu F_n}$).

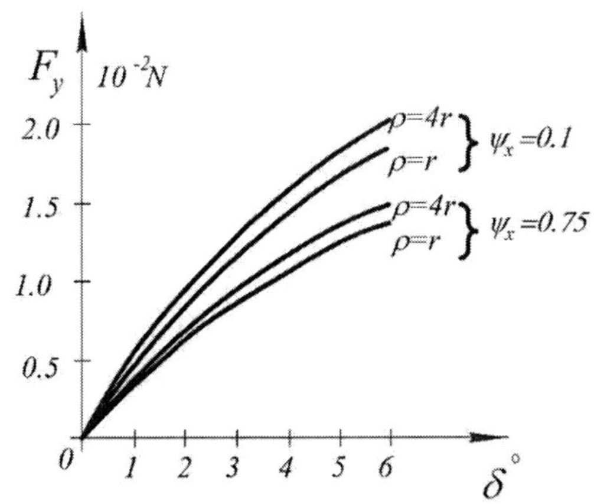

Figure 2. The relationship between the lateral force and the slip angle for various values of the curvature of the wheel treadmill and the traction coefficient (longitudinal force)

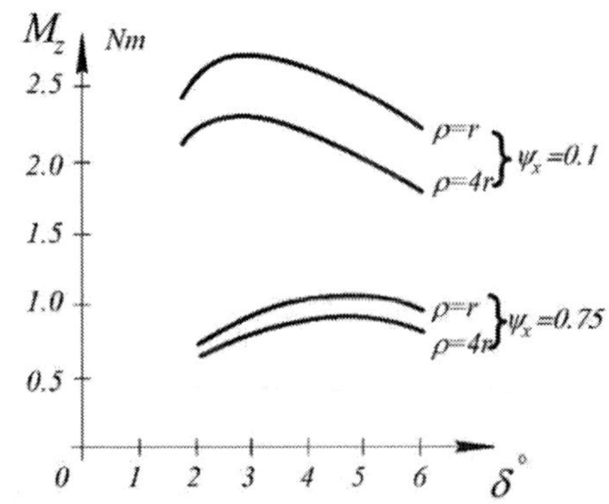

Figure 3. The relationship between the moment of rotation resistance and the slip angle for various values of the curvature of the wheel treadmill and the traction coefficient (longitudinal force)

To assess the wear intensity across the width of the treadmill of a toroidal elastic wheel rolling with a slip on a rigid base, we use the expression for the linear friction power in the contact area:

$$P_{fr}^{lin.} = \int_{-a}^{x_0} \mu q_n V^{sl} dx \int_{-a}^{x_0} \mu q_n V^{sl} dx = \frac{a - x_0}{a + x_0} \sqrt{\xi_x^2 + \delta^2} \left(a^2 x_0 - \frac{x_0^3}{3} + \frac{2}{3} a^3 \right). \tag{22}$$

The total friction power, which gives an integrated estimate of the treadmill wear intensity, in turn, can be found as:

$$P_{fr} = \int_{-b}^{+b} P_{fr}^{lin.} dy. \quad (23)$$

4. Results and Conclusions

The obtained relationships make it possible to calculate the lateral force and stabilizing moment acting in the contact patch of the toroidal wheel rolling with a slip, and the wear intensity of the wheel treadmill at various values of the slip angle δ, realized in the contact of the longitudinal tangential force F_x and the radius of curvature of the wheel ρ in the transverse plane. This, in turn, allows one to determine analytically the coefficient of lateral slip taking into account the acting longitudinal force F_x [9].

As seen in the above graphs, the influence of the curvature of the treadmill in the transverse direction on the magnitude of the lateral force is not so great. Along with this, the influence of this curvature on the value of the stabilizing moment at small longitudinal tangential forces in the contact area at $\rho < (2 \dots 3)r$ is evident.

For a longitudinally elongated contact patch $\left(\frac{\rho}{r} < 1\right)$, the lateral force is smaller and the stabilizing moment is greater than for a contact patch elongated in the transverse direction (all else being equal).

The power of friction losses in contact area during wheel rolling with a slip slightly depends on the toroidality of the wheel.

In further works [10–13] the authors also considered other questions of the mechanics of wheel rolling, including along a curved path.

References

[1] Virabov R V and Mamaev A N 1983 The effect of toroidal elastic wheels on the uneven wear along the width of the treadmill *Proc. Universities, Mechanical Engineering* **9** 94–7

[2] Virabov R V and Mamaev A N 1978 Determination of the power of friction losses in the contact area of the friction pair "wheel with pneumatic tire – rigid base" *Stepless adjustable gears: Proc. interuniversity* **3** 61–7

[3] Virabov R V and Mamaev A N 1980 Analysis of kinematic and power relationships when rolling an elastic wheel on a rigid base *Mechanics of machines* **57** 101–6

[4] Virabov R V, Mamaev A N and Balabina T A 2010 General issues of the interaction of an elastic wheel with a rigid supporting surface *Abstracts of the International Scientific and Practical Conference AAI "Automobile and Tractor Engineering"* (Moscow: MSTU "MAMI").

[5] Balabina T A and Mamaev A N 2014 The rolling mechanics of an elastic wheel on a rigid supporting surface *Collection: Engineering: trends, prospects and development technologies* 20–5

[6] Virabov R V and Mamaev A N 1987 To the question of the smallest value of the coefficient of rolling resistance of an elastic wheel on a rigid horizontal surface *Proc. Universities, Mechanical Engineering* **10** 85–8

[7] Mamaev A N 1982 Determination of the tangential elasticity coefficient of a wheel with a toroidal treadmill shape *Proc. Universities, Mechanical Engineering* **10** 80–6

[8] Mamaev A N, Vukolova G S and Dmitrieva L N 1999 The influence of the form of the adopted law of the distribution of normal pressures in the contact patch of the wheel with a rigid base on the calculated power and kinematic characteristics of the wheel *Collection of scientific papers dedicated to the 60th anniversary of the reconstruction of MAMI (*Moscow: Publ. MSTU MAMI) 112–8

[9] Virabov R V, Mamaev A N, Marinkin A P and Yuryev Yu M 1986 The influence of the rolling mode of the elastic wheel on the value of lateral force during lateral slip *Bulletin of mechanical engineering* **1** 33–35

[10] Virabov R V and Mamaev A N 1980 Analysis of power relations when rolling a driven elastic cylindrical wheel along a curved path *Mechanics of machines* **57** 105–12
[11] Virabov R V and Mamaev A N 1980 Determination of forces and moments acting on a toroidal wheel during curvilinear rolling *Proc. Universities. Engineering* **3** 30–4
[12] Mamaev A N, Sazanov I V and Nazarov Yu P 1990 Determination of the power characteristics of an elastic wheel when rolling by withdrawal along a curved path *Materials of the II All-Union Symposium "Problems of tires and rubber-cord materials. Strength and Durability"* (Moscow: NIISHP) 50–6
[13] Mamaev A N The influence of the working conditions of tires on the wear of their treadmill *Truck*, **9**

Influence of rolling condition and geometry of tire tread on its wear intensity

M Yu Karelina[1,3], T A Balabina[1], V V Filatov[1], A N Mamaev[2] and A S Vorontsov[3]

[1]Department of Machinery Parts and Theory of Mechanisms, Moscow Automobile and Road State Technical University (MADI), 64, Leningradsky Ave., Moscow, 125319, Russia
[2]Moscow Polytechnic University, 38, B. Semyonovskaya Street, Moscow, 107023, Russia
[3]Department of Materials Science and Resource-Saving Technologies, Yanka Kupala State University of Grodno, 22, Ozheshko str, Grodno, 230023, Belarus

E-mail: karelinamu@mail.ru

Abstract. Wheel tire wear annually takes hundreds of thousands of rubber tons. In this regard, the issue of reducing this negative phenomenon is urgent, for which it is necessary to know what and how affects tire wear. Factors affecting the wear of elastic wheels can be different: recipe of the rubber compound, the amount of traction (or braking) force in contact with the road, the type of movement is straight or cornering, the installation angles of the wheels (camber and toe-in angles), tire design, driving mode - uniform or with acceleration (braking). In this paper, we consider the influence on tire wear of such parameters as the value of the longitudinal force acting in the contact, the camber and toe-in angles, the radius of the motion trajectory and such a structural parameter as the convexity of the tread. When assessing the wear intensity, as is accepted by most researchers, we proceed from the fact that wear is proportional to the friction power in the contact of two interacting bodies.

1. Introduction

Wheel tire wear annually takes hundreds of thousands of rubber tons. In this regard, the issue of reducing this negative phenomenon is urgent, for which it is necessary to know what and how affects tire wear. Factors affecting the wear of elastic wheels can be different: recipe of the rubber compound, the amount of traction (or braking) force in contact with the road, the type of movement is straight or cornering, the installation angles of the wheels (camber and toe-in angles), tire design, driving mode - uniform or with acceleration (braking). Let us consider the influence on the tire wear of such parameters as the longitudinal force acting in the contact, the camber and toe-in angles, the radius of the motion trajectory and such a structural parameter as the convexity of the tread.

2. Problem statement

The solution of these problems is possible from consideration of the wheel interaction mechanics with the supporting surface. This was done, in particular, in [1–8], in which there are dependencies that allow one to determine the force and kinematic characteristics including the power of friction losses in the wheel contact with the supporting surface.

The wear rate is proportional to the work or power $P_{fr}=F\cdot\xi\cdot V$ [1, 2], spent on friction in the contact. Since the relative loss of speed (slip) of the wheel ξ depends on the tangential force F realized in the contact, the dependence $P_{fr}=f(F)$ has the form of a parabola of n-order. Moreover, with relatively small tangential forces ($F<$ (0.2 ... 0.3) μF_n, where μ – coefficient of sliding friction, F_n – normal load on the wheel), the friction loss power in the contact $P_{fr}\sim F^2$.

An increase in the saturation of the treadmill pattern increases its wear resistance both due to the larger abrasion area of the treadmill and due to a decrease in the tangential elasticity and a decrease in the relative loss of speed and power of friction losses in the contact, respectively.

At the same time, the question of the positive effect of reducing the tangential elasticity coefficient of the wheel is not so clear, since from the point of view of reducing the powertrain load when the axles of an all-wheel drive vehicle are blocked, it is desirable to have tires with greater tangential elasticity, which reduces the uneven distribution of tangential forces over the axles, and therefore and uneven tire wear on different axles.

Another structural parameter of elastic wheels that affects their wear (although to a much lesser extent) is convexity, i.e. curvature of the treadmill of the wheel in the transverse direction (toroidality).

The physical picture of the phenomena occurring during rolling of a toroidal wheel is described in [3]. It also gives the derivation of equations that quantitatively describe the force and kinematic dependences for toroidal wheels and, in particular, equations that allow one to calculate *the linear (i.e., in longitudinal section)* and *the total* power of friction losses in the contact, and, therefore, evaluate the unevenness of the wear rate using the width of the treadmill of the toroidal wheel and give an integrated assessment of the wear rate.

Figure 1 shows the distribution of linear friction power in contact across the width of the treadmill for a wheel with a massive rubber tire having a curvature radius of treadmill in the transverse plane $\rho=2r=260$ mm (the height of the rubber layer $h=24$ mm, width B = 47 mm, normal load F_n = 3000 N) for various braking torque.

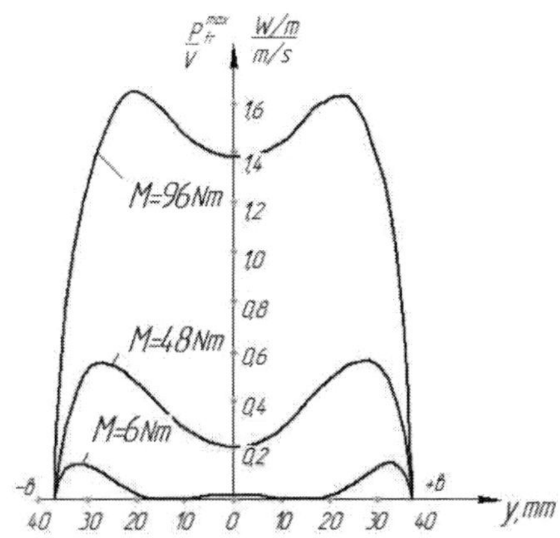

Figure 1. Distribution of linear friction power in contact across the width of the treadmill of the wheel with the convex shape of the treadmill

The decrease of the treadmill curvature radius ρ in the transverse plane increases the unevenness and intensity of wear along its width, especially when $\rho<$ (2 ... 3)r; the greatest wear of the driven wheel should take place near the edges of the contact spot, which leads to a decrease in the radius ρ, and, consequently, to an increase in the intensity of wear, i.e. to progressive treadmill wear. With an increase in the moment of rolling resistance (or torque), the distribution of the linear power of the friction losses in the contact in different longitudinal sections is equalized, and the total wear rate increases.

Figure 2 shows the dependence of the total frictional power loss in contact on the curvature radius of the elastic massive wheels in the transverse plane, and Figure 3 shows the influence ("significance") of the treadmill convexity on the total friction loss in the contact.

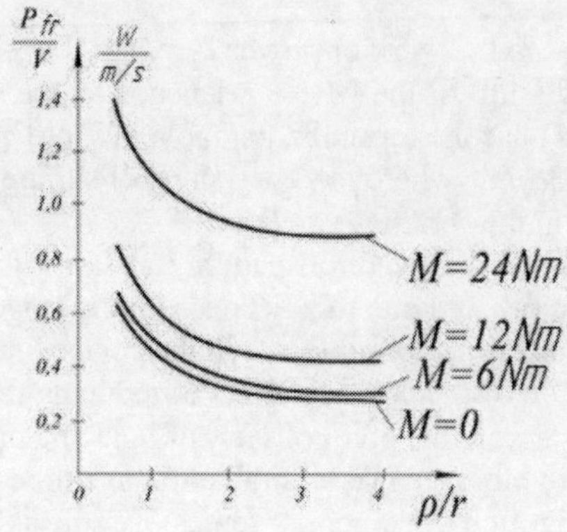

Figure 2. Dependence of the friction power losses in contact on the wheel toroidality (radius of wheel treadmill curvature in the transverse plane)

As follows from these figures, power losses increase with wheel treadmill curvature in the transverse direction increasing, and most intensively at $\rho < (2 \ldots 3)r$; with torque increasing (braking or traction) on the wheel. The fraction of friction losses in the contact due to the toroidality of the wheel decreases many times and is practically insignificant at high values of the torque.

Tire wear to a large extent depends on the amount of the angle of the drive with which the wheel rolls [4–6]. Its influence is illustrated by the graphical dependencies shown in Figures 4–5. The friction power losses in the contact of the wheel rolling with the wheel withdrawal increases sharply with an increase in the withdrawal angle; the type of this dependence has the form of a parabolic curve.

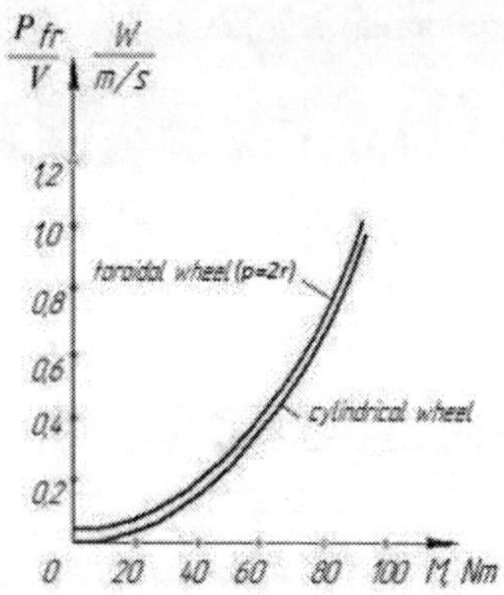

Figure 3. Dependence of the friction power losses in the contact on the transmitted moment for cylindrical and toroidal ($\rho = 2r$) wheels

If we compare the influence on the tire wear intensity of the camber and toe-in angles with which the wheel is mounted on the car, we can conclude that the role of the toe-in angle is many times greater than the influence of the camber angle.

Among the poorly studied is the question of the elastic wheel wear during its curvilinear movement, including rolling with a draw [7, 8].

Figure 5 shows the graphical dependences of the friction loss power in the contact characterizing the wear intensity on the withdrawal angle and the traction force realized during the curved motion of the wheel with a turning radius of one meter.

The performed studies show [9] that, with a decrease in the curvature radius of the trajectory, the wear intensity of the wheel increases, especially sharply at small radii. Moreover, the wear of the treadmill along its width is uneven: for driven and braking wheels, the outer edges (with respect to the center of rotation) are subjected to the greatest wear, while the leading ones have internal ones. However, this unevenness appears noticeably only at small turning radii and for wide wheels (and for twin wheels).

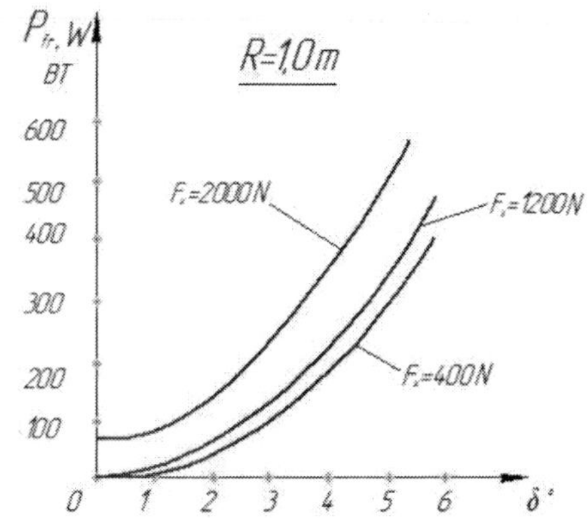

Figure 4. Dependence of the friction power losses in the contact on the withdrawal angle for various tangential forces in the contact

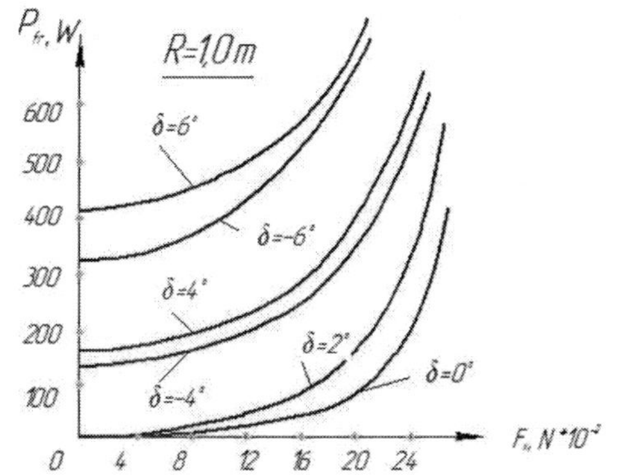

Figure 5. Dependence of the friction power losses in contact on the withdrawal angle and the realized traction force during curvilinear motion of the wheel with a turning radius of one meter

The rolling of a wheel with a draw along a curved trajectory increases the friction power losses in the contact, and its value depends on the sign of the withdrawal angle (Figure 5).

When rolling with the draw along both a curved and a straight trajectory, the friction power losses in the contact, and hence the wear intensity, does not depend much on the toroidality of the wheel.

3. Conclusion

The materials presented in this work allow us to more reasonably analyze the influence of rolling conditions and tire design on the performance indicators of both the tires themselves and the wheeled vehicle as a whole. This, in turn, improves the quality of tire and vehicle design.

References

[1] Virabov R V and Mamaev A N 1978 Determination of the friction power losses in the contact of the friction pair "wheel with pneumatic width – rigid base" *Interuniversity collection of scientific papers "Continuously variable gears"* (Yaroslavl) pp 61–7

[2] Virabov R V and Mamaev A N 1980 Analysis of kinematic and power relations when rolling a wheel on a rigid base *Machine mechanics "Science"* **57** pp 101–6

[3] Mamaev A N 1980 *Study of the influence of the geometry of the elastic wheel and the curvature of the trajectory of its movement on rolling resistance* Cand. dissertation (Moscow: MAMI) 204 p
[4] Balabina T A, Mamaev A N and Chepurnoy S I 2013 Determination of the camber and toe angles of elastic wheels providing the least rolling resistance *Tidings MSTU "MAMI"* **1(15)**
[5] Balabina T A, Mamaev A N and Chepurnoy S I 2014 Force characteristics of the wheel when rolling with camber and toe *Fundamental problems of technical sciences* Coll. of articles of the Int. sci.-pract. Conf. (Ufa, 19 February)
[6] Mamaev A N 1989 Features of elastic toroidal wheels rolling with the withdrawal *Problems of Tires and Rubber-cord Composites* Mater. First All-Union Conf. (Moscow, 17–19 October 1989) pp 22–8
[7] Virabov R V and Mamaev A N 1980 Analysis of force relations when rolling a cylindrical wheel along a curved trajectory *Machine mechanics "Science"* **57** 106–12
[8] Mamaev A N, Sazanov I V and Nazarov Yu P 1990 Determination of the power characteristics of an elastic wheel during rolling with withdrawal along a curved trajectory *Problems of Tires and Rubber-cord Materials. Strength and Durability* The Second All-Union Symp. Abstracts of reports (Moscow, 23–25 October 1990) pp 192–8
[9] Mamaev A N 1999 Influence of tire working conditions on the wear of their treadmill *Sci. J. "Truck"* **9**

Investigation of changes in the properties of engine oil depending on the sulfur content in gasoline

A Khaziev and A Laushkin

Moscow Automobile and Road Construction State Technical University (MADI), 64, Leningradsky ave., Moscow, 125319, Russia

E-mail: madi-chim@mail.ru

Abstract. In exploitation of modern cars, there are cases of power unit failure due to a sharp deterioration in the properties of engine oil and loss of performance of the lubricant. The article describes the causes of failures of the engines of modern cars associated with the loss of the resource of oil in the engine and the processes occurring in the lubricant. They are increase of kinematic viscosity, decrease of Total Base Number, increase of the acid number, decrease of flash point, increase of water content, decrease of additives content, increase of wear products. Decrease in properties and motor oils lifetime was caused primarily by the deviation of the physicochemical parameters of applied motor gasoline beside regulatory petrol. The process and chemical reaction of sulfurous acid formation during engine operation are described. We have established a pattern of changes in physico-chemical parameters of motor oils in automotive gasoline sulfur compounds. Made quantification is produced by the combustion of 1 kg of gasoline sulfur acid. The low values of the loss of the alkaline number of motor oil, caused by the low sulfur content of the ecological class 5 gasoline, are a prerequisite for an increase periodicity of oil change in the operation of modern cars.

Keywords: car, engine, motor oil, operating conditions, sulfur content, sulfur acid, sulfur dioxide, additive package, base number, the neutralization of the acid.

1. Introduction

In exploitation of modern cars, there are cases of power unit failure due to a sharp deterioration in the properties of engine oil and loss of performance of the lubricant. These processes occur in the normal operation of cars, most often in the winter at low ambient temperatures and are not accompanied by extreme loads on the vehicle's engine.

Tests of samples of motor oils, taken from the lubrication system of failed power units and tested in the MADI-CHIM laboratory, allow us to conclude that there are significant deviations of oil properties:

- increase of kinematic viscosity;
- decrease of TBN;
- increase acid number;
- decrease of flash point;
- increase of water content;
- decrease of additives content;
- increase of wear products.

Result of experimental studies of the MADI-CHIM test laboratory is decrease properties and motor oils lifetime was caused primarily by the deviation of the physicochemical parameters of applied motor gasoline beside regulatory petrol [1]. A significant increase in the viscosity of the engine oil, accompanied by a loss of fluidity of the lubricant in the positive temperature range is the main cause of oil starvation and a reduction (or loss) of the engine life and occurs due to the ingress of motor gasoline into the oil [2].

These processes are accompanied by a decrease of lubricant viscosity, oxidation of oil additives, the evaporation of gasoline from the oil, the accumulation of low-temperature sludge on engine parts, which is formed in conditions of low temperatures of the power unit in the interaction of crankcase gases containing gasoline, water etc. and oil residues rapid response of additives, deterioration of the physico-chemical properties of the lubricant, indicated by TBN decreasing and the growth of TAN, increased density, viscosity and freezing point of the engine oil.

TBN decreasing is usually associated with the ingress of sulfuric acid produced at the combustion of sulfur compounds contained in gasoline. Oil produced in the Russian Federation is rich in sulfur compounds. For example, sulfur content for the Volga and Ural fields reaches 6 % [3].

Most often sulfur is contained in oil and oil products in compounds [4]:

- elemental sulfur (*S*);
- hydrogen sulphide (H_2S);
- mercaptans (*R-SH*);
- sulfides (*R'-S-R*);
- disulphides (*R'-S-S-R*);
- thiophene derivatives (C_4H_4S);
- high-molecular sulfur compounds;
- complex compounds containing also oxygen, nitrogen, etc.

The presence of sulfur affects the quality of commercial petroleum products and requires additional technological operations to remove it from motor gasolines.

So, the combustion of sulfur contained in motor gasoline generates sulfur dioxide (SO2), which is very toxic, can cause human poisoning, affect the environment and lead to the formation of acid rain [5–8].

Technical regulations of the Customs Union TP TC 013/2011 "The requirements for automobile and aviation gasoline, diesel and marine gasoline, jet fuel and fuel oil" [9] establishes a gradual reduction of the sulfur content for motor gasoline. Since July 1, 2016 in the Russian Federation, there should be motor gasoline of ecological class K5 with sulfur concentration less than 10 ppm in circulation.

2. Formation of sulfuric acid during engine operation

In the operating of the internal combustion engine, its fuel cylinders are supplied with a certain ratio of automobile fuel and air. At the same time, a working mixture is formed which ignites from the spark of the spark plug and performs useful work during combustion.

Along with automobile gasoline, sulfur compounds are also burned in the engine.

The reaction of sulfur combustion is as follows:

$$S + O_2 = SO_2, \tag{1}$$

where S is a sulfur;

O_2 is oxygen;

SO_2 – sulfur oxide (sulfur dioxide, sulfurous anhydride) [10].

Let us calculate the amount of sulfur dioxide formed by burning 1 kg of motor gasoline of ecological class 5.

We will describe the molar masses of the compounds participating in the combustion reaction:

$$M_S = 32 \text{ g/mol},$$

$$MSO_2 = 32 + 16 \cdot 2 = 64 \text{ g/mol}.$$

Let's compose the proportion:

$$M_S = 32 \text{ g/mol} \; - \; M_{SO2} = 64 \text{ g/mol}$$

$$m_S = 10 \text{ mg} \quad - \quad m_{SO2} = X \text{ mg},$$

then X = (10 × 64)/32 = 20 mg, that is, when 1 kg of motor gasoline of ecological class 5 burns, 20 mg of SO_2 is formed.

Sulfuric anhydride, when reacted with water, forms sulphurous acid:

$$SO_2 + H_2O = H_2SO_3. \quad (2)$$

Taking into account the fact that both sulfuric anhydride and water compounds participate in the chemical reaction represented, water is formed in a large amount at burning gasoline [11], then we calculate the amount of acid formed as a result of combustion of gasoline by sulfur dioxide:

$$M_{SO2} = 64 \text{ g/mol} \quad - \quad M_{H2SO3} = 1 \cdot 2 + 32 + 16 \cdot 3 = 82 \text{ g/mol}$$

$$m_{SO2} = 20 \text{ mg} \quad - \quad m_{H2SO3} = Y \text{ mg}$$

$$Y = 20 \cdot 82/64 = 25.625 \text{ m}.$$

Thus, when burning 1 kg of automobile gasoline (ecological class 5), 25.625 mg of sulfuric acid is formed.

3. Neutralization of the acid formed during the combustion of motor gasoline

Sulfuric acid, formed during the combustion of gasoline, enters the engine oil and is neutralized with an additive package. Various additive packages will react with the acid in different ways. To assess the neutralizing ability of motor oils, the indicator "neutralization number" or otherwise "TBN" is used [12]. When it is determined in laboratory conditions, the oil is oxidized with hydrochloric acid HCl and the amount of acid spent on titration is recalculated into an equivalent alkali – KOH [13].

By neutralization reaction $H_2SO_3 + 2KOH = K_2SO_3 + 2H_2O$ we calculate the amount of equivalent alkali needed to neutralize 1 g of sulfuric acid:

$$M_{H2SO3} = 82 \text{ g/mol} \quad - \quad 2\,M_{KOH} = 2 \cdot (39 + 16 + 1) = 112 \text{ g/mol}$$

$$m_{H2SO3} = 1 \text{ g} \quad - \quad m_{KOH} = Z \text{ g},$$

then to neutralize 1 g of conditionally taken H_2SO_3 it will be required:

$$Z = 1 \cdot 112/82 = 1.366 \text{ g of potassium hydroxide.}$$

4. Neutralization of sulfuric acid during engine operation

Not all acid, which is formed during the combustion of motor gasoline, causes its oxidation. Most of it is taken out with the exhausted gases into the engine neutralization system. According to the results of previous studies at maximum engine revolutions under load, with a gasoline consumption of 20 liters per hour, 16 m^3 of exhaust gas are emitted during the combustion of one liter of gasoline. Breakdown of gases into the crankcase of the engine depends on many factors and is from 16 to 28 l/min [14]. As a percentage, this is 0.3...0.53 %. In the same ratio, sulfuric acid will enter the crankcase.

We calculate the amount of sulfuric acid that enters the engine oil when the engine is running. The consumption of gasoline with a modern car is 5l / 100 km. For the service interval, which is usually in the Russian Federation 15000 km, the car will consume 750 liters (≈555 kg) of gasoline of ecological class 5. In this case, 555·25.625 = 14221.9 mg or about 14.2 g of sulfuric acid will be released. And the crankcase will get: (14.2·(0.3...0.53))/100 = (0.0426...0.0753) g of sulfuric acid.

To neutralize sulfuric acid, it will be necessary to do the following: (0.0426 ... 0.0753) · 1.366 = (0.058 ... 0.103) g KOH.

In the car engine's carter, volume is 3.5 liters (≈3 kg) of motor oil, the total alkaline value is 8 mg KOH/1 g, then there will be 20,4 grams of equivalent alkali in the crankcase.

Under normal vehicle operating conditions (automobile gasoline corresponding to GOST 32513-2013 [1], engine oil recommended by the manufacturer, technically sound engine of the car, observance of service intervals of maintenance), only 0.3–0.5 % of the total alkaline number goes to neutralize the acid formed. Taking into account that mainly acidic detergents have the property of neutralizing acids, the resulting acid has practically no effect on the detergent and other properties of the motor oil [15].

5. Conclusions

The low values of the loss of the alkaline number of motor oil, caused by the low sulfur content of the ecological class 5 gasoline, are a prerequisite for an increase periodicity of oil change in the operation of modern cars.

Using of motor gasoline in cars that does not comply with the normative and technical document [1, 9] in terms of sulfur content accelerates the process of triggering additives and reduces the life of engine oil, which negatively affects engine reliability and its resource.

To ensure the declared services intervals for the lubricant, the operating conditions of the cars should be monitored: use automotive materials that meet the manufacturer's requirements automotive equipment, follow the instructions for maintenance intervals, monitor the level and condition of engine oil in the engine. In the event of a change in the operating conditions of the car, it is necessary to adjust the frequency of the engine oil change.

References

[1] GOST 32513-2013 *Motor fuels. Gasoline unleaded. Technical conditions* 2014 (Moscow: Standartinform)

[2] Khaziev A A, Laushkin A V and Gorina E B 2013 The reasons for changing the properties of engine oil *Grusovik* **6** 15–6

[3] Ogorodnikov S K 1978 *Petrochemist handbook* (Leningrad: Chimia)

[4] Kalechits I V 1973 *Cimestry of hydrogenation processes* (Moscow: Chimia)

[5] Tian H 2014 Advances in the study on endogenous sulfur dioxide in the cardiovascular system *Chin. Med. J.* **127(21)** 3803–7

[6] Liu D 2010 Sulfur dioxide: a novel gaseous signal in the regulation of cardiovascular functions *Mini-Reviews in Med. Chem.* **11** 1039–45

[7] Kowalski B I, Abasin D D and Petrov O N 2016 Method of Control the Effect of Temperature for the Oxidation Process of Partially Synthetic Motor Oils *Procedia Engineer.* **150** 480–5

[8] Izrael Yu A, Nazarov I M, Pressman A Ya et al 1989 *Acid rain* (Leningrad)

[9] TR CU 013/2011 2014 Technical regulations of the customs union *"Requirements for automobile and aviation gasoline, diesel and marine fuel, jet fuel and fuel oil"* (Moscow: Standartinform)

[10] Harlampidi H E 2000 Seraorganic compounds of oil, methods of purification and modification *Sorosovsky Educational magazine* **6(7)** 42–6

[11] Laushkin A V and Khaziev A A 2012 Causes of watering of motor oil in exploitation *Bull. of MADI* **1(28)** 63–7

[12] Vasileva L S 2004 *Automobile Exploitation Materials* (Moscow: Nauka-Press)

[13] GOST 11362-96 *Petroleum products and lubricants. Neutralization number, Potentiometric titration method* (Minsk)

[14] Kuznetsov E S et al 2001 *Technical exploitation of automobiles: a textbook for universities* ed. E S Kuznetsov 4rd ed Pererab. and additional (Moscow: Nauka)

[15] Laushkin A V and Khaziev A A 2014 Theoretical aspects of the change in the alkaline number of engine oil during the operation of the power plant *Coll. Sci. papers on the mater. of the int. sci. and pract. Conf. "Modernization and scientific research in the transport complex"* (Perm, PIDU) pp 140–4

IOP Conf. Series: Materials Science and Engineering **832** (2020) 012081 doi:10.1088/1757-899X/832/1/012081

Analysis of operating modes of a traction electric drive

A V Klimov[1], V V Filatov[1], I K Maslennikov [2], B K Ospanbekov[1] and V Ch Belash[3]

[1]Moscow Automobile and Road State Technical University (MADI), 64, Leningradsky Ave., Moscow, 125319, Russia
[2]FSUE The Central Scientific Research Automobile and Automative Engines Institute (NAMI), 2 Avtomotornaya St., Moscow, 125438 Russia
[3]Department of Materials Science and Resource-Saving Technologies, Yanka Kupala State University of Grodno, 22, Ozheshko str, Grodno, 230023, Belarus

E-mail: klimmanen@mail.ru

Abstract. Nowadays the modern society thinks more about the adverse impact of transport on the environment. Therefore development engineers and manufacturers of automobile and bus facilities have recently started to take more interest in creation of transport facilities with a reduced impact on the environment. These transport facilities comprise an energy storage unit as a traction battery that has a key impact on their cost and characteristics, such as like a carrying capacity and a distance on one charge. Development engineers must get all the necessary technical specifications of system components, as well as rapidly and flexibly evaluate an impact of their changes on service characteristics. It allows applying a simulation approach by using the Matlab/Simulink software suite while analysing motion modes under service conditions. This article contains the results of such an analysis for motion modes of passenger vehicles with high capacity depending on their routes. This analysis was executed by recording speed rates with subsequent statistical processing. It shows that average running speeds don't exceed 15 km/h. And the vehicle doesn't reach any maximum speed values. It has been established that the electric engine working points relate to the first zone of the external characteristic. The most typical speed values do not exceed 1/3 from the maximum speed, and the values of a torsion tractive effort torque and a brake torque are within the limits and do not exceed 35 from the maximum short-time torque.

1. Introduction

Stricter environmental regulations applicable to transport environmental impacts, as well as technology consumer requirements to reduce the total cost of ownershipб demand from development engineers to solve problems concerning energy efficiency and environmental compatibility of automotive vehicles. These problems can be solved by using drive gears and electric energy sources with higher efficiency as compared with the ones widely used here and now. This problem is particularly actual in the city area where there is a higher level of air pollution. The key issue of using such vehicles involves an efficiency of the conversion of energy to be stored in an energy storage unit which is very expensive and has no appropriate technical specifications contributing to widespread introduction of such vehicles. The cost for such an energy storage unit makes up to half the cost for the whole vehicle. At the same time in order to provide necessary operational kilometersб engineers have to increase their capacities which brings increase in weight and reduction in useful capacity of vehicles. Therefore, this is an actual problem to determine optimal technical specifications for traction electric engines contributing to reduction in energy consumption subject to service conditions.

2. Analysis of motion modes of transport facilities

It is necessary to determine the most characteristic operating modes for vehicles en route, but without determining any optimal technical specifications necessary to minimize energy consumption for vehicle moving. Therefore, it is necessary to analyze motion modes. Routes of passenger vehicles of the city of Moscow were analyzed for that purpose.

Motion processes on routes of passenger vehicles with high capacities have been chosen as a study object because today electric buses with increased ecological properties are being introduced in this segment. Table 1 shows their technical specifications.

Table 1. Technical specifications of transport facilities with high capacity

Spec	Identification	Value
Gross weight, kg	m_p	18000
Unladen weight, kg	m_s	12260
Height, m	H	3.26
Length, m	B	2.54
Tyres		225/70R22.5
Full speed, km/h	v	75

Time dependences of speeds on a number of intracity routes where electric buses had already worked or planned to work were received to determine operating characteristics by using GPS-tracker. Table 2 shows the results of statistical analysis.

Table 2. Essential operating characteristics

Route No.	V_{max}, km/h	V_{av}, km/h	S, km
107	51.99	14.7	15.73
14	51.05	16.56	16.8
17	49.88	11.66	22.84
22	44.99	13.36	28.6
24	54.2	15.49	12.15
34	44.35	11.09	10.06
36	40.48	13.7	24.92
41	46.92	14.15	16.18
42	45.04	12.83	9.61
73	56.1	13.9	22.5
76	63.49	12.66	26.61
83	64	13.17	8.39
832	49.5	12.13	20.32
908	39.29	10.44	18.7
Average	50.09	13.27	18.01

The maximum speed values in these routes did not exceed 50.09 km/h, the average operating speed was within 10–15 km/h.

The analysis of a vertical road alignment in the route shows that on the average, there are ascents of no more than 4 % and their maximum distance does not exceed 5% from the route length or no more than 0.9–1 km as a whole.

IOP Conf. Series: Materials Science and Engineering **832** (2020) 012081 doi:10.1088/1757-899X/832/1/012081

3. Simulation of operating modes for electric engines as an electric bus component

The review [1–6] shows that in order to analyze essential operating characteristics while moving with a reasonable degree of accuracy it is sufficient to apply a simulation model for a vehicle in-line motion to be described as follows (1)–(5):

$$mg\delta = F_{\text{T}} - F_k - F_w - F_\alpha \tag{1}$$

$$F_{\text{T}} = M_T i \eta r_k \tag{2}$$

$$F_k = mgf\cos\alpha \tag{3}$$

$$F_{\text{C.B}} = 0.5 c_x \rho_w S(v - v_w) \tag{4}$$

$$F_\alpha = mg\sin\alpha \tag{5}$$

where m is a vehicle weight, δ is a coefficient of rotating masses, F_{T} is a traction power, F_k is a rolling resistance force, F_w is an air resistance, F_α is a resistance force when moving upward, M_T is a torsion torque on the traction engine shaft, i is a shaft line transmission ratio, η is an efficiency factor of a wheel drive mechanical part, r_K is a rolling radius, f is a rolling resistance coefficient, α is a roadway slope, c_x is an air shape resistance coefficient, ρ_w is a density of air, S is a frontal area, v is a vehicle speed, v_w is a wind velocity.

The analysis of the most probable operating modes should be executed by using a simulation model of a vehicle because it allows taking into account both its aspects and those of service conditions. The simulation model is realized by using the MATLAB/Simulink math modeling (Figure 1).

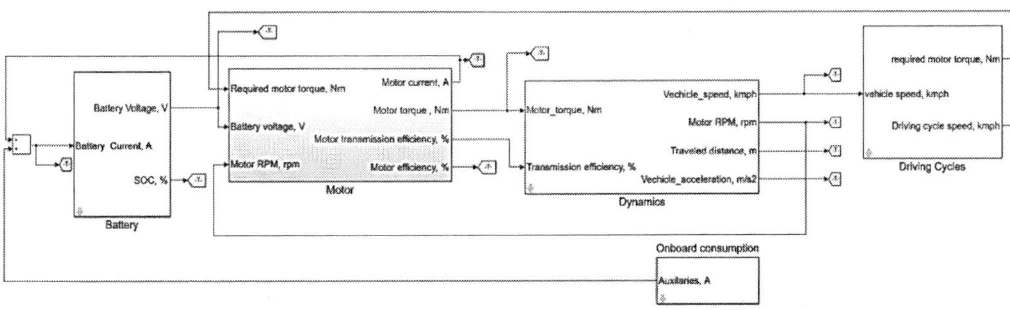

Figure 1. Simulation model to calculate operating characteristics of a traction voltage system of an electric bus

The simulation model includes units which describe processes in the motor, a battery, a unit describing driving cycles and a motion resistance force unit.

As far as this case doesn't require an analysis of transition processes in electrical circuits of a voltage changer and a drive motor. The simulation model of an electric traction drive is made on the basis of characteristics of a traction drive and $M_m = f(N_m)$dependences of an efficiency factor from a torsion torque and a rotational speed$\eta_m = f(M_m; N_m)$ and is described as follows (6, 7):

$$P_m = \omega_m M_m \tag{6}$$

$$P_e = \frac{P_m}{\eta_m}, \qquad P_e = P_m \eta_m \tag{7}$$

where P_m. P_e is a mechanical and electrical capacity of the electric engine, ω_m is a rotational speed of an electric engine rotor shaft, η_m is an efficiency factor.

The results which show the classification of the electric engine working points according to their characteristics received in the simulation modelling of a vehicle motion en routes and indicated in Table 1 in compliance with the received motion modes are presented in Figure 2.

IOP Conf. Series: Materials Science and Engineering **832** (2020) 012081 doi:10.1088/1757-899X/832/1/012081

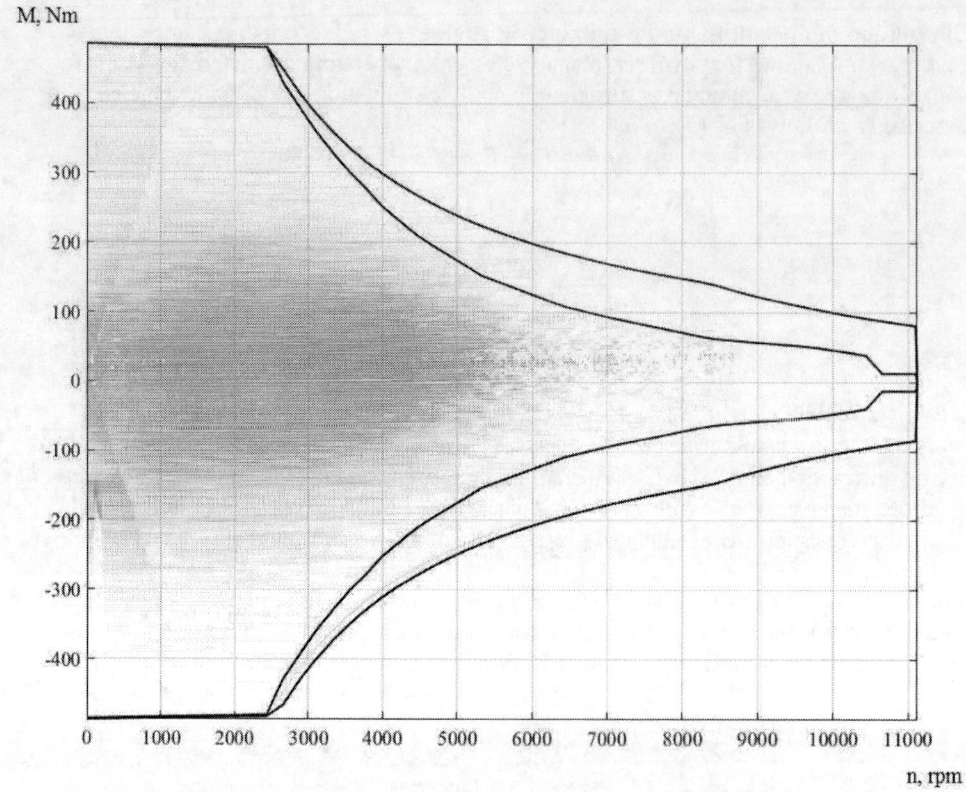

Figure 2. Classification of the electric engine working points

Based on the data received it can be seen that the electric engine working points relate to the first zone of the external characteristic. The most typical speed values do not exceed 1/3 from the maximum speed, and the values of a torsion tractive effort torque and a brake torque are within the limits and don't exceed 45 % from the maximum short-time torque. It is worthwhile to choose a nominal design mode to engineer an electric engine or to choose an engine corresponding to the indicated recommendations. A maximum efficiency coefficient should be reached at low rotor shaft speeds.

But if we analyze electrical processes in an electric engine which can be described by the equation (8), through the example of one phase [7, 8]:

$$U = \frac{d\psi}{dt} + iR \tag{8}$$

where U means a loading voltage, ψ means a phase magnetic linkage, t means time, i means a phase current, R means a phase resistance.

It is seen that high mechanical characteristics of an electric engine can be received when increasing a number of phase winding circuits, but in this case the phase winding resistance also increases; and the more the circuits are, the less the copper slot space factor becomes due to a great number of insulation. The phase resistance increases faster than a magnetic flux linkage derivative. The second method to increase a magnetic flux linkage in time is to reduce a polar pitch, but at the same time the number of slots for placing a phase winding grows and it results in a phase resistance increase. Both methods result in an increase in weight of active materials and it involves an increase in weight and dimensional characteristics of an electric engine and an increase in cost.

The third method involves an increase in a magnetic flux linkage derivative due to an increase in speed and a change of the magnetic field towards the stator. For that purpose it is necessary to use interim mechanical transmissions. This approach allows reducing weight of active materials, dimensions and cost.

4. Summary

The analysis of operating modes for the traction electric engine as a component of a passenger vehicle due to service conditions has shown that these operating modes don't exceed 45 % values from the maximum value according to the rotational rate and are within 35 % limits according to the torsion and brake torque. Development engineers who create traction electric engines should choose a nominal ultimate moment to reduce energy consumption when operating within stated limits. In order to reduce cost and weight and dimensional characteristics it is needed to increase the speed for the magnetic field change in an electric engine by using mechanical transmissions.

References

[1] Litvinov A S and Farobin Ya E 1989 Automobile: Theory of operating characteristics (Moscow: Mashinostroenie Publ.)
[2] Maamria D, Gillet K, Colin G, Chamaillard Y and Nouillant C 2016 On the use of Dynamic Programming in eco-driving cycle computation for electric vehicles *2016 IEEE Conference on Control Applications* (Buenos Aires: IEEE) pp 1288–93
[3] Jiquan Wang 2016 *Battery electric vehicle energy consumption modelling. Testing and prediction: a practical case study* (Eindhoven: Technische Universiteit Eindhoven)
[4] Rios-Torres J, Sauras-Perez P, Alfaro R, Taiber J et al 2015 Eco-Driving System for Energy Efficient Driving of an Electric Bus *SAE International Journal of Passenger Cars – Electronic and Electrical Systems* **8(1)** 79–89
[5] Kang H, Shen A and Sarker 2017 Velocity optimization of pure electric vehicles with traffic dynamics consideration *IEEE 37th International Conference on Distributed Computing Systems* (Atlanta: IEEE) pp 2206–12
[6] Lin X, Gorges D and Liu S 2014 Eco-driving assistance system for electric vehicles based on speed profile optimization *2014 IEEE Conference on Control Applications* (Juan Les Antibes: IEEE) pp 629–34
[7] Vinogradov N V, Goryainov F A and Sergeev P S 1969 *Engineering of electrical machines* (Moscow: Energia)
[8] Ivanov-Smolensky A V 1969 *Electromagnetic fields and processes in electrical machines and their physical modelling* (Moscow: Energia)

Features of calculation and design of pavement enduring prolonged static load

A V Korochkin

Moscow Automobile and Road State Technical University (MADI), 64, Leningradsky ave., Moscow, 125319, Russia

E-mail: andrey_korochkin@mail.ru

Abstract. The main calculation in the design of pavement is the calculation of the dynamic load. Calculation of static load is recommended in some cases specified in regulatory documents. It is necessary to prove that the calculation of the static load is required for any calculations of pavement. Using the calculation of pavement as an example, it is proved that a design that can withstand dynamic load does not always withstand static load. For the first time, a recommendation is given on the need to calculate the static load for any calculation of pavement, and not only in the cases indicated in the regulatory documents. It is proved that calculations for static load must be performed for any calculations of pavement. It is noted that different regulatory documents give different parameters for the calculation; therefore, before the start of the calculation, it is recommended to decide on which regulatory documents the calculation will be made.

Keywords: pavement, static load, standard load, dynamic coefficient

1. Introduction

In the process of a car moving along the road, an interaction occurs between the wheel of the car and pavement [2]. Moreover, depending on the road conditions and driving mode, this interaction has differences associated primarily with such parameters as:

- the nature of the load that the car has on the cover;
- load application time;
- frequency of load application.

The magnitude of the load depends on the carrying capacity of the car, the number of axles and their layout and is characterized by specific pressure in the area of contact of the car wheel with the pavement [8].

The load application time depends on the speed of the car on the road. The frequency of application of the load is directly related to the intensity of road traffic and its distribution by the hours of the day [13]. These parameters are taken into account when designing and calculating pavement.

Pavements on the stretches of roads are calculated for short-term repeatedly applied (dynamic) load [10]. At the same time, quite often a single calculation for a short-term load is not enough and a check for long-term (with a loading duration of at least 10 minutes) is required. Moreover, in some cases, the calculation of pavement is carried out only for long-term (static) load [3].

According to existing norms and rules, the calculation for long-term load is the main one for calculating pavement for:

- sidewalks, pedestrian streets, park alleys, squares of cities, etc.;
- car parking in parking lots, roadside service areas, rest areas, etc.

Additionally, it is necessary to check pavement for static load when designing:

- transitional high-speed lanes and ramps at intersections at the same and different levels;
- on roadsides and stopping strips;
- at public transport stops;
- at collection points in the case of toll roads;
- at regulated intersections;
- level crossings, etc.

Figure 1. Traffic jams on the waterfront

Testing for the long-term impact of the load is especially relevant when designing streets and city roads, where there are a large number of public transport stops, there are pedestrian walkways and pedestrian streets, and intra-quarter streets and driveways are often used as temporary and sometimes permanent parking for vehicles [6]. And where the phenomenon of traffic jams is widespread, cars are idle for much more than 10 minutes.

This article discusses the features of the calculation and design of non-rigid pavement under short-term and long-term loading and the effect of static load on the structure of pavement. The aim of the study is to prove the necessity of calculating the static load in all cases without exception when designing pavements.

2. General design characteristics of the impact of the car on pavement and the procedure for calculating non-rigid pavements depending on the duration of the load

In the General case, the calculation of pavement is carried out on the load determined by regulatory documents in force in the Russian Federation [1].

IOP Conf. Series: Materials Science and Engineering **832** (2020) 012082 doi:10.1088/1757-899X/832/1/012082

The magnitude of the regular load is determined by GOST 32960-2014, GOST R 52748-2007 and SP 34.13330.2012.

Depending on the category of the road and the type of coating on it (capital, lightweight, etc.), it is 115 or 110 kN per vehicle axis. When calculating pavement, the pressure of one wheel on the coating is uniformly distributed over the area of the wheel imprint (*D*). Moreover, it should be remembered that the diameters of the circles are equal to the track imprint of the wheel (*D*) in the static and dynamic positions are not equal [12]. The value of the parameter *D* varies depending on the regulatory document.

So, according to GOST GOST R 52748-2007 in the static position D = 0.34 m, in the dynamic D = 0.39 m, regardless of the category of the road, ODM 218.046-01 recommends the following *D* values, depending on the standard load:

- at a load of 100 kN for a fixed wheel D = 33 cm, for a moving wheel D = 37 cm;
- at a load of 110 kN for a fixed wheel D = 34 cm, for a moving wheel D = 39 cm;
- at a load of 130 kN for a fixed wheel D = 37 cm, for a moving wheel D = 42 cm.

GOST 32960-2014 does not regulate the diameter of the imprint of the wheel (D), but recommends the value of the pressure of the wheel on the coating (*p*): 0.8 MPa calculation of pavement of capital type; 0.6 MPa for the calculation of lightweight pavement and transitional types.

PNST 265-2018 offers the following regulatory parameters:

The standard pressure *p*:

- 0.8 MPa – for roads with capital pavements;
- 0.6 MPa – for roads with lightweight and transitional pavements.

And the standard load P_{st}:

Table 1. Standard loads

Load calculation group	Standard static load P_{st}, kN		Calculated load parameters	
	Axle	Wheel	Pavement pressure *p*, MPa	Diameter D_d/D_{st} , cm
A-10	100	50,0	0,6	37/33
A-11,5	115	57,5	0,8	34/30

Note – The numerator shows the diameter D_d – for a moving wheel, in the denominator D_{st} – for a fixed wheel.

In the general case, the diameters of the prints of the fixed wheel D_{st} and the moving wheel. D_d is determined by the following formulas:

$$D_{st} = \sqrt{\frac{40P_{st}}{\pi p}};$$

$$D_d = \sqrt{k_{dyn}}D_{st},$$

where k_{dyn} is the dynamic coefficient.

Thus, when designing and calculating pavements for short-term multiple loads, parameters other than those used to calculate long-term one-time loads are used. Also, the calculations of pavement for long and short-term load also differ.

Currently, there are several methods for calculating non-rigid pavement type. The most commonly used calculation is according to ODN 218.046-01 and PNST 265-2018. The calculation principles are the same in both of these documents, some calculated coefficients and individual parameters differ.

According to ODN 218.046-01 and PNST 265-2018, the calculation of the strength of non-rigid road pavements on road sections with short-term load is carried out according to the following criteria: elastic deflection, shear resistance of molded layers, resistance of monolithic layers to bending fatigue stress. In addition, if necessary, a check for frost resistance is performed.

In this case, the strength and deformation characteristics of materials and soils are used with repeated application of a load of 0.1 s duration.

Road clothes at stops and stopping lanes located on roadsides, intersections, at ramps and approaches to intersections with railroad tracks should be additionally checked for a single load lasting at least 10 minutes using static values of the calculated parameters [4].

For pavements of reinforced roadside lanes, parking lots, as well as sidewalks, pedestrian streets and squares in cities, the calculation is performed only for a single load lasting at least 10 minutes. The calculation is performed according to the condition of shear stability.

If one or several calculations for dynamic and static loads do not satisfy the required conditions, adjustments are made to the pavement design until all conditions are met [15].

3. An example of the calculation of pavement for short and long loads

We shall consider the features of the calculation and design of pavement for short and long-term loading and shall also analyze the effect of static load on the structure of pavement. As an example, we take the calculation of pavement for a highway of the second category in the Krasnodar Territory. We will calculate according to PNST 265-2018.

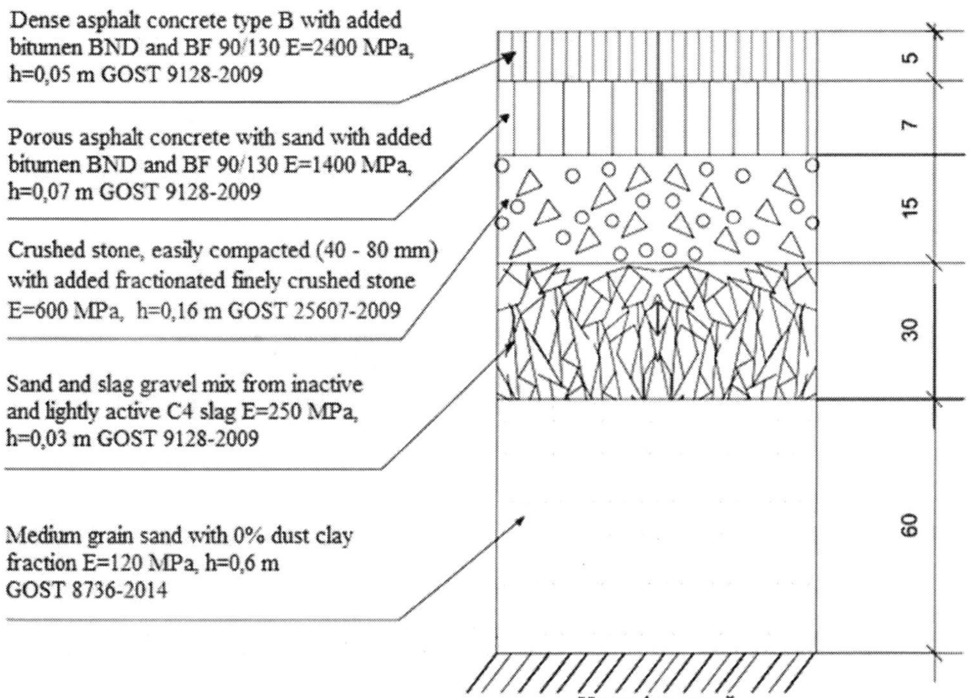

Figure 2. Road surface elements

Initial data for the calculation of pavement
Design area – Krasnodar Territory;
Category of the projected road – II;
Road climatic zone – IV;
Type of terrain for humidification – 1;
The specified reliability – K_r = 0.95 (by agreement with the customer);
Type of pavement – capital;
Soil of a working layer of a subgrade – Soil heavy loam;
Groundwater level, counting from the bottom of the pavement – 10 m;
Subgrade compaction coefficient K_{com} = 1.01 – 0.98.

IOP Conf. Series: Materials Science and Engineering **832** (2020) 012082 doi:10.1088/1757-899X/832/1/012082

Determination of the total estimated number of applications of the estimated load for the service life.
Estimated load – A11.5 (at $P = 0.8$ MPa);
The service life of pavement, years – 12;
Design Load Parameters:
Wheel load $Q = 57.5$ kN;
Tire pressure $P = 0.8$ MPa;
Diameter of wheel imprint $D_d = 34$ cm;
Diameter of wheel imprint from static loading $D_{st} = 30$ cm.
The total estimated number of load applications is set initially and is 91964 aW.

The estimated shear characteristics (elastic modulus and shear characteristics) of soil and sand are shown in table 2:

Table 2. Calculated shear characteristics

Layer material	E, MPa	Angle of internal friction, degrees	**Angle of internal friction (static), degrees**	Coupling, MPa	**Coupling (static), MPa**
Sand of medium grain with a dusty clay fraction of 0 %	120	28.1786	**32**	0.003	**0.004**
Heavy loamy soil	46.8554	7.7126	**19.9518**	0.0098	**0.0223**

The calculated characteristics of the layers of pavement are shown in table 3:

Table 3. Design characteristics of pavement layers

Layer material	Thick ness, cm	Elastic modulus of elastic deflection, MPa	Shear modulus, MPa	Bending modulus, MPa	Bending tensile strength R_0, MPa	m	α
Dense asphalt concrete type B with added bitumen 90/130	5	2400	550	3600	9.5	5	6.3
Porous asphalt concrete with sand with added bitumen 90/130	7	1400	380	2200	7.8	4	7.6
Crushed stone, easily compacted (40–80 mm) with added fractionated finely crushed stone	15	450	450	450	–	–	–
Sand and slag gravel mix from inactive and lightly active C4 slag (40mm maximum grain)	30	250	250	250	–	–	–
Sand of medium grain with a dusty clay fraction of 0 %	60	120	120	120	–	–	–

The total thickness of the pavement is 117 cm.
Calculation of frost resistance.
The design is considered frost-resistant if the condition is met:

$$l_b \leq l_{don}$$

The calculation is made using the nomogram. With a pavement thickness of 117 cm, this condition is met. Frost resistance is provided.

Calculation of elastic deflection. Strength criterion has the form:

$$E_{rev} > E_{min} K_{pr}^{res}$$

where E_{rev} is the total calculated modulus of elasticity of the structure, determined by the nomogram;

E_{min} – the minimum required modulus of elasticity of the structure, determined by the empirical formula;

K_{pr}^{res} – the required coefficient of strength according to the criterion of elastic deflection (for roads of the II category $K_{pr}^{res} = 1,2$);

$$E_{min} = \sqrt{\frac{p}{0,6}} 98,65[lg \sum N_p - c]$$

where p is the calculated pressure on the coating;

$\sum N_p$ – the number of applications of the estimated lane load for the standard service life;

c is an empirical parameter equal to the calculated load: A-10 – 3.55; A-11.5 – 3.20.

As a result of the calculation, we obtain:

$E_{min} = 325\ MPa.$

$E_{rev} = 394{,}25\ MPa.$

The structural strength coefficient obtained by calculation is equal to:

$$K_{pr} = \frac{E_{rev}}{E_{min}} = \frac{394{,}25}{325} = 1{,}21$$

The required strength factor K_{pr}^{res} is 1.2.

Durability is ensured.

Calculation of the design of non-rigid pavement under the condition of shear stability of underlying soil and loosely coupled structural layers.

Inadmissible shear deformations in the structure will not accumulate if the condition is provided in the soil of the subgrade and in loosely coupled (sandy) layers:

$$T \leq \frac{T_{pr}}{K_{pr}^{res}}$$

where T is the active shear stress from the current short-term or long-term load;

T_{pr} – ultimate shear stress, exceeding which causes a violation of shear strength;

K_{pr}^{res} – the required coefficient of strength (for roads of the II category $K_{pr}^{res} = 1{,}0$);

The calculation is carried out separately on the ground (heavy loam) and on the sand

Active shear stresses are determined by the nomogram. The calculated elastic moduli for the layers of the pavement are assigned from the condition of short-term loading.

According to the nomogram and formulas, we determine the calculated coefficient of strength K_{pr}.

As a result of calculating the shear in the soil, we obtain $K_{pr} = 1.41$

Required strength factor $K_{pr}^{res} = 1.0$

Durability is ensured.

When calculating the shear in the sand, we obtain $K_{pr} = 1.1$

Required strength factor $K_{pr}^{res} = 1.0$

Durability is ensured.

Design calculation for the resistance of monolithic layers to fatigue fracture from tensile bending.

The calculation is based on the condition:

$$\sigma_r \leq \frac{R_N}{K_{pr}^{res}}$$

where K_{pr}^{res} – the required coefficient of strength taking into account the given reliability (for roads of the II category $K_{pr}^{res} = 1.0$);

IOP Conf. Series: Materials Science and Engineering **832** (2020) 012082 doi:10.1088/1757-899X/832/1/012082

R_N is the tensile strength of the layer material in bending taking into account fatigue phenomena;

σ_r is the largest tensile stress in the monolithic layer.

σ_r is determined using the nomogram, leading the real construction to a two-layer model. All asphalt concrete layers are referred to the upper layer of the model.

As a result of the calculation, we obtain:

R_N = 2.26 MPa;

σ_r = 1.0676 MPa

The structural strength coefficient obtained by calculation is equal to:

$$K_{pr} = \frac{R_N}{\sigma_r} = \frac{2,26}{1,0676} = 2,12$$

The required strength factor is 1.

Durability is ensured.

Thus, the presented pavement meets all calculation conditions when exposed to a short-term (dynamic) repeating load.

Now we check the same pavement for the perception of a long load. The initial parameters for the calculation are the same, but for a static state.

Design characteristics of pavement layers.

The calculated characteristics for soil, sand, crushed stone and slag-crushed stone bases remain the same as for dynamic calculation. The calculated characteristics of monolithic layers of pavement for static calculation are shown in Table 4.

Table 4. Design characteristics of monolithic layers of pavement for static calculation

Layer material	Thickness, cm	Shear modulus, MPa
Dense asphalt concrete type B with added bitumen 90/130	5	300
Porous asphalt concrete with sand with added bitumen 90/130	7	280

The total thickness of the pavement is 117 cm.

Calculation of the long-term load on shear stability in weakly connected layers.

The calculation of shear stability under static load is carried out as well as for dynamic, but taking into account static parameters.

Accounting for shear resistance is carried out under the condition:

$$T \leq \frac{T_{pr}}{K_{pr}^{res}}$$

where T is the active shear stress from the current short-term or long-term load;

T_{pr} – ultimate shear stress, exceeding which causes a violation of shear strength;

K_{pr}^{res} – the required coefficient of strength (for roads of the II category K_{pr}^{res} = 1,0);

The calculation is carried out separately on the ground (heavy loam) and on the sand

Active shear stresses are determined by the nomogram. The calculated elastic moduli for the layers of the pavement are assigned from the condition of short-term loading.

According to the nomogram and formulas, we determine the calculated coefficient of strength K_{pr}

As a result of calculating the shear in the soil, we obtain K_{pr} = 2,97

Required strength factor K_{pr}^{res} = 1,0

Durability provided.

When calculating the shear in the sand, we obtain K_{pr} = 0,82

Required strength factor K_{pr}^{res} = 1,0

Durability is not ensured.

Thus, the calculation clearly shows that the pavement fully meets the regulatory conditions when exposed to short-term (dynamic) loads, when exposed to a static load, it is not able to provide the required strength. This design of pavement requires adjustment, after which it must again be calculated before ensuring strength according to all criteria [5].

4. The need to calculate pavement for static load

The current regulatory documents (ODN 218.046-01, PNST 265-2018) indicate the requirements according to which it is necessary to check the design of pavement for long-term (static) load. It also lists the cases in which such a check is necessary. However, one should not forget that the probability of stopping vehicles for a long period of time (more than 10 minutes) exists not only in the cases specified in ODN 218.046-01 and PNST 265-2018.

This primarily concerns city streets and roads where there are intersections with traffic lights, the carriageway is often used for stops and parking, and the phenomenon of traffic jams is also widespread [9]. That is, in the city there are practically no places where the car for one reason or another could not be stopped for 10 minutes or more.

Therefore, in urban conditions, the calculation of the static load is required.

Also, carriageways of suburban highways should not be discounted, which are not as much as in cities, but also have a static effect on the carriageway [14]. We consider several such cases:

1. Traffic accidents.

In this case, whole traffic flows stop and the static effect on the coating is quite significant [7].

2. Traffic jams.

It is no secret that traffic jams are not a rare occurrence on suburban highways, especially during the summer season, and in "narrow" sections cars stand idle for much more than 10 minutes.

3. Low speed in the territories of industrial enterprises, shopping centers, parking lots, etc. Calculation of short-term load is made from the condition that the time the impact of the wheel on the road is 0.1 sec. However, at low speed this condition is not satisfied. Therefore, a check for static impact in this case is necessary [11].

5. Findings

Currently, there are several methods for calculating non-rigid structures of pavement. The main, most popular of them are:

- Calculation methodology according to ODN 218.046-01 "Designing non-rigid road pavement", approved and enforced by the Order of the State Road Service "Rosavtodor" of the Ministry of Transport of the Russian Federation dated December 20, 2000 No. OC-35-P.
- Calculation procedure according to PNST 265-2018 "General automobile roads. Designing non-rigid pavements", approved by the order of Rosstandart of 04/11/2018 No. 3-PNST.

In addition to these documents for the design and calculation of pavement, the data of other applicable regulatory documents are used: GOST R 52748-2007 "Regulatory loads, design load schemes and approximation dimensions"; GOST 32960-2014 "General automobile roads. Regulatory loads, design load schemes", SP 34.13330.2012 "Roads", etc. All of them have the status of "active" and are actively used in the Russian Federation for the design of roads. Therefore, before starting the design and calculation of pavement, it is necessary to decide what regulatory documents to use in specific conditions.

An example of calculating pavement on the dynamic and static effect of a car's wheel on a coating showed that pavement that fully meets regulatory conditions when exposed to a short-term (dynamic) load is not able to provide the required strength when exposed to a static load.

An analysis of the current conditions of road transport in urban conditions and on country roads convincingly proves that the probability of stopping transport on any part of the road is very high, and therefore the statistical effect on the coating of a car's wheel cannot be discounted.

Thus, in order to increase the service life of road pavement, the calculation for long loads must be performed under any traffic conditions of vehicles, not limited to the recommendations of regulatory documents.

References

[1] Apestin V K, Shak A M and Yakovlev Yu M 1977 *Testing and evaluation of the strength of non-rigid road clothes* (Moscow: Transport) 102 p

IOP Conf. Series: Materials Science and Engineering **832** (2020) 012082 doi:10.1088/1757-899X/832/1/012082

[2] Babkov V F A review of experimental work on measuring stress in the ground *Studies of methods for calculating the thickness of road surfaces*: *Proceedings of DorNII. Edition of Gushosdor* vol I pp 156–231

[3] Vasiliev A P 2004 Reference encyclopedia of the road builder *Repair and maintenance of roads in 2 volumes.* Moscow

[4] Vasiliev A P 2010 *Operation of highways:* in 2 volumes, vol 1 (Moscow: Publ. Center "Academy") 320 p

[5] Ivanov N N 1973 *Design and calculation of non-rigid pavement* (Moscow: Transport) 328 p

[6] Korochkin A V 2012 The study of hard road pavement with asphalt concrete pavement in urban conditions *Transport construction* **7** 8–11

[7] Korochkin A V 2011 Calculation of rigid pavement taking into account the impact of a moving vehicle *Science and technology in the road industry* **2** 8–10

[8] Korochkin A V 2011 Studying the impact of a moving vehicle on the construction of pavement *Building Materials* **1** 28–9

[9] Korochkin A V 2015 On the dynamic effect of a vehicle on the construction of road clothes *Roads and Bridges* **33/1** 30–3

[10] Krasikov O A 2006 *Strength assessment and calculation of reinforcement of non-rigid pavement* (Almaty: KazgosINTI) 308 p

[11] Nemchinov M V 1985 *Coupling qualities of road surfaces and traffic safety of cars* (Moscow: Transport) 231 p

[12] Silyanov V V and Domke E R 2008 Transport and operational qualities of highways and city streets: a textbook for students. higher textbook. institutions (Moscow: Publishing Center "Academy") 352 p

[13] Savchenko E T and Maksin M O 2016 Analysis of the feasibility of the construction of asphalt concrete and cement concrete road pavements *Young Scientis* **21** 204–7

[14] Radovsky B S and Ilyev E B 1976 Modern trends in the design of pavement on roads with heavy traffic *Construction and operation of roads* **4** (Moscow: TSBNTI Minavtodor of the RSFSR) 39 p

[15] Rukavishnikova E E, Lubkina K A and Skvortsov A V 2015 *Road Design at IndorPavement* (Tomsk: Tomsk University Press)

Effect of pressure boost on the diesel performance and gaseous emission with fuel spray pressure up to 3000 bar

Mikhail G. Shatrov[1], Andrey Y Dunin[1] and Thin Quynh Nguyen[2]

[1]Moscow Automobile and Road State Technical University (MADI), 64, Leningradsky ave., Moscow, 125319, Russia

[2]University of transport and communications, 3 Cau Giay Strs., Ha Noi, 122000, Vietnam

E-mail: a.u.dunin@yandex.ru

Abstract. The study presents the results of a single-cylinder engine modeling that assesses the effect of turbochargers on diesel performance and gaseous emission with the fuel injection pressure up to 3000 bar. 3D modeling is performed in Fire ESE Diesel. The results show that indicator power, soot, HC increase, and BSFC, NO_x decrease when using turbochargers. The boost pressure increases from 1 to 3.5 bar, the indicator power and BSFC vary tremendous, while the boost pressure increases from 3.5 to 4.5 bar, the indicator power, and BSFC change slowly. Indicator power grows by 15 %, BSFC reduces 16.6 %, NO_x value goes down by 85 % and soot value goes up by 92.9 %, the maximum value of HC goes up by 51.6 % with a boost pressure of 4.5 bar and SOI at 17° BTDC. When the SOI falls, BSFC reduces and the maximum value of this reduction is 6.4 %, while indicator power increases by 5.6 % with the boost pressure of 3.5 bar. NO_x value goes down by 70.3 % and soot value goes up by 88.6 % with the boost pressure of 4.5 bar. With the SOI of 5° BTDC, the turbocharging pressure from 2.5 to 4.5 bar, gaseous emissions meet the EURO V standard without the use of exhaust gas after-treatment systems.

Keyword: heavy-duty diesel, NO_x, soot, HC, combustion, fuel injection pressure 3000 bar, turbocharge.

1. Introduction

The exhaust gas after-treatment technology of modern diesel engines is researched and applied to reduce emissions and meet the emission standards that are becoming strictly. EURO emission standards are shown in table 1 [1].

Table 1. EU emission standards for heavy-duty CI (diesel) engines: steady-state testing

Stage	Date	Test	CO (g/kWh)	HC (g/kWh)	NO_x (g/kWh)	PM (g/kWh)
Euro III	1999.10 EEV only	ESC & ELC	1.5	0.25	2.0	0.02
	2000.10		2.1	0.66	5.0	0.10[a]
Euro IV	2005.10		1.5	0.46	3.5	0.02
Euro V	2008.10		1.5	0.46	2.0	0.02
Euro VI	2013.01	WHSC	1.5	0.13	0.4	0.01

[a] PM = 0.13 g/kWh for engines < 0.75 dm^3 swept volume per cylinder and a rated power speed > 3000 min^{-1}

IOP Conf. Series: Materials Science and Engineering **832** (2020) 012083 doi:10.1088/1757-899X/832/1/012083

In order to meet these emissions levels, there are three groups of solutions, namely, biofuels and additives in lubricating oil, exhaust gas after-treatment, and engine design. Additionally, operating mode and controlling technologies play an important role in reducing emission and improving the productivity of the engine and vehicle [1, 2]. In the group of engine design, improvement of the fuel system can be used, which includes increasing fuel injection pressure, changing the nozzle diameter, etc. When increasing the fuel injection pressure with other modifications, including optimized nozzle geometries, engine design and other engine advances, fuel efficiency could improve 4 % [3, 4]. Improving the pressure booster, high fuel injection pressure, adjustment of start of injection (SOI) angle and using EGR can meet EURO VI standards for the engine without the use of exhaust gas after-treatment.

Increasing of injection pressure is the basic development trend of the common rail fuel system [5, 6] which ensures multiple injection with the desired shape of the front edge of the basic injection rate [7, 8] and organization of fuel distribution in the combustion chamber [9, 10].

The desired fuel injection law at any operation mode of the engine is formed by variation of the control impulse duration [7, 8, 11, 12] and pressure in the common rail. It also depends on the wave phenomenon originating in the high-pressure line and affects considerably the fuel injection process in case of a multistage injection [13].

Currently, four engine simulation packages are used in the automotive industry, including Ricardo, GT-Suite, AVL-Suite and Lotus Engine Simulation (LESoft), which helps to reduce the time for engine design and costs. AVL-ESE Diesel is a part of the AVL-Fire package. It is used to set up, analyze, simulate and optimize combustion and emission formation in Diesel engines [14, 15]. In this research, the effect of rising the pressure booster is studied with fuel injection pressure up to 3000 bar and adjustment of fuel injection start angle. The results of modeling will give overview of increasing pressure booster on diesel engine performance and gaseous emissions

2. Simulation model

The sub-models, engine specifications, initial and boundary conditions are shown in tables 2, 3 and 4. The single-cylinder has a displacement of approximately 1.47 (l) and compression ratio of 15.4:1.

Table 2. AVL-FIRE Sub-Models

Spray model	Wave
Spray wall interaction model	Walljet1
Droplet Evaporation model	Dukowicz
Turbulence model	k-zeta-f
Ignition model	Auto - ignition
Combustion model	ECFM-3Z model
NO formation	Extended Zeldovich
Soot formation	Kinetic

Table 3. Engine Specifications [16]

Bore D, mm	120
Stroke S, mm	130
Connecting rod length, mm	224
Number of cylinders	1
Displacement, l	1.47
Compression ratio (-)	15.4
Number of valves	4
Spray angle	160°
Number of nozzle holes	8
Nozzle hole diameter, d_c (mm)	0.24

Table 4. Initial and boundary conditions

Engine speed, (rpm)	1400
Intake air temperature (K)	300
Intake air pressure (bar)	3.5
Cylinder head temperature (K)	550.15
Piston top temperature (K)	575.15
Fuel injection temperature (K)	330.15
Cylinder wall temperature (K)	475.15

IOP Conf. Series: Materials Science and Engineering **832** (2020) 012083 doi:10.1088/1757-899X/832/1/012083

The geometry of the piston and the combustion chamber are shown in figure 1. The computational mesh is shown in figure 2 with 57504 elements. Figure 3 illustrates the rate of fuel injection which is used in the model. The rate of injection is determined from the experiment signal.

Figure 1. Geometry of piston and combustion chamber

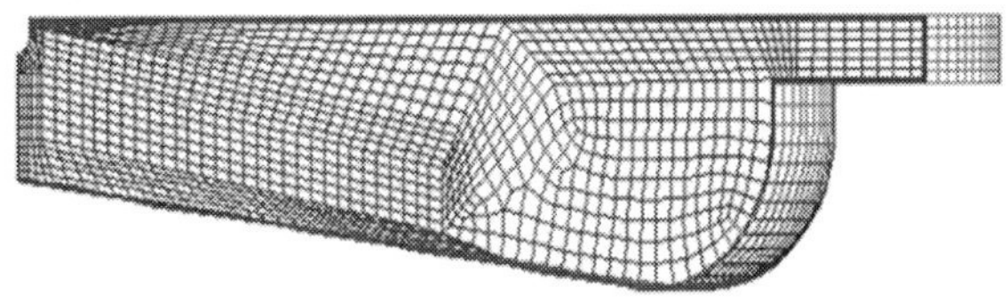

Figure 2. Computational mesh

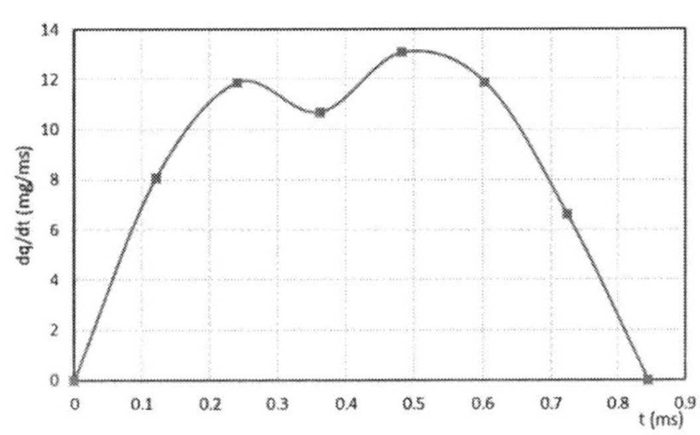

Figure 3. Rate of fuel injection

3. Simulation results

3.1. Combustion characteristics

Figures 4, 5 and 6 demonstrate the change of mean pressure, rate of heat release and mean temperature in the cylinder when the pressure increases.

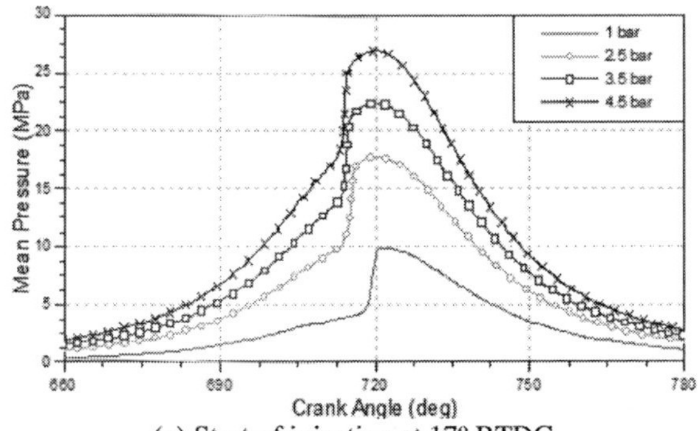

(a) Start of injection at 17° BTDC

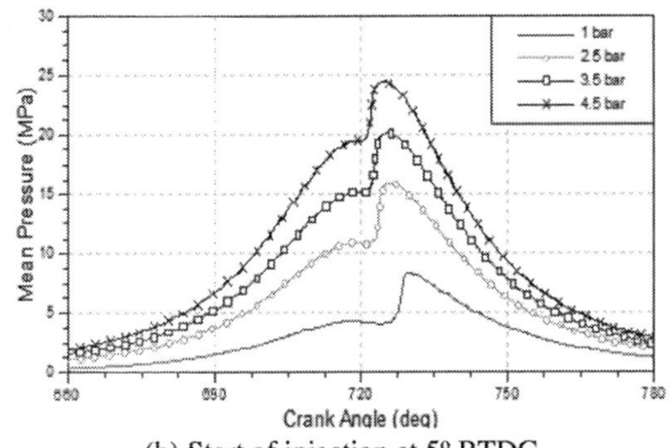

(b) Start of injection at 5° BTDC

Figure 4. Mean pressures in the combustion chamber with increasing of boost pressure

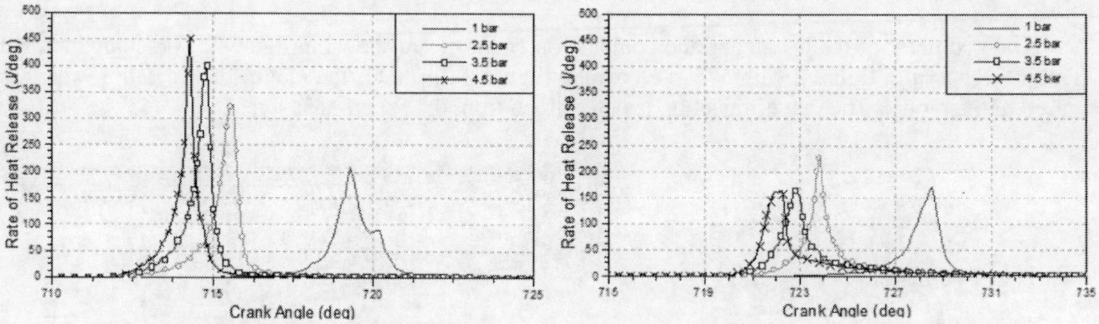

(a) Start of injection at 17° BTDC (b) Start of injection at 5° BTDC

Figure 5. Rate of heat release with increasing of boost pressure

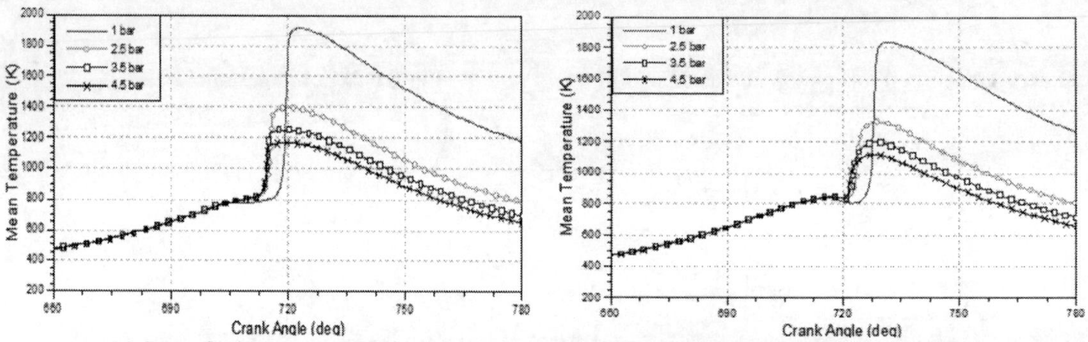

(a) Start of injection at 17° BTDC (b) Start of injection at 5° BTDC

Figure 6. Mean temperature in the combustion chamber with increasing of boost pressure

As the boost pressure increases, the mean pressure also raises. When the SOI is 17° before top dead center (BTDC), the mean pressure in the combustion chamber increases with increasing boost pressure and reaches the maximum value at the top dead center (TDC). The maximum value of the mean pressure rises to 2.75 times with boost pressure of 4.5 bar (figure 4a). The SOI at 5° BTDC, the time of achieving the maximum value of mean pressure varies from 9.5 to 4.7° after top dead center (ATDC) that corresponds to the boost pressure from 1 to 4.5 bar. The maximum value of the mean pressure goes up 2.9 times with boost pressure of 4.5 bar (figure 4b).

When using turbochargers, the maximum rate of heat release will move closer to the SOI. The rate of heat release will go up when boost pressure rises and the SOI is 17° BTDC. The maximum value of the rate of heat release climbs 2.9 times with boost pressure of 4.5 bar (figure 5a). While the SOI at 5° BTDC, the rate of heat release increases when the boost pressure increases from 1 to 2.5 bar, however it decreases when the boost pressure goes up from 2.5 to 3.5 bar and slightly changes when the boost pressure rises from 3.5 to 4.5 bar (figure 5b).

In contrast, the mean temperature goes down with turbocharging. It is because the boost pressure is small, the temperature at the end of the compression period decreases, the air density is low, the combustion process extends so the mean pressure cannot leap (figure 6). Therefore, the mean temperature in the combustion chamber will increase as the boost pressure decreases

3.2. BSFC and indicator power

The results of BSFC, the indicator power and increasing of the boost pressure are shown in fig. 7.

It can be noticed that the indicator power engine goes up and BSFC decreases with increasing of boost pressure. When the boost pressure rises from 1 to 3.5 bar the indicator power and BSFC vary tremendous, while the boost pressure increases from 3.5 to 4.5 bar, the indicator power and BSFC change slowly. The boost pressure at 3.5 bar, indicator power increases by 13.6 %, BSFC goes down by 15 %. And indicator power grows by 15 %, BSFC reduces by 16.6 % with a boost pressure of 4.5 bar.

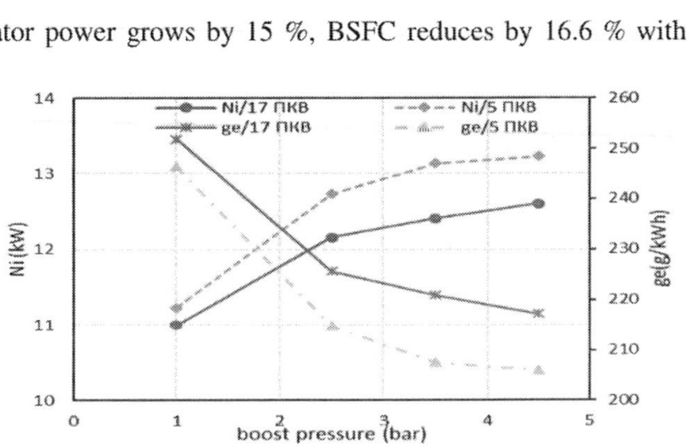

Figure 7. BSFC and indicator power when boost pressure rises

When the SOI reduces, the indicator power increases and the BSFC decreases. From the results in Figure 8, it is found that the BSFC is the smallest, the indicator power is the largest with the turbocharging pressure of 4.5 bar, the SOI of 5° BTDC. The boost pressure at 3.5 bar, the maximum value of BSFC reduction is 6.4 %, while indicator power increases by 5.6 %.

3.3. Gaseous emission

NO_x value decreases and soot increase when boost pressure goes up. The result is shown in figure 8a. NO_x is formed by the reaction of oxygen and nitrogen at high temperatures. It is because the mean temperature falls, at the same time, the contact area between the air and nitrogen reduces, so NO_x goes down with increasing boost pressure. The SOI at 17° BTDC, the soot value is the opposite of NO_x. NO_x value goes down by 85 % and soot value goes up by 92.9 %. While the SOI at 5° BTDC, NO_x value declines by 70.3 % and soot value raises by 88.6 %.

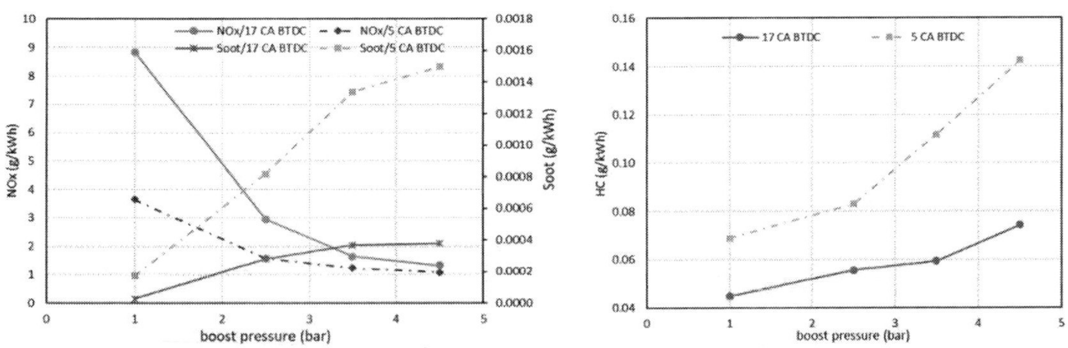

(a) Changing of NO_x and soot (b) HC emission

Figure 8. Gaseous emission when boost pressure rise

HC volume is the unburnt fuel. The boost pressure increases, the contact area between the air and the fuel decreases, at the same time, the burning process is quickly so the HC value rises. The result is shown in fig. 8b.

The results of spray length and temperature field in the combustion chamber are shown in fig. 9–12.

From the results, can be see that spray length and spray cone diameter decrease, the combustion area also goes down when the boost pressure increases. Because air density, temperature in the combustion chamber increase, the combustion process is quick, so the indicator power upraises and gaseous emissions reduces.

It is found that the SOI at 5° BTDC, the turbocharging pressure from 2.5 to 4.5 bar, gaseous emissions meet the EURO V standard without the use of exhaust gas after-treatment systems. To ensure the economy when using the booster, the boost pressure should use less than 3.5 bar.

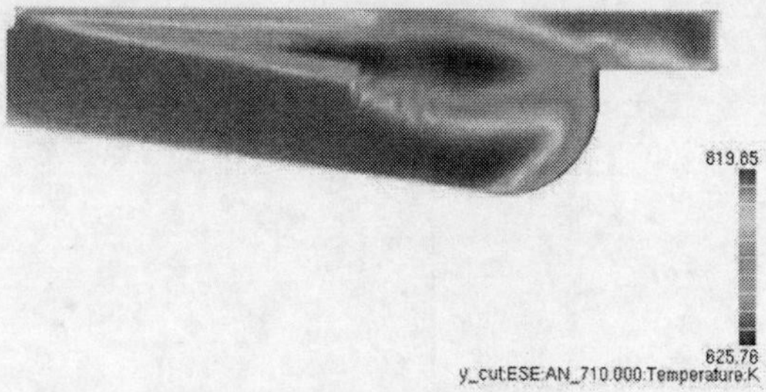

(a) 1 bar boost pressure

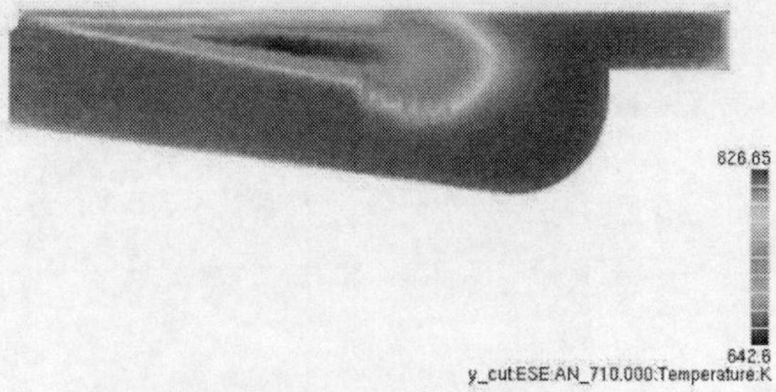

(b) 2.5 bar boost pressure

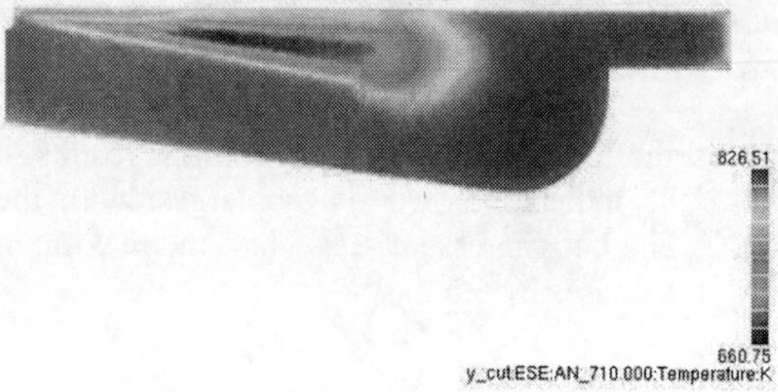

(c) 3.5 bar boost pressure

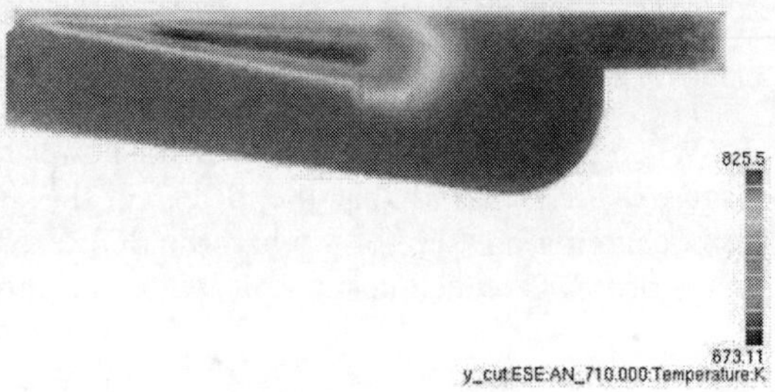

(d) 4.5 bar boost pressure

Figure 9. Temperature field in the combustion chamber with the piston at 10° BTDC, start of injection 17° BTDC.

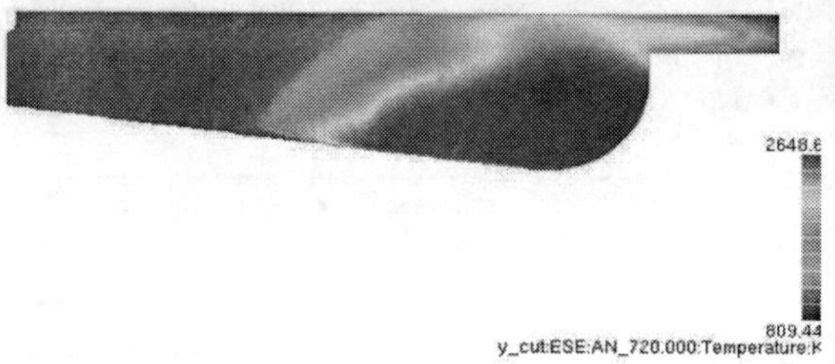

(a) 1 bar boost pressure

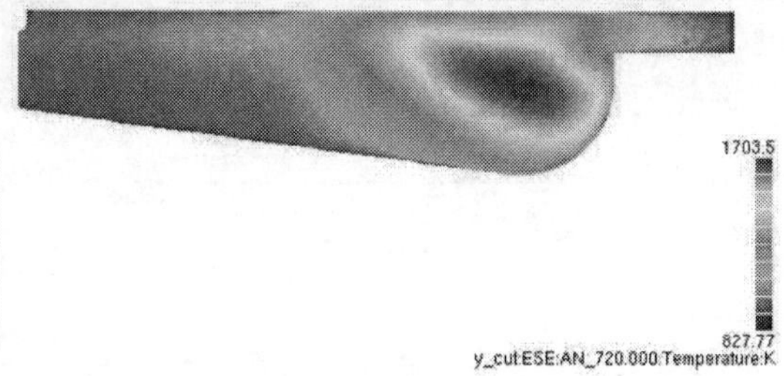

(b) 2.5 bar boost pressure

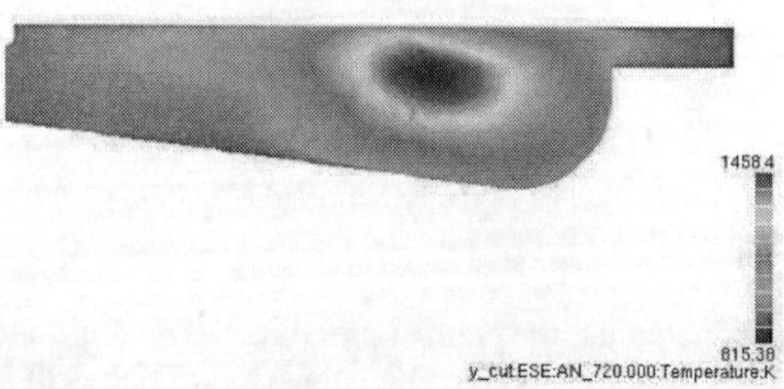

(c) 3.5 bar boost pressure

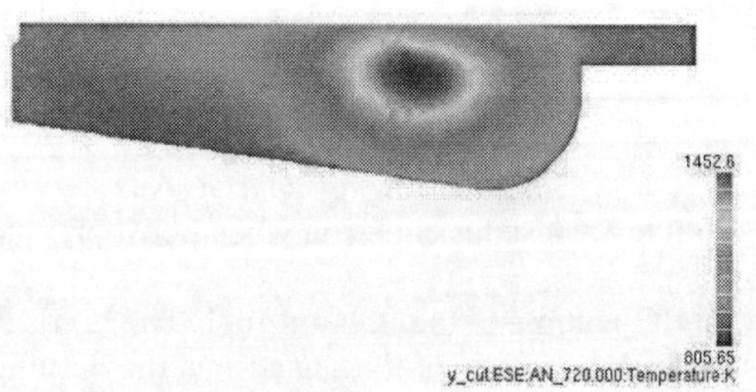

(d) 4.5 bar boost pressure

Figure 10. Temperature field in the combustion chamber with the piston at TDC and SOI of 17° BTDC

IOP Conf. Series: Materials Science and Engineering **832** (2020) 012083 doi:10.1088/1757-899X/832/1/012083

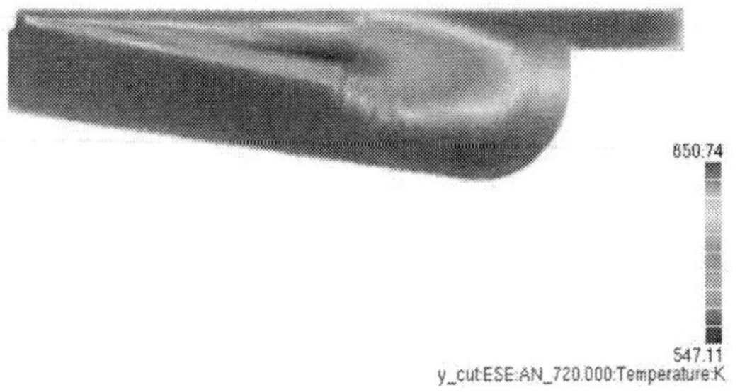

(a) 1 bar boost pressure

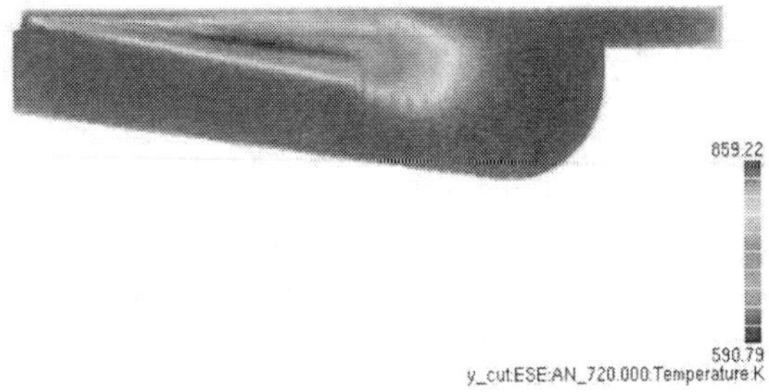

(b) 2.5 bar boost pressure

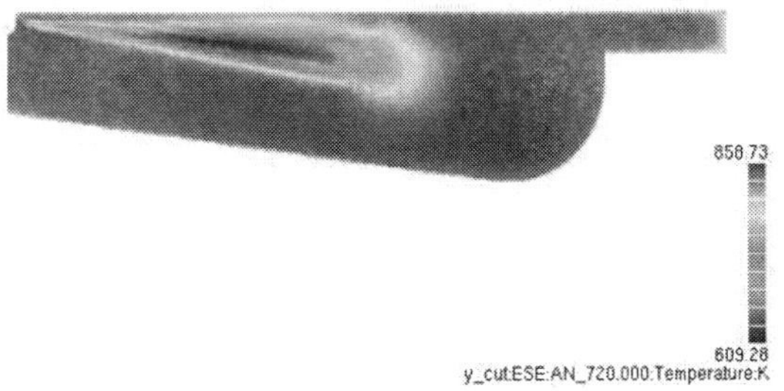

(c) 3.5 bar boost pressure

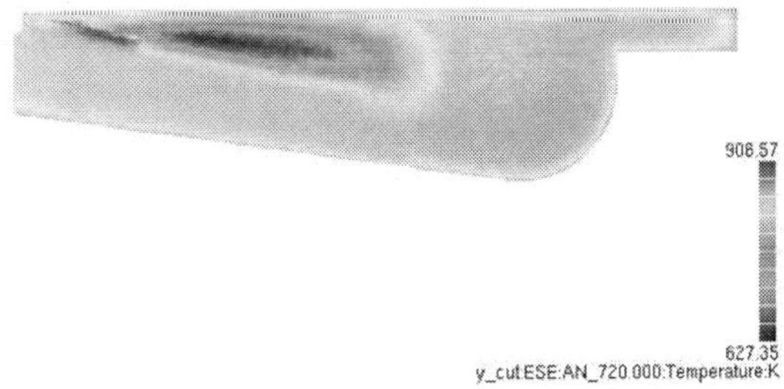

(d) 4.5 bar boost pressure

Figure 11. Temperature field in the combustion chamber with the piston at TDC and SOI of 5° BTDC

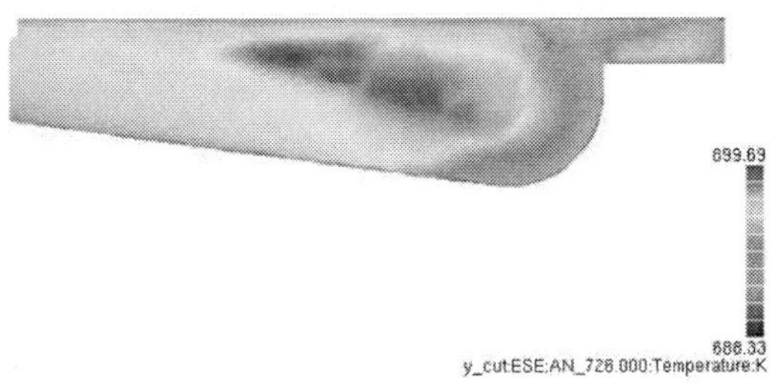

(a) 1 bar boost pressure

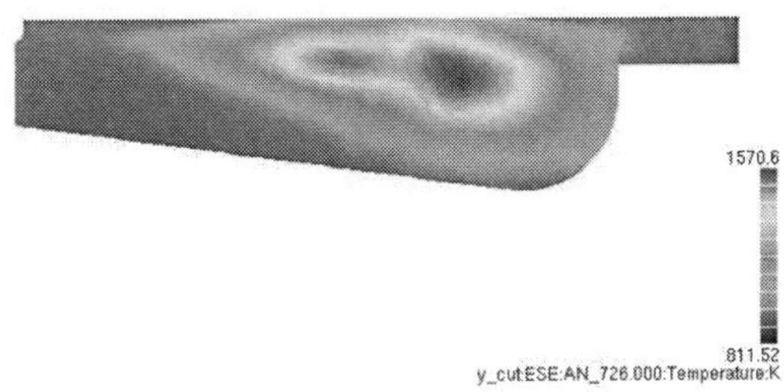

(b) 2.5 bar boost pressure

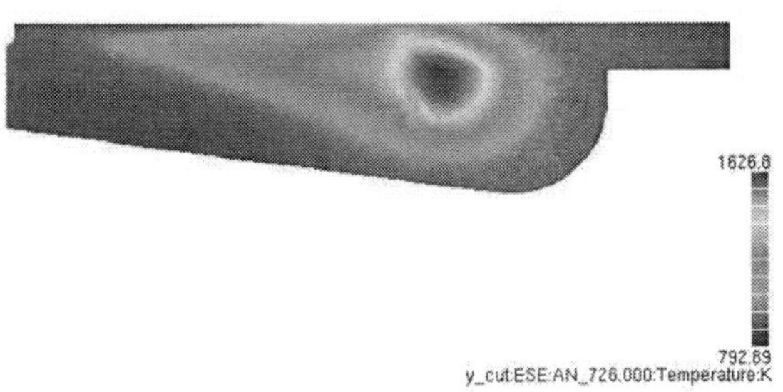

(c) 3.5 bar boost pressure

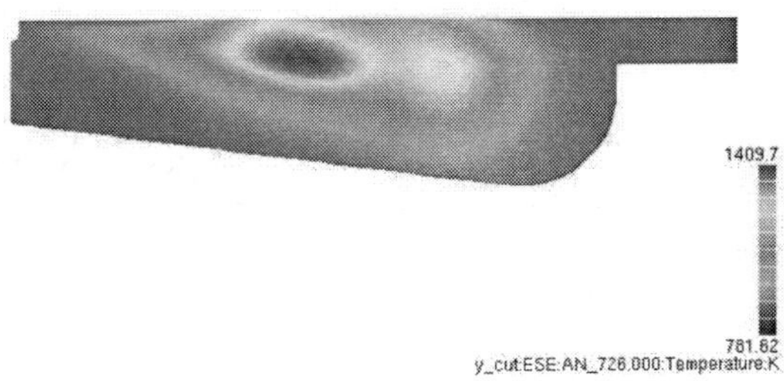

(d) 4.5 bar boost pressure

Figure 12. Temperature field in the combustion chamber with the piston at 6° ATDC and SOI of 5° BTDC

4. Conclusions

In this study, the effect of boost pressure on diesel performance, emissions gaseous is simulated by AVL Fire software. The summary results are as follows:

- As the air pressure increases, the combustion of fuel take place is earlier and moves toward the SOI;
- With 3000 bar of fuel injection pressure, it is necessary to combine turbocharging with adjustment of SOI for increasing indicator power, reducing BSFC and improving emissions;
- As the boost pressure rises, the indicator power of engine goes up, BSFC decreases. The boost pressure increases from 1 to 3.5 bar, the indicator power and BSFC vary tremendous, however the boost pressure raises from 3.5 to 4.5 bar, the indicator power and BSFC change slowly. Indicator power grows by 15 %, BSFC reduces 16.6 % with a boost pressure of 4.5 bar. Reducing SOI, the maximum value of BSFC reduction is 6.4 %, while indicator power increases by 5.6 % with the boost pressure at 3.5 bar;
- NOx goes down with increasing boost pressure. The SOI at 17° BTDC, NOx value reduces by 85 % and soot value goes up to 92.9 %, while the SOI at 5° BTDC, NOx value falls to 70.3 % and soot value raises to 88.6 %;
- When the boost pressure grows, the HC value rises. The maximum value of HC increase is 51.6 %;
- The SOI at 5° BTDC, the turbocharging pressure from 2.5 to 4.5 bar, gaseous emissions meet the EURO V standard without the use of exhaust gas after-treatment systems. To ensure the economy when using the booster, the boost pressure should use less than 3.5 bar.

References

[1] *Engine Emission Control* Retrieved from: https://www.dieselnet.com/tech/engine_emission-control.php

[2] Rajamani V K, Schoenfeld S and Dhongde A 2012 Parametric Analysis of Piston Bowl Geometry and Injection Nozzle Configuration using 3D CFD and DoE *SAE Technical Papers* 12

[3] Shatrov M G, Golubkov L N, Dunin A U, Yakovenko A L and Dushkin P V 2015 Influence of high injection pressure on fuel injection perfomances and diesel engine worcking process *Thermal Sci.* **19(6)** 2245–53

[4] Shatrov M G, Golubkov L N, Dunin A U, Yakovenko A L and Dushkin P V 2015 Research of the injection pressure 2000 bar and more on diesel engine parameters *Int. J. of Applied Res.* **10(20)** 41098–102

[5] Shatrov M G, Golubkov L N, Dunin A Yu, Dushkin P V and Yakovenko A L 2017 A method of control of injection rate shape by acting upon electromagnetic control valve of common rail injector *Int. J. of Mechan. Engineer. and Technol.* **8(11)** 676–90

[6] Shatrov M G, Golubkov L N, Dunin A U, Dushkin P V and Yakovenko A L 2017 The new generation of common rail fuel injection system for Russian locomotive diesel engines *Pollution Res.*, **36(3)** 678–84

[7] Shatrov M G, Malchuk V I, Dunin A U and Yakovenko A L 2016 The influence of location of input edges of injection holes on hydraulic characteristics of injector the diesel fuel system *Int. J. of Applied Engineer. Res.* **11(20)** 10267–73

[8] Shatrov M G, Malchuk V I, Dunin A Y, Shishlov I G and Sinyavski V V 2018 A control method of fuel distribution by combustion chamber zones and its dependence on injection conditions *Thermal Sci.* **22(5)** 1425–14

[9] Shatrov M G, Sinyavski V V, Dunin A Y, Shishlov I G and Vakulenko A V 2017 Method of conversion of high- and middle-speed diesel engines into gas diesel engines *Facta Universitatis. Ser. Mechan. Engineer.* **15(3)** 383–95

[10] Shatrov M G, Sinyavski V V, Dunin A Yu, Shishlov I G, Vakulenko A V and Yakovenko A L 2018 Using simulation for development of the systems of automobile gas diesel engine and

its operation control *Int. J. of Engineer. and Technol.* **7(2.28)** 288–95

[11] Shatrov M G, Golubkov L N, Dunin A U, Yakovenko A L and Dushkin P V 2016 Experimental research of hydrodynamic effects in common rail fuel system in case of multiple injection *Int. J. of Applied Engineer. Res.* **11(10)** 6949–53

[12] Wloka J.A and Pflaum S Wachtmeister 2010 Potential and Challenges of a 3000 Bar Common-Rail Injection System Considering Engine Behavior and Emission Level *SAE Techn. Papers* 13

[13] Johnson J E, Yoon S H, Naber J D, Lee S-Y, Hunter G, Truemner R and Harcombe T 2012 Characteristics of 3000 bar Diesel Spray Injection under Non-Vaporizing and Vaporizing Conditions *ICLASS 2012, 12th Triennial Int. Conf. on Liquid Atomization and Spray Systems* (Heidelberg, German)

[14] *AVL-FIRE User Guides* 2014

[15] Firat M, Altun Ş and Varol Y 2016 A numerical study of no and soot formation in an automotive diesel engine fueled with diesel-biodiesel blends *J. of FCE – sci. paper* **4**

[16] Alekseev A B 2010 *Improving the performance of truck diesel engines by choosing a combustion chamber and optimizing fuel equipment* PhD dissertation thesis (Moscow)

IOP Conf. Series: Materials Science and Engineering **832** (2020) 012084 doi:10.1088/1757-899X/832/1/012084

Influence of temperature and geometric parameters of elements in a turbocompressor seal assembly on its operability

D A Nikitin[1], A R Asoyan[2], P D Nikitin[1] and C A Turoverov[1]

[1]Saratov State Agrarian University named after N.I. Vavilov, Russia
[2]Moscow Automobile and Road Construction State Technical University, 64, Leningradsky ave., Moscow, 125319, Russia

E-mail: asoyan.ar@mail.ru , nilppr@rambler.ru

Abstract. Currently, most power units using internal combustion engines are equipped with turbocharging and that is explained by the requirements for efficiency, environmental safety and specific power characteristics. At the same time, the turbocompressor (TCR) is one of the resource-determining elements of power units, which accounts for a significant number of failures. Studies conducted towards the analysis of turbocompressor failures have shown that the main reason for it lies in a malfunction of the seal assembly, which is a consequence of extremely unfavorable conditions for its operation, consisting in a high (up to 800 °C) exhaust temperature, aggressive environment, a high (up to 120,000 min^{-1}) rotor speed, the sliding speeds of the mating parts, up to 150 m/s. Oil for lubrication and cooling of the TCR is supplied through floating plain bearings, while the cavities of the turbine and compressor are isolated by means of the seal assemblies. Violation of their operability leads to the ingress of oil in a cavity of the impellers, to the imbalance of a rotor and, ultimately, to a turbocompressor failure. The determining factor affecting the operability of the seal assembly is a heat density of its parts. This factor is especially manifested in operating under forced conditions and engine shutdown, when cooling of a turbine volute is stopped by blown air, and a bearing assembly is stopped by circulating oil. In this case, a large part of the thermal energy is transferred from a heated turbine volute to the seal assembly and to the bearing assembly. The focus is on the temperature influence on performance of the turbocompressors' sealing rings.

1. General information about turbocompressors usage

At present, the production of atmospheric diesels comprises not more than 40 % from the total volume of all production and the steady trend is being observed to its reduction at the world market. For trucks, building and agricultural machinery, the most widely used diesel engines are Kamins, Caterpillar (USA); Volvo, Scania (Sweden); Deutz, MAN (Germany). In Russia, the main diesel producers of this group are KAMAZ and YaMZ. At the same time, the turbocompressor (TCR), being the main unit of the boost system, significantly affects the equipment downtime, a cost of spare parts, repair and reliability of a power unit as a whole. In this regard, ensuring the reliability of a turbocmpressor and developing the cost-effective repair technologies are the important tasks.

2. Problem and hypothesis statement

TCR of diesel engines [1, 2] (Fig. 1) operate under extremely adverse conditions, which include high (up to 800 °C) exhaust gas temperatures, aggressive environment, and a high (up to 120,000 min^{-1}) rotor speed, sliding speeds of the mating parts, up to 150 m/s.

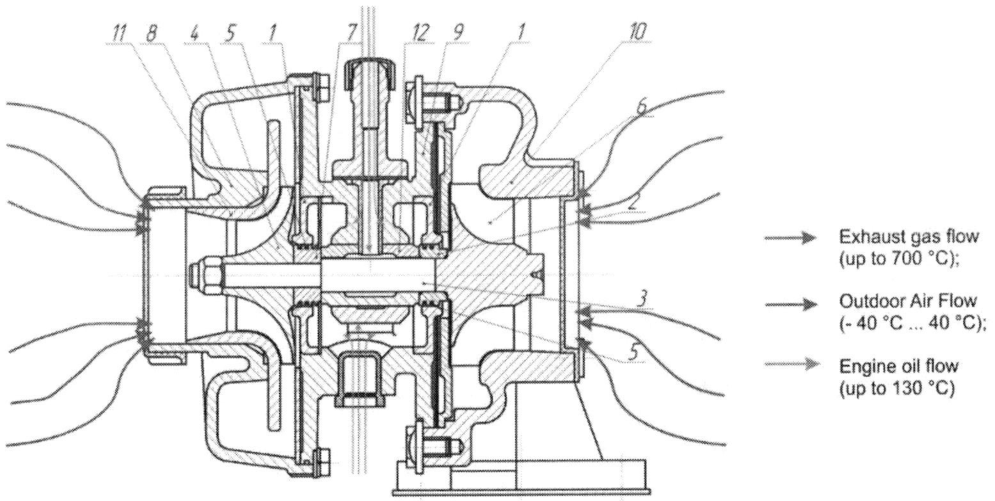

Figure 1. General view of a turbocompressor: 1 – a seal cover; 2 – a sleeve of a rotor shaft; 3 – a rotor shaft; 4 – a compressor wheel; 5 – sealing rings; 6 – a turbine wheel; 7 – an oil deflector; 8 – compressor housing; 9 – the medium body; 10 – turbine housing; 11 – an insert compressor; 12 – a bearing

One of the critical components of the TCR, affecting its operating cycle, is the lubrication system seal assembly. Severe temperature operating conditions of the seal assembly (especially from the turbine side) significantly affect the reliability of the turbocompressor and the power unit as a whole, where failures due to the TCR reach to 25.6 % (Fig. 2). At the same time, subject to operation requirements and timely performance of scheduled maintenance of the power unit, new TCRs, as a rule, satisfactorily fulfill the envisaged resource. The TCRs that underwent repair impacts, often even in the initial period of operation, demonstrate increased oil consumption, and the mean time between failures, as a rule, does not exceed 50 % of the operating time of a new turbocompressor.

A detailed analysis of failure causes of both new and restored TCRs is presented in [1]. The studies results reflecting the TCR failures are presented in Figure 3. As can be seen from the diagram, the main cause of failures is increased oil consumption, which is a consequence of the seal assembly poor operation. The reason for the wear of the turbine wheels and the compressor, as a rule, is a violation of the balancing of these parts due to carbon formation (in case of loss of tightness of the seal) and the presence of some solid particles from outside. The same reason leads to a break in the weld of the rotor shaft. Thus, up to 88 %, the reason for all turbocompressor failures is poor operation of the seal assembly.

To understand the reasons for this, we examine the seal assembly in detail. The oil seal of most modern turbocompressor of automotive engines is combined, which unites in itself the contact and labyrinth seals, formed respectively by the outer surface of the sealing ring and the housing, the ends of the seal and the walls of the groove (Figure 4). At the same time, the key role in this seal is played by elastic split metal rings of rectangular cross-section, installed in the grooves of the rotor bushings on the compressor and turbine sides. The effectiveness of this seal is most affected by the amount of clearance in the ring-groove mate, the elasticity of the seal ring, and the amount of thermal clearance in the seal ring.

IOP Conf. Series: Materials Science and Engineering **832** (2020) 012084 doi:10.1088/1757-899X/832/1/012084

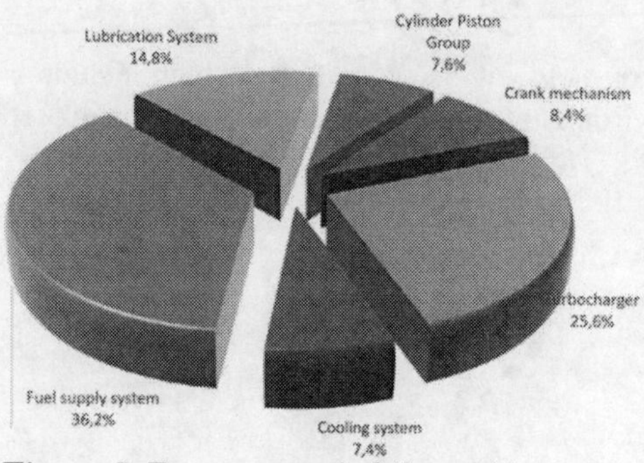

Figure 2. For power unit failures, % attributable to various systems

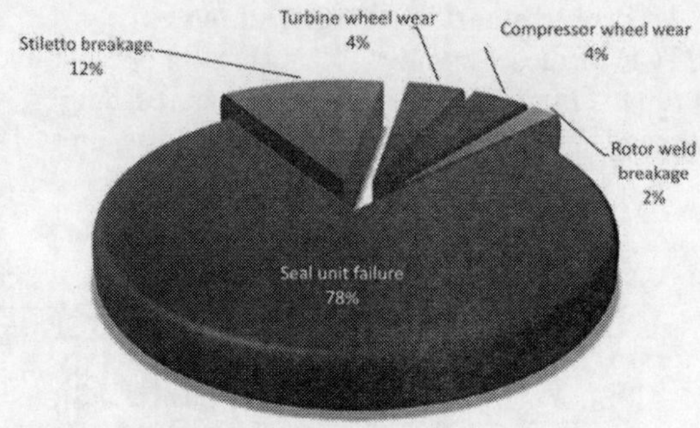

Figure 3. Reasons for failure of reconditioned turbocompressors

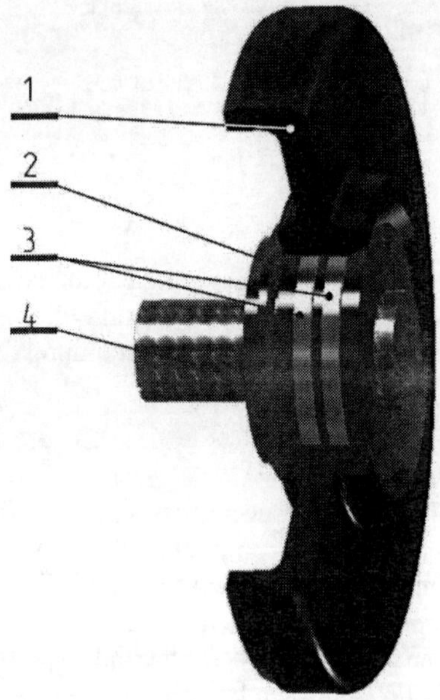

Figure 4. TCR sealing assembly:
1 – housing part (with cut-out);
2 – a sleeve;
3 – split sealing rings;
4 – a fragment of rotor

Based on the analysis of TCR failures, the design of the seal assembly and its elements, and also taking into account that the majority of TCR failures occur suddenly, bypassing the stage of increasing manifestations of the malfunction, the authors made the assumption that the malfunction occurs due to the loss of elasticity of the sealing rings [3] and contact with a female part. Accordingly, a working hypothesis has been put forward and it states the following: the performance of the seal assembly is violated when extreme temperature conditions occur as a result of violations of the operating requirements, and can be caused by deformations of the sealing rings that extend beyond the elasticity range.

3. Mathematical apparatus for modeling temperature effect of on details of seal assembly

In the framework of the hypothesis that has been put forward, the temperature fields of the extreme ring on the turbine side of the sealing ring were first theoretically determined as being most susceptible to temperature effects from the exhaust gases. To determine the temperature fields in this ring, we propose the following technique.

The exhaust gases in the turbine cavity have the property of internal friction, the tangential stress of which obeys the following law:

$$\tau = \mu\left(\frac{du}{dn}\right), \tag{1}$$

where μ is the dynamic coefficient of viscosity; $\frac{du}{dn}$ – a normal flow velocity gradient.

As you know, any gas flow (both laminar and turbulent) along the channel walls has a boundary layer, the flow rate in which tends to 0 as it approaches the walls. A similar situation arises in the temperature (thermal) boundary layer. In this situation, the temperature in this layer varies from the temperature of the boundary (wall) to the temperature of the main gas stream. The thickness of this layer in a turbulent flow is determined by the thermophysical properties of gas [4]. The laminar motion in the general case is described by the Navier-Stokes equation:

$$\left.\begin{aligned} \frac{Du_x}{Dt} &= -\frac{1}{\rho}\cdot\frac{\partial P}{\partial x} + \nu\cdot\left(\frac{\partial^2 u_x}{\partial x^2} + \frac{\partial^2 u_x}{\partial y^2} + \frac{\partial^2 u_x}{\partial z^2}\right) \\ \frac{Du_y}{Dt} &= -\frac{1}{\rho}\cdot\frac{\partial P}{\partial y} + \nu\cdot\left(\frac{\partial^2 u_y}{\partial x^2} + \frac{\partial^2 u_y}{\partial y^2} + \frac{\partial^2 u_y}{\partial z^2}\right) \\ \frac{Du_z}{Dt} &= -\frac{1}{\rho}\cdot\frac{\partial P}{\partial z} + \nu\cdot\left(\frac{\partial^2 u_z}{\partial x^2} + \frac{\partial^2 u_z}{\partial y^2} + \frac{\partial^2 u_z}{\partial z^2}\right) \end{aligned}\right\}, \tag{2}$$

where: $\frac{Du_x}{Dt}, \frac{Du_y}{Dt}, \frac{Du_z}{Dt}$ are the substantial derivatives of the velocity with respect to the coordinates of the space x, y, z; p is the density of the liquid or gas in the stream; P is the pressure; $\nu = \frac{\mu}{\rho}$ – the kinematic coefficient of viscosity of liquid (gas).

The substantial derivative of the velocity for any of the coordinates is the sum of the local and convective derivatives. On X axis, the substantial derivative of the velocity is described by the following relations:

$$\frac{Du_x}{Dt} = \frac{\partial u_x}{\partial t} + u_x\frac{\partial u_x}{\partial x} + u_y\frac{\partial u_y}{\partial y} + u_z\frac{\partial u_z}{\partial z} \tag{3}$$

where $\frac{\partial u_x}{\partial t}$ is the local derivative of speed, which determines the change in speed at a given point in space in time; $u_x\frac{\partial u_x}{\partial x} + u_y\frac{\partial u_y}{\partial y} + u_z\frac{\partial u_z}{\partial z}$ is the convective derivative of velocity, which determines the rate of change in velocity at a given time in the vicinity of a given point in space.

If there are no viscosity forces in the flow, then the terms of equation 3 containing ν are equal to 0 – this irrotational motion of the working fluid is called potential (in this case, friction creates vortex formation).

The system of equations 2, which describes the distribution of the flow velocities of a working fluid (gas or liquid), is reduced (i.e., without viscous terms) to the system of the Euler equations of potential motion.

It should be noted that turbulent gas motion in heat conduction problems can be considered as potential precisely because of its turbulence.

The temperature distribution is described by the Fourier-Kirchhoff equation:

$$\frac{DT}{Dt} = \alpha\cdot\left(\frac{\partial^2 T}{\partial x^2} + \frac{\partial^2 T}{\partial y^2} + \frac{\partial^2 T}{\partial z^2}\right) + \frac{\mu}{\rho\cdot c_p}\cdot\Phi, \tag{4}$$

where *DT/Dt* – substantial derivative temperature; Wed – heat capacity of the working fluid (gas); Φ is the Rayleigh dissipative function.

Since for most gases and liquids at moderate speeds the dispersive Rayleigh function is small, the energy equation 4 reduces to the following form:

$$\frac{DT}{Dt} = \alpha\cdot\left(\frac{\partial^2 T}{\partial x^2} + \frac{\partial^2 T}{\partial y^2} + \frac{\partial^2 T}{\partial z^2}\right) \tag{5}$$

In case of stability loss, the laminar boundary layer can become turbulent. The condition for the existence of laminar motion on the walls is

$$\frac{d}{D_z} \cdot \left(\mu \cdot \frac{du_x}{dz}\right)_{z=0} < 0 \tag{6}$$

where $\left(\frac{du_x}{dz}\right)_{z=0}$ is the velocity gradient of the layer on the wall.

In this case, the stability of the layer is determined from the following expression:

It is obvious that the layer is stable at

$$\left.\frac{\partial^2 u_x}{\partial z^2}\right|_{z=0} < 0 \text{ , i. e. } \frac{dT}{dz} > 0 \text{ at } \left.\frac{\partial^2 u_x}{\partial z^2}\right|_{z=0} > 0. \tag{7}$$

Thus, if the gas in the turbine is cooled (it gives energy to the walls of the casing, to the parts of the rotor, including the rings and the turbine), then a laminar boundary layer is stored on the walls of these parts.

Using equations 6 and 7 and formulating the boundary conditions, we can find the distribution of gas flow velocities in the laminar (boundary) layer of the working fluid flow and calculate the temperature distribution in it, which, in turn, will allow, based on Fourier hypothesis, to determine the heat flux density given (perceived) by the walls of the parts from the working fluid (in our case, exhaust gas):

$$q = -\lambda_f \cdot \left(\frac{dT}{dn}\right)_{n=0}, \tag{8}$$

where λ_f is the coefficient of thermal conductivity of the working fluid (gas); $\left(\frac{dT}{dn}\right)_{n=0}$ – temperature gradient in the boundary layer at the boundary with the wall.

Knowing the heat flux q and the temperatures T_ω and the working fluid (exhaust gases) outside the boundary layer T_f allows to find the heat transfer coefficient introduced by I. Newton:

$$\alpha = \frac{q}{T_f - T_\omega}. \tag{9}$$

The heat transfer coefficient, being a scalar value, is not a physical parameter; it is very convenient for generalizing empirical and calculated dependencies.

Given that the problem of determining, ultimately, the temperatures of specific parts of a turbocompressor is quasistatic, which allows to neglect the local derivatives of speed and temperature, and the boundary conditions (temperatures of the working fluid and body parts and gas pressure in the separated cavities are known from experiments), the solution for the task to be stated seems quite possible.

As a result of these considerations, we have obtained a mathematical model of the temperature distribution over the volume of parts, which allows to carry out some computational experiments with the operation of turbocompressor parts in various modes.

4. Computational experiment results

Calculations of the outermost ring temperature from the turbine side have shown (Fig. 5) that a significant temperature of the material of the body of the ring during operation is in the range 120...620^{about}C. The highest temperatures are observed from the turbine side and along the surface of the ring lock.

Such an uneven temperature distribution can not but cause the ring deformation, the calculation results are presented in Fig. 5.

The analysis on the calculation results addressed the temperature deformations, which are presented in Figure 5, shows that the ring deformations reach 0.017 mm, which must be taken into account when designing the TCR geometric elements.

Taking into account the temperature deformation of the ring, in order to prevent "setting" and rotation of the rings with the sleeve, it is necessary to provide a gap of at least 0.1 mm across the width of the ring in the groove. In addition, for some manufacturers, the minimum value of the thermal gap in the lock of the ring is 0 with a positive upper deviation of, (Fig. 5.d) leads to a joint, which causes the additional force in this zone. The results relating to the stress calculations arising from this interaction are presented in Fig. 6, and the results of calculating the safety factor for yield strength are shown in Fig. 7.

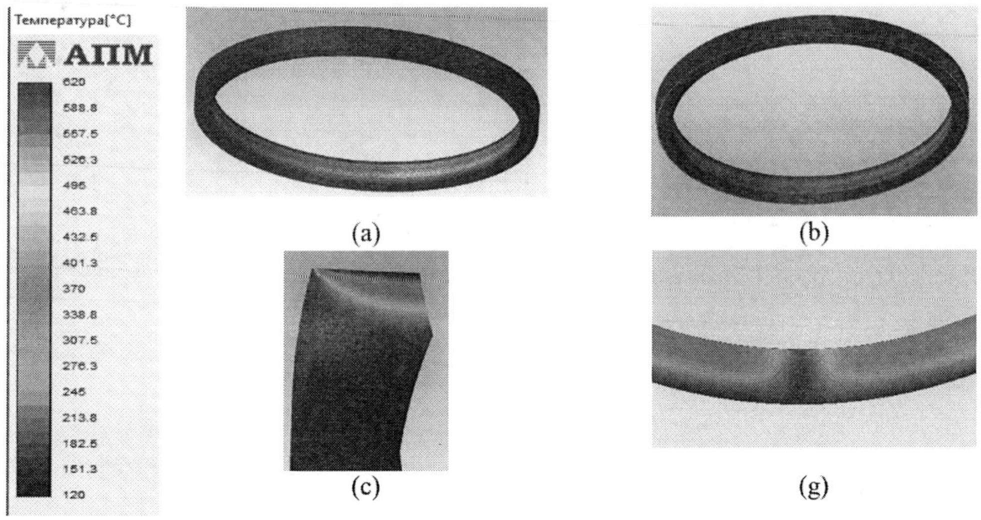

Figure 5. Temperature distribution in the sealing ring, the extreme one from the turbine side: (a) – view from the turbine side; (b) – view from the side of the central bearing; (c) – cross section of the ring; (d) – the section along the ring

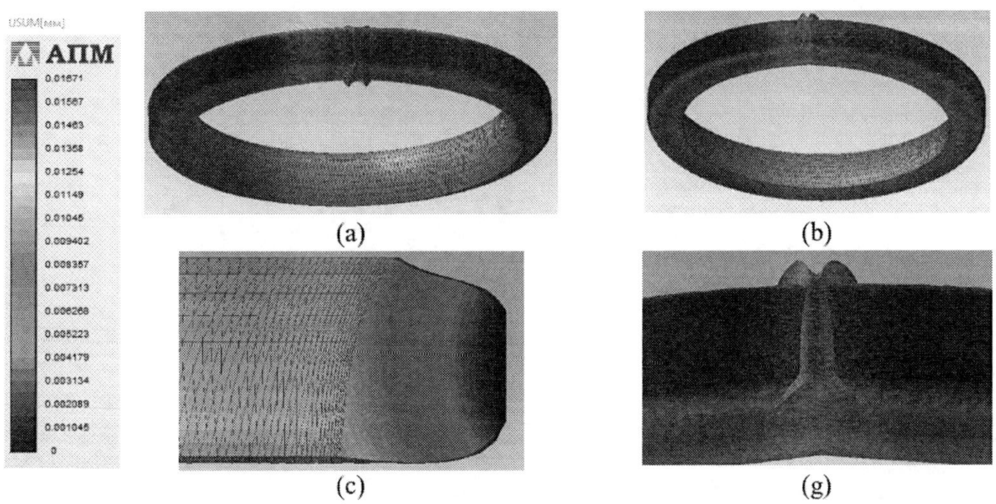

Figure 6. Temperature distribution in the sealing ring, the extreme one from the turbine side: (a) – view from the turbine side; (b) – view from the side of the central bearing; (c) – cross section of the ring; (d) – the section along the ring

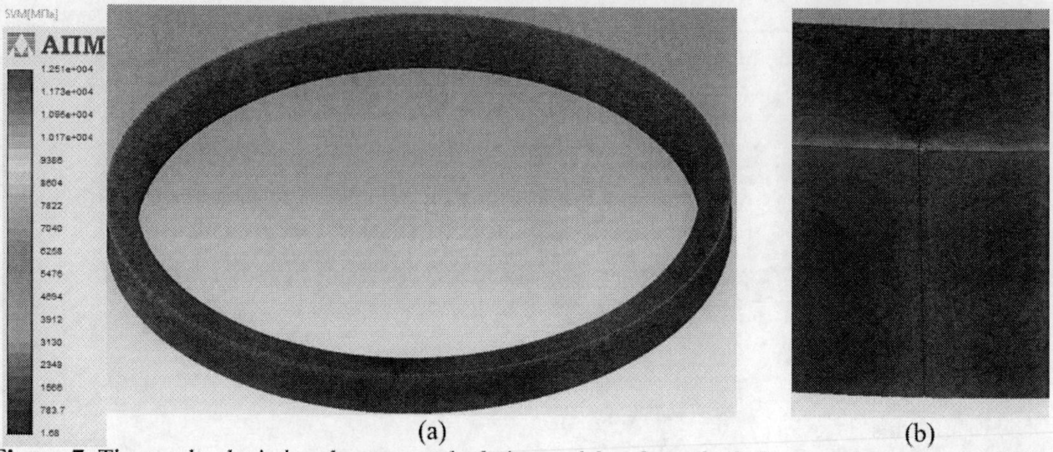

Figure 7. The results depicting the stress calculations arising from the influence of temperature: (a) in the compactor ring as a whole, (b) in the lock's zone

The stress data in the lock's zone, as suggested in the hypothesis, go beyond the yield strength, the safety factor for this indicator is significantly less than 1 (Fig. 7), which leads to the loss of operability of at least the extreme sealing ring on the turbine side.

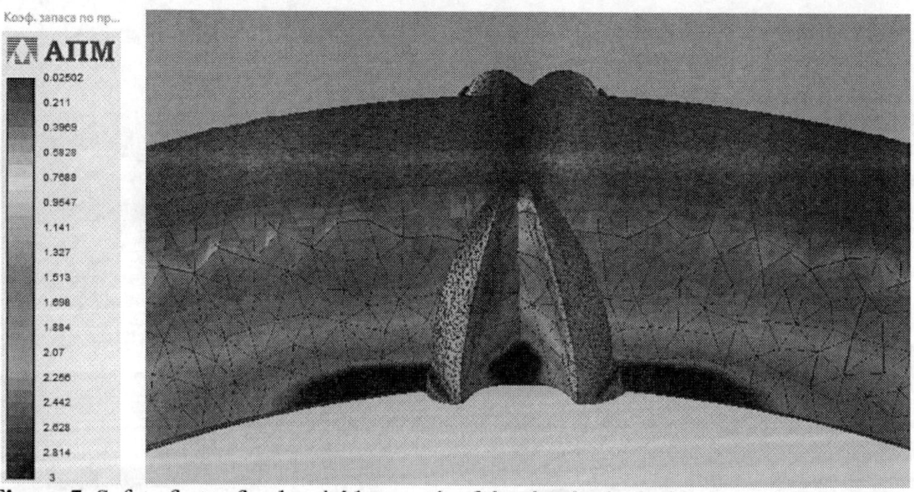

Figure 7. Safety factor for the yield strength of the ring in the lock's area

5. Analysis of calculation results, recommendations problem solution and conclusion

As can be seen from Figures 5–7, the main reason for the malfunction of the seal assembly is the absence of a thermal gap in the ring lock under the influence of temperature, which leads to unnormalized excess stress and deformations beyond the yield strength of the material. As one of the ways to solve this problem, the authors propose to significantly increase the thermal gap in the lock to a value exceeding the value of thermal deformation, i.e., according to the above calculations, to a value of 0.15 mm for the TCR under consideration, at least for a ring extreme side of the turbine.

The calculation results towards deformations, stress, and safety factor for yield strength are presented in Fig. 8.

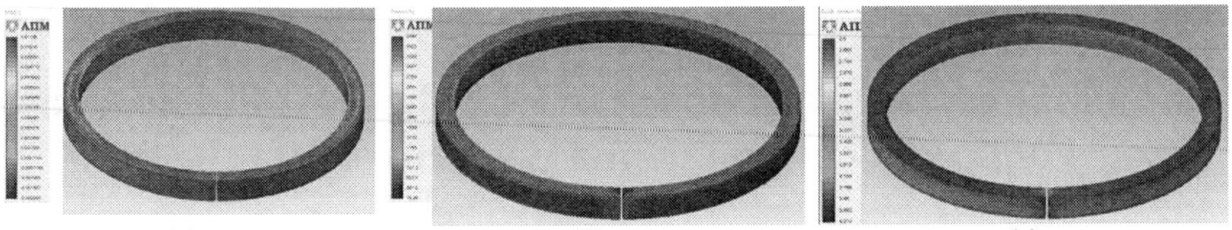

(a) (b) (c)

Figure 8. The calculation results (a) – deformations; (b) – volatages; (c) – tensile strength by yield strength

The calculation results on deformations, voltages and the safety factor for the yield strength of the TCR-7 turbocompressor, O-ring of the extreme side of the turbine, having a thermal gap of 0.15 mm, showed that this thermal gap allows avoiding a "joint" even under the most adverse conditions of the TCR operation in the lock, which, in turn, eliminates the deformation of the ring beyond the yield strength, and, therefore, allows to maintain the health of the sealing ring and the seal assembly as a whole.

Given that an increase in the lock in the free state entails an increase in oil consumption, it is advisable to use rings with the size of the lock in the free state proposed by the authors only for the extreme rings from the turbine side. At the same time, it is possible not to manufacture special rings, but to use selective assembly when picking the TCP.

References

[1] Nikitin D A, Mezhetskiy G D, Denisov A S, Asoyan A R, Nikitin P D and Chekmarev V V 2017 Reasons for turbocompressors' failure and recommendations for repairing sealing components *Agricultural Scientific J.* **10** 50–4

[2] Denisov A S, Asoyan A R, Nikitin D A, Orlov N I, Vedeneev A A and Shirokova S A 2018 In the book: *Improvement of road systems and service technologies. Influence of thermal stress of turbocharger elements on its resource* (Saratov: Saratov State Technical University Press.) pp 188–92

[3] Mezhetskiy G D, Nikitin D A, Chekmarev V V and Mezhetskiy D V 2015 Effect of temperature on the thermal fatigue strength of its parts *Scientific review* **8** 121–5

[4] Bargende M and Putter R G 1986 Ermitttung der Ladungs bewegung in motorischen Brennraumen durch Messung instationarer Oberflachen temperatureuraturverlante *MTZ* **12** 491–2

IOP Conf. Series: Materials Science and Engineering **832** (2020) 012085 doi:10.1088/1757-899X/832/1/012085

The use of reusable fuel injection with the aim of improving the economic and environmental performance of the diesel engine

E A Savastenko[1], A A Savastenko[1], E V Gorbachevskiy[1] and V A Markov[2]

[1]Moscow Automobile and Road State Technical University (MADI), 64, Leningradsky ave., Moscow, 125319, Russia
[2]Head of the Piston Engines Department, Bauman Moscow State Technical University, 5, 2-a Baumanskaya str., Moscow, 105005, Russia

E-mail: e.d.u.a.r.d@inbox.ru

Abstract. The current trend in the development of motor-and-tractor diesel engines imposes tough requirements on fuel equipment and the fuel supply process. The implementation of it is difficult to combine with the requirements of reducing heat generation at the initial stage of combustion and the rate of pressure increment to reduce exhaust emissions and engine noise. A decrease in the duration of the injection process and an increase in the volumetric feed rate lead to a growth in the cycle dynamism factor and an increase in the operation noise. This scientific article discusses the use of multiple injection of fuel in a diesel engine to reduce the emission of nitrogen oxides, smoke and particulate matter by improving the heat transfer processes, the mixture formation and, as a result, heat generation. Low exhaust gases toxicity and smokiness values are achieved.

1. Introduction

The development of engines with promising energy and environmental parameters requires the development of new options for the fuel supply system for various diesel models [1, 2]. In this regard, important tasks are to ensure injection pressure up to 200MPa and higher [3, 4]; as well as regulating the shape of the leading edge of the injection characteristics [5, 6] and organizing the distribution of fuel in the combustion chamber [7, 8]. The required law of fuel injection for any engine operation mode is formed by changing the control pulse duration and the pressure in the fuel accumulator. It also depends on the wave phenomenon that occurs in the high-pressure line and has a significant effect on the fuel injection during multi-stage injection [9–12].

These materials are based on experimental data obtained by masters and graduate students of RUDN University and according to a mathematical model developed at Moscow State Automobile and Road Technical University (MADI) [1, 13]. The experiments were carried out on a single-cylinder diesel engine equipped with battery-powered fuel equipment (CR). In the course of research, tests were carried out on various methods of supplying fuel to the diesel combustion chamber. In particular, several preliminary injections of fuel were used in relation to the main one with varying injection angles, injection pressures and angular intervals between pre-injections.

2. Experimental study of reusable fuel injection

The processing of experimental data consisted in analysis of the dependence of economic, toxic, noise indicators of the diesel engine, as well as smokiness and their compliance with modern requirements for exhaust emissions. A diagram of the multi-stage injection characteristic is shown in Fig. 1.

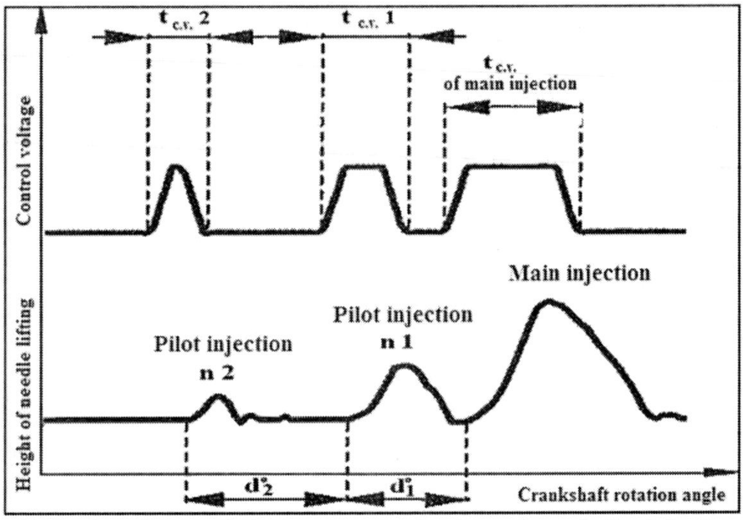

Figure 1. Stepped injection characteristic ($t_{c.v.2}$ – control voltage time of pilot injection n_2; $t_{c.v.1}$ - control voltage time of pilot injection n_1; d°$_1$ – delay between pilot injection n_1 and basic (main) injection; d°$_2$ – delay between pilot injection n_2 and pilot injection n_1) [14]

Five injections were studied with different amounts of fuel (different times of the control voltage impulse ($t_{c.v.}$)) and different delays between the pilot and main injections for illustrating the effect of pre-injection (pilot injection) on the indexes of the diesel engine exhaust gases emissions. The experiment was carried out at a constant frequency of the crankshaft (n = 2000 rpm) and injection pressure (P_{inj} = 55 MPa).

Emission indexes of the diesel engine exhaust gases are shown in table 1. Graphs of nitric oxides (NO_x) emissions, specific effective fuel consumption (g_e), soot and noise characteristics of the diesel engine are shown in figure 2.

Table 1. Emissions indexes from engine exhaust gases with factory settings

g_e (g/kW·h)	NO_x (g/kW·h)	Soot (g/kW·h)	HC (g/kW·h)	CO (g/kW·h)	Noise (dB)
305,5	1,5	0,72	0,71	4,8	84,9

Figure 2 shows that the optimal delay between the pilot injection n1 and the main injection is d° 1 = 6° of crankshaft rotation. The results of the first test showed both positive and negative points: soot indicators decreased by almost 40 %, HC indicators reduced by almost 16 %, CO indicators decreased by almost 17 %, specific effective fuel consumption reduced by 3 %. But negative phenomena appeared too: NO_x roused by 12 %, the noise of the combustion process increased slightly by 2 %.

This allows making an electronic control unit with a readout board that reads diesel engine data in real time and changes certain parameters set by the operator in the program. The pressure in the cylinder of the diesel was recorded by a piezoelectric sensor. Bosch ESAS250 gas analyzer was used for gas analysis, as well as equipment for determining the noise indexes of engines.

We made some changes: P_{inj} = 67.5 MPa and injection angle of the main portion of fuel (φ = 3.5 crankshaft rotation after TDC) for emissions reduction of solid particles HC, soot and NOx.

Figure 3 shows that the results of tests conducted with the changed injection pressure and the injection angle of the main portion of the fuel can be concluded that there were no large changes and the change in the injection angle of the main portion of the fuel did not give results. The indexes of CO and solid particles of CH remained unchanged, and some indicators increased: NO_x increased by almost 20 %. This is mainly due to the fact that because of the increased injection pressure, the fuel oxidizes and burns better, since the dispersion of fuel atomization has increased, which in turn leads to bigger speed of pressure rise rate in the cylinder. As a result of the injection grows, as could be expected, the smoke level decreased by almost 100 %, to 0.2 g/kW·h, which is a significant advantage of the fact that the injection pressure is increased.

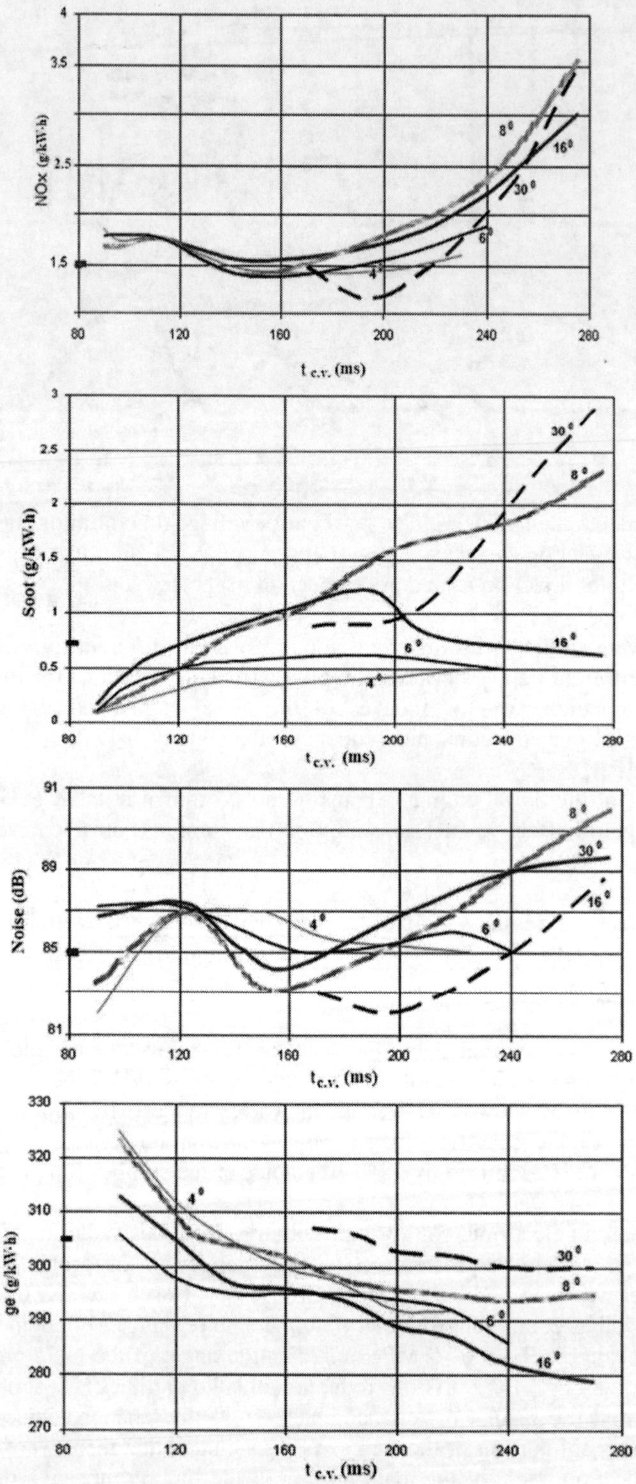

Figure 2. One pilot injection n_1 with various delays of the crankshaft rotation angle relatively to the main injection. 4^0, 6^0, 8^0, 16^0, 30^0 – delay angles between pilot and main injections in angles of the crankshaft rotation

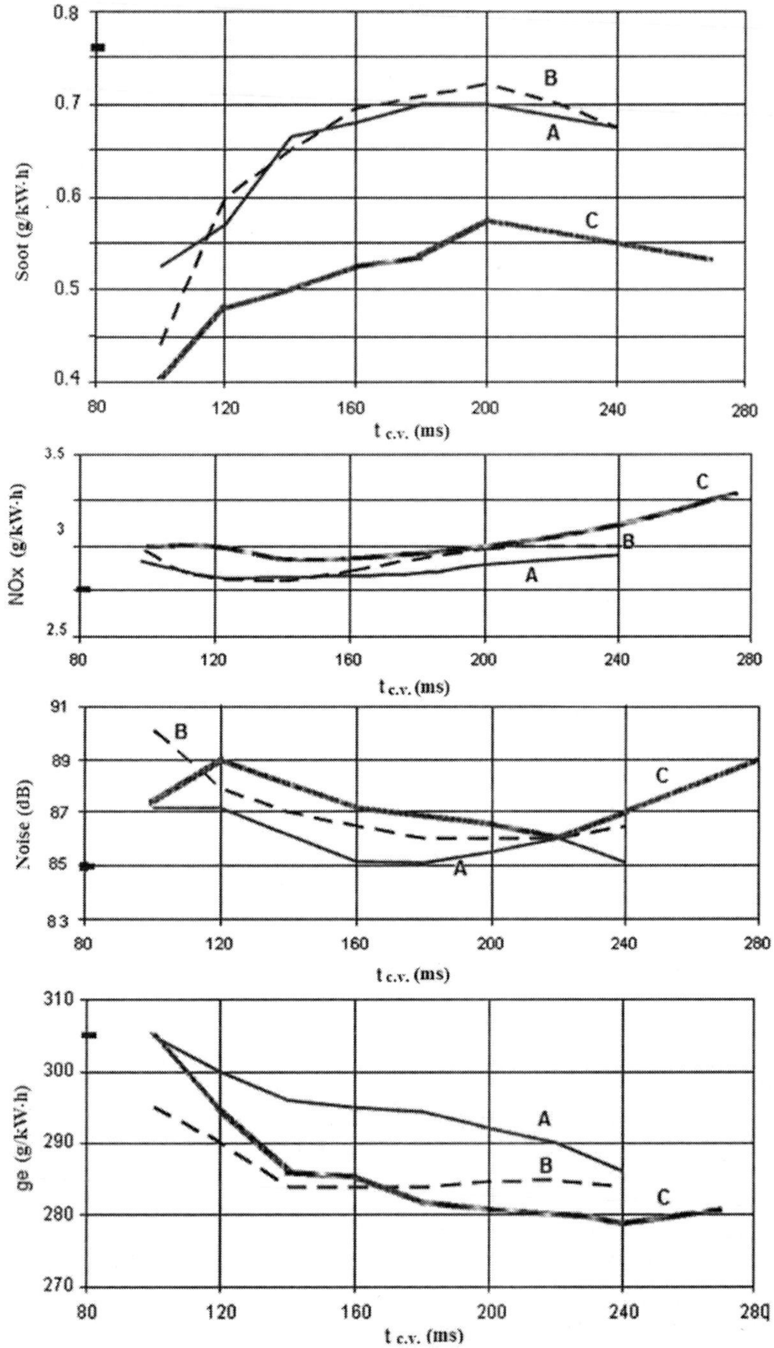

Figure 3. Influence of $t_{c.v}$ and delay time between the pilot injection n1 and the main injection on ecological, economic and noise indicators of the engine: **A** – Delay between the pilot and the main injection is 6° of crankshaft rotation. Fuel pressure in the battery is 55 MPa; Start of the main injection is 3.5° of crankshaft rotation. **B** – Delay between the pilot and the main injection is 6° of crankshaft rotation. Fuel pressure in the battery is 55 MPa; Start of the main injection is 1.5° of crankshaft rotation. **C** – Delay between the pilot and the main injection is 6° of crankshaft rotation. Fuel pressure in the battery is 67.5 MPa; Start of the main injection is 3.5° of crankshaft rotation

Due to the high NO_x emissions, the exhaust gas recirculation rate was increased by 8 %. NO_x rates have declined, but soot emissions have increased. Thus, the optimal settings were chosen: the delay between the pilot injection and the main one, the control voltage time (figure 4), where the highlighted line is the optimal dependence of the pressure in the cylinder on the nozzle needle lift; the angle of injection of the main fuel charge. The characteristic of the pressure change in the cylinder was chosen on the basis of the following considerations: the characteristic of the pressure change in the cylinder should be as flat as possible, without sudden changes in pressure. Also Figure 4 shows the optimal dependence of the heat generation rate on the nozzle needle lift, where two peaks are clearly traced, which characterize the pilot injection and the main one, respectively.

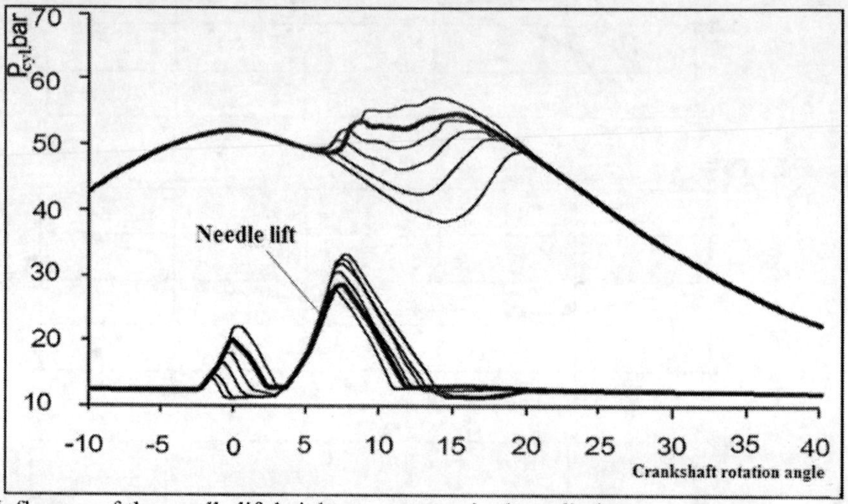

Figure 4. Influence of the needle lift height on pressure in the cylinder.
– Delay between the pilot and the main injection is 6° of crankshaft rotation. Fuel pressure in the battery is 67.5 MPa; Start of the main injection is 3.5° of crankshaft rotation after TDS $n = 2000$ rpm

3. Conclusions

The use of the multiphase injection characteristic in the diesel engine reduces indicators of emissions with exhaust gases due to more improved mixture formation process and, as a result, heat generation. An increase in the injection pressure leads to a decrease in emissions of soot and solid particles of hydrocarbons, more dispersed atomization of the fuel, which reduces the ignition delay time. The characteristics of heat generation and pressure in the cylinder become gentler, which allows increasing power, due to reduction of time for mixture formation. The angular intervals between the pilot and the main injection are determined by the optimal indicators of toxicity and noise of the diesel engine. The efficiency of using the battery system of high pressure fuel supply of Common Rail type to reduce emissions from exhaust gases increases with the use of the exhaust gas recirculation system.

Reference

[1] Shatrov M G, Golubkov L N, Dunin A U, Dushkin P V and Yakovenko A L 2017 The new generation of common rail fuel injection system for Russian locomotive diesel engines *Pollution Research* **36(3)** 678–84

[2] Shatrov M G, Sinyavski V V, Dunin A Y, Shishlov I G and Vakulenko A V 2017 Method of conversion of high- and middle-speed diesel engines into gas diesel engines *Facta Universit. Ser. Mechan. Engineer.* **15(3)** 383–95

[3] Shatrov M G, Golubkov L N, Dunin A U, Yakovenko A L and Dushkin P V 2015 Influence of high injection pressure on fuel injection perfomances and diesel engine worcking process *Thermal Sci.* **19(6)** 2245–53

[4] Shatrov M G, Golubkov L N, Dunin A U, Yakovenko A L and Dushkin P V 2015 Research of the injection pressure 2000 bar and more on diesel engine parameters *Int. J. of Applied Res.* **10(20)** 41098–102

[5] Shatrov M G, Golubkov L N, Dunin A Yu, Dushkin P V and Yakovenko A L 2017 A method of control of injection rate shape by acting upon electromagnetic control valve of common rail injector *Int. J. of Mechan. Engineer.and Technol.* **8(11)** 676–90

[6] Shatrov M G, Sinyavski V V, Dunin A Yu, Shishlov I G, Vakulenko A V and Yakovenko A L 2018 Using simulation for development of the systems of automobile gas diesel engine and its operation control *Int. J. of Engineer. and Technol.* **7(2.28** 288–95

[7] Shatrov M G, Malchuk V I, Dunin A U and Yakovenko A L 2016 The influence of location of input edges of injection holes on hydraulic characteristics of injector the diesel fuel system *Int. J. of Applied Engineer. Res.*, **11(20)** 10267–73

[8] Shatrov M G, Malchuk V I, Dunin A Y, Shishlov I G and Sinyavski V V 2018 A control method of fuel distribution by combustion chamber zones and its dependence on injection conditions *Thermal Sci.* **22(5)** 1425–34

[9] Shatrov M G, Golubkov L N, Dunin A U, Yakovenko A L and Dushkin P V 2016 Experimental research of hydrodynamic effects in common rail fuel system in case of multiple injection *Int. J. of Applied Engineer. Res.* **11(10)** 6949–53

[10] Savastenko A A and Stepanova M E 2010 *Features of the fuel equipment operation of a diesel engine with a two-stage system of fuel supply into the cylinder* (Moscow: RUDN Publ) 107 p

[11] Savastenko A A, Kuzyakov V V, Devyanin S N and Starosvetsky V V 2003 *Study of the staged fuel supply of fuel injection pump assembly of the company "Nippon Denso"* (Moscow: RUDN Publ) 74 p

[12] Savastenko, A A, Stepanova M E and Melnik I S 2010 *Hydrodynamic processes in fuel equipment of a separated type with double-spring nozzles* (Moscow: RUDN) 317 p

[13] Markov V A, Devyanin S N and Malchuk V I 2007 *Injection and Atomization of Fuel in Diesel Engines* (Moscow: BMSTU) 360 p

[14] Golubkov L N, Grishin A B and Emelyanov L A 2003 *The results of calculation studies and optimization of the battery fuel system with electrically-controlled injectors Piston engines and fuels in the XXI century* (Moscow: MADI) p 37

The regression model of automated control of timely replacement of air diesel air filter

V S Aslamova, T Taiwan, E A Rush

Irkutsk State Transport University, 15 Chernyshevsky Street, Irkutsk, 664074, Russia

E-mail: irgups-journal@yandex.ru

Abstract. The article analyzes the literature sources and experience of four-year operation of the diesel locomotive of series 2TE116UM No. 016 on the Ulan-Bator railway (Mongolia). The results of the analysis show that the turbocharger does not exhaust its service life until regular preventive maintenance, due to the onset of such a phenomenon as surging, stability of its operation is disrupted. It has been established that the main causes of surge emergence are: flow stall from the diffuser walls, and/or from the guide vanes, and / or from the impeller blades and others. These deviations from the operating conditions are caused by various factors, the main among which is the contamination of the air filter. The dynamics of the onset of surging, the inspection and the replacement of the air filter are analyzed. An empirical dependence of the air flow on the degree of pressure increase in the turbocharger is obtained. In order to automate the control of the turbocharger's unattended operation in the diesel locomotive operation mode set by the controller, the dependences of the pumping distance on the ambient temperature, the degree of pressure increase and the pressure at the inlet of the turbocharger, depending on the degree of air filter contamination, were investigated. The source data from the controller was processed in the Statgraphics Plus package and adequate regression equations were obtained to determine the running period magnitude before the surge start from each parameter individually and from the combined effect of all parameters. The coefficients of determination of the regression dependencies obtained are 92-99%.

1. Introduction

The mathematical analysis of experimental data with their subsequent approximation is of great importance for the development of automated processes of controlling the functioning of complex systems. The approximation with simultaneous statistical data processing is often provided by regression analysis methods [1]. Out of the entire set of functions depending on the input variables and external influences, only the regression function has a minimum forecast error of the output variable of the object under study. For that reason, regression models are widely used in various industrial, economic and educational systems [1-8]. The development of regression models helps to avoid an exhaustive search of options in the design and selection of optimal operating modes of equipment and processing procedures through the use of information technologies. A good enough approximation to the real process requires the use of a more complex multivariate regression model, which we will consider below.

2. Turbocharger surge dynamics, inspection and filter replacement graphs

During the operation of diesel locomotives, diesel air filters are gradually contaminated. With that, their passage area decreases, which leads to a decrease in air flow through the diesel [9-10]. The dis-charge characteristic shifts to the lower flow zone, the temperature of the exhaust gases rises. An in-crease in gas temperature can not compensate for a decrease in gas air flow; therefore, the speed of the turbine and compressor is reduced. As a result of in-creasing the boost pressure, the air consumption de-creases, and the joint operation mode of the turbo-charger and the engine moves to a point at another hydraulic characteristic located closer to the surge limit.

The change in air flow through the engine leads to a change in the indicator efficiency and engine shaft speed. With that, the efficiency of the turbocharger also changes, which affects the efficiency of the engine cycle.

If to maintain a constant frequency of rotation of the diesel shaft by increasing the fuel supply when the air filters are contaminated, this will result in an even greater increase in the temperature of the exhaust gases. In this case, the flow characteristics of the diesel engine and the line of operating modes of the turbocharger will be located closer to the surge limit [11].

Works [12-16] evaluated the impact of reducing the boost pressure on the output parameters of a diesel locomotive engine, as well as the effect of the degree of pollution of air filters on the constancy of performance of a diesel locomotive turbocharger.

The results of the operational data analysis showed that the turbocharger service life does not become exhausted until regular preventive maintenance, because stability of its operation is disrupted due to the onset of surging.

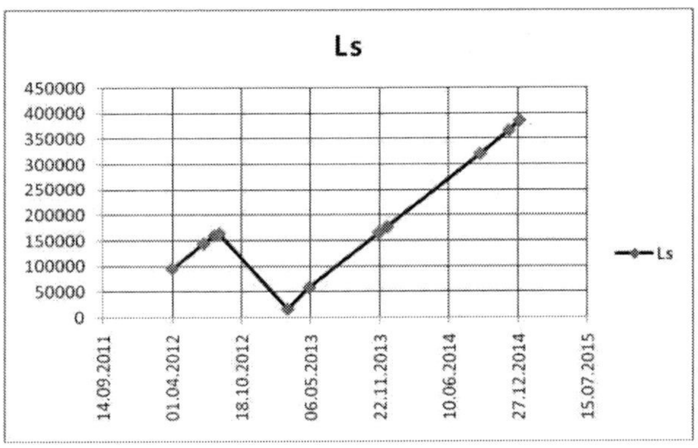

Figure 1. The dynamics of the onset of surging during the run *Ls*.

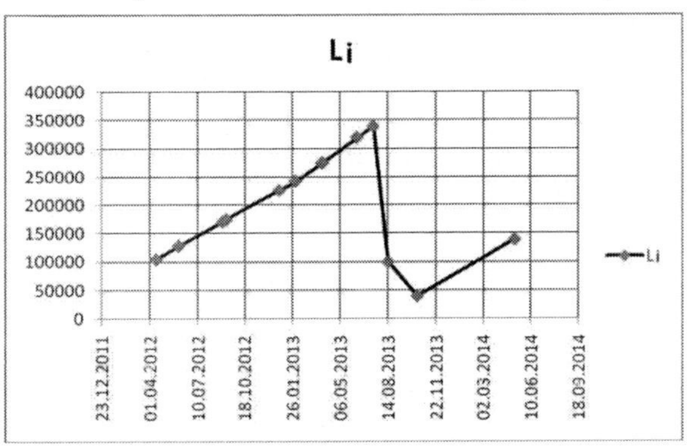

Figure 2. The dynamics of carrying out the inspection of air filter during the running period of the diesel locomotive *Li*.

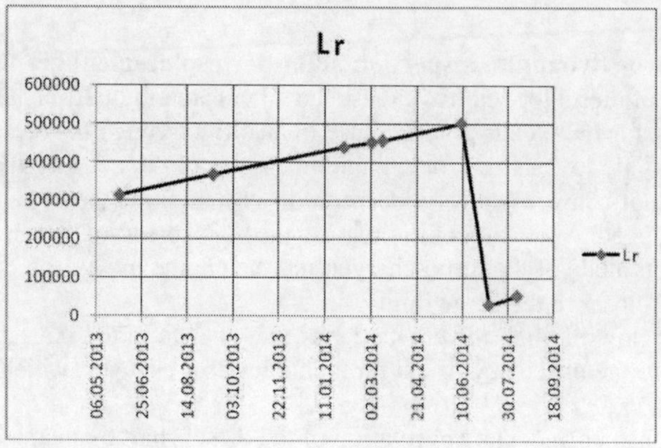

Figure 3. The dynamics of replacing the air filter.

It is established that the main causes of surge occurrence are: flow stall from the walls of the diffuser, and / or from the guide vanes, and / or from the vanes of the impeller; as well as instability of the flow in a vaneless space; self-oscillations of air in the "compressor - blow-off-and-boosting receiver" system. These deviations from the operating conditions are caused by various factors, the main among which is the contamination of the air filter.

As can be seen from Fig. 1, surging occurred in 0.5-3 months after the inspection of the air filter, in 4.5-6.5 months after replacing the filter.

To prevent the occurrence of surging, the workers of the Ulan-Bator Railway tried to carry out the inspection as often as possible (Fig. 2), or to replace the air filter of a locomotive, which led to heavy material expenditures.

In order to identify the quantitative effect of the technological parameters of a diesel locomotive's turbocharger on the running period at which surging occurs, and to develop empirical models for the automated control of safe running period, statistical processing of the experimental data has been performed using the Statgraphics Plus package.

Let us denote the following: *T* is the ambient air temperature, °C; *pz* is the pressure in front of the turbocharger, Pa; πк is the degree of pressure increase; *G* is the air consumption, kg/s. With increasing running period *L* of the diesel locomotive, the hydraulic resistance of the air filter increases due to contamination and the value of *pz* increases. To prevent surging, the filter is to be replaced or inspected, during which it is subjected to cleaning. This results in a decrease in the value of *pz*. Therefore, the distance of run of the locomotive L_{rp}, at which surge occurred, was calculated by the formula

$$L_{rp} = L - L_f \tag{1}$$

where *L* - from the beginning of operation, km, L_f is the running period during which the filter was inspected or replaced.

The air flow depends on πк and is described by linear regression (2), see Fig. 4. The regression validity coefficients (2): the coefficients of determination R^2 and of Darbin-Watson *DW*, mean square σ and absolute Δ errors are listed in the table.

$$G = -0.9188 + 2.2345\pi_k \tag{2}$$

Figure 5 shows a comparison of the experimental data of G_e with the calculation results of G_c by formula (2). The regression validity coefficients (2) are listed in the table.

Studying the influence of each parameter separately on the value of the safe running period, we obtained one-dimensional regressions, the reliability coefficients of which are given in the table.

The dependence of L_{rp} on *T* is quadratic (see Fig. 6).

IOP Conf. Series: Materials Science and Engineering **832** (2020) 012086 doi:10.1088/1757-899X/832/1/012086

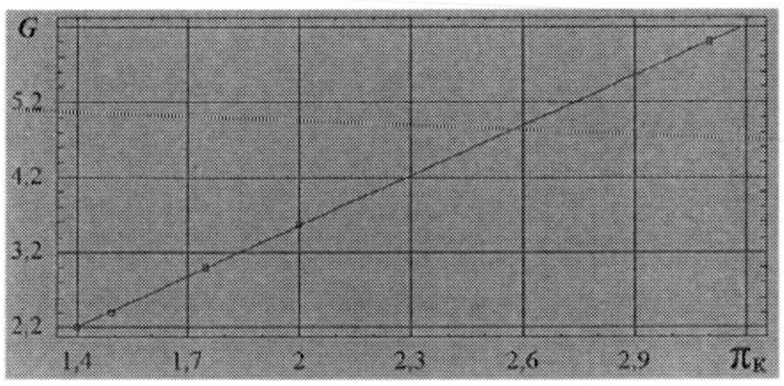

Figure 4. Dependence of the air flow G on the degree of pressure increase π_K in the turbocharger.

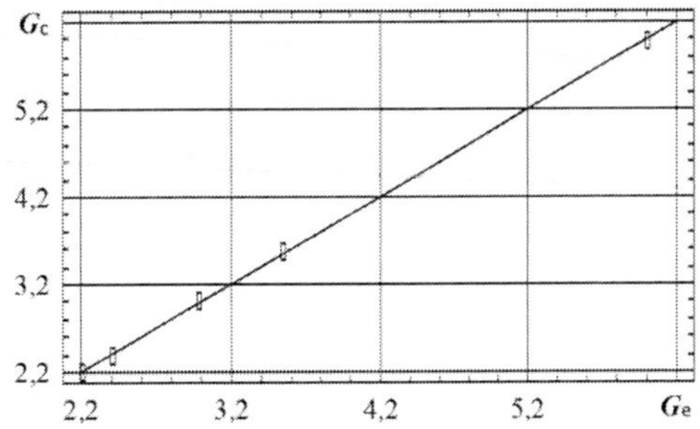

Figure 5. Comparison of values calculated by formula (1) G_c with experimental values G_e.

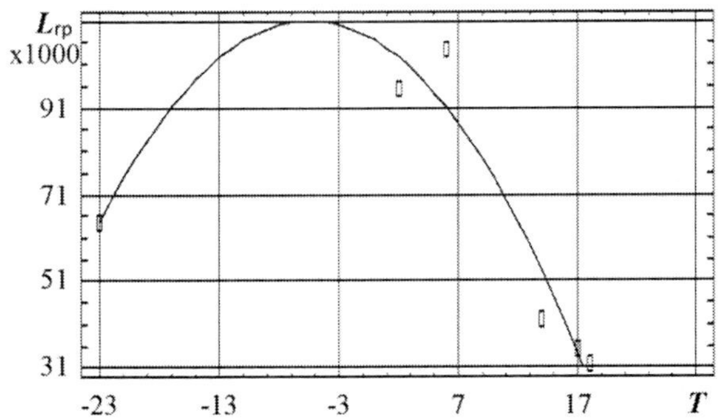

Figure 6. Dependence of surge running period on air temperature.

The greatest value of L_{rp} can be found from the necessary condition for the maximum function:

$$\frac{\partial L_{rp}}{\partial T} = 1674.98 - 2 \cdot 152.46 \cdot T = 0 \rightarrow \max L_{rp}$$

At $T = \dfrac{1674.98}{304.92} \approx 5.49\ ^{\circ}\mathrm{C}$

The regression validity coefficients (3) are shown in the table. The accuracy of the regression found can be judged by Fig. 7, which compares the running period values calculated by formula (3) L_{crp} (abscissa axis) with experimental L_{rp} (ordinate axis).

$$L_{crp} = 106674.0 - 1674.98T - 152.46T^2 \qquad (3)$$

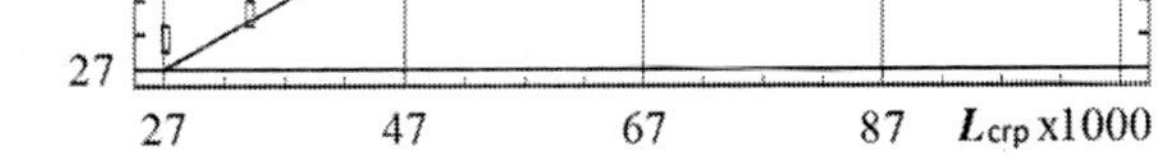

Figure 7. Comparison of the values of L_{crp} calculated by formula (3) with experimental L_{rp}.

The dependence of L_{rp} on π_K is parabolic (see Fig. 8). The regression validity coefficients (4) are presented in the table.

$$L_{rp} = -5.130 \cdot 10^6 + 3.810 \cdot 10^6 - 693361.0\pi_k^{\ 2} \qquad (4)$$

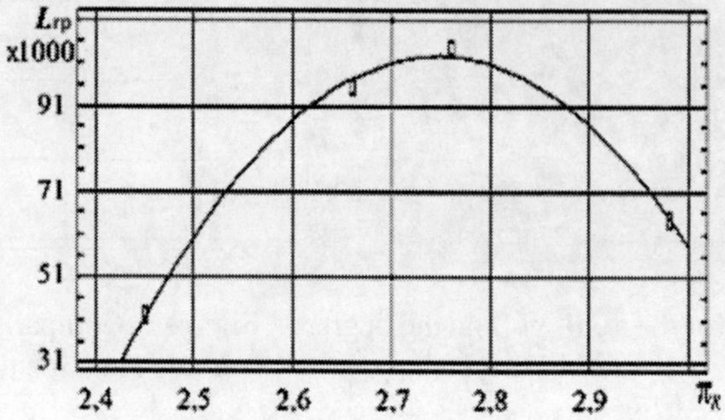

Figure 8. The dependence of L_{rp} on π_K.

From the necessary condition for the existence of an extremum of the function, we find that the greatest value of L_{rp} is achieved when $\pi_K = 2.747$, because

$$\frac{\partial L_{rp}}{\partial \pi_k} = 3809610.0 - 2 \cdot 693361 \cdot \pi_k = 0 \rightarrow \max L_{rp}$$

At $\pi_k = \dfrac{3809610}{1386722} \approx 2.747\ ^{\circ}\mathrm{C}$

The accuracy of the regression found can be judged by Fig. 9, which compares the running period values calculated by formula (4) L_{crp} (abscissa axis) with experimental L_{rp} (ordinate axis).

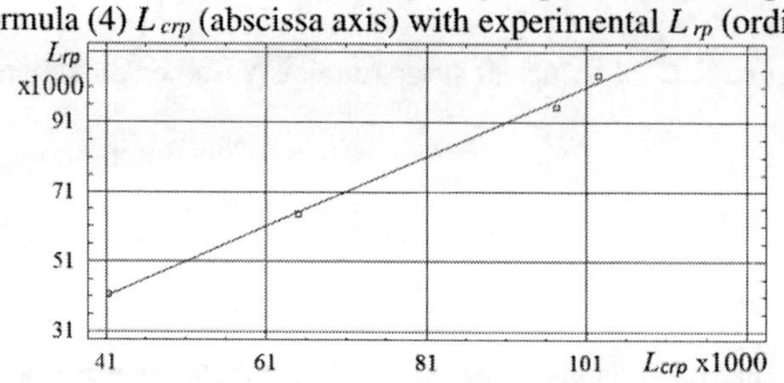

Figure 9. Comparison of the values of L_{crp} calculated by formula (4) with experimental L_{rp}.

The regression dependence of L_{rp} x1000 on the parameter p_z is quadratic (see Fig. 10):

$$L_{rp} = -2.91352 \cdot 10^6 + 39.29 \cdot pz - 0.00013pz^2 \qquad (5)$$

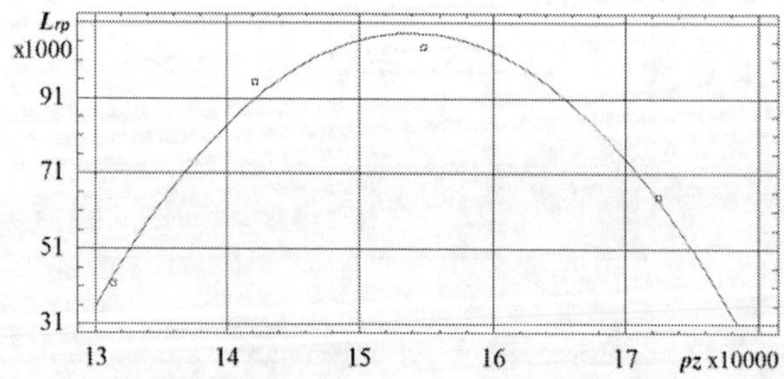

Figure 10. Dependence of L_{rp} on pz.

The regression validity coefficients (5) are listed in the table. Fig. 11 shows a comparison of the running period values calculated by formula (5) L_{crp} (abscissa axis) with experimental data L_{rp} (ordinate axis).

The fourth compatibility condition for the valve SJ:

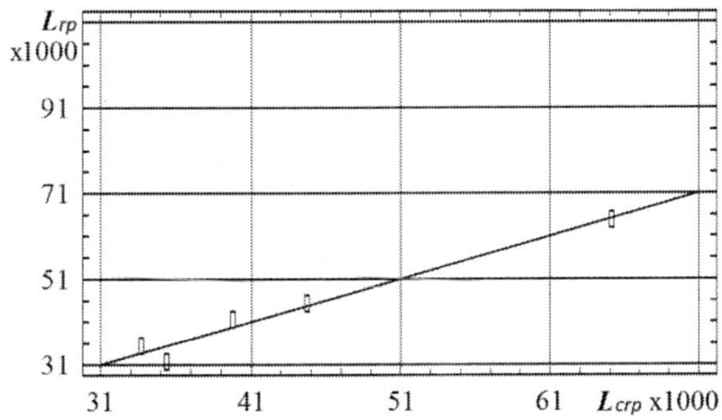

Figure 11. Comparison of the values of L_{crp} calculated by formula (5) with experimental L_{rp}.

A characteristic feature of the modern stage of railway transport management is the automation of production and management processes. It is achieved by creating automated and automatic systems of machines, robotic complexes, creating automated workplaces for specialists and managers using information technologies to analyze processes and machines, and choosing automation options [15-20]. Computer technology will make it possible to effectively use modeling and software systems in the process of locomotive control.

The multi-parameter linear regression equation describing the combined influence of the parameters under consideration is written as

$$L_{rp} = -43418.4 + 43607.8 \cdot \pi_k^2 - 1.172 p_z - 154.837 T^2 \quad (6)$$

It can be seen from the formula that with an increase in the parameters p_z and T, surging occurs faster (L_{rp} decreases), and with the increase of the parameter $\pi_к$ the value of L_{rp} increases. The regression validity coefficients (6) are given in the table.

Table 1. Regression validity coefficients.

Formula No.	R^2, %	DW	σ	Δ
(2)	99.99	2.74	0.0156	0.0113
(3)	92.16	1.54	11392.60	10859.30
(4)	99.67	1.45	2917.77	21744.0
(5)	98.46	1.45	6181.34	19119.70
(6)	96.99	3.26	7231.95	4331.31

The obtained equation (6) will allow controlling the amount of safe running period during operation of the locomotive, which should not exceed the value of L_{rp} obtained by formula (6) and accurately describes 99.26% of the initial data.

The accuracy of the regression found can be judged by Fig. 9, which compares the running period values calculated by formula (6) L_{crp} (abscissa axis) with experimental L_{rp} (ordinate axis).

Thus, the obtained multi-parameter linear regression was used in the development of an automated software module [17, 21-29] that monitors the running period of a diesel locomotive and prevents surging as a result of timely replacement of the air filter.

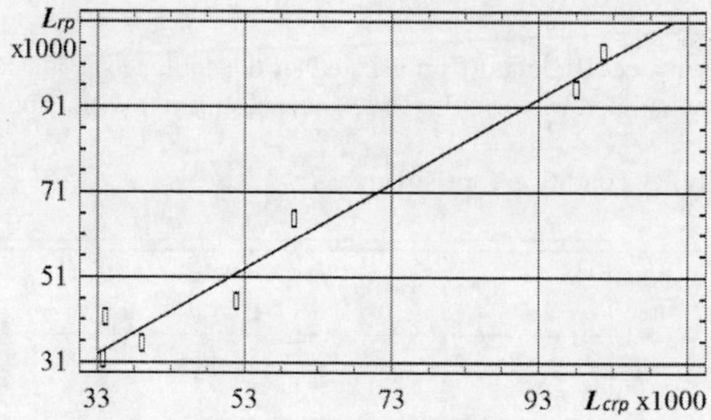

Figure 12. Comparison of the values of L_{crp} calculated by formula (6) with experimental L_{rp}.

3. Conclusion

The use of automated calculation systems can significantly reduce the time to solve the problem. The programming functions built into the system make it possible to create simple program modules necessary for repeated calculations, for example, to find the coordinate $x_{max.o}$ and the radius $r_{max.p}$, used in further calculations as the initial approximation for the *Given – minimize* computational unit.

The built-in Given block allows solving the problems of exploring functions for an extremum (*Given – minimize, – maximize, – minerr*, etc.), however, it is limited by initial approximations, which in most cases leads to incorrect results. The main advantage of the Given block, in this case, is the ability to calculate rational geometric parameters of the seat (thickness) by solving an optimization problem with constraints on allowable stresses.

The presented example of solving the problem of optimizing the geometric dimensions of a thin-walled seat under dynamic and static loading in PTC MathCAD, its automation using the built-in Given block can also be supplemented with data obtained from modeling in various systems (MSC.vN4W, APM WinMachine, etc.).

References

[1] Prokofiev A B, Shakhmatov E V 2005 *Using regression analysis to study complex systems management processes: method. instructions* (Samara, SSAU Publ.) p 23

[2] Kolosov A D, Gozbenko V E, Shtayger M G, Kargapoltsev S K, Balanovskiy A E, Karlina A I, Sivtsov A V, Nebogin S A 2019 Comparative evaluation of austenite grain in high-strength rail steel during welding, thermal processing and plasma surface hardening *IOP Conference Series: Materials Science and Engineering* **560** 012185

[3] Khomenko A P, Gozbenko V E, Kargapoltsev S K, Minaev N V, Karlina A I 2017 *International Journal of Applied Engineering Research* T. 12 **23** pp 13773-13778

[4] Kargapoltsev S K, Khomenko A P, Gozbenko V E, Minaev N V, Karlina A I 2017 Development of new lubricants for reducing the wear of the elements of the path and running parts of rolling stock *International Journal of Applied Engineering Research* Vol. 12 **22** pp 12362-12368

[5] Konstantinova M V, Balanovskiy A E, Gozbenko V E, Kargapoltsev S K, Karlina A I, Shtayger M G, Guseva E A, Kuznetsov B O 2019 Application of plasma surface quenching to reduce rail side wear *IOP Conference Series: Materials Science and Engineering* **560** 012146

[6] Grechneva M V, Balanovskiy A E, Gozbenko V E, Kargapoltsev S K, Karlina A I, Shtayger M G, Karlina Yu I, Govorkov A S 2019 Quality and reliability improvement of "tube-tube plate" welded joints during welding by pulse pressure *IOP Conference Series: Materials Science and Engineering* **560** 012142

[7] Guseva E A, Kargapoltsev S K, Balanovskiy A E, Karlina A I, Shtayger M G, Gozbenko V E, Konstantinova M V, Sivtsov A V 2019 Comparative evaluation of corrosion resistance of wheel and rail steels in various media *IOP Conference Series: Materials Science and Engineering* **560** 012181

[8] Shtayger M G, Balanovskiy A E, Kargapoltsev S K, Gozbenko V E, Karlina A I, Karlina Yu I, Govorkov A S, Kuznetsov B O 2019 Investigation of macro and micro structures of compounds of high-strength rails implemented by contact butt welding using burning-off *IOP Conference Series: Materials Science and Engineering* **560** 012190
[9] Mezheritsky A D 1971 *Turbochargers of marine diesel engines* (Leningrad: Sudostroenie Publ.) p 192
[10] Skovorodnikov E I, Anisimov A S, Grishina Yu B 2012 Evaluation of the effect of reducing the boost pressure on the output parameters of a diesel engine *The Journal of Transib Railway Studies. Omsk State Transport Un-ty* **1 (9)** pp 39-42
[11] Rush E A, Taiwan Tungalag 2013 Evaluation of the efficiency of the air cleaning system of a diesel engine of a diesel locomotive of the 2TE116UM series in the conditions of Mongolia *Transport infrastructure of the Siberian region: Materials of the IV international scientific-practical conference* **1** pp 221-224
[12] Rush E A, Tungalag Taiwan 2015 Analysis of the degree of influence of air filter pollution on the stability of the turbocharger of a diesel engine 2TE116UM *Transport infrastructure of the Siberian region: Materials of the IV international scientific-practical conference* **1** pp 285-290
[13] Balanovskiy A E, Shtayger M G, Karlina A I, Kargapoltsev S K, Gozbenko V E, Karlina Yu I, Govorkov A S, Kuznetsov B O 2019 Surface hardening of structural steel by cathode spot of welding arc *IOP Conference Series: Materials Science and Engineering* **560** 012138
[14] Vdovin K N, Feoktistov N A, Sinitskii E V, Gorlenko D A, Durov N A2015 Production of high-manganese steel in arc furnaces. Part 1 *Steel in Translation* Vol. 45 **10** pp 729-732
[15] Vdovin K N, Gorlenko D A, Feoktistov N A 2016 Characteristics of excess phase in cast high-manganese steel *Steel in Translation* Vol. 46 **7** pp 484-488
[16] Vdovin K N, Feoktistov N A, Gorlenko D A 2016 The effect of the cast high-manganese steel primary structure on its properties *Materials Science Forum* **870** pp 339-344
[17] Rush E A, Tungalag Taiwan 2015 Reliability analysis of the turbocharger of a diesel locomotive 2TE116UM *Transport infrastructure of the Siberian region: Materials of the IV international scientific-practical conference* **1** pp 295-300
[18] Lombardi P, Styczynski Z, Sokolnikova T, Suslov K 2014 Use of energy storage in Isolated Micro Grids *Proceedings - 2014 Power Systems Computation Conference, PSCC* 7038361
[19] Suslov K V, Stepanov V S, Solonina N N 2013 Smart grid: Effect of high harmonics on electricity consumers in distribution networks *IEEE International Symposium on Electromagnetic Compatibility* 6653416 pp 841-845
[20] Lombardi P, Sokolnikova T, Suslov K V, Styczynski Z 2012 Optimal storage capacity within an autonomous micro grid with a high penetration of Renewable Energy Sources *IEEE PES Innovative Smart Grid Technologies Conference Europe* 6465706
[21] Certificate of state registration of a computer program 2016 "Determining the surge distance value depending on the technological parameters of a diesel locomotive turbocompressor" No. 2016611489 was registered in the Computer Program Register on April 13, 2016
[22] Sivtsov A V, Sheshukov O Y, Tsymbalist M M, Nekrasov I V, Makhnutin A V, Egiazar'yan D K, Orlov P P 2019 Steel Semiproduct Melting Intensification in Electric Arc Furnaces Using Coordinated Control of Electric and Gas Conditions: II. On-Line Control of the State of the Charge and Melt Zones in Electric Arc Furnaces *Russian Metallurgy (Metally)* Vol. 2019 **6** pp 565-569
[23] Sivtsov A V, Sheshukov O Y, Tsymbalist M M, Nekrasov I V, Makhnutin A V, Egiazar'yan D K, Orlov P P 2018 Steel Semiproduct Melting Intensification in Electric Arc Furnaces Using Coordinated Control of Electric and Gas Conditions: I. Heat Exchange and Structure of the Electric Arc Furnace Laboratory *Russian Metallurgy (Metally)* Vol. 2018 **12** pp 1108-1113
[24] Nekrasov I V, Sheshukov O Y, Metelkin A A, Sivtsov A V, Tsymbalist M M 2016 Slag conditions in electrosmelting: A review *Steel in Translation* Vol. 46 **6** pp 435-442

[25] Sivtsov A V, Sheshukov O Y, Tsymbalist M M, Nekrasov I V, Egiazar'yan D K 2015 The Valve Effect of an Electric ARC and Problems in Controlling Electric-ARC Furnaces *Metallurgist* Vol. 59 **5-6** pp 380-385
[26] Nekrasov I V, Sheshukov O Y, Metelkin A A, Sivtsov A V, Tsymbalist M M, Egiazar'yan D K 2015 Ensuring Consistent Foamability in Electric-Furnace Slags *Metallurgist* Vol. 59 **3-4** pp 300-304
[27] Sheshukov O Y, Nekrasov I V, Sivtsov A V, Tsymbalist M M, Egiazar'yan D K, Metelkin A A 2014 Dynamic monitoring of slag oxidation and thickness in the ladle-furnace unit *Steel in Translation* Vol. 44 **1** pp 43-46
[28] Gozbenko V E, Khomenko A P, Kargapoltsev S K, Minaev N V, Karlina A I 2017 Creating of the alternative lubricants and practice of their use *International Journal of Applied Engineering Research* Vol. 12 **22** pp 12369-12372

IOP Conf. Series: Materials Science and Engineering **832** (2020) 012087 doi:10.1088/1757-899X/832/1/012087

The reduced wear of wheel pair flanges of freight cars

V E Gozbenko[1, 2], **B Tuvshintur**[1], S K Kargapoltsev

[1] Irkutsk State Transport University, 15 Chernyshevsky Street, Irkutsk, 664074, Russia
[2] Angarsk State Technical University, 60 Chaykovskogo Street, Angarsk, 665835, Russia

E-mail: irgups-journal@yandex.ru

Abstract. Reduced wear of rails and wheels of rolling stock is a very important task of railway transport. Solving this problem not only in-creases the lifespan of wheels and rails, but often leads to a positive reduction of fuel consumption for train traction, reduces noise when the train moves in curves. The intensity of wear of rails and wheels of rolling stock depends on many factors. The practice of Ulan Bator railway shows that since 1992, the flange of the wheels has started to wear out intensively, thereby increasing operating costs to repair and reducing service life of the railcar wheel pairs. In order to determine comprehensive measures for the rotation of wheel pairs as well as to determine the intensity of wear of the flange thickness of railcar wheel pairs, experiments have been con-ducted on the wear and turning of wheel pairs of freight cars. To substantiate comprehensive measures for the rotation of wheel pairs and determine the intensity of wear of the flange thickness of the railcar wheel pairs, experimental tests on a freight car after the overhaul with the 33-mm flange thickness have been carried out throughout the year.

1. Introduction

The track length of the Ulan Bator railway is 2430 km, of which 1814.13 km are the main track, 379.39 km are station tracks, 211.61 km are approach lines. The main route from the border of the Russian Federation to the border of the People's Republic of China is 1113 km. In the main trunk line, 350 km are occupied by P65 type rails and 1,435.74 km are occupied by P50 type rails. The reduced wear of rails is up to 3 mm. The length of the curve sections of the track with a radius of less than 600 m is 551 km, i.e. 30% of the main track run. 272 km of track are operated on reinforced concrete sleepers, and 842 km of track on wooden sleepers. There are 1528 pieces of construction equipment in the railway in total, including 141 pieces with R65 rails and 1124 pieces with P50 rails. All point switchings are operated using wooden blocks. Over 900 km of the main route, the ballast is crushed quarry, and it is the sand and gravel mixture in the remaining sections. About 30% of the track with a crushed quarry ballast have a layer thickness under the sleeper of less than 15 cm with a contamination of more than 20%.

Reducing the wear of rails and wheels of rolling stock is a crucial task in railway transport. Solving this problem not only increases the lifespan of wheels and rails, but often results in to a positive reduction of fuel consumption for train traction and reduces noise when the train moves in curves.

The intensity of wear of rails and wheels of rolling stock depends on many factors. For the last 10 years, there have been significant changes in the construction of the track and the rolling stock caused by technical advancements in the railway transport [1-4]. The experience of the Ulan Bator Railways shows that since 1992 the wheel flange has been wearing out intensively, as a result of which, the running costs for repair are increasing, and the service life of the railcar wheel pairs is decreasing.

To reduce the wear of wheel pairs, as well as to increase the service life of wheels on the Ulan Bator railway, there has been a thorough experimental research related to ensuring guaranteed safety. According to the analysis of the research, the following proposals were made: 1) to reduce the permissible size of the flange thickness; 2) to grind thin-rimmed wheel pairs with a different profile of the taping line and run them into operation; 3) to insert the template DSH-4 on the working surfaces of the wheel pairs in order to determine the sharp flanges; 4) to set the minimum size of the wheel pair flange thickness to 24 mm during operation; 5) to introduce the lubrication of rails [5-11].

2. Experimental studies Experiment No. 1

To carry out complex measures to rotate wheel pairs and determine the wear intensity of the thickness of the flange of wheel pairs of railcars, there was an experimental study to determine the wear of freight railcars with 33-mm thickness of the flange over a year. As a result of these studies, the data obtained are listed in Table 1.

The results of experimental studies of the flange thickness are listed in Table 2.

When carrying out the experiment, we took in-to account the calculated wear of the thickness of the flange of wheel pairs of railcars according to the formula:

$$H = L \times H \tag{1}$$

where L is the railcar run (km), H= 0.00008 is the wear of the flange for 1 km of the railcar run (mm).

Table 1. Experimental studies Experiment No.1

Date	Relative size	811935							
		wheel pair 1		wheel pair 2		wheel pair 3		wheel pair 4	
		side A	side B	side A	side B	side A	side B	side A	side B
		Initial flange thickness							
1/V	Flange thickness	31.7	32.1	32.1	32.2	32.7	32	32.6	31.7
		After the 4947-km run							
1/VI	Flange thickness	30.2	31.4	31.4	31.5	31.1	31.4	31.6	30.9
	Wear	1.5	0.7	0.7	0.7	1.6	0.6	1	0.8
	Wear after 100 km	0.011	0.005	0.005	0.005	0.012	0.004	0.007	0.006
		After the 12449-km run							
1/VIII	Flange thickness	29.1	30.3	30.6	30.1	29.1	30.9	30.9	30
	Wear	1.1	1.1	0.8	1.4	2	0.5	0.7	0.9
	Wear after 100 km	0.008	0.008	0.006	0.011	0.016	0.004	0.005	0.007
		After the 12775-km run							
1/X	Flange thickness	28.4	29.4	30	29.7	28.4	30.4	30.6	29.6
	Wear	0.7	0.9	0.6	0.4	0.7	0.5	0.3	0.4
	Wear after 100 km	0.014	0.018	0.012	0.008	0.014	0.010	0.006	0.008
		After the 30171-km run in total							
	Wear	3.3	2.7	2.1	2.5	4.3	1.6	2	2.1
	Wear after 100 km	0.010	0.008	0.006	0.008	0.014	0.005	0.006	0.006
	Wear intensity	2.0 times	1.6 times	1.2 times	1.6 times	1.75 times	minimal	1.2 times	1.2 times

Table 2. The results of experimental studies of the flange thickness

Operation time	Flange thickness, mm		Notes (measures taken)
	The thickness of the A-side flange	The thickness of the B-side flange	
0	33	33	
After 6 months	28	32	Rotation of wheel pairs
	32	28	Repeated rotation of wheel pairs
After 12 months	27	26	
After 16 months	25	25	Wheel pair machining
	22 months in total		

When calculating the thickness of the flange of wheel pairs of railcars, it is necessary to take into account the location of the wheel pair (see Fig. 1) under the railcars (K_1), temperature of the environment (K_2) and the flange ovality (K_3). These coefficients are given in Table 3-5 [12-15].

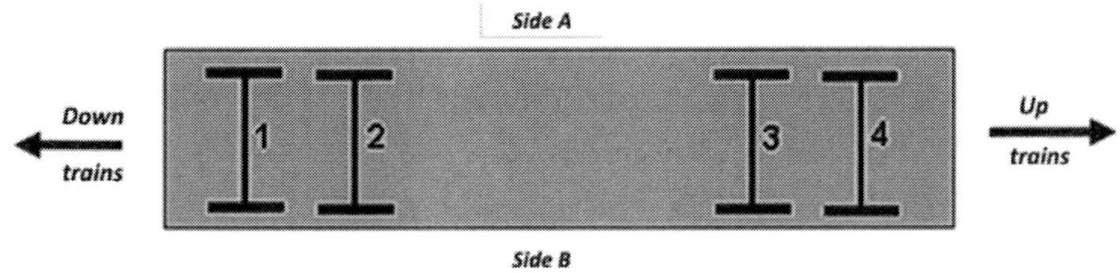

Figure 1. Location of wheel pairs.

Table 3. The location of the wheel pair under the railcars.

Location of a wheel pair	wheel pair 1		wheel pair 2		wheel pair 3		wheel pair 4	
	A	*B*	*A*	*B*	*A*	*B*	*A*	*B*
K_1	1.16	0.96	1.0	1.12	1.45	1.09	1.0	1.2

Table 4. The values of the coefficient K2 (medium temperature).

Month	2: 3: 4: 9: 10	5: 6: 7: 8	11: 12: 1
K_2	1.0	1.3	0.95

The diagram of the wear of the thickness of the flange at a coefficient of K_2 equal to 1 is shown in Fig. 2.

When carrying out the experiment, the wear of the thickness of the flange of wheel pairs of railcars was calculated according to the refined formula:

$$H_1 = L_1 \times 0.00008 \times K_1 \times K_2 \times K_3 \qquad (2)$$

where 0.00008 is the wear of the flange for 1 km of the railcar run (km); L is the run of the railcar (km); K_1 is the coefficient of location of wheel pairs; K_2 is the temperature coefficient of the environment; K_3 is the coefficient of ovality of the flange.

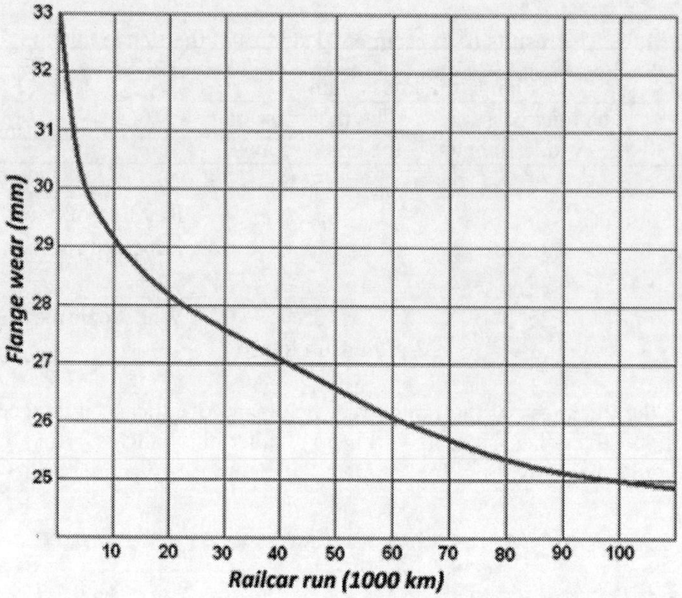

Figure 2. The diagram of the flange thickness wear.

Table 5. The values of the coefficient K3

Run L_1 (km)	Flange wear (mm)	The wear of the flange with a run of 1 km (mm)	coefficient K_3
Up to 10000	33 – 29	0.0004	5.0
10.000 – 20.000	29 –28	0.0001	1.25
20.000 – 60.000	28 – 26	0.00005	0.63
60.000 – 100.000	26 – 25	0.000025	0.31

The wear of the flange thickness can be calculated by the formula:

$$\sum H = H_1 + H_2 + ... + H_n \tag{3}$$

Since January 2014, 4 test gondola cars were selected after the repair in the depot, formed into one train and used to transport coal along the Ulan-Bator-Baganuur route. There were certain methods developed for the study and the following railcars were tested: two railcars, according to the current instructions for machining wheel pairs of freight cars, were used with the flange thickness of 33 mm. With the next two railcars, the machining was performed with the flange thickness of about 30 mm. On a trip in April and July 2014, the thicknesses of the flanges of wheel pairs for all four railcars under study were measured, with the measurements compared [16-23].

3. Experiment No. 2

To study the profile of the flange in order to determine the thickness with which it is necessary to machine the flanges of railcar wheel pairs.

It is accepted that the working surface of the flanges in the place of contact with the rail top should be released at an angle of 60°.

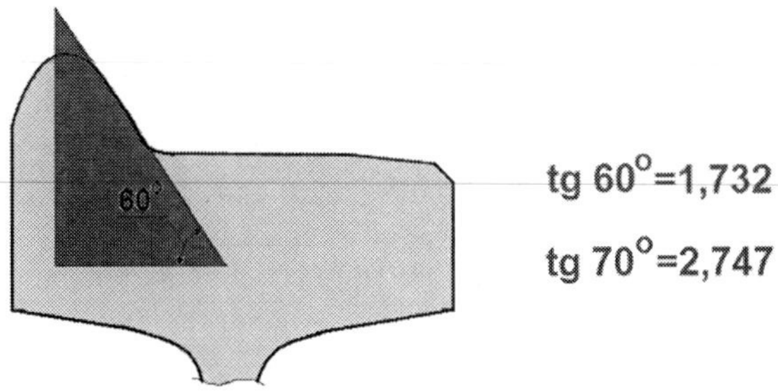

Figure 3. Determination of the profile of the railcar wheel pair flanges.

During operation, the flanges often change shape; therefore, the working surfaces are inclined at an angle of approximately 71^0. This type of railway fault is called a sharp flange, and the wheelset with such a defect is prohibited from being brought under the wagon. On the basis of changes in the thickness of the crests and the study of geometrical data of the wheel, you can approximately determine how the thickness and other parameters of the bandage after turning the wheel are changed (see Table 6).

Table 6 makes it possible to determine in advance what kind of repair it is and for which railcars we need to bring the wheel pairs after machining (see Table 7).

Using the above data in Tables 6 and 7, we will determine the economic indicators after the wheel machining, with parameters with 24-mm thickness and 40-mm thickness.

If you machine the wheels with the flange thickness of 33 mm, then the bandage becomes 24.4 mm in size and such a wheel pair can already be brought under the railcar, but only with a current repair. If machining is carried out with the flange thickness of approximately 30 mm, then the thickness of the bandage becomes 29.6 mm and this wheel pair can already be brought under the railcar not only with a repair at the depot, but also with a current repair.

Table 6. Machining data for wheel pairs of railcars with the 24-mm wear of flanges.

Flange thickness, mm	Reducing the thickness of the bandage (mm), with the working surface being at an angle of	
	60^0	70^0
33	15.6	24.7
32	13.8	22.0
31	12.0	19.2
30	10.4	16.5
29	8.7	13.7
28	7.0	11.0
27	5.2	8.3

Table 7. Sizes of the studied wheel pairs during repair.

Type of repair	Wheel pair size	
	Flange thickness (mm)	Bandage thickness (mm)
Complete overhauling	30-33	not less than 35
At the depot	28-33	not less than 27
Current	26-33	not less than 24

If you machine the wheels with the flange thickness of 33 mm, then the bandage becomes 24.4 mm in size and such a wheel pair can already be brought under the railcar, but only with a current repair. If machining is carried out with the flange thickness of approximately 30 mm, then the thickness of the

IOP Conf. Series: Materials Science and Engineering **832** (2020) 012087 doi:10.1088/1757-899X/832/1/012087

bandage becomes 29.6 mm and this wheel pair can already be brought under the railcar not only with a repair at the depot, but also with a current repair.

4. Experiment No. 3

When investigating the wear of the flange, we were able to determine that if the flange thickness can wear out to 29-26 mm at the working edge of the flange, the galling of metal into the top of the head of the flange should necessarily appear. Because of such faults in Ulan-Bator Railway, more than 500 railcars had to be uncoupled and repaired, which led to large economical costs.

Figure 4 shows the studied position of the pointed edge of the flange in the upper part of the flange (about 2 mm lower than the top of the flange) before rolling the wheel onto the switch point, which butts against the frame rail with a gap of 3 mm.

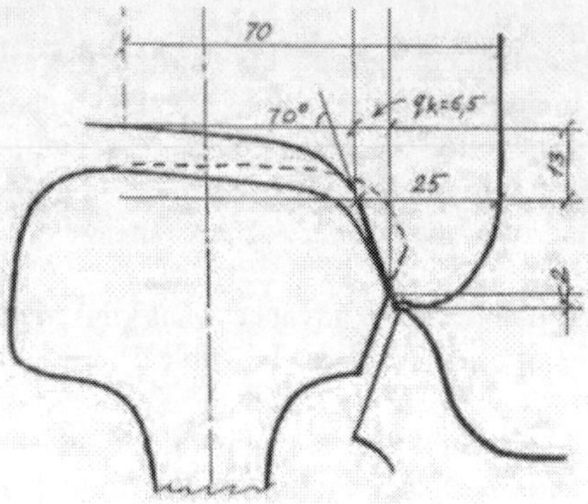

Figure 4. The sharp flange during the wheel rolling onto the switch point.

The rejection criterion, which characterizes the critical shape of the wear of the flange, can be represented as a horizontal projection forming the worn surface of the flange from a point located 2 mm below its top and of a point that is 13 mm higher than the average taping line.

According to the accepted terminology, the sharp flange is the galling of metal to the top of the flange on the edge of the transition of the worn-out part to the unworn one. This galling most often occurs under the action of friction forces during two-point contact of the wheel with the rail. This, as a rule, occurs in the curves at the moment when the contact moving along the flange, in contrast to the normal contact, does not open, and the flange sliding along the lateral surface of the rail is directed towards the top [24-28].

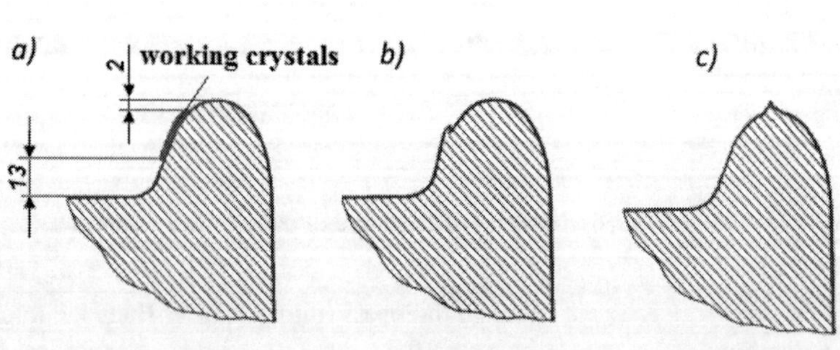

Figure 5. The main faults of the flange of wheel sets of railcars – a) a method for eliminating a fault; b) a fault on the working surface of the flange; c) on a non-working surface.

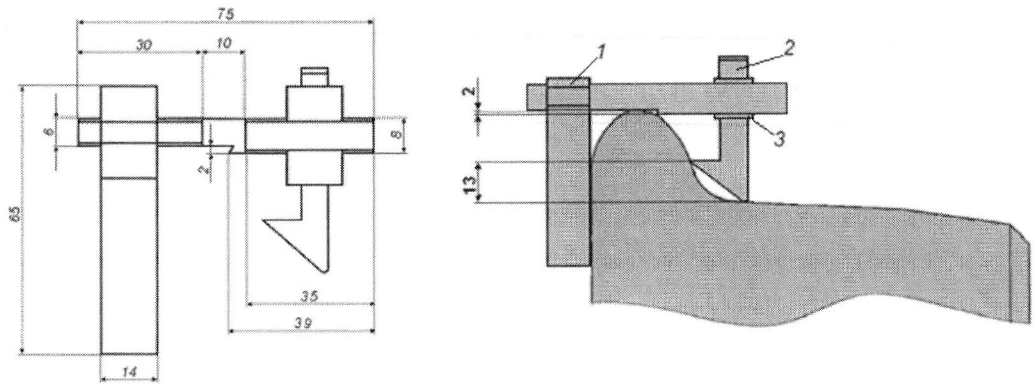

Figure 6. The diagram of the template DSh-4.

5. Conclusion

Based on a wide range of studies, the DSh-4 template was put into operation. As a result of these actions, the number of uncoupled railcars for the considered wheel pair faults with a sharp flange decreased by 80% and, accordingly, the cost savings were reduced.

References

[1] Lysyuk V C 2005 Interaction of rails and wheels in curves *Track and track facilities* **10** pp 11-17

[2] Bulatov Y N, Kryukov A V, Suslov K V 2017 Multi-agent technologies for control of distributed generation plants in the isolated power systems *Far East Journal of Electronics and Communications* Vol. 17 **5** pp 1197-1210

[3] Solodusha S, Suslov K, Gerasimov D 2016 Applicability of Volterra integral polynomials in the control systems of electric power facilities *International Conference Stability and Oscillations of Nonlinear Control Systems (Pyatnitskiy's Conference)* 7541227

[4] Suslov K V, Solonina N N, Smirnov A S 2011 Smart Grid: A new way of receiving primary information on electric power system state *IEEE PES Innovative Smart Grid Technologies Conference Europe* 6162654

[5] Galiev I I, Nekhaev V A, Nikolaev V A 2012 The safety of the movement of freight trains and the dynamic properties of the carriage of the wagon. *Journal of Transsib Railway Studies* **1** pp 107-112

[6] Sivtsov A V, Sheshukov O Y, Tsymbalist M M, Nekrasov I V, Makhnutin A V, Egiazar'yan D K, Orlov P P 2019 Steel Semiproduct Melting Intensification in Electric Arc Furnaces Using Coordinated Control of Electric and Gas Conditions: II. On-Line Control of the State of the Charge and Melt Zones in Electric Arc Furnaces *Russian Metallurgy (Metally)* **6** pp 565-569

[7] Sivtsov A V, Sheshukov O Y, Tsymbalist M M, Nekrasov I V, Makhnutin A V, Egiazar'yan D K, Orlov P P 2018 Steel Semiproduct Melting Intensification in Electric Arc Furnaces Using Coordinated Control of Electric and Gas Conditions: I. Heat Exchange and Structure of the Electric Arc Furnace Laboratory *Russian Metallurgy (Metally)* **12** pp 1108-1113

[8] Nekrasov I V, Sheshukov O Y, Metelkin A A, Sivtsov A V, Tsymbalist M M 2016 Slag conditions in electrosmelting: A review *Steel in Translation* Vol. 46 **6** pp 435-442

[9] Sivtsov A V, Sheshukov O Y, Tsymbalist M M, Nekrasov I V, Egiazar'yan D K 2015 The Valve Effect of an Electric ARC and Problems in Controlling Electric-ARC Furnaces *Metallurgist* Vol. 59 **5-6** pp 380-385

[10] Nekrasov I V, Sheshukov O Y, Metelkin A A, Sivtsov A V, Tsymbalist M M, Egiazar'yan D K 2015 Ensuring Consistent Foamability in Electric-Furnace Slags *Metallurgist* Vol. 59 **3-4** pp

300-304
[11] Sheshukov O Y, Nekrasov I V, Sivtsov A V, Tsymbalist M M, Egiazar'yan D K, Metelkin A A 2014 Dynamic monitoring of slag oxidation and thickness in the ladle-furnace unit *Steel in Translation* Vol. 44 **1** pp 43-46
[12] Monakhov I K 2008 Width. Elevation. Lubrica-tion *Track and the track facilities* **11** pp 20-24
[13] Yolkin K S, Yolkin D K, Kolosov A D, Ivanov N A, Shtayger M G 2018 Technologies, which allow to reduce an impact of metal silicon production on the environment *IOP Conference Series: Materials Science and Engineering* **411** 012028
[14] Kolosov A D, Gozbenko V E, Shtayger M G, Kargapoltsev S K, Balanovskiy A E, Karlina A I, Sivtsov A V, Nebogin S A 2019 Comparative evaluation of austenite grain in high-strength rail steel during welding, thermal processing and plasma surface hardening *IOP Conference Series: Materials Science and Engineering* **560** 012185
[15] Konstantinova M V, Balanovskiy A E, Gozbenko V E, Kargapoltsev S K, Karlina A I, Shtayger M G, Guseva E A, Kuznetsov B O 2019 Application of plasma surface quenching to reduce rail side wear *IOP Conference Series: Materials Science and Engineering* **560** 012146
[16] Guseva E A, Kargapoltsev S K, Balanovskiy A E, Karlina A I, Shtayger M G, Gozbenko V E, Konstantinova M V, Sivtsov A V 2019 Comparative evaluation of corrosion resistance of wheel and rail steels in various media *IOP Conference Series: Materials Science and Engineering* **560** 012181
[17] Vdovin K N, Feoktistov N A, Gorlenko D A 2016 The effect of the cast high-manganese steel primary structure on its properties *Materials Science Forum* Vol. 870 pp 339-344
[18] Nazarov N S, Yakimova G A, Gozbenko V E et al. 2003 The use of industrial waste in compositions for lubricating rails (Irkutsk: Ir-kut. un-ty Publ.)
[19] Karpushenko N I, Velichko D V, Bobovnikova N A 2010 Influence of gauge and condition of running gears of the rolling stock on wear in-tensity *The transport of the Russian Federation* **3 (28)** pp 10-14
[20] Balanovskiy A E, Shtayger M G, Karlina A I, Kargapoltsev S K, Gozbenko V E, Karlina Yu I, Govorkov A S, Kuznetsov B O 2019 Surface hardening of structural steel by cathode spot of welding arc *IOP Conference Series: Materials Science and Engineering* **560** 012138
[21] Balanovsky A E, Shtayger M G, Kondra'Ev V V, Karlina A I, Govorkov A S 2018 Comparative analysis of structural state of welded joints rails using method of Barkhausen effect and ultrasound *Journal of Physics: Conference Series* **1118** 012006
[22] Ivanchik N N, Balanovsky A E, Shtayger M G, Sysoev I A, Karlina A I 2018 Capability enhancement of production of activating fluxes for arc welding using ultradispersed products of silicon waste processing *IOP Conference Series: Materials Science and Engineering* **411** 012035
[23] Balanovsky A E, Shtayger M G, Grechneva M V, Kondrat'Ev V V, Karlina A I 2018 Comparative metallographic analysis of the structure of St3 steel after being exposed to different ways of work-hardening *IOP Conference Series: Materials Science and Engineering* **411** 012012
[24] Kuznetsov V M 2000 About the sharp flanges on the wheel pairs *Track and track facilities* **9** pp 16-19
[25] Balanovsky A E, Shtayger M G, Kondrat'Ev V V, Van Huy V, Karlina A I 2018 Plasma-arc surface modification of metals in a liquid medium *IOP Conference Series: Materials Science and Engineering* **411** 012013
[26] Balanovsky A E, Shtayger M G, Kondrat'Ev V V, Nebogin S A, Karlina A I 2018 Complex metallographic researches of 110G13L steel after heat treatment *IOP Conference Series: Materials Science and Engineering* **411** 012014
[27] Vdovin K N, Feoktistov N A, Sinitskii E V, Gorlenko D A, Durov N A 2015 Production of high-manganese steel in arc furnaces. Part 1 *Steel in Translation* Vol. 45 **10** pp 729-732

[28] Vdovin K N, Gorlenko D A, Feoktistov N A 2016 Characteristics of excess phase in cast high-manganese steel *Steel in Translation* Vol. 46 **7** pp 484-488

Statistical study of the wear of wheel pair flanges of freight cars

V E Gozbenko[1, 2], B Tuvshintur[1], S K Kargapoltsev

[1] Irkutsk State Transport University, 15 Chernyshevsky Street, Irkutsk, 664074, Russia
[2] Angarsk State Technical University, 60 Chaykovskogo Street, Angarsk, 665835, Russia

E-mail: irgups-journal@yandex.ru

Abstract. Reduced wear of rails and wheels of rolling stock is a very important task of railway transport. The flange profile studies were conducted to determine the machining sizes for the flange of railcar wheel pairs. The studies made it possible to determine that the rejection criterion characterizing dangerous form of the flange wear is a horizontal projection forming the worn-out surface of the flange between the points located at about 2 mm below its top and 13 mm above the average taping line. There are calculations made on the wear of the wheel pairs of gondola cars of the EC70 model manufactured in China after three years of operation.

1. Introduction

According to the reports of the VU-53 form of the freight car depot VChD-1 of the Dzunkhara station of the Ulan Bator Railway, 3,740 wheels with different faults were machined in 2014, but most of the wheels turned out to be with worn flanges [1-5].

Analysis of the wheel pairs of railcars and their faults showed that the main ones are: 1) No. 14 (thin flange) - 41%; 2) No. 22 (shelled treads) - 36%; 3) No. 20 (slid flats) - 21%. There were no other faults, for example, No. 73 (the difference between the diameters of the wheels) detected in the car depot, but the depot data say that it reaches 19% [6-11].

Earlier, with a 1524 mm gauge, the high flange in the taping line increased first, and then the wear of the flange began to increase more slowly [12-19]. After the gauging was done, the situation changed - the flange began to wear out faster. Within 5 months, June - October 1995, in the wheel pairs of freight cars (Ulan Bator Railway), the thickness of the wheel pair flanges reached the limit, i.e., less than 23 mm. High flanges began to increase slowly, so the corresponding machining was minimized. Therefore, the service life of the wheel pairs has significantly decreased. As a result of these measures, the service life of the wheel pairs began to decrease. To increase the time of service of wheel pairs of railcars on the Ulan Bator railway, a series of measures were undertaken in 1995-1998 [20-24]. Upon reaching the flange thickness 25-26 mm, railcar wheel pairs were rotated. Since January 1998, the technology of building-up the flanges has been introduced. The wheel pairs were rotated alternately when the desired flange thickness was reached. The first wheel pair was replaced by the third, the second wheel pair was replaced by the fourth pair, etc. To carry out comprehensive measures for the rotation of wheel pairs and to determine the intensity of wear of the flange thickness of railcar wheel pairs, experiments have been conducted on determining the wear and machining of wheel pairs of freight railcars.

2. Research Methods

The wear of the flanges of wheel pairs of gondola cars of model EC70 made in China after 3-year operation has been calculated. With that, the railcar mileage ranges from 60-70 thousand km, the main cargos are iron ores at the sections between the Eroo – Dzamyn-Ude stations of the Ulan-Bator Railway.

The statistical data for the calculation were obtained from 248 railcars after the repair in the freight car depot of the Dzunkhara of the Ulan Bator Railway. As is known, the release thickness of the flange of wheel pairs is 33 mm. When operating these railcars for 3 years after the release, a change in the thickness of the flange of wheel pairs of railcars ranges within 0-7 mm. Sample size $n = 496 \div 1984$, depending on the wheel pair of the railcar. The level of significance is accepted $\alpha = 0.005$ reliability $\gamma = 0.95$.

We will carry out a statistical study for wheel pairs in 248 railcars. The calculation is based on the results of the data obtained on the first wheel pair, the second wheel pair, the third wheel pair, the fourth wheel pair, the first bogie, the second bogie, the left wheels, the right wheels and the data on all wheel pairs.

Let us make a series of the distribution of the sample [13, 29-31], dividing by 7 parts and finding the length of the partial interval $h = \frac{7-0}{7} = 1$.

3. Research Summary

A number of distributions of the frequencies of the sample and the distribution of the relative frequencies are given in Tables 1 and 2.

Table 1. A number of distributions of the frequencies of the sample

Numeric interval	0-1	1.5-2	2.5-3	3.5-4	4.5-5	5.5-6	6.5-7
n_i	1	11	113	192	135	42	2
n_i/n	0.002	0.022	0.228	0.387	0.272	0.085	0.004

Table 2. A number of the distribution of the relative frequencies

x_i	0.5	1.75	2.75	3.75	4.75	5.75	6.75
n_i/nh	0.002	0.022	0.228	0.387	0.272	0.085	0.004

We find selected numeric characteristics according to the formulas:

$$\overline{x}_b = \frac{\sum_{i=1}^{k} x_i n_i}{n}, D_b = \frac{\sum_{i=1}^{k} x_i^2 n_i}{n} - (\overline{x}_b)^2, \sigma = \sqrt{D_b} \quad (1)$$

where k is the number of intervals;
The results of the calculations are listed in Table 3.

Table 3. The results of the calculations

i	x_i	n_i	u_i	$u_i n_i$	$u_i^2 n_i$	$(u_i+1)^2$
1	0.5	1	0.5	0.5	0.25	2.25
2	1.75	11	1.75	19.25	33.69	83.19
3	2.75	113	2.75	310.75	854.56	1589.06
4	3.75	192	3.75	720	2700.00	4332.00
5	4.75	135	4.75	641.25	3045.94	4463.44
6	5.75	42	5.75	241.5	1388.63	1913.63
7	6.75	2	6.75	13.5	91.13	120.13
				1946.75	8114.19	12503.69
				3.925	0.95	

Let us construct a histogram of the density of relative frequencies (see Fig. 1).

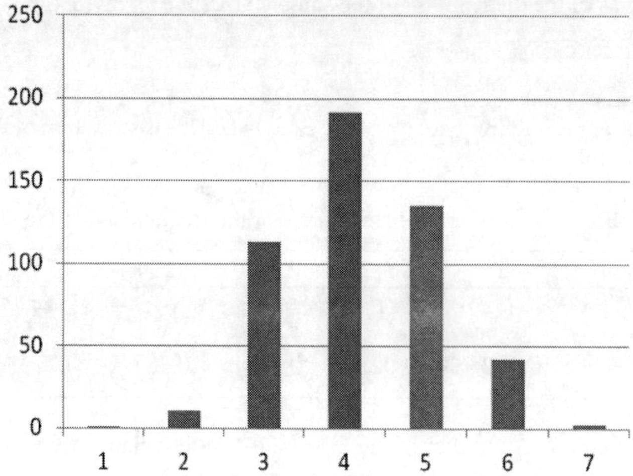

Figure 1. The histogram of the density of relative frequencies for the first wheel pair.

Table 4. The results of the calculations

The first wheel pair	**1**	**11**	**113**	**192**	**135**	**42**	**2**
x_i	0.002	0.022	0.228	0.387	0.272	0.085	0.004

The density function of normal distribution law:

$$f(x) = \frac{1}{\sigma\sqrt{2\pi}} e^{\frac{-(x-a)^2}{2\sigma^2}}, \qquad (2)$$

where $a = \bar{x}_b, \sigma_b \sqrt{\frac{n-1}{n}} = \sigma = S$.

Next, we find the confidence interval for the mathematical expectation *M(x)* and the mean square deviation $\sigma(x)$ with reliability $\gamma = 0.95$

$$\bar{x}_b - t_\gamma \frac{S_b}{\sqrt{n}} < M(x) < \bar{x}_b + t_\gamma \frac{S_b}{\sqrt{n}},$$

$S(1-q) < \sigma < S(1+q)$, at $q < 1$,

$0 < \sigma < S(1+q)$, at $q < 1$.

Find q using [15]

To verify the validity of the hypothesis put forward, we use Pearson's criterion at a significance level of $\alpha = 0.05$. The observed criterion value is calculated according to the formula:

$$x_{obs}^2 = \sum_{i=1}^{k} \frac{(n_i - n_i)^2}{n_i}, \qquad (3)$$

where n_i – are statistical frequencies, n_i – are theoretical frequencies, defined in the case of the normal law of distribution according to the formulas:

$$n_i = \frac{nh}{S} \times \varphi(z_i); \; z_i = \frac{x_i - \overline{x_b}}{S}; \; \varphi(z_i) = \frac{1}{2\pi} \times e^{\frac{-z_i^2}{2}} \qquad (4)$$

Find the values of $\varphi(z_i)$ with [4]

After calculating the value of the criterion, we conclude that the hypothesis proposed is true.

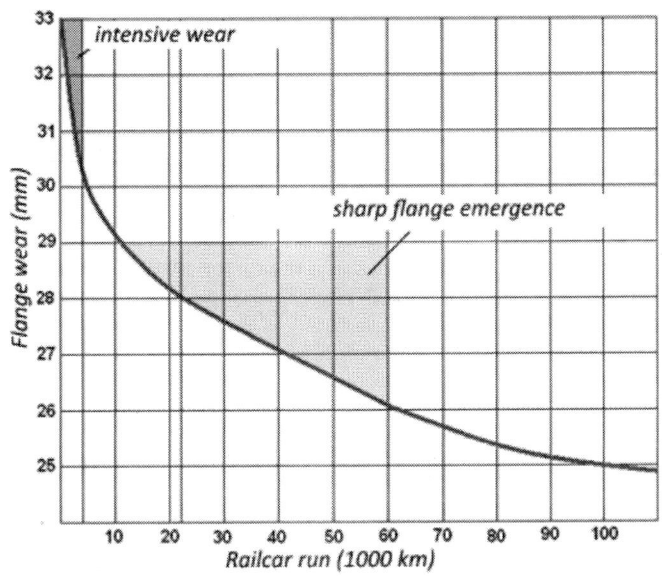

Figure 2. The intensive wear of the flange after the machining and to increase the life service of the wheel pair

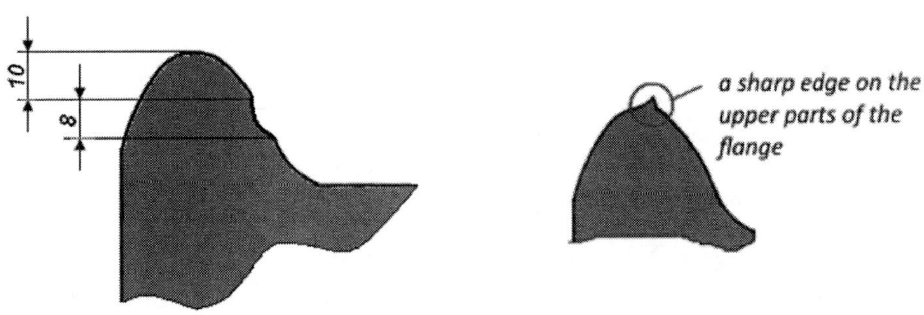

IOP Conf. Series: Materials Science and Engineering **832** (2020) 012088 doi:10.1088/1757-899X/832/1/012088

Figure 3. The flange for wheel pairs without the flange building-up

Figure 4. The working part of the flange is the surface located within 2 mm from the top - 13 mm from the wheel thread

4. Conclusion

1. When comparing the thickness of the 30-mm and 33-mm flange, it was revealed that wheel pairs with the 30- mm thickness have the ability to avoid intense wear of the flange during a run of 120-130 thousand km i.e., approximately when worn down to 27 mm. Therefore, it is proposed to release wheel pairs with an initial thickness of 30 mm for the utuilization with a depot and complete overhaul repairs. In this case, it becomes possible to increase the service life of the wheel pair for 44 months.
2. We recommend that the DSh-4 template be put into operation on the railway.
3. With a bandage thickness of up to 33-36 mm, machining is not recommended. For maximum use of the wheel pair resource, it is necessary to machine the flanges to 27 mm and then use them under the railcars that have undergone the current overhaul repair. It is necessary to introduce changes in the instructions for inspection and examination of wheel pairs of railcars.
4. It is necessary to set the schedule for the rotation of the wheel pair under the railcars, coming from the data on the railcar runs. This will result in savings in the measurement of the wheel pairs of the railcars during inspection.
5. To set the maximum thickness of the wheel flange when machining. This makes it possible to avoid the intensive wear of the flange after the machining and to increase the life service of the wheel pair (see Fig. 2).
6. According to the instructions for inspection and examination of wheel pairs of railcars, when machining the wheels, it is necessary to leave pittings on the flange with a depth of not more than 2 mm, located not more than 10 to 18 mm from the top of the flange for wheel pairs without the flange building-up, so that the rim will be ground 3-4 mm less (see Fig. 3).
7. Observance of technical requirements according to the "Reminder for railcar inspectors No. 724-2009 PKB TsV" according to clause 4.2.1 "Sharp edge on the upper parts of the flange, without cutting, is not subject to rejection. The working part of the flange is the surface located within 2 mm from the top - 13 mm from the wheel thread (see Fig. 4).

References

[1] Lysyuk V C 2005 Interaction of rails and wheels in curves *Track and track facilities* **10** pp 11-17

[2] Bulatov Y N, Kryukov A V, Suslov K V 2017 Multi-agent technologies for control of distributed generation plants in the isolated power systems *Far East Journal of Electronics and Communications* Vol. 17 **5** pp 1197-1210

[3] Solodusha S, Suslov K, Gerasimov D 2016 Applicability of Volterra integral polynomials in the control systems of electric power facilities *International Conference Stability and Oscillations of Nonlinear Control Systems (Pyatnitskiy's Conference)* 7541227

[4] Suslov K V, Solonina N N, Smirnov A S 2011 Smart Grid: A new way of receiving primary information on electric power system state *IEEE PES Innovative Smart Grid Technologies Conference Europe* 6162654

[5] Galiev I I, Nekhaev V A, Nikolaev V A 2012 The safety of the movement of freight trains and the dynamic properties of the carriage of the wagon. *Journal of Transsib Railway Studies* **1** pp 107-112

[6] Sivtsov A V, Sheshukov O Y, Tsymbalist M M, Nekrasov I V, Makhnutin A V, Egiazar'yan D K, Orlov P P 2019 Steel Semiproduct Melting Intensification in Electric Arc Furnaces Using Coordinated Control of Electric and Gas Conditions: II. On-Line Control of the State of the Charge and Melt Zones in Electric Arc Furnaces *Russian Metallurgy (Metally)* **6** pp 565-569

[7] Sivtsov A V, Sheshukov O Y, Tsymbalist M M, Nekrasov I V, Makhnutin A V, Egiazar'yan D K, Orlov P P 2018 Steel Semiproduct Melting Intensification in Electric Arc Furnaces Using

Coordinated Control of Electric and Gas Conditions: I. Heat Exchange and Structure of the Electric Arc Furnace Laboratory *Russian Metallurgy (Metally)* **12** pp 1108-1113
[8] Nekrasov I V, Sheshukov O Y, Metelkin A A, Sivtsov A V, Tsymbalist M M 2016 Slag conditions in electrosmelting: A review *Steel in Translation* Vol. 46 **6** pp 435-442
[9] Sivtsov A V, Sheshukov O Y, Tsymbalist M M, Nekrasov I V, Egiazar'yan D K 2015 The Valve Effect of an Electric ARC and Problems in Controlling Electric-ARC Furnaces *Metallurgist* Vol. 59 **5-6** pp 380-385
[10] Nekrasov I V, Sheshukov O Y, Metelkin A A, Sivtsov A V, Tsymbalist M M, Egiazar'yan D K 2015 Ensuring Consistent Foamability in Electric-Furnace Slags *Metallurgist* Vol. 59 **3-4** pp 300-304
[11] Sheshukov O Y, Nekrasov I V, Sivtsov A V, Tsymbalist M M, Egiazar'yan D K, Metelkin A A 2014 Dynamic monitoring of slag oxidation and thickness in the ladle-furnace unit *Steel in Translation* Vol. 44 **1** pp 43-46
[12] Monakhov I K 2008 Width. Elevation. Lubrica-tion *Track and the track facilities* **11** pp 20-24
[13] Yolkin K S, Yolkin D K, Kolosov A D, Ivanov N A, Shtayger M G 2018 Technologies, which allow to reduce an impact of metal silicon production on the environment *IOP Conference Series: Materials Science and Engineering* **411** 012028
[14] Kolosov A D, Gozbenko V E, Shtayger M G, Kargapoltsev S K, Balanovskiy A E, Karlina A I, Sivtsov A V, Nebogin S A 2019 Comparative evaluation of austenite grain in high-strength rail steel during welding, thermal processing and plasma surface hardening *IOP Conference Series: Materials Science and Engineering* **560** 012185
[15] Konstantinova M V, Balanovskiy A E, Gozbenko V E, Kargapoltsev S K, Karlina A I, Shtayger M G, Guseva E A, Kuznetsov B O 2019 Application of plasma surface quenching to reduce rail side wear *IOP Conference Series: Materials Science and Engineering* **560** 012146
[16] Guseva E A, Kargapoltsev S K, Balanovskiy A E, Karlina A I, Shtayger M G, Gozbenko V E, Konstantinova M V, Sivtsov A V 2019 Comparative evaluation of corrosion resistance of wheel and rail steels in various media *IOP Conference Series: Materials Science and Engineering* **560** 012181
[17] Vdovin K N, Feoktistov N A, Gorlenko D A 2016 The effect of the cast high-manganese steel primary structure on its properties *Materials Science Forum* Vol. 870 pp 339-344
[18] Nazarov N S, Yakimova G A, Gozbenko V E et al. 2003 The use of industrial waste in compositions for lubricating rails (Irkutsk: Ir-kut. un-ty Publ.)
[19] Karpushenko N I, Velichko D V, Bobovnikova N A 2010 Influence of gauge and condition of running gears of the rolling stock on wear in-tensity *The transport of the Russian Federation* **3 (28)** pp 10-14
[20] Balanovskiy A E, Shtayger M G, Karlina A I, Kargapoltsev S K, Gozbenko V E, Karlina Yu I, Govorkov A S, Kuznetsov B O 2019 Surface hardening of structural steel by cathode spot of welding arc *IOP Conference Series: Materials Science and Engineering* **560** 012138
[21] Balanovsky A E, Shtayger M G, Kondra'Ev V V, Karlina A I, Govorkov A S 2018 Comparative analysis of structural state of welded joints rails using method of Barkhausen effect and ultrasound *Journal of Physics: Conference Series* **1118** 012006
[22] Ivanchik N N, Balanovsky A E, Shtayger M G, Sysoev I A, Karlina A I 2018 Capability enhancement of production of activating fluxes for arc welding using ultradispersed products of silicon waste processing *IOP Conference Series: Materials Science and Engineering* **411** 012035
[23] Balanovsky A E, Shtayger M G, Grechneva M V, Kondrat'Ev V V, Karlina A I 2018 Comparative metallographic analysis of the structure of St3 steel after being exposed to different ways of work-hardening *IOP Conference Series: Materials Science and Engineering* **411** 012012
[24] Kuznetsov V M 2000 About the sharp flanges on the wheel pairs *Track and track facilities* **9** pp 16-19
[25] Balanovsky A E, Shtayger M G, Kondrat'Ev V V, Van Huy V, Karlina A I 2018 Plasma-arc

surface modification of metals in a liquid medium *IOP Conference Series: Materials Science and Engineering* **411** 012013

[26] Balanovsky A E, Shtayger M G, Kondrat'Ev V V, Nebogin S A, Karlina A I 2018 Complex metallographic researches of 110G13L steel after heat treatment *IOP Conference Series: Materials Science and Engineering* **411** 012014

[27] Korneev S A, Krupnikov I V, Polyakov S N, Shalai V V 2006 Calculation and experi-mental method for determination of material parameters of elastic-plastic materials on the paths of active deformation of small curva-ture *Omsk Scientific Bulletin* **4 (38)** pp 86-90

[28] Vdovin K N, Feoktistov N A, Sinitskii E V, Gorlenko D A, Durov N A 2015 Production of high-manganese steel in arc furnaces. Part 1 *Steel in Translation* Vol. 45 **10** pp 729-732

[29] Dokin V N, Kuzmin O V 2010 On the randomi-zation of B-distributions *Combinatorial and probable problems of discrete mathematics: Coll. ofsc. papers* (Irkutsk: Irkut. un-ty Publ.) pp 34-46

[30] Vdovin K N, Gorlenko D A, Feoktistov N A 2016 Characteristics of excess phase in cast high-manganese steel *Steel in Translation* Vol. 46 **7** pp 484-488

[31] Gmurman V E 1979 *A guide to solving problems according to the theory of probabilities and mathematical statistics* (Moscow: Vysshaya Shkola Publ.)

IOP Conf. Series: Materials Science and Engineering **832** (2020) 012089 doi:10.1088/1757-899X/832/1/012089

Removal of burrs from small-size high-precise parts for coaxial radio components

Yu I Karlina[1], S K Kargapoltsev[2], V E Gozbenko[2, 3], A I Karlina[1]

[1] Irkutsk National Research Technical University, 83 Lermontov Street, Irkutsk, 664074, Russia
[2] Irkutsk State Transport University, 15 Chernyshevsky Street, Irkutsk, 664074, Russia
[3] Angarsk State Technical University, 60 Chaykovsky Street, Angarsk, 665835, Russia

E-mail: karlinat@mail.ru

Abstract. The urgency of the problem of removing burrs from parts of coaxial radio components is due to the fact that, according to technical requirements, burrs are not allowed on these parts, and on materials beryllium bronze, brass, alloy 29 NK used for their production, it is not possible to exclude the appearance of burrs during processing by cutting. The analysis of the influence of technological heredity of the number and size of burrs obtained on the CNC automatic turning operation to the choice of the method of deburring. The parameters of the burrs that affect the choice of the method of deburring, and the degree of their influence on the choice of a particular method are determined. On the CITIZEN bar lathes, an experimental selection of the cutting tool and processing modes was carried out in order to minimize the number and size of burrs on parts made of beryllium bronze, brass, alloy 29 NK. The installation of selected processing modes and tools will ensure the production of parts of coaxial radio components at the turning stage with a minimum number and size of burrs, which will reduce the laboriousness and time of deburring.

1. Introduction

The production of coaxial radio components, which are part of the electronic component base, is one of the areas of import substitution. In modern electronic equipment there is a tendency to increase the frequency of the signal and, accordingly, there is an increasing demand for signal transmission from the microstrip line without loss and distortion, filtering interference [1]. Therefore, the details of connectors, super high frequency and low frequency inputs, racks and noise filters have special technical requirements.

The main parts of the radiofrequency connectors are the inner and outer conductors and the insulator between them. Signal transmission in coaxial radio components occurs along the inner side of the outer conductor and the outer side of the inner conductor. The higher the frequency and the greater the electrical conductivity and magnetic permeability of the metal in a thin surface layer, the smaller the thickness of this layer [1].

The inner and outer conductors must have exact dimensions, a surface without burrs and sinks, a uniform galvanic coating with a highly conductive layer. During assembly, a snug fit of the mating surfaces must be ensured; the conductive layer must not be disturbed.

Internal and external conductors for leaky connectors are released from beryllium bronze, less often from brass. The choice of material is due to the requirement to ensure the density of mating surfaces during assembly. According to GOST 18175-78, the BrB2 beryllium-bronze alloy has the property of very good deformability in the hardened state and is intended for the manufacture of spring-loaded parts for responsible appointment [2].

IOP Conf. Series: Materials Science and Engineering **832** (2020) 012089 doi:10.1088/1757-899X/832/1/012089

After thermal hardening, beryllium bronze becomes more durable, elastic and ductile. This allows compressing the lamellas of the housing during assembly of the connector.

Internal and external conductors for hermetic connectors are produced from 29 NK alloy or 15Kh25T steel to obtain by the agreed thermal expansion coefficients of junctions with glass C52-1 that are consistent in magnitude. In this case, the glass is an insulator, the tightness of the connector is provided by the junction of the insulator with the conductors.

2. Problems of ensuring the quality of parts of coaxial radio components during cutting edge machining

The main problems that arise when ensuring the quality of parts of coaxial radio components are caused by the use of hard-to-handle materials, small size and high precision and surface quality requirements.

BrB2 beryllium bronze material has high strength and hardness, 29NK alloy material has high hardness. When these materials are processed by cutting, hard burrs that are firmly mated to the surface appear on the surface.

The determining size for the parts of coaxial radio components is the diameter. Outer conductors are produced from blanks with a diameter of up to 10 mm, for example, the Housing for one of the leaky radiofrequency connectors is produced from the BrB2 bar with a diameter of 5 mm, and the produced for one of the sealed radiofrequency connectors from the circle 29 NK 6 mm in diameter. Internal conductors, current outputs, etc. are produced from wire and have diameters less than 1 mm. The small dimensions of parts often require the use of cutting in one pass, since the allowances for processing are already minimally acceptable.

The accuracy of the details of coaxial radio components is high and reaches 6 accuracy levels. High precision manufacturing of small parts is ensured by the use of modern longitudinal turning lathes with CNC CITIZEN CINCOM K16E-VII, CITIZEN CINCOM K16-VI, CITIZEN CINCOM R07-VI, as well as precision cutting tools. These machines provide high rigidity of the technical system and have additional advantages, for example, the function of wave turning, which facilitates the implementation of interrupted turning.

According to GOST 4GO.070.014-79, burrs, ragged and sharp edges are not allowed on metal parts of electronic equipment made of metal [3]. When cutting billets of beryllium bronze, brass, alloy 29 NK, burrs occur on the surface of the parts and on the edges. It is not possible to completely eliminate the appearance of agnails during cutting in the size of a batch of parts. Actual is the problem of minimizing the number and size of burrs. Technological heredity in the number and size of burrs obtained in a turning operation determines the choice of further methods for removing burrs [4].

3. Selection of tools and cutting modes for parts of coaxial radio components

The selection of tools and cutting modes was carried out for parts of coaxial radio components made of beryllium bronze, brass, 29 NK alloy, produced on longitudinal turning CNC machines CITIZEN CINCOM in order to minimize the number and size of burrs. The best results were obtained when processing using precision cutting tools from Swiss companies Utilis, Fraisa, IFANGER. The cutting tool of these manufacturers is designed for micromechanics, made of hard alloy and coated with TIALN + Al2O3 (ZrCN), has high wear resistance. Of particular importance is the presence of a special coating. ZrCN zirconium carbonide is known for its unique performance characteristics and is actively used for processing non-ferrous metals and aluminum alloys. The coating prevents sticking of the processed material to the surface of the tool, which reduces the build-up during cutting and facilitates chip removal. Zirconium carbonitride has such properties as resistance to abrasion and corrosion processes, a high level of hardness and impact strength, low friction coefficient, tribological properties, and inertness to many types of acids. A high level of hardness prevents the formation of defects on the cutting surface, namely, in the area of these defects, the heating of the treated surface increases. Low coefficient of friction also reduces the level of heating of the treated surface. Due to overheating, the number and size of burrs increases. The Al2O3 coating (corundum) also possesses the properties necessary for lowering the temperature in the cutting area – high hardness and thermal conductivity. Corundum is second only to superhard materials based on diamond and boron nitrite in these properties. Due to the high thermal conductivity, redistribution of heat generated during cutting occurs. The amount of heat leaving the tool increases, reducing the amount of heat leaving the workpiece and chips. The removal of heat from the treated surface reduces its heating. Tool holders and collets provide reliable fastening of the tool, support the declared rigidity of the system device - tool - part for turning bar machines, such as CITIZEN CINCOM.

When selecting the types and processing parameters, the need was taken into account for chip removal, planned tool wear resistance, early access to the design dimensions and control of the size of the allowances at each stage of processing, surface quality requirements. The selection of cutting modes was carried out experimentally for each cutter, plate and drill from a typical set of tools for these parts. As a result of the experiment, processing modes were selected that satisfy the above requirements, in which the number and size of burrs are minimal. It was found that the processing parameters obtained experimentally differ from the parameters recommended by the manufacturer. The manufacturer sets either the range of permissible values of the parameter or its exact value [5, 6]. However, minimizing the number and size of burrs imposes additional restrictions on the value of the processing parameter.

The following parameters were experimentally selected for each type of tool and diameter range:

- feed F, mm / min;
- number of spindle revolutions n, rpm:
- cutting depth f, mm.

The experimentally selected values of the processing parameters are within the range of recommended values or slightly lower. It was found that when setting the minimum feed rate recommended by the manufacturer and the maximum value from the cutting speed range recommended by the manufacturer, the cutters overheat, when setting the minimum values from the supply range and speed recommended by the manufacturer, the necessary chip-breaking space and chip discharge are not provided, and at the maximum values of the feed range and speed range recommended by the manufacturer results in large chips and the tool burns. The experiment was carried out on a new tool, so the cause of combustion may be overheating due to poor heat removal of the chips and due to work with low feeds.

A slight decrease in feed, speed and depth of cut compared to the recommended values allows ensuring the planned wear resistance of the tool and reducing the number and size of burrs. After installing a new tool at the phase of the lapping tool, the first 5 - 7 parts of beryllium bronze and brass are obtained without burrs. Details of alloy 29 NK can be obtained without burrs during 12 hours of work on a new tool. After passing through the phase of the lapping tool, minor defects, scratches, etc. appear on the tool blade. At this phase, the tool is not worn out, but burrs already appear on the parts, the number and size of which are stable until the moment of wear of the tool. After this phase the number and size of burrs increases significantly and the tool must be replaced. Thus, on parts made of alloy 29 NK, it was possible to eliminate the appearance of burrs when working with new tools during 1 working shift, then it is necessary to replace the tool and use it for turning parts from softer materials, but this is often impractical, since there is a loss of time for readjustment, and the volumes of production of parts from softer materials do not allow the effective use of the entire tool remaining after each working shift of turning parts from alloy 29 NK. On the details of beryllium bronze and brass it was impossible to eliminate the appearance of burrs by selecting tools and processing modes.

Complete deburring cannot be achieved by installing additional tools on the machine - steel or ceramic brushes. The installation of abrasive wheels on a rubber bond showed that the wheels are destroyed under the influence of an oil cutting fluid.

For each type of tool, the optimal processing conditions were selected according to the above criteria, on the basis of which it is recommended to create references for further setting processing parameters when compiling control programs and launching them without additional experiments on machines. The creation of standard templates of control programs also helps to reduce the cost of production preparation.

4. Changing burr parameters depending on the tool and processing conditions

According to the international standard, [7] a burr is a plastically deformable material formed on the edge of a part as a result of machining or cutting. Other protruding defects formed during cutting are also considered as burrs. The main burr parameters affecting the method and time of their removal are:

1. dimensions (height, thickness, length);
2. hardness;
3. cross-sectional shape;
4. configuration in the longitudinal direction;
5. location (accessibility) [9, 10].

IOP Conf. Series: Materials Science and Engineering **832** (2020) 012089 doi:10.1088/1757-899X/832/1/012089

The base or root of the burr is the part adjacent to the surface to be treated, and the apex of the burr is its end, which is removed from the surface to be processed by the length of the burr. The burr thickness t_b is the burr root thickness, defined in its cross section along the surface to be treated. The height of the burr h_b is the size determined perpendicular to the surface being machined from the root to the part of the burr farthest from the surface. The length of the burr is the size characterizing the length of the burr along the surface to be treated. The burr hardness H_b is the burr root hardness measured on the theoretical line of the surface to be treated. Figure 1 shows the main geometric parameters of the burr.

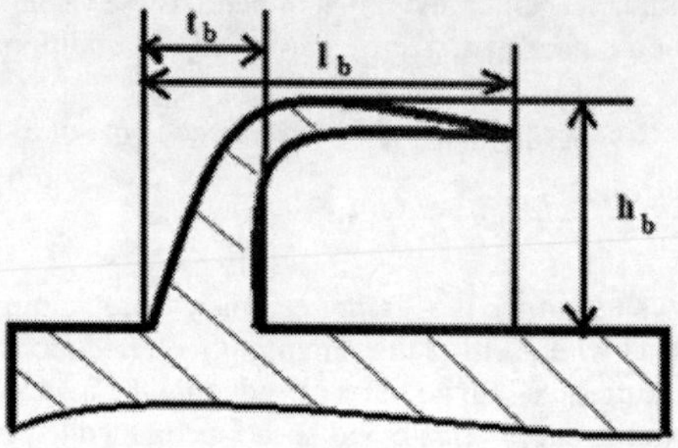

Figure 1. Main geometric parameters of the burr: h_b is the height of the burr, l_b is the length of the burr, t_b is the thickness of the burr root.

The process of burr formation at the exit in the feed direction with longitudinal turning, as well as in the direction of the main movement with free orthogonal cutting, is affected by the physical and mechanical properties of the processed materials, as well as technological parameters, such as:

- the degree of wear of the blade;
- depth of cut;
- rake angle of the blade and the angle of the edge of the workpiece;
- slice thickness;
- cutting speed;
- feed;
- the main angle in the plan;
- edge angle [10];
- feed;
- processing conditions - the use of cutting fluid;
- rigidity of the system machine - fixture - tool - part.

The cross section of the burrs on the details of beryllium bronze, brass and alloy 29 NK, formed during cutting, has a shape close to round. They have a reinforced base, are uniformly strong along the entire length, and break off with difficulty. These are long stretching burrs, often consisting of one root and several peaks. With a decrease in feed, speed and depth of cut, the burrs shorten, become drop-shaped, and the root thickness decreases.

The degree of significance of each of the parameters of the burrs depends on the influence on the choice of method and the laboriousness of removing burrs. The most important thing is the location (accessibility) of burr. The burrs on the outer surfaces of the parts are easily removed with a tumbling, which does not require large investments and laboriousness. In the presence of hard-to-reach burrs on parts having internal surfaces, the burr root thickness becomes significant, and burr hardness and its length become important for various processing methods.

IOP Conf. Series: Materials Science and Engineering **832** (2020) 012089 doi:10.1088/1757-899X/832/1/012089

5. Deburring methods after cutting edge machining

The choice of the deburring method after cutting edge machining depends on the geometry of the part, the material of the part, the quantity and size of the burrs. Depending on the geometry, parts can be divided into two groups: parts that do not have internal surfaces and parts that have internal hard-to-reach surfaces (grooves, pockets, etc.). Parts that do not have internal surfaces are subject to abrasive treatment in a tumbling drum. Protective covers must be worn on the external surfaces of parts not subject to abrasive treatment, for example, containing threads. For parts having internal surfaces, various deburring methods are possible. The choice of the method depends on the technological heredity after processing by cutting in the form of the number and size of burrs, as well as on the cost and loading of work centers implementing a suitable method for removing burrs.

When removing burrs manually by an additional plumbing operation, deburring from the outer surfaces is done with a tumbling, and manually deburring only from the inner surfaces. Locksmith operation is a stripping with a scraper, a drill, abrasive wheels on a rubber bond. The laboriousness and quality of the plumbing operation depends on the number, thickness of the root and the strength of the burrs.

Deburring in a thermal pulse installation is possible for parts made of beryllium bronze, brass, 29 NK alloy, provided that the burr root thickness is at least 0.1 mm and four times less than the minimum thickness of the part. Otherwise, thin-walled parts may be damaged. The batch size of the parts loaded into the thermal pulse installation should be within the limits recommended for the selected processing mode. A significant change in the size of the batch of the load leads to the need to adjust the processing program and change the chamber volume of the thermal pulse installation.

Deburring with robots for stripping is only possible with large volumes of parts. The cost of automating the process of deburring is quite high, and is cost-effective only with two or three shift work of machines that produce such parts. The effectiveness of this method of deburring is independent of the number and parameters of the burrs.

The chemical etching deburring experiment showed that only thin burrs with a root thickness of up to 0.02 mm are removed, and burrs with a thick root are not removed, the root remains. The root thickness of some of the burrs obtained in the selected cutting conditions exceeds the thickness of 0.02 mm.

6. Conclusion

Due to technical requirements, burrs are not allowed on the parts of coaxial radio components. These parts are made of hard-to-work materials - beryllium bronze, brass, alloy 29 NK, and when these materials are processed by cutting, burrs appear on the surface of the parts and on the edges. It is not possible to eliminate the appearance of burrs for the period of the tool's operability by selecting tools and processing modes. The geometry of the part and technological heredity according to the parameters of the burrs obtained on a CNC automatic lathe operation determines the choice of methods for further deburring. Minimizing the size and number of burrs obtained by cutting allows them to be removed at the lowest cost. For parts having internal surfaces, the following deburring methods are preferable:

- deburring from the outer surfaces with tumbling, deburring from the inner surfaces with manual bench work, the complexity and quality of the work depends on the quantity, hardness and thickness of the burrs root;
- deburring in a thermal pulse installation is possible if the condition of the ratio of the thickness of the root of the burr and the thickness of the part is not less than 1: 4 and the thickness of the root of the burr not more than 0.1 mm, it is also necessary to ensure the recommended batch size of the parts;
- deburring by a robot for stripping is cost-effective only for large volumes of parts.

The choice of deburring method is influenced by such parameters of the burrs as quantity, root thickness, hardness. The minimum number and size of burrs is necessary for CNC automatic lathe operations by installing specially selected tools and processing modes. According to the results of an experiment on the selection of tools and processing modes in order to minimize the number and size of burrs, it is necessary to create a directory to the types of processing for each type of tool, it is advisable to use standard templates of control programs with the established processing modes according to the directory.

IOP Conf. Series: Materials Science and Engineering **832** (2020) 012089 doi:10.1088/1757-899X/832/1/012089

Acknowledgments

The reported study was funded by RFBR, project number 19-38-901184.

References

[1] Dzhurinsky K B 2006 *Miniature coaxial radio components for SHF microelectronics. Connectors, coaxial microstrip transducers, adapters, SHF-inputs, glands, low-frequency inpits, insulating stands, interference filters* (Moscow: Tekhnosfera Publ.) p 216

[2] GOST 18175-78 2004 *Interstate standard. Tinless bronzes, pressure treated / Non-ferrous metals. Bronze. Technical conditions Stamps* (Moskow. IPK Publishing House of Standards) p 148

[3] OST 4GO.070.014-79 1992 Details of electronic equipment (Moscow: LLC Radio Standard-TsNIIRES) p 15

[4] Zverev E A, Skeeba V Yu, Martyushev N V, Skeeba P Yu 2017 Integrated quality ensuring technique of plasma wear resistant coatings *Key Engineering Materials* vol. 736 pp. 132-137.

[5] *Unilis Multidec 2018 Tools Catalog.* [Electronic resource] URL: https://drive.google.com/file/d/16RklcSx8hpMf1kysP6b0e5IJNu9172Oc/view?usp=sharing. Access date 09/09/2019.

[6] *Fraisa tool catalog.* [Electronic resource] URL: https://drive.google.com/file/d/18-Zug4pifm1PE5D6INBhCWsPSBoOlL5Z/view?usp=sharing. Access date 09/09/2019.

[7] STD-01. WBTC. 1996 *draft. –DeburringTechnologyInternational* (Inc., KansasCity, MO)

[8] Sergiev A P, Logachev V N 2006 *On the issue of burr formation* Education, science, production and management: (Stary Oskol: LLC TNT) pp 399-400.

[9] Proskurin A A 2016 *The development of the device and the study of the modes of reverse cutting down of parts from sheet metal without burrs* The dissertation for the degree of candidate of technical sciences. (Moscow National ResearchTtechnological University) p 172

[10] Toropov A A 1999 *Prediction and minimization of burrs during cutting, modeling of the processes of their formation* The dissertation for the degree of candidate of technical sciences (St. Petersburg State Technical University) p 235

IOP Conf. Series: Materials Science and Engineering **832** (2020) 012090 doi:10.1088/1757-899X/832/1/012090

The use of intelligent systems when regulating road traffic

I E Agureev[1], K P Andreev[2,3], E V Ionov[1], A Ya Svistunova[1] and V V Terentyev[2]

[1]Polytechnical Institute, FSBEI HE TSU, 92, Lenin avenue, Tula, 300012, Russia
[2]Road Transport Department, FSBEI HE RSATU, 1, Kostychev Str., Ryazan, 390044, Russia

E-mail: kosta066@yandex.ru

Abstract. Statistics show that approximately half of all fatal accidents occur at unregulated and inadequately lit pedestrian crossings. To minimize the severity of the consequences, it is necessary to develop a set of organizational and technical measures to inform traffic participants about a safe pedestrian crossing. One of the ways to solve this problem may be to introduce the development of leading design organizations aimed at the use of intelligent technical systems to ensure the process of safe crossing of highways by pedestrians, i.e. smart pedestrian crossing. The article discusses the possibility of introducing an intelligent system into the traffic control process in sections of the road network with heavy traffic of cars and pedestrians. The operating principle of this system is based on the recognition of moving vehicles and timely warning pedestrians, as well as determining the location of pedestrians at the crossing and warning drivers about people on the road.

1. Introduction

The rapid pace of life in the modern city places certain demands on the organization of the transport process. First, municipal authorities pay attention to the possibility of the road network to provide the required traffic capacity without reducing the speed of the traffic flow in general. To solve this problem, various models of the organization of traffic in existing sections of the road network (RN) are developed and analyzed, new transport interchanges are designed and built (often at different levels), and measures are taken to improve the transport infrastructure as a whole. The implementation of the whole complex of the above measures ultimately leads to the desired result and traffic on the city streets becomes more comfortable and safe for car drivers, but there is a pedestrian on the road as well. As practice shows, the interests of the latter are not always taken into account when developing organizational and technical measures to improve the transport process, which leads to an increase in road traffic accidents with pedestrians in RN sections which seem perfectly designed [1, 2].

2. Target setting

The problem of road safety at the present stage of social development is gaining priority. The United Nations declared the period of 2011–2020 as the "Decade of Action for Road Safety" [3]. In the White Paper of 2011, the European Commission proposed a new approach to a "zero vision" of road safety, in order to exclude cases of death and serious injuries occurring on roads. The Government of the Russian Federation approved the Road Safety Strategy in the Russian Federation for 2018–2024 by Decree No. 1-r dated January 8, 2018, which states that in order to increase the safety of pedestrians in

IOP Conf. Series: Materials Science and Engineering **832** (2020) 012090 doi:10.1088/1757-899X/832/1/012090

road traffic accidents and their consequences it is necessary to equip pedestrian crossings with a set of technical traffic management tools using innovative methods and technologies [4].

According to the data presented on the official website of the State Inspectorate for Road Traffic Safety of the Russian Federation for 2016–2018, about 511 thousand accidents occurred on roads of the Russian Federation. This is 3-4 times more than in Europe. For three years, 57.610 people died and 651.367 people were injured as a result of road traffic accidents [5]. According to the NHTSA's National Center for Statistics and Analysis in the United States for the same period, these figures included 18.340 dead and more than 220 thousand people were injured. It is obvious that the figures of road traffic injuries among pedestrians in the USA as well as in Europe are much lower than in Russia.

Road traffic accidents involving pedestrians make up a significant part of the total number of accidents – approximately 30 % for 2016–2018 (Table 1) [5].

Table 1. Statistical information on road traffic accidents involving pedestrians

Year	Total number of road traffic accidents	Number of pedestrian accidents				
		Total	% of the total	caused by drivers	caused by pedestrians	in the dark
2018	168.099	48.814	29.04	32.966	18.987	19.791
2017	169.432	51.839	30.59	31.234	20.861	22.401
2016	173.694	52.001	29.93	29.933	21.747	21.665

An analysis of the statistical data presented in Table 1 makes it possible to conclude that in the majority of cases accidents with pedestrians are caused by drivers (approximately 1.5 times more often than pedestrians), and a high percentage of pedestrian accidents happen in the dark (40 % of the total number of accidents involving pedestrians in 2018). Statistics show that every third victim crossed the road at a pedestrian crossing.

The great number of car collisions with pedestrians in the zone of pedestrian crossing is due to the following reasons:

- insufficient visibility of the crossing in the dark and during rainfall (heavy rain, snow, fog) due to poor lighting;
- poorly visible road marking (high degree of wear of the zebra marking, especially after the autumn-winter period);
- low discipline of road users (both drivers and pedestrians).

3. Discussion

Leading scientific and design organizations conduct research aimed at introducing intelligent systems into the process of pedestrians' safe crossing highways, i.e. developing a safe pedestrian crossing [6].

In many cities around the world, pedestrian crossings are equipped with traffic light systems and pedestrians have to press a button to cross the roadway. Such a traffic light is always green for vehicles when there is no pedestrian who wants to cross the road. To indicate the intention of crossing, pedestrians inform the system of their presence by pressing a button. With such a traffic light control scheme, various problems arise. For example, older pedestrians prefer to wait until the car passes, rather than press a button. In addition, teens play with buttons and press them before and after crossing, which leads to traffic delays and affects the behavior of vehicle drivers. This approach is also a problem for pedestrians with disabilities, such as blind people and people who have difficulty to touch a button [7].

Employees of the Transport Institute of Tyumen State Oil and Gas University under the direction of I.A. Anisimov [8] developed an intelligent pedestrian crossing system, including a motion sensor. The principle of operation is as follows. A pedestrian approaches the carriageways and enters the range of the motion sensor. Moreover, to increase the reliability of the device, a button is provided that one can press and cross the road. The motion sensor includes yellow warning lights located above the roadway, and any driver approaching the pedestrian crossing will be warned of the imminent appearance of a pedestrian on the zebra. In the dark, the pedestrian crossing area is additionally illuminated by bright LED light. The sensor determines the response zone exclusively for movement across the roadway. The disadvantage of this system is the operation of a motion sensor per person taller than 100–110 cm, which will not allow to recognize a child or an animal.

Scientists from the Republic of Bulgaria [9] have proposed an improved system for ensuring traffic safety at pedestrian crossings by the use of a range of organizational and technical means. The technical solution is based on recognition of pedestrians in the waiting area. The recognition method is based on a background pattern that is identified by the camera. When pedestrians enter the waiting area, the picture changes, as they are located between the camera and the background, which is a signal to inform the traffic control system. The study also presents the algorithm of the system modeling and controlling the traffic light system. According to the authors, the use of this method of traffic regulation will make it possible to reduce traffic delays caused by automatic traffic light control, as well as eliminate risks for traffic users when using buttons. The proposed approach will improve traffic safety in areas with pedestrian crossings and will be useful for people with disabilities who are not always able to press the button. The negative aspects when using this system include the lack of visibility of the background pattern during heavy rainfall, which can lead to false triggering of automatic control, and the possibility of unauthorized painting the background picture or drawing a silhouette (shadow) of a person on it by outsiders should also be taken into account.

In order to improve road safety, Dorokhin S.V. [10] proposed to equip the pedestrian crossing by highlighting a safe area in front of the stop line and the zebra marking, and monitoring the actions of traffic participants. The length of the safe zone was also determined. The mechanism of the sequence of movement of cars and pedestrians within the crosswalk was described. Suggestions for changing some sections of the RF Traffic Rules and state industry standards regarding the actions of pedestrians and the arrangement of pedestrian crossings were presented. The concepts of a dashed stop line and a special regulation sign were introduced, indicating the zone of the pedestrian crossing. A method for monitoring the actions of road users was proposed, which did not exclude administrative punishment for violation of the requirements of transport legislation. Creating a safe area between the car and the pedestrian will reduce the risk of traffic accidents and will increase road safety. The arrangement of the safe zone requires significant capital investments, the expansion of the carriageway and the introduction of amendments to certain sections of the RF Traffic Rules and state industry standards.

4. Materials and methods

In this work, it is proposed to consider the possibility of introducing an intelligent system based on the effect of early recognition of moving vehicles and timely warning pedestrians, as well as determining the presence of pedestrians at the crossing and warning drivers about the possibility of finding people in the way (Figure 1) [11].

A distinctive feature of the intelligent transport system under consideration from existing projects is the ability to visualize the presence of a person at a pedestrian crossing for the driver by automatic turning on additional lights located directly above the "zebra" that vertically illuminate the markings and traffic signs. In view of this, the system is equipped with video cameras that record the presence of a person in the waiting area and allow monitoring the start and end of the transition process.

The algorithm for informing pedestrians about approaching a car at high speed is as follows:

- a radar detector mounted on an illumination pole detects the speed of vehicles approaching the pedestrian crossing;
- when capturing the excess of the allowed speed, the information board installed in the pedestrian's visual contact zone is triggered;
- a warning message is displayed on the board indicating that the vehicle is approaching at high speed. The board is equipped with an additional sound signal for visually impaired pedestrians.
- Consider the principle of the system for informing drivers about the presence of pedestrians on the roadway:
- a motion sensor located in the waiting area detects a person at a pedestrian crossing;

- when determining a pedestrian, an electrical impulse from the motion sensor is transmitted to an information board mounted on a special support 50–70 meters before the crossing and designed to inform drivers about the presence of people on the roadway. The board can work continuously, and to enhance the visual effect for the driver, additionally blink or change color from yellow to red when people are on the way.

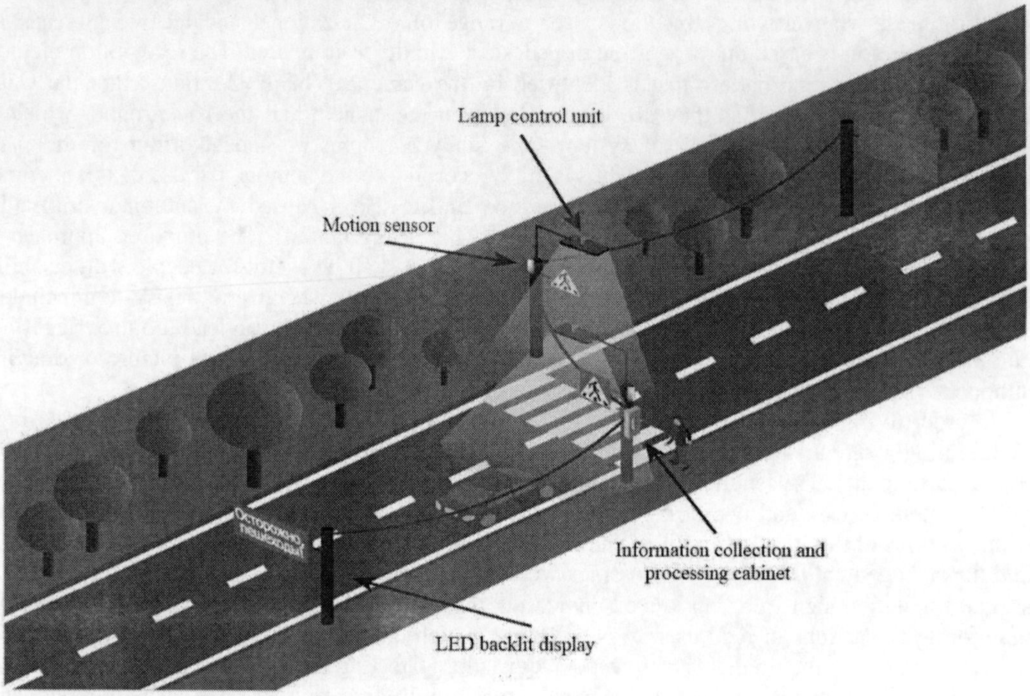

Figure 1. The principle of the intelligent pedestrian crossing

Consider the option of installing a warning system that excludes the algorithm for determining the presence of a person in the waiting area of a pedestrian crossing. In this case, the principle of operation of the intelligent system is based solely on radar data on the speed of the approaching vehicle. If a violation of the speed limit is recorded on the information board, a warning message about the presence of a pedestrian on the roadway is displayed for drivers. At the same time, the radar sends a pulsed signal to the receiver of the information board for pedestrians, on which accordingly a warning message appears about the car approaching and its actual speed. Speed information will allow pedestrians to estimate the time interval for a safe crossing correctly.

The possibility of using modern innovative developments in the intelligent transport system under consideration expands the scope of its application. Special supports with an installed lighting console and projections will provide a high level of visualization of the pedestrian crossing by all road users, which will positively affect the safety of pedestrians crossing the roadway, including in the dark [12].

The possibility of warning drivers about pedestrians by navigation applications, and later on transmitting a signal about the need to reduce speed and stop before the pedestrian crossing is considered for the future.

Currently, authorities in a number of regions are implementing similar projects of the intellectual pedestrian crossing. In particular, in the city of Ryazan, such a transition has been established and is successfully functioning (Figure 2).

The practice of regions where similar safety systems are already used when crossing the carriageways shows a significant decrease in mortality and the number of accidents.

Figure 2. Pedestrian crossing on Mikhailovskiy highway in Ryazan

5. Conclusion

The practice of implementing intelligent systems in the organization of traffic shows the high efficiency of their application. The range of artificial intelligence in the road sector is very wide. The work to optimize the traffic in the RN sections using intelligent pedestrian crossings should be carried out in a complex and take into account the needs of all participants in this process. Only in this case it will be possible to achieve greater discipline for both drivers and pedestrians, which will ultimately lead to a decrease in the number of accidents and reduce the severity of their consequences.

References

[1] Kiselev V A, Shemyakin A V, Polischuk S D, Terentyev V V, Andreev K P and Churilov D G 2018 Optimization of urban transport infrastructure *Transport business in Russia* **5** 138–40
[2] Dorokhin S V, Terentyev V V and Andreev K P 2017 Road safety: problems and solutions *World of transport and technological machines* **2 (57)** 67–73
[3] UN General Assembly *Decade of Action for Road Safety 2011–2020* 2010
[4] Government of the Russian Federation *Road Safety Strategy in the Russian Federation for 2018–2024* 2018
[5] Retrieved from: https://gibdd.rf/
[6] Agureev I E, Kretov A Yu and Matsur I Yu 2013 The study of traffic light control algorithms for the intersection with various traffic flow parameters *Bulletin of Tula State University Technical science* **7-2** 54–61
[7] Anciaes P R and Jones P 2018 Estimating preferences for different types of pedestrian crossing *Facilities Transportation Research Part F* **52** 222–37
[8] Anisimov I A 2014 The use of autonomous intelligent signalling systems to reduce accidents in the area of unregulated pedestrian crossings *Organization and traffic safety* (the VII All-Russian Scientific and Practical Conference) pp 19–22
[9] Evstatiev B, Balbuzanov T, Beloev I and Pencheva V 2019 Intelligent system for improved safety of pedestrian traffic lights *Transport Problems* vol 14 **1** 35–43
[10] Dorokhin S V, Zelikov V A and Denisov G A 2018 Improvement of road traffic safety in the zone of unsignaled pedestrian crossings *Transportation Research Procedia* 122–128
[11] Stepanischeva E S and Rybakov A V 2015 Intelligent crosswalk *Research of young scientists is a contribution to the innovative development of Russia* (Youth research and innovation competition ("UMNIK"). Comp. by Lozovskaya M V and Badelin A G) pp 237–238
[12] Polosukhina M V and Krasnikova D A 2016 Intelligent street lighting systems to enhance traffic safety at pedestrian crossings *Actual directions of research of the XXI century: theory and practice* **4(5-3(25-3))** 133–7

Methods for determining transport delays at simple intersections

M K Bedanokov, Y H Guketlev, A Z Udzhukhu, A S Labutin

Engineering Faculty, MSTU, 191, Pervomayskaya str., Maikop, 385000, Russia

E-mail: bedan64@mail.ru

Abstract. The article analyzes the flow of traffic to the city of Maikop. Existing methods for determining transport delays at intersections have been considered. It is established that the assessment of transport delays at intersections with a railway crossing is important from the point of view of traffic safety. The subject of the study is the intersection of Pushkin and Zheleznodorozhnaya streets in Maikop. The object of the study is the transport delay at this intersection. A new approach for transport delay calculating has been proposed. Preliminary results of the study have been presented; the author's formula for calculating transport delays has been given. The proposed method for transport delay calculating at another intersection has been verified.

1. Introduction

The modern development of society leads to the fact that motor vehicle fleet has been growing continuously and constantly over the years, that is, the growth rate of the number of cars remains stably high [1]. At the same time, the road network remains virtually unchanged. The consequences of this imbalance can be observed in almost any city in Russia: it is either constant traffic jams in large cities, or traffic jams during rush hours in small towns. In this regard, transport delays increase significantly.

In our opinion, traffic delay is one of the main characteristics of traffic, which can most fully characterize the interaction of the organization of traffic, the traffic capacity of intersections (section of a road) and the traffic flow intensity. In recent years, works have begun to appear related to transport delays at traffic lights [2, 3] and at pedestrian crossings [4, 5].

The increase in transport delays has a number of negative consequences, which include increased emissions, economic losses, reduced speed, etc. Besides, traffic delays contribute to psychophysiological fatigue of drivers, which leads to an increase in the number of conflict situations and accidents. One of the factors for increasing emergency situations is that drivers tend to drive through this intersection as quickly as possible, and most often they violate traffic rules. This is especially evident at railway crossings. Despite the fact that most railway crossings are equipped with barriers, the statistics of accidents at railway crossings remains stably high in Russia [6] (Figure 1).

Figure 1 shows that over the past four years, on average, at least one person has been injured or hurt in every accident, and every third accident has caused death. These statistics and a significant number of injured people and deaths at railway crossings show that improving road safety at railway crossings is an important task.

Consequently, determination of transport delays in an area of intersections of highways near railway crossings as one of the factors of road safety will allow us to assess the state of traffic organization at this intersection. Adoption of necessary measures for the optimal organization of traffic is a prerequisite for reducing accidents at road intersections.

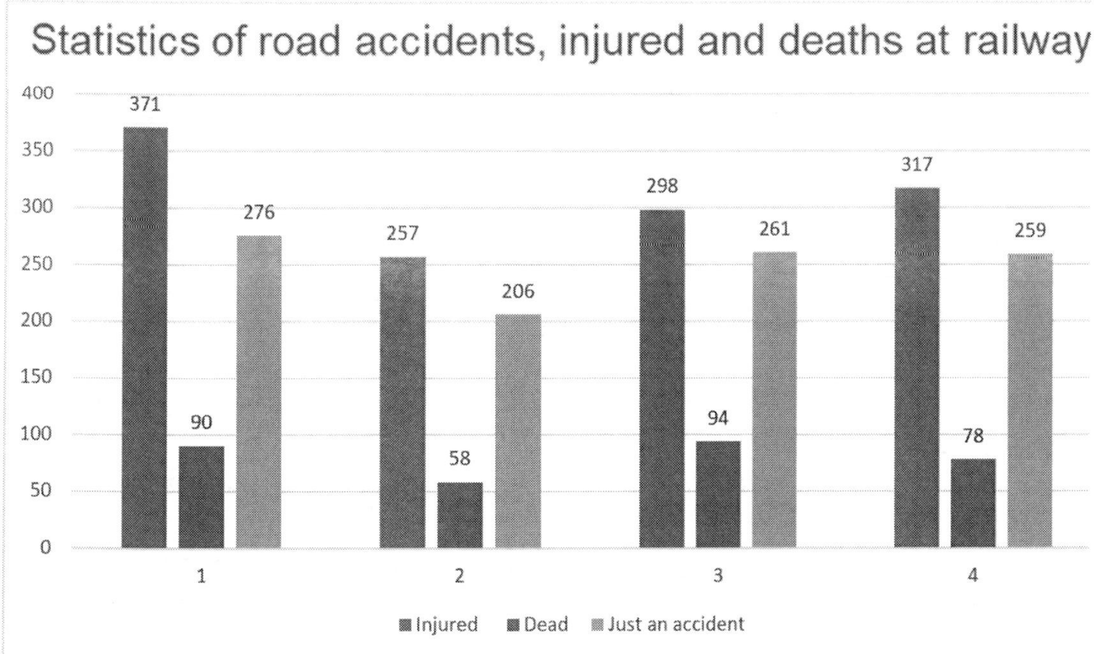

Figure 1. Road accident statistics at railway crossings

2. Analysis of the flow of traffic to the city of Maikop

The city of Maikop is located in the southern Russia and it is the capital of the Republic of Adygeya. The road network through Maikop is the shortest way to a number of large cities in the southern Russia, the tourist cluster and the Black Sea coast. A multiple increase in transit flow through Maikop is expected with the setting the bridge across the Kerch Strait into operation.

The city of Maikop has mostly a rectangular layout (Figure 2). Entrance and exit of road transport to and out of the city of Maikop is carried out in 5 directions. Three of these directions have intersections (No. 7–9; Figure 2) located next to the railway. As a rule, the largest traffic flow at these intersections comes from the secondary direction. This leads to the fact that there are significant automobile queues in the secondary directions that can reach 300 meters or more (Figure 3, 4).

Due to a high traffic load at the crossing on Hakurate Street (No. 8, Figure 2) this crossing is closed for reconstruction, since a two -level interchange is being constructed at this intersection.

Analysis of traffic flows by the example of the city of Maikop shows that the largest load at simple intersections is observed at intersections located near the railway. Significant queues can be observed in secondary directions at relatively low traffic intensities at such crossings.

3. Existing methods for determining transport delays

In our opinion, all methods for determining transport delays can be divided into two groups: field methods and analytical ones. Each of these methods has its advantages and disadvantages. Let us consider each of these methods briefly.

All methods for determining transport delays during direct work with the traffic flow at an intersection are referred to the field methods. These methods are described in a number of works [9–11]. The advantage of the field method is that it helps to obtain the most reliable results. It should

IOP Conf. Series: Materials Science and Engineering **832** (2020) 012091 doi:10.1088/1757-899X/832/1/012091

also be noted that only field observations provide initial data for all other methods of traffic research. The disadvantage of this method is that it is labor-consuming and requires participation of several people to carry out measurements.

It should be noted that modern means of video recording and digital data processing can significantly simplify the complexity of field methods.

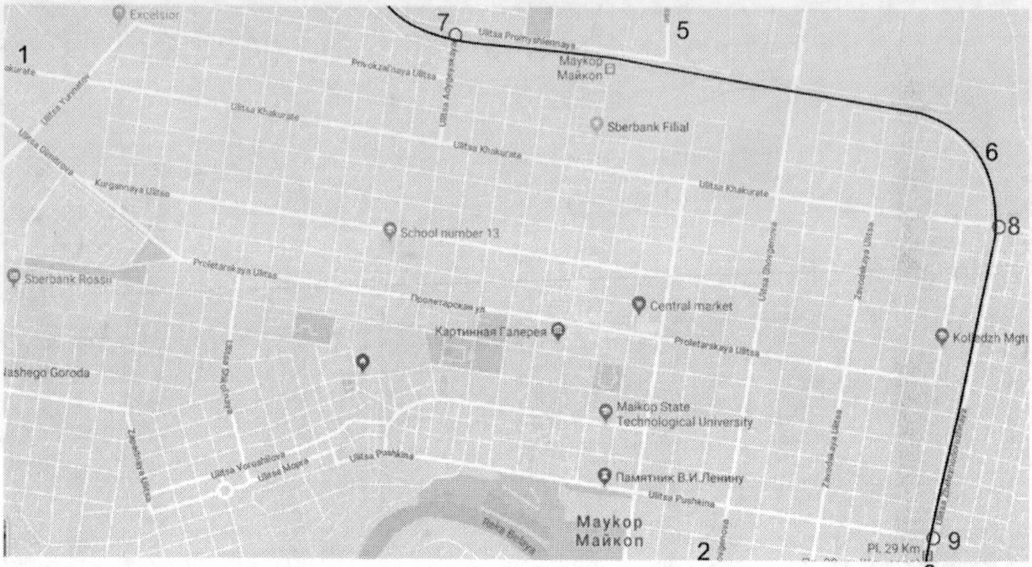

Figure 2. The scheme of traffic flows to Maikop. 1, 2, 3, 4, 5 – directions of entry and exit of motor vehicles in the city of Maikop; 6 – railway; 7, 8, 9 – intersections of highways located near the railway.

Figure 3. The car queue at the intersection of Adygeyskaya and Promyshlennaya streets

Figure 4. The queue of cars at the intersection of Pushkin and Zheleznodorozhnaya streets

In our opinion, analytical methods include both calculation formulas and results of traffic modeling at intersections. The most popular approach for calculating transport delays was proposed by Professor Yu.A. Kremenets [9]. This approach consists in the following: components of transport losses, even at constant traffic intensities on intersecting roads, vary over large ranges and are different for each vehicle.

The time spent is estimated by an average delay of one vehicle $t_{\Delta H}$, which is calculated considering the influence of many different indicators, if there are some indicators. Below there is a general view of an average delay:

$$t_{\Delta_H} = t_{\Delta_{H1}} + t_{\Delta_{H2}} + t_{\Delta_{H3}} \tag{1}$$

where $t_{\Delta H1}$ is average time spent while waiting for an acceptable interval, c;

$t_{\Delta H2}$ and $t_{\Delta H3}$ indicate average time that a car spends in a queue on the secondary road, and the vehicle deceleration time before the intersection, c.

In the case of constant deceleration and acceleration in the process of changing speed and the exponential distribution of time intervals between vehicles on the main road, the average delay of the vehicle in this direction of secondary junction is:

$$t_{\Delta_H} = \frac{e^{N_m \cdot t_b} - N_m \cdot t_b - 1}{N_m - N_s \cdot \left(e^{N_m \cdot t_b} - N_m \cdot t_b - 1\right)} + \frac{V_a}{7.2} \cdot \left(\frac{1}{a_T} + \frac{1}{a_P}\right) \tag{2}$$

where e is Napierian base;

N_m is traffic flow intensity on the main road in both directions, v/ s;
N_s is intensity per average secondary road lane in the considered traffic direction, v/s;
V_a is vehicle speed on the main road, km / h;
t_b is boundary interval, s;
a_T and a_P are deceleration and acceleration respectively, m/s^2.

The average vehicle delay $t_{\Delta H}$ at the intersection is defined as the weighted average value of the delays for all directions (approaches to an intersection) of the secondary road, calculated according to the formula (2) [9]:

$$\bar{t}_{\Delta_H} = \frac{\sum_1^n \left(t_{\Delta_{Hj}} \cdot N_j\right)}{\sum_1^n N_j} \tag{3}$$

where Nj is traffic intensity on the j-direction of the secondary road, v / h; n is a number of directions (approaches to the intersection) of the secondary road.

The disadvantage of the above approach is that the accuracy of calculations is not always acceptable.

4. Authors' research to determine transport delays

At present video fixation devices are becoming increasingly important in road traffic organization. The cameras provide information on traffic flow in a continuous mode and thereby contribute to road safety.

4.1. Characteristics of intersection no. 9

To obtain data on traffic delays at intersection 9, we used a video camera to record traffic flows. The traffic management diagram at this intersection is shown in Figure 5.

The traffic management diagram at intersection 9 of Pushkin Street and Zheleznodorozhnaya Street is shown in Figure 5. Both Pushkin Street and Zheleznodorozhnaya Street have two lanes.

This crossing serves as a link between the city center and the Lago-Naki and Guzeriple tourism cluster.

A number of pedestrians crossing the railway in the crossing area is insignificant. Public transport passes through the crossing.

The crossing is equipped with a traffic light and an automatic barrier. The traffic direction at the intersection is shown in Figure 5; approximate traffic intensity is shown in Figure 6.

The largest lines of automobile transport in length and time are observed from direction 4 (Figure 6). From this direction, the motor vehicle turns left. This direction makes the main contribution to the average vehicle delay (3).

4.2. Regression equations for transport delay determination

The transport delay at the intersection was determined by the queue length. We fixed the initial time. At given time, the queue length was counted from direction 4. It was considered constant until the first car in the queue turned left. The time and the length of the queue were recorded again. If the queue increased, then it was assumed that the queue increased at the moment when the first vehicle drove away. Similarly, the time and length of the queue were recorded after each vehicle that turned left. Thus, you can get delay for each vehicle in the queue for the entire observation time.

According to the results of the research, we received queueing time for each car. Using these data we constructed a graph of the dependence of a vehicle idle time on the queue length (Figure 8). The sum of a vehicle idle time gave a transport delay in this direction.

IOP Conf. Series: Materials Science and Engineering **832** (2020) 012091 doi:10.1088/1757-899X/832/1/012091

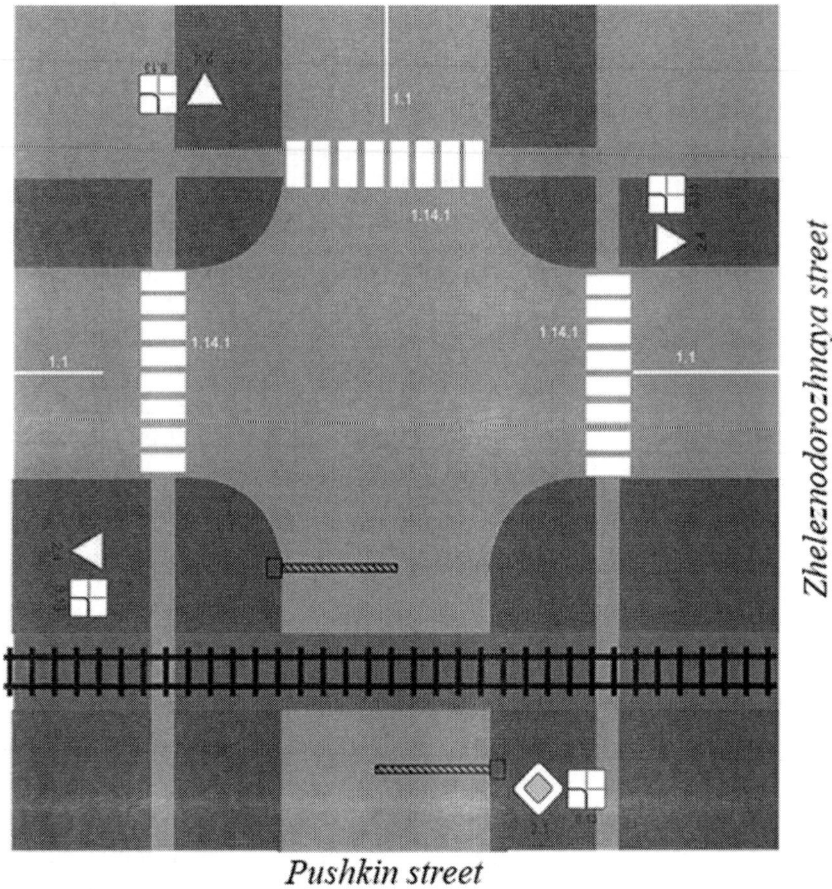

Figure 5. The intersection of Pushkin and Zheleznodorozhnaya streets

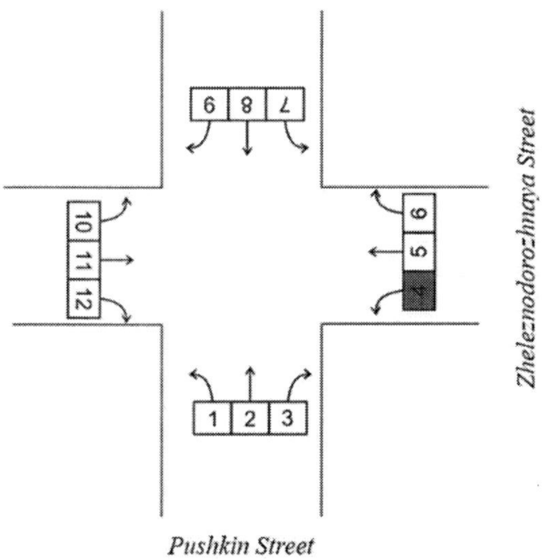

Figure 6. Traffic directions at the intersection of Pushkin and Zheleznodorozhnaya Streets

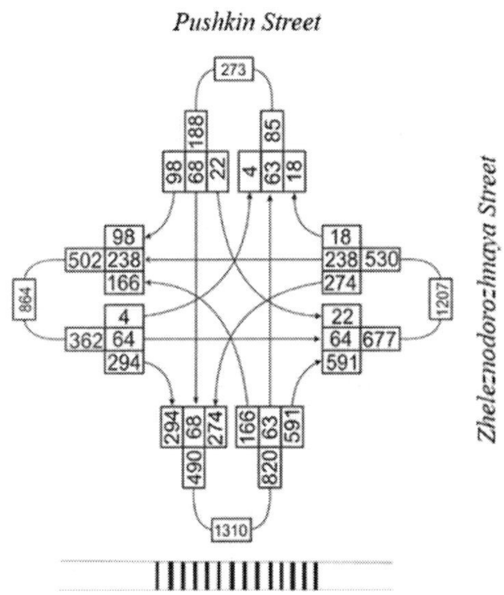

Figure 7. Traffic intensity at the intersection of Pushkin and Zheleznodorozhnaya streets

IOP Conf. Series: Materials Science and Engineering **832** (2020) 012091 doi:10.1088/1757-899X/832/1/012091

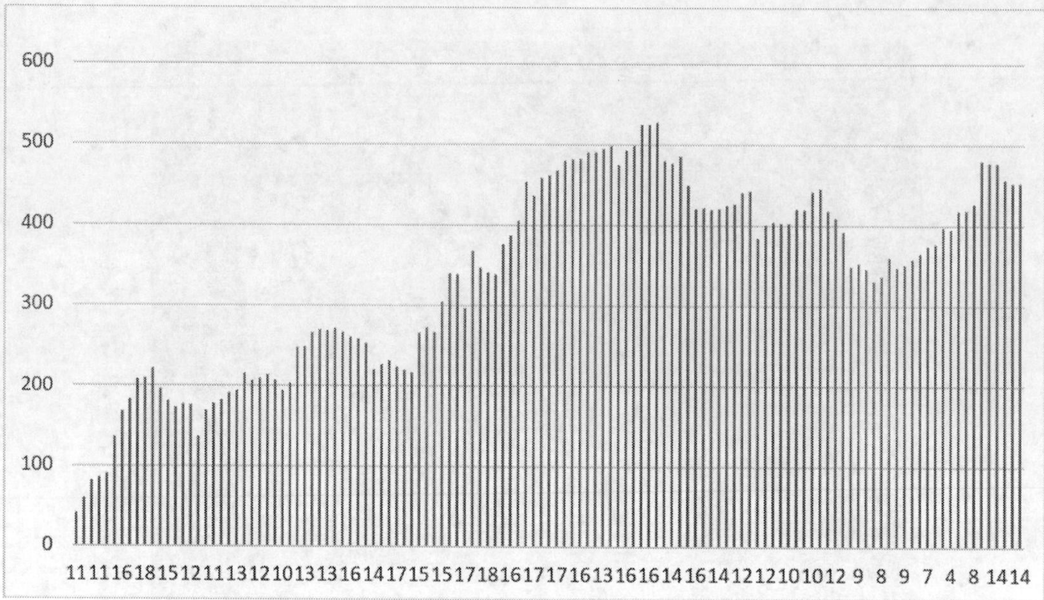

a

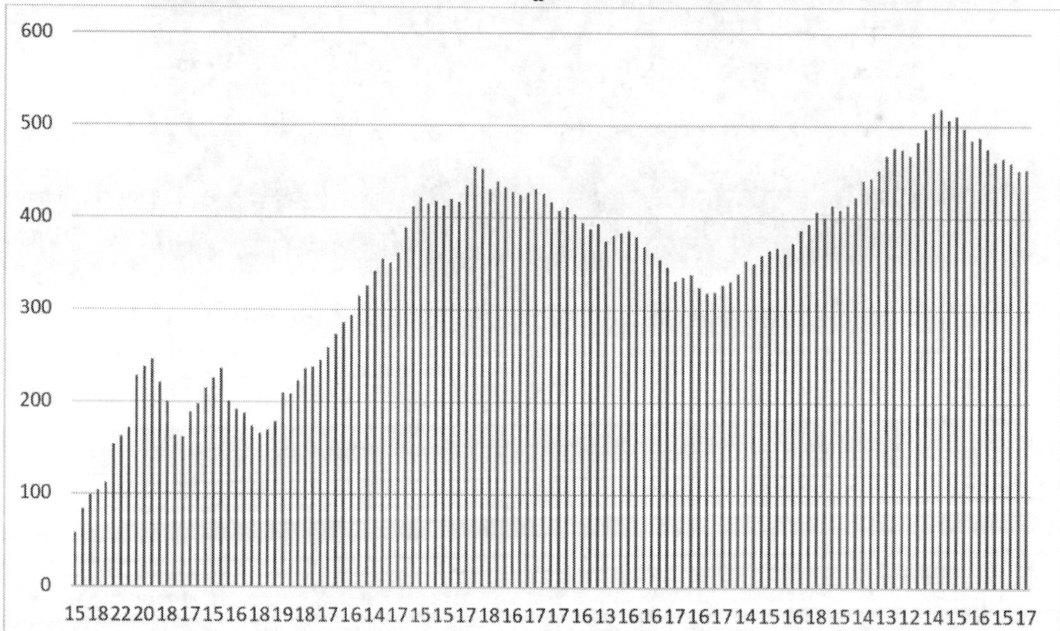

b

Figure 8. The graph of the dependence of the transport delay of cars (vertical) on the queue (horizontal); a – for crossing no. 9, b – for crossing no. 7

4.3. Development of a model for determining transport delays

We propose to consider the following approach to accounting traffic intensity: turning to the left from direction 4 (Figure 6) is prevented by the movement of vehicles from directions 1, 2, 7, 8, 11, 12. These directions are considered to be the main directions relatively to direction 4. The other directions do not interfere with the movement of vehicles from direction 4. The following factors affect the amount of transport delay (or the length of vehicle queue) from direction 4: traffic intensity from direction 4 and traffic intensities from directions 1, 2, 7, 8, 11, 12. Obviously, we can consider the sum of the intensities of movements from directions 1, 2, 7, 8, 11, 12, because the movement of vehicles from these directions is an obstacle for vehicles from direction 4.

We can assume that the magnitude of the transport delay at the investigated intersection depends linearly on the above parameters. Let's denote the total intensity of movement in the directions 1, 2, 7, 8, 11, 12 by X_1 and traffic intensity in direction 4 a by X_2. Then the transport delay can be represented as the following expression:

$$Y = k_1 \cdot X_1 + k_2 \cdot X_2 + C \quad (4)$$

where Y is transport delay, k_1 and k_2 are arbitrary coefficients, C is an intercept term.

Thus, to determine the transport delay, it is necessary to investigate the interaction of two factors that is to conduct a two-factor analysis.

After the two-factor analysis, the following regression equation is obtained:

$$Y = 51.4476 + 1.1171X_1 + 0.2081X_2. \quad (5)$$

4.4. Comparison of transport delay determination results

To verify formulas (4, 5, 7) calculations of the transport delay at the considered intersection have been made; the results of hourly observations at different times and on different days have been taken as initial data. Some results on measurements of one hour are presented in Table 1.

Table 1. The results of the transport delay calculations at intersection No. 9 according to the formulas (2, 5)

No.	The number of passing cars in the main directions (1. 2. 7. 8. 11. 12)	A number of passing cars in a secondary direction (4)	Secondary direction Queue length (vehicles)	Experimental transport delay in the secondary direction . s.	Transport delay (formula 4).	Inaccuracy. %	Transport delay (formula 5).	Inaccuracy. %
1	47	36	15	92	0.61	111.44	111.44	21.13
2	46	35	15	102	0.45	110.12	110.12	7.96
3	45	32	13	92	–0.05	108.38	108.38	17.80
4	44	33	14	126	0.09	107.47	107.47	–14.71
5	49	27	13	129	–0.97	111.80	111.80	–13.33
6	50	28	15	119	–0.71	113.13	113.13	–4.93
7	51	25	12	124	–1.43	113.62	113.62	–8.37
8	52	24	11	117	–1.70	114.53	114.53	–2.11
9	61	29	13	121	–0.30	125.63	125.63	3.82
10	62	30	13	118	–0.10	126.95	126.95	7.59
11	52	32	9	107	0.11	116.20	116.20	8.59
12	51	31	10	103	–0.08	114.87	114.87	11.53
13	57	27	16	98	–0.78	120.74	120.74	23.21
14	57	26	14	119	–1.02	120.53	120.53	1.29
15	49	25	13	120	–1.50	111.39	111.39	–7.18
16	50	25	13	97	–1.47	112.51	112.51	15.98
17	54	24	12	99	–1.64	116.77	116.77	17.94
18	53	24	12	91	–1.67	115.65	115.65	27.09
19	49	29	13	104	–0.52	112.22	112.22	7.90
20	50	29	12	135	–0.49	113.34	113.34	–16.05
21	50	31	12	118	–0.10	113.75	113.75	–3.60
22	51	30	13	134	–0.27	114.66	114.66	–14.43
23	55	28	13	125	–0.59	118.71	118.71	–5.03
24	55	27	13	123	–0.82	118.51	118.51	–3.65
	Total			2713	–45.62		2763	

The analysis of the above table shows that the maximum inaccuracy of the transport delay is 27% using formula (5). However, the total estimated hourly transport delay (2763 s) differs from the experimental one (2713 s) by 1.8%.

4.5. Verification of the obtained equation at intersection No. 7 of Adygeyskaya and Promyshlennaya streets
To verify the obtained formula in the form of a regression equation (5), the transport delay in the secondary direction at intersection No. 7 of Adygeyskaya and Promyshlennaya streets has been calculated. The obtained transport delay values have been compared with experimental data. The intersection pattern is similar to the intersection of Pushkina Street and Zheleznodorozhnaya Street (Figure 4). The calculation results are presented in the form of Table 2. The results of the calculations of the transport delay according to formula (2) are also shown in this table.

Table 2. The results of the transport delay calculations at intersection No. 7 according to the formulas (2, 5)

№	The number of passing cars in the main directions (1. 2. 7. 8. 11. 12)	A number of passing cars in a secondary direction (4)	Secondary direction Queue length (vehicles)	Experimental transport delay in the secondary direction , s,	Transport delay (formula 2),	Transport delay (formula 5),	Inaccuracy, %
1	58	36	15	112	0.76	123.73	10,47
2	57	35	15	135	0.62	122.41	-9,33
3	63	32	13	139	0.25	128.48	-7,57
4	63	33	14	119	0.40	128.69	8,14
5	69	27	13	125	–0.62	134.15	7,32
6	68	28	15	129	–0.42	133.24	3,28
7	62	25	12	140	–1.18	125.91	-10,06
8	61	24	11	143	–1.47	124.59	-12,88
9	59	29	13	132	–0.32	123.39	-6,52
10	60	30	13	129	–0.12	124.72	-3,32
11	63	32	9	116	0.25	128.48	10,76
12	64	31	10	119	0.10	129.39	8,73
13	57	27	16	101	–0.78	120.74	19,55
14	58	26	14	133	–1.00	121.65	-8,53
15	51	25	13	120	–1.43	113.62	-5,31
16	52	25	13	137	–1.40	114.74	-16,25
17	54	24	12	141	–1.64	116.77	-17,19
18	53	24	12	135	–1.67	115.65	-14,33
19	57	29	13	129	–0.35	121.16	-6,08
20	58	29	12	119	–0.34	122.27	2,75
21	55	31	12	126	–0.01	119.34	-5,29
22	55	30	13	123	–0.19	119.13	-3,15
23	54	28	13	122	–0.61	117.60	-3,61
24	53	27	13	128	–0.86	116.27	-9,16
	Total			3052	–12.05	2946.12	

Table 2 shows that the maximum accuracy in calculating the transport delay by the formula (5) does not exceed ± 20 %.

The estimated transport hour delay at this intersection (2946 s) differs from the experimental one (3052 s) by 3.5 %. Thus, the preliminary obtained result using formula (5) for determining the transport delay shows a high level of correlation.

Obviously, further research is needed for the development of the proposed approach. A significant advantage of the proposed approach, the result of which is expression 5, is its simplicity of calculation, in contrast to expression (2).

5. Conclusion

The following work has been carried out in the research:

- Experimental data in the form of video files have been collected.
- The experimental data have been processed, as a result, a graph of the dependence of the transport delay on the queue length has been obtained (Figure 8).
- A new approach has been proposed for determining transport delays. The equation (5) has been obtained, that shows the dependence between transport delay and traffic intensity.
- The initial data obtained as a result of the experiment for the investigated intersection and the results calculated by the formula (5) are presented in Table 1.
- The transport delay has been calculated using the formula (5) at another intersection (No. 7), and Table 2 has been constructed on the basis of the results. A rather high accuracy of determining the transport delay has been obtained.

References

[1] Akhunova I B and Udzhukhu A Z 2017 On some additional directions to the tasks in the Federal target program "Improving Road Safety" *Bulletin of transport information. Information and practical journal* **3(261)** 33–5
[2] Andronov R V and Leverenz E E 2017 Calculation of vehicle delays at an isolated regulated intersection during its operation at high loading levels using Monte Carlo method *Bulletin of Civil Engineers* **1(60)** 221–6
[3] Boyarsky S N 2015 Improving the efficiency of highway intersection with a high value of the traffic load factor *The dissertation for the degree of Candidate of Technical sciences* (Yekaterinburg)
[4] Saha A, Ghosh I and Chandra S 2017 Delay at Signalized Intersections under Mixed Traffic Conditions *Journal of Transportation Engineering* **143(8)**
[5] Preethi P, AbyVarghese and Ashalath R 2016 Modelling Delay at Signalized Intersections under Heterogeneous Traffic Conditions *Transportation Research Procedia* **17** 529–38
[6] Naumova N A 2018 Method for determining vehicle delays at intersections, taking into account pedestrian traffic *Modern high-tech technologies* **8** 116–20
[7] Chikalin E N 2012 Analysis of transport delays at unregulated pedestrian crossings *Bulletin of Irkutsk State Technical University* **9(68)** 168–74
[8] *Road safety indicators.* Retrieved from: http://stat.gibdd.ru
[9] Kremenets Yu A, Pechersky M P and Afanasyev M B 2005 *Technical means of traffic management* (Moscow, "Academkniga" ICC)
[10] Bulavina L V 2009 An experimental study of the characteristics of transport and pedestrian traffic (Yekaterinburg)
[11] *TranspoVolume.* Retrieved from: http://www.transpovolume.ru/ranvols-878-1.html

IOP Conf. Series: Materials Science and Engineering **832** (2020) 012092 doi:10.1088/1757-899X/832/1/012092

Development of an approach to the selection of the necessary method of arrangement of the left turn in the area of controlled intersection

S V Dorokhin[1], A N Novikov[2], V A Zhulai[1] and D V Likhachev[1]

[1]Voronezh State University of Forestry and Technologies named after G.F. Morozov, 8, Timiryazeva str., Voronezh, 394087, Russia
[2]Orel State University named after I.S. Turgenev, 95, Komsomolskaya str., Orel, 302026, Russia

E-mail: dsvvrn@yandex.ru

Abstract. In this article the analysis of the main ways of the organization of the left turn in a zone of the regulated intersection with use of foreign and domestic scientific sources is carried out. In the analysis the authors determined that today in the choice of the necessary way of organizing the left turn option uses a minimal set of data, namely, intensity of traffic, causing a traffic controlled junctions may not be effective. Studies of changes in the main characteristics of the traffic flow as a result of conflict analysis "left turn-direct conflicting flow" allowed the authors to develop a new approach to the selection of the necessary method of arrangement of the left turn, taking into account changes in the main characteristics of the traffic flow – delay, speed and queue length. The experiment showed the feasibility of using the proposed approach.

1. Introduction

Daily due to the rapid growth of level of automobilization [1-3] in the world and on the roads of the Russian Federation there are many negative factors, such as the loss of time, the formation of queues and congestion, unnecessary fuel consumption, wear of components, wear of road surface, environmental degradation, and another important factor is the occurrence of accidents, injury and death, leading to substantial losses for the state. These problems are particularly acute in the regulated areas, in this regard, it is necessary to develop a new approach to the selection of the optimal control scheme, namely the introduction of an additional phase of regulation.

When introducing an additional control phase on regulated areas, it is necessary to analyze conflicting directions and consider the traffic process using additional characteristics of the traffic flow in order to determine the rational use of the additional time of the traffic light cycle. In this regard, the main objective of the study – the rationale for entering the additional phase regulation and development approach to the choice of rational schemes of organization of traffic left turn, taking into account changes of characteristics of the traffic flow.

2. Analysis of existing approaches to the organization of the left turn

An analysis of the world experience in introducing an additional left-turn band was carried out using many sources [4–8], including research reports, state and federal design guidelines, and reference books.

Based on the analysis, the existing foreign approaches to the specialization of the left-turning lane are divided depending on the parameters and traffic conditions:

1. Approaches based on the accident rate of the regulated area; these approaches, despite the use of the characteristics of the traffic flow, are mainly based on the number of road accidents.

According to the approaches assigned to the first group, there are basic parameters necessary when taking into account the specialization of the left-turning band:

- Incoming intensity (N_{inc}) is the total hourly peak traffic on the main road, cars move in the same direction together with the left-hand flow.
- Left-turn flow intensity (N_{Left}) is the part of the incoming intensity that turns left at the intersection.
- Left rotation percentage ($\%_{Left}$) – percentage of incoming intensity that turns left:

$$\%_{Left} = [N_{Left} \div N_{inc}] \times 100. \quad (1)$$

- Forward flow rate (N_{dir}) is the part of the incoming intensity that goes directly through the intersection:

$$N_{dir} = N_{Left} + N_{inc}. \quad (2)$$

- Counter intensity (N_{count}) – the total peak hourly intensity of vehicles moving towards the incoming intensity.

Despite the basic parameters of the traffic flow, the fundamental factor will be the number of accidents at the intersection associated with the left turn, a total of four or more such accidents in 12 months or six or more in 24 months will be the basis for the specialization of the left turn.

2. Approaches based on the analysis of the main characteristics of the traffic flow.

The scientists of Park, B. and Lee, J. [6] have improved the main parameters used in the specialization of the left-turning band: 1. a critical interval of 5.5 seconds (instead of 5.0 seconds); 2. time to turn left 4.3 s (instead of 3.0 s); 3. time to clear the strip 3.2 s (instead of 1.9 s).

In addition, depending on the speed of movement and the volume of the left turn from the direct flow, critical values were obtained for entering a specialized band [8].

According to studies of the approaches assigned to the second group, the rationale for introducing an additional regulatory phase is to compare the intensity of the conflicting flow and the percentage of left rotation.

In domestic practice, to determine the algorithm for entering the additional regulation phase, it is initially necessary to consider the structure of the cycle and the sequence of calculation of its main elements. In more detail, the domestic experience was determined by the authors in a scientific study [9,10], however, to date, the use of an additional left-hand lane is associated with the introduction of a separate third phase in the traffic control cycle if the intensity of the left turning flow is more than 120 aut./h [11]. The main calculation will depend on the geometry of the regulated section in the area of calculating the saturation flux of such a band and, accordingly, the duration of the regulation phase [12].

As a result of using a minimal data set, namely, knowing only the intensity of the left turn, the organization of traffic at an adjustable intersection is possibly inefficient. In order to expand the field of knowledge in this matter, the necessary measure will be the development of a sound scientific approach using previously unused characteristics of the transport stream to develop an approach to the introduction of a specialized left-handed lane and regulation phase. This event will further improve the efficiency of regulated intersections. During the study of scientific literature, the main approaches to the introduction of an additional left-handed control phase were established. It was determined that when entering the extended regulation phase, the main parameters of the traffic flow, such as the length of the queue, the intensity of the conflicting flow, speed and delays of the vehicles, are not taken into account. In this regard, the necessary measure is the development of a scientifically based approach to the introduction of a specialized regulation phase at controlled intersections, taking into account the main characteristics of the traffic flow.

3. Crossroads research

An analysis of the main regulated intersections in Voronezh showed that today there are several ways to organize left-hand traffic (these methods are typical for many cities of the Russian Federation), which are as follows:

1. Organization of the movement of the left turn, together with the direct movement from one lane in one phase of regulation («method 1»).
2. Organization of the movement of the left turn from a specialized lane in one phase of regulation («method 2»).
3. Organization of a left turn from a specialized lane together with the use of an «elongated regulation phase» («method 3»).
4. Organization of a left turn from a specialized lane in a specialized phase of regulation («method 4»).

The presented methods of organizing the movement of the left turn are quite different and require the development of a reasonable description of the input of each of them. In the presented scientific sources there are no strictly regulated requirements for the selection of each of them.

A study of the main regulated intersections in Voronezh showed that with various ways of organizing traffic, the main parameters of the traffic flow, namely, speed, queue length and vehicle delays, change. In order to develop a reasonable method for choosing the necessary organization of the left turn, it is necessary to assess the change in the main parameters of the traffic flow.

To evaluate changes in these parameters (speed, delay, queue length), an analysis of their change is necessary for various combinations of left-turning N_1 traffic flow and oncoming conflicting N_2. During model experiments, changes in speed, delays and maximum queue length for a conflicting stream were determined (Fig. 1–3).

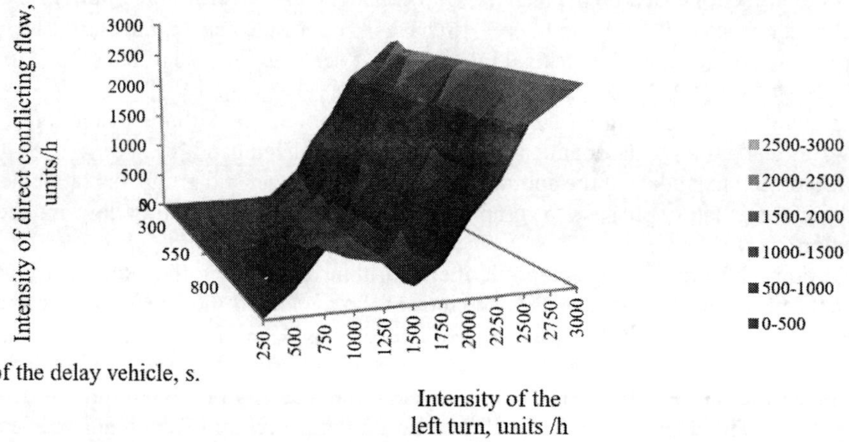

Figure 1. Change in delays with the ratio of intensities N_1 to N_2

The minimum value for vehicle delays is 2.77 s with a ratio of N_1 = 50 units/h to N_2 = 250 units/h, the maximum value is 2644.97 s. with a ratio of N_1 = 50 units/h to N_2 = 3000 units/h, and the average delay is 918.19 s. In most ratios of intensities N_1 and N_2, delays in the model are directly dependent and have a pronounced upward trend with downward correction and a local maximum of delays of 756.98 s with a ratio of N_1 = 1000 units/h to N_2 = 750 units/h.

The minimum value of the queue length of vehicles is 1 aut. with a ratio of N_1 = 50 units/h to N_2 = 250 units/h, the maximum value is 1087 aut. with a ratio of N_1 = 1000 units/h to N_2 = 3000 units/h, and the average value of the length of the vehicle queue is 918.19 aut. In most ratios of intensities N_1 and N_2, the queue length in the model is directly dependent and has a pronounced upward trend.

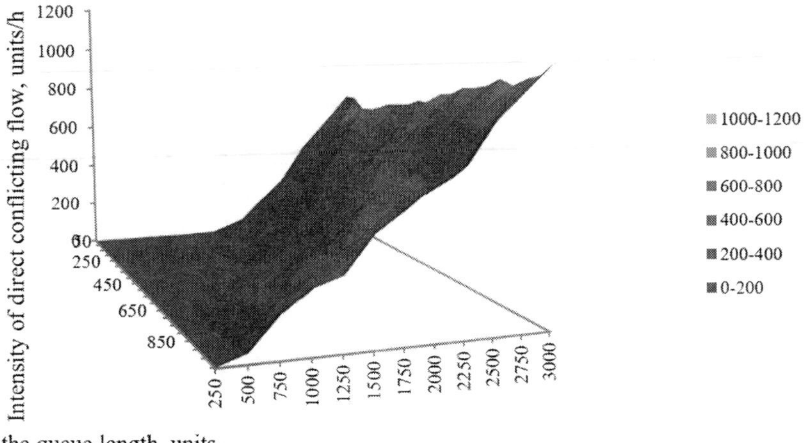

Figure 2. Change in queue length with intensity ratio N_1 to N_2

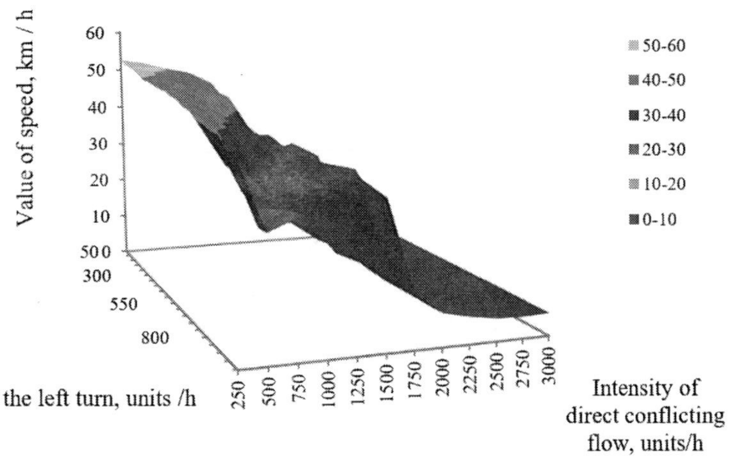

Figure 3. Change in speed with intensity ratio N_1 to N_2

The minimum vehicle speed is 4.75 km/h with a ratio of N_1 = 750 units/h to N_2 = 2750 units/h, the maximum value is 52.55 km/h with a ratio of N_1 = 1000 units/h to N_2 = 3000 units/h, and the average vehicle speed is –23.35 km/h. In most ratios of intensities N_1 and N_2, speed in the model is inversely related and have a pronounced downward trend with upward correction and a local minimum of 26.7 km/h with a ratio of N_1 = 950 units/h to N_2 = 500 units/h.

As a result of determining the main parameters of the traffic flow with a combination of the intensities of the left-hand flow and the direct conflicting flow for one and two lanes, it became possible to determine the possible ranges of changes in speed, delay and queue length, which are fundamental parameters in determining the necessary way of organizing the movement of the left-turn flow.

4. Approach to the organization of the left turn in the area of the regulated intersection

In order to determine the parameter values for each possible combination, the authors proposed to switch to a system of ratio coefficients, each of which will be interpreted by certain values of the considered parameters (queue length, speed, delay). The obtained values of the coefficients are presented in table 1.

$$K_1 = \frac{N_1}{N_2} = n = \{t_n = x_n;\ l_n = y_n;\ v_n = z_n\} \tag{3}$$

where K_1 – the coefficient of the ratio of N_1 to N_2, N_1 – intensity of movement of the left turn, units/h; N_2 – the intensity of the movement of a direct conflicting stream moving in one lane, units h; n – the value of the ratio coefficient (0.02 ... 4.0); t_n – delay time of the vehicle, s.; x_n – the value of the delay time, for a given ratio of intensities, s.; l_n – vehicle queue length, units; y_n – is the value of the queue length, for a given intensity ratio, units; v_n – speed, km/h; z_n – the value of the speed of movement, for a given ratio of intensity, km/h.

Table 1. Value of the ratio N_1 and N_2

N_2 \ N_1	50	100	150	200	250	300	350	400	450	500	550	600	650	700	750	800	850	900	950	1000
250	0.20	0.40	0.60	0.80	1.00	1.20	1.40	1.60	1.80	2.00	2.20	2.40	2.60	2.80	3.00	3.20	3.40	3.60	3.80	4.00
500	0.10	0.20	0.30	0.40	0.50	0.60	0.70	0.80	0.90	1.00	1.10	1.20	1.30	1.40	1.50	1.60	1.70	1.80	1.90	2.00
750	0.07	0.13	0.20	0.27	0.33	0.40	0.47	0.53	0.60	0.67	0.73	0.80	0.87	0.93	1.00	1.07	1.13	1.20	1.27	1.33
1000	0.05	0.10	0.15	0.20	0.25	0.30	0.35	0.40	0.45	0.50	0.55	0.60	0.65	0.70	0.75	0.80	0.85	0.90	0.95	1.00
1250	0.04	0.08	0.12	0.16	0.20	0.24	0.28	0.32	0.36	0.40	0.44	0.48	0.52	0.56	0.60	0.64	0.68	0.72	0.76	0.80
1500	0.03	0.07	0.10	0.13	0.17	0.20	0.23	0.27	0.30	0.33	0.37	0.40	0.43	0.47	0.50	0.53	0.57	0.60	0.63	0.67
1750	0.03	0.06	0.09	0.11	0.14	0.17	0.20	0.23	0.26	0.29	0.31	0.34	0.37	0.40	0.43	0.46	0.49	0.51	0.54	0.57
2000	0.03	0.05	0.08	0.10	0.13	0.15	0.18	0.20	0.23	0.25	0.28	0.30	0.33	0.35	0.38	0.40	0.43	0.45	0.48	0.50
2250	0.02	0.04	0.07	0.09	0.11	0.13	0.16	0.18	0.20	0.22	0.24	0.27	0.29	0.31	0.33	0.36	0.38	0.40	0.42	0.44
2500	0.02	0.04	0.06	0.08	0.10	0.12	0.14	0.16	0.18	0.20	0.22	0.24	0.26	0.28	0.30	0.32	0.34	0.36	0.38	0.40
2750	0.02	0.04	0.05	0.07	0.09	0.11	0.13	0.15	0.16	0.18	0.20	0.22	0.24	0.25	0.27	0.29	0.31	0.33	0.35	0.36
3000	0.02	0.03	0.05	0.07	0.08	0.10	0.12	0.13	0.15	0.17	0.18	0.20	0.22	0.23	0.25	0.27	0.28	0.30	0.32	0.33

To determine the necessary way of organizing traffic at a regulated intersection from the four previously considered it is necessary to interpret the obtained values of vehicle delays with the existing service levels of the regulated intersection.

As a result of the mathematical analysis of the data, the main limits were determined by combining the critical intensities for the left turn and the direct conflicting flow, characterized by the ratio coefficient. Each coefficient obtained can be described by the corresponding value of the characteristics of the traffic flow (queue length, delay and speed). The study of changes in these indicators when using traffic light control and comparing the obtained data with the value of the delay characteristic for the corresponding level of service, during the course of the study made it possible to determine the corresponding areas characterized by the initial parameters – the intensity – and determine the corresponding areas for each level. For each service level, a mathematical model and its main parameters are determined by the result of mathematical analysis.

Based on the analysis of changes in the main values of the traffic flow during the study of the «left turn-direct conflicting flow» conflict, such as speed, queue length and delay value, it is proposed to use the ratio coefficient, which will serve as the main value when choosing the necessary method of organizing the movement of the left turn at an adjustable intersection. Each value of the coefficient will be characterized by the corresponding value of the main values that will be observed at the intersection. By connecting the obtained coefficient value with the value of the delay value of cars, it is possible to determine the level of service at the intersection and thereby determine the necessary way of organizing traffic, which is shown in detail in table 2.

As a result of the mathematical analysis of the conflict in question, an algorithm was developed for choosing the necessary method of organizing movement in an adjustable area.

Table 2. Methods of organizing the movement of the left-hand flow by service levels

Type	Thresholds, units/h	Coefficient	Delays, s.	Queue length, aut	Speed, km/h	Service level
method 1	1) $N_2 = 500$, $N_1 = 50–400$; 2) $N_2 = 1000$, $N_1 = 50–200$; 3) $N_2 = 1500$, $N_1 = 50–100$; 4) $N_2 = 2000$, $N_1 = 50$	1) 0,1–0,8; 2) 0,05–0,2; 3) 0,03–0,07; 4) 0,03	1) 2,69–9,3; 2) 4,14–9,45; 3) 5,92–7,39; 4) 8,73	1) 0,18–0,82; 2) 0,85–0,9; 3) 1,42–1,58; 4) 2,33	1) 52,7–48,7; 2) 51,93–49,57; 3) 50,81–50,17; 4) 49,06	A (≤ 10 s.)
method 2	1) $N_2 = 500$, $N_1 = 450–700$; 2) $N_2 = 1000$, $N_1 = 250–300$; 3) $N_2 = 1500$, $N_1 = 150–250$; 4) $N_2 = 2000$, $N_1 = 100$; 5) $N_2 = 2500$, $N_1 = 50$	1) 0,9–1,4; 2) 0,25–0,3; 3) 0,1–0,17; 4) 0,05; 5) 0,02	1) 11,46–19,16; 2) 10,3–16,29; 3) 11,52–19,28; 4) 10,67; 5) 14,56	1) 0,97–1,37; 2) 0,97–1,12; 3) 1,5–1,6; 4) 2,58; 5) 3,93	1) 47,75–44,72; 2) 49,07–47,02; 3) 48,5–46,42; 4) 48,48; 5) 46,15	B (10-20 s.)
method 3	1) $N_2 = 500$, $N_1 = 750–800$; 2) $N_2 = 1000$, $N_1 = 350–500$; 3) $N_2 = 2000$, $N_1 = 150$	1) 1,5–1,6; 2) 0,35–0,5; 3) 0,08	1) 24,4–29,28; 2) 21,18–30,69; 3) 20,97	1) 1,8–2,05; 2) 1,1–1,35; 3) 2,48	1) 42,95–41,62; 2) 45,96–43,62; 3) 45,73	C (20–35 s.)
method 4	1) $N_2 = 500$, $N_1 = 850–900$; 2) $N_2 = 2500$, $N_1 = 100$; 3) $N_2 = 3000$, $N_1 = 50$	1) 1,7–1,8; 2) 0,04; 3) 0,02	1) 38,53–38,93; 2) 41,39; 3) 51,21	1) 1,9–2,18; 2) 5,37; 3) 11,38	1) 39,91–39,47; 2) 40,53; 3) 37,24	D (35–55 s.) E (55–80 s.) F (>80 s.)

5. Experiment

To assess the effectiveness of the proposed approach, a model experiment was performed at one of the regulated intersections of Voronezh using the Aimsun simulation product (fig. 4, 5).

Figure 4. Satellite image of the object of study

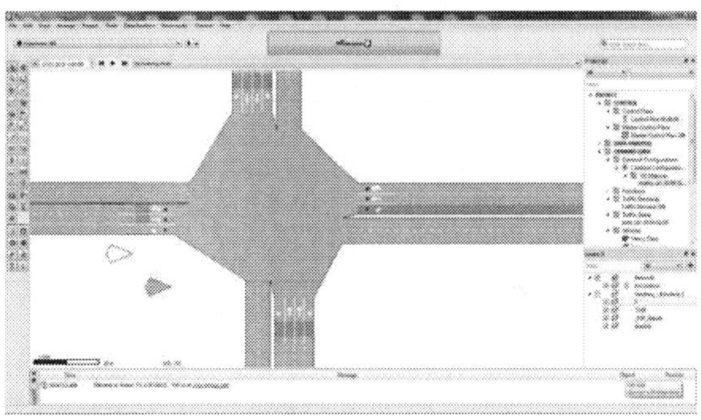

Figure 5. Model of the object of study

The organization of the left turn at the object of study occurs according to «method 2». As a result of the simulation, it was found that the average value of the delays is 160.19 s, and the average value of the vehicle speed is 25.54 km/h.

Field studies allowed us to determine the intensity of the left turn $N_1 = 87$ units/h, the intensity of the oncoming flow $N_2 = 800$ units/h.

Using the approach proposed for work (table 2), it follows that for these combinations of intensity, which are included in the threshold values of service level A, it is advisable to use the «method 1» of organizing the left turn (the movement of the left-turn flow together with the direct flow from one lane to single phase).

As a result of the simulation, after implementing this method for these intensities, the average value of the delays decreased by 19 %, and the average value of the speed increased by 16 % (table 3).

Table 3. Indicators of traffic flow before and after the reorganization of the existing traffic management scheme

Traffic characteristic	Indicators of traffic flow with the existing traffic management scheme	Indicators of traffic flow after the reorganization of the existing traffic management scheme
Delays, s.	160,19	129,84
Speed, km/h	25,54	29,81

It is established that when introducing a new method of organizing a left turn, the situation in the analyzed object of study improves.

6. Conclusion

As a result of the study, it was found that to date, schemes for organizing the movement of the left turn are not effective, in connection with which congestion situations are observed at intersections and many transport characteristics are of increased and reduced value. In order to improve the transport situation and the smooth distribution of the traffic flow, namely, the harmonious movement of vehicles, a new approach was proposed to equip the left turn in the area of the regulated intersection. The performed model experiment showed the feasibility and effectiveness of using the approach proposed to develop based on the study of the «left turn – direct conflicting flow» conflict and determining the values of the main characteristics of the traffic flow – speed, queue length, and delays. As a result of testing the developed approach in the studied section of the road network, it was possible to reduce the average delay by 19 % and increase the speed by 16 %.

References

[1] Blinkin M Ya and Reshetova E M 2013 *Road Safety Background, International Experience, Basic Institutions* (Moscow: National Research University Higher School of Economics)

[2] Jacobs G, Aeron-Thomas A and Astrop A 2000 *Estimating Global Road Fatalities* TRL Report 445 (U.K.: Transport Research Laboratory. Crowthorne)

[3] Kopits E and Cropper M 2005 Traffic Fatalities and Economic Growth *Accident Analysis and Prevention* **37** 169–78

[4] Gifford J, Pelletiere D and Collura J 2001 Stakeholder Requirements for Traffic Signal Preemption and Priority in the Washington D.C. Region *Transportation Research Record* **1748** 1–7

[5] Shelby S G, Bullock D M and Gettman D 2006 Transition Methods in Traffic Signal Control *Transport. Res. Record J. of the Transport. Res. Board* **1978** 130–40

[6] Luyanda F, Gettman D, Head L, Shelby S, Bullock D and Mirchandani P 2003 ACS-Lite algorithmic architecture: Applying adaptive control system technology to closed-loop traffic signal control systems *Transport. Res. Record: J. of the Transport. Res. Board* **1856** 175–184

IOP Conf. Series: Materials Science and Engineering **832** (2020) 012092 doi:10.1088/1757-899X/832/1/012092

[7] Park B, Santra P, Yun I and Lee D 2001 Optimization of time-of-day breakpoints for better traffic signal control *Transport. Res. Record: J. of the Transport. Res. Board* **1867** 217–213
[8] Park B and Lee J 2008 Procedure for determining time-of-day break points for coordinated actuated traffic signal systems *KSCE J. of Civil Engineer.* **12(1)** 37–44
[9] Novikov A, Katunin A, Novikov I and Shevtsova A 2017 Research of Influence of Dynamic Characteristics for Options Controlled Intersection *Procedia Engineer.* **187** 664–71
[10] Dorokhin S V, Zelikov V A, Strukov Y V, Likhachev D V, Novikov A N, Novikov I A and Shevtsova A G 2018 Investigation of methods for calculating duration of light signal regulation cycle *J. of Physics: Conf. Ser.* **1015** 032128
[11] National standard of the Russian Federation GOST R 52289-2004 "*Technical means of traffic management. Rules for the use of road signs, markings, traffic lights, road fences and guiding devices*" (approved by order of the Federal Agency for Technical Regulation and Metrology of 15 December 2004 no 120-st)
[12] Levashev A G, Mikhailov A Yu and Golovnykh I M 2007 *Design of adjustable intersections* (Irkutsk: Publishing house of Ir State Technical University)

Analytical model of the laminar traffic phase states and its use for calculating the level of service indicators for road network

D B Efimenko, V M Vlasov, V N Bogumil, V A Demin and S A Filatov

Moscow Automobile and Road State Technical University (MADI), 64, Leningradsky ave., Moscow, 125319, Russia

E-mail: td@madi.ru

Abstract. The article is dedicated to the description and discussion of theoretical models that may be used for calculating the level of service indicators for road network. Authors developed the approach for calculating the level of service indicators for urban road network sections with signalized intersections. The core of the proposed models of assessing the level of service is based on the method, developed by American experts for main roads and adapted by the authors for the city roads. The approach to assessing the level of service is based on the advanced analytical model of the fundamental diagram, describing various phases of traffic flow. Herein authors introduce the concept of a laminar traffic flow in the sections of the road network, which is implemented for the level of service calculating.

1. Introduction

Modern society needs high-quality transport services. At the same time, the demands for transportation of goods and passengers are constantly growing. The growing demand for passenger transport has led to an increase in the level of motorization and to the aggravation of related problems, including a high level of accidents on the roads, the occurrence of chronic congestion and environmental pollution.

In Russia the most typical example is Moscow. In Moscow every day occur about 650 traffic jam situations, each of which is idle an average of 500–600 cars. The average speed within central part of the City is about 18 km/h [1]. Similar conditions are observed in many other megacities of the world [2–5].

The most effective means of dealing with the consequences of uncontrolled motorization are intelligent transport systems (ITS). The ITS project should include three basic complex subsystems [6, 7]:

1) Subsystems that solve the problem of optimizing the transport system performance, focused on management of transport flows.

2) Subsystems of dispatching management of transport (public, municipal, special, freight) united by a single automated system of information for analytical support of Executive bodies.

3) Subsystems, providing safety and security of transport processes.

Exploitation of ITS is very expensive affair for a city administration. Therefore, an important task is a comprehensive assessment of the ITS functioning quality. From the point of view of the road users, the ability to move through the urban streets and city roads network is considered as a service.

In this case, assessment of situation must be based on evaluation of actual parameters of traffic flow on sections of the road network [5, 8].

A scientific basis for the assessment of traffic flow parameters (speed, density, intensity) was laid in the works of American scientists. In the second half of the last century they had formulated a number of indicators under the title "Level of service". They determined six levels of service designated A, B, C, D. E, F, where the "A" level indicates the lightest condition of traffic flow. Indicators for assessing these levels are set out in the Highway Capacity Manual (HCM), fundamental American document, which has been repeatedly supplemented and republished [9]. Each level of service depends on relation of average traffic flow speed and free flow speed for the highway. Similar approach was realized in Russia road branch methodical document ODM 218.2.020-2012 "Guidelines on the assessment of road capacity" [10].

However, the analysis of limit values of indicators in both these documents indicates that they are applicable only for highways and not designed to assess the level of service on urban roads. The reason for this conclusion is that the characteristics of traffic flow on city roads are strongly influenced by speed limits, prescribed by city authorities. These limits are significantly different in value from the speed limits for highways. This, in turn, significantly affects the boundary values of indicators assessing level of service. This article deals with the evaluation of the indicator "level of service" using the fundamental diagram [11, 12], which is considered as an analytical dependence of the average speed on the average density at different phase states of the traffic flow, and provides a critical analysis of an alternative approach to the description of the fundamental diagram based on the use of a single approximating analytic formula. When assessing the indicator "level of service" we take into account the speed limit, prescribed for the road sector.

2. Two types of car following models, used for describing traffic flow parameters

2.1. Continuous model of Van Aerde

The mathematical theory of traffic flow, started with fundamental diagram, was developed mainly by mathematicians and physicists. There were two branches in the development of the mathematical theory of traffic flow, using "car following model" The first one is approximation theory, which try to construct mathematical model, using statistical data of science experiments. The well-known model was continuous model of Van Aerde [13]. This model described the gap (h) between two vehicles: lider and follower, as function of current velocity (v) and velocity of free flow state (v_0) as follows:

$$h = c_1 + \frac{c_2}{v_0 - v} + c_3 v\,, \qquad (1)$$

where c_1, c_2, c_3 – constants.

The inverse value is density (ρ). Therefore, we have:

$$\rho = \frac{1}{c_1 + \dfrac{c_2}{v_0 - v} + c_3 v}\,. \qquad (2)$$

The graph of the Van Aerde continuous "Speed – Density" model represented in fig. 1.

The approximation approach has its own advantages and disadvantages. Advantages include simplicity of the analytical model and good approximation to real situation, gained by choice of coefficients values. The analytical model of the "speed-density" shows that dependence is non-linear and decreasing function of the traffic flow density. When the traffic flow density changes from zero to a maximum value, the curve is changed so that in total it represent some different sections, which correspond to different phase states of the traffic flow.

Disadvantages are:

1) the model doesn't explain cause of traffic flow phase state changes;

2) the character of curve for numeric interval of small values of density is in contradiction with real situation. Indeed, the curve slows down immediately after starting point of density value, equal to zero. It means that average speed of traffic flow drops at initial section of the curve, corresponding to situation of almost empty road. This is contradiction with the real situation on the road and drivers behavior. Really, within some interval near the zero density value, average speed of the traffic flow doesn't drop down and stay constant.

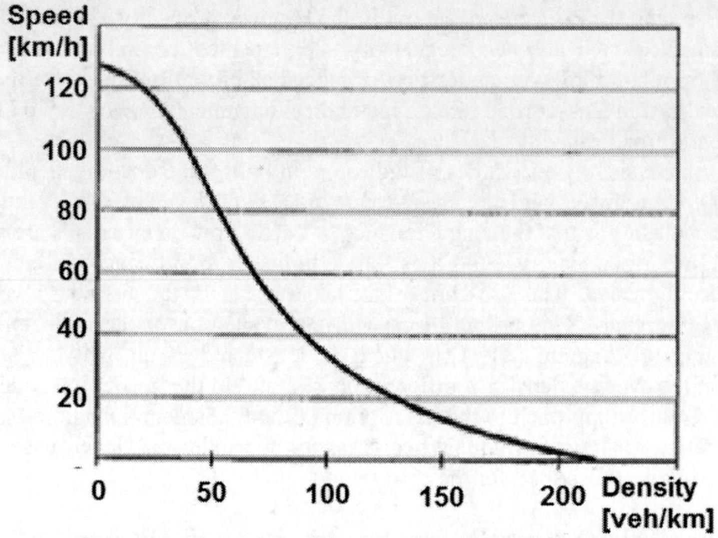

Figure 1. The graph of the Van Aerde continuous model

2.2. Methods of engineering psychology, used to derive the model of traffic flow

In the models of the follow-the-leader system "driver-vehicle" treated as a closed system with feedback, methods of engineering psychology are used to identify the mechanism by which the driver selects a certain position in relation to other cars and the road in order to describe the processes of information processing and decision making in the management of the vehicle [14].

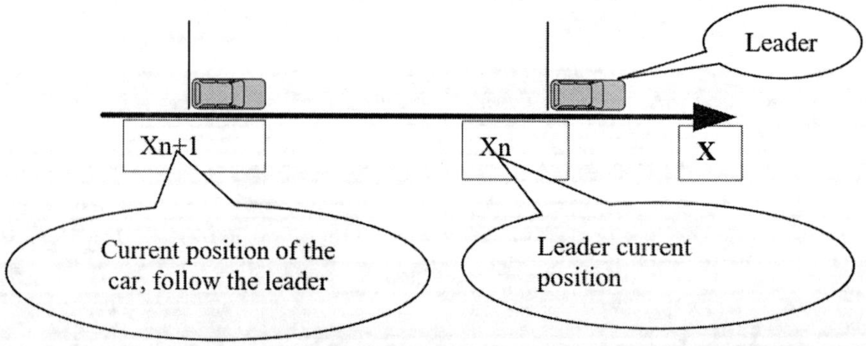

Figure 2. Position of the leader and the car following the leader

Engineering-psychological approach is used to examine the unique sensory and motor characteristics of the driver and restrictions: on this basis, strive to minimize the number of situations in which the driver cannot properly manage vehicle, as well as to regulate the absolute and relative speed [15–17]. The scheme of movement of the car following the "leader" is shown in fig. 2. Here "x" is the current position of the leader, and "x_{n+1}" is the current position of the car following the leader.

The advantage of the leader models, in comparison with the continuous model of Van Aerde, is a detailed account of the human factor, in this case – a detailed account of driver behavior and its impact on phase state of traffic flow. In accordance with this approach, the traffic flow on the roads can be considered as a complex closed system "driver – vehicle – traffic conditions – means of regulation". In this system, a control element influencing the speed of movement is the driver. He, relative to the speed of movement, keeps in mind states of other components of the system. The person (driver) is the main link in the system "driver – vehicle–road–environment".

3. Four phase states of the "laminar" traffic flow

3.1. Explanation of "laminar traffic flow" term and reason to use it.

When we analyze the situation with traffic flow which occurs on long section of town's arterial roads with few or without intersections, we can note that state of the flow changes slightly during the day time, without hops. In that case the *average speed* of the traffic flow doesn't fall down to zero at any time and at any traffic conditions if there are no obstacles. So we may use an analogy with the term "laminar", which is taken from hydrodynamics theory and means the way in which the liquid layers move without agitation and pulsation (that is, without messy rapid changes of velocity and pressure). The meaning of the term "laminar" is consistent with the cases of traffic flow when traffic is moving relatively freely, the number of "bottle necks" is small and the changes of flow parameters is slow.

The main reason for the term "laminar traffic flow" is that in this state it behavior precisely described with "follow the leader" models. Therefore we shall formulate our definition of this term.

For definition construction we use analogy with the term "laminar", which is taken from hydrodynamics theory and means the way in which the liquid layers move without agitation and pulsation (that is, without messy rapid changes of velocity and pressure). The meaning of the term "laminar" is consistent with the cases of traffic flow when traffic is moving relatively freely, the number of overtaking is small and the change of flow parameters is slow.

Strictly speaking, in the case of free flow, when the cars move freely from one lane to another, the analogy with laminar flow is not quite correct. The cars move with different velocities, moving freely within the drive way, from right lane to left and back. However, in this case, we believe the main condition is the absence of obstacles to the flow. That is the main condition for laminar traffic flow occur (at any value of flow density). The second important condition for the existence of the laminar traffic flow is sustained flow condition close to the stationary, characterized by the absence of sharp changes. The third important condition, it is possible to consider the traffic flow in this state without interruption, even when the value of the flow density, close to the maximum density value. Though, the concept of "maximum density" is rather arbitrary, since it is impossible to tell the exact meaning of this value but only a certain range of these values, as mentioned above.

<u>Definition</u>: *laminar traffic flow* is the traffic flow which meets no artificial obstacles and bottlenecks in its path, not experiencing the sharp fluctuations of it velocity along the path and moving without stopping at any value of it average density [18–19].

Thus, we assume the traffic flow is laminar at any phase state if the following conditions take place:

1) Absence of obstacles to the flow of the vehicles.
2) Stability condition, expressed in the absence of sharp fluctuations of velocity and flow density.

When these conditions are met, the term laminar traffic flow is applicable for all its phase states.

Thus, we can assume laminar traffic flow under the above conditions arising in the areas of urban road network with uninterrupted motion.

Violation of the stability condition occurs when the transport streams convergence or spit, when traffic streams have to go around a parked car or accident scene, at signalized intersection.

In conclusion, we can say that laminar traffic flow condition is important because it guaranties correctness of "follow the leader" models implementation for describing traffic flow behavior.

3.2. Analytical model of congested flow

IOP Conf. Series: Materials Science and Engineering **832** (2020) 012093 doi:10.1088/1757-899X/832/1/012093

Let us analyze the literature mathematical description of the motion of stationary traffic flow for different parts of the range of variation of traffic flow density. For this aim we can see analytical expressions of the function "speed-density".

As a basis we have taken a model of "follow the leader" for "congested flow" [15, 16]. Basic assumptions upon which the analytical model of a congested flow stream designed, when the density of the flow, tending to the limiting values $\rho = \rho_{max}$, are as follows:

- the cars move on the lane, not overtaking each other;
- the movement of each car following its leading vehicle (leader), described by certain differential equations, which reflects the increased sensitivity of the driver at small changes of distance between the cars;
- the derivation of the analytical dependence of the average values of flow "speed - density" assumes the existence of an equilibrium state of the stream;
- when traffic flow density reaches the maximum value average flow speed reaches the minimum value, but different from zero.

It is proved that the behavior of cars in heavy traffic on a separate strip adequately described by the model of follow the leader, which is represented by the differential equation:

$$x(t)_{n+1} = Cx(t)_n \frac{x(t)_n - x(t)_{n+1}}{(x(t)_n - x(t)_{n+1})}, \tag{3}$$

where

$x(t)_n$ = current leader position in time t;
$x(t)_{n+1}$ = current position of car, following the leader in time t;
$x''(t)_{n+1}$ = acceleration of the car, following the leader in time t;
$x'(t)_{n+1}$ = velocity of car, following the leader in time t;
$x'(t)_n$ = velocity of leader car in time t;
C = the proportionality coefficient.

It has been shown (Haight, 1963) that the equation (3) resolved and shows the equation for average speed stationary traffic flow (V_{flow}) in the form:

$$V_{flow} = C_0 e^{-C_1 \rho}, \tag{4}$$

where

C_0, C_1 = coefficients;
ρ = average traffic density.

The equation (4) shows that when the maximum density of a laminar traffic flow is reached, the flow velocity is minimal, but not equal to zero. This corresponds to the fact that cars in real traffic flow at a maximum density are not stop in the state of "bumper to bumper" corresponding to maximum density, as was suggested by the author of the first analytical model of traffic flow (Greenshields, 1934), and traffic flow continue to move at a low average speed keeping the smallest possible secure interval between cars. This corresponds to the real situation on the arterials and highway section when there are no obstacles for the traffic. We need to note that in this case "maximum density" term is idealization of the real situation.

3.3. Analytical model of "synchronized flow"

When density of traffic flow became less then it correspond to "congested" flow the average speed rise up to approximately 30 km/h, traffic flow phase state changes for "synchronized flow" and equation (4) is not adequate for this speed range.

The relation between the basic parameters of traffic flow in phase of it "synchronized motion" can be described by a rule formulated by the competent authorities of the State of California Department of Transport: "The minimum distance between two moving one behind the other cars, must be proportional to the velocity of the slave vehicle" [16].

To formulate analytical description of this state, let us denote $x'_{n+1}(t)$ – current speed of the vehicle, following the leader at time t. Then this rule can be written in analytical form as follows:

$$x(t)_{n+1} = c(x_n - x_{n+1}) + k \tag{5}$$

where

$x(t)_n$ = current leader position in time t;

$x(t)_{n+1}$ = current following the leader car position in time t;

C, k = coefficients.

As noted above, the task of obtaining the desired ratios of the average values of speed and density of flow is solved in the assumption of the existence of the equilibrium state of the flow. That is, you must accept the assumption that there is a very stable equilibrium state in each phase. Note that this condition is just performed when "laminar traffic flow" conditions occur.

For the state of traffic flow equilibrium, described by model (5) we have the average distance between vehicles, which is inversely proportional to the average traffic flow density. Therefore, for average traffic flow speed (V_{flow}) we obtain the expression:

$$V_{flow} = \frac{c}{(\rho - l)} + k, \tag{6}$$

where

c, k, l = constants.

We can see that in the "synchronized flow" phase state may be described by a hyperbolic function. The coefficient l in equation (6) is a value of flow density (ρ_0), which corresponds to the vertical asymptote of the hyperbola whose equation is $\rho = \rho_v$. With this in mind, equation (6) can be rewritten in the form:

$$V_{flow} = \frac{c}{(\rho - \rho_v)} + k, \tag{7}$$

3.4. Analytical model of free flow state

When conditions of free flow occur each driver chooses the most convenient speed depending on the speed limit prescribed for the road section. In free flow conditions the increase in traffic flow density within certain limits does not reduce the average traffic flow velocity (V_{flow}) within some narrow range of density near the zero value [18, 19]. Hence, the equation linking average speed of free and density of free-flowing traffic flow looks like this:

$$V_{flow} = V_{max}, \quad 0 < \rho < \rho_0, \tag{8}$$

where V_{max} =constant; maximum value of average speed of traffic flow when free flow conditions occur.

The Graph of the function "speed-density" for "free flow" phase state (when traffic flow density changes within the range of $0 < \rho < \rho_0$) has the form of a straight line parallel to the ρ-axis.

3.5. Proof of the existence of the traffic flow fourth phase

When the density of the traffic flow exceed value of $\rho = \rho_0$, the traffic "free flow" phase state transform into another phase state.

Here we outline theoretical justification for the existence of a fourth phase, – the phase, which is transitioning traffic flow state from the state of "free flow" to the state of "synchronized flow".

The formal proof of existence of "transition state" phase is based on the following considerations.

Since the traffic flow is orderly movement of motor vehicles of large mass, according to Newton's laws of movement, "speed-density" function for different phase states should be smooth. This means that each value of argument (density) should not be a value of discontinuity of the function itself and its first derivative. As we can see above, different parts of "speed-density" function are described with different equation. The conditions of smoothness at the points of conjugation of the function different part could analytically be written as follows:

$$\lim_{\rho \to \rho_{ij}-0} V_i(\rho_{ij}) = \lim_{\rho \to \rho_{ij}+0} V_j(\rho_{ij}), \qquad (9)$$

where

V_i, V_j – the value of the function "speed-density" to the left and to the right of the conjugation point;

ρ_{ij} = the conjugation point of the i-th and j-th sections of this function.

Similarly, for the first derivation of the function we have:

$$\lim_{\rho \to \rho_{ij}-0} V_i'(\rho_{ij}) = \lim_{\rho \to \rho_{ij}+0} V_j'(\rho_{ij}), \qquad (10)$$

where V_i', V_j' – the value of the first derivative of the function "speed-density" to the left and right of the conjugation point.

As shown above, the influence of density absents and average speed remains constant within some interval $[0, \rho_0]$ for the function section (8) Therefore, the value of the first derivative at any point of interval $[0, \rho_0]$ equal to zero.

From other side, synchronized traffic flow is represented by hyperbola, which is decreasing function. Therefore, the value of the first derivative at any value of density for this section is less than zero. If after "free flow" phase state, the next state is "synchronized flow" phase state, then condition (10) could not be fulfilled. That is impossible. Conclusion– there must be a "transition phase", smoothly connecting "free flow" phase state and "synchronized flow" phase state. So, formally, the need for the existence of "transition phase" is justified. It means that laminar traffic flow has distinct four phase states.

3.6. Formal derivation of analytical expression for "transition phase" of laminar traffic flow

We now describe the analytical expressions derivation of the "speed – density" function for "transition phase". To do this, we must note that this phase state appears when the vehicle moves at the high speed and the driver began to understand the dangerous situation, created by close distance from leader. Therefore, we perform the construction of differential equations, reflecting the behavior of the driver, following the leader in the "transition phase", on the basis of the theoretical analysis of the behavior of vehicle drivers in emergency situations. We considered a theoretical analysis developed by the Russian researchers A. Utkin and V. Chvanov. Driving a car at the maximum allowable speed was regarded by them as an emergency situation.

A. Utkin detailed the features of the driver behavior and visual perception of the situation in case of dangerous situations. He took the time interval to possible collision as the measure of transport risk. He developed a model of driver behavior in the transport stream based on the description of the interaction of two or three cars moving sequentially one behind the other [20].

V. Chvanov performed analysis of the main patterns of drivers' behavior, using the theory of engineering psychology concerning the activities of operators in the special conditions, to which Chvanov was assigned the driving of vehicles at maximum allowed speed [21].

We can conclude that the A. Utkin and V. Chvanov results adequately describe the behavior of the driver during the "transition phase" driving. Than we chose Chvanov's verbal model "Danger - speed" as the basis for analytical model construction. The main content of this model is as follows: "A sense of the danger of the driver depends on the selected speed. The growth of danger sense can be compensated by lowering car speed, which allows driver to maintain the emotional stress at acceptable level" [21].

We comment on this model as follows. When driving the car in the free flow condition, the main danger source is the high kinetic energy of the car, which moving at very high speed regarding to driving conditions. This factor creates the driver a feeling of great danger. So, minimum distance to the leader is estimated by the driver on the basis of the analysis of, distance, road conditions, vehicle brake system capacity and other factors. As a result, driver should change the car speed if the distance reduced so as to avoid a collision if a sudden and sharp deterioration of the situation will occur. Simultaneously, he must take in mind the car limited maneuvering and braking capabilities at high speed.

This implies that driver behavior should be based on the possibility of reducing the vehicle speed when approaching a leader to maintain high emotional stress at an acceptable level. In this case the negative acceleration during convergence with the leader should be gradually increased, not exceeding a certain value, in order to sustain the high speed movement of the car on the road. These terms in a differential equation describing the negative acceleration of the driven vehicle $(x''(t)_{n+1})$ can be represented as follows:

$$x''(t)_{n+1} = \frac{C_1}{(x_n - x_{n+1})}, \tag{11}$$

where

C_1 = constant;

$(x_n - x_{n+1})$ = distance value, which forced driver, following the leader, to react during convergence.

The expression (11) defines the value of negative acceleration when approaching to the leader during the "transition phase". After integration equation (11) for following the leader car speed $x'(t)_{n+1})$ we have the equation:

$$x'(t)_{n+1} = \frac{C_1}{(x_n - x_{n+1})^2} + C_0, \tag{12}$$

where

C_0, C_1 = constants.

For the equilibrium state of the flow in the "transition phase", given the fact that the average distance between vehicles is inversely proportional to the average flow density and $C_0 = V_{max}$, we get from (12) for the average flow speed (V_{flow}):

$$V_{flow} = V_{max} - C(\rho - \rho_0)^2, \quad \rho > \rho_0 \tag{13}$$

V_{max} = maximum value of average flow speed during "free flow" phase; ρ_0 – maximum value of average flow density, when fare flow state occurs;

C_1 = constant.

Analysis of equation (13) shows that analytical expression of transition phase state of laminar traffic flow is one branch of parabola, which has maximum value V_{max} when density equal to ρ_0. The conditions (9) and (10) are fulfilled in this point.

Analytical expressions of phase states of laminar traffic flow are shown in Table 1.

Table 1. Analytical expressions to calculate the speed and intensity of traffic flow in different parts of the curve "speed-density" density change of the above thread

Curve section	Traffic flow phase	Traffic flow density (vehicles/km) interval		"Speed - density" equation for the curve section
Section 1	Free flow phase	0	ρ_0	$V_{flow} = V_{max}$
Section 2	Transition phase	ρ_0	ρ_1	$V_{flow} = V_{max} - C(\rho - \rho_0)^2$
Section 3	Synchronized flow phase	ρ_1	ρ_2	$V_{flow} = \frac{C_2}{(\rho - \rho_v)} + k$
Section 4	Congested flow phase	ρ_2	ρ_{max}	$V_{flow} = V_o e^{-C\rho}$

4. The results of experimental studies conducted to confirm the adequacy of the analytical models of the laminar flow phase states

4.1. Objects of experimental studies and technology for gathering initial data

Objects of experimental studies were:

- gathering of measured speed, capacity data of traffic flow in the different phase states;

initial calculation of gathered data and estimating of analytical expressions parameters for "speed-density" functions for different phase state of traffic flow and fitting different part of the function to fulfill conditions (9), (10).

The data was gathered for street and arterials sections of Moscow road network with different speed limits (110 km/h, 80 km/h and so on). Each session lasted 15 or 30 minutes.

For measurements, a road section with a dry, flat surface was selected. Measurements were carried out in the years 2008–2011 [18]. A total of about 20 sessions were held in order to obtain baseline data. However, part of the data had to be discarded due to inappropriate conditions in which they were collected. Each session lasted 15 or 30 minutes, during which the transit time of a pre-selected and measured section of each vehicle in the selected lane was measured. The total number of vehicles passing through the lane during the session was calculated.

Processing the data collected was to calculate the traffic flow average speed capacity and density per lane. Then parameters of analytical expressions were calculated.

In Table 2, as an example, given the parametric model of those functions with numeric parameter values and ranges of flow density for different phase states on the main road with an established speed limit of 110 km/h [18]. Density concept refers to the flow, which is equivalent to the flow of homogenous flow of passenger's cars.

Table 2. Parametric "speed-density" function model for various phase states of traffic flow on the left lane of the main road with limited speed of 110 km/h

Number of state	Traffic flow phase	"Speed - density" equation for the curve section	Flow density (vehicles/km) interval
1.	Free flow phase	$V_1(\rho) = 110$	From 0 to 9,1
2.	Transition phase	$V_2(\rho) = 110 - 0{,}17(\rho - 9{,}1)^2$	From 9,1 to 22,2
3.	Synchronized flow	$V_3(\rho) = 5 + \frac{799}{0{,}7\rho - 5}$	From 22,2 to 49,0
4.	Congested flow	$V_4 = 86e^{-0{,}02}$	From 49,0 to ~160

Figure 3 shows the graph of the stage "Speed-density" function, representing analytical equations of traffic phase states, represented in table 2.

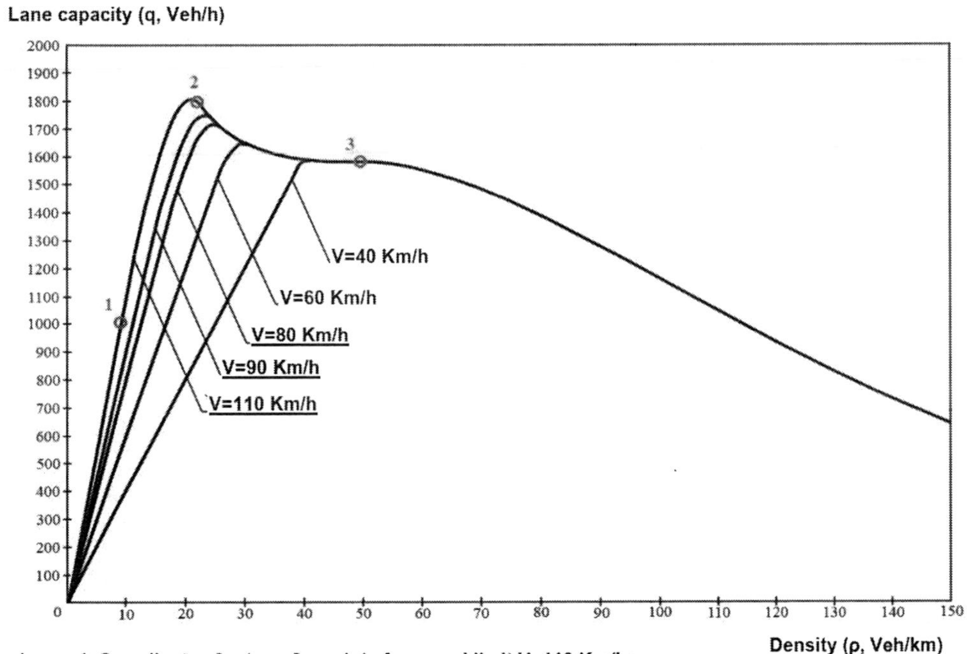

Legend: Coordinates for transfer points for speed limit V=110 Km/h:
1. Transfer "Free flow" phase state to "Transition" phase state (q=1001 Veh/h, ρ=9.1 Veh/km)
2. Transfer "Transition" phase state to "Synchronized flow" phase state (q=1794 Veh/h, ρ=22.2 Veh/km)
3. Transfer "Synchronized flow" phase state to "Congested flow" phase state (q=1583 Veh/h, ρ=49.0 Veh/km)
4. Point 3 is point of indistinguishability for each "Capacity-Density" function. Their values become statistically indistinguishable when average density of traffic flow exceed density value, corresponding to point 3.

Figure 3. The graph for the stage "Speed-density" function, representing analytical equations of traffic phase states, represented in table 2

5. Calculating the level of service indicators for road network, keeping in mind speed limits for urban arterials and streets segments

The figure 3 represents 5 different graphs of "Capacity-Density" relationship corresponding to different traffic speed limits for urban arterials and streets segments.

We suppose that if average speed of traffic flow is close to established speed limit for a streets segments, then conditions correspond to "Free flow" phase state. We can see that initial parts of graphs are represented by straight-line segment, which corresponds to different constant average speed of traffic flow for "Free flow" phase state. Different slope of each initial part of graphs reflects different speed limits. Therefore transfer point coordinates inside "Capacity-Density" phase space for "Free flow" phase state depend on speed limit established for street or arterial segment. Further we can see that after some density value "Capacity-Density" relationship is statistically indistinguishable for each street or arterial section, and not depends on initial speed limit for a street or arterial segment. Then we can see that for small speed limit established, traffic flow has only three phase states. These conclusions give us opportunity to recalculate parameters of phase state of traffic flow. The results of this recalculation are represented in table 3.

Table 3. Phase states boundaries for laminar traffic flow on freeway flow sections with different values of speed limits

Value of speed limit, km/h	Traffic flow phase state	Density variation interval, vehicles/km
$V_{max} = 60$ km/h	Free flow state	$0 < \rho \leq 24{,}5$
	Synchronized flow state	$24{,}5 < \rho \leq 60$
	Congested flow state	$60 < \rho \leq \sim 140$
$V_{max} = 80$ km/h	Free flow state	$0 < \rho \leq 16{,}8$
	Transition state	$16{,}8 < \rho \leq 25{,}6$
	Synchronized flow state	$25{,}6 < \rho \leq 60{,}0$
	Congested flow state	$60 < \rho \leq \sim 140$
$V_{max} = 90$ km/h	Free flow state	$0 < \rho \leq 13{,}9$
	Transition state	$13{,}9 < \rho \leq 25{,}6$
	Synchronized flow state	$25{,}6 < \rho \leq 60{,}0$
	Congested flow state	$60 < \rho \leq \sim 140$
$V_{max} = 110$	Free flow state	$0 < \rho \leq 9{,}1$
km/h	Transition state	$9{,}1 < \rho \leq 22{,}2$
	Synchronized flow state	$22{,}2 < \rho \leq 60$
	Congested flow state	$60 < \rho \leq \sim 140$

6. Conclusions

Analysis of traffic flow shows that there are some conditions when specific traffic behavior arises. This occurs when the traffic flow meets no artificial obstacles and bottlenecks in its path, not experiencing the sharp fluctuations of velocity along the path and moving without stopping at any value of it average density. This traffic behavior was named "laminar traffic flow". We assume the traffic flow is laminar at any phase state if the conditions, mentioned above, take place.

Theoretical investigations results prove four phase stage of traffic flow and each phase stage can be described with different "Capacity-Density" relationship analytical expression. When "Free flow" phase stage arises average flow speed remains constant inside some density value interval.

A speed limits for traffic, established by local authorities for streets and arterials sections of urban roads network influences phase state of traffic, particularly changes density values interval for free flow state. This, in turn, influences estimations of traffic flow phase state.

When some average density value of traffic flow is exceeded, "Capacity-Density" relationship is statistically indistinguishable for each street or arterial section, and does not depend on initial speed limit of traffic flow for the sections.

References

[1] Bogumil V N and Efimenko D B 2013 *Urban transport dispatch control system helps to increase Intelligent Transport Systems effectiveness* Proc. of the 11th Europ. Transp. Congr. (Prague, September 19–20) pp 20–5

[2] Foss T 2014 Safe and Secure Intelligent Transport Systems (ITS). Proceedings of Transport Research Area (TRA) *5th Conference on Transport at Solutions from Research to Deployment* (Paris, France, April 2014)

IOP Conf. Series: Materials Science and Engineering **832** (2020) 012093 doi:10.1088/1757-899X/832/1/012093

[3] Torok A 2016 Statistical Analysis of a Multi-Criteria Assessment of Intelligent Traffic Systems for the Improvement of Road Safety *J. of Finance and Econ.* **4(5)** 127–135 Retrieved from: http://pubs.sciepub.com/jfe/4/5

[4] Van De Ven T, Long J and Wedlock M 2013 ITS ACTION PLAN *Free Road Safety Traffic Information* Retrieved from: http://ec.europa.eu/transport/themes/its/studies/its_en.htm.

[5] Raiyn J 2016 Speed Adaptation in Urban Road Network Management *Transp. and Telecommunicat.* **17(2)** 111–21

[6] Vlasov V M et al 2011 *Intelligent Transportation System in Automobile and Road Complex* (Moscow: Mailer) 487 p

[7] Yatskiv I, Kabashkin I and Savrasov M 2015 Development of TTI Research Strategy and Main Directions of Research Activities Abstr. of the 15th Int. Conf. *Reliability and Statistics in Transportation and Communication (RelStat'15)* ed. I.V. Kabashkin and I.V. Yatskiv pp 7–8

[8] Hughes B P, Newstead S, Anund A, Shu C.C and Felkmer T 2015 A review of models relevant to road safety *Accident Analysis & Prevention* **74** 250–70 DOI:10.1016/j.aap.2014.06.003.

[9] *Highway Capacity Manual* 2010 (TRB, Washington, DC) 1134 p

[10] ODM 218.2.020-2012 *Guidelines on the assessment of road capacity*

[11] Greenshields B D 1934 A study of traffic capacity *Proc. (US) highway res. Board* **14** 448–494

[12] Greenberg H 1959 An analysis of traffic flow *Operat. Res.* **7** 79–85.

[13] Chantler R E, Herman R and Montroll E W 1958 Traffic Dynamics: Studies in Car Following *Operat. Res.* **6** 165–184

[14] Lighthill M J and Whitham G B 1955 On kinetic waves II. A theory of traffic flow on crowded roads *Proc. of the Royal Society Ser. A 1178* **229** 317–45

[15] Paips L A 1953 A Mathematical analysis of Traffic Dynamics *J. Appl. Phys.* **24** 274–81

[16] Haight F A 1963 *Mathematical Theories of Traffic Flow* (Elsevier, Academic Press Inc.) 255 p

[17] Drew D 1968 *Traffic Flow Theory and Control* (New York: McGraw Hill)

[18] Bogumil V N 2011 *Evaluation of Basic Parameters of Traffic Flows on the Road Network of the City, Based on the Processing of the Navigation Data of Urban Passenger Transport* PhD. Dissertation (Moscow) 196 p

[19] Bogumil V N and Vlasov V M 2016 *Intelligent Transportation Systems for road users and public transport* Saarbrucken, Germany: LAP LAMBERT Acad. Publ.) 157 p

[20] Utkin A 2008 *The determination of indicators of traffic safety based on an assessment of the drivers of interactions in the transport flow* PhD Dissertation (Moscow)

[21] Chvanov V V 2010 *Methods to assess and improve road safety subject to the conditions of the driver*(Moscow) 415 p

Smart Transport in road transport infrastructure

S A Evtiukov, E V Kurakina and S S Evtiukov

Department of Ground Transport and Processing Machines, Saint Petersburg State University of Architecture and Civil Engineering, St. Petersburg, 190005, Russia

E-mail: elvl_86@mail.ru

Abstract. Introduction of new technologies in cities through effective interaction between people and modern IT management tools has introduced the concept of «smart city» able to independently manage its own resources, energy, space and information to improve life quality. One of the main components of a «smart city» is «smart transport» using latest communication technologies for efficient movement, monitoring of location, interaction between vehicles and other traffic elements, general traffic safety improvement resulting in a decrease in the accident number (tendency towards zero mortality). One of the ways to expand this field of UAV application is their use for primary photofixation of the accident site, which will both increase the objectivity of the investigation and help to reduce time required for the investigation. Solving problems related to traffic safety in a «smart city» using IT technologies and UAVs — smart transport components results in fast collection of information, including that from hard-to-reach zones, the ability to take pictures from small heights and near objects to obtain high-resolution images. These are a possibility of application in emergency zones without unnecessary risk to life and health of personnel; improvement of the objectivity of primary information from the accident site; the possibility of automatic processing of the primary information with its transformation into the necessary types of formalized documents.

1. Introduction

Modern stages of construction, reconstruction, repair, monitoring or audit of road traffic safety cannot be imagined without the integrated use of terrain data in a geo-information mode, plans and graphs, a set of processes, methods of searching, storing and processing, presenting and distributing information, that is, without modern IT-technologies. IT-technology includes the photogrammetric method, which allows determining the displacement of the vertical and horizontal points of the object under investigation in two and three coordinates. The photogrammetric method consists in measuring the difference between the coordinates of object's points found from photographs for various stages of observation. The initial observation period is conventionally considered a zero cycle. The photographs obtained in this cycle fix the initial state of the object. The increase in deformations of buildings or structures (or deformations of vehicles in a road traffic accident (RTA)) in the process of observing them are determined by the displacements of the observed points in the photographs of the so-called deformation cycle.

The use of graphic descriptions in schemes and plans is one of the fixation methods during the inspection of the site of a road traffic accident (RTA). The protocol of the accident site inspection contains a description and characteristics of all elements of the accident site. To study the accident site it is necessary to apply the photogrammetric method.

IOP Conf. Series: Materials Science and Engineering **832** (2020) 012094 doi:10.1088/1757-899X/832/1/012094

Analysis of the scientific and methodological support to the research in the field of photogrammetry in case of an accident shows that this area has been little studied. And the existing scientific literature describing the scanning process and processing of scan results for the purposes of construction, reconstruction and repair of highways does not provide a complete understanding and solution of practical problems.

2. Statement of a problem

However, the analysis of scientific and methodological support allowed us to identify the results of scientific research of the authors working in the related fields of research. So, A.V. Komissarov in his thesis «Theory and technology of laser scanning for spatial modeling of territories» offers «a system solution to the problem of matching the speed and accuracy of technological operations of laser imaging to the requirements of various branches of the country's economy, by means of developing a theory and technology of laser scanning to collect geospatial data» [1–3, 6, 8, 9, 14].

Scientific research by A.A. Kochneva is aimed at the design of highways based on airborne laser scanning. In her thesis «Development of modified digital terrain models according to airborne laser scanning for road design», all aspects of the application of airborne laser scanning technology for constructing digital terrain models for the design of highways are described in detail [7, 10, 11]. Based on the research, an algorithm for interpolating an array of airborne laser scanning data has been developed, the features of which are low computational complexity and a high degree of rarefaction of arrays of points forming flat horizontal and inclined surface areas. The work done, described in the thesis is impressive, but the accuracy of airborne laser scanning, although it meets the requirements, is far from the accuracy obtained by classical shooting methods. The author considers shooting with unmanned aerial vehicles (UAVs) to be a more rational technology, which provides high-quality images, an accurate cloud of points (up to 3 cm in a plan view and height) and a more favorable price / quality ratio.

D.V. Beregovoy in his thesis «Creating topographic plans based on survey data from an unmanned aerial vehicle and automating the process of interpretation» considers the following situation: «at the moment there are a lot of modern technologies that allow you to create a topographic plan. The most popular and effective ones are a tachometric survey and a survey using GNSS. With high productivity of office work they show low field efficiency [1, 7, 11]. However, it is important to understand that in order to increase the productivity it is much easier to automate office work than what is done directly on site. Therefore, the implementation of an effective method of creating a topographic plan based on technologies such as laser scanning and aerial photography (including using UAVs) is more promising in comparison with other methods».

The results of the patent search confirmed increased attention of specialists over the past 10 years to the development and improvement of UAV structures intended for aerial photography of the terrain [3, 14].

3. Materials and methods

Investigating the accuracy and errors of various methods of ground-based laser scanning application, the author delves into mathematical analysis, on the basis of which he proposes a number of "methodological, technical and technological solutions for laser scanning", which improve the quality of the measurement processes using ground-based laser scanners and scan processing. Based on the mathematically based conclusions on the optimal method, it is difficult to imagine real time savings when working with ground-based laser scanning when shooting highways.

The technical solutions proposed in the studied inventions and utility models are aimed at increasing the stability of flights of unmanned aerial vehicles (UAVs), improving their control systems, including via secure communication channels, ensuring flight safety.

Inventions and utility models in the field of aerial photography have a very specific, narrowly professional nature. They can be used primarily to provide areas such as aerial mapping and aerial reconnaissance.

The solutions proposed by the authors suggest, among others, improving the quality of aerial photography from the standpoint of distinguishing small objects and obtaining 3D images in photographs.

4. Results

The authors propose a technique for the use of UAVs in road inspection. Let us present the main provisions of the methodology and the algorithm for its implementation. In accordance with the requirements of the regulations, the time of arrival of the road patrol crew to the accident site with victims should not exceed 20 minutes from the moment of notification of the accident. In accordance with the current governing documents, the patrol crew of the traffic police on the site of an accident with victims must take investigative actions together with the investigator of the Investigation Department of Accidents with Victims and Safety of Roads of the Main Investigation Department of the Main Directorate of the Ministry of Internal Affairs of the Russian Federation and the expert on whose territory the incident occurred. The tasks of the patrol crew are the localization of the accident site, its initial inspection, calling for emergency medical assistance if necessary, traffic management in the area of the accident, participation in the preparation of the traffic accident layout plan. The listed range of tasks testifies the absolute necessity of the presence of the patrol crew at the accident site. In accordance with such prerequisites, a phased algorithm for the use of UAVs for photo-recording of the accident site is presented as follows (Table 1).

Table 1. Stages of the implementation of the methodology for the use of UAVs in road inspection to photograph the accident site

Stage	Stage characteristics
Informing about the accident	The traffic police operational service receives a signal about the accident; it is transmitted to the nearest traffic police station and the nearest traffic police crew
Video monitoring of the situation on the road section from the traffic police station to the accident site	The traffic police station, equipped with the unmanned aerial system, sends a UAV to the accident site, which moves to the determined destination point by including in its automated control system the coordinates of the accident site The flight is carried out at an altitude of up to 500 meters, ensuring reliable video monitoring of the situation on the road section from the traffic police station to the accident site Video information from UAV surveillance cameras in real time is transmitted to the monitors at the flight control center of the traffic police station and to the video monitor installed in the traffic police car connected to the flight control center of the traffic police station
Traffic management correction in areas close to the accident site	According to the results of video surveillance, the traffic police and patrol car crews make decisions and correct traffic control already in the areas close to the accident site, and based on the information received, the crew can optimize the route of their advancement to the accident site taking into account the inevitable compaction of traffic flows on the road sections adjacent to the accident site
UAV photo survey	The UAV after arrival to the accident site with a given coordinate point is guided in the manual mode from the flight control center for orthogonal hovering over the accident site at an altitude of up to 100 meters After the patrol crew arrives at the accident site and the first priority actions are taken, they organize the localization of the accident site and coordinate actions with the UAV flight control center for photographing The manually controlled UAV drops to a shooting height (approximately 30 meters), is coordinated around the aerial photography center point and takes pictures
Receiving and processing the results of the UAV photo survey	Photographs from the UAV are transmitted via communication channels to the flight control center in the form of an orthophotomap of the accident section. The orthophotomap is converted to a formalized document type – a traffic accident layout plan -using a special computer program in the flight control center The received accident layout plan is sent via communication channels to the computer of the traffic police car, where it is printed and accepted for use in further investigation
UAV return to the place of permanent deployment	Upon completion of the UAV photo survey mode, it returns to the place of its permanent deployment in the automatic control mode; in the process of returning it can conduct video monitoring of changes in the traffic situation on the road sections coinciding with its flight route

IOP Conf. Series: Materials Science and Engineering **832** (2020) 012094 doi:10.1088/1757-899X/832/1/012094

In this regard, the use of UAVs to photograph the accident site is considered from the standpoint of automating the initial inspection of this place, increasing the objectivity of fixing all trace factors and the speed of drawing up the necessary formalized documents, the main one of which is the layout plan of the accident.

Solving the tasks of ensuring road safety in a «smart city» through the use of IT technologies and the UAV component of the intelligent transport is based on increasing the objectivity of primary information from an accident site, an opportunity to automate the processing of this primary information and transform it into the necessary types of formalized documents. In this regard, the methodology for the use of UAVs in road inspection is proposed. The practical implementation of the methodology was carried out on the road network of the ring road (the Ring Road) of St. Petersburg. The study of the locations of traffic police stations, locations of accidents concentrations, and typical conditions of their occurrence will allow to formulate the key requirements for UAVs intended for aerial photography of accident sites on this route, namely the normative range, duration and altitude of flights. To substantiate these parameters, it is important to have information about the relative position of traffic police stations on the Ring Road and locations of traffic accidents concentrations. The location of traffic police stations on the Ring Road is fixed; it is known and can be mapped. To determine the concentration of accidents it is supposed to use statistical information on all accidents that have occurred on the Ring Road for the past three years.

To process this information, a special methodology has been developed for statistical processing of data on accidents at the Ring Road of St. Petersburg.

The methodology is based on the division of the entire Ring Road into areas 1 km long with subsequent sampling for each accident site with their characteristics at the time of commission, causes of occurrence and severity of consequences for their participants. In total, about 910 cases of accidents with victims are to be processed.

Then this information is visualized on a detailed large-scale map of the Ring Road and the areas of accidents concentrations are identified.

At the next stage, the distances from the traffic police stations to the locations of accidents concentrations adjacent to them are determined and the probabilistic-statistical estimation of the distances from the corresponding stations to the borders of the concentration zones is carried out. According to the obtained results, the necessary distance of the continuous UAV flight and its time in flight are determined.

The altitude flight regulations are based on the study of data on the presence and characteristics of man-made structures and natural obstacles in areas adjacent to the intended flight routes of the UAVs.

Testing and approbation of the methodology were performed using the example of the Ring Road section between 62nd and 70th km along the inner ring and between 71st and 79th km along the outer ring. In accordance with the above proposals, the flight zones are limited to the ranges presented in Table 2.

Table 2. The flight zone ranges in the Ring Road of St. Petersburg

	Flight zone range	
No.	Along the inner ring, km	Along the outer ring, km
1	57.7…68.7	73.2…84.2
2	74.4…68.7	57.1…67.6
3	83.1…94.1	47.8…58.8
4	94.7…105.5	36.5…47.2
5	105.3…115.8	26.1…36.7
6	9.5…19.8	122.3…132.3
7	63.8…72.6	67.4…78.1

Generalized information on the state of road safety in the flight zones, confirming their emergency danger is presented in Table 3, and the scheme of assigning flight zones to the parts of the Ring Road is shown in Figure 2.

Experimental studies of the UAV flight mode for carrying out aerial photography of the accident site have been obtained. The studies included flight tests of a quadcopter to estimate the error of transformation of a photo-video survey into a layout plan, depending on the geolocation point of the quadcopter in the GLONASS system; determining the height of the quadcopter hovering for shooting an accident site; working out the process of transforming aerial photography into a layout plan.

During the flight tests, DJIPHANTOM 3 ADVANCED quadcopter equipped with a video camera was used (Table 3). Technical characteristics of the UAV are given in Table 4.

Table 3 Technical characteristics of DJIPHANTOM 3 ADVANCED quadcopter

Parameter	Value
Weight, grams	1280
Maximum lifting speed, m/s	5
Maximum speed of descent, m/s	3
Maximum speed, m/s	16 (no wind)
Maximum flight altitude, m	6000
Maximum radius of action, m	2000
Maximum flight time, min	23
Operating temperature range, °C	from 0 to 40
GPS mode	GPS/GLONASS

Table 4 Technical characteristics of DJIPHANTOM 3 ADVANCED photo-video camera

Parameter	Value
Sensor	Sony EXMOR 1/2.3" Pixels: 12.76 M
Lens	FOV 94°20 mm
ISO range	100-3200 (video) 100-1600 (photo)
Shutter speed	8 s – 1/8000 s
Maximum size of the photo	4000×3000 mm
Photo modes	Single photo Continuous shooting: 3/5/7 photos. Auto exposure bracketing (AEB): 3/5 Photos with an exposure offset of 0.7 EV. Time-lapse
Supported card types	MicroSD Maximum capacity: 64 GB
Video recording modes	UHD: 4096×2160p 24/25, 3840×2160p 24/25/30 FHD: 1920×1080p 24/25/30/48/50/60 HD: 1280×720p 24/25/30/48/50/60
Supported file formats	FAT32/ex FAT Photo: JPEG, DNG Video: MP4/MOV (MPEG-4 AVC/H.264)
Operating temperature range	from 0°C to 40°C

A general view of DJIPHANTOM 3 ADVANCED quadcopter is shown in Figure 1.

Traditional software for UAV orientation in flight space provides for its topographic adjustment in the GLONASS system with the launch point coordinates. In our case, this is the point of the permanent UAV deployment (usually a stationary traffic police station). Taking into account the distance from the topographic adjustment point to the accident site of up to 5 km, there is an assumption about the occurrence of a significant error in determining the coordinates of the location of

IOP Conf. Series: Materials Science and Engineering **832** (2020) 012094 doi:10.1088/1757-899X/832/1/012094

characteristic objects involved in the accident at the place of its occurrence. This assumption was verified experimentally [5, 13, 14].

The test flights of the UAV were carried out in the waters of the Gulf of Finland along the coastline of a sandy beach near Solnechny settlement in the Leningrad region at a visual communication distance (up to 2 km) at altitudes up to 10–30 meters.

a) b)

Figure 1. DJI PHANTOM 3 ADVANCED quadcopter: a) general view, b) general view in flight

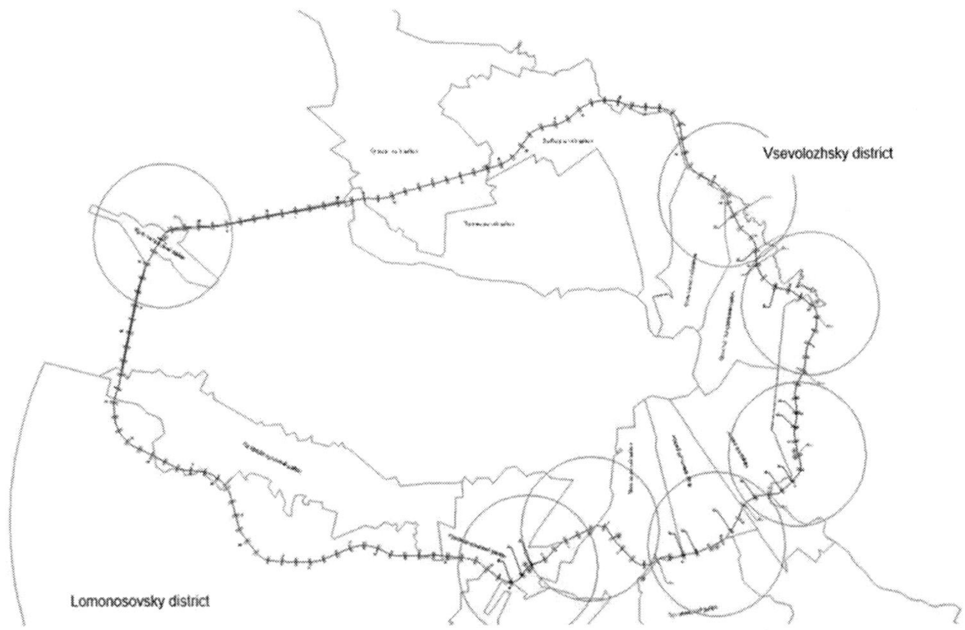

Figure 2. The reference diagram of flight zones BLAH for the participants in the KAD of Saint-Petersburg

The UAV topographic adjustment was carried out at the place of take-off, and the estimation of the error of the program for transforming photography into a layout plan was carried out at a distance of 2 km in relation to objects ranging in size from 0.1 m to 7.0 m. The experiment showed that the transformation error occurs due to the large distance from the launch site to the accident site but it is not systemic and depends on random deviations of the flight path of the UAV and can be up to 15 % [4].

To exclude this negative phenomenon, at the time of shooting the UAV was topographically adjusted not to the place of its launch, but to the characteristic objects located at the area of shooting and the subsequent photo. Such an approach made it possible to reduce the error in the transformation of photography into a layout plan by up to 1.5 %. The tests were carried out on the territory of the Faculty of Automobile and Road Building of SPSUACE at the parking lot of the vehicle inspection station. For shooting, a traffic situation was simulated and the full-scale samples of passenger cars were used. As a result of the study, it was established that to fully cover the accident site with a camera, it is advisable to follow the following recommendations:

– in the event of an accident, accompanied by scattering of vehicles after the collision, full coverage of the accident site in the photograph is achieved when shooting is performed from a height of about 30 m;

– in the event of an accident that is not accompanied by vehicle scattering – from a height of about 10 m. Figure 3 shows the photo of the place imitating an accident from a height of 10 m.

Clarification of the aerial photography transformation algorithm into a road accident layout plan is as follows: a UAV is sent to the accident site, upon arrival at the site the altitude is selected, photographs are taken and the video is shot, all photo and video materials are reflected and stored on the flash card, after the fixation of the accident site the UAV returns to the traffic police station; upon arrival of the UAV the official transfers all information to a personal computer; knowing at least one value, you can use the special software to set all the dimensions and coordinates of the accident site that are used for the layout plan (Fig. 3).

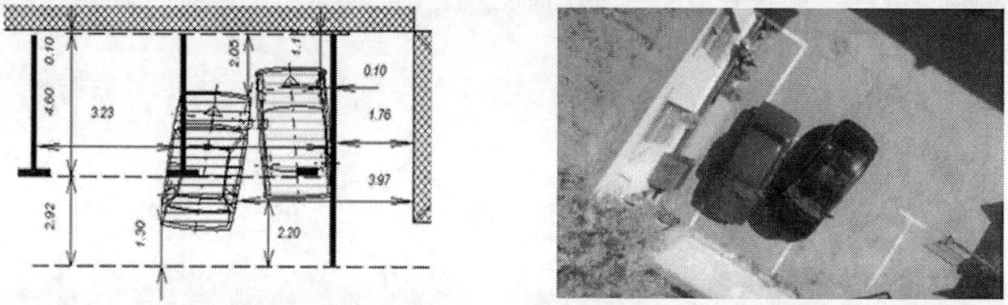

a) b)

Figure 3. Imitation of an accident site: a) the photo of the place imitating an accident from a height of 10 m, b) the layout plan of the imitated accident

Application of intellectual vehicles in the "smart city" presupposes the introduction of IT-technologies in road inspection, which allows obtaining more detailed and accurate information about the state of the roadway and the environment. It helps to make the right decisions at all stages of the life cycle of highways and is aimed at improving safety. The results exclude gross errors associated with the "human factor", as a part of the process is automatic. It is possible to assess in more detail both the geometric flaws of the longitudinal and transverse profiles, and the defects of the coating itself.

The article presents experimental approbation of the method of primary inspection of the accident site using aerial photography from an unmanned aerial vehicle.

5. Conclusion

The flight mode of an unmanned aerial vehicle (altitude, speed, trajectory) for aerial photography of the accident site has been substantiated. An algorithm has been developed for transforming aerial photography from an accident site into a traffic accident layout plan; a technique for using UAVs in road inspection has been developed, in particular, processing aerial photography materials in the interests of developing a traffic accident layout plan.

Acknowledgements

According to the research topic presented in the article, the authors carry out scientific and practical activities confirming their contribution to the development of intelligent transport systems in a smart city, in particular the use of IT-technologies in road inspection in order to improve road safety, striving for zero mortality.

References

[1] Beregovoy D V 2018 Creation of topographic plans based on survey data from UAVs and automation of the interpretation process (St. Petersburg: St. Petersburg Mining University) 136 p.

[2] Brannolte U, Pavel Pribyl P and Silyanov V (2017). Simulation of Regional Mortality Rate in Road Accidents *Transportation Research Procedia* **20** 112–24. DOI: 10.1016/j.trpro.2017.01.032.

[3] Briese C 2012 Rdioametric calibration of multi-wavelength airborne laser scanning data *ISPRS Annals of the Photogrammetry, Remote Sensing and Spatial Information Sciences* **1–7**, 335–340. DOI https://doi.org/10.1007/978-94-017-8663-8.

[4] Dobromirov V N, Evtiukov S S and Kurakina E V 2017 Improving safety assessment methods of traffic on high-speed roads *World Transport and Technological Machines* **1(56)** 94–100

[5] Evtiukov S, Kurakina E, Lukinskiy V and Ushakov A 2017. Methods of accident reconstruction and investigation given the parameters of vehicle condition and road environment *Transportation Research Procedia* **20** 185–92. DOI: 10.1016/j.trpro.2017.01.049.

[6] Evtiukov S S, Kurakina E V and Perevalov N V 2019 IT-technologies in road inspection (St. Petersburg, Petropolis Publishing House) 132 p.

[7] Kochneva A A 2018 *Development of modified digital terrain models according to airborne laser scanning for road design*, Thesis for a Candidate degree in technical sciences. St. Petersburg, 144 p.

[8] Komissarov A V 2015 System presentation of laser scanning *Geodesy and Cartography* **7** 18–23, DOI: 10.22389 / 0016-7126-2015-901-7-18-23.

[9] Komissarov A V 2015 *Theory and technology of laser scanning for spatial modeling of territories*, Thesis for a Doctoral degree in technical sciences. Novosibirsk, 278 p.

[10] Kurakina E, Evtiukov S and Rajczyk J 2018 Forecasting of road accident in the DVRE system *Transportation Research Procedia* **36** 380–5. https://doi.org/10.1016/j.trpro.2018.12.111.

[11] Kurakina E V, Evtiukov S S and Golov E V 2017 Reconstruction of road accidents (St. Petersburg: Petropolis Publishing House) 204 p.

[12] Pribyl P, Novikov A, Vasileva V and Katunin A 2017 ITS Control of Highways Capacity *Transportation Research Procedia* **20** 468–73. DOI: 10.1016/j.trpro.2017.01.076.

[13] Rajczyk P, Kurakina E and Knapiński M 2018 The influence of surface topography on the safety of road and utility surfaces *Transportation Research Procedia* **36** 640–48. Retrieved from: https://doi.org/10.1016/j.trpro.2018.12.139.

[14] Rajczyk P and Evtiukov S 2018 New technological solutions in improving road safety, Transportation Research Procedia, **36**, 649–53. Retrieved from: https://doi.org/10.1016/j.trpro.2018.12.133.

Agent modeling of vehicle traffic at a regulated intersection

K N Mezentsev

Moscow Automobile and Road Construction State Technical University (MADI), 64, Leningradsky ave., Moscow, 125319, Russia

E-mail: mknpost@rambler.ru

Abstract. The article contains information about the rules of agent-based modeling of the movement of a vehicle in a stream. The vehicle is considered as a homogeneous agent with reactive behavior. The reactive behavior of the agent allows you to control the distance with respect to the vehicle in front and react to the traffic light signal at an adjustable intersection. Testing the reactive behavior of the agents was performed in the NetLogo multi-agent programming environment. The article provides NetLogo code for a fragment of the agent model responsible for the movement of vehicles at the intersection. Obtained using the model, the graph of traffic intensity corresponds to the theoretical graph.

1. Introduction

The flow of vehicles can be considered as a combination of homogeneous or heterogeneous agents. When modeling the movement of vehicles, the most difficult option is modeling the behavior of heterogeneous agents. A simpler solution can be obtained if the flow of vehicles is considered as a collection of moving homogeneous agents. Such agents have a single scenario of behavior during changes in their environment. We list the main features of the behavior of a homogeneous vehicle agent.

The agent must control the distance to the front of the moving agent. The agent performs distance control by fixing the speed of the agent moving in front. If there is no agent at a safe distance, then the agent moves at a speed equal to the speed of the agent in front, which it increases by a constant value. If an agent is detected at a safe distance in front, it is necessary to reduce the speed by a constant value. In this case, the speed of the agent must vary within specified limits.

The agent must respond to a traffic signal. We will accept for the motion model that the traffic light can find in two states: supply of the enable signal, supply of the inhibitory signal of movement. Signals are green and red. When the red signal is fixed, the agent starts braking.

The model interface is shown in figure 1. NetLogo multi-agent modeling environment was used as a model development tool. Programming agent behavior is performed in an embedded language, which was created on the basis of programming languages such as Logo and StarLogo [1, 2].

The model works in discrete mode. The track consists of two segments. Agents move from left to right, after passing segments of the route the agent is destroyed. When the setup button is clicked, the numbers-of-cars agents are distributed on the main route. The number of agents numbers-of-cars = {10,...,100} change step 1. In the process of model work, new TPC agents appear on the left. The appearance of agents is a multiple of: f-input = {0,...,100} step 5.

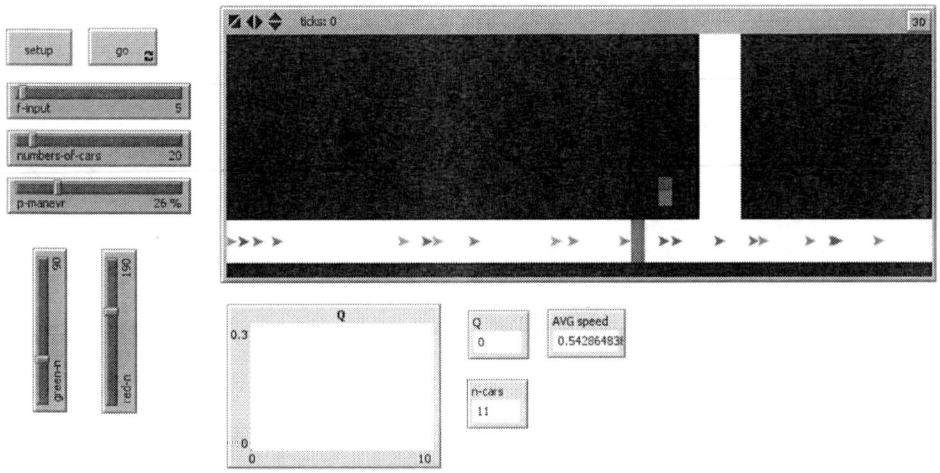

Figure 1. Model interface; Q is the intensity of movement; AVG speed – average speed on a segment, n-cars – the number of cars on a segment

The green-n and red-n values specify the multiplicity when switching the green and red phase signals. These values vary from 1 to 300 in increments of 1.

TPC agents accelerate movement by an acceleration value equal to 0.0045 units, if there is no TPC ahead of you at a given distance equal to one; otherwise TPC slows down the movement by a deceleration value equal to 0.026 units.

At the fork, a TPC is selected with a probability of p-manevr for turning to the side segment. The value ranges from 0 to 100 percent. The step of change is 1.

In the process, a graph of the change in the intensity Q of movement on the segment of the route to the traffic light is displayed. The traffic intensity is calculated by formula 1:

$$Q = K \cdot V \ (1)$$

where K is the density of the traffic flow, V is the average speed of the TPC on the segment of the route, formula 2.

$$K = \frac{N}{L} \ (2)$$

N is the number of cars on the segment of the route, L is the length of the segment of the route.
A homogeneous agent has the following parameters:
speed – TPC speed. The initial speed value is set randomly from a range of 0.1 to 0.9 units;
speed-limit – speed limit;
speed-min – minimum speed;
manevr? – a sign of maneuver. Rotate to branch, true, otherwise false.
R is the radius of view of the environmental agent.

2. Agent Behavior

Programming discrete models in NetLogo requires a basic procedure, a code that defines the behavior of the entire model [3]:

```
to go
        ;Departure of vehicles
        if (f-input != 0) and (ticks >= 100) [
          if  ticks mod f-input = 0 [ put-car ]
```

IOP Conf. Series: Materials Science and Engineering **832** (2020) 012095 doi:10.1088/1757-899X/832/1/012095

```
        ]
        ; Traffic light control
        set-lighter
        ; Driving a vehicle
        ask turtles [
                ; Traffic light tracking, red signal fixation
                let list-ag patches in-radius R
                let red? any? list-ag with [ pcolor = red]
                ;Red signal and stop line?
                ifelse red? and (pcolor = green )
                [ set speed  0 ] ;Yes
                [ go-rbase   ] ;No traffic allowed
        ]
        calc-q ;Calculate Q
        tick ;Discrete step
end
```

The put-car command is used to put the agent on the track. Agents appear on the left and move along the segment of the route towards the traffic light. The agent in the process of moving inspect the environment in the radius of the viewing radius R. In the model, this value is assumed to be equal to 10 units. If a red signal is detected and the agent is on the stop line, then the agent speed is reset to zero. This behavior of the agent in combination with distance control forces agents to carry out a "tackle" in the flow of traffic to the agent traveling in front, when a red signal is given by the traffic light. The go-rbase command forces the agent to move along the track with distance control.

Set-lighter command is responsible for switching the traffic signal. Switching is performed multiple of the values red-n, green-n according to formula 3.

$$t \bmod phase = 0 \quad (3)$$

where t is the values of the current discrete step, phase is the multiplicity of the signal switching phase.

Calculation of the current value of the load intensity of the stream Q is performed by calling the calc-q command. The calculation of the intensity occurs on the segment of the route, which begins from the point of start of movement to the stop-line. To determine the segment length, you need to calculate the difference between the stop line coordinate and the coordinate of the starting point of the trace segment along the x axis.

The agent movement code is specified in the go-rbase procedure.

```
;Distance control
;Search ahead of the traveling agent
let car-ahead one-of turtles-on patch-ahead 1
ifelse  car-ahead != nobody [
set speed  [speed] of car-ahead
slow-down-car ]  ;Ahead there is an agent, braking
[ speed-up-car; ] ;No agent, increase speed
;Speed control
if speed < speed-min [set speed  speed-min]
if speed > speed-max [set speed speed-max]
; Deciding to turn to the side segment
if random 100 < p-manevr [
        if (pxcor >= 10 ) and (pxcor <= 11) [
        set manevr? true ;Decision is made
        set heading 360
        set xcor 10 ]
]
;Removing Agents completed movement
```

ifelse not manevr?
[if xcor + 1 > max-pxcor [die]]
[if ycor + 1 > max-pycor [die]]
fd speed ;Driving on the highway
The slow-down-car and speed-up-car commands decrease or increase the speed of the agent. The speed-max agent speed limit is set to 1, and the minimum speed-min agent speed is zero.

3. Model testing

The model in action is shown in figure 2. The traffic light gives a green signal that permits movement. After passing the stop line, two agents decided to make a maneuver, one continues to move along the main highway.

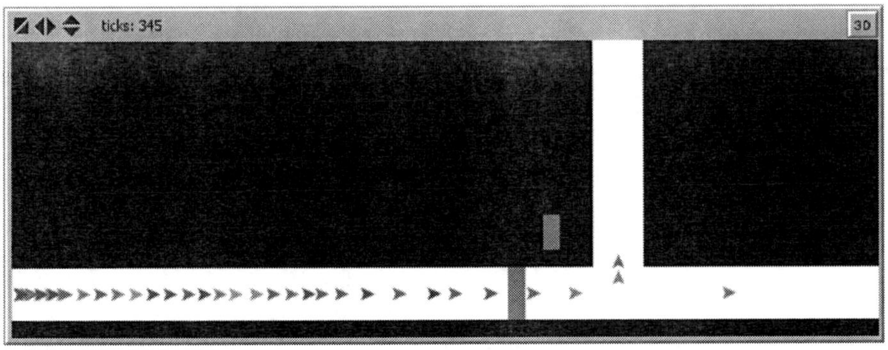

Figure 2. f-input =5, numbers-of-cars=20, p-manevr=26;green-n=90,red-n=190

Figure 3 shows a graph of the movement intensity, it can be seen that on the controlled segment in front of the traffic light there are fluctuations in traffic intensity. The nature of the graph changes is close to a parabola, which corresponds to the mode of motion in the stream [4].

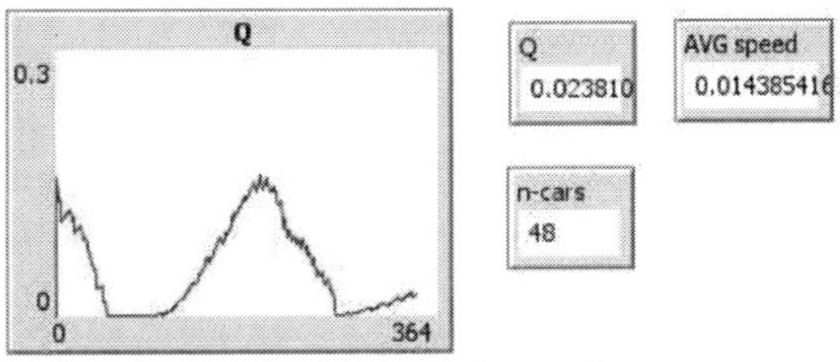

Figure. 3. Traffic intensity on the segment

4. Conclusion

The considered algorithm of the behavior of vehicles in the stream allows you to get a traffic model close to the real situation on the tracks. It can be concluded that it is advisable to use a model based on the concept of reactive homogeneous agents as a model of the behavior of a vehicle driver. This model of behavior is advisable to use when modeling the movement of unmanned vehicles.

References

[1] Mezentsev K N 2017 *Multiagent systems. Lecture course* (Moscow: MADI)
[2] Mezentsev K N 2014 *Multi-agent modeling in NetLogo* (Moscow: MADI)
[3] Mezentsev K N 2015 *Multi-agent modeling in NetLogo. Tutorial* (St. Petersburg: Lan Publ. House)
[4] Gorev A E 2010 *Fundamentals of the theory of transport systems* (St. Petersburg: SPbGASU)

IOP Conf. Series: Materials Science and Engineering **832** (2020) 012096 doi:10.1088/1757-899X/832/1/012096

Development of a methodology for optimizing the urban route in the metropolitan area

A N Novikov[1], S V Eremin[1] and A G Shevtsova[2]

[1]Orel State University named after I.S. Turgenev, 95, Komsomolskaya St., Orel, 302026, Russia
[2]Belgorod State Technological University named after V.G. Shukhov, 46, Kostyukova St., Belgorod, 308012, Russia

E-mail: shevtsova-anastasiya@mail.ru

Abstract. One of the urgent tasks is to determine the composition of the bus fleet to serve each route. At the same time, it is necessary to be able to vary the type and number of buses that have different technical characteristics, in this regard, in the framework of the study, the issues of optimizing bus movement modes on city routes were considered. Basically, on the city route network, buses stop at all stopping points. However, quite often there is a situation when there is a sufficiently large unevenness of passenger traffic at various stopping points. In this regard, design schemes for analyzing the effectiveness of combined and shortened routes are proposed. Key relationships include empirical estimates of passenger flows. The parameters of the models are the volume of the bus fleet, the irregular production of buses, the probability of a denied boarding at individual stopping points.

1. Introduction

As part of the quality of service, the question arises of assessments of denied boarding and passenger waiting times [1–3]. Basically, the type of bus is determined by the presence of this type in the transport park of the city. In solving this problem, it is necessary to solve the problems of accounting for the stochastic nature of the route network and obtaining statistics on passenger flows.

The choice of type and number of buses is determined by many factors, including:

- economic (associated with the availability of material and technical base for their storage, repair, etc.);
- social (associated with the transport mobility of the population, given the assessment of the loss of time waiting for all passengers and the conditions of transportation comfort, etc.);
- technical (condition of the roadway and related structures, UDS throughput, permissible speeds, etc.);
- operational (climatic conditions, etc.);
- regulatory (environmental requirements, road safety requirements, maximum permissible bus occupancy, etc.).

The listed factors are the initial premises for solving optimization problems, and some of the listed factors can act as criteria, and some as limitations.

As a result of the occurrence of these factors and the presentation of special requirements for them, city routes are often not effective. To solve this problem, especially when developing and reorganizing routes in urban agglomerations, it is necessary to develop methods for optimizing urban routes.

2. Analysis of existing methods for optimizing urban routes according to the criterion of the park

In the general case, the task of choosing a bus depot for the maintenance of the route is multi-criteria, in which the main criteria, namely, the waiting time for passengers and the costs of the enterprise for the maintenance of the bus fleet, come into a compromise. It is necessary to minimize both the total waiting time (T_{exp}) of all passengers on the selected route and the expenses of the enterprise (taking into account the profit) R_{ent}:

$$T_{exp} \to min \quad R_{ent} \to min \tag{1}$$

One of the approaches to solving such a problem is the formation of a generalized additive criterion [4, 5]:

$$k_{opt} = \alpha_T T_{exp} + \alpha_R R_{ent} \tag{2}$$

The values of the coefficients can be obtained from some social standards; however, the approach to the formation of their values does not have a formal setting. It is necessary to understand the cost losses in expectation in terms of the general economic task of improving the standard of living. Moreover, the multicriteria problem reduces to scalar optimization and when solving it there are no conceptual problems, only technical difficulties may arise (if a sufficiently adequate simulation model is used):

$$k_{opt} \to min \tag{3}$$

For different values of the coefficients and knowing that the relationship between T_{exp} and R_{ent} is inversely proportional, you can build this generalized dependence of the loss function (in a simplified version of simulation, as a function of the number of buses on the route).

Based on the analysis of existing methods, it was found that they propose to jointly use two models, namely, simulation and analytical, based on the use of relations (1) - (3) to form a generalized parameter that reduces to scalar optimization [6,7]:

$$F = \sum_{i=1}^{l} \left[60 t_i C\left(\sum_{j=1}^{m} T_{exp\,ji} S_{ji}\right)\right] + R_{ent} \to min \tag{4}$$

where S_{ji} – determines the volume of passenger traffic on the corresponding interval of the route; T_{expji} – the average waiting time; m – the total number of stops, regardless of direction.

However, this model involves the simultaneous use of buses of only one capacity. In the general case, it is necessary to optimize the fleet of vehicles, taking into account the possibility of using buses of various capacities (from a certain number of passenger capacities set in advance for other reasons).

As a result of the analysis of existing optimization methods, it was found that they are based on two criteria – the waiting time and costs of the enterprise without performing an analysis of the capacity of roads, which is an extremely important task when optimizing city routes, especially in urban areas. In this regard, the next stage of the study is the development of methods for optimizing the urban route, taking into account previously unused data.

3. Development of a methodology for optimizing the urban route

For the main objective of the study - the optimization of the urban route, taking into account an expanded set of criteria, it is proposed to use a heuristic algorithm that involves less severe restrictions:

1. There is a set of capacities $\{W_i\}$, $i = 1..m$, which may have buses.
2. The minimum number of A_{min} buses can be set either by attracting the maximum capacity to the route, or by an acceptable interval.
3. The maximum number of A_{max} buses can be set based on the available:

- the number of drivers in the enterprises;
- the number of buses of the selected capacity;
- the number of buses with a minimum capacity of W_{min};
- the smallest capacity of the streets and stopping points of the route Ω_{min};

4. The quality of service is determined by the total number of passenger seats in all the bus in a given time interval, and it should be no less than the average traffic volume in the most stressful section for the same time interval [8].

This problem is an integer programming problem and can be formally presented as follows:

$$\{W_i\}, i = 1 \dots m, W_i \in [W_{min}, W_{max}], \{A_i\}, i = 1 \dots m, A_i \in [A_{min}, A_{max}] \tag{5}$$

$$\sum_{i=1}^{m} A_i W_i = W^{exp} \geq max(\lambda_i T_{exp}); \sum_{i=1}^{m} A_i \leq \Omega_{min} \tag{6}$$

where A_i is the number of buses with W_i capacity indicator W^{exp} - the total number of passenger seats in all buses.

The optimization function is a general function, and does not have continuity properties. Constraints in solving the optimization problem can lead to ambiguous solutions of the tuple $\{Ai\}$, $i = 1..m$. As a solution to the problem, methods of integer multi-criteria search are used using search optimization procedures in MatLab.

The procedure contains the following steps:

1. For all W_i, subject to restrictions, we find A_i.
2. The problem of bandwidth limitation is separately solved Ω_{min}.
3. $\forall \omega_1$, the permissible A_{imin} is calculated and the value of the objective function $F\ (A_{imin})$ is determined.
4. Increase the value of A_{1min} by 1.
5. We calculate the objective function $F\ (A_{1min+1})$.
6. If $F\ (A_{1min+1}) \leq F\ (A_{1min})$, a return is made to step 4 for the iterative repetition of the algorithm to find the minimum of the functional. Otherwise, go to step 7.
7. We consider 1 bus with a given capacity of ω_2 and go to step 4.
8. Subsequent steps 4–7 are repeated until the optimal tuple sets A = {Ai} i = 1..m .. are determined.

One of the variants of the algorithm for solving a similar problem [9-10] involves algorithmization in MatLab and gives an initial approximation of the distribution problem.

One option for the development of the method is to include the possibility at different time intervals (peak hours, day, late evening, etc.) to choose the composition of the buses in the necessary, but different quantity and different capacity.

As a result of the analysis of the algorithm, it was found that an increase in capacity reduces the probability of failure. At the same time, both the capacity and the number of buses directly proportionally affect the reduced costs of the enterprise. It is also shown that a small increase in the costs of the enterprise can give significant savings in reduced costs [11–13].

When considering a selected route, the following options are possible: a certain reduction in the costs of the enterprise with a simultaneous increase in passenger costs for waiting; Both company expenses and waiting times are reduced. Reduction of expenses of the enterprise does not lead to a change in the waiting time; the reduction in waiting time does not increase the costs of the enterprise [14–15].

4. Experiment

To perform the experiment and evaluate the effectiveness of the algorithm proposed for development, the team of authors performed an analysis of two routes, namely No. 85 (fig. 1), which is the first in popularity, and No. 22, as the least loaded. Route number 85 passes through the historical center of the city of Krasnoyarsk, connecting it from west to east, passing through the October bridge.

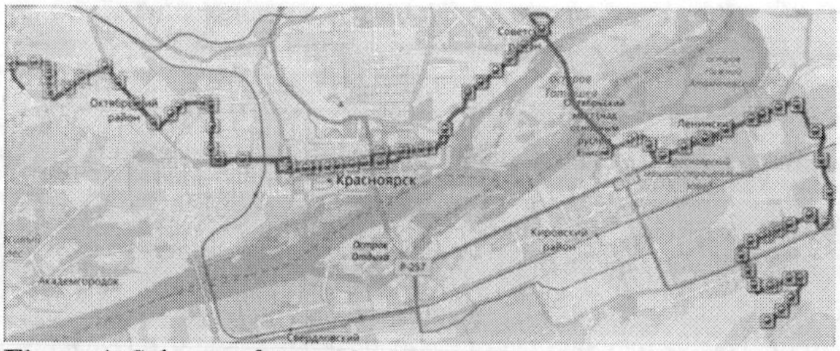

Figure 1. Scheme of route No. 85

For the selected set of routes, the main characteristics of passenger flows are calculated, in particular, the unevenness coefficients for the hours of the day and others. So, it is clear that the structure of unevenness can be significantly different (Fig. 2).

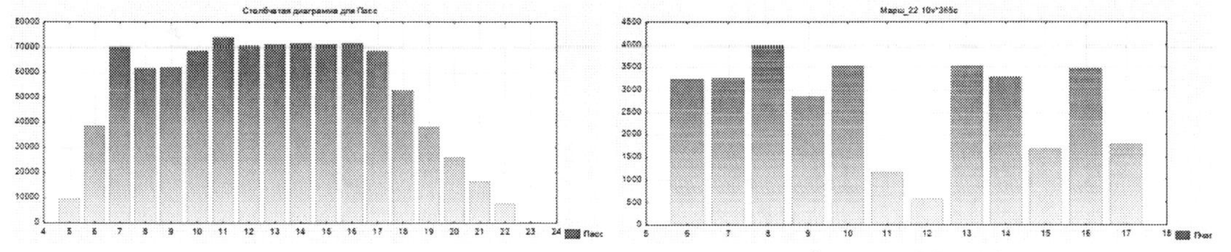

a) route No. 85 b) route No. 85

Figure 2. The structure of intraday unevenness of passenger flow

As a result of the analysis of certain routes and testing of the developed methodology, it was found that the most significant characteristic of the random process is the autocovariance function, which is defined as $R(n)=M\xi_{n+k}\xi_k$, and as its estimate in the analysis of the time series $(x_1, \ldots, x_N)$ of length N, the value is used:

$$\hat{R}_N(n,x)=\frac{1}{N-n}\sum_{k=0}^{N-n-1}x_{n+k}x_k \tag{7}$$

This estimate is unbiased, i.e. on various routes, schedules of series of traffic volumes were constructed, including route No. 85 (Fig. 3)

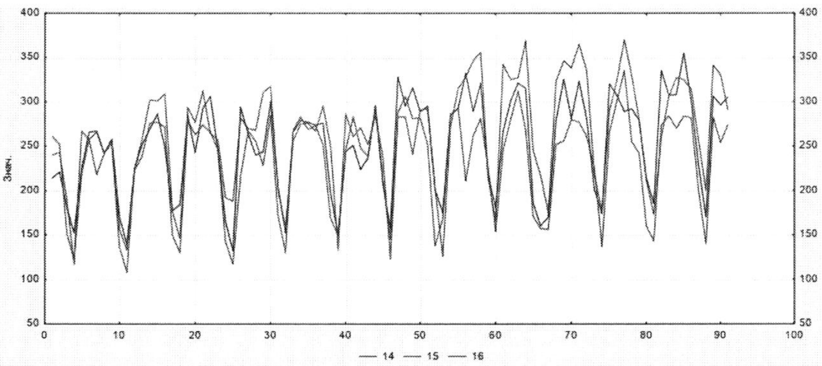

Figure 3. Time series of passenger flow of route No. 85

The graph clearly shows the cyclical nature of the passenger flow, which is also superimposed on a linear trend, which makes the process unsteady in two respects. As part of the tasks of assessing the seasonality of passenger traffic, spectral analysis estimates were used, which assumed an assessment of the periodogram which is an estimate of spectral density.

For passenger flows of the chosen route, autocorrelation functions and periodogram were constructed (fig. 4).

To generate interconnected passenger flows (with cross-correlations similar to those obtained by statistical analysis, we propose a preliminary linear transformation of the elements of time series of passenger flows with one time index. So, if the random vector has the characteristics $M\xi=m_\xi$ *u* $D\xi=||cov(\xi_i, \xi_j)||$, then for the random vector $\eta=L\xi+C$, the mathematical expectation vector of the vector η will be $M\eta=Lm_\xi+C$, and the dispersion matrix will be η - $D\eta=L{\cdot}D\xi{\cdot}L^T$ D. Predetermined dispersion matrix of the resulting vector pre solve the system of nonlinear equations to find the matrix L. As a result, the proposed set of time series generating models of passenger covers the requirements for establishing an adequate simulation model parametrization streams obtained based on statistical analysis.

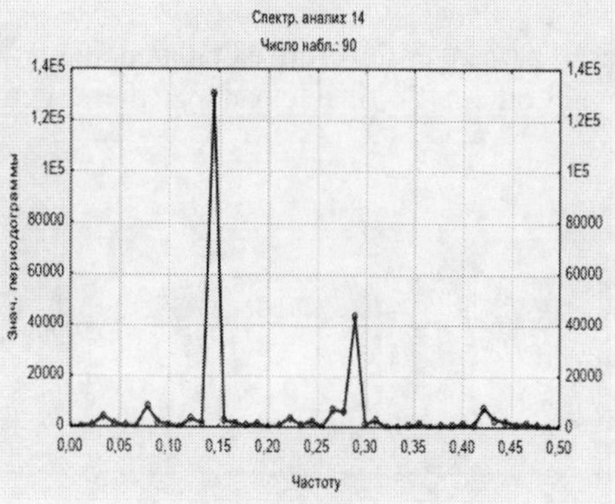

a) function

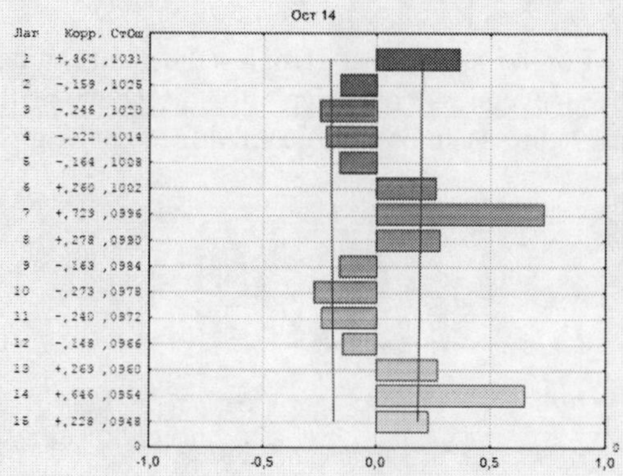

b) periodogram

Figure 4. Characteristics of passenger traffic on route No. 85

Thus, the introduction of progressive forms of organization of bus movement on the route contributes to the improvement of public transport services. At the same time, there is often a redistribution of passengers between routes (in particular, the attraction of new passengers to high-speed and: express lines), as a result of which the parameters of passenger flows Q, Q_j, R_j change. These changes should be considered when planning and organizing transportation.

As a result of the experiment on the existing city route, it was found that the application of the proposed methodology allows to increase the efficiency of the selection of the necessary composition, taking into account the redistribution of passenger flow in the city metropolitan area

5. Conclusion

Based on the analysis of the developed algorithm, it is shown in the work that an increase in capacity reduces the probability of failure. At the same time, both the capacity and the number of buses directly proportionally affect the reduced costs of the enterprise. It is also shown that a small increase in the costs of the enterprise can give significant savings in reduced costs.

When considering a certain selected route, the following options are possible: a certain reduction in costs with a simultaneous increase in passenger costs for waiting; both company expenses and waiting times are reduced. Cost reduction does not change the waiting time; reduced time waiting does not increase costs.

As a result of the study, the problem of optimizing the composition of the fleet for servicing routes was solved, which is formally presented in the form of a multi-criteria choice between the cost of maintaining the fleet and the time spent by passengers on travel. To solve the problem, a joint use of the analytical and simulation models is proposed. A solution to the problem of distributing buses along the routes of the city's transport network is proposed.

References

[1] Pattnaik S B, Mohan S and Tom V M 1998 Urban bus transit route network design using genetic algorithmю *J. of Transport. Engineer.* **124(4)** 368–75
[2] Toner J 2001 Organization of a tender for buses in London *Principles and Practice. Report at the 7th Int. Conf. on Competition and Property* (Molde, Norway: Public Transport)
[3] Tahmassebpour M 2016 Performance evaluation and scalability of IP-based and heuristic-based job scheduling algorithm backup systems *Indian J. Sci. Technol.* **9** 17–24
[4] Bennett D, Xiao N and Armstrong M 2003 Using genetic algorithms to create multicriteria class intervals for choropleth maps *Annals of the Associat. of American Geograph.* **93(3)** 595–623
[5] Nepal K and Park D 2003 Routing algorithms for transportation systems and service improvement projects in urban transportation networks *Tech. rep., Department of Civil Engineering, Tokyo Institute of Technology* (Tokyo, Japan: Ookayama, Meguroku)

[6] Mesjasz-Lech A 2014 Development of public transport in the city – a challenge for urban logistics in terms of sustainable development *Forum Sci. Econo.* **2(4)** 63–75
[7] Paulley N 2006 The demand for public transport: The effects of fares, quality of service, income and car ownership *Transport Policy* **13(4)** 295–306
[8] Mirotin L, Ignatenko A and Marunich V 2011 Logistic look at passenger traffic *Logistics* **4** 31–3
[9] Regirer S A and Shapovalov D S 2003 Filling passengers with space in public transport *Automat. and Remote Control* **8** 111–21
[10] Semenova O S 2009 *Mathematical modeling in problems of optimizing the movement of urban passenger transport, taking into account the imposition of route schemes* (Novokuznetsk)
[11] Novikov A N, Kulev A V and Peshekhonov M V 2013 Comparison of location systems and their application in intelligent transport systems *World of transport and technolog. machines* **2(41)** 109–13
[12] Korchagin V A, Novikov A N, Lyapin S A and Rizaeva Yu N 2016 Complex self-developing transport systems *World of transport and technolog. machines* **2(53)** 110–6
[13] Kulev A V, Novikov A N, Kulev M V, and Kuleva N S 2016 Increasing the efficiency of urban passenger transport functioning *In the coll. Inform. Technol. and Innovat. in Transport mater. of the 2nd Int. Sci. and Pract. Conf.* ed. by A N Novikov pp 378–82
[14] Zhankaziev S V, Novikov A N, Vorobyev A I, Kulev A V and Morozov D Y 2017 Efficiency of operation and functioning of the system of an indirect transport flow regulation and control *Int. J. of Applied Engineer.* **12(13)** 3645–52
[15] Fernandez R 2010 Modelling public transport stops by microscopic simulation *Transport. Rese. Part C: Emerging Technolog.* **18(6)** 856–68

IOP Conf. Series: Materials Science and Engineering **832** (2020) 012097 doi:10.1088/1757-899X/832/1/012097

Results of parking survey in the central part of Moscow for new tariff determination

P I Pospelov, D S Martyakhin, V V Rudakova and D M Strokov

Moscow Automobile and Road State Technical University (MADI), 64, Leningradsky Ave., Moscow, 125319, Russia

E-mail: pospelov@madi.ru

Abstract. The need to conduct a study of the parking space in the center of Moscow was caused by a significant overload of paid parking lots and, thus, traffic jams due to car mileage in search of free parking spaces in the central part of the city, increased noise and air pollution. Excessive loading of parking lots also leads to numerous violations of parking rules. The demand for parking within the Boulevard Ring is determined by the concentration of historical monuments, government agencies, museums, exhibition halls, shops, leisure facilities and other facilities that are extremely attractive to residents and visitors of the city. The results of one of the survey stages included the following: the inventory of existing car parking within the territory in question and the indicators of the current level of the parking space system development in the territory under consideration. The inventory included the analysis of parking place location with the allocation of parking lots on the street-road network, in underground and elevated capital structures, and in courtyards. Simultaneously with the monitoring, they carried out mass surveys of drivers parking cars in paid parking lots of the road network.

1. Introduction

The rational use of the parking space of any city always implies the lack of influence on the throughput of the street-road network, an efficient use of the city territory and, as a basic requirement, the provision of the transport system easy use, of which parking is an integral part [1, 2, 5, 7, 8]. The introduction of paid parking is urgently needed in cases where the above conditions are not met. The high pace of motorization in Moscow, the underdeveloped network of streets and roads, the lack of opportunities for the construction of spacious modern parking lots force us to take measures constantly to improve the transport system of the city. Only the improvement of this system can provide a balance of interests between individual users. The development of the public transport system in combination with all components of the transport infrastructure will solve the transport problems of the city.

2. Inspection of the parking space of Moscow

The demand for parking is mainly influenced by the planning features of Moscow, the historical tendency for the development of residential areas in the peripheral areas of the city and the concentration of employment places in its central part.

The studies included the inventory with the allocation of parking lots on the road network, in underground and elevated capital structures, with the record of the following data for each parking space: parking address, its type, capacity, number of vehicles, number of violations and monitoring with the record of car parking filling on weekdays during the following periods: morning hours from 6:30 to 7:30; daytime hours from 12:00 to 16:00; evening hours from 21:00 to 23:00. For the convenience of inspection and control, the studied area was divided into 40 polygons in accordance with the planning boundaries of the plots with an average area of about 15 hectares.

Thus, the total number (17239) of parking spaces was determined with a breakdown into the territory of the road network and the territory outside it, as well as the number of parked cars in the morning and evening hours. The ratio of parking lot number in the courtyard territories and capital structures to the number of parking lots located on the road network determines the possibility of parking space provision for residents permanently residing in this territory. Of the total number of parking lots (17239) there are 7435 lots (43.1 %) on the road network and 9804 lots (56.9 %) outside it. The ratio by polygons is significantly different and ranges from 1 to 39 %.

The number of parking lots for disabled people on SRN is 2.46 % and, taking into account the number of lots in the courtyard areas, it complies with the standards.

The distribution of the number of parking lots by the criterion of charging a fee: 8322 lots (48.3 % of the total) – paid parking, 8917 lots (51.7 %) – free parking.

Minimum tariffs contribute to the maximum use of parking space, but the search time for free space will be maximum. The maximum tariff corresponds to private underground parking, but their occupancy rarely exceeds 40 % even during peak hours. In general, it is necessary to strive for parking space occupancy at the level of 85 % – the figure generally accepted throughout the world. With this load, the time to search for free space will be minimal and accessibility of social facilities (pharmacies, government agencies, hospitals, etc.) will be maximum for car owners.

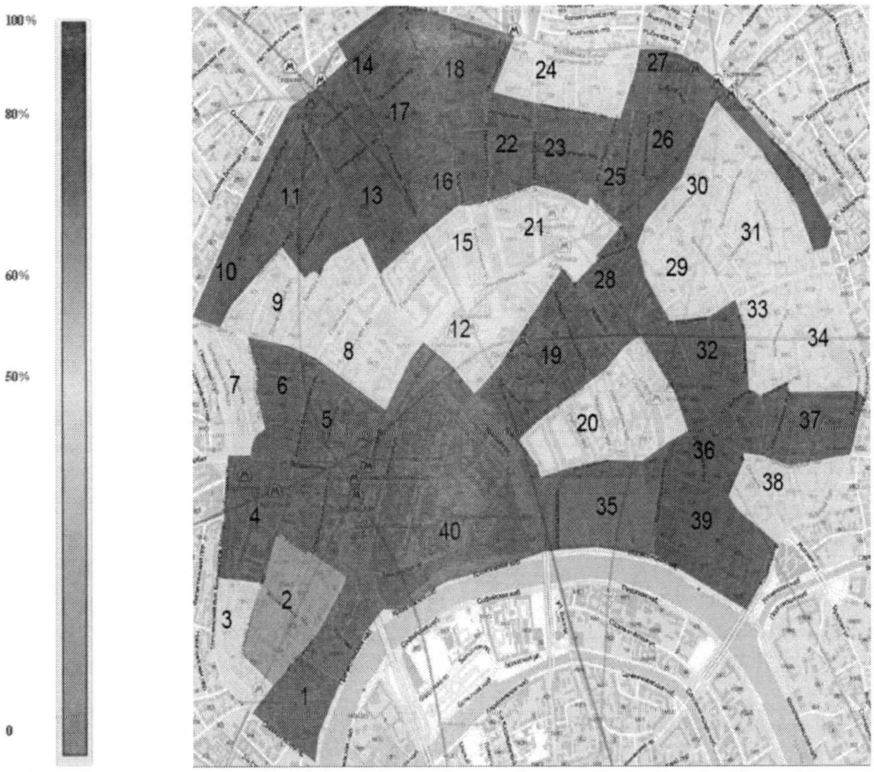

Figure 1. Occupancy of parking lots on the road network during daytime, %

The greatest occupancy of parking lots was noted during the daytime (Fig. 1). This indirectly confirms that the bulk of trips to the central part of the city are labor trips involving a significant period of parking space use. With an average occupancy of the parking space of the territory in question during the daytime equal to 78 %, the occupancy values for different sites vary from 27 % to 148 %. In the morning and evening (night) hours, the occupancy makes from 10 to 100 %, with an average occupancy of 46 %. Traditionally, a large occupancy of parking lots was noted at the borders of the territory bounded by the streets with high traffic intensity.

The violations of parking rules are caused not only by the attempts to refuse to pay for the service, but can also be caused by the urgent need for short-term parking in the conditions of a significant shortage of parking spaces. Confirmation of this circumstance can be found during the analysis of violations when cars are parked at each of 40 allocated sites. The main conclusion: the busier the parking lot is, the more cars are parked with violations.

With an average relative number of cars parked with violations – 20.2 % (daytime hours) and 11.8 % (morning and evening hours), the number of cars parked with violations varies from 0.0 to 91.8 % (daytime hours) and from 0.0 to 26.9 % (evening hours) (Fig. 2).

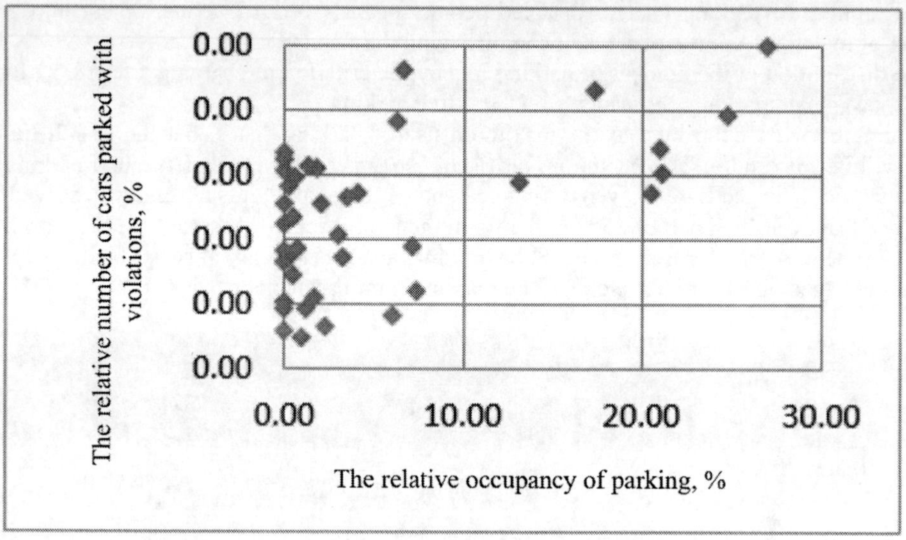

Figure 2. Dependence of the relative number of cars parked with violations (%) on the relative occupancy of parking in the morning and evening hours (%)

Mass surveys of drivers parking cars at paid parking lots of the street road network were carried out simultaneously. The purpose of the questionnaire was to study the motivation of drivers during a parking lot selection and the difficulties this choice causes.

The questionnaire contained the following questions: frequency of paid parking lot use at the SRN; the purpose of the trip; the location of departure and destination points; time to search for a parking spot; the assessment of tariff level; source and amount of financial costs for parking.

To achieve statistical reliability of the survey results, the number of respondents amounted to more than 2500 people, from 50 to 70 questionnaires at each of the sites on average.

The analysis of driver questionnaire results on the issue of paid parking use frequency (Fig. 3) indirectly indicates that 58 % of respondents accurately use parking for business trips (weekdays and almost every day in the questionnaire), although other categories of drivers may also use parking for business trips. Quite a lot of respondents (17 %) use paid parking lots during visits no more than once a month.

The vast majority of respondents (70 %) answered the following to the question about the purpose of the trip to the central part of the city – the working one (Fig. 4). A working trip, as a rule, implies that a car occupies a parking space for a long time, which is not released for a long time and the number of parking users during the day is negligible. Such a parking lot is only vacated in the evening hours. This state of affairs is unfavorable for other users and leads to the fact that the share of trips for tourism (4 % of respondents), cultural (13 %), sports (2 %) and other purposes (11 %) is small in the central historical part of the city. This is also due to the fact that the drivers who use parking for business trips occupy it earlier than drivers who use parking for other purposes.

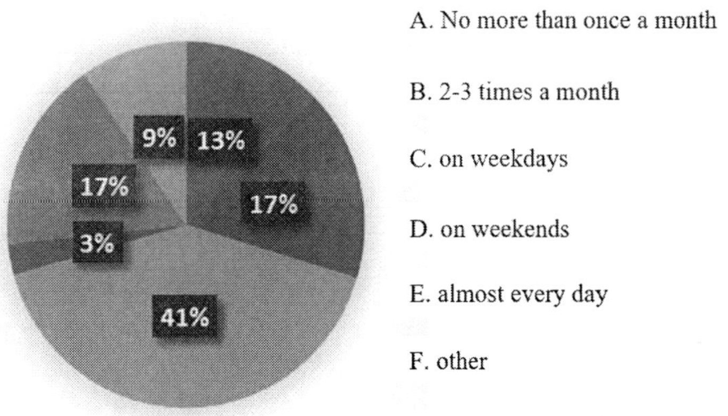

Figure 3. Frequency of paid parking lot use (% of respondents)

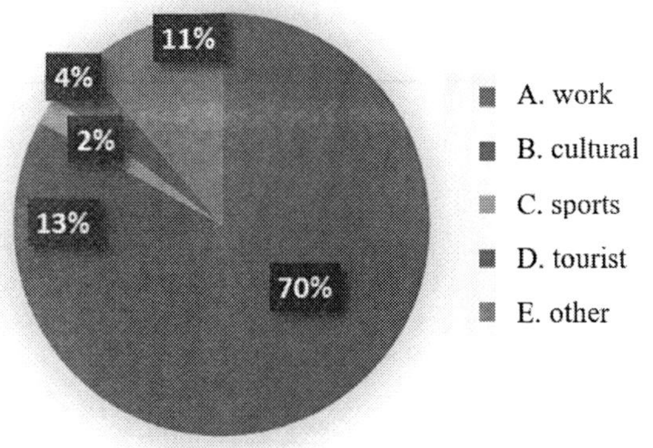

Figure 4. Distribution of trips by purpose during paid parking application

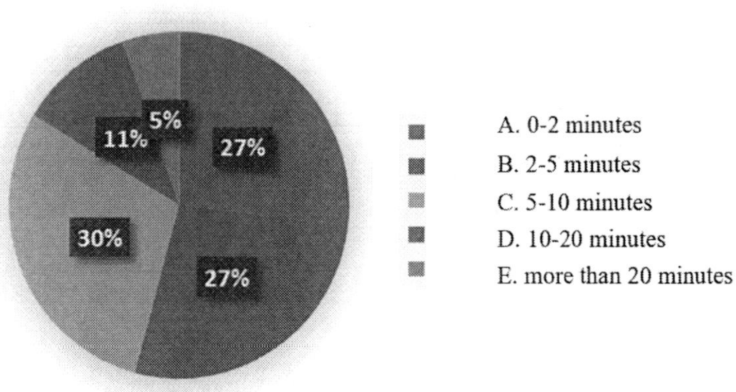

Figure 5. Parking lot search time

One of the main characteristics of parking space use is the time spent by drivers to find a free parking lot. Among the respondents, 69 % of drivers spent up to 10 minutes searching for a parking spot (Fig. 5). Due to the fact that the parking space is mainly occupied for a long time by parking lots associated with business trips, the question arose about the source of payment for parking. 39 % of surveyed drivers said that their parking lot is paid by the employer.

The key point during paid parking consideration is the ratio of real (non-potential) users to the level of tariffs for parking. 47 % of surveyed drivers rated the existing tariffs as high and 45 % as satisfactory. Only 5 % of respondents consider the tariff level to be low.

3. Conclusions

Based on the foregoing, the following conclusions can be made:

1. The results of the parking space inspection with a constant number of parking spaces in the central part of the city can be considered stable enough to carry out forecasts on the establishment of new tariffs for parking lot paying.
2. The main purpose of the trip to the city center is a working one, with a considerable duration associated with a low parking payment rate and the fact that up to 40 % of parking is paid by the employer. Thus, the frequency of travel is directly related to the goal.
3. The duration for a parking place search in the vast majority of cases does not exceed 10 minutes, and in about half of cases – no more than 5 minutes.
4. About 50 % of surveyed drivers rated paid parking rates as high and 5 % as low. From 10 to 12 % of surveyed drivers rate the tariffs as low – these are the drivers who have purchased subscriptions and use parking lots only on weekends.

4. Summary

Based on the study, the following proposals were made to change the payment rates for parking on the road network within the Boulevard Ring of Moscow:

- to increase the cost of parking;
- to limit parking time to 2 hours;
- to apply a progressive scale of payment for parking with a simultaneous increase of tariff for the second and subsequent hours of parking.

During preparation of the proposals, we proceeded from the fact that the use of paid parking should remain accessible to most residents of the city, and everyone should be able to come to the city center by car freely and park it when parking is really necessary.

References

[1] Shup D 2014 *The high cost of free parking* (Moscow: OOO "Media Car") 733 p
[2] Duvanova I A 2015 *Stroitel'stvo unikal'nykh zdaniy i sooruzheniy* **12(39)** 43–56
[3] Andrews J 2000 Don`t Park Here *Planning* **66(10)** 20–3
[4] Brown J, Daniel H and Donald Shoup 2001 Unlimited Access *Transportation* **28(3)** 233–67
[5] Calthrop E, Steff P and Kurt Van Dender 2000 Harking Policies and Road Pricing *Urban Studies* **37(1)** 63–76
[6] El-Fadel M and Hayssam S 2001 Parking Facilities in Urban Areas Air and Noise Impacts *Journal of Urban Plann. and Developm.* **127(1)** 16–33
[7] Gould C 2003 Parking: When Less is More *Transportat. Plann.* **27(1)** 3–11
[8] Amy H, Fisher D and Collura J 2002 Drivers' Parking Decisions: Advanced Parking Management Systems *J. of Transportat. Engineer.* **128(1)** 49–57

IOP Conf. Series: Materials Science and Engineering **832** (2020) 012098 doi:10.1088/1757-899X/832/1/012098

Methodology for determining the required length and number of traffic lanes, other parameters, taking into account the passenger city transport motion

L V Zhivoglyadova[1], V G Zhivoglyadov[2] and V V Siliyanov[3]

[1]KVVU, 4, Krasina str., Krasnodar, 350063, Russia
[2]Academy of Science and Technology and social progress, 4, Kalinina str., Krasnodar, Russia
[3]Moscow Automobile and Road State Technical University (MADI), 64, Leningradsky ave., Moscow, 125319, Russia

E-mail: Zhivogladova @mail.ru, jivogljadov_vg@mail.ru

Abstract. The article presents methods for determining the maximum severity and speed, high-speed and regulated roads, as well as other parameters taking into account the movement of passenger urban transport. The introduction of meta-dynamics will reduce conflict situations to 35–40 %, and, accordingly, the number of places in cities and between them. The lack of published methods for determining the necessary length and number of movements and the principles of their development adversely affects road traffic in cities, in particular, the establishment of the causes of congestion. Traffic safety cannot be ensured without decision making.

Zones resulting from traffic shortages, as well as secondary streets, due to the mismatch of traffic sizes to the number of lanes, primarily the construction of roads [1]. For each car, it is necessary to calculate the interval between the front bumpers ($\gamma = 1000\ V/\lambda_{max}$) in terms of ensuring road safety; at the place of residence for night and day storage; for short-term storage (markets, stations, stadiums, hospitals, shops, etc.) [2–4]. Due to the lack of the necessary space for short-term (daily) car storage outside these areas, drivers are forced to use half sons for these purposes, including at least a 50 percent lane deficit in cities and creating this necessary road safety [1, 2, 4].

Road traffic is made possible by calculation and estimation methods. This means that the tools cannot be implemented in a timely manner with the aim of assessing the existing length of the lane, including in the case of highways, and in general in cities, places, regions and the Russian Federation. For each car (on average), 320 square meters (80x4 square meters) of lane are required, 35 square meters – for night garage storage and 25 square meters for short-term storage outside the carriageway, i.e. only 380 sq.m. The proposed methods for calculating and assessing the sufficiency, insufficiency of movement in the territorial formations of cities, with study [1, 3], have the form:

- express roads (highways):

$$L_{CD} = (N_{z.\,avt.} + N_{Tranz.}) \cdot L_{d} \cdot h_{TG} \cdot h_{в} \cdot K_{TO} /1000, \text{ km/city}; \quad (1)$$

- adjustable roads:

$$L_{MRD} = (N_{z.\,avt.} + N_{Tranz.}) \cdot V_{a} \cdot \tau_j \cdot C \cdot h_{тг} \cdot h_{в} \cdot K_{то} / t_{зел}\ 1000. \quad (2)$$

In particular, methods for operational use are presented.

$$L_{MRD} = (N_{z.\,avt.} + N_{Tranz.}) \cdot V_{a} \cdot \tau_j \cdot C \cdot h_{TG} \cdot h_{в} \cdot K_{то} / t_{zel} \cdot 3600. \quad (3)$$

For any situations in the object of study:

$$L = (N_{z.\,avt.} + N_{Tranz.}) \cdot V_a \cdot h_{TG} \cdot h_в \cdot K_{TO} / \lambda, \text{ km/citi}. \quad (4)$$

Knowing that λ = q · V_a [5], we obtain:

$$L = (N_{z.\,avt.} + N_{Tranz.}) \cdot h_{TG} \cdot h_в \cdot K_{TO} / q, \text{ km/city}. \quad (5)$$

The spatial interval between the front bumpers of the vehicle is γ =1000V/λ and λ = 1000V/γ, we obtain

$$L = (N_{z.\,avt.} + N_{Tranz.}) \cdot \gamma \cdot h_{TG} \cdot h_в \cdot K_{TO} / 1000, \text{ km/city}, \quad (6)$$

where $N_{s\ author}$ – the number of registered vehicle fleet in the control object (OU), TE/mountains; $N_{tranzit}$ – the number of transit fleet daily (on average), TE / city; L_d is the length of the dynamic longitudinal dimension [6], *m*/ TE; – traffic intensity (average for OU), TE/*h*:

$$\bar{\lambda} = \sum_{j=1}^{4} \sum_{k=1}^{K} \sum_{m=1}^{M} (\lambda_{jkm} / 4 \cdot K \cdot M), \quad (7)$$

λ_{max} =3600 ·t_{zel} / (τ_j ·C) – lane potential, TE / h; j = 1, 2, 3, 4, (1, 3 – odd), (2, 4 – clear) – traffic numbers of traffic flows (TP); k = 1, 2, ..., K – lane numbers in directions, j – through; m = 1,2,...,M – numbers of roads (streets) in the OU, in the directions.

Each street will be counted twice, since on each street there are no clear values; Va is the speed of movement (in formula (2) in m/s) in the rest of the equations of a single measurement, *km/h*; τ_j is the time interval between the front bumpers of vehicles when driving, stop lines with a traffic-resolution signal, s/TE; C is the duration of the traffic light control cycle [1, 3], s/cycle; q, r – dynamic characteristics (harmonic grand ranks) of traffic flows in odd, clear directions, and other maneuvers, respectively [1, 3];

$$t_{zel} = \alpha\,(C - T_o) = q/(q + r)(C - T_o), \qquad t_{kr} = (1 - \alpha) \cdot (C - T_o), \text{ s/cycle};$$

1000 – the number of meters per kilometer, m/km; 3600 – number of seconds per hour, s/h;

q – electric traffic, TE/km; α is the share of the enabling measure in the sum of the main measures (allowing, prohibiting); h_{TG} – coefficient of technical efficiency of the fleet; h in – probability of departure, i.e. receive vehicles to the lanes of the road network, [1, 5];

K_{TO} is the coefficient of transport hazard, K_{TO} = 1.15 ... 1.3 [4].

t_r – reaction of the first car of the queue to t_{zel} [1];

4 – the number of directions of movement.

For the operational solution of problems of assessing the sufficiency or insufficiency of the length and number of lanes in key and other directions of traffic flows, the most acceptable formulas are (3)–(7).

In the calculations of cycles and main measures, adequate traffic intensities are taken λ_j, λ_{j+1}, λ_{j+2} and λ_{j+3}, t. e. $q = r$, (0,46 + 0,46 = 0,92). Inequalities:

$$(0 \le q + r \le 1), \qquad 0 \le max\,(\lambda_j \tau_j; \lambda_{j+2} \tau_{j+2}) + max\,(\lambda_{j+1} \tau_{j+1}; \lambda_{j+3} \tau_{j+3}) \le 1, \quad (8)$$

characterizing the rational operation of traffic light control modes, can take place in real conditions:

$$max\,(\lambda_j; \lambda_{j+2}) + max\,(\lambda_{j+1}; \lambda_{j+3}) \le 1656\ TE/h, \quad (9)$$

where the terms can have dimensions:

$$0 \le max\,(\lambda_j; \lambda_{j+2}) \le 1656\ TE/h, \quad 0 \le max\,(\lambda_{j+1}; \lambda_{j+3}) \le 1656\ TE/h, \quad (10)$$

but in total, the intersecting flow directions should not go beyond 1656 *TE/h*.

The duration of the intermediate ticks is calculated according to the method described in [1], which allows one to obtain optimal traffic light control modes.

Determination of the number of lanes in the meridional l_m and latitudinal l_w measurements are carried out by the following expressions:

$$K_M = L_{MRD}\,(l_M / (l_M + l_{Ch}))/ l_M = L_{MRD} / (l_M + l_{Ch}), \text{ bands/merid}; \quad (11)$$

$$K_{Ch} = L_{MRD}\,(l_{Ch} / (l_M + l_{Ch}))/ l_{Ch} = L_{MRD} / (l_M + l_{Ch}), \text{ bands/latitudes}. \quad (12)$$

In large cities and in inter-city traffic – between large cities it is advisable to build a hundred-year-long perspective – 16-lane motorways with eight lanes in each direction:

$$M_M = K_M /16, \text{ motorway/merid.}; \quad M_{Ch} = K_{Ch} /16, \text{ motorway/lat.}$$

Determination of the area Smag occupied by lanes (highways) is carried out by the following equations:

$$S_{PDG} = L_{MRD} \cdot B_p, \text{ sq. km /city,} \tag{13}$$

where Bp and L_{MRD} are measured in km.

The density coefficient of motorways (lanes) is determined by the following expression:

$$K_{PPDG} = S_{PDG} / S_G, \text{ harmonic quantity;} \tag{14}$$

$$K_{PPDG} = (S_{PDG} / S_{cit.})\ 100, \%. \tag{15}$$

There is a linear relationship between the coefficient of transport hazard and the product of the speed and density of movement, and the proportional dependence is inverse between the intensity of movement, etc.:

$$K_{TO} = V \cdot q / \lambda \cdot h \approx 1{,}315; \tag{16}$$

$$K_{TO} = q \cdot \gamma / 1000 \cdot h \approx 1{,}315; \tag{17}$$

$$K_{TO} = \lambda \cdot \gamma / 1000 \cdot V \cdot h \approx 1{,}315, \tag{18}$$

where ≈ 0.76; λ = 0.23 TE/s; T = 6.3 s/TE; γ = 1000 · V / λ; after substituting γ into formula (18), we obtain:

$$K_{TO} = 1/h = 1/0{,}76 = 1.315.$$

The traffic safety coefficient is the ratio of the product of the density and speed (q · V) to the product of the coefficient of transport hazard and the traffic intensity (K_{TO} · λ) or the product of the density of traffic and the space interval between the front bumpers of vehicles (q · γ) the product of the coefficient of transport hazard multiplied by 1000 (K_{TO} ·1000), and finally, the inverse of the coefficient of transport hazard:

$$K_{BDD} = (q \cdot V) / (K_{TO} \cdot \lambda), \text{ relative value.} \tag{19}$$

Check: (TE·*km·h/km·h·b/r*·TE) = 1; the formula is correct.

$$K_{BDD} = (q \cdot \gamma) / (K_{TO} \cdot 1000), \text{ relative value.} \tag{20}$$

Check: (TE·*m·km/km*·TE·*b/r·m*) = 1; the formula is correct.

$$K_{BDD} = (\lambda\ V) / (V\ K_{TO}\ \lambda) = 1/K_{TO} \approx 0.76 \text{ on average.} \tag{21}$$

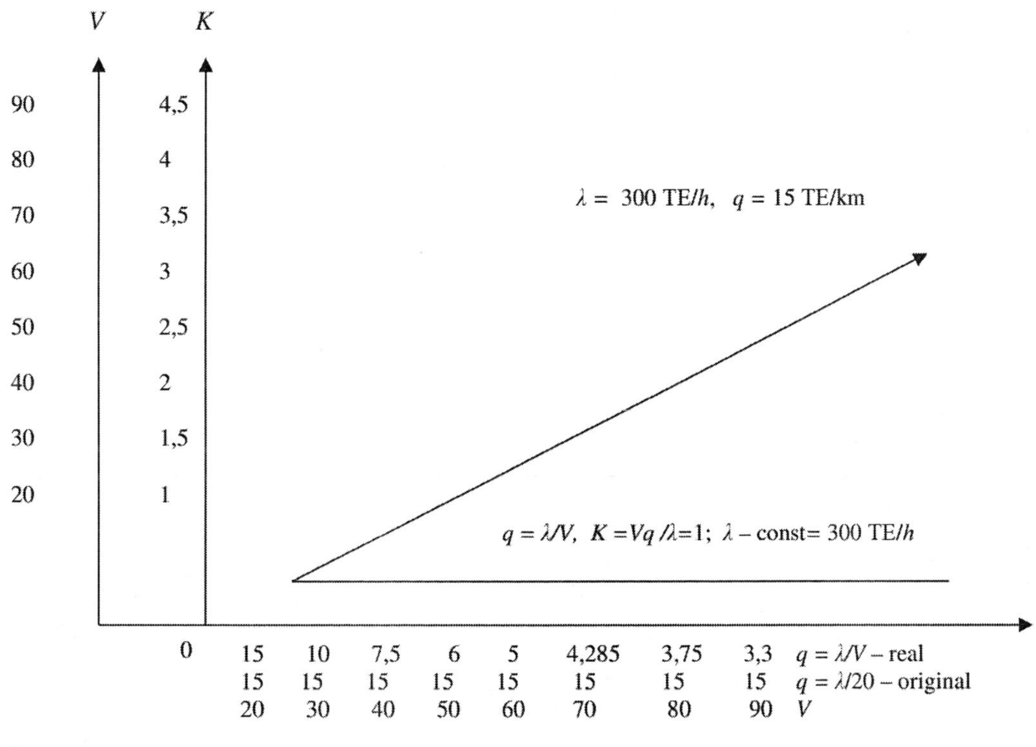

Figure. The dynamics of the road safety coefficient

Conclusion

In conclusion, the following should be noted:

- the new formulas presented in the article to determine the necessary length, unlike [3], taking into account the coefficient of transport hazard and the probable size of the traffic and the number of lanes, will allow us to estimate the size of the provision of urban and rural lanes educational institutions not only in districts, territories, regions and republics, but also, integrally, in the whole of the Russian Federation.
- The use of the transport hazard coefficient increases the length of lanes by about 26 %, but at the same time helps to create conditions for ensuring traffic safety by at least 30–35 %, relative to the existing state of traffic [4, 7].
- Before reconstructing regulated road networks, it is necessary to assess the degree of provision of street-road networks with lanes, then, if not only sufficient, but also if there is a reserve of 30–40 %, you can engage in the implementation of modern adaptive traffic light control tools for sequentially skipping flows (this is one of the last options for improving O and UDD on networks at the same level).
- At the same time, it should be borne in mind that 3- and 4-phase hard traffic lights regulation, as well as the passage of left-hand traffic or pedestrian flows according to the cycles allowed for movement in cycles, significantly increase delays [1–3], which are adequate in time – the maintenance of a motor vehicle depot in 406 cars (only at one 4-way intersection with two lanes in each direction, 828 TE/h in each lane) during the year [1, 3]. This is, in fact, an economic justification for the construction of multi-level interchanges for the passage of both traffic and pedestrian flows, linearly related to ensuring safety, both road traffic and environmental traffic, due to a sharp increase in the speed of communication of the public transit passenger transport and bandwidth lanes.
- Road safety depends on the intensity of the conflict between traffic and pedestrian flows, the latter on the frequency, size and complexity of hazardous areas, the density of the information field, and on the isolation of flows in maneuvers and directions, including pedestrian flows and ultimately as a result, from the costs of eliminating these dangerous sections and creating traffic isolation.
- There is a linear relationship between road safety and the cost of creating a road environment that excludes conflict situations not only at intersections, but also on road sections. The quicker the safe road conditions are created, the faster the moment of trouble-free operation of vehicles will come.
- The construction of 16-lane motorways and express roads with the specialization of lanes for passengers, freight and other vehicles will significantly increase the throughput and speed relative to existing regulated road networks, and eliminates conflicting traffic situations, which means creating safe conditions for traffic and pedestrian flows.

References

[1] Zhivoglyadov V G 2005 *The theory of traffic and pedestrians* (Rostov-on- Don: Ed. Universities North-Caucasus. region Technical sciences) 1082 p
[2] Silyanov V V 1977 *The theory of traffic flows in road design and traffic management* (Moscow: Transport) 303 p
[3] Zhivoglyadov V G 2009 *The methodology of increasing the effectiveness of traffic management* (St. Petersburg Abstract for the degree of Doctor of Technical Sciences) 35 p
[4] Zhivoglyadova L V 2005 *Conflict theory of transport and pedestrian flows* (Rostov-on-Don: Izv. University of the North Caucasus. region, Engineering) 275 p
[5] Haight F 1966 *The mathematical theory of transport flows* (Moscow: Mir) 226 p
[6] Klinkovstein G I and Afanasyev M B 2001 *Organization of road traffi* (Moscow: Transport) 247 p
[7] Zhivoglyadova L V 2005 *Traffic safety assessment based on modeling of conflicting traffic intersections* (Moscow: Rotaprint MADI (GTU)) 20 p

AUTHOR INDEX